Advances in Computer Vision and Pattern Recognition

Titles in this series now included in the Thomson Reuters Book Citation Index!

Advances in Computer Vision and Pattern Recognition is a series of books which brings together current developments in this multi-disciplinary area. It covers both theoretical and applied aspects of computer vision, and provides texts for students and senior researchers in topics including, but not limited to:

- Deep learning for vision applications
- Computational photography
- Biological vision
- Image and video processing
- Document analysis and character recognition
- Biometrics
- Multimedia
- Virtual and augmented reality
- Vision for graphics
- Vision and language
- Robotics

Sébastien Marcel · Julian Fierrez · Nicholas Evans
Editors

Handbook of Biometric Anti-Spoofing

Presentation Attack Detection and Vulnerability Assessment

Third Edition

Editors
Sébastien Marcel
Idiap Research Institute
Martigny, Switzerland

Julian Fierrez
Universidad Autonoma de Madrid
Madrid, Spain

Nicholas Evans
EURECOM
Biot Sophia Antipolis, France

ISSN 2191-6586 ISSN 2191-6594 (electronic)
Advances in Computer Vision and Pattern Recognition
ISBN 978-981-19-5290-6 ISBN 978-981-19-5288-3 (eBook)
https://doi.org/10.1007/978-981-19-5288-3

This Springer imprint is published by the registered company Springer Nature Singapore Pte Ltd.
The registered company address is: 152 Beach Road, #21-01/04 Gateway East, Singapore 189721, Singapore

Foreword

It is a pleasure and an honour both to write the Foreword for a leading and prestigious book. This is the Third Edition of the *Handbook of Biometrics Anti-Spoofing* and I was co-editor for the first two Editions, and now I have been promoted. (More realistically, I have not worked in Anti-Spoofing for some time now.) So I admit bias, and I am taking the opportunity to look at the book from the outside. Essentially, I shall aim to substantiate my claim that the Handbook is indeed leading and prestigious, and I find it not hard to do.

The first evidence I can offer is that the Handbook was the first text to concentrate on anti-spoofing. There appear to be no other books yet which concentrate on this important topic, though it is among the content of some and naturally within those describing the security domain. I have been privileged to be involved in biometrics from its earliest days, in my case in the 1980s before the word biometrics was coined. Even then we knew that if we were working or were ever to work properly, then we needed to countenance the possibility that someone would try and break the system. In modern terminology, we did not consider the possibility that there might be spoofing or presentation attacks. In the First Edition, I, Sébastien, and Stan Li (the Editors then) argued at length over whether the book should be titled "Spoofing" or "Anti-spoofing"—and that is the nature of pioneering work: one does not even know how best to describe it. So the Handbook's uniqueness and initiative strongly underpin a claim that the text is leading and prestigious.

Secondly, the evidence concerns the people involved: those who have edited this Handbook, those who have authored its Chapters, and those who have reviewed them. Sébastien Marcel is well known for his work in multi-modality anti-spoofing, Nicholas Evans is well known for his works on speech and other biometrics, and Julian Fierrez is well known for his prize-winning works on face and signature. All have contributed to the infrastructure this fascinating subject requires and are due many accolades for their other contributions. This is a great team of Editors. Then we find a cohort of authors who are leading lights in this topic area, and like the Editors, they too have contributed to the infrastructure and the building of this topical material. That the contributors derive from many leading institutions around the world further underpins the suggestion that the pedigree of those involved emphasizes the claim that the text is leading and prestigious.

The third evidence is that the book's coverage is highly appropriate to lead the way in studies of anti-spoofing. The coverage includes major modalities, like Finger, Face, Iris, and Voice (noting that voice/speaker is often—unfortunately—considered within the signal/speech domain rather than within biometrics). There are naturally other biometrics. In previous editions, these have included signature, gait, and vein and two of these maintain in the new Edition. There are also methodological contributions and there is updated material on standards and legal issues. The latter two are of enormous importance as identity verification and recognition continue to pervade society and enable convenient, practicable, and secure access. So the comprehensiveness and the insight afforded by the Handbook's structure reinforce the claim to leadership and prestige.

There are many other metrics that further contribute to this evidence. The Handbook has been highly cited in its previous two Editions, and naturally, the Chapters within it have their own citations too. My own students have enjoyed the book, and many others do. The book is published within Springer's excellent series on biometrics. Then there are sales and library penetration, neither of which I can assess. That a book reaches its third Edition assures one of the market penetration since publishers will not consider furthering texts that have not proved substantially their value. Even in my biased view, I find it easy to find evidence that this book is the leading and prestigious book and a pioneering text in biometric anti-spoofing. By its content, presentation, ethos, and style, the Third Edition of the *Handbook of Biometrics Anti-Spoofing* makes an excellent addition to this compelling series. My congratulations to all who have contributed to this.

March 2022

Mark S. Nixon
Professor Emeritus
School of Electronics and Computer Science
University of Southampton
Southampton, UK

Preface

The study of Presentation Attack Detection (PAD) is now an established field of research across the biometrics community. Increasing awareness of vulnerabilities to presentation attacks has continued to fuel the growing impetus to develop countermeasures to prevent the manipulation and to protect the security and reliability of biometric recognition systems. Progress has been rapid, with PAD nowadays being a key component to almost any deployment of biometric recognition technology and now standardized by the ISO and the IEC.

The tremendous advances since the publication of the second edition in 2019 are the principal reason why we decided to compile a third edition of the *Handbook of Biometric Anti-Spoofing*. As for the second edition, we published an open call for expressions of interest and invited updates to previous chapters in addition to entirely new contributions. The third edition is arranged in seven different parts dedicated to PAD in fingerprint recognition, iris recognition, face recognition, and voice recognition, with two more covering other and multi-biometrics, legal aspects, and standards. Like the second edition, the first four parts all start with an introduction to PAD that is specific to each biometric characteristic. We are delighted that all four also contain a review of PAD competitions (the last being combined with the respective introduction), all updated to provide an overview of progress and the latest results. New chapters feature throughout and complement updates with coverage of new detection approaches and models, multi-spectral/multi-channel techniques, multiple biometric PAD databases and solutions, and legal perspectives. The third edition comprises 21 chapters. We are extremely grateful to the authors and also the reviewers who have helped to ensure quality. Both authors as well as reviewers are listed separately herein.

We wish also to thank the team at Springer for their support and especially *Celine Chang* and *Vinothini Elango* who helped significantly towards the preparation of this third edition. Our thanks also to Prof. Mark Nixon. Mark served as co-editor for both the first and second editions but, following his recent retirement, stepped down from this role for the third edition. We refused to let him off the hook completely, however, and are delighted that he so kindly agreed to provide the foreword. Thank you Mark!

We are confident that the third edition serves as a timely update and defacto reference to biometric presentation attack detection for researchers, students, engineers, and technology consultants alike. Like the second edition, a number of chapters are accompanied by open-source software, scripts, and tools that can be used by others to reproduce some of the results presented in the book. Additional resources can be downloaded from https://gitlab.idiap.ch/biometric-resources. In almost all cases, the experimental work reported was performed using standard or common databases and protocols which we hope will allow and inspire others to join the PAD efforts. PAD will surely remain at the forefront of research in our fields for years to come. While tremendous progress has been made, new threats are emerging and call for greater investments to ensure that biometrics recognition technology remains both reliable and secure.

Martigny, Switzerland — Sébastien Marcel
Madrid, Spain — Julian Fierrez
Biot Sophia Antipolis, France — Nicholas Evans
March 2022

List of Reviewers

Zahid Akhtar INRS-EMT, University of Quebec, Canada
José Luis Alba Castro Universidad de Vigo, Spain
André Anjos Idiap Research Institute, Switzerland
Sushil Bhattacharjee Idiap Research Institute, Switzerland
Christophe Champod University of Lausanne, Switzerland
Xingliang Cheng Tsinghua University, China
Adam Czajka Research and Academic Computer Network (NASK), Poland and University of Notre Dame, USA
Héctor Delgado Nuance Communications, Spain
Nesli Erdogmus Izmir Institute of Technology, Turkey
Nicholas Evans EURECOM, Department of Digital Security, France
Jiangjiang Feng Tsinghua University, China
Julian Fierrez Biometrics and Data Pattern Analytics—BiDA Lab, Universidad Autonoma de Madrid, Spain
Javier Galbally European Commission, Joint Research Centre, Italy
Anjith George Idiap Research Institute, Switzerland
Luca Ghiani University of Cagliari, Department of Electrical and Electronic Engineering, Italy
Marta Gomez-Barrero da/sec Biometrics and Internet Security Research Group, Hochschule Darmstadt, Germany
Abdenour Hadid University of Oulu, Finland
Guillaume Heusch Idiap Research Institute, Switzerland
Ivan Himawan Queensland University of Technology, Brisbane, Australia
Els Kindt KU Leuven, Belgium
Tomi Kinnunen School of Computing, University of Eastern Finland, Finland
Jukka Komulainen Center for Machine Vision and Signal Analysis, University of Oulu, Finland
Ketan Kotwal, Idiap Research Institute, Switzerland
Vedrana Krivokuća Hahn, Idiap Research Institute, Switzerland
Stan Z. Li Chinese Academy of Sciences, China
Sébastien Marcel Idiap Research Institute, Switzerland

Gian Luca Marcialis University of Cagliari, Department of Electrical and Electronic Engineering, Italy
Jeanne Pia Mifsud Bonnici University of Groningen, The Netherlands
Amir Mohammadi Idiap Research Institute, Switzerland
Aythami Morales Biometrics and Data Pattern Analytics—BiDA Lab, Universidad Autonoma de Madrid, Spain
Mark S. Nixon University of Southampton, UK
Jonathan Phillips NIST, USA
Hugo Proenca University of Beira Interior, Instituto de Telecomunica coes, Portugal
Kiran B. Raja Norwegian Biometrics Laboratory, Norwegian University of Science and Technology (NTNU), Norway
Raghavendra Ramachandra Norwegian Biometrics Laboratory, Norwegian University of Science and Technology (NTNU), Norway
Arun Ross Michigan State University, USA
Md Sahidullah MULTISPEECH, Inria, France
Richa Singh IIIT-Delhi, India
Juan Eduardo Tapia Farias, Hochschule Darmstadt, Germany
Massimiliano Todisco Department of Digital Security, EURECOM, France
Ruben Tolosana Biometrics and Data Pattern Analytics—BiDA Lab, Universidad Autonoma de Madrid, Spain
You Zhang University of Rochester, USA

Contents

Contributors

(Andy) Fang Zhaoyuan University of Notre Dame, Notre Dame, IN, USA

Becker Benedict University of Notre Dame, Notre Dame, USA

Boutros Fadi Fraunhofer Institute for Computer Graphics Research IGD, Darmstadt, Germany

Bowyer Kevin Notre Dame University, Notre Dame, IN, USA

Boyd Aidan University of Notre Dame, Notre Dame, IN, USA

Busch Christoph da/sec - Biometrics and Internet-Security Research Group, Hochschule Darmstadt and ATHENE (National Research Center for Applied Cybersecurity), Darmstadt, Germany;
Biometrics Laboratory, Norwegian University of Science and Technology (NTNU), Trondheim, Norway

Cappelli Raffaele Università di Bologna, Cesena, Italy

Casula Roberto University of Cagliari, Cagliari, Italy

Chen Xinhui Lehigh University, Bethlehem, USA

Cheng Xingliang Tsinghua University, Beijing, China

Czajka Adam University of Notre Dame, Notre Dame, IN, USA

Damer Naser Fraunhofer Institute for Computer Graphics Research IGD, Darmstadt, Germany

Das Priyanka Clarkson University, Clarkson, NY, USA

Delgado Héctor Nuance Communications, Madrid, Spain

Duan Zhiyao University of Rochester, Rochester, USA

Engelsma Joshua J. Michigan State University, East Lansing, USA

Evans Nicholas Department of Digital Security, EURECOM (France), Biot, France

Fang Meiling Fraunhofer Institute for Computer Graphics Research IGD, Darmstadt, Germany

Fierrez Julian School of Engineering, Universidad Autonoma de Madrid, Madrid, Spain;
Biometrics and Data Pattern Analytics—BiDA Lab, Escuela Politecnica Superior, Universidad Autonoma de Madrid, Madrid, Spain

Fontanillo López Cesar Augusto KU Leuven – Law Faculty – Citip – iMec, Leuven, Belgium

Galbally Javier eu-LISA / European Union Agency for the Operational Management of Large-Scale IT Systems in the Area of Freedom, Security and Justice, Tallinn, Estonia;
European Commission, Joint Research Centre, Ispra, Italy

George Anjith Idiap Research Institute, Martigny, Switzerland

Gomez-Barrero Marta Hochschule Ansbach, Ansbach, Germany

Gonzalez-Garcia Carlos Biometrics and Data Pattern Analytics—BiDA Lab, Escuela Politecnica Superior, Universidad Autonoma de Madrid, Madrid, Spain

Gonzalez-Soler Lazaro Janier da/sec - Biometrics and Internet-Security Research Group, Hochschule Darmstadt, Darmstadt, Germany

Hernandez-Ortega Javier School of Engineering, Universidad Autonoma de Madrid, Madrid, Spain

Jiang Fei Tencent Technology Co., Ltd., Shenzhen, China

Johnson Alan University of Notre Dame, Notre Dame, USA

Kamble Madhu EURECOM, Biot, France

Kindt Els J. KU Leuven – Law Faculty – Citip – iMec, Leuven, Belgium

Kinnunen Tomi School of Computing, University of Eastern Finland (Finland), Joensuu, Finland

Kohli Naman West Virginia University, Morgantown, WV, USA

Komulainen Jukka University of Oulu, Oulu, Finland;
Visidon Ltd, Oulu, Finland

Lee Kong-Aik Institute for Infocomm Research, A*STAR (Singapore), Singapore, Singapore

Li Lantian Tsinghua University, Beijing, China

Li Xiaobai Center for Machine Vision and Signal Analysis, University of Oulu, Oulu, Finland

Linortner Michael Department of Computer Science, University of Salzburg, Salzburg, Austria

Liu Si-Qi Department of Computer Science, Hong Kong Baptist University, Kowloon, Hong Kong

Maciejewicz Piotr Medical University of Warsaw, Warsaw, Poland

Marcel Sébastien IDIAP Research Institute, Martigny, Switzerland

Marcialis Gian Luca Università di Cagliari, Cagliari, Italy

McGrath Joseph University of Notre Dame, Notre Dame, IN, USA

Micheletto Marco University of Cagliari, Cagliari, Italy

Mohammadi Amir IDIAP Research Institute, Martigny, Switzerland

Morales Aythami School of Engineering, Universidad Autonoma de Madrid, Madrid, Spain

Namboodiri Anoop IIIT Hyderabad, Hyderabad, India

Nautsch Andreas Department of Digital Security, EURECOM (France), Biot, France

Noore Afzel Texas A &M University-Kingsville, Kingsville, TX, USA

Orrù Giulia University of Cagliari, Cagliari, Italy

Ortega-Garcia Javier Biometrics and Data Pattern Analytics—BiDA Lab, Escuela Politecnica Superior, Universidad Autonoma de Madrid, Madrid, Spain

Patino Jose EURECOM, Biot, France

Poiret Eric IDEMIA, Paris, France

Popli Additya IIIT Hyderabad, Hyderabad, India

Raja Kiran Biometrics Laboratory, Norwegian University of Science and Technology (NTNU), Trondheim, Norway

Ramachandra Raghavendra Biometrics Laboratory, Norwegian University of Science and Technology (NTNU), Trondheim, Norway

Rathgeb Christian da/sec Biometrics and Internet Security Research Group, Hochschule Darmstadt, Darmstadt, Germany

Sahidullah Md Université de Lorraine, CNRS, Inria, LORIA, Nancy, France

Schuckers Stephanie Clarkson University, Clarkson, NY, USA

Schuiki Johannes Department of Computer Science, University of Salzburg, Salzburg, Austria

Singh Richa Department of CSE, IIT Jodhpur, Jodhpur, RJ, India

Stokkenes Martin Norwegian University of Science and Technology, Gjøvik, Norway

Tandon Saraansh IIIT Hyderabad, Hyderabad, India

Todisco Massimiliano Department of Digital Security, EURECOM (France), Biot, France

Tolosana Ruben Biometrics and Data Pattern Analytics—BiDA Lab, Escuela Politecnica Superior, Universidad Autonoma de Madrid, Madrid, Spain

Trokielewicz Mateusz Warsaw University of Technology, Warsaw, Poland

Uhl Andreas Department of Computer Science, University of Salzburg, Salzburg, Austria

Vatsa Mayank Department of CSE, IIT Jodhpur, Jodhpur, RJ, India

Venkatesh Sushma Norwegian University of Science and Technology, Gjøvik, Norway;
Biometrics Laboratory, Norwegian University of Science and Technology (NTNU), Trondheim, Norway;
da/sec Biometrics and Internet Security Research Group, Hochschule Darmstadt, Darmstadt, Germany;
Hochschule Ansbach, Ansbach, Germany

Vera-Rodriguez Ruben Biometrics and Data Pattern Analytics—BiDA Lab, Escuela Politecnica Superior, Universidad Autonoma de Madrid, Madrid, Spain

Wang Dong Tsinghua University, Beijing, China

Wang Xin National Institute of Informatics (Japan), Tokyo, Japan

Wasnik Pankaj Norwegian University of Science and Technology, Gjøvik, Norway

Wimmer Georg Department of Computer Science, University of Salzburg, Salzburg, Austria

Xu Mingxing Tsinghua University, Beijing, China

Yadav Daksha West Virginia University, Morgantown, WV, USA

Yamagishi Junichi National Institute of Informatics (Japan), Tokyo, Japan

Yambay David Clarkson University, Clarkson, NY, USA

Yu Zitong Center for Machine Vision and Signal Analysis, University of Oulu, Oulu, Finland

Yuen Pong C. Department of Computer Science, Hong Kong Baptist University, Kowloon, Hong Kong

Zhang You University of Rochester, Rochester, USA

Zhao Guoying Center for Machine Vision and Signal Analysis, University of Oulu, Oulu, Finland

Zheng Thomas Fang Tsinghua University, Beijing, China

Zhu Ge University of Rochester, Rochester, USA

Part I
Fingerprint Biometrics

Chapter 1
Introduction to Presentation Attack Detection in Fingerprint Biometrics

Javier Galbally, Julian Fierrez, Raffaele Cappelli, and Gian Luca Marcialis

Abstract This chapter provides an introduction to Presentation Attack Detection (PAD) in fingerprint biometrics, also coined as anti-spoofing, describes early developments in this field, and briefly summarizes recent trends and open issues.

1.1 Introduction

"*Fingerprints cannot lie, but liars can make fingerprints*". Unfortunately, this paraphrase of an old quote attributed to Mark Twain[1] has been proven right in many occasions now.

As the deployment of fingerprint systems keeps growing year after year in such different environments as airports, laptops, or mobile phones, people are also becoming more familiar to their use in everyday life and, as a result, the security weaknesses of fingerprint sensors are becoming better known to the general public. Nowadays it is not difficult to find websites or even tutorial videos, which give detailed guidance on how to create fake fingerprints which may be used for spoofing biometric systems.

[1] Figures do not lie, but liars do figure.

J. Galbally
European Commission, Joint Research Centre, Ispra, Italy
e-mail: javier.galbally@ec.europa.eu

J. Fierrez
Biometrics and Data Pattern Analytics-BiDA Lab, Universidad Autonoma de Madrid, Madrid, Spain
e-mail: julian.fierrez@uam.es

R. Cappelli
Università di Bologna, Cesena, Italy
e-mail: raffaele.cappelli@unibo.it

G. L. Marcialis (✉)
Università di Cagliari, Cagliari, Italy
e-mail: marcialis@unica.it

S. Marcel et al. (eds.), *Handbook of Biometric Anti-Spoofing*, Advances in Computer Vision and Pattern Recognition, https://doi.org/10.1007/978-981-19-5288-3_1

As a consequence, the fingerprint stands out as one of the biometric traits which has arisen the most attention not only from researchers and vendors, but also from the media and users, regarding its vulnerabilities to Presentation Attacks (PAs, aka spoofing), as the attempt to impersonate someone else by submitting an artifact or Presentation Attack Instrument. This increasing interest of the biometric community in the security evaluation of fingerprint recognition systems against presentation attacks has led to the creation of numerous and very diverse initiatives in this field: the publication of many research works disclosing and evaluating different fingerprint presentation attack approaches [1–4]; the proposal of new countermeasures to spoofing, namely, novel presentation attack detection methods [5–7]; related book chapters [8, 9]; Ph.D. and MSc Thesis which propose and analyze different fingerprint PA and PAD techniques [10–13]; several patented fingerprint PAD mechanisms both for touch-based and contactless systems [14–18]; the publication of Supporting Documents and Protection Profiles in the framework of the security evaluation standard Common Criteria for the objective assessment of fingerprint-based commercial systems [19, 20]; the organization of competitions focused on vulnerability assessment to fingerprint presentation attacks [21–23]; the acquisition of specific datasets for the evaluation of fingerprint protection methods against direct attacks [24–26], the creation of groups and laboratories which have the evaluation of fingerprint security as one of their major tasks [27–29]; or the acceptance of several European Projects on fingerprint PAD as one of their main research interests [30, 31].

The aforementioned initiatives and other analogue studies have shown the importance given by all parties involved in the development of fingerprint-based biometrics to the improvement of the systems security and the necessity to propose and develop specific protection methods against PAs in order to bring this rapidly emerging technology into practical use. This way, researchers have focused on the design of specific countermeasures that enable fingerprint recognition systems to detect fake samples and reject them, improving this way the robustness of the applications.

In the fingerprint field, besides other PAD approaches such as the use of multibiometrics or challenge-response methods, special attention has been paid by researchers and industry to the so-called *liveness detection* techniques. These algorithms use different physiological properties to distinguish between real and fake traits. Liveness assessment methods represent a challenging engineering problem as they have to satisfy certain demanding requirements [32]: (i) non-invasive, the technique should in no case be harmful to the individual or require an excessive contact with the user; (ii) user friendly, people should not be reluctant to use it; (iii) fast, results have to be produced in a very reduced interval as the user cannot be asked to interact with the sensor for a long period of time; (iv) low cost, a wide use cannot be expected if the cost is excessively high; and (v) performance, in addition to having a good fake detection rate, the protection scheme should not degrade the recognition performance (i.e., false rejection) of the biometric system.

Liveness detection methods are usually classified into one of two groups: (i) *Hardware-based* techniques, which add some specific device to the sensor in order to detect particular properties of a living trait (e.g., fingerprint sweat, blood pressure, or odor); (*ii*) *Software-based* techniques, in this case the fake trait is detected once the sample has been acquired with a standard sensor (i.e., features used to distinguish between real and fake traits are extracted from the biometric sample, and not from the trait itself).

The two types of methods present certain advantages and drawbacks over the other and, in general, a combination of both would be the most desirable protection approach to increase the security of biometric systems. As a coarse comparison, hardware-based schemes usually present a higher fake detection rate, while software-based techniques are in general less expensive (as no extra device is needed) and less intrusive since their implementation is transparent to the user. Furthermore, as they operate directly on the acquired sample (and not on the biometric trait itself), software-based techniques may be embedded in the feature extractor module which makes them potentially capable of detecting other types of illegal break-in attempts not necessarily classified as presentation attacks. For instance, software-based methods can protect the system against the injection of reconstructed or synthetic samples into the communication channel between the sensor and the feature extractor [33, 34].

Although, as shown above, a great amount of work has been done in the field of fingerprint PAD and big advances have been reached over the last two decades, the attacking methodologies have also evolved and become more and more sophisticated. This way, while many commercial fingerprint readers claim to have some degree of PAD embedded, many of them are still vulnerable to presentation attack attempts using different artificial fingerprint samples. Therefore, there are still big challenges to be faced in the detection of fingerprint direct attacks.[2]

This chapter represents an introduction to the problem of fingerprint PAD [35, 36]. More comprehensive and up-to-date surveys of recent advances can be found elsewhere [37–40]. The rest of the chapter is structured as follows. An overview into early works in the field of fingerprint PAD is given is Sect. 1.2, while Sect. 1.3 provides a summary of recent trends and main open issues. A brief description of large and publicly available fingerprint spoofing databases is presented in Sect. 1.4. Conclusions are finally drawn in Sect. 1.5.

1.2 Early Works in Fingerprint Presentation Attack Detection

The history of fingerprint forgery in the forensic field is probably almost as old as that of fingerprint development and classification itself. In fact, the question of whether or not fingerprints could be forged was positively answered [41] several years before it was officially posed in a research publication [42].

Regarding modern automatic fingerprint recognition systems, although other types of attacks with dead [43] or altered [44] fingers have been reported, almost

[2] https://www.iarpa.gov/index.php/research-programs/odin/.

all the available vulnerability studies regarding presentations attacks are carried out either by taking advantage of the residual fingerprint left behind on the sensor surface, or by using some type of gummy fingertip (or even complete prosthetic fingers) manufactured with different materials (e.g., silicone, gelatin, plastic, clay, dental molding material, or glycerin). In general, these fake fingerprints may be generated with the cooperation of the user, from a latent fingerprint or even from a fingerprint image reconstructed from the original minutiae template [1–3, 24, 45–49].

These very valuable works and other analogue studies have highlighted the necessity to develop efficient protection methods against presentation attacks. One of the first efforts in fingerprint PAD initiated a research line based on the analysis of the skin perspiration pattern which is very difficult to be faked in an artificial finger [5, 50]. These pioneer studies, which considered the periodicity of sweat and the sweat diffusion pattern, were later extended and improved in two successive works applying a wavelet-based algorithm and adding intensity-based perspiration features [51, 52]. These techniques were finally consolidated and strictly validated on a large database of real, fake, and dead fingerprints acquired under different conditions in [25]. Recently, a novel region-based liveness detection approach also based on perspiration parameters and another technique analyzing the valley noise have been proposed by the same group [53, 54]. Part of these approaches have been implemented in commercial products [55] and have also been combined with other morphological features [56, 57] in order to improve the presentation attack detection rates [58].

A second group of fingerprint liveness detection techniques has appeared as an application of the different fingerprint distortion models described in the literature [59–61]. These models have led to the development of a number of liveness detection techniques based on the flexibility properties of the skin [6, 62–64]. In most of these works, the user is required to move his finger while pressing it against the scanner surface, thus deliberately exaggerating the skin distortion. When a real finger moves on a scanner surface, it produces a significant amount of distortion, which can be observed to be quite different from that produced by fake fingers which are usually more rigid than skin. Even if highly elastic materials are used, it seems very difficult to precisely emulate the specific way a real finger is distorted, because the behavior is related to the way the external skin is anchored to the underlying derma and influenced by the position and shape of the finger bone.

Other liveness detection approaches for fake fingerprint detection include the combination of both perspiration and elasticity-related features in fingerprint image sequences [65]; fingerprint-specific quality-related features [7, 66]; the combination of the local ridge frequency with other multiresolution texture parameters [56]; techniques which, following the perspiration-related trend, analyze the skin sweat pores visible in high-definition images [67, 68]; the use of electric properties of the skin [69]; using several image processing tools for the analysis of the fingertip surface texture such as wavelets [70], or three very related works using Gabor filters [71], ridgelets [72], and curvelets [73]; and analyzing different characteristics of the Fourier spectrum of real and fake fingerprint images [74–78].

A critical review of some of these solutions for fingerprint liveness detection was presented in [79]. In a subsequent work [80], the same authors gave a comparative analysis of the PAD methods efficiency. In this last work, we can find an estimation of some of the best performing static (i.e., measured on one image) and dynamic (i.e., measured on a sequence of images) features for liveness detection, that were later used together with some fake-finger specific features in [78] with very good results. Different static features are also combined in [81], significantly improving the results of the individual parameters. Other comparative results of different fingerprint PAD techniques are available in the results of the 2009 and 2011 Fingerprint Liveness Detection Competitions (LivDet 2009 and LivDet 2011) [82, 83].

In addition, some very interesting hardware-based solutions have been proposed in the literature applying multispectral imaging [84, 85], an electrotactile sensor [86], pulse oximetry [87], detection of the blood flow [14], odor detection using a chemical sensor [88], or a currently very active research trend based on Near Infrared (NIR) illumination and Optical Coherence Tomography (OCT) [89–94].

More recently, the third type of protection methods which fall out of the traditional two-type classification software- and hardware-based approaches have been started to be analyzed in the field of fingerprint PAD. These protection techniques focus on the study of biometric systems under direct attacks at the *score level*, in order to propose and build more robust matchers and fusion strategies that increase the resistance of the systems against presentation attack attempts [95–99].

Outside the research community, some companies have also proposed different methods for fingerprint liveness detection such as the ones based on ultrasounds [100, 101], light measurements [102], or a patented combination of different unimodal experts [103]. A comparative study of the PAD capabilities of different commercial fingerprint sensors may be found in [104].

Although the vast majority of the efforts dedicated by the biometric community in the field of fingerprint presentation attacks and PAD are focused on touch-based systems, some works have also been conducted to study the vulnerabilities of contactless fingerprint systems against direct attacks, and some protection methods to enhance their security level have been proposed [17, 50, 105].

The approaches mentioned above represent the main historical developments in fingerprint PAD until ca. 2012-2013. For a survey of more recent and advanced methods in the last 10 years, we refer the reader to [37–40] and the ODIN program.[3]

1.3 A Brief View on Where We Are

In the next chapters of the book, the reader will be able to find information about the most recent advances in fingerprint presentation attack detection. This section merely summarizes some ongoing trends in the development of PADs and some of the main open issues.

As stated in the previous Section, independent and general-purpose descriptors were proposed for feature extraction since from 2013 [38]. In general, these features

[3] https://www.iarpa.gov/index.php/research-programs/odin/.

looked for minute details of the fake image which are added or deleted, impossible to catch by the human eye. This was typically done by appropriate banks of filters aimed at deriving a set of possible patterns. The related feature sets can be adopted to distinguish live from fake fingerprints by machine learning methods.

"Textural features" above looked as the most promising until the advent of deep learning approaches [39, 40]. These, thanks to the increased availability of datasets, allowed the design of a novel generation of fingerprint PAD [26, 106, 107] which exploited the concept of "patch", a very small portion of the fingerprint image to be processed instead of taking the image as a whole input to the network. However, textural features have not yet been left behind because of their expressive power and the fact that they explicitly rely on the patch definition [108, 109].

Among the main challenges to be faced with in the near future, it is important to mention the following[110]:

- assessing the robustness of anti-spoofing methods against novel presentation attacks in terms of fabrication strategy, adopted materials, and sensor technology; for instance, in [111], it has been shown that the PAD error rates of software-based approaches can show a three-fold increase when tested on PA materials not seen during training;
- designing effective methods to embed PAD in fingerprint verification systems [112], including the need for computationally efficient PAD techniques, to be used on low-resources systems such as embedded devices and low-cost smartphones;
- improving explainability of PAD systems; the use of CNNs is providing great benefits to fingerprint PAD performance, but such solutions are usually considered as "black boxes" shedding little light on how and why they actually work. It is important to gain insights into the features that CNNs learn, so that system designers and maintainers can understand why a decision is made and tune the system parameters if needed.

1.4 Fingerprint Spoofing Databases

The availability of public datasets comprising real and fake fingerprint samples and of associated common evaluation protocols is basic for the development and improvement of fingerprint PAD methods.

However, in spite of the large amount of works addressing the challenging problem of fingerprint protection against direct attacks (as shown in Sect. 1.2), in the great majority of them, experiments are carried out on proprietary databases which are not distributed to the research community.

Currently, the two largest fingerprint spoofing databases publicly available for researchers to test their PAD algorithms are as follows:

- LivDet DBs (LivDet 2009–2021 DBs) [21–23]: These datasets, which share the acquisition protocols and part of the samples, are available from 2009 to 2021

Fingerprint Liveness Detection Competitions websites[4, 5] and are divided into the same train and test sets used in the official evaluations. Over seven editions, LivDet shared with the research community over 20,000 fake fingerprint images made up of a large set of materials (play doh, silicone, gelatine, latex...) on a wide brands of optical and solid-state sensors. Over years, LivDet competitions also proposed challenges as the evaluation of embedding fingerprint PAD and matching [22, 23] and of a novel approach to provide spoofs called "Screenspoof" directly from the user's smartphone screen [22]. The LivDet datasets are available for researchers by signing the license agreement.
- ATVS-Fake Fingerprint DB (ATVS-FFp DB) [24]: This database is available from the Biometrics group at UAM.[6] It contains over 3,000 real and fake fingerprint samples coming from 68 different fingers acquired using a flat optical sensor, a flat capacitive sensor, and a thermal sweeping sensor. The gummy fingers were generated with and without the cooperation of the user (i.e., recovered from a latent fingerprint) using modeling silicone.

1.5 Conclusions

The study of the vulnerabilities of biometric systems against presentation attacks has been a very active field of research in recent years [113]. This interest has led to big advances in the field of security-enhancing technologies for fingerprint-based applications. However, in spite of this noticeable improvement, the development of efficient protection methods against known threats (usually based on some type of self-manufactured gummy finger) has proven to be a challenging task.

Simple visual inspection of an image of a real fingerprint and its corresponding fake sample shows that the two images can be very similar and even the human eye may find it difficult to make a distinction between them after a short inspection. Yet, some disparities between the real and fake images may become evident once the images are translated into a proper feature space. These differences come from the fact that fingerprints, as 3-D objects, have their own optical qualities (absorption, reflection, scattering, and refraction), which other materials (silicone, gelatin, and glycerin) or synthetically produced samples do not possess. Furthermore, fingerprint acquisition devices are designed to provide good quality samples when they interact, in a normal operation environment, with a real 3-D trait. If this scenario is changed, or if the trait presented to the scanner is an unexpected fake artifact, the characteristics of the captured image may significantly vary.

In this context, it is reasonable to assume that the image quality properties of real accesses and fraudulent attacks will be different and therefore image-based

[4] http://livdet.diee.unica.it.

[5] http://people.clarkson.edu/projects/biosal/fingerprint/index.php.

[6] http://biometrics.eps.uam.es/.

presentation attack detection in fingerprint biometrics would be feasible. Key early works in this regard have been summarized in the present chapter.

Overall, the chapter provided a general overview of the progress which was initially made in the field of fingerprint PAD and a brief summary about current achievements, trends, and open issues, which will be further developed in the next chapters.

Acknowledgements This work was mostly done (2nd Edition of the book) in the context of TABULA RASA: Trusted Biometrics under Spoofing Attacks, and BEAT: Biometrics Evaluation and Testing projects funded under the 7th Framework Programme of EU. The 3rd Edition update has been made in the context of EU H2020 projects PRIMA and TRESPASS-ETN. This work was also partially supported by the Spanish project BIBECA (RTI2018-101248-B-I00 MINECO/FEDER).

References

1. van der Putte T, Keuning J (2000) Biometrical fingerprint recognition: don't get your fingers burned. In: Proceedings IFIP conference on smart card research and advanced applications, pp 289–303
2. Matsumoto T, Matsumoto H, Yamada K, Hoshino S (2002) Impact of artificial gummy fingers on fingerprint systems. In: Proceedings of SPIE optical security and counterfeit deterrence techniques IV, vol 4677, pp 275–289
3. Thalheim L, Krissler J (2002) Body check: biometric access protection devices and their programs put to the test. ct magazine, pp 114–121
4. Sousedik C, Busch C (2014) Presentation attack detection methods for fingerprint recognition systems: a survey. IET Biom 3:219–233(14). http://digital-library.theiet.org/content/journals/10.1049/iet-bmt.2013.0020
5. Derakhshani R, Schuckers S, Hornak L, O'Gorman L (2003) Determination of vitality from non-invasive biomedical measurement for use in fingerprint scanners. Pattern Recognit 36:383–396
6. Antonelli A, Capelli R, Maio D, Maltoni D (2006) Fake finger detection by skin distortion analysis. IEEE Trans. Inf Forensics Secur 1:360–373
7. Galbally J, Alonso-Fernandez F, Fierrez J, Ortega-Garcia J (2012) A high performance fingerprint liveness detection method based on quality related features. Futur Gener Comput Syst 28:311–321
8. Franco A, Maltoni D (2008) Fingerprint synthesis and spoof detection. In: Ratha NK, Govindaraju V (eds) Advances in biometrics: sensors, algorithms and systems. Springer, pp 385–406
9. Li SZ (ed) (2009) Encyclopedia of biometrics. Springer
10. Coli P (2008) Vitality detection in personal authentication systems using fingerprints. PhD thesis, Universita di Cagliari
11. Sandstrom M (2004) Liveness detection in fingerprint recognition systems. Master's thesis, Linkoping University
12. Lane M, Lordan L (2005) Practical techniques for defeating biometric devices. Master's thesis, Dublin City University
13. Blomme J (2003) Evaluation of biometric security systems against artificial fingers. Master's thesis, Linkoping University
14. Lapsley P, Less J, Pare D, Hoffman N (1998) Anti-fraud biometric sensor that accurately detects blood flow 5(737):439
15. Setlak DR (1999) Fingerprint sensor having spoof reduction features and related methods 5(953):441

16. Kallo I, Kiss A, Podmaniczky JT (2001) Detector for recognizing the living character of a finger in a fingerprint recognizing apparatus 6(175):64
17. Diaz-Santana E, Parziale G (2008) Liveness detection method. EP1872719
18. Kim, J., Choi, H., Lee, W.: Spoof detection method for touchless fingerprint acquisition apparatus (2011). 1054314
19. Centro Criptologico Nacional (CCN) (2011) Characterizing attacks to fingerprint verification mechanisms CAFVM v3.0. Common Criteria Portal
20. Bundesamt fur Sicherheit in der Informationstechnik (BSI) (2008) Fingerprint spoof detection protection profile FSDPP v1.8. Common Criteria Portal
21. Ghiani L, Yambay DA, Mura V, GLM, Roli F, Schuckers S (2017) Review of the fingerprint liveness detection (livdet) competition series: 2009 to 2015. Image Vis Comput 58:110–128. https://doi.org/10.1016/j.imavis.2016.07.002
22. Casula R, Micheletto M, Orrù G, Delussu R, Concas S, Panzino A, Marcialis G (2021) Livdet 2021 fingerprint liveness detection competition – into the unknown. In: Proceedings of international joint conference on biometrics (IJCB 2021). https://doi.org/10.1109/IJCB52358.2021.9484399
23. Orrù G, Tuveri P, Casula R, Bazzoni C, Dessalvi G, Micheletto M, Ghiani L, Marcialis G (2019) Livdet2019 – fingerprint liveness detection competition in action 2019. In: Proceedings of IEEE/IAPR international conference on biometrics (ICB 2019). https://doi.org/10.1109/ICB45273.2019.8987281
24. Galbally J, Fierrez J, Alonso-Fernandez F, Martinez-Diaz M (2011) Evaluation of direct attacks to fingerprint verification systems. J Telecommun Syst, Special Issue of Biom Syst Appl 47:243–254
25. Abhyankar A, Schuckers S (2009) Integrating a wavelet based perspiration liveness check with fingerprint recognition. Pattern Recognit 42:452–464
26. Spinoulas L, Mirzaalian H, Hussein ME, AbdAlmageed W (2021) Multi-modal fingerprint presentation attack detection: evaluation on a new dataset. IEEE Trans Biom Behav Identity Sci 3:347–364. https://doi.org/10.1109/TBIOM.2021.3072325
27. Biometrics Institute (2011) Biometric Vulnerability Assessment Expert Group. http://www.biometricsinstitute.org/pages/biometric-vulnerability-assessment-expert-group-bvaeg.html
28. NPL (2010) National Physical Laboratory: Biometrics. http://www.npl.co.uk/biometrics
29. CESG (2001) Communications-Electronics Security Group - Biometric Working Group (BWG). https://www.cesg.gov.uk/policyguidance/biometrics/Pages/index.aspx
30. BEAT (2012) BEAT: Biometrices evaluation and testing. http://www.beat-eu.org/
31. Tabula Rasa (2010) Trusted biometrics under spoofing attacks (tabula rasa). http://www.tabularasa-euproject.org/
32. Maltoni D, Maio D, Jain A, Prabhakar S (2009) Handbook of fingerprint recognition. Springer
33. Cappelli R, Maio D, Lumini A, Maltoni D (2007) Fingerprint image reconstruction from standard templates. IEEE Trans Pattern Anal Mach Intell 29:1489–1503
34. Cappelli R (2009) Synthetic fingerprint generation. In: Handbook of fingerprint recognition. Springer, pp 270–302
35. Galbally J, Marcel S, Fierrez J (2014) Image quality assessment for fake biometric detection: application to iris, fingerprint and face recognition. IEEE Trans. Image Process 23(2):710–724. https://doi.org/10.1109/TIP.2013.2292332
36. Hadid A, Evans N, Marcel S, Fierrez J (2015) Biometrics systems under spoofing attack: an evaluation methodology and lessons learned. IEEE Signal Process Mag 32(5):20–30. https://doi.org/10.1109/MSP.2015.2437652
37. Marasco E, Ross A (2014) A survey on antispoofing schemes for fingerprint recognition systems. ACM Comput Surv 47(2). https://doi.org/10.1145/2617756
38. Sousedik C, Busch C (2014) Presentation attack detection methods for fingerprint recognition systems: a survey. IET Biom 3(4):219–233. https://doi.org/10.1049/iet-bmt.2013.0020
39. Karampidis K, Rousouliotis M, Linardos E, Kavallieratou E (2021) A comprehensive survey of fingerprint presentation attack detection. J Surveill Secur Saf 2:117–61

40. Singh JM, Madhun A, Li G, Ramachandra R (2021) A survey on unknown presentation attack detection for fingerprint. In: Yildirim Yayilgan S, Bajwa IS, Sanfilippo F (eds) Intelligent technologies and applications. Springer International Publishing, Cham, pp 189–202
41. Wehde A, Beffel JN (1924) Fingerprints can be forged. Tremonia Publish Co
42. de Water MV (1936) Can fingerprints be forged? Sci News-Lett 29:90–92
43. Sengottuvelan P, Wahi A (2007) Analysis of living and dead finger impressions identification for biometric applications. In: Proceedings of international conference on computational intelligence and multimedia applications
44. Yoon S, Feng J, Jain AK (2012) Altered fingerprints: analysis and detection. IEEE Trans Pattern Anal Mach Intell 34:451–464
45. Willis D, Lee M (1998) Biometrics under our thumb. Netw Comput (1998). http://www.networkcomputing.com/
46. Sten A, Kaseva A, Virtanen T (2003) Fooling fingerprint scanners - biometric vulnerabilities of the precise biometrics 100 SC scanner. In: Proceedings of Australian information warfare and IT security conference
47. Wiehe A, Sondrol T, Olsen K, Skarderud F (2004) Attacking fingerprint sensors. NISlab, Gjovik University College, Technical report
48. Galbally J, Cappelli R, Lumini A, de Rivera GG, Maltoni D, Fierrez J, Ortega-Garcia J, Maio D (2010) An evaluation of direct and indirect attacks using fake fingers generated from ISO templates. Pattern Recognit Lett 31:725–732
49. Barral C, Tria A (2009) Fake fingers in fingerprint recognition: glycerin supersedes gelatin. In: Formal to practical security, LNCS 5458, pp 57–69
50. Parthasaradhi S, Derakhshani R, Hornak L, Schuckers S (2005) Time-series detection of perspiration as a liveness test in fingerprint devices. IEEE Trans Syst Man Cybernect - Part C: Appl Rev 35:335–343
51. Schuckers S, Abhyankar A (2004) A wavelet based approach to detecting liveness in fingerprint scanners. In: Proceedings of biometric authentication workshop (BioAW), LNCS-5404. Springer, pp 278–386
52. Tan B, Schuckers S (2006) Comparison of ridge- and intensity-based perspiration liveness detection methods in fingerprint scanners. In: Proceedings of SPIE biometric technology for human identification III (BTHI III), vol 6202, p 62020A
53. Tan B, Schuckers S (2008) A new approach for liveness detection in fingerprint scanners based on valley noise analysis. J Electron Imaging 17:011,009
54. DeCann B, Tan B, Schuckers S (2009) A novel region based liveness detection approach for fingerprint scanners. In: Proceedings of IAPR/IEEE international conference on biometrics, LNCS-5558. Springer, pp 627–636
55. NexIDBiometrics (2012). http://nexidbiometrics.com/
56. Abhyankar A, Schuckers S (2006) Fingerprint liveness detection using local ridge frequencies and multiresolution texture analysis techniques. In: Proceedings of IEEE international conference on image processing (ICIP)
57. Marasco E, Sansone C (2010) An anti-spoofing technique using multiple textural features in fingerprint scanners. In: Proceedings of IEEE workshop on biometric measurements and systems for security and medical applications (BIOMS), pp 8–14
58. Marasco E, Sansone C (2012) Combining perspiration- and morphology-based static features for fingerprint liveness detection. Pattern Recognit Lett 33:1148–1156
59. Cappelli R, Maio D, Maltoni D (2001) Modelling plastic distortion in fingerprint images. In: Proceedings of international conference on advances in pattern recognition (ICAPR), LNCS-2013. Springer, pp 369–376
60. Bazen AM, Gerez SH (2003) Fingerprint matching by thin-plate spline modelling of elastic deformations. Pattern Recognit 36:1859–1867
61. Chen Y, Dass S, Ross A, Jain AK (2005) Fingerprint deformation models using minutiae locations and orientations. In: Proceedings of IEEE workshop on applications of computer vision (WACV), pp 150–156

62. Chen Y, Jain AK (2005) Fingerprint deformation for spoof detection. In: Proceedings of IEEE biometric symposium (BSym), pp 19–21
63. Zhang Y, Tian J, Chen X, Yang X, Shi P (2007) Fake finger detection based on thin-plate spline distortion model. In: Proceedings of IAPR international conference on biometrics, LNCS-4642. Springer, pp 742–749
64. Yau WY, Tran HT, Teoh EK, Wang JG (2007) Fake finger detection by finger color change analysis. In: Proceedings of international conference on biometrics (ICB), LNCS 4642. Springer, pp 888–896D
65. Jia J, Cai L (2007) Fake finger detection based on time-series fingerprint image analysis. In: Proceedings of IEEE international conference on intelligent computing (ICIC), LNCS-4681. Springer, pp 1140–1150
66. Uchida K (2004) Image-based approach to fingerprint acceptability assessment. In: Proceedings of international conference on biometric authentication, LNCS 3072. Springer, pp 194–300
67. Marcialis GL, Roli F, Tidu A (2010) Analysis of fingerprint pores for vitality detection. In: Proceedings of IEEE international conference on pattern recognition (ICPR), pp 1289–1292
68. Memon S, Manivannan N, Balachandran W (2011) Active pore detection for liveness in fingerprint identification system. In: Proceedings of IEEE telecommunications forum (TelFor), pp 619–622
69. Martinsen OG, Clausen S, Nysather JB, Grimmes S (2007) Utilizing characteristic electrical properties of the epidermal skin layers to detect fake fingers in biometric fingerprint systems-a pilot study. IEEE Trans Biomed Eng 54:891–894
70. Moon YS, Chen JS, Chan KC, So K, Woo KC (2005) Wavelet based fingerprint liveness detection. Electron Lett 41
71. Nikam SB, Agarwal S (2009) Feature fusion using Gabor filters and cooccrrence probabilities for fingerprint antispoofing. Int J Intell Syst Technol Appl 7:296–315
72. Nikam SB, Argawal S (2009) Ridgelet-based fake fingerprint detection. Neurocomputing 72:2491–2506
73. Nikam S, Argawal S (2010) Curvelet-based fingerprint anti-spoofing. SIViP 4:75–87
74. Coli P, Marcialis GL, Roli F (2007) Power spectrum-based fingerprint vitality detection. In: Proceedings of IEEE workshop on automatic identification advanced technologies (AutoID), pp 169–173
75. Jin C, Kim H, Elliott S (2007) Liveness detection of fingerprint based on band-selective Fourier spectrum. In: Proceedings of international conference on information security and cryptology (ICISC), LNCS-4817. Springer, pp 168–179
76. Jin S, Bae Y, Maeng H, Lee H (2010) Fakc fingerprint detection based on image analysis. In: Proceedings of SPIE 7536, sensors, cameras, and systems for industrial/scientific applications XI, p 75360C
77. Lee H, Maeng H, Bae Y (2009) Fake finger detection using the fractional Fourier transform. In: Proceedings of biometric ID management and multimodal communication (BioID), LNCS 5707. Springer, pp 318–324
78. Marcialis GL, Coli P, Roli F (2012) Fingerprint liveness detection based on fake finger characteristics. Int J Digit Crime Forensics 4:1–19
79. Coli P, Marcialis GL, Roli F (2007) Vitality detection from fingerprint images: a critical survey. In: Proceedings of international conference on biometrics (ICB), LNCS 4642. Springer, pp 722–731
80. Coli P, Marcialis GL, Roli F (2008) Fingerprint silicon replicas: static and dynamic features for vitality detection using an optical capture device. Int. J Image Graph 495–512
81. Choi H, Kang R, Choi K, Jin ATB, Kim J (2009) Fake-fingerprint detection using multiple static features. Opt Eng 48:047,202
82. Marcialis GL, Lewicke A, Tan B, Coli P, Grimberg D, Congiu A, Tidu A, Roli F, Schuckers S (2009) First international fingerprint liveness detection competition – livdet 2009. In: Proceedings of IAPR international conference on image analysis and processing (ICIAP), LNCS-5716, pp 12–23

83. Yambay D, Ghiani L, Denti P, Marcialis GL, Roli F, Schuckers S (2011) LivDet 2011 - fingerprint liveness detection competition 2011. In: Proceedings of international joint conference on biometrics (IJCB)
84. Nixon KA, Rowe RK (2005) Multispectral fingerprint imaging for spoof detection. In: Proceedings of SPIE 5779, biometric technology for human identification II (BTHI), pp 214–225
85. Rowe RK, Nixon KA, Butler PW (2008) Multispectral fingerprint image acquisition. In: Ratha N, Govindaraju V (eds) Advances in biometrics: sensors, algorithms and systems. Springer, pp 3–23
86. Yau WY, Tran HL, Teoh EK (2008) Fake finger detection using an electrotactile display system. In: Proceedings of international conference on control, automation, robotics and vision (ICARCV), pp 17–20
87. Reddy PV, Kumar A, Rahman SM, Mundra TS (2008) A new antispoofing approach for biometric devices. IEEE Trans Biomed Circuits Syst 2:328–337
88. Baldiserra D, Franco A, Maio D, Maltoni D (2006) Fake fingerprint detection by odor analysis. In: Proceedings of IAPR international conference on biometrics (ICB), LNCS-3832. Springer, pp 265–272
89. Cheng Y, Larin KV (2006) Artificial fingerprint recognition using optical coherence tomography with autocorrelation analysis. Appl Opt 45:9238–9245
90. Manapuram RK, Ghosn M, Larin KV (2006) Identification of artificial fingerprints using optical coherence tomography technique. Asian J Phys 15:15–27
91. Cheng Y, Larin KV (2007) In vivo two- and three-dimensional imaging of artificial and real fingerprints with optical coherence tomography. IEEE Photon Technol Lett 19:1634–1636
92. Larin KV, Cheng Y (2008) Three-dimensional imaging of artificial fingerprint by optical coherence tomography. In: Proceedings of SPIE biometric technology for human identification (BTHI), vol 6944, p 69440M
93. Chang S, Larin KV, Mao Y, Almuhtadi W, Flueraru C (2011) Fingerprint spoof detection using near infrared optical analysis. In: Wang J (ed.) State of the art in biometrics. Intechopen, pp 57–84
94. Nasiri-Avanaki MR, Meadway A, Bradu A, Khoshki RM, Hojjatoleslami A, Podoleanu AG (2011) Anti-spoof reliable biometry of fingerprints using en-face optical coherence tomography. Opt Photon J 1:91–96
95. Rattani A, Poh N, Ross A (2012) Analysis of user-specific score characteristics for spoof biometric attacks. In: Proceedings of of IEEE computer society workshop on biometrics at the international conference on computer vision and pattern recognition (CVPR), pp 124–129
96. Marasco E, Ding Y, Ross A (2012) Combining match scores with liveness values in a fingerprint verification system. In: Proceedings of IEEE international conference on biometrics: theory, applications and systems (BTAS), pp 418–425
97. Hariri M, Shokouhi SB (2011) Possibility of spoof attack against robustness of multibiometric authentication systems. SPIE J Opt Eng 50:079,001
98. Akhtar Z, Fumera G, Marcialis GL, Roli F (2011) Robustness analysis of likelihood ratio score fusion rule for multi-modal biometric systems under spoof attacks. In: Proceedings of IEEE international carnahan conference on security technology (ICSST), pp 237–244
99. Akhtar Z, Fumera G, Marcialis GL, Roli F (2012) Evaluation of serial and parallel multibiometric systems under spoofing attacks. In: Proceedings of international conference on biometrics: theory, applications and systems (BTAS)
100. Ultra-Scan (2012). http://www.ultra-scan.com/
101. Optel (2012). http://www.optel.pl/
102. PosID (2012). http://www.posid.co.uk/
103. VirdiTech (2012). http://www.virditech.com/
104. Kang H, Lee B, Kim H, Shin D, Kim J (2003) A study on performance evaluation of the liveness detection for various fingerprint sensor modules. In: Proceedings of international conference on knowledge-based intelligent information and engineering systems (KES), LNAI-2774. Springer, pp 1245–1253

105. Wang L, El-Maksoud RA, Sasian JM, William Kuhn P, Gee K, VSV (2009) A novel contactless aliveness-testing fingerprint sensor. In: Proceedings of SPIE novel optical systems design and optimization XII, vol 7429, p 742915
106. Chugh T, Cao K, Jain AK (2018) Fingerprint spoof buster: use of minutiae-centered patches. IEEE Trans Inf Forensics Secur 13:2190–2202. https://doi.org/10.1109/TIFS.2018.2812193
107. Park E, Cui X, Nguyen THB, Kim H (2019) Presentation attack detection using a tiny fully convolutional network. IEEE Trans Inf Forensics Secur 14:3016–3025. https://doi.org/10.1109/TIFS.2019.2907184
108. Xia Z, Yuan C, Lv R, Sun X, Xiong NN, Shi Y (2020) A novel weber local binary descriptor for fingerprint liveness detection. IEEE Trans Syst Man Cybernet: Syst 50:1526–1536. https://doi.org/10.1109/TSMC.2018.2874281
109. Agarwal S, Rattani A, Chowdary CR (2021) A comparative study on handcrafted features v/s deep features for open-set fingerprint liveness detection. Pattern Recognit Lett 147:34–40. https://doi.org/10.1016/j.patrec.2021.03.032
110. Jain AK, Deb D, Engelsma JJ (2021) Biometrics: trust, but verify
111. Chugh T, Jain AK (2021) Fingerprint spoof detector generalization. IEEE Trans Inf Forensics Secur 16:42–55. https://doi.org/10.1109/TIFS.2020.2990789
112. Micheletto M, Marcialis GL, Orrù G, Roli F (2021) Fingerprint recognition with embedded presentation attacks detection: are we ready? IEEE Trans Inf Forensics Secur 16:5338–5351. https://doi.org/10.1109/TIFS.2021.3121201
113. Nixon KA, Aimale V, Rowe RK (2008) Spoof detection schemes. In: Jain AK, Flynn P, Ross A (eds) Handbook of biometrics. Springer, pp 403–423

Chapter 2
Vision Transformers for Fingerprint Presentation Attack Detection

Kiran Raja, Raghavendra Ramachandra, Sushma Venkatesh, Marta Gomez-Barrero, Christian Rathgeb, and Christoph Busch

Abstract Automated fingerprint recognition systems, while widely used, are still vulnerable to presentation attacks (PAs). The attacks can employ a wide range of presentation attack species (i.e., artifacts), varying from low-cost artifacts to sophisticated materials. A number of presentation attack detection (PAD) approaches have been specifically designed to detect and counteract presentation attacks on fingerprint systems. In this chapter, we study and analyze the well-employed Convolutional Neural Networks (CNN) with different architectures for fingerprint PAD by providing an extensive analysis of 23 different architectures in CNNs. In addition, this chapter presents a new approach introducing vision transformers for fingerprint PAD and validates it on two different public datasets, LivDet2015 and LivDet2019, used for fingerprint PAD. With the analysis of vision transformer-based F-PAD, this chapter covers both spectrum of CNNs and vision transformers to provide the reader with a one-place reference for understanding the performance of various architectures. Vision transformers provide at par results for the fingerprint PAD compared to CNNs with more extensive training duration suggesting its promising nature. In addition, the chapter presents the results for a partial open-set protocol and a true

K. Raja (✉) · R. Ramachandra (✉) · S. Venkatesh (✉) · C. Busch (✉)
Biometrics Laboratory, Norwegian University of Science and Technology (NTNU), Trondheim, Norway
e-mail: kiran.raja@ntnu.no

R. Ramachandra
e-mail: raghavendra.ramachandra@ntnu.no

C. Busch
e-mail: christoph.busch@ntnu.no

S. Venkatesh · C. Rathgeb (✉)
da/sec Biometrics and Internet Security Research Group, Hochschule Darmstadt, Darmstadt, Germany
e-mail: christian.rathgeb@h-da.de

S. Venkatesh · M. Gomez-Barrero (✉)
Hochschule Ansbach, Ansbach, Germany
e-mail: marta.gomez-barrero@hs-ansbach.de

S. Marcel et al. (eds.), *Handbook of Biometric Anti-Spoofing*, Advances in Computer Vision and Pattern Recognition, https://doi.org/10.1007/978-981-19-5288-3_2

open-set protocol analysis where neither the capture sensor nor the material in the testing set is known at the training phase. With the true open-set protocol analysis, this chapter presents the weakness of both CNN architectures and vision transformers in scaling up to unknown test data, i.e., generalizability challenges.

2.1 Introduction

Automated fingerprint recognition systems, like any other biometric system, are vulnerable to presentation attacks (PAs). The widespread use of fingerprint recognition in unsupervised access control scenarios is often challenged by low-cost presentation attack species (i.e., artefacts). Willis and Lee demonstrated the weakness of fingerprint sensors when presented with artificial fingers on six different fingerprint scanners [1], where four of them accepted the artificial fingers. The former work inspired many of the subsequent studies on investigating the vulnerability of the fingerprint systems [2–5]. Along with studying the vulnerability, necessary defense mechanisms were developed to detect and counteract presentation attacks, popularly termed presentation attack detection (PAD).

With the increased use of deep learning for various tasks, fingerprint PAD (F-PAD) has also progressed from classical machine learning-based approaches to deep learning approaches. This chapter is designed to study and analyze the well-employed Convolutional Neural Networks (CNNs) under different architectures for fingerprint PAD. The chapter presents an extensive analysis of 23 different architectures in CNNs to demonstrate their applicability for F-PAD and analyze their strength and weaknesses. In addition, this chapter aims to provide a comparative study along with introducing vision transformers for F-PAD. Although a number of different approaches and networks have been proposed, we restrain the scope of the chapter to standard off-the-shelf approaches available widely for any reader to reproduce the results.

The analysis consists of three different scenarios where the first analysis is performed using the weights from a pre-trained ImageNet, which are directly employed and fine-tuned for the task of F-PAD under a transfer learning scenario. The second and third analyses examine the ability of all 23 architectures for F-PAD when weights are learnt from scratch using two well-employed activation functions in the last fully connected layer, specifically, Softmax and Sigmoid activations. We include DenseNet121 [6], DenseNet169 [6], DenseNet201 [6], EfficientNetB0 [7], EfficientNetB1 [7], EfficientNetB2 [7], EfficientNetB3 [7], EfficientNetB4 [7], EfficientNetB5 [7], EfficientNetB6 [7], EfficientNetB7 [7], InceptionResNetV2 [8], InceptionV3 [9], MobileNet [10], ResNet101 [11], ResNet101V2 [12], ResNet152 [11], ResNet152V2 [12], ResNet50 [11], ResNet50V2 [12], VGG16 [13], VGG19 [13], and Xception [14] architectures in all the analysis conducted in this work. While the first analysis is targeted to cover several CNN published up to 2020, this chapter also introduces the most recently proposed vision transformers for F-PAD. While CNNs have tedious and computationally intensive training, vision transformers are advocated to be efficient in training and resources. Motivated by such arguments, this chapter also benchmarks their performance for F-PAD for the first time in the literature. With the analysis of vision transformer-based F-PAD, this chapter covers both spectrum of CNNs and vision transformers to provide the reader with a one-place reference for understanding the performance of various architectures. The chapter adopts an unbiased protocol for each network by training them up to 100 epochs for CNNs and 30 epochs for vision transformers considering the previous results indicating the ability of vision transformers to achieve the state-of-the-art results against the CNNs with few epochs of training.

All of the analysis is further conducted on a publicly available challenge set of LivDet 2015 and LivDet 2019 datasets, which has a variety of capture devices and a wide range of artefacts. Further, to demonstrate the advantages and constraints of various networks, this chapter presents experiments in three different settings. The first analysis includes studying the CNN architectures when the training set consists of multiple species of artefacts (i.e., materials used for creating artefacts), which corresponds to a partial open-set protocol. The second analysis assumes that the testing data is partially known while the capture device remains the same. This protocol also presents the partial open-set protocols in line with the previous protocol. Within this analysis, the chapter presents the results obtained on various combinations of sensors available in the LivDet 2019 dataset. Finally, in the third analysis, a true open-set protocol is employed where neither the capture sensor nor the material in the testing set is known at the training phase. The true open-set protocols present the weakness of both CNN architectures and vision transformers in scaling up to unknown test data.

In the rest of the chapter, a set of background works on F-PAD is presented in Sect. 2.2 where details of classical approaches are discussed, and further, a set of works employing deep learning is presented in Sect. 2.3.2. A brief overview of vision transformers and the rationale behind using them for F-PAD is presented in Sect. 2.4. Section 2.5 presents the details of the LivDet 2015 and LivDet 2019 datasets for the convenience of the reader. In Sect. 2.6, the results are analyzed, and a few observations are noted from the obtained results. Finally, Sect. 2.7 provides discussions, conclusions, and future directions for advancing F-PAD to provide a complete overview of F-PAD.

2.2 Related Works

Most of the works proposed for fingerprint PAD can be classified under solutions based on hardware, software, or a combination of both [4, 15, 16]. We provide a brief listing of the existing related works on F-PAD for the benefit of the reader and limit the scope of the chapter not to include a detailed discussion; however, we acknowledge that an extensive set of works can be found on the topic of F-PAD. Readers are referred to the following works for a detailed overview of approaches at different points of time in literature [17–20].

Hardware-based methods either use some additional sensors or different sensing technologies to determine the liveness of the presented fingerprint [21, 22]. The new generation of sensing technologies for detecting PAs in fingerprints includes solutions such as Optical Coherence Tomography (OCT) [23–26] and Multi-Spectral Imaging (MSI) [27, 28]. OCT can capture the subsurface fingerprints and thus can easily detect overlay artefacts, when positioned on the skin surface. The same can be used when detecting fingerprint artefacts that do not have the same internal and external fingerprints. MSI for detecting artefacts uses the signature of images in different spectral bands to classify bona fide fingerprints versus artefacts that typically differ in characteristics. Short Wave Infrared Imaging (SWIR) has been recently proposed detecting the attacks on fingerprint systems [29]. Laser speckle contrast imaging has been proposed as another new sensing technology to detect artefacts efficiently in fingerprint systems [30]. Impedance-based F-PAD has been recently proposed [31]. Additionally, new generation sensors capable of capturing fingerprint and finger-vein simultaneously have been proposed to detect the artefacts in which the finger-vein characteristics are proposed to detect the liveness through blood flow for bona fide presentations [32].

2.3 Software-Based Solutions for F-PAD

Several software-based PAD approaches found can also be classified as static methods and dynamic methods [33]. For instance, dynamic methods for PAD are based on characteristic changes as a result of interaction with the capture device, such as the increase in humidity on a fingertip. Such characteristics have been investigated by Derakhshani et al. [34] who examined the change in perspiration from the pores of the fingertip in a fixed interval of a few seconds to determine the presentation as bona fide or attack. Inspired by this work, a variant of the same was proposed by employing wet saturation percentage change and dry saturation percentage change by Parthasaradhi et al. [35]. At the same time, the deformation of fingerprint as a model of skin elasticity was also proposed as a mode for detecting presentation attack by subsequent works [2, 8, 31]. These approaches inherently use the characteristics of sensors through customized algorithmic processing to classify the artefacts. Similar other approaches include using pulse-oximetry [36] and measuring blood pressure [37], odor [21], and skin-impedance [38] for detecting the F-PAD.

2.3.1 Hand-crafted Texture-Based Solutions for F-PAD

A number of classical hand-crafted feature extraction techniques have been used in different works for F-PAD that include features like Local Binary Patters (LBP), Multi-scale LBP (MSLBP), Uniform-LBP, Local Phase Quantization (LPQ), Local Coherence Pattern (LCP), Binarized Statistical Independent Features (BSIF), Weber

Local Descriptor (WLD), Local Binary Gradient Orientation (LBGO), Speeded-Up Robust Features (SURF), and Scale Invariant Feature Transformation (SIFT) all of which have been used with different classifiers.

Wavelet analysis was employed for PAD by analyzing the surface using wavelet transform [39], and later, LBP on wavelets was proposed for F-PAD [40]. In a similar direction, multi-scale LBP (MSLBP) for F-PAD was proposed considering the complementarity of LBP features from different scales [41]. LPQ was introduced by capturing the effects of blurring in fingerprint acquisition to detect the PAD [42]. BSIF was further introduced for F-PAD, arguing the independent responses obtained from different statistically learnt filters [43, 44]. WLD was proposed for PAD based on the histogram of differential excitation and orientation for bona fide versus artefacts [45], and Weber Local Binary Descriptor (WLBD) was further proposed to take into account the shortcomings of WLD [46]. Coherence of the signals was also employed for detecting liveness in fingerprints by employing LCP where the difference of the dispersion gradient field was used to differentiate bona fide and artefacts [47]. Guided filtering and hybrid image analysis have been explored to detect the F-PAD additionally [48]

In an ensemble approach, multiple textural features such as pore features and residual noise first-order statistics were proposed for F-PAD [49]. SURF, PHOG, and texture features extracted from the Gabor wavelet were proposed for F-PAD making use of low-level features and shape analysis in another ensemble approach [50]. In a similar direction to the ensemble approach, multiple Support Vector Machines (SVM) were trained using LBP and alike features for detecting the artefacts [51]. Yet another ensemble approach using features such as Fisher Vector (FV), Vector of Locally Aggregated Descriptors (VLAD), and Bag of Words (BoW) was proposed for F-PAD [52]. Recently another ensemble approach using LBP, Complete LBP, Local Adaptive Binary Pattern (LABP), and BSIF was jointly used for detecting artefacts in a method referred to as HyFiPAD [17]. In addition to texture-based approaches alone, quality features were explored to detect the artefacts considering the quality of bona fide fingerprint to be different from artefact fingerprint [53, 54].

2.3.2 Deep-Learning-Based Solutions for F-PAD

Multiple works have employed transfer learning to extract the features from fingerprint-based on networks trained on ImageNet dataset [55]. Initially, features from Deep-Convolutional Neural Networks (D-CNN) and LBP were investigated for detecting artefacts [56]. An extension of the previous approach was further proposed using transfer learning by basing the approach on AlexNet and VGG [57]. Transfer learning was further extended by taking random patches for extracting features and making a decision using multiple features extracted from pre-trained networks [58].

A dedicated network with multiple voting was proposed as Finger-Net to detect the F-PAD [59], and later, another network ConvNet was proposed for F-PAD [60]. A lightweight dedicated CNN called SlimResCNN was introduced by expanding the convolutional filters to extract contrastive features from F-PAD [61]. Using the standards D-CNN model based on ResNet and adjusting the learning rate with scale equalization in the fully concatenation layer and the last convolution layer was investigated for F-PAD [62]. Extending the previous work and considering the limitations of Global Average Pooling (GAP), another new dedicated network called FLDNet is introduced based on DenseNet with changes in dense block structure where the residual path was proposed for F-PADspscitezhang2020fldnet. The scores from Slim-ResCNN were further combined with scores from local feature structure extracted by Octantal Nearest-Neighborhood Structure (ONNS) for F-PAD later in somewhat an ensemble manner [63]. A cycle consistency-based De-Folding task is recently proposed for F-PAD specifically capturing the hierarchical features within the fingerprint structure [64].

Siamese networks have been explored for F-PAD by combining CaffeNet and GoogLeNet architectures [65]. Other paradigms such as Deep Belief Networks have been used for F-PAD [66] including the use of triplet embeddings has been explored for F-PAD using custom-made networks [67]. Adversarial learning and Adversarial Representational Learning (ARL)-based F-PAD has been proposed in another set of works [68, 69]. Genetic optimization was investigated to optimize D-CNNs for fingerprint detection using DenseNet architecture [70]. A recent approach using Universal Material Generator (UMG) has been recently proposed using the paradigm of adversarial learning for creating a generalizable F-PAD [64, 71]

2.4 Data-Efficient Image Transformers (DeiT) for F-PAD

The DeiT model [72] extends the idea of Vision Transformer (ViT) [73] that has been shown to outperform existing convolutional neural networks using a Transformer encoder. While the CNNs use convolutions to extract the features, ViTs do not use any convolutions but apply a standard Transformer directly to images. The images in the ViT approach are split into patches and converted into equivalent tokens in Natural Language Processing (NLP). While such an approach extracts the features from patches, the invariance to image translation and scaling is not accounted for as in CNNs. Thus, the first generation of transformers could not generalize well when trained on insufficient amounts of data. This limitation was mitigated by employing larger datasets consisting of millions of images (14M-300M images) [73] which are performed at par with state-of-the-art CNNs. Motivated by such success, ViTs have been well explored for several tasks in image classification.

However, the ViT models require training on expensive infrastructure for a longer duration. DeiT (data-efficient image transformers), on the other hand, are proven to be more efficiently trained transformers for image classification tasks with lesser computing resources and shorter training duration as compared to the original ViT models. Unlike ViTs, DeiTs employ a different training strategy and introduce the distillation token without using the Multi-Layer-Perceptron (MLP) head for the pre-training but only a linear classifier. Knowledge Distillation (KD) refers to the training paradigm in which a student model leverages "soft" labels from a strong teacher network. The output vector of the teacher's softmax function rather provides a "hard" label improving the performance of the student model. A number of variants for DeiT have been proposed using strong teacher models using KD.

DeiTs add a new token obtained through the distillation to the initial embeddings (patches of images and class token). The tokens are used similarly to the class token, where they interact with other embeddings through self-attention, which is further output by the network after the last layer. The DeiTs with distillation employ the target objective as the distillation component of the loss. The distillation embedding thus allows the DeiT model to learn from the output of the teacher, as in a regular distillation, and retains the complementarity to the class embedding [72]. Although a detailed discussion would benefit the reader, we refer the reader to original works on DeiT model [72, 74] and Vision Transformer (ViT) [73] for a detailed discussion considering the scope of this chapter.

Motivated by the remarkable performance of DeiTs [72] using limited data and much lesser computational resources for training, we explore DeiTs for fingerprint PAD. This is the first attempt to use them for F-PAD to the best of our knowledge. Within the scope of this chapter, we restrict ourselves to employing DeiT-small distilled,[1] DeiT-base distilled using an image of size 224×224,[2] and DeiT-base distilled using an image of size 384×384[3] in a transfer learning manner. The choice is specifically made based on performances noted on the ImageNet dataset, where the top performances can be noted. DeiT-small distilled (referred to in the rest of the article as DeiT-small-distilled-224) achieves an accuracy of 81.2% top-1 accuracy on ImageNet, while DeiT-base distilled (referred to in the rest of the article as DeiT-base-distilled-224) obtains 82.9% and DeiT-base distilled with input size of 384×384 (referred to in the rest of the article as DeiT-base-distilled-384) obtains 85.2% accuracy. While the authors report better performance with higher resolution (384×384) in fine-tuning, it is also advocated to increase the model size for achieving better performance [72] depending on the computational resources. Further, for all the analysis in this chapter, we simply resort to patch size of 16×16[4] with DeiT configured to Gaussian Error Linear Unit (GELU) as activation function and the parameters described in Table 2.1 for DeiT-base-distilled-224, DeiT-small-distilled-224, and DeiT-base-distilled-384.

2.5 Databases

We provide brief details of databases employed to evaluate different CNN architectures and DeiTs for analyzing results. First, we present the LivDet 2015 dataset to

[1] Obtained from https://huggingface.co/facebook/deit-small-distilled-patch16-224.

[2] Obtained from https://huggingface.co/facebook/deit-base-distilled-patch16-224.

[3] Obtained from https://huggingface.co/facebook/deit-base-distilled-patch16-384.

[4] Code will be made publicly available for readers to reproduce the results.

Table 2.1 Parameter configuration in DeiT for DeiT-base-distilled-224, DeiT-small-distilled-224 and DeiT-base-distilled-384 employed in this chapter

Parameter	DeiT-base-distilled-224	DeiT-small-distilled-224	DeiT-base-distilled-384
Hidden activation	GELU	GELU	GELU
Hidden dropout probability	0.0	0.0	0.0
Hidden size	384	384	768
Image size	224	224	384
Initializer range	0.02	0.02	0.02
Intermediate size	3072	1536	3072
Layer normalization factor	1e-12	1e-12	1e-12
Number of attention head	6	6	12
Number of hidden layers	12	12	12
Patch size	16	16	16

illustrate the effectiveness of various deep learning networks and further provide a comparative evaluation against proposed vision transformers in a closed-set protocol. Similarly, we also provide a detailed analysis on the LivDet 2019 dataset for both deep learning paradigms in closed-set protocols. Further, to complement the detailed study, we also present the experiments in open-set settings where the artefact species of the test set is completely unknown for various combinations as detailed further.

2.5.1 LivDet 2015 Database

LivDet 2015 dataset [20] is captured using four optical devices such as Green Bit, Biometrika, Digital Persona, and CrossMatch. The details of the capture devices used in the dataset are presented in Table 2.2 and the choice of the dataset is based on the following factors: (1) availability of both disjoint training and testing set, (2) availability of corresponding data captured from multiple capture devices, (3) varying resolution of sensors and capture conditions, and (4) the testing set involving presentation artefacts from unknown materials to study generalization capabilities.

Table 2.2 Fingerprint capture devices and resolution of images captures

Scanner	Model	Resolution [dpi]	Image Size [px]	Format
Green bit	DactyScan26	500	500 × 500	PNG
Biometrika	HiScan-PRO	1000	1000 × 1000	BMP
Digital persona	U.are.U 5160	500	252 × 324	PNG
CrossMatch	L Scan Guardian	500	640 × 480	BMP

Table 2.3 Details of LivDet 2015 dataset. * indicates unknown artefact species

Dataset	Bona fide	Presentation attack species					
		Ecoflex	Gelatine	Latex	WoodGlue	Liquid ecoflex	RTV
Known training and testing artefact species							
Green Bit	1000	250	250	250	250	250*	250*
Biometrika	1000	250	250	250	250	250*	250*
Digital persona	1000	250	250	250	250	250*	250*
CrossMatch	1500	270	–	–	–	–	–
Unknown testing artefact species							
				Gelatine	Body double	Playdoh	OOMOO
CrossMatch	–	–	–	300	300	281	297

2.5.2 *LivDet 2019 Database*

Bona fide images are captured in multiple acquisitions of all fingers from different subjects. Bona fide acquisitions further include normal mode, with wet and dry fingers and with high and low pressure mimicking real-life data acquisition scenarios. The presentation attack artefacts were further collected cooperatively with different types of species that include Ecoflex, gelatine, latex, wood glue, a liquid Ecoflex, and RTV (a two-component silicone rubber) as listed in Table 2.3. Unlike the known training and testing sets captured from Green Bit, the Biometrika, and the Digital Persona datasets, the dataset captured from CrossMatch capture device includes Playdoh, Body Double, Ecoflex, OOMOO (a silicone rubber), and a novel form of Gelatine from which the artefacts are captured.

We further adapt the original protocols of the LivDet 2015 dataset evaluation, where the original dataset is divided into two parts by using images from different subjects: a training set and a testing set. The training set is further subdivided into 80% for training and 20% for validation of the training. The testing sets include both known and unknown artefacts species, i.e., materials not included in the training set. The unknown materials are liquid Ecoflex and RTV for Green Bit, Biometrika, and Digital Persona datasets, and OOMOO and Gelatine for Crossmatch dataset.

Table 2.4 Device characteristics for LivDet 2019 datasets

Scanner	Model	Resolution [dpi]	Image Size [px]	Format	Type
Green bit	DactyScan84C	500	500×500	BMP	Optical
Orcanthus	Certis2 Image	500	$300 \times n$	PNG	Thermal swipe
Digital persona	U.are.U 5160	500	252×324	PNG	Optical

Table 2.5 Composition of the LivDet 2019 dataset

	Train						Test			
Dataset	Live	Wood Glue	Ecoflex	Body Double	Latex	Gelatine	Live	Mix 1	Mix 2	Liquid Ecoflex
Green bit	1000	400	400	400	–	–	1020	408	408	408
Orcanthus	1000	400	400	400	–	–	990	384	308	396
Digital persona	1000	250	250	–	250	250	1019	408	408	408

This practice has been adopted to assess the reliability of algorithms under attack by unknown materials.

2.5.3 *LivDet 2019 Database*

LivDet 2019 consists of sub-sets of data captured using Orcanthus Certis2 device and Green Bit DactyScan84C device from LivDet 2017 and in addition to Digital Persona U.are.U 5160 images from LivDet 2015 train set. The characteristics of different capture devices are provided in Table 2.4. Further, the number of images captured for training and testing set in LivDet 2019 is also presented in Table 2.5 and a set of bona fide samples along with artefacts of different species are presented in Fig. 2.1.

The LivDet 2019 presentation artefacts were collected using the cooperative method. While bona fide images correspond to multiple acquisitions of at least six fingers of different subjects, the artefacts are both cooperative and noncooperative. Mold is created through user collaboration in cooperative mode and the same is created using latent fingerprint in noncooperative mode.

However, it has to be noted that the artefacts in the test set in LivDet 2019 were created using materials different from those in the train set. This set consists of 6400 images, while the test set contains 6565 images. The number of samples for each scanner is shown in Table 2.5. In contrast to the LivDet 2015 dataset, two new spoof material mixes were employed in LivDet 2019 providing a fully unknown/unseen/never-seen-before material at the training phase to test the capability of the PAD algorithm.

We choose to present the results on LivDet 2019 to provide an analysis on the performance of the algorithms for various acquisition surfaces and images. As images

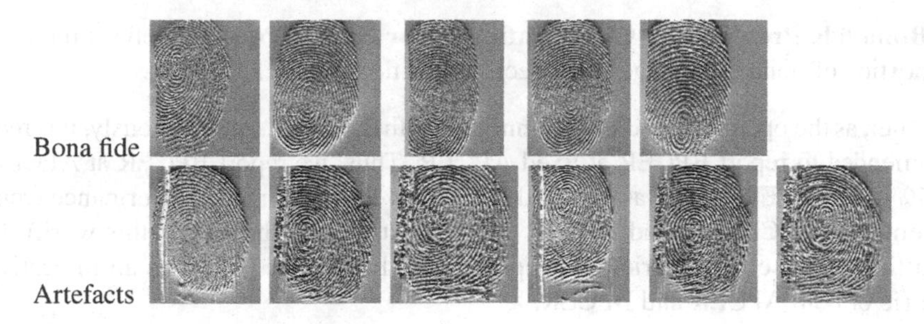

Fig. 2.1 Sample images captured from Orcanthus capture device. The top row presents data captured in normal settings referred to as Normal-I, Normal-II, Normal-III, Normal-IV, and Normal-V from different sessions while the bottom row presents the artefacts created using Latex, Liquid Ecoflex, and Mix in two different sessions in the order of mentioning

from different sensors exhibit different characteristics. Such change in characteristics is expected to impact the generalization of PAD. Further, the dataset is created by different persons reflecting a real-life attack scenario where each attacker can have different skills. Testing the PAD algorithms on the unknown class of attacks further can provide insights into the generalization of the algorithms.

2.6 Experiments and Results

We present the results on two different datasets mentioned previously. We first report the results when multiple artefact species (classes) are known (combined set) during training, and a few unknown artefacts classes are present in the testing. We then evaluate the impact of limited seen classes during training and device interoperability using a combination of the various training set and testing set under different combinations. We employ 23 different CNN architectures such as DenseNet121, DenseNet169, DenseNet201 EfficientNetB0, EfficientNetB1, EfficientNetB2, EfficientNetB3, EfficientNetB4, EfficientNetB5, EfficientNetB6, EfficientNetB7, InceptionResNetV2, InceptionV3, MobileNet, ResNet101, ResNet101V2, ResNet152, ResNet152V2, ResNet50, ResNet50V2, VGG16, VGG19, and Xception. We study the applicability of the networks in three different settings such as transfer learning by using the weights of the networks learnt on ImageNet dataset (reported as Transfer Learning), training the network from scratch using Softmax activation, and training the network from scratch using Sigmoid activation. Further, we present a comparison against results obtained from three variants of DeiT. All the results are presented in terms of Attack Presentation Classification Error Rate (APCER) and Bona fide Presentation Classification Error Rate (BPCER) as defined in the International Standard ISO/IEC 30107-3 [75] where

- **Attack Presentation Classification Error Rate (APCER)**: Defines the proportion of attack samples incorrectly classified as bona fide face images.

- **Bona fide Presentation Classification Error Rate (BPCER)**: Defines the proportion of bona fide images incorrectly classified as attack samples.

Further, as the operational settings cannot fix both the errors simultaneously, it is recommended to report BPCER at fixed APCER. Thus, we report BPCER at APCER = 5% and APCER = 10% as it is widely employed to report the performance (represented as BPCER_20 and BPCER_10, respectively, in the rest of this work). In addition to these two metrics, we report Equal Error Rate (EER) as an indicative metric of both APCER and BPCER.

2.6.1 Evaluation on LivDet 2015 Dataset with Multiple Known Classes (Combined Set) Training and Few Unknown Classes in Testing

We first present the results obtained on the LivDet 2015 dataset, where the training is conducted on the training set provided by the authors. The training set consists of bona fide images captured by Green Bit, Biometrika, Digital Persona, and CrossMatch and, correspondingly, the artefacts created using Ecoflex Gelatine Latex and WoodGlue. Unlike the training, the set consists of images from Liquid Ecoflex and RTV. As noted from Table 2.6, the networks tend to achieve better results as compared to transfer learning in most of the cases. Finally, we make a certain set of observations from this set of experiments:

- The EER decreases for all the networks when Softmax is employed as the classification layer's activation function. While similar trends can be observed for Sigmoid activation function, a small drop in performance can be observed (Fig. 2.2).
- Further, training the networks with Sigmoid activation decreases average EER in most cases except ResNet50, ResNet50V2, VGG16, and VGG19, where the EER tends to increase both for the trained network from scratch using Softmax and even the transfer learning approach.
- We further note that the EfficientNet architectures provide lower error rates than the rest of the architectures, with EfficientNetB4 providing the lowest EER (5.70% with Softmax, 6.21% with Sigmoid activation) and lowest BPCER at APCER=5% (6.18% with Softmax, 6.85% with Sigmoid activation).
- When the decisions from various networks are fused, one can note the drop in error rates and a significant drop in BPCER for both APCER = 5% and APCER = 10%. Choosing Top-5 and Top-10 performing networks based on the lowest BPCER at APCER=5% further decreases the EER rates but provides slightly varying BPCER.
- Fusion of scores from various networks tends to improve the results for fusion from all networks, top 5 and top 10 networks, and decreases the error rates in all training cases from scratch or transfer learning setting. Such a performance gain indicates the complementarity of the networks in making the decisions (Fig. 2.3).

Table 2.6 Results on LivDet 2015. Trained on the combined dataset and tested on a few unknown artefacts in testing set

	Transfer learning			Retrained—softmax			Retrained—sigmoid		
Network	EER	BPCER_20	BPCER_10	EER	BPCER_20	BPCER_10	EER	BPCER_20	BPCER_10
DenseNet121	15.03	36.22	22.80	6.76	8.25	4.92	6.96	9.74	5.47
DenseNet169	13.46	33.90	19.18	8.61	15.84	6.79	6.78	8.92	5.11
DenseNet201	13.24	33.50	17.93	6.73	8.92	5.12	6.99	8.83	5.25
EfficientNetB0	**7.80**	**12.34**	**5.96**	6.47	7.99	4.32	6.79	8.19	5.31
EfficientNetB1	8.77	15.04	7.68	6.85	9.21	4.56	6.87	8.25	5.01
EfficientNetB2	8.49	13.88	7.03	7.23	8.83	6.14	6.94	8.34	4.96
EfficientNetB3	9.23	14.73	8.03	5.78	6.27	**3.53**	6.81	8.92	4.84
EfficientNetB4	8.81	14.68	7.83	**5.69**	**6.18**	4.14	**6.20**	**6.85**	4.96
EfficientNetB5	9.37	15.62	8.85	6.92	8.43	5.27	6.61	8.23	4.42
EfficientNetB6	9.30	16.00	8.43	7.07	9.66	4.23	8.61	13.08	7.67
EfficientNetB7	9.64	17.35	9.10	13.03	28.08	16.42	6.37	8.05	4.62
InceptionResNetV2	39.35	61.78	61.78	7.60	11.70	5.96	6.26	6.99	4.24
InceptionV3	22.01	62.58	44.92	8.73	16.57	7.36	6.50	8.36	**4.18**
MobileNet	15.22	43.52	24.20	6.05	6.85	4.32	6.18	7.39	4.71
ResNet101	9.63	19.35	9.23	7.16	9.39	5.70	7.45	10.39	6.27
ResNet101V2	17.53	21.56	21.56	6.25	7.72	3.89	8.04	12.15	6.76
ResNet152	9.65	16.82	9.30	6.79	9.50	4.09	10.59	21.60	11.34
ResNet152V2	17.77	15.57	15.57	12.41	24.45	15.17	11.52	24.67	13.37
ResNet50	8.88	14.59	7.92	13.53	28.16	17.95	10.97	18.89	11.61
ResNet50V2	15.39	33.84	23.52	12.59	25.94	16.51	13.21	27.96	17.08
VGG16	12.22	28.10	14.93	13.67	27.14	17.02	10.06	16.55	10.07
VGG19	12.59	27.52	15.19	7.68	12.24	5.76	10.37	19.60	10.67
Xception	17.15	51.81	30.97	9.33	15.71	8.61	9.33	15.77	8.92
Fusion	6.34	7.68	4.32	5.72	6.30	3.18	5.72	6.36	3.53
Fusion—top 5	6.20	7.19	4.45	4.79	4.65	2.40	4.98	4.96	2.89
Fusion—top 10	6.18	7.30	4.23	5.03	5.09	2.54	5.14	5.34	2.73

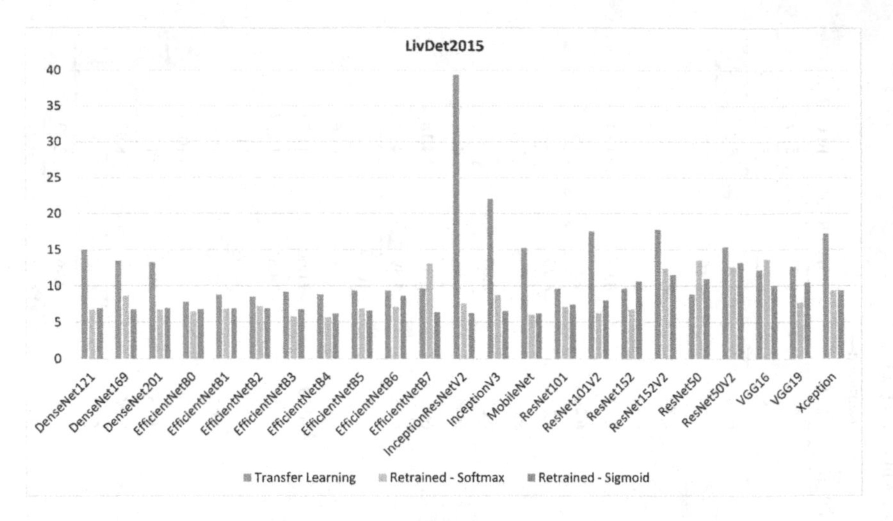

Fig. 2.2 Decrease in EER as a result of training from scratch on LivDet 2015 dataset

Table 2.7 Results on LivDet 2015 dataset using data-efficient transformers in three different configuration

Configuration	EER	BPCER_20	BPCER_10
DeiT-base-distilled-224	4.82	4.69	2.55
DeiT-small-distilled-224	6.77	8.38	4.84
DeiT-base-distilled-384	**4.79**	**4.53**	**2.06**

We further evaluate the proposed approach using vision transformers for fingerprint PAD, and the results are presented in Table 2.7. As it can be noted, all the three configurations employed in this chapter compete against state-of-the-art CNNs presented in Table 2.7. We note a gain of over 2% in EER in two of the three chosen configurations and a lower BPCER at APCER at 5 and 10%. Specifically, the lowest BPCER at APCER = 5% of 4.54% obtained using DeiT-base-distilled-384 configuration is 1.64% lower than the best performing CNN of EfficientNetB4 architecture. The promising results indicate the applicability of DeiT for F-PAD, albeit small limitation in performance. The drop in performance can be argued due to the data itself, which is not well aligned according to the occurrence of minutia. We hypothesize this limitation can be easily addressed in future works to achieve near ideal performance, without computational overhead as seen in training CNNs. As it can be noted from Detection Error Trade-off (DET) curve in Fig. 2.3, the APCER and BPCER simultaneously decrease for DeiT-based PAD as against other networks indicating the promising nature of DeiT-based F-PAD.

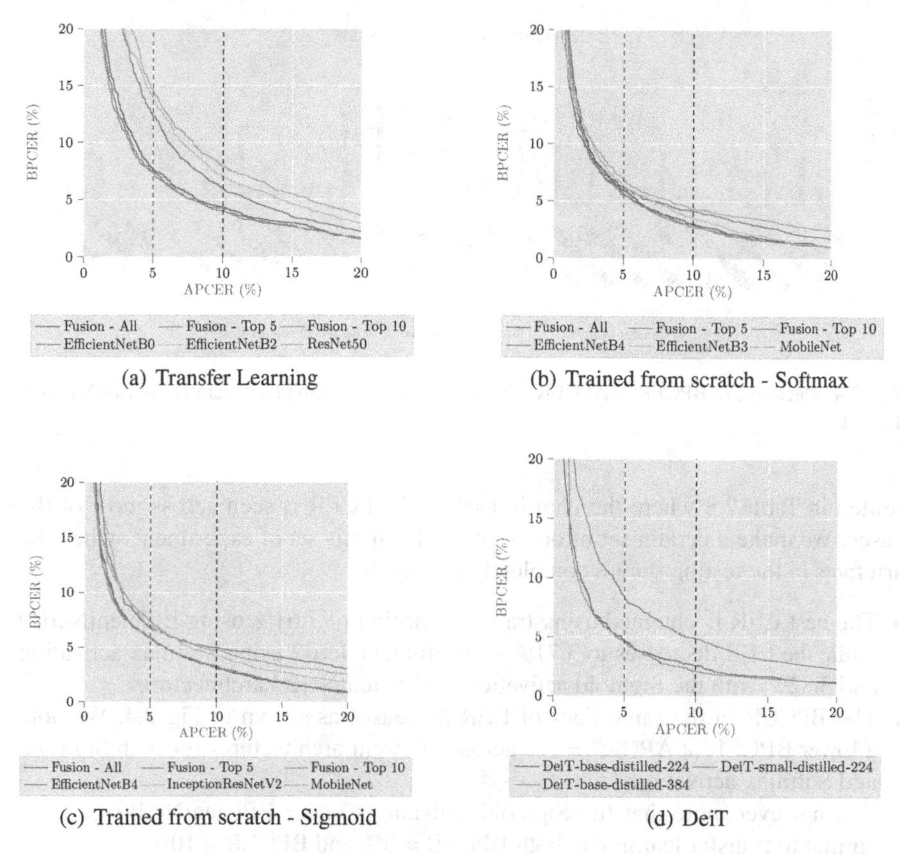

(a) Transfer Learning

(b) Trained from scratch - Softmax

(c) Trained from scratch - Sigmoid

(d) DeiT

Fig. 2.3 DET curves obtained on LivDet 2015 dataset

2.6.2 Evaluation on LivDet 2019 Dataset with Multiple Known Classes (combined Set) in Training and All Unknown Classes in Testing

In the lines of the previous set of experiments, we evaluate 23 different CNN architectures on the LivDet 2019 dataset. The training dataset includes multiple known artefacts such as Live, Wood Glue, Ecoflex, Body Double, Latex, and Gelatine. At the same time, the test images consist of live images and the artefacts referred to as Mix 1, Mix 2, and Liquid Ecoflex. Thus, the experiments correspond to true open-set experiments except for the live images that correspond in both training and testing sets. Similar to the previous set of experiments, we study the applicability of various architectures in three different settings of transfer learning by using the weights of the networks learnt on the ImageNet dataset (reported as Transfer Learning), training the network from scratch using Softmax activation, and training the network from scratch using Sigmoid activation. The results from this set of experiments are pre-

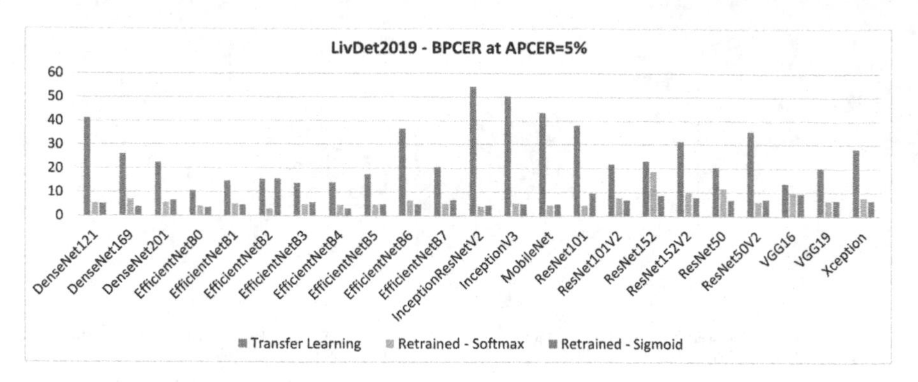

Fig. 2.4 Decrease in BPCER at APCER = 5% as a result of training from scratch on LivDet 2019 dataset

sented in Table 2.8 where the gain in EER and BPCER is seen across most of the cases. We make a certain set of observations from this set of experiments when the artefacts in the testing set are completely unknown:

- The best EER is obtained using transfer learning of 7.61% using EfficientNetB0 while the EER decreases to 3.71% with EfficientNetB2 using Softmax activation and 3.82% with the Sigmoid activation in EfficientNetB4 architecture.
- The BPCER in the same lines of EER decreases as shown in Fig. 2.4. We note a lower BPCER at APCER = 5% across different architectures for both Sigmoid and Softmax activations.
- We, however, note that the Sigmoid activations under EfficientNetB2 perform similar to transfer learning at both BPCER = 5% and BPCER = 10%.
- Unlike the results obtained on the LivDet 2015 dataset, the results on the LivDet 2019 dataset tend to improve by a large margin when trained from scratch, indicating the ability of networks to deal with unknown artefacts.
- In the case of fusion, the decrease in error rates can be clearly seen from Table 2.8 providing better results both when scores are fused from all networks or top-performing networks (5 and 10). While the fused networks decrease the EER to 5.22% for transfer learning, the EER goes further down to 2.84 and 1.98% when trained from scratch using Softmax and Sigmoid, respectively.
- The results indicate the benefit of training the networks for the dedicated task of F-PAD irrespective of the architecture as seen with other tasks in computer vision applications.

The same dataset, when evaluated with the approach using vision transformers, performs better than the state-of-the-art CNNs as reported in Table 2.9. The lowest EER of 3.26% obtained is better than the best performing CNN of EfficientNetB2 with 3.71%. While the improvement seems marginal, it must be noted that the EER is significant at a 95% confidence interval computed in a bootstrapped manner. Further, a clear improvement at BPCER at APCER = 5% and APCER = 10% can be seen for base-patch16-384 configuration, indicating the promising nature of using DeiT

Table 2.8 Results on LivDet 2019 dataset. Combined dataset

	Transfer learning			Retrained—softmax			Retrained—sigmoid		
Network	EER	BPCER_20	BPCER_10	EER	BPCER_20	BPCER_10	EER	BPCER_20	BPCER_10
DenseNet121	16.89	41.17	28.00	5.11	5.42	3.20	5.03	5.12	3.17
DenseNet169	12.91	26.08	16.44	5.87	6.94	3.80	4.43	3.80	2.18
DenseNet201	11.92	22.46	13.71	5.45	5.65	2.94	5.72	6.51	3.93
EfficientNetB0	**7.61**	**10.49**	**5.57**	4.39	4.10	2.51	4.24	3.53	2.31
EfficientNetB1	7.99	14.63	6.11	5.03	5.05	3.20	4.77	4.49	2.87
EfficientNetB2	8.78	15.32	7.89	**3.71**	**3.03**	**1.81**	8.75	15.52	7.16
EfficientNetB3	8.48	13.77	7.07	4.81	4.85	2.77	5.41	5.68	3.66
EfficientNetB4	8.82	14.00	7.69	4.66	4.59	2.54	**3.82**	**3.10**	**1.91**
EfficientNetB5	9.13	17.40	8.39	4.88	4.66	2.67	4.85	4.82	2.54
EfficientNetB6	13.10	36.51	19.35	5.76	6.44	4.13	4.92	4.82	2.51
EfficientNetB7	9.99	20.38	9.97	5.03	5.12	2.71	5.72	6.61	3.63
InceptionResNetV2	33.74	54.34	54.34	4.28	3.99	2.21	4.66	4.33	2.41
InceptionV3	21.04	50.30	37.12	5.07	5.32	3.24	4.96	4.92	2.54
MobileNet	18.10	43.35	28.66	4.73	4.52	2.74	4.96	4.92	2.74
ResNet101	15.56	38.03	23.97	4.51	4.46	3.07	8.10	9.61	7.10
ResNet101V2	10.75	21.80	11.46	6.32	7.56	4.42	5.94	6.64	4.06
ResNet152	11.32	22.99	12.81	11.05	18.76	11.62	7.04	8.49	5.71
ResNet152V2	12.30	31.24	15.39	7.42	10.07	5.22	6.25	7.73	3.70
ResNet50	20.86	20.44	20.44	8.36	11.59	7.69	5.98	6.67	4.26
ResNet50V2	13.40	35.47	19.35	5.41	5.91	3.04	6.21	6.94	3.27
VGG16	36.08	13.63	13.63	7.38	9.74	5.55	6.81	9.38	4.79
VGG19	10.53	20.14	11.03	5.64	6.41	2.87	5.87	6.51	4.00
Xception	15.52	28.24	22.75	6.36	7.76	3.93	5.49	6.44	3.70
Fusion	6.37	7.96	3.93	3.26	2.58	1.85	3.07	2.21	1.65
Fusion—top 5	6.10	6.93	3.73	3.14	2.48	1.55	2.71	1.98	1.35
Fusion—top 10	5.22	5.38	3.00	2.84	2.01	1.29	2.61	1.85	1.22

Table 2.9 Results on LivDet 2019 dataset—DeiT vision transformer

Network	EER	BPCER_20	BPCER_10
DeiT-base-distilled-224	4.08	3.33	1.94
DeiT-small-distilled-224	3.74	3.07	1.84
DeiT-base-distilled-384	**3.25**	**2.37**	**1.55**

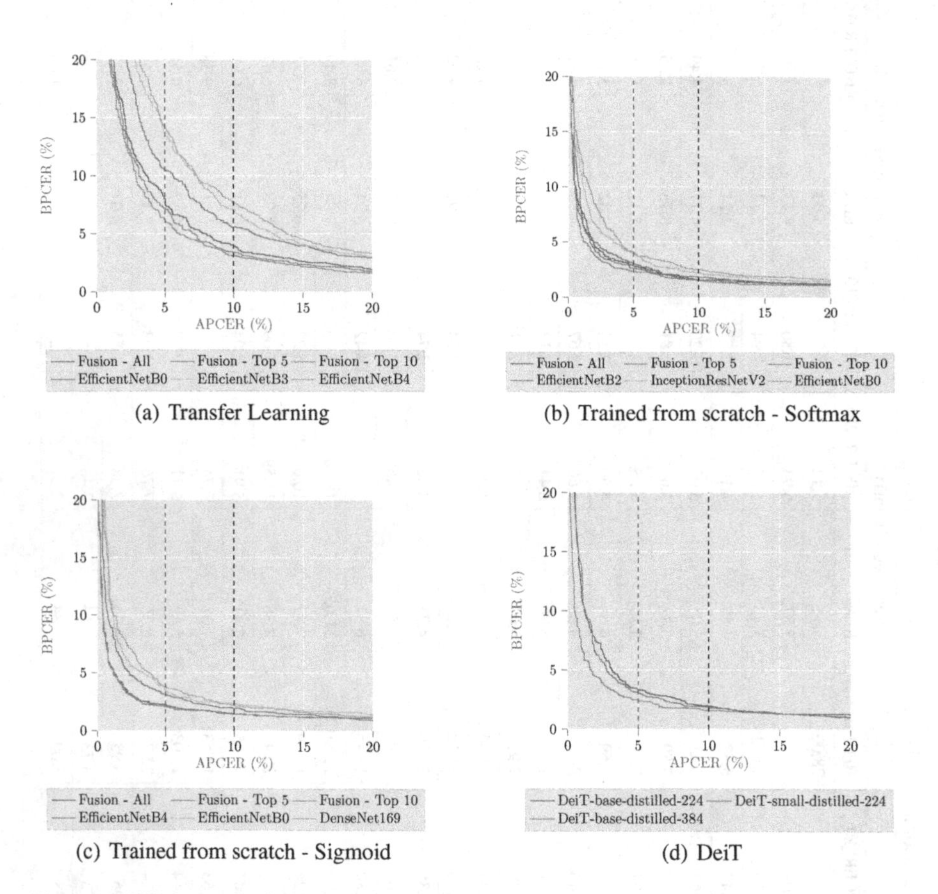

Fig. 2.5 DET curves obtained on LivDet 2019 dataset

for future works. However, it has to be noted that the performance of the DeiT (all configurations) is slightly lower than the fusion approach, nonetheless limited complexity as against training multiple networks for obtaining limited performance gain. Further, it can be noted that the APCER and BPCER simultaneously decrease for DeiT-based PAD as against other networks as shown in DET curve in Fig. 2.5.

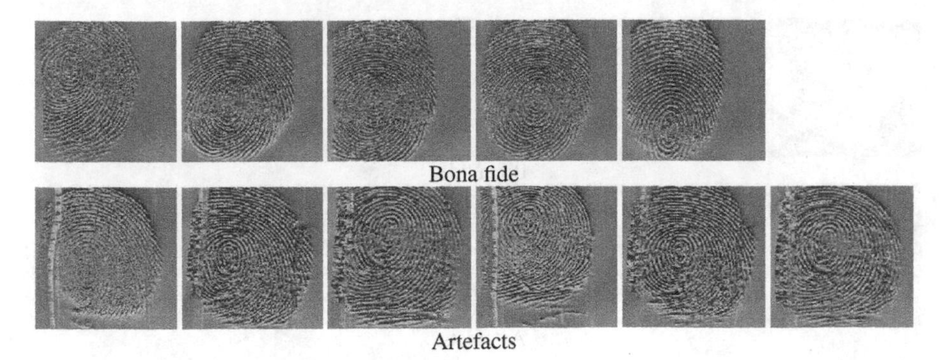

Fig. 2.6 LivDet 2019—visualization of explainable features obtained using EfficientNetB4 in transfer learning setting trained on combined set on images shown in Fig. 2.1. One can note from the figure that the attention of the network appears constant across the image correlating to the low performance of the network noted in Table 2.8

2.6.3 Analysis of Explainability

Intrigued by the performance of different networks and DeiT, we also analyze to visualize the regions focused by images. Due to the limitations of the page, we present the analysis using EfficientNetB4 architecture for the sample images noted in Fig. 2.1 when the combined set is used for the training. The visualization of features is considered by EfficientNetB4 in transfer learning setting and training from scratch using Softmax and Sigmoid activation in Figs. 2.6, 2.7, and 2.8, respectively. All the visualizations are presented by overlaying the features using a color-map indicating different attention of image regions. While the attention appears to be uniform across the image when transfer learning is employed as seen in Fig. 2.6, one can notice the different attention across different regions of fingerprint images in determining the artefacts against bona fide in networks trained from scratch as noted in Figs. 2.7 and 2.8.

In a similar manner, we also analyze the attention for DeiT in all three configurations. While there are 12 layers available in the configurations employed in this work, we choose the penultimate layer (Layer-11) to present our analysis as shown in Figs. 2.9, 2.10, and 2.11 corresponding to DeiT-base-distilled-224, DeiT-small-distilled-224, and DeiT-base-distilled-384, respectively. One can note that the best performing configuration of DeiT-base-distilled-384 has attention spread across different regions in the fingerprint image.

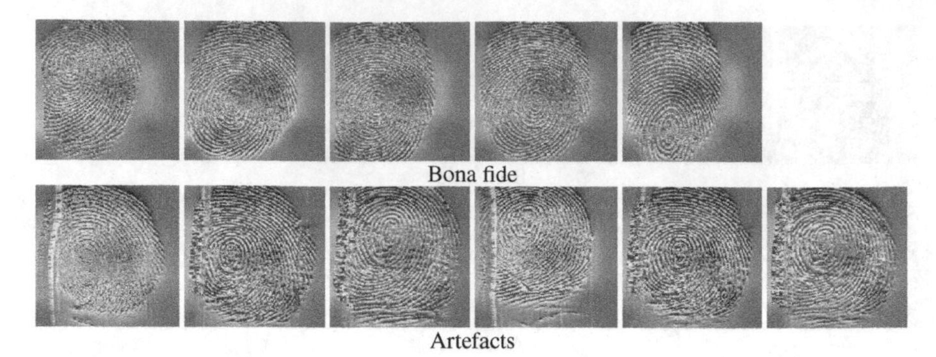

Fig. 2.7 LivDet 2019—visualization of explainable features obtained using EfficientNetB4 trained on combined set on images shown in Fig. 2.1 (Softmax activation). Clear variation of the attention of the network can be seen; however, it must be noted that the network does not capture all visible artefacts from attack images leading to minor performance degradations as seen in Table 2.8

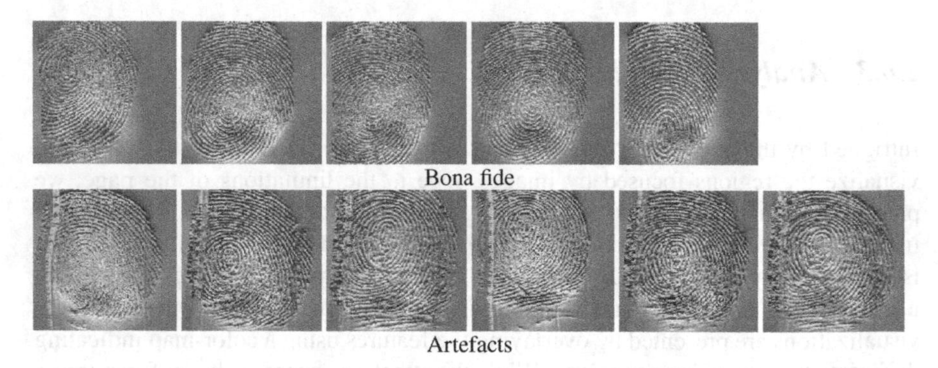

Fig. 2.8 LivDet 2019—visualization of explainable features obtained using EfficientNetB4 trained on combined set on images shown in Fig. 2.1 (Sigmoid activation). The attention of the network can be seen to differ across the image but obvious artefacts in the images in the bottom row are not well captured (left-most image for instance). A potential hypothesis explaining such a behavior is the presence of similar lines appearing in bona fide capture due to degradation of the sensor across captures making the network to ignore artefacts

2.6.4 LivDet 2019—Impact of Limited Seen Classes During Training and Sensor Interoperability

We further investigate the robustness and applicability of various networks against the proposed approach using vision transformers when the training set consists of limited artefacts. In order to study these, we perform two sets of experiments where the first set of experiments consists of training and testing set images captured from the same sensor and the second set of experiments consists of bona fide and artefacts captured from two different sensors. We conduct the same analysis as in previous sec-

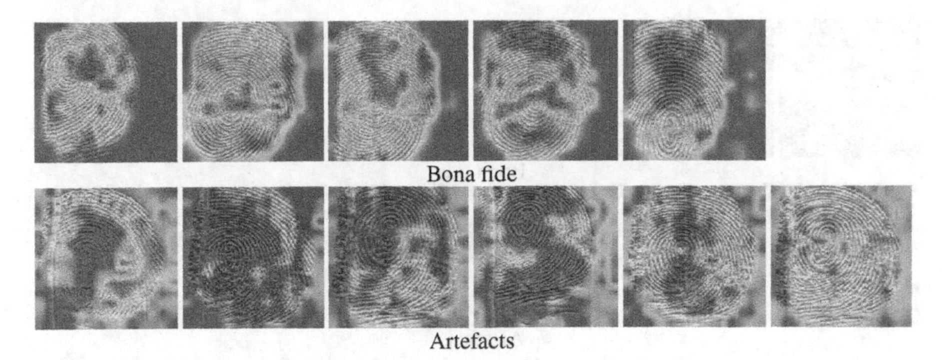

Fig. 2.9 LivDet 2019—visualization of explainable features obtained using DeiT-base-distilled-224 (Layer-10) trained on combined set on images shown in Fig. 2.1. Compared to the attention of other networks, DeiTs appear to capture artefacts to a better extent and this can be evidently seen by comparing the results shown in top row (bona fide) versus bottom row (artefacts)

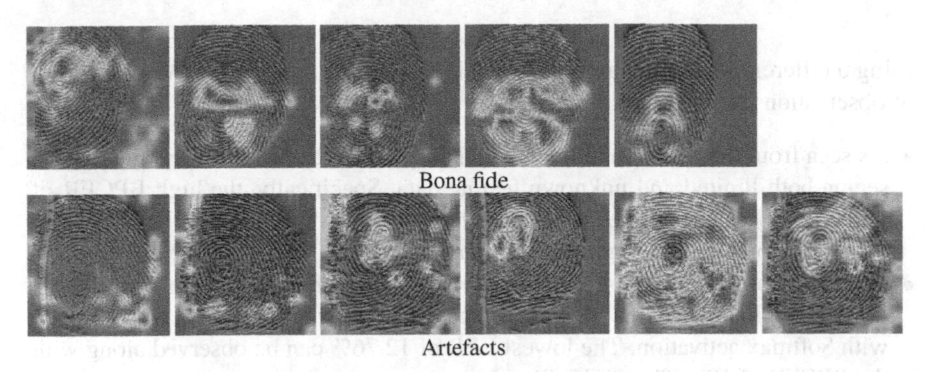

Fig. 2.10 LivDet 2019—visualization of explainable features obtained using DeiT-small-distilled-224 (Layer-10) trained on combined set on images shown in Fig. 2.1. Compared to attention of DeiT-base-distilled-224 (Layer-10) in Fig. 2.9, DeiT-small-distilled-224 (Layer-10) appears to focus on other regions providing an indication of complementarity of DeiT configurations

tions of studying transfer learning, training from scratch with Softmax and Sigmoid activation.

2.6.4.1 Livdet 2019—Unknown Testing Set

In the next section, we present the results obtained under an unknown testing set before presenting the results that correspond purely to device interoperability. We first train the networks with bona fides and artefacts captured with DigitalPersona and tested with artefacts captured using the same capture device (i.e., DigitalPersona) in Table 2.10. To recap, it must be noted that the artefacts in LivDet 2019 are created

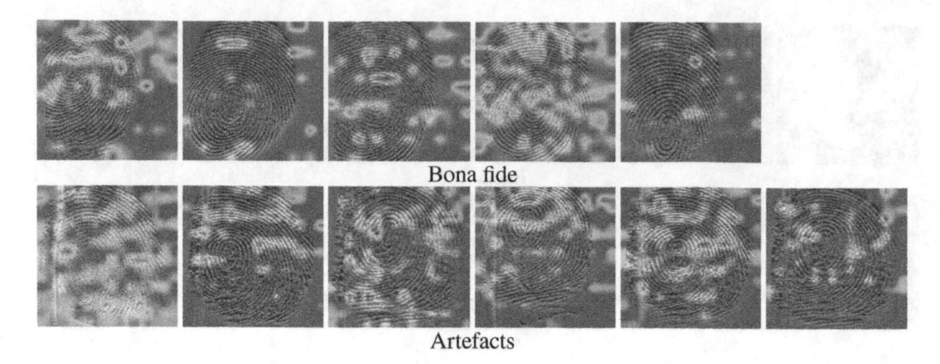

Fig. 2.11 LivDet 2019—visualization of explainable features obtained using DeiT-base-distilled-384 (Layer-10) trained on combined set on images shown in Fig. 2.1. Compared to attention of DeiT-base-distilled-224 (Layer-10) in Figs. 2.9 and 2.10, the DeiT configuration extracts fully complementary features justifying the low EER and providing new approach in future works

using a different material than the materials used in the training set. We make a set of observations as noted below:

- As seen from the results in Table 2.10, the networks tend to overfit in the case of seeing both limited and unknown testing data. Specifically, the high BPCER of 100% in the case of ResNet is a clear indication of this fact, and similar trends can be seen despite the training from scratch for ResNet101v2 and ResNet152.
- In line with previous results, one can note the effectiveness of EfficientNet architectures generalizing to a better extent as against all other architectures when trained with Softmax activation. The lowest EER of 12.76% can be observed along with the BPCER of 19.05% at APCER = 5%.
- In sharp contrast to previous observations of ResNet101v2 failing with Softmax, the network trained from scratch with Sigmoid activation results in a reasonable BPCER of 18.06% at APCER = 5%. These results indicate the complementarity of the networks in making a decision that can provide a basis for the fusion of networks trained with Sigmoid and Softmax activations.
- It has to be, however, noted with caution that the training with Sigmoid activation increases the error rates generally as against the Softmax-based activation.

While Table 2.10 indicates the clear challenge of existing networks in failing to detect unknown attacks captured from DigitalPersona capture device, Table 2.11 presents the same results obtained using DeiT. As it can be noted, the proposed approach using DeiT obtains the lowest BPCER at the given APCER and EER, indicating the promising directions for future works using vision transformer-based approaches for detecting unknown attacks. Further, the networks provide inconsistent results, with some networks providing lower EER but higher BPCER in corresponding settings, and one would need to identify the networks for different operating thresholds (for instance, APCER = 5%); DeiT provides consistently similar results under one configuration, specifically DeiT-base-distilled-384 with EER = 11.55%,

Table 2.10 Results on LivDet 2019 dataset—train data from Digital Persona and test data from Digital Persona. *Note—artefacts in training and testing are from different materials and different creation processes

	Transfer learning			Retrained—softmax			Retrained—sigmoid		
Network	EER	BPCER_20	BPCER_10	EER	BPCER_20	BPCER_10	EER	BPCER_20	BPCER_10
DenseNet121	31.99	66.63	55.35	13.68	25.12	16.19	19.15	43.57	31.01
DenseNet169	29.44	73.11	56.82	20.61	45.34	35.43	18.24	38.08	27.58
DenseNet201	25.53	67.81	52.70	23.55	48.58	39.94	18.54	38.57	28.46
EfficientNetB0	16.68	34.15	23.55	24.92	62.22	44.75	15.70	35.53	25.71
EfficientNetB1	17.33	45.93	29.83	16.00	38.47	24.44	16.41	41.51	26.10
EfficientNetB2	16.49	**32.67**	24.93	17.86	28.95	23.16	17.96	39.55	31.50
EfficientNetB3	22.28	55.35	40.63	17.47	42.79	31.70	26.69	72.13	53.09
EfficientNetB4	**15.21**	35.23	23.85	16.00	32.38	25.02	15.81	38.96	28.95
EfficientNetB5	16.98	48.18	28.85	17.66	38.57	28.95	15.01	34.05	22.96
EfficientNetB6	18.06	42.39	27.38	**12.75**	**19.03**	**15.21**	**13.98**	24.04	17.47
EfficientNetB7	15.41	34.94	22.47	18.94	64.57	29.05	14.59	25.91	19.63
InceptionResNetV2	37.39	74.09	71.25	16.41	45.24	30.62	18.24	46.71	29.54
InceptionV3	32.83	81.45	67.32	18.94	38.76	28.66	19.15	50.44	31.21
MobileNet	33.43	76.84	65.55	24.62	53.29	43.18	24.32	64.47	44.16
ResNet101	31.70	74.98	68.30	14.29	21.69	16.78	17.02	50.44	33.46
ResNet101V2	19.53	39.84	31.11	24.76	100.00	100.00	16.83	**18.05**	**18.05**
ResNet152	24.44	67.22	50.15	28.13	100.00	100.00	19.76	34.35	26.89
ResNet152V2	19.43	44.36	34.54	16.11	33.76	22.47	19.63	50.15	37.39
ResNet50	43.54	100.00	100.00	22.77	52.11	41.12	20.90	56.92	37.10
ResNet50V2	22.67	56.13	44.85	17.86	37.49	25.12	21.58	42.00	31.50
VGG16	40.00	100.00	100.00	19.04	51.62	32.38	13.07	19.23	15.01
VGG19	22.80	51.62	36.60	17.93	32.48	23.85	18.54	41.12	27.28
Xception	33.13	50.05	50.05	18.84	42.00	27.67	18.45	32.78	25.22
Fusion	18.65	41.61	29.54	13.05	22.37	15.90	13.64	26.40	16.09
Fusion—top 5	14.92	25.71	19.73	12.46	22.08	15.41	12.27	21.59	13.74
Fusion—top 10	14.03	28.16	17.86	11.78	22.57	12.95	12.27	21.59	13.74

Table 2.11 Results on LivDet 2019 dataset—train data from Digital Persona and test data from Digital Persona. *Note—artefacts in training and testing are from different materials and different creation processes

Network	EER	BPCER_20	BPCER_10
DeiT-base-distilled-224	15.15	22.69	18.95
DeiT-small-distilled-224	18.78	29.66	24.26
DeiT-base-distilled-384	**11.55**	**17.36**	**12.26**

BPCER = 17.37%, and BPCER = 12.27% at APCER = 5% and APCER = 10% ,respectively. While the fusion approach can provide better BPCER at APCER = 5% as against DeiT-based approach, one can easily note the lower BPCER at all other lower APCER as against other networks and fusion-based decisions as shown in Fig. 2.5. Specifically, the DET for DeiT-base-distilled-384-based PAD demonstrates the better scalability for PAD at lower APCER as illustrated in Fig. 2.12d.

However, the results obtained on retraining the networks using both Softmax and Sigmoid activation for data captured from GreenBit and Orcanthus capture device are more promising as noted in Tables 2.12 and 2.14, respectively. Taking a closer look at Table 2.12, one can notice that the EER drops from 6.30% in the transfer learning case to 1.57 and 1.18% in networks trained from scratch using Softmax and Sigmoid, respectively. While for the same set of experiments using a vision transformer provides an EER of 2.45%, it can be noted that the CNNs perform better in dealing with data from GreenBit capture device. Motivated by this observation, we inspected the images captured in the Orcanthus capture device as illustrated in Fig. 2.1 and we note that the images inherently present different characteristics making the networks learn the differences between bona fide and artefacts easily. Specifically, one can note the enhanced artefacts in the captured images for attacks on the left side of the image. Similar observations can be made to the images captured with the GreenBit capture device. Nonetheless, DeiT provides comparable results at both APCER = 5% and APCER = 10% (Tables 2.13 and 2.15).

2.6.4.2 LivDet 2019—True Unknown Artefacts and Capture Device Interoperability

While the previous set of results discussed the observations when the training set and testing came from the same capture device, despite different materials, we present the results obtained in the true open-set setting where the artefacts and the capture device can be unknown during training. In order to present such an analysis, we consider the data captured with one device in the training set and data captured from a different device in the testing set. Such a case inherently presents artefacts that are unknown during the training phase. For instance, we consider the data captured using DigitalPersona device in training and data captured in GreenBit for testing as shown in Table 2.16. In such a setting, the artefacts are captured using Latex, WoodGlue,

Table 2.12 Results on LivDet 2019 dataset—train data from GreenBit and test data from GreenBit. *Note—artefacts in training and testing are from different materials and different creation processes

	Transfer learning			Retrained—softmax			Retrained—sigmoid		
Network	EER	BPCER_20	BPCER_10	EER	BPCER_20	BPCER_10	EER	BPCER_20	BPCER_10
DenseNet169	7.75	12.35	5.98	3.14	2.06	1.47	2.70	1.96	0.78
DenseNet201	9.08	16.37	8.04	2.86	0.98	0.29	4.17	3.43	1.57
EfficientNetB0	7.94	11.76	6.96	2.45	1.76	0.78	4.82	4.41	1.76
EfficientNetB1	6.86	**8.13**	**4.21**	2.29	1.47	0.49	2.35	1.27	0.88
EfficientNetB2	7.06	10.39	4.51	3.60	2.35	1.08	1.47	0.88	0.78
EfficientNetB3	9.80	16.08	9.71	3.35	1.96	0.88	4.09	2.94	0.88
EfficientNetB4	8.50	14.51	7.75	3.52	2.75	1.37	2.78	1.67	0.88
EfficientNetB5	6.86	9.61	5.10	1.86	0.78	0.69	2.35	1.76	1.37
EfficientNetB6	**6.29**	8.43	3.73	**1.56**	**0.68**	**0.58**	**1.176**	**0.68**	**0.49**
EfficientNetB7	8.83	15.78	7.75	3.43	1.57	0.78	1.96	1.08	0.78
InceptionResNetV2	27.77	66.18	53.63	5.00	5.00	2.35	2.86	2.25	1.57
InceptionV3	17.99	42.25	31.57	3.63	3.14	2.25	3.03	1.57	0.78
MobileNet	12.45	24.71	15.20	4.91	4.80	2.55	2.45	1.08	0.49
ResNet101	10.79	19.12	11.08	4.14	2.55	1.27	3.04	1.86	0.98
ResNet101V2	7.69	12.55	4.80	3.60	2.84	1.08	2.94	1.57	0.98
ResNet152	7.84	11.67	6.27	4.51	3.92	2.25	4.12	3.14	1.86
ResNet152V2	8.09	12.25	7.06	4.17	3.73	1.67	4.09	3.33	1.37
ResNet50	13.43	22.25	17.45	4.02	3.53	1.67	5.81	6.86	3.73
ResNet50V2	8.53	14.90	7.84	4.42	3.73	2.06	5.81	7.06	2.35
VGG16	14.80	16.86	16.86	3.82	3.53	2.35	3.92	3.33	1.76
VGG19	8.14	13.14	6.27	5.72	6.96	3.24	3.92	3.63	2.16
Xception	13.65	28.82	17.75	2.78	1.96	1.08	2.35	1.37	0.59
Fusion	3.84	3.24	1.57	1.57	0.69	0.49	1.27	0.69	0.49
Fusion—top 5	3.76	3.24	1.76	1.23	0.69	0.49	0.82	0.69	0.49
Fusion—top 10	3.63	2.65	1.57	1.18	0.49	0.49	0.90	0.49	0.49

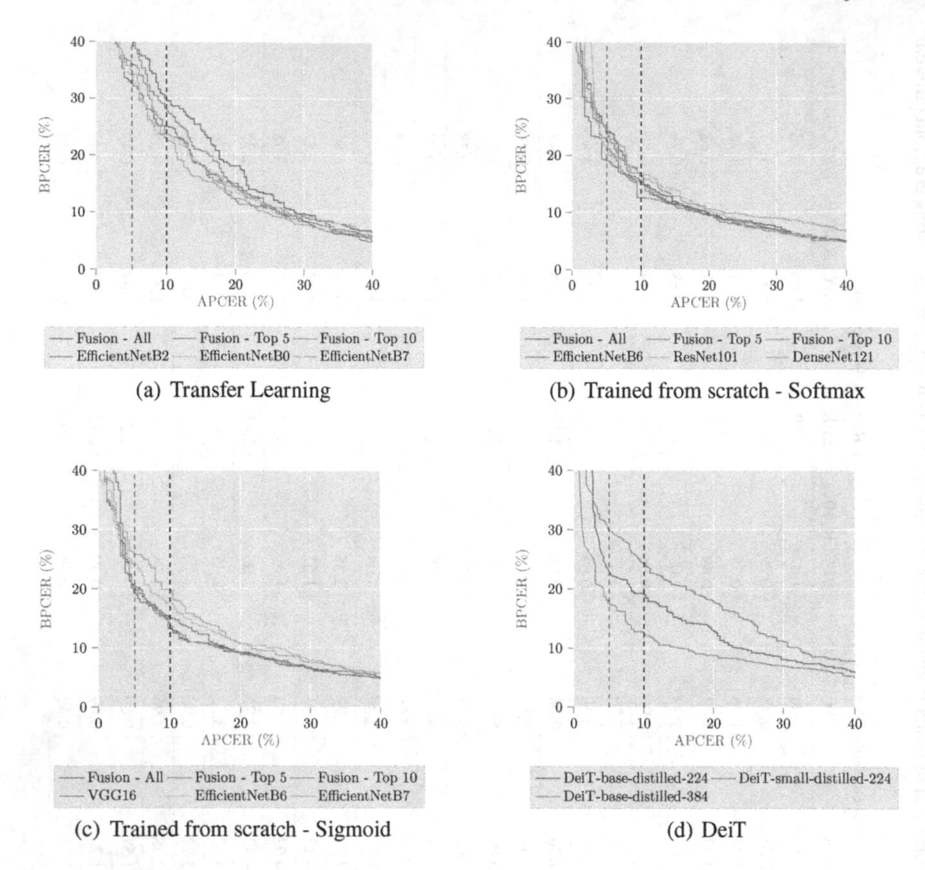

(a) Transfer Learning

(b) Trained from scratch - Softmax

(c) Trained from scratch - Sigmoid

(d) DeiT

Fig. 2.12 DET curves for results on LivDet 2019 dataset—train data from Digital Persona and test data from Digital Persona. *Note—artefacts in training and testing are from different materials and different creation processes

Table 2.13 Results on LivDet 2019 dataset—train data from GreenBit and test data from GreenBit. *Note—artefacts in training and testing are from different materials and different creation processes

Network	EER	BPCER_20	BPCER_10
DeiT-base-distilled-224	2.85	1.86	1.177
DeiT-small-distilled-224	3.27	2.15	0.88
DeiT-base-distilled-384	**2.45**	**1.176**	**0.68**

Table 2.14 Results on LivDet 2019 dataset—train data from Orcanthus and test data from Orcathus. *Note—artefacts in training and testing are from different materials and different creation processes

	Transfer learning			Retrained—softmax			Retrained—sigmoid		
Network	EER	BPCER_20	BPCER_10	EER	BPCER_20	BPCER_10	EER	BPCER_20	BPCER_10
DenseNet121	1.93	1.01	1.01	1.93	1.01	1.01	10.58	16.16	11.62
DenseNet169	17.94	43.03	27.98	17.94	43.03	27.98	4.78	4.65	2.53
DenseNet201	3.64	2.53	1.41	3.64	2.53	1.41	4.95	4.75	2.42
EfficientNetB0	5.66	6.26	3.74	5.66	6.26	3.74	2.83	**1.31**	0.51
EfficientNetB1	3.68	2.53	1.31	3.68	2.53	1.31	6.53	8.89	4.85
EfficientNetB2	2.39	1.21	0.91	2.39	1.21	0.91	3.31	2.22	1.72
EfficientNetB3	6.46	6.97	4.14	6.46	6.97	4.14	6.26	7.98	3.64
EfficientNetB4	4.95	4.85	2.12	4.95	4.85	2.12	5.86	6.97	3.64
EfficientNetB5	4.69	4.14	2.02	4.69	4.14	2.02	8.18	16.87	6.26
EfficientNetB6	5.43	6.06	2.42	5.43	6.06	2.42	8.19	12.73	6.26
EfficientNetB7	5.70	6.26	3.23	5.70	6.26	3.23	3.13	2.12	1.31
InceptionResNetV2	**2.39**	**0.90**	**0.60**	**2.39**	**0.90**	**0.60**	3.50	2.12	**0.20**
InceptionV3	5.66	6.46	3.13	5.66	6.46	3.13	4.60	4.34	2.73
MobileNet	4.65	4.44	2.93	4.65	4.44	2.93	**2.22**	1.72	1.31
ResNet101	5.27	4.04	2.63	5.27	4.04	2.63	3.54	2.93	1.62
ResNet101V2	4.04	3.84	2.42	4.04	3.84	2.42	4.55	4.44	3.13
ResNet152	10.86	17.78	11.31	10.86	17.78	11.31	5.76	6.46	2.53
ResNet152V2	7.98	11.62	5.15	7.98	11.62	5.15	10.86	16.77	11.52
ResNet50	7.98	13.54	5.96	7.98	13.54	5.96	7.27	10.51	4.95
ResNet50V2	7.54	9.70	5.45	7.54	9.70	5.45	6.77	8.59	4.65
VGG16	3.74	2.63	1.41	3.74	2.63	1.41	8.08	10.51	5.86
VGG19	5.06	5.35	2.42	5.06	5.35	2.42	3.13	2.12	1.41
Xception	7.91	11.01	5.15	7.91	11.01	5.15	7.88	11.41	5.76
Fusion	4.85	4.75	2.22	2.48	1.41	0.81	2.48	1.41	0.81
Fusion—top 5	3.68	2.42	1.21	1.52	1.01	0.61	1.93	0.81	0.30
Fusion—top 10	4.05	3.33	1.52	1.72	1.11	0.71	2.02	0.91	0.40

Table 2.15 Results on LivDet 2019 dataset—train data from Orcanthus and test data from Orcathus. *Note—artefacts in training and testing are from different materials and different creation processes

Network	EER	BPCER_20	BPCER_10
DeiT-base-distilled-224	2.94	2.022	1.31
DeiT-small-distilled-224	3.95	3.53	1.51
DeiT-base-distilled-384	**2.02**	**1.41**	**1.01**

Ecoflex 00–50, and Gelatine-based material using DigitalPersona device, and the testing set consists of artefacts consisting of Latex 1, Latex 2, Mix 2, Mix 1, Liquid Ecoflex 1, and Liquid Ecoflex 2—which are not present in the training set. While different CNN architectures presented promising and competitive results, LivDet 2019 is created using a different material than the materials used in the training set. We make a set of observations as noted below:

- Training the network from scratch using EfficientNet architecture obtains better results than the results obtained using the vision transformer-based approach as noted in Table 2.17. Specifically, EfficientNetB6 architecture provides an EER of 20.78 and 20.20% for networks trained from scratch using Softmax and Sigmoid activation. In contrast, the vision transformer-based approach provides an EER of 22.22%.
- The obtained results using vision transformers provides a clear advantage over several CNN architectures and achieve slightly worse results as compared to EfficientNet. We refer to the earlier works where a similar observation was made in certain cases and conclude this as a benefit of the highly optimized architecture of EfficientNet [72].
- Motivated by such an observation, we also study other combinations of training and testing in unseen data settings. An indicative set of results is presented in Table 2.18 for different CNN architectures. The corresponding results from the vision transformer are presented in Table 2.19. In the set of results presented in those two tables indicate the advantage of vision transformer-based PAD where it achieves an EER of 5.67% and a BPCER of 2.73% at APCER = 10%. At the same time, the best performing CNN (EfficientNetB2) obtains an EER of 11.01% and a BPCER of 11.92% at APCER = 10%, and the same CNN architecture achieves a BPCER of 9.12% and a BPCER of 8.78% at APCER = 10% when trained with Sigmoid activation.
- Similar benefits of the vision transformer-based fingerprint PAD for unseen data can be seen across different train-test combinations as reported further in Table 2.20 and we omit the full comparison due to constraints of the page limit.

Table 2.16 Results on LivDet 2019 dataset—trained on Latex, WoodGlue, Ecoflex 00–50, and Gelatine-based material captured using Digital Persona capture device and tested on artefacts consisting of Latex 1, Latex 2, Mix 2, Mix 1, Liquid Ecoflex 1, and Liquid Ecoflex 2 captured from GreenBit capture device

	Transfer learning			Retrained—softmax			Retrained—sigmoid		
Network	EER	BPCER_20	BPCER_10	EER	BPCER_20	BPCER_10	EER	BPCER_20	BPCER_10
DenseNet121	46.03	93.63	84.41	41.98	100.00	100.00	43.04	83.53	75.20
DenseNet169	46.68	100.00	100.00	39.37	100.00	69.22	43.92	100.00	100.00
DenseNet201	50.29	100.00	100.00	46.47	100.00	100.00	41.86	76.96	68.43
EfficientNetB0	27.72	67.45	51.86	28.02	100.00	100.00	35.24	74.31	64.61
EfficientNetB1	32.95	82.45	69.41	27.02	100.00	100.00	18.89	100.00	34.71
EfficientNetB2	27.47	72.94	55.10	25.29	100.00	38.14	31.86	75.00	61.47
EfficientNetB3	34.18	87.65	72.75	22.89	52.94	40.39	31.57	68.73	55.69
EfficientNetB4	34.80	82.06	68.43	28.23	100.00	100.00	21.47	66.67	47.45
EfficientNetB5	25.02	**59.60**	**46.07**	26.66	100.00	48.63	25.98	59.71	45.88
EfficientNetB6	27.23	66.37	54.31	**20.77**	**50.68**	**35.39**	**20.19**	**41.66**	**32.35**
EfficientNetB7	**23.62**	67.16	46.76	30.59	70.49	54.80	31.18	68.24	55.98
InceptionResNetV2	29.84	100.00	100.00	40.29	80.69	68.33	38.24	74.90	67.06
InceptionV3	39.90	100.00	100.00	35.39	81.27	69.31	41.70	88.04	82.16
MobileNet	43.14	86.37	78.04	38.08	100.00	100.00	37.53	100.00	100.00
ResNet101	32.45	78.33	65.98	23.54	100.00	100.00	27.39	70.39	52.45
ResNet101V2	36.27	83.53	74.02	35.02	100.00	100.00	27.55	100.00	100.00
ResNet152	38.51	94.02	83.82	27.69	100.00	100.00	27.47	63.82	47.25
ResNet152V2	41.76	79.41	69.41	42.94	100.00	100.00	43.42	100.00	68.92
ResNet50	30.42	100.00	100.00	47.06	100.00	100.00	27.35	65.29	52.45
ResNet50V2	37.94	70.10	61.37	26.00	100.00	47.45	45.00	81.18	75.20
VGG16	45.40	100.00	100.00	28.33	67.45	53.04	33.77	70.59	61.47
VGG19	36.67	77.45	66.57	43.82	100.00	61.27	44.81	84.71	76.96
Xception	36.55	100.00	100.00	43.82	100.00	100.00	42.68	86.47	81.27
Fusion	26.41	57.25	41.86	34.12	57.75	49.80	34.71	55.69	48.14
Fusion—top 5	24.80	57.94	46.76	29.80	60.00	48.14	21.34	54.61	36.67
Fusion—top 10	22.35	60.39	44.12	24.71	46.08	35.98	19.71	43.14	30.39

Table 2.17 Results on LivDet 2019 dataset—trained on Latex, WoodGlue, Ecoflex 00–50, and Gelatine-based material captured using Digital Persona capture device and tested on artefacts consisting of Latex 1, Latex 2, Mix 2, Mix 1, Liquid Ecoflex 1, and Liquid Ecoflex 2 captured from GreenBit capture device

Network	EER	BPCER_20	BPCER_10
DeiT-base-distilled-224	29.49	66.43	56.62
DeiT-small-distilled-224	31.45	67.51	53.68
DeiT-base-distilled-384	22.22	57.80	42.88

2.6.5 Analysis of Explainability in True Unknown Data Setting

We further analyze the explainability of the features learnt using various approaches as in the previous section for true unknown data settings where the training and testing sets are captured from different materials and capture devices. Similar to the previous analysis, we choose EfficientNetB4 to provide comparable results, and specifically, we choose network trained using data captured from Digital Persona and tested on data captured using Orcanthus. Figures 2.13, 2.14, and 2.15 present the regions used for making decisions using the networks trained in transfer learning setting and trained from scratch setting with Softmax and Sigmoid activation, respectively, on the images shown in Fig. 2.1.

As noted from the results in Table 2.18, the explainability analysis shows the activation maps which are uniform in Figs. 2.13, 2.14, and 2.15 supporting the low performance in true open-set scenario. While on the other hand, it can be noted that the Figs. 2.16 and 2.17 for DeiT-base-distilled-224 and DeiT-small-distilled-224 present varying attention maps supporting error rates observed in Table 2.19. While the attention maps for DeiT-base-distilled-224 and DeiT-small-distilled-224 do not necessarily look into specific regions of image to obtain low error rates, activation maps in Fig. 2.18 present activations maps that differ significantly for bona fide images as compared to the artefacts supporting the low error rates observed in Table 2.19.

Table 2.18 Results on LivDet 2019 dataset—trained on Latex, WoodGlue, Ecoflex 00–50, and Gelatine-based material captured using Digital Persona capture device and tested on artefacts consisting of Mix, Liquid Ecoflex, and Latex with two different processes captured from Orcanthus capture device

	Transfer learning			Retrained—softmax			Retrained—Sigmoid		
Network	EER	BPCER_20	BPCER_10	EER	BPCER_20	BPCER_10	EER	BPCER_20	BPCER_10
DenseNet121	46.03	93.63	84.41	41.98	100.00	100.00	43.04	83.53	75.20
DenseNet169	46.68	100.00	100.00	39.37	100.00	69.22	43.92	100.00	100.00
DenseNet201	50.29	100.00	100.00	46.47	100.00	100.00	41.86	76.96	68.43
EfficientNetB0	27.72	67.45	51.86	28.02	100.00	100.00	35.24	74.31	64.61
EfficientNetB1	32.95	82.45	69.41	27.02	100.00	100.00	18.89	100.00	34.71
EfficientNetB2	27.47	72.94	55.10	25.29	100.00	38.14	31.86	75.00	61.47
EfficientNetB3	34.18	87.65	72.75	22.89	52.94	40.39	31.57	68.73	55.69
EfficientNetB4	34.80	82.06	68.43	28.23	100.00	100.00	21.47	66.67	47.45
EfficientNetB5	25.02	**59.60**	**46.07**	26.66	100.00	48.63	25.98	59.71	45.88
EfficientNetB6	27.23	66.37	54.31	**20.77**	**50.68**	**35.39**	**20.19**	**41.66**	**32.35**
EfficientNetB7	**23.62**	67.16	46.76	30.59	70.49	54.80	31.18	68.24	55.98
InceptionResNetV2	29.84	100.00	100.00	40.29	80.69	68.33	38.24	74.90	67.06
InceptionV3	39.90	100.00	100.00	35.39	81.27	69.31	41.70	88.04	82.16
MobileNet	43.14	86.37	78.04	38.08	100.00	100.00	37.53	100.00	100.00
ResNet101	32.45	78.33	65.98	23.54	100.00	100.00	27.39	70.39	52.45
ResNet101V2	36.27	83.53	74.02	35.02	100.00	100.00	27.55	100.00	100.00
ResNet152	38.51	94.02	83.82	27.69	100.00	100.00	27.47	63.82	47.25
ResNet152V2	41.76	79.41	69.41	42.94	100.00	100.00	43.42	100.00	68.92
ResNet50	30.42	100.00	100.00	47.06	100.00	100.00	27.35	65.29	52.45
ResNet50V2	37.94	70.10	61.37	26.00	100.00	47.45	45.00	81.18	75.20
VGG16	45.40	100.00	100.00	28.33	67.45	53.04	33.77	70.59	61.47
VGG19	36.67	77.45	66.57	43.82	100.00	61.27	44.81	84.71	76.96
Xception	36.55	100.00	100.00	43.82	100.00	100.00	42.68	86.47	81.27
Fusion	19.41	46.26	32.63	18.08	54.24	32.93	14.55	33.43	20.30
Fusion—top 5	17.30	34.44	23.74	10.86	23.03	12.02	10.58	17.98	11.21
Fusion—top 10	14.24	31.01	19.29	10.86	23.03	12.02	11.13	20.91	11.82

Table 2.19 Results on LivDet 2019 dataset obtained using DeiT—trained on Latex, WoodGlue, Ecoflex 00–50, and Gelatine-based material captured using Digital Persona sensor and tested on artefacts consisting of Mix, Liquid Ecoflex, and Latex with two different processes captured from Orcanthus sensor

Network	EER	BPCER_20	BPCER_10
DeiT-base-distilled-224	18.85	67.97	44.24
DeiT-small-distilled-224	21.06	72.22	51.91
DeiT-base-distilled-384	5.69	6.875	2.73

Table 2.20 Results on LivDet 2019 dataset—Unseen attack results obtained using DeiT-based PAD

Network	EER	BPCER_20	BPCER_10
Train GreenBit :: Test Digital Persona			
DeiT-base-distilled-224	8.51	12.75	6.96
DeiT-small-distilled-224	8.78	12.37	8.84
DeiT-base-distilled-384	6.36	7.56	3.53
Train GreenBit :: Test Orcathus			
DeiT-base-distilled-224	14.44	26.06	18.28
DeiT-small-distilled-224	20.33	33.33	28.08
DeiT-base-distilled-384	14.25	23.93	17.67
Train Orcanthus :: Test DigitalPersona			
DeiT-base-distilled-224	18.84	37.58	28.36
DeiT-small-distilled-224	19.39	49.11	37.13
DeiT-base-distilled-384	14.54	28.38	17.97
Train Orcanthus :: Test GreenBit			
DeiT-base-distilled-224	19.36	64.08	42.39
DeiT-small-distilled-224	15.86	43.13	26.66
DeiT-base-distilled-384	20.01	41.60	30.22

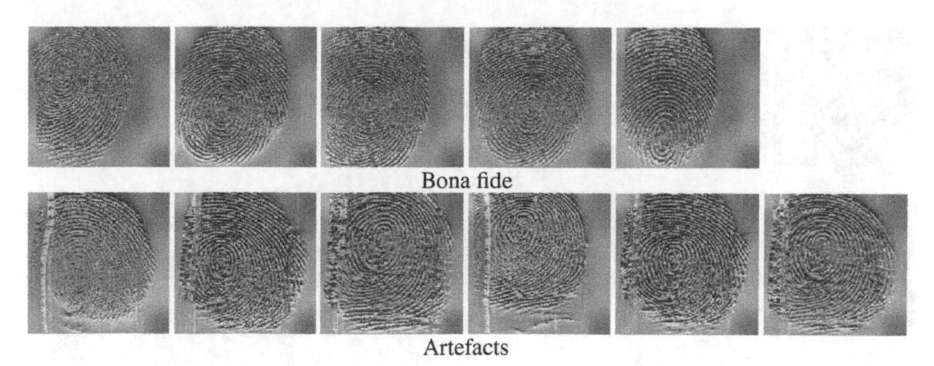

Fig. 2.13 Visualization of explainable features obtained using EfficientNetB4 in transfer learning setting trained on data captured from Digital Persona sensor and tested on data captured from Orcanthus sensor for images shown in Fig. 2.1. As noted from the figure, the activation of the network is very sparse which correlates to the low performance shown in Table 2.18. The network fails to generalize when trained on Latex, WoodGlue, Ecoflex 00–50, and Gelatine-based material captured using Digital Persona capture device and tested on artefacts consisting of Mix, Liquid Ecoflex, and Latex with two different processes captured from Orcanthus capture device

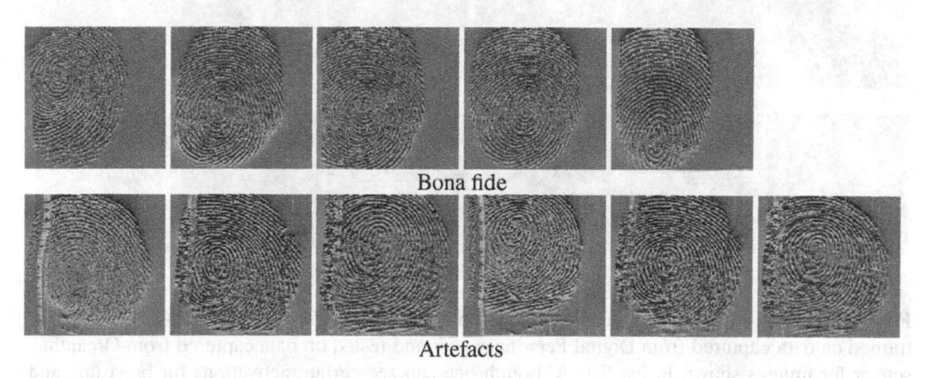

Fig. 2.14 Visualization of explainable features obtained using EfficientNetB4 trained on data captured from Digital Persona sensor and tested on data captured from Orcanthus sensor for images shown in Fig. 2.1 (Softmax activation). The activation of the network is very sparse correlating to the low performance shown in Table 2.18 failing to generalize across different materials and capture sensors similar to Fig. 2.13

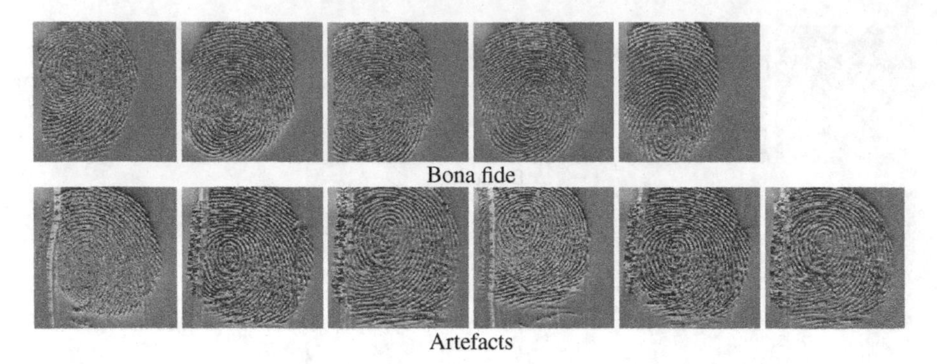

Fig. 2.15 Visualization of explainable features obtained using EfficientNetB4 trained on data captured from Digital Persona sensor and tested on data captured from Orcanthus sensor for images shown in Fig. 2.1 (Sigmoid activation). Similar to Figs. 2.13 and 2.14, one can observe very limited activations corresponding to low performance reported in Table 2.18

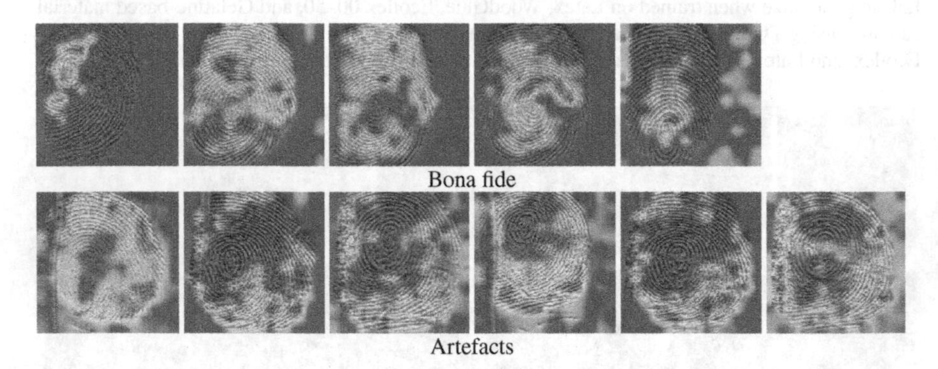

Fig. 2.16 Visualization of explainable features obtained using DeiT-base-distilled-224 (Layer-10) trained on data captured from Digital Persona sensor and tested on data captured from Orcanthus sensor for images shown in Fig. 2.1. Although one can see certain activations for bona fide and artefacts, they do not appear very different confirming the low performance of this configuration for F-PAD as seen in Table 2.19

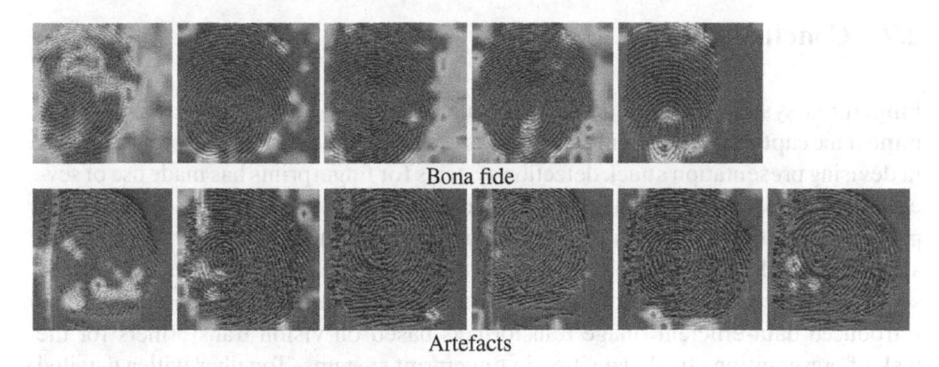

Fig. 2.17 Visualization of explainable features obtained using DeiT-small-distilled-224 (Layer-10) trained on data captured from Digital Persona sensor and tested on data captured from Orcanthus sensor for images shown in Fig. 2.1. Low and very similar activations for both bona fide and artefacts in this figure corroborate with low performance reported in Table 2.19 under this configuration

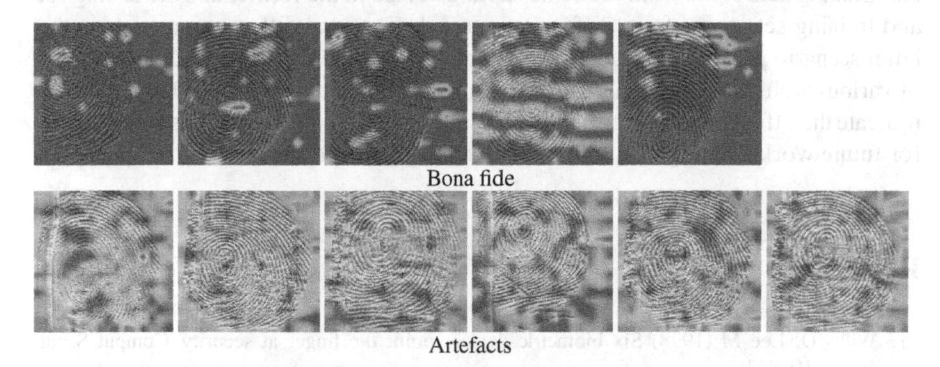

Fig. 2.18 Visualization of explainable features obtained using DeiT-small-distilled-384 (Layer-10) trained on data captured from Digital Persona sensor and tested on data captured from Orcanthus sensor for images shown in Fig. 2.1. One can note significantly different activations in this figure as compared to Figures 2.16 and 2.17 indicating differences in bona fide and artefacts. Such differences lead to higher attack detection rate as noted in Table 2.19 achieving best performance under this configuration

2.7 Conclusion

Fingerprint systems need to be enabled with presentation attack detection to determine if the captured fingerprint corresponds to a bona fide or artefact. Recent progress in devising presentation attack detection systems for fingerprints has made use of several different deep learning approaches, including dedicated networks. This chapter presented a detailed analysis of 23 different CNN architectures for fingerprint presentation attack detection for the benefit of the reader. The analysis is expected to serve as a reference point for choosing the network. In addition, the chapter also introduced data-efficient image transformers based on vision transformers for the task of presentation attack detection in fingerprint systems. Together with a detailed analysis of the error rates, we have presented the explainability analysis to provide the reader with an understanding of attention to different approaches. Further, the experiments conducted on the closed-set and open-set protocols where the testing set and training sets come from the same capture device in the former and the testing set and training sets come from different capture devices and different materials in the latter scenario provide the reader with a complete overview of benefits and weakness of various architectures. Finally, the experiments conducted on two different datasets indicate the effectiveness of data-efficient image transformers, a promising direction for future works on fingerprint presentation attack detection.

References

1. Willis D, Lee M (1998) Six biometric devices point the finger at security. Comput Secur 5(17):410–411
2. Van der Putte T, Keuning J (2000) Biometrical fingerprint recognition: don't get your fingers burned. In: Smart card research and advanced applications. Springer, pp 289–303
3. Matsumoto T, Matsumoto H, Yamada K, Hoshino S (2002) Impact of artificial "gummy" fingers on fingerprint systems. In: Optical security and counterfeit deterrence techniques IV, vol 4677. International Society for Optics and Photonics, pp 275–289
4. Schuckers S (2002) Spoofing and anti-spoofing measures. Inf Secur Tech Rep 7(4):56–62
5. Zwiesele A, Munde A, Busch C, Daum H (2000) BioIS study—comparative study of biometric identification systems. In: 34th annual 2000 IEEE international Carnahan conference on security technology (CCST). IEEE Computer Society, pp 60–63
6. Huang G, Liu Z, Van Der Maaten L, Weinberger KQ (2017) Densely connected convolutional networks. In: Proceedings of the IEEE conference on computer vision and pattern recognition, pp 4700–4708
7. Tan M, Le Q (2019) Efficientnet: rethinking model scaling for convolutional neural networks. In: International conference on machine learning (ICML), pp 6105–6114
8. Szegedy C, Ioffe S, Vanhoucke V, Alemi AA (2017) Inception-v4, inception-ResNet and the impact of residual connections on learning. In: AAAI conference on artificial intelligence (AAAI)
9. Szegedy C, Vanhoucke V, Ioffe S, Shlens J, Wojna Z (2016) Rethinking the inception architecture for computer vision. In: IEEE conference on computer vision and pattern recognition (CVPR), pp 2818–2826

10. Howard AG, Zhu M, Chen B, Kalenichenko D, Wang W, Weyand T, Andreetto M, Adam H (2017) Mobilenets: efficient convolutional neural networks for mobile vision applications. arXiv:1704.04861
11. He K, Zhang X, Ren S, Sun J (2016) Deep residual learning for image recognition. In: Proceedings of the IEEE conference on computer vision and pattern recognition, pp 770–778
12. He K, Zhang X, Ren S, Sun J (2016) Identity mappings in deep residual networks. In: European conference on computer vision. Springer, pp 630–645
13. Simonyan K, Zisserman A (2014) Very deep convolutional networks for large-scale image recognition. arXiv:1409.1556
14. Chollet F (2017) Xception: deep learning with depthwise separable convolutions. In: Proceedings of the IEEE conference on computer vision and pattern recognition, pp 1251–1258
15. Grosz SA, Chugh T, Jain AK (2020) Fingerprint presentation attack detection: a sensor and material agnostic approach. In: 2020 IEEE international joint conference on biometrics (IJCB), pp 1–10. https://doi.org/10.1109/IJCB48548.2020.9304863
16. Micheletto M, Marcialis GL, Orrù G, Roli F (2021) Fingerprint recognition with embedded presentation attacks detection: are we ready? IEEE Trans Inf Forensics Secur
17. Sharma D, Selwal A (2021) Hyfipad: a hybrid approach for fingerprint presentation attack detection using local and adaptive image features. Vis Comput 1–27
18. Sousedik C, Busch C (2014) Presentation attack detection methods for fingerprint recognition systems: a survey. IET Biom 3(4):219–233
19. Raja K, Raghavendra R, Venkatesh S, Gomez-Barrero M, Rathgeb C, Busch C (2019) A study of hand-crafted and naturally learned features for fingerprint presentation attack detection. In: Handbook of biometric anti-spoofing. Springer, pp 33–48
20. Ghiani L, Yambay DA, Mura V, Marcialis GL, Roli F, Schuckers S (2017) Review of the fingerprint liveness detection (livdet) competition series: 2009–2015. Image Vis Comput 58:110–128
21. Baldisserra D, Franco A, Maio D, Maltoni D (2006) Fake fingerprint detection by odor analysis. In: International conference on biometrics. Springer, pp 265–272
22. Engelsma JJ, Cao K, Jain AK (2018) Raspireader: open source fingerprint reader. IEEE Trans Pattern Anal Mach Intell 41(10):2511–2524
23. Auksorius E, Boccara AC (2015) Fingerprint imaging from the inside of a finger with full-field optical coherence tomography. Biomed Opt Express 6(11)
24. Bicz A, Bicz W (2016) Development of ultrasonic finger reader based on ultrasonic holography having sensor area with 80 mm diameter. In: 2016 International Conference of the biometrics special interest group (BIOSIG). IEEE, pp 1–6
25. Raja K, Auksorius E, Raghavendra R, Boccara AC, Busch C (2017) Robust verification with subsurface fingerprint recognition using full field optical coherence tomography. In: Proceedings of the IEEE conference on computer vision and pattern recognition workshops, pp 144–152
26. Sousedik C, Breithaupt R, Busch C (2013) Volumetric fingerprint data analysis using optical coherence tomography. In: 2013 International Conference of the biometrics special interest group (BIOSIG). IEEE, pp 1–6
27. Yu X, Xiong Q, Luo Y, Wang N, Wang L, Tey HL, Liu L (2016) Contrast enhanced subsurface fingerprint detection using high-speed optical coherence tomography. IEEE Photonics Technol Lett 29(1):70–73
28. Harms F, Dalimier E, Boccara AC (2014) En-face full-field optical coherence tomography for fast and efficient fingerprints acquisition. In: SPIE defense+ security. International Society for Optics and Photonics, pp 90,750E–90,750E
29. Tolosana R, Gomez-Barrero M, Kolberg J, Morales A, Busch C, Ortega-Garcia J (2018) Towards fingerprint presentation attack detection based on convolutional neural networks and short wave infrared imaging. In: 2018 international conference of the biometrics special interest group (BIOSIG). IEEE, pp 1–5
30. Mirzaalian H, Hussein M, Abd-Almageed W (2019) On the effectiveness of laser speckle contrast imaging and deep neural networks for detecting known and unknown fingerprint presentation attacks. In: 2019 international conference on biometrics (ICB). IEEE, pp 1–8

31. Kolberg J, Gläsner D, Breithaupt R, Gomez-Barrero M, Reinhold J, von Twickel A, Busch C (2021) On the effectiveness of impedance-based fingerprint presentation attack detection. Sensors 21(17):5686
32. Ramachandra R, Raja K, Venkatesh SK, Busch C (2019) Design and development of low-cost sensor to capture ventral and dorsal finger vein for biometric authentication. IEEE Sens J 19(15):6102–6111
33. Tuveri P, Ghiani L, Zurutuza M, Mura V, Marcialis GL (2019) Interoperability among capture devices for fingerprint presentation attacks detection. In: Handbook of biometric anti-spoofing. Springer, pp 71–108
34. Derakhshani R, Schuckers S, Hornak LA, O'Gorman L (2003) Determination of vitality from a non-invasive biomedical measurement for use in fingerprint scanners. Pattern Recogn 36(2):383–396
35. Parthasaradhi ST, Derakhshani R, Hornak LA, Schuckers S (2005) Time-series detection of perspiration as a liveness test in fingerprint devices. IEEE Trans Syst, Man, Cybern, Part C (Appl Rev) 35(3), 335–343
36. Reddy PV, Kumar A, Rahman S, Mundra TS (2008) A new antispoofing approach for biometric devices. IEEE Trans Biomed Circuits Syst 2(4):328–337
37. Drahansky M, Notzel R, Funk W (2006) Liveness detection based on fine movements of the fingertip surface. In: 2006 IEEE information assurance workshop. IEEE, pp 42–47
38. Martinsen OG, Clausen S, Nysæther JB, Grimnes S (2007) Utilizing characteristic electrical properties of the epidermal skin layers to detect fake fingers in biometric fingerprint systems–a pilot study. IEEE Trans Biomed Eng 54(5):891–894
39. Moon YS, Chen J, Chan K, So K, Woo K (2005) Wavelet based fingerprint liveness detection. Electron Lett 41(20):1112–1113
40. Zhang Y, Fang S, Xie Y, Xu T (2014) Fake fingerprint detection based on wavelet analysis and local binary pattern. In: Chinese conference on biometric recognition. Springer, pp 191–198
41. Jia X, Yang X, Cao K, Zang Y, Zhang N, Dai R, Zhu X, Tian J (2014) Multi-scale local binary pattern with filters for spoof fingerprint detection. Inf Sci 268:91–102
42. Ghiani L, Marcialis GL, Roli F (2012) Fingerprint liveness detection by local phase quantization. In: 2012 21st international conference on pattern recognition (ICPR). IEEE, pp 537–540
43. Ghiani L, Hadid A, Marcialis GL, Roli F (2013) Fingerprint liveness detection using binarized statistical image features. In: 2013 IEEE sixth international conference on biometrics: theory, applications and systems (BTAS). IEEE, pp 1–6
44. Ghiani L, Hadid A, Marcialis GL, Roli F (2017) Fingerprint liveness detection using local texture features. IET Biom 6(3):224–231
45. Gragnaniello D, Poggi G, Sansone C, Verdoliva L (2013) Fingerprint liveness detection based on weber local image descriptor. In: 2013 IEEE workshop on biometric measurements and systems for security and medical applications (BIOMS). IEEE, pp 46–50
46. Xia Z, Yuan C, Lv R, Sun X, Xiong NN, Shi YQ (2018) A novel weber local binary descriptor for fingerprint liveness detection. IEEE Trans Syst, Man, Cybern: Syst 50(4):1526–1536
47. Kim W (2016) Fingerprint liveness detection using local coherence patterns. IEEE Signal Process Lett 24(1):51–55
48. Tan G, Zhang Q, Hu H, Zhu X, Wu X (2020) Fingerprint liveness detection based on guided filtering and hybrid image analysis. IET Image Proc 14(9):1710–1715
49. Marasco E, Sansone C (2010) An anti-spoofing technique using multiple textural features in fingerprint scanners. In: 2010 IEEE workshop on biometric measurements and systems for security and medical applications. IEEE, pp 8–14
50. Dubey RK, Goh J, Thing VL (2016) Fingerprint liveness detection from single image using low-level features and shape analysis. IEEE Trans Inf Forensics Secur 11(7):1461–1475
51. Agarwal S, Chowdary CR (2020) A-stacking and a-bagging: adaptive versions of ensemble learning algorithms for spoof fingerprint detection. Expert Syst Appl 146:113,160
52. González-Soler LJ, Gomez-Barrero M, Chang L, Pérez-Suárez A, Busch C (2021) Fingerprint presentation attack detection based on local features encoding for unknown attacks. IEEE Access 9:5806–5820

53. Galbally J, Alonso-Fernandez F, Fierrez J, Ortega-Garcia J (2012) A high performance fingerprint liveness detection method based on quality related features. Futur Gener Comput Syst 28(1):311–321
54. Galbally J, Alonso-Fernandez F, Fierrez J, Ortega-Garcia J (2009) Fingerprint liveness detection based on quality measures. In: 2009 international conference on biometrics, identity and security (BIdS). IEEE, pp 1–8
55. Krizhevsky A, Sutskever I, Hinton GE (2012) Imagenet classification with deep convolutional neural networks. Adv Neural Inf Process Syst 25:1097–1105
56. Nogueira RF, de Alencar Lotufo R, Machado RC (2014) Evaluating software-based fingerprint liveness detection using convolutional networks and local binary patterns. In: 2014 IEEE Workshop on biometric measurements and systems for security and medical applications (BIOMS) proceedings. IEEE, pp 22–29
57. Nogueira RF, de Alencar Lotufo R, Machado RC (2016) Fingerprint liveness detection using convolutional neural networks. IEEE Trans Inf Forensics Secur 11(6):1206–1213
58. Park E, Kim W, Li Q, Kim J, Kim H (2016) Fingerprint liveness detection using CNN features of random sample patches. In: 2016 international conference of the biometrics special interest group (BIOSIG). IEEE, pp 1–4
59. Wang C, Li K, Wu Z, Zhao Q (2015) A DCNN based fingerprint liveness detection algorithm with voting strategy. In: Chinese conference on biometric recognition. Springer, pp 241–249
60. Jung H, Heo Y (2018) Fingerprint liveness map construction using convolutional neural network. Electron Lett 54(9):564–566
61. Zhang Y, Shi D, Zhan X, Cao D, Zhu K, Li Z (2019) Slim-ResCNN: a deep residual convolutional neural network for fingerprint liveness detection. IEEE Access 7:91476–91487
62. Yuan C, Xia Z, Jiang L, Cao Y, Wu QJ, Sun X (2019) Fingerprint liveness detection using an improved CNN with image scale equalization. IEEE Access 7:26953–26966
63. Zhang Y, Gao C, Pan S, Li Z, Xu Y, Qiu H (2020) A score-level fusion of fingerprint matching with fingerprint liveness detection. IEEE Access 8:183,391–183,400
64. Liu H, Kong Z, Ramachandra R, Liu F, Shen L, Busch C (2021) Taming self-supervised learning for presentation attack detection: In-image de-folding and out-of-image de-mixing. arXiv:2109.04100
65. Marasco E, Wild P, Cukic B (2016) Robust and interoperable fingerprint spoof detection via convolutional neural networks. In: 2016 IEEE symposium on technologies for homeland security (HST), pp 1–6. https://doi.org/10.1109/THS.2016.7568925
66. Kim S, Park B, Song BS, Yang S (2016) Deep belief network based statistical feature learning for fingerprint liveness detection. Pattern Recognit Lett 77:58–65. https://doi.org/10.1016/j.patrec.2016.03.015. http://www.sciencedirect.com/science/article/pii/S0167865516300198
67. Pala F, Bhanu B (2017) Deep triplet embedding representations for liveness detection. In: Deep learning for biometrics. Springer, pp 287–307
68. Fei J, Xia Z, Yu P, Xiao F (2020) Adversarial attacks on fingerprint liveness detection. EURASIP J Image Video Process 2020(1):1–11
69. Chugh T, Jain AK (2020) Fingerprint spoof detector generalization. IEEE Trans Inf Forensics Secur 16:42–55
70. Jian W, Zhou Y, Liu H (2020) Densely connected convolutional network optimized by genetic algorithm for fingerprint liveness detection. IEEE Access 9:2229–2243
71. Grosz SA, Chugh T, Jain AK (2020) Fingerprint presentation attack detection: a sensor and material agnostic approach. In: 2020 IEEE international joint conference on biometrics (IJCB). IEEE, pp 1–10
72. Touvron H, Cord M, Douze M, Massa F, Sablayrolles A, Jégou H (2021) Training data-efficient image transformers & distillation through attention
73. Dosovitskiy A, Beyer L, Kolesnikov A, Weissenborn D, Zhai X, Unterthiner T, Dehghani M, Minderer M, Heigold G, Gelly S, Uszkoreit J, Houlsby N (2020) An image is worth 16 × 16 words: transformers for image recognition at scale
74. Wu B, Xu C, Dai X, Wan A, Zhang P, Yan Z, Tomizuka M, Gonzalez J, Keutzer K, Vajda P (2020) Visual transformers: token-based image representation and processing for computer vision

75. ISO/IEC JTC1 SC37 Biometrics: ISO/IEC 30107-3 (2017) Information Technology—Biometric presentation attack detection - Part 3: Testing and Reporting. International Organization for Standardization

Chapter 3
Review of the Fingerprint Liveness Detection (LivDet) Competition Series: From 2009 to 2021

Marco Micheletto, Giulia Orrù, Roberto Casula, David Yambay, Gian Luca Marcialis, and Stephanie Schuckers

Abstract Fingerprint authentication systems are highly vulnerable to fingerprint artificial reproductions, called fingerprint presentation attacks. Detecting presentation attacks is not trivial because attackers refine their replication techniques from year to year. The International Fingerprint liveness Detection Competition (LivDet), an open and well-acknowledged meeting point of academies and private companies that deal with the problem of presentation attack detection, has the goal to assess the performance of fingerprint presentation attack detection (FPAD) algorithms by using standard experimental protocols and data sets. Each LivDet edition, held biannually since 2009, is characterized by a different set of challenges to which the competitors must provide solutions The continuous increase of competitors and the noticeable decrease in error rates across competitions demonstrate a growing interest in the topic. This paper reviews the LivDet editions from 2009 to 2021 and points out their evolution over the years.

M. Micheletto (✉) · G. Orrù · R. Casula · G. L. Marcialis
University of Cagliari, Cagliari, Italy
e-mail: marco.micheletto@unica.it

G. Orrù
e-mail: giulia.orru@unica.it

R. Casula
e-mail: roberto.casula@unica.it

G. L. Marcialis
e-mail: marcialis@unica.it

D. Yambay · S. Schuckers
Clarkson University, Potsdam, USA
e-mail: yambayda@clarkson.edu

S. Schuckers
e-mail: sschucke@clarkson.edu

S. Marcel et al. (eds.), *Handbook of Biometric Anti-Spoofing*, Advances in Computer Vision and Pattern Recognition, https://doi.org/10.1007/978-981-19-5288-3_3

3.1 Introduction

In recent years, fingerprint-based biometric systems have achieved a significant degree of accuracy. This biometry can reasonably be considered the most mature from both an academic and an industrial point of view. Nowadays, it is one of the most suitable solutions in every application that requires a high level of security. The reason for this success is mainly due to its universality, duration, and individuality properties. Nevertheless, it has been repeatedly shown that artificial fingerprints can fool a fingerprint-based personal recognition system [1, 2]. As a matter of fact, most standard scanners on the market cannot distinguish images of a real fingerprint from the so-called "spoofs". For this purpose, a fingerprint presentation attack detector (FPAD) is designed to automatically decide during usage about the liveness of the fingerprint presented to the sensor. In these systems, additional information is usually exploited to check whether the fingerprint is authentic or not. Liveness detection systems are therefore included in two categories: (1) hardware-based, which add specific sensors to the scanner to detect specific traits that ensure vivacity (e.g., sweat, blood pressure, etc.); (2) software-based, which process the sample acquired by a standard sensor to verify the liveness, by extracting distinctive features from the image and not from the finger itself.

Among the several initiatives, the *International Fingerprint Liveness Detection Competition* (LivDet) is a biennial meeting aimed at making the point on the limits and perspectives of presentation attacks detectors (PADs). The competition occurred and evolved in seven editions between 2009 [3] and 2021 [4], proposing new challenging falsification methods, kind of scanner, and spoofs' materials.

This paper examines all the editions of the LivDet competition, analyzing the progress of the liveness algorithms in twelve years.

Section 3.2 describes the background of spoofing and liveness detection. Section 3.3 describes the methods used in the competition test sets. Section 3.4 examines the results in terms of system accuracy. Section 3.5 discusses the analysis of the fingerprint's image quality. Section 3.6 concludes the paper.

3.2 Fingerprint Presentation Attack Detection

Since 1998, the year in which the vulnerability of personal recognition systems to artificial fingerprint replicas was demonstrated [5], the scientific community has continued to study and propose new methods to protect such systems with the so-called Presentation Attack Detectors (PADs). In the same way, the ability of attackers continues to evolve over the years and the study of attack techniques is the basis of research on Presentation Attack detection. It is therefore important to highlight the main techniques for creating a fake fingerprint, which can be consensual or non-consensual. The consensual approach is regarded as the worst-case scenario, as the user's collaboration permits the creation of a high-quality fake. This category includes the basic *molding and casting* method [1] and sophisticated 2D and 3D printing techniques [6]. Three phases are involved in the *molding and casting* process. In the initial stage, the volunteer presses a finger into a silicone material, leaving a negative impression of her or his biometric trait on a mold. After that, the mold is filled with a casting substance, such as latex, liquid ecoflex, or glue. The solidified material is released from the mold and represents an exact replica of the genuine fingerprint, allowing a presentation attack against a fingerprint recognition system. The non-consensual method involves greater risks for the user as it is more realistic in terms of attacks on a biometric system.

Unlike the casts created in the consensual method with the collaboration of a volunteer, the latent imprint can be detected by an impostor through various techniques [7]. By applying magnetic powders on a smooth or non-porous surface, it is possible to highlight the impression so that it can be photographed or scanned [8]. This will be subsequently processed and the negative will be created and printed on a sheet. Another way to create the fake is to engrave the negative on a printed circuit in order to drip the material.

To defend against these types of attacks, various software solutions for liveness detection have been implemented over the years [9]. As with other biometrics, presentation attack detection techniques have also undergone an evolution passing from the analysis of ridges and valleys, to local hand-crafted methods based on morphology, color and texture analysis such as BSIF and LBP [24–26], and to more modern deep-learning techniques [27, 28].

3.3 The Fingerprint Liveness Detection Competition

The Fingerprint Liveness Detection Competition was established with the goal of developing a standard for evaluating PAD technologies.

Born in 2009 [3] through the collaboration of the University of Cagliari and Clarkson University, the number of participants and algorithms has steadily increased over the editions. Using a competition rather than merely sharing datasets might ensure free, third-party testing. Because of a conflict of interest, Clarkson and Cagliari have never competed in LivDet.

The Algorithms part of the LivDet competition provides a common experimental protocol and data sets and evaluates the performance of FPAD systems with different edition-by-edition challenges that simulate open issues of the state of the art. Each competitor receives a set of images (training set), trains their FPADs with them and submits the solution. The organizers test the algorithms on a set of new images (test set) and calculate the performances and the final ranking.

The interest in the Algorithms part of the LivDet competition and the number of participants and algorithms grows from edition to edition as shown in Table 3.1.

The Systems part of the LivDet competition originally began with one competition, but in 2017 a second competition was added. For both competitions, each competitor submits a fingerprint sensor trained on their own FPAD data, with some

Table 3.1 Number of competitors and algorithms submitted to the seven editions of the Fingerprint Liveness Detection Competition—Algorithm part

LivDet edition	2009	2011	2013	2015	2017	2019	2021
# Participants	4	3	9	10	12	9	13
# Algorithms	4	3	11	12	17	12	23

FPAD recipes supplied to competitors. The organizers then recruit live subjects and create PAIs to test the system based on which competition to which the system is submitted. The first competition is a PAD only module for systems, which detects if the input is live or spoof. The second competition is the incorporation of presentation attack detection with matching to determine overall performance of a system. For this portion, a live person enrolls and a spoof is created of the same fingerprint of the enrolled person. A successful attack bypasses the PAD module and matches the enrolled fingerprint.

3.4 Methods and Dataset

3.4.1 Algorithms part

This section will analyze the composition and the methodology for creating each dataset of all the Algorithms part competitions from 2009 to 2021. In these seven editions, a total of twenty-five datasets were created, using fourteen distinct scanners, the majority of which were optical kind, as shown in Table 3.2; numerous materials have been tested throughout the years in order to determine the best ones for fabricating fingerprint replicas. In our experience, the ease with which a successful spoof can be produced depends on subjective factors such as casting the material on the mold or the difficulty of detaching the replica from the mold without ruining one or both, and objective characteristics of composition and behavior. As a matter of fact, certain materials have been shown to be less suited than others. Several of them cannot duplicate the ridges and valleys pattern without introducing visible artifacts such as bubbles and alterations of ridge edges. Fig. 3.1 reports the best and most "potentially harmful" materials employed in each competition. The final choice was made based on the best trade-off between spoof effectiveness and the fabrication process.

In the first edition of 2009, the three optical scanners Crossmatch, Identix, and Biometrika were used. The fingerprint images are created via the consensual method, utilizing gelatine, silicone, and play-doh as spoof materials (Fig. 3.2a, c).

LivDet 2011 presents four datasets related to four optical sensors: Biometrika, Digital Persona, ItalData, and Sagem. Consensual spoofs were fabricated with gelatin, latex, ecoflex, play-doh, silicone, and woodglue (Fig. 3.2d, f).

Table 3.2 Device characteristics for all LivDet datasets. The only kinds of scanner employed in all competitions are optical (O) and thermal swipe (TS)

Scanner	Model	Res. [dpi]	Image size [px]	Type	Edition
Biometrika	FX2000	569	315 × 372	O	2009, 2011
	HiScan-PRO	1000	1000 × 1000	O	2015
Crossmatch	Verifier 300 LC	500	480 × 640	O	2009
	L Scan Guardian	500	800 × 750	O	2013, 2015
Dermalog	LF10	500	500 × 500	O	2021
DigitalPersona	U.are.U 5160	500	252 × 324	O	2015, 2017, 2019
	4000B	500	355 × 391	O	2011
GreenBit	DactyScan26	500	500 × 500	O	2015
	DactyScan84C	500	500 × 500	O	2017, 2019, 2021
Identix	DFR2100	686	720 × 720	O	2009
ItalData	ET10	500	640 × 480	O	2011, 2013
Orcanthus	Certis2 Image	500	300 × N	TS	2017, 2019
Sagem	MSO300	500	352 × 384	O	2011
Swipe	–	96	208 × 1500	TS	2013

The 2013 dataset consists of images from four devices: Biometrika, Crossmatch, Italdata, and Swipe, the first non-optical scanner introduced in the competition. However, this edition stands out for another relevant novelty: the non-cooperative method in producing the spoof images of the Biometrika and Italdata datasets. In contrast, the Crossmatch and Swipe datasets were created with the consensual approach. Except for silicone, the adopted materials are the same as the previous edition, with the addition of body double and modasil (Fig. 3.2g, i).

LivDet 2015 is again characterized by four optical scanners: GreenBit, Biometrika, Digital Persona, and Crossmatch. The fake fingerprints were all created through the cooperative technique. For the Green Bit, Biometrika, and Digital Persona datasets, spoof materials included ecoflex, gelatine, latex, woodglue, a liquid ecoflex, and RTV (a two-component silicone rubber); for the Crossmatch dataset, play-doh, Body Double, Ecoflex, OOMOO (a silicone rubber), and a novel form of gelatin (Fig. 3.2j, l). In this edition, the testing sets included spoof images of unknown materials, i.e., materials which were not included in the training set. This strategy has been used to determine the robustness of algorithms in facing unknown materials. The results obtained prompted the organizers to mantain the same protocol for subsequent editions.

LivDet 2017 consisted of data from three fingerprint sensors: Green Bit DactyScan84C, Digital Persona U.are.U 5160 configured on an Android device, and a new thermal swipe sensor called Orcanthus. The materials adopted to generate the training set are wood glue, ecoflex, body double, whereas gelatine, latex, and liq-

		2009	2011	2013	2015		2017		2019		2021	
Material	**Type**	-	-	-	TR	TS	TR	TS	TR	TS	TR	TS
Body Double	Siliconic		•	•	•	•	•		•			•
Ecoflex	Siliconic		•	•	•	•	•		•			
GLS20	Siliconic											•
Modasil	Siliconic			•								
OOMOO	Siliconic					•						
Play-Doh	Siliconic	•		•	•	•						
RFast30	Siliconic											•
RProFast	Siliconic										•	
RTV	Siliconic					•						
Silicone	Siliconic	•	•									
Latex	Rubber		•	•	•	•		•	•		•	
Gelatine	Hybrid	•	•	•		•		•	•			
Liquid Ecoflex	Hybrid					•		•		•		
Mix1	Hybrid									•		•
Elmers Glue	Glue											•
Mix2	Glue									•		
Woodglue	Glue		•	•	•	•	•		•			

Fig. 3.1 Materials characteristics and frequency over the seven LivDet editions. The train and test materials were completely separated from 2017 to examine the PADs' resilience against "never-seen-before" materials

uid ecoflex compose the test set, namely, it is wholly made up of never-seen-before materials;(Fig. 3.2m, o). It's worth mentioning that this LivDet version also features new characteristics: two sets of persons with varying degrees of ability were involved in manufacturing spoofs and gathering associated images. In this manner, we may simulate a real-world scenario in which the operator's capacity to build the train set differs from that of potential attackers.

The scanners in LivDet 2019 are the same as in previous editions, albeit all three were tested on Windows OS. This edition presents for the first time multi-material combinations, both of different consistency and of different nature. The others were the same of the previous edition (Fig. 3.2p, r).

The latest edition features only two scanners, GreenBit, and Dermalog. For each of them, two datasets were created: the first through the consensual approach and the second through the new pseudo-consensual method, called ScreenSpoof [10]. Except for latex, body double, and "mix1" (already present in previous competitions), new materials have been chosen for the fakes, such as RProFast, Elmers glue, gls20, and RPro30 (Fig. 3.2s, u).

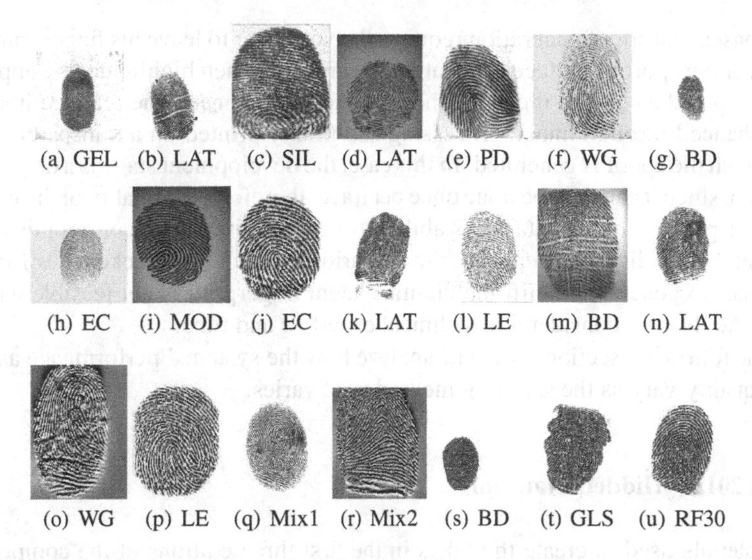

Fig. 3.2 Examples of successful spoof attacks with various materials

3.4.1.1 Specific Challenges

With the passing of the editions, the organizers have made several innovations, both in terms of the production of fakes and in terms of the difficulties that participants must address: LivDet 2013 [11] introduced a dataset with spoofs created by latent fingerprints; LivDet 2015 [12] included some "unknown" materials within the dataset used for detectors' evaluation; in LivDet 2017 [13], train and test sets have been collected by different skilled operators, and they were not completely separated since some users are present in both of them, in order to test the so-called "user-specific" effect [14]; in LivDet 2019 [15] participants were asked to submit a complete algorithm that could output not only the image's liveness probability, but also an integrated score which included the probability of fingerprint matching, and finally LivDet 2021 [4] presented two datasets created using the new and effective ScreenSpoof method. The most significant novelties introduced in the competitions will be discussed thoroughly in the following sections.

LivDet 2013—Consensual and Semi-consensual

In the cooperative artifact fabrication scenario, the subject is requested to press his finger on a moldable and stable substance to get the negative of his fingerprint. The actual fingerprint can be obtained by casting the spoof material on the negative mold. This method is usually considered the "worst-case" scenario since it allows to fabricate high-quality spoofs, although this type of attack has a low probability of being carried out since it requires the victim's collaboration. On the other hand, semi-consensual spoof generation requires the volunteer to leave his fingerprint on a smooth or non-porous surface. The latent fingerprint is then highlighted by applying magnetic powders so that it can be photographed or scanned. The scanned image is then enhanced through image processing and finally printed on a transparent sheet from which the spoof is generated. In this case, the development stage is a destructive operation, since it can only be done once per trace; therefore, the quality of the mold is entirely dependent on the attacker's ability to complete it without compromising the latent mark's quality. Furthermore, the scenario in which the attacker uses forensic techniques to generate and lift the victim's latent fingerprint is not feasible without the attacker having considerable technical expertise and time.

In the following sections, we will analyze how the systems' performance and the fakes' quality vary as the spoofing method used varies.

LivDet 2015—Hidden Materials

The materials used to create the fakes in the first three editions of the competition were essentially the same. This constant may have led competitors to enter parameters capable of recognizing the characteristics of a specific material and, therefore, of easily distinguishing a live fingerprint from a fake. To try to manage any type of material, some materials never used before were included in the LivDet 2015 [12] test set: liquid ecoflex and RTV for the Green Bit, Biometrika, and Digital Persona sensors OOMOO and a new gelatine for Crossmatch. As a matter of fact, in a realistic scenario, it is not possible to determine whether the material used for the attack is present in the dataset used to train the system. Therefore, the competitors' results explain not only the system's accuracy but also its cross-material reliability. This novelty was maintained for all subsequent editions: from 2017 onwards the train and test materials were different to evaluate the robustness of the PADs to the so-called "*never-seen-before*" *materials*.

LivDet 2017—User-Specific Effect and Operator Skill

The 2017 [13] edition of Livdet was focused on determining how much the data composition and collection techniques influence the PAD systems. Although the aim of the competition was clear from the beginning, the following aspects have been analyzed in a later work [16] and only marginally in the LivDet 2017 report for the sake of space: (1) the presence of operators with different skills in fabricating the replicas, one with a considerable degree of expertise (high skilled) and one composed of novice forgers (low skilled) ; (2) the presence of some users in both the training and testing parts of the three datasets. As a matter of fact, in previous editions, the train and test sets were completely separated as none of the users were present in both. Nevertheless, it has been demonstrated [14] that having different acquisitions of the same fingerprint in both parts of the dataset resulted in substantially higher classification results when compared to a system with distinct users in train and test.

The additional information coming from the enrolled user was called "*user-specific effect*", whereas PAD systems are often based on "generic user" techniques because the final user population is unknowable. User-specific information is essential when creating a fingerprint presentation attacks detector that must be incorporated into a fingerprint verification system. It is assumed that the system will have some knowledge of the users in this scenario, which may be used to develop a more robust software module for presentation attacks when authenticating subjects. It is worth noting how the focus was starting to move from stand-alone PAD to *integrated systems* in this edition. This tendency will lead to the next edition "LivDet in action" challenge.

LivDet 2019—LivDet in Action

In the early 2000s, most competitions focused on the matching process, such as the Fingerprint Verification Competition [17].

Since its inception, the Liveness Detection Competition series has been running to provide a standard for assessing PAD algorithms in the same way that matching performance was measured. The subsequent LivDet editions were therefore focused on evaluating the performance of a stand-alone fingerprint liveness detector. Nevertheless, in real-world applications, the FPAD system is generally used in conjunction with a recognition system to determine whether the input is from a genuine user or an impostor attempt to deceive the system [18]. For this purpose, in LivDet 2019 [15], we invited all the competitors to submit a complete algorithm capable of outcoming not only the classic image's liveness probability but also an integrated match score (IMS) that includes the degree of similarity between the given sample and the reference template. Based on state-of-the-art methods in such fields, our investigation provided a clear overview of how integrating a standard verification system with a PAD may affect final performances. This challenge, considering its importance in a real application context, was also maintained for the next LivDet2021 edition.

LivDet 2021—ScreenSpoof Method

In comparison to previous editions, the latest represents a turning point. Firstly, two distinct challenges were designed to test the competitor algorithms: the first one, called "Liveness Detection in Action" in the wake of the previous edition, investigated the impact of combining a PAD system with a fingerprint verification system; the second task, named "Fingerprint representation", was designed to test the compactness and representativeness achieved by feature sets adopted, which are significant factors to ensure reliable performances in terms of accuracy and speed in current biometric systems. Moreover, in this edition, two different methods for creating spoofs were used, as in LivDet 2013. We took inspiration from our recent work [10], where we have shown that it is possible to create fine-quality replicas by properly pre-processing the snapshots of the latent fingerprints left on the smartphone screen surface. We called this technique "*ScreenSpoof*". Although it is a semi-consensual fabrication method, our findings revealed a threat level comparable to that of attacks using spoofs fabricated with the complete consensus of the victim. For this reason, this acquisition protocol was repeated in the last edition. Unlike LivDet 2013, the results obtained for this new method are as effective as the consensual method and, in some specific cases, even more successful.

3.4.2 LivDet Systems

The LivDet Systems competition does not produce the publically available datasets seen in the Algorithms competition; however, with each competition, the dataset collected on a submitted system is supplied to the respective competitors to assist them in improving their systems. Each submission includes a fully packaged fingerprint system with software installation. The systems were equipped with only a liveness module until the 2017 edition [19], in which we also accepted integrated systems, with enrollment, matching and liveness modules. A system outputs a score between 0 and 100 representing liveness or 0–100 for match scores. In particular, in the PAD only competition, a score greater than 50 indicates a live sample, while in the integrated systems competition, a score greater than 50 means that a sample is both a match and live. Each competition contained several PAIs, with a portion of them known to the competitor ahead of time and the remainder unknown. The 2017 systems competition also included the concept of unknown mold types with the use of 3D printed molds [20] for some PAIs being kept from the competitors.

3.4.3 Performance Evaluation

Another significant step forward of the latest edition regards the nomenclature adopted for the performance evaluation, which has been updated in favor of ISO/IEC 30107-1:2016 terminology [21]. Table 3.3 reports the related terms and their meaning. In particular, APCER and BPCER evaluate the performance of the FPADs, whereas FNMR, FMR, and IAPMR assess the performance of an integrated system. Additional metrics often employed and deriving from those previously mentioned are:

- *Liveness Accuracy*: percentages of samples correctly classified by the PAD, that is the inverse of the weighted average of APCER and BPCER.
- *Integrated Matching (IM) Accuracy*: percentages of samples correctly classified by the integrated system, that is the inverse of the weighted average of FNMR, FMR, and IAPMR.

Table 3.3 Correspondence between the ISO terms and the terms used in older Livdet editions (2009–2019)

ISO/IEC 30107-1 term	Old LivDet term	Meaning
Attack Presentation Classification Error Rate (APCER)	FerrFake	Rate of misclassified fake fingerprints.
Bona fide Presentation Classification Error Rate (BPCER)	FerrLive	Rate of misclassified live fingerprints
False non-match rate (FNMR)	100-IMG_accuracy 100-Genuine Acceptance Rate (GAR) False Reject Rate (FRR)	Rate of incorrectly rejected genuine live fingerprints
False match rate (FMR)	False Accept Rate (FAR)	Rate of incorrectly accepted zero-effort impostors
Impostor Attack Presentation Match Rate (IAPMR)	100-IMI_accuracy 100-SGAR	Rate of incorrectly classified impostor live or genuine fake fingerprints

3.5 Examination of Results

Different challenges, different types of PADs, and different materials make the comparison between LivDet editions non-trivial. The purpose of this review is to compare the editions from different points of view to highlight the differences between acquisition sensors, spoofing techniques, and the nature of PADs.

A first comparison is reported in Table 3.4 which shows the mean and standard deviation of APCER and BPCER of the three best algorithms of each edition. Although it is necessary to weigh the differences between edition and edition, there is an evident downward trend in errors in both the classification of live and fake samples.

This trend was interrupted only in the last edition regarding the APCER, which has very high values. This increase is partially explained by the presence of a new forgery methodology, the ScreenSpoof method [10]. This aspect is discussed in more detail in Sect. 3.5.1. The APCER values relating to the consensual acquisition method

Table 3.4 Mean and standard deviation BPCER and APCER values for each LivDet competition

	LivDet 2009	LivDet 2011	LivDet 2013	LivDet 2015	LivDet 2017	LivDet 2019	LivDet 2021
BPCER	22.03 (±9.94)	23.52 (±13.85)	6.26 (±9.18)	5.94 (±3.21)	4,99 (±0.89)	4,15 (±2.65)	1,1 (±0.65)
APCER	12.47 (±6.12)	26.97 (±18.24)	21.55 (±37.21)	6.89 (±3.83)	4,97 (±0.62)	4,75 (±5.71)	27,02 (±23.31)

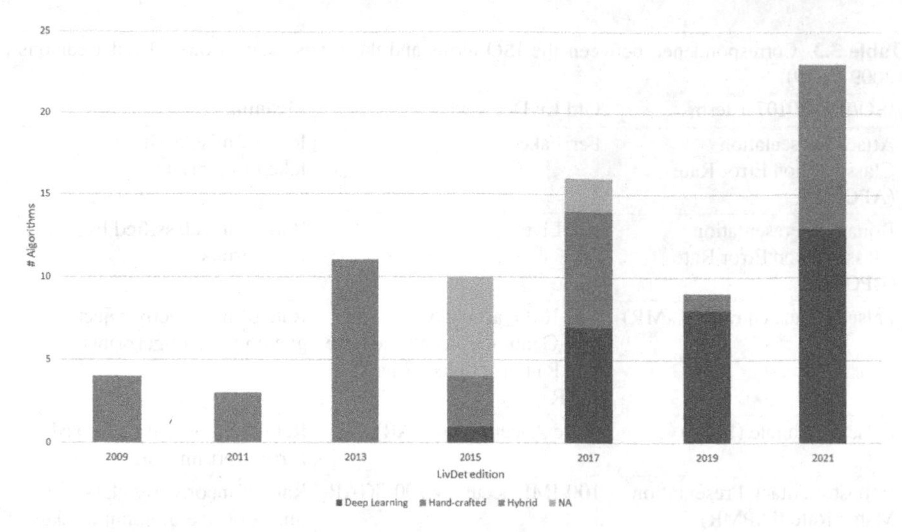

Fig. 3.3 Types of PAD submitted to LivDet—algorithms part

alone are, however, very high: the three best methods obtained an average APCER of 17.13% on the LivDet 2021 dataset acquired through the consensual technique. Another reason for the drop in performance lies in the materials used to manufacture the spoofs: in fact, the 2021 training set is the only one that does not contain any glue-like material; for this reason, we hypothesize that materials of this nature in the test set are very insidious for PADs. The analysis of the materials of the last three editions is reported in Sect. 3.5.2.

The combined examination of Table 3.4 and Fig. 3.3, showing the types of PADs of the various editions, underlines the contribution of deep-learning techniques in liveness detection starting from the 2015 edition, whose winner [22] has presented a CNN-based solution. In the following competitions, submissions based on deep-learning have increased, up to become the most numerous since 2019. In the last edition, several hybrid solutions were presented.

It is important to underline that the difficulty of LivDet datasets has increased since 2015, with the introduction of "never-seen-before" materials in the test sets. Therefore, the leap in the precision of the PADs from 2013 to 2015 and subsequent editions is to be considered more significant.

Figure 3.4 summarizes the average liveness accuracy sensor-by-sensor for the datasets created by consensual method of each LivDet edition. In each bin the range between maximum and minimum accuracy is shown by whiskers. It is easy to observe that within each edition there is a great variability between sensors, demonstrating the great influence of the technical specifications of each of them. Obviously, this graph is to be read in light of the considerations of the composition of the datasets (e.g., consensual and non-consensual) and the challenges of each edition.

Nevertheless, the joint analysis of liveness accuracy and IM accuracy and IAMPR, respectively in the Figs. 3.4, 3.5, highlights the following behavior: a higher IM

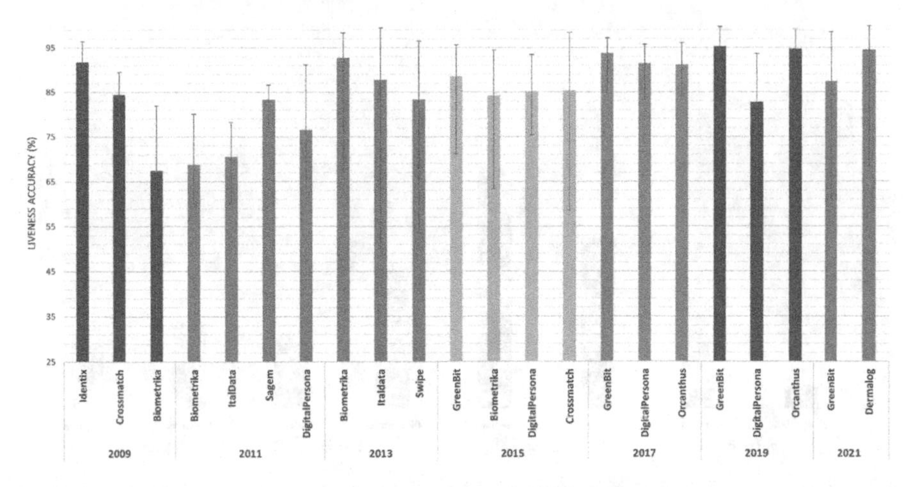

Fig. 3.4 Liveness accuracy for each competition dataset from 2009 to 2021

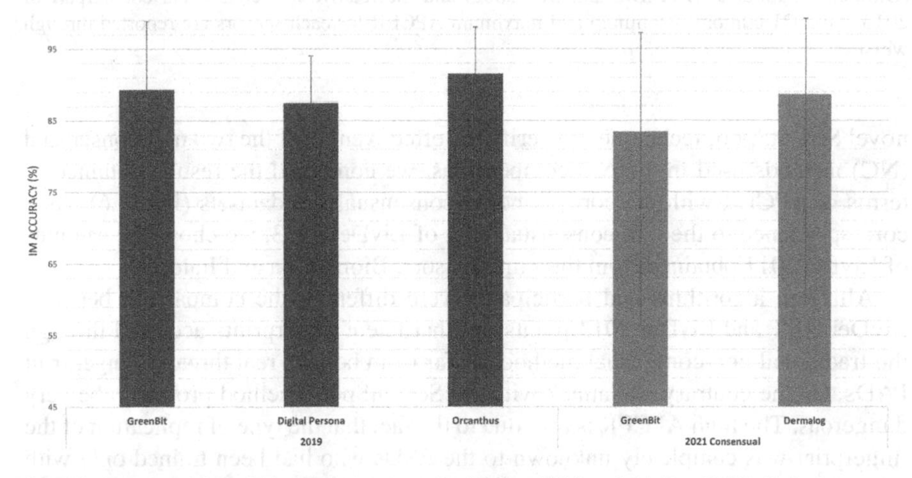

Fig. 3.5 Integrated Matching (IM) accuracy for LivDet 2019 and LivDet 2021. Only these two editions have evaluated the performance of the integrated PAD and matcher systems

accuracy always corresponds to a better liveness accuracy, independent of the dataset. This suggests that the PAD system has a significant impact on the performance of the integrated system, regardless of the fusion rules and matching algorithms used.

3.5.1 Non Consensual Verses Consensual data

In the 2013 and 2021 editions, in addition to the consensual datasets, two non-consensual datasets were created, using *latent fingerprints*. The LivDet 2011 subset contains latents obtained through magnetic powder, while LivDet 2021 uses the

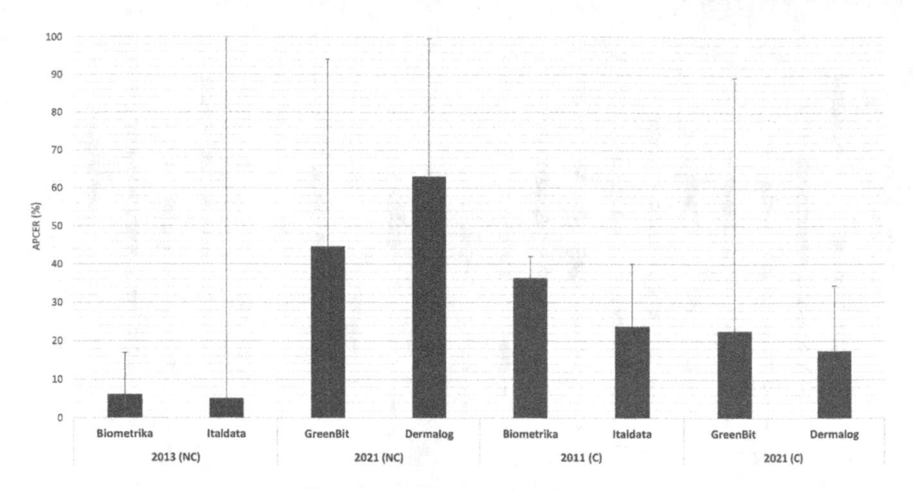

Fig. 3.6 Comparison between APCER for consensual datasets of LivDet 2011 and 2021 (Biometrika, Italdata, GreenBit, and Dermalog) and the relative non-consensual counterpart of 2013 and 2021 editions. Minimum and maximum APCER for each sensors are reported through whiskers

novel ScreenSpoof technique. To verify the effectiveness of the two non-consensual (NC) methods used in the two competitions, we compared the results obtained in terms of APCER with the corresponding consensual (C) datasets (Fig. 3.6). As a correspondence to the non-consensual data of LivDet 2013, we chose the samples of LivDet 2011 obtained from the same sensors, Biometrika and Italdata.

Although algorithms and participants were different, the comparison between LivDet 2011 and LivDet 2013 points out that latent fingerprints acquired through the traditional non-consensual method are far from being a real threat to fingerprint PADs. On the contrary, the attack with the ScreenSpoof method proved to be very dangerous. The high APCER is also due to the fact that the type of replication of the fingerprint was completely unknown to the PADs who had been trained only with fake obtained by the consensual method. The higher error rate, the lower effort to manufacture the fingerprint artifacts, and the complete unawareness of the user make this attack much more realistic than the classic consensual method and highlight the need to make future PADs "aware" of this eventuality.

3.5.2 *Materials Analysis*

The type of material used to make the spoof highly influences the success of an attack on a PAD [23]. To evaluate the generalization capacity of the PADs and simulate a realistic and dangerous eventuality, the test sets of the last three editions of LivDet contained only materials "never seen before", therefore different from those of the

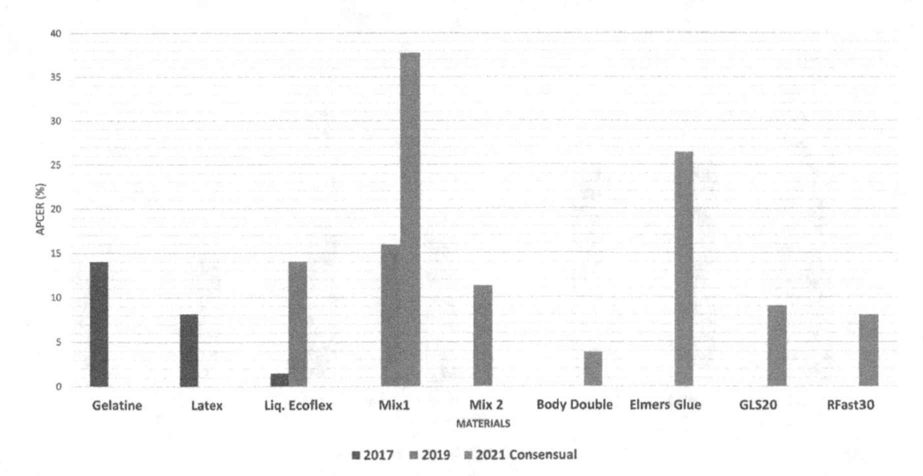

Fig. 3.7 Comparison of APCER for test set materials from the last three editions of LivDet (consensual datasets only)

training sets. Figure 3.7 analyzes the rates of misclassified fake fingerprints for each material used in the consensual test datasets of the last three LivDet editions.

The most immediate comparison can be made between the test materials of LivDet 2017 and LivDet 2019 that share the training set. As always, a certain variability due to the different PADs submitted must be taken into account.

However, the behavior of the liquid ecoflex deserves attention as it turns out to be very easy to recognize in the 2017 edition while it is on average with the other materials in 2019. It is worth noting that in 2017 two characteristics helped the PADs in the classification. The first is the presence of users in the test shared with the training set to analyze the so-called user-specific effect. The second is that an inexperienced operator built part of the 2017 test set to evaluate the difference between the attackers' skills. We, therefore, believe that these two details and the difference in PADs lead to the marked difference in APCER of the two editions for this material.

On the other hand, they do not explain the significant performance disparity for the other material in common across the two editions. Mix 1 is present in the test set of LivDet 2019 and LivDet 2021. These editions present a very different training set. In particular, the training set of LivDet2021 contains only silicone and rubbers materials (Fig. 3.1). PADs trained with these data compositions do not learn the characteristics of the gluey materials and cannot distinguish them from live samples. In support of this thesis, we can evidence how the second most challenging material to classify is Elmer's glue. In contrast, the other three silicone materials have a lower error rate.

To understand if the fakes made up of these two materials are actually a risk to the security of the fingerprint recognition system, we also report the results on the integrated system. Figure 3.8 highlights how the spoofs made with the two glue-like

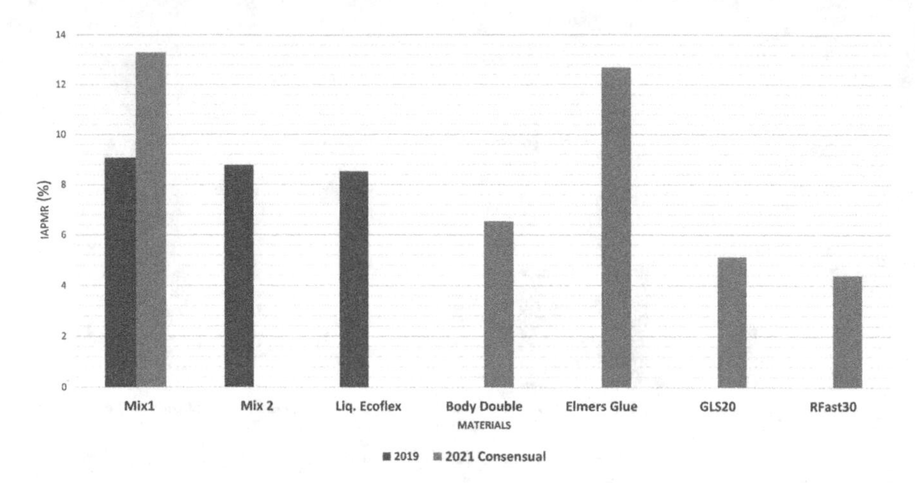

Fig. 3.8 Comparison of IAPMRs for test set materials from the last three editions of LivDet (consensual datasets only)

materials are dangerous not only from the liveness detection point of view but also manage to cheat the matcher and therefore report a high IAPMR. Another evidence that emerges from the materials analysis is related to the Body Double: although the PADs easily recognize it, it achieves a high IAMPR. This demonstrates the high quality of spoofs that keep information relating to the individual (minutiae).

3.5.3 *LivDet Systems Results*

The number of submitted systems remained low for the LivDet Systems competition until the 2017 competition which saw a drastic rise in the amount of competitors (Table 3.5). Starting from the two presented in LivDet 2011, in 2017, there were a total of 14 submitted system to the two sub-competition: nine systems submitted (with a 10th unofficial submission) for the PAD only competition and five for the integrated PAD and matcher systems competition [19].

Table 3.5 Number of systems submitted to the LivDet Systems competition

LivDet systems edition	2011	2013	2015	2017 PAD Only	2017 Integrated systems
# Participants	2	2	1	9(10)	5

Table 3.6 Error rates by competitor for LivDet Systems 2017 PAD only competition

System	APCER	BPCER	ACE
F1	2.06	4.45	3.25
ZF1	1.62	1.33	1.475
ZF1+	2.61	0.91	1.76
LF10	3.23	0.82	2.025
GB	10.08	1.89	5.985
Anon1	7.62	22.6	15.11
Anon2	10.63	10.19	10.41
Anon3	8.17	20.18	14.175
Anon4	0.4	18.4	9.4

Across these competitions it can be seen that the general trend of error rates has decreased with one exception. The winning system in Livdet 2013 had the best overall results from all of the Livdet Systems competitions.

In 2011, Dermalog performed at a BPCER of 42.5% and a APCER of 0.8%. GreenBit performed at a BPCER of 38.8% and a APCER of 39.47%. Both systems had high BPCER scores. The 2013 edition produced much better results since Dermalog performed at a BPCER of 11.8% and a APCER of 0.6%. Anonymous1 performed at a BPCER of 1.4% and a APCER of 0.0%. Both systems had low APCER rates. Anonymous1 received a perfect score of 0.0% error, successfully determining every spoof finger presented as a spoof. Anonymous2, in 2015, scored a BPCER of 14.95% and a APCER of 6.29% at the (given) threshold of 50 showing an improvement over the general results seen in LivDet 2011, however, the anonymous system did not perform as well as what was seen in LivDet 2013. For LivDet Systems 2017 (Table 3.6), from the nine submissions for the PAD only competition, Dermalog ZF1 performed the best with an APCER of 1.62% and a BPCER of 1.33%. The second place system is the Dermalog ZF1+ with an APCER of 2.61% and a BPCER of 0.91% and the third place system is the Dermalog LF10 with an APCER of 3.23% and a BPCER of 0.82%. In both Livdet Systems 2015 and Livdet 2017, the most common problem with submitted systems was play-doh. In LivDet 2015, the play-doh color contributed significantly to the system's issues with spoofs, resulting in a 28% APCER and thus accounting for a substantial portion of the total 6.29% APCER. In LivDet 2017, even though the Dermalog ZF1 had an APCER of 1.62 percent, it had a 10% error rate for play-doh spoofs.

Overall, LivDet Systems 2017 showed a tremendous improvement in the error rates recorded from the systems. Increased in the strength of algorithms for detecting PAIs has improved throughout constant testing of systems and there is the expectation that a future LivDet Systems competition will show even more growth.

3.6 Conclusion

The LivDet competition series is becoming increasingly crucial in bringing together the most recent solutions and assessing the state of the art in fingerprint PAD. The continuous increase of competitors over the years clearly indicates the growing interest in this research topic. This chapter presented the design and results of the seven Liveness Detection International Competitions on software-based fingerprint liveness detection methods and fingerprint systems with artifact detection capabilities. Each competition has presented new challenges to the research community, resulting in unique solutions and new insights, proving to be essential in taking fingerprint liveness detection to the next level. However, this heterogeneity has made the comparison of the results more challenging. Overall, the liveness detection algorithms and systems have improved significantly over time, even under harsh conditions such as including never-before-seen materials in the test set or introducing the new Screen-Spoof fabrication method. On the other hand, the investigation of integrated systems performed in the last two editions revealed that PAD algorithms achieved an average accuracy of around 90%. This implies that a typical PAD would miss 10% of both presentation attacks and legitimate presentations. Obviously, the acceptability of this error rate depends on the application, but in general, PAD technology in this context does not appear to be fully mature. Furthermore, the reported spoof material analysis confirmed two points: (1) the operator's skill to create and exploit counterfeit replicas has a significant impact on performance, as seen by a comparison of the 2017 and 2019 test sets; (2) since PAD systems rely mostly on the training-by-example method, the representativeness of the training set has a serious influence on their accuracy. This was especially evident in the 2021 edition, where the absence of gluey material in the training set resulted in an unexpected increase in the error rate for such materials. In conclusion, we believe that the findings presented in this chapter will be beneficial to academics and companies working on presentation attack detection, encouraging them to improve existing solutions and uncover new ones. For our part, we will continue on this path and keep posing always new challenges to help fight this still, unfortunately, open problem.

References

1. Matsumoto T, Matsumoto H, Yamada K, Hoshino S (2002) Impact of artificial "gummy" fingers on fingerprint systems. In: van Renesse RL (ed) Optical security and counterfeit deterrence techniques IV, vol 4677. International Society for Optics and Photonics, SPIE, pp 275 – 289. https://doi.org/10.1117/12.462719
2. Marcel S, Nixon MS, Li SZ (2014) Handbook of biometric anti-spoofing: trusted biometrics under spoofing attacks. Springer Publishing Company, Incorporated
3. Marcialis GL, Lewicke A, Tan B, Coli P, Grimberg D, Congiu A, Tidu A, Roli F, Schuckers S, (2009) First international fingerprint liveness detection competition-livdet, (2009). In: Foggia P, Sansone C, Vento M (eds) Image analysis and processing - ICIAP 2009. Springer, Berlin, Heidelberg, pp 12–23

4. Casula R, Micheletto M, Orrù G, Delussu R, Concas S, Panzino A, Marcialis GL (2021) Livdet 2021 fingerprint liveness detection competition - into the unknown. In: 2021 IEEE international joint conference on biometrics (IJCB), pp 1–6. https://doi.org/10.1109/IJCB52358.2021.9484399
5. Willis D, Lee M (1998) Six biometric devices point the finger at security. Netw Comput 9(10):84–96
6. Arora SS, Jain AK, Paulter NG (2017) Gold fingers: 3d targets for evaluating capacitive readers. IEEE Trans Inf Forensics Secur 12(9):2067–2077. https://doi.org/10.1109/TIFS.2017.2695166
7. Jain AK, Feng J, Nagar A, Nandakumar K (2008) On matching latent fingerprints. In: 2008 IEEE computer society conference on computer vision and pattern recognition workshops, pp 1–8. https://doi.org/10.1109/CVPRW.2008.4563117
8. Sodhi G, Kaur J (2001) Powder method for detecting latent fingerprints: a review. Forensic Sci Int 120(3):172–176. https://doi.org/10.1016/S0379-0738(00)00465-5
9. Marasco E, Ross A (2014) A survey on antispoofing schemes for fingerprint recognition systems. ACM Comput Surv 47(2):28:1–28:36. https://doi.org/10.1145/2617756
10. Casula R, Orrù G, Angioni D, Feng X, Marcialis GL, Roli F (2021) Are spoofs from latent fingerprints a real threat for the best state-of-art liveness detectors? In: 2020 25th international conference on pattern recognition (ICPR). IEEE, pp 3412–3418
11. Ghiani L, Yambay D, Mura V, Tocco S, Marcialis GL, Roli F, Schuckers S (2013) Livdet 2013 fingerprint liveness detection competition 2013. In: 2013 international conference on biometrics (ICB), pp 1–6. https://doi.org/10.1109/ICB.2013.6613027
12. Mura V, Ghiani L, Marcialis GL, Roli F, Yambay DA, Schuckers S (2015) Livdet 2015 fingerprint liveness detection competition 2015. In: 2015 IEEE 7th international conference on biometrics theory, applications and systems (BTAS), pp 1–6. https://doi.org/10.1109/BTAS.2015.7358776
13. Mura V, Orrù G, Casula R, Sibiriu A, Loi G, Tuveri P, Ghiani L, Marcialis GL (2018) Livdet 2017 fingerprint liveness detection competition 2017. In: 2018 international conference on biometrics (ICB), pp 297–302. https://doi.org/10.1109/ICB2018.2018.00052
14. Ghiani L, Marcialis GL, Roli F, Tuveri P (2016) User-specific effects in fingerprint presentation attacks detection: insights for future research. In: 2016 international conference on biometrics (ICB). IEEE,, pp 1–6
15. Orrù G, Casula R, Tuveri P, Bazzoni C, Dessalvi G, Micheletto M, Ghiani L, Marcialis GL (2019) Livdet in action - fingerprint liveness detection competition 2019. In: 2019 international conference on biometrics (ICB), pp 1–6 (2019). https://doi.org/10.1109/ICB45273.2019.8987281
16. Orrù G, Tuveri P, Ghiani L, Marcialis GL (2019) Analysis of "user-specific effect" and impact of operator skills on fingerprint pad systems. In: International conference on image analysis and processing. Springer, pp 48–56
17. Cappelli R, Ferrara M, Franco A, Maltoni D (2007) Fingerprint verification competition 2006. Biom Technol Today 15(7–8):7–9
18. Chingovska I, Anjos A, Marcel S (2013) Anti-spoofing in action: Joint operation with a verification system. In: 2013 IEEE conference on computer vision and pattern recognition workshops, pp 98–104. https://doi.org/10.1109/CVPRW.2013.22
19. Yambay D, Schuckers S, Denning S, Sandmann C, Bachurinski A, Hogan J (2018) Livdet 2017 - fingerprint systems liveness detection competition. In: 2018 IEEE 9th international conference on biometrics theory, applications and systems (BTAS), pp 1–9. https://doi.org/10.1109/BTAS.2018.8698578
20. Engelsma JJ, Arora SS, Jain AK, Paulter NG (2018) Universal 3d wearable fingerprint targets: advancing fingerprint reader evaluations. IEEE Trans Inf Forensics Secur 13(6):1564–1578. https://doi.org/10.1109/TIFS.2018.2797000
21. ISO/IEC 30107-3 (2016) Information technology-Biometric presentation attack detection-Part 3: testing and reporting
22. Nogueira RF, de Alencar LR, Campos Machado R (2016) Fingerprint liveness detection using convolutional neural networks. IEEE Trans Inf Forensics Secur 11(6):1206–1213. https://doi.org/10.1109/TIFS.2016.2520880

23. Chugh T, Jain AK (2021) Fingerprint spoof detector generalization. IEEE Trans Inf Forensics Secur 16:42–55. https://doi.org/10.1109/TIFS.2020.2990789
24. Zhang Y, Fang S, Xie Y, Xu T (2014) Fake fingerprint detection based on wavelet analysis and local binary pattern. In: Sun Z, Shan S, Sang H, Zhou J, Wang Y, Yuan W (eds) Biometric recognition. Springer International Publishing, Cham, pp 191–198
25. González-Soler LJ, Gomez-Barrero M, Chang L, Pérez-Suárez A, Busch C (2021) Fingerprint presentation attack detection based on local features encoding for unknown attacks. IEEE Access 9:5806–5820. https://doi.org/10.1109/ACCESS.2020.3048756
26. Keilbach P, Kolberg J, Gomez-Barrero M, Busch C, Langweg H (2018) Fingerprint presentation attack detection using laser speckle contrast imaging. In: 2018 international conference of the biometrics special interest group (BIOSIG), pp 1–6 (2018). https://doi.org/10.23919/BIOSIG.2018.8552931
27. Hussein ME, Spinoulas L, Xiong F, Abd-Almageed W (2018) Fingerprint presentation attack detection using a novel multi-spectral capture device and patch-based convolutional neural networks. In: 2018 IEEE international workshop on information forensics and security (WIFS), pp 1–8. https://doi.org/10.1109/WIFS.2018.8630773
28. Tolosana R, Gomez-Barrero M, Kolberg J, Morales A, Busch C, Ortega-Garcia J (2018) Towards fingerprint presentation attack detection based on convolutional neural networks and short wave infrared imaging. In: 2018 international conference of the biometrics special interest group (BIOSIG), pp 1–5. https://doi.org/10.23919/BIOSIG.2018.8553413

Chapter 4
A Unified Model for Fingerprint Authentication and Presentation Attack Detection

Aditya Popli, Saraansh Tandon, Joshua J. Engelsma, and Anoop Namboodiri

Abstract Typical fingerprint recognition systems are comprised of a spoof detection module and a subsequent recognition module, running one after the other. In this study, we reformulate the workings of a typical fingerprint recognition system. We show that both spoof detection and fingerprint recognition are correlated tasks, and therefore, rather than performing these two tasks separately, a joint model capable of performing both spoof detection and matching simultaneously can be used without compromising the accuracy of either task. We demonstrate the capability of our joint model to obtain an authentication accuracy (1:1 matching) of TAR = 100% @ FAR = 0.1% on the FVC 2006 DB2A dataset while achieving a spoof detection ACE of 1.44% on the LivDet 2015 dataset, both maintaining the performance of standalone methods. In practice, this reduces the time and memory requirements of the fingerprint recognition system by 50 and 40%, respectively; a significant advantage for recognition systems running on resource constrained devices and communication channels.

Additya Popli and Saraansh Tandon—These authors have contributed equally.

A. Popli (✉) · S. Tandon (✉) · A. Namboodiri (✉)
IIIT Hyderabad, Hyderabad, India
e-mail: additya.popli@research.iiit.ac.in

S. Tandon
e-mail: saraansh.tandon@research.iiit.ac.in

A. Namboodiri
e-mail: anoop@iiit.ac.in

J. J. Engelsma
Michigan State University, East Lansing, USA
e-mail: engelsm7@msu.edu

S. Marcel et al. (eds.), *Handbook of Biometric Anti-Spoofing*, Advances in Computer Vision and Pattern Recognition, https://doi.org/10.1007/978-981-19-5288-3_4

4.1 Introduction

Due to their widespread usage in many different applications, fingerprint recognition systems are a prime target for attackers. One of the most widely known methods of attack is known as a presentation attack (PA), which can be realized through the use of commonly available materials like gelatin, play-doh, and silicone or more expensive and sophisticated 3D printing techniques (these subsets of presentation attacks are also known as spoof attacks). To counter these attacks, various fingerprint presentation attack detection (FPAD) approaches to automatically detect and flag spoof attacks prior to performing authentication have been proposed [1].

The typical pipeline of a fingerprint recognition system equipped with such a spoof detector is illustrated in Fig. 4.1a. First, a fingerprint image acquired by the fingerprint reader is passed to a spoof detector, where a decision is made as to whether the image comes from a live (bonafide) fingerprint or from a fake (spoof) fingerprint. If the image is determined to be bonafide, it is passed on to the authentication module for matching. In such a system, the spoof detector can run in parallel with the authentication module (only releasing the decision if the spoof detector decides bonafide), or the spoof detector can run in series with the authentication module (as a pre-filter before ever running authentication). Both modules running in parallel requires significant memory, especially for recent spoof detection and matching algorithms that leverage deep convolutional neural networks (CNNs). If instead the algorithms were to run in series, memory can be saved, but the time needed to release a decision to the user is increased. Both of these limitations are most prominently manifested on resource constrained devices and weak communication channels. Given these limitations, we propose a reformulation to the typical workings of a fingerprint recognition

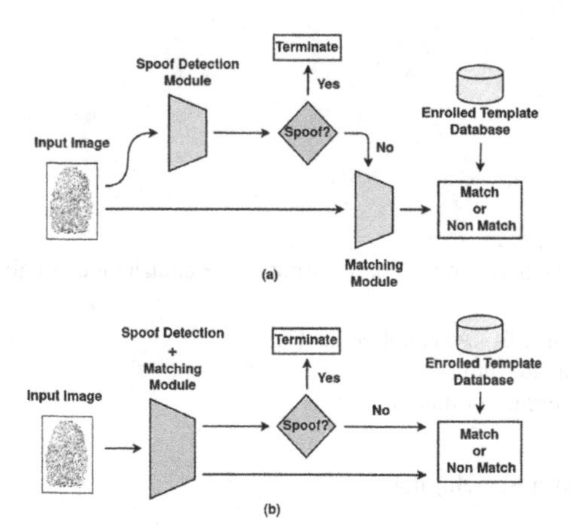

Fig. 4.1 **a** Standard fingerprint recognition pipeline with separate networks running either in parallel (larger memory requirement) or series (increased time) for fingerprint spoof detection and matching **b** Proposed pipeline with a common network for both the tasks that reduce both time and memory consumption without significantly affecting the accuracy of either task

system (Fig. 4.1a). We posit that FPAD and fingerprint matching are correlated tasks and propose to train a model that is able to jointly perform both functions. Our system design is shown in Fig. 4.1b. Our joint model maintains the accuracy of published stand-alone FPAD and matching modules, while requiring 50% and 40% less time and memory, respectively.

Our motivation for coupling FPAD and fingerprint matching into a single model and our intuition for doing so are based upon the following observation. Many FPAD algorithms and most fingerprint matching algorithms rely heavily on Level 1 (ridge-flow), Level 2 (minutiae), and Level 3 (pores) features. For example, minutiae points have been shown to provide significant utility in both fingerprint matching [2], and FPAD [3] systems. Fingerprint pores have also been used for both tasks [4, 5].

Given the possibility of correlation between FPAD and fingerprint matching tasks (i.e., they both benefit by designing algorithms around similar feature sets), we conduct a study to closely examine the relationship between these two tasks with the practical benefit of reducing the memory and time consumption of fingerprint recognition systems. Our work is also motivated by similar work in the face recognition and face PAD domains where the correlation of these two tasks was demonstrated via a single deep network for both tasks [6]. However, to the best of our knowledge, this is the first such work to investigate and show that PAD and recognition are related in the fingerprint domain where different feature sets are exploited than in the face domain.

More concisely, our contributions are as follows:

- A study to examine the relationship between fingerprint matching and FPAD. We show that features extracted from a state-of-the-art fingerprint matcher ([7]) can also be used for spoof detection. This serves as the motivation to build a joint model for both the tasks.
- The first model capable of simultaneously performing FPAD and fingerprint matching. Figure 4.2 shows that the embeddings extracted from this model for bonafide and spoof images are well separated while keeping the distance between embeddings extracted from different impressions of the same fingerprint together.
- Experimental results demonstrating matching accuracy[1] of TAR = 100% @ FAR = 0.1% on FVC 2006 and fingerprint presentation attack detection ACE of 1.44% on LiveDet 2015, both similar to the performance of individual published methods.
- A reduction in time and memory requirements for a fingerprint recognition system of 50% and 40%, respectively, without sacrificing significant system accuracy. Our algorithm has significant advantages for resource constrained fingerprint recognition systems, such as those running on smartphone devices.

[1] The terms authentication, verification and matching have been used interchangeably to refer to a 1:1 match surpassing a threshold.

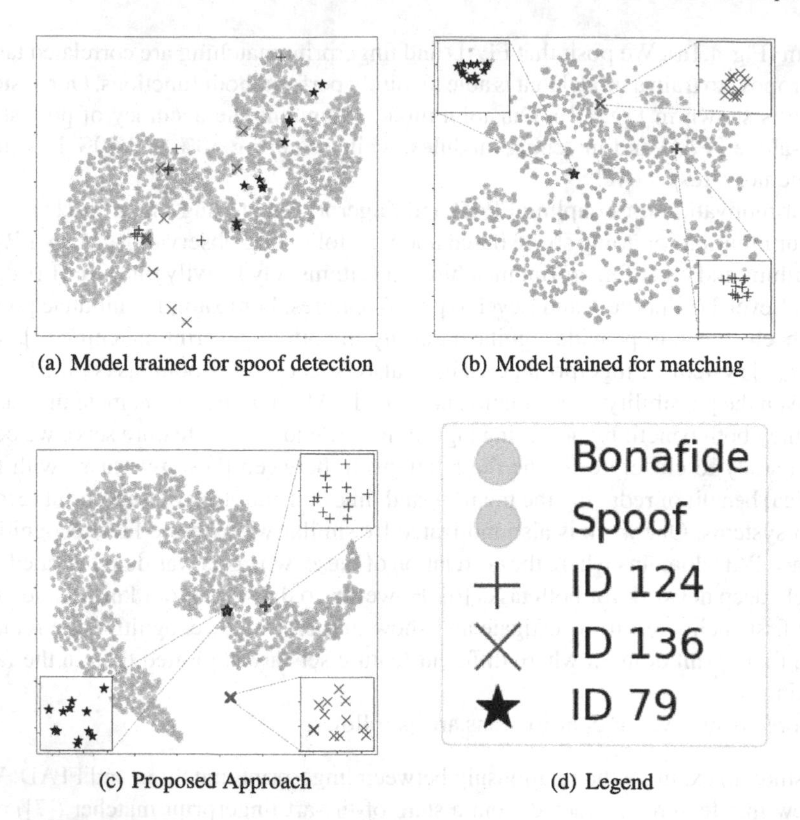

Fig. 4.2 t-SNE visualization of the embeddings generated by models trained for spoof detection and fingerprint matching. Unlike **a** and **b** which only extract embeddings useful for their respective task (i.e., spoof detection, or matching), our model **c** is able to extract embeddings which can be used for both tasks

4.2 Related Work

4.2.1 Fingerprint Spoof Detection

Fingerprint spoof detection approaches that have been proposed in the literature can be broadly classified into hardware-based and software-based solutions [1]. While hardware-based solutions rely upon detecting the physical characteristics of a human finger with the help of additional sensor(s), software-based approaches extract features from fingerprint images already captured for the purpose of fingerprint matching and do not require any additional sensors [1]. Early software-based approaches relied on hand-crafted or engineered features extracted from the fingerprint images to classify them as either bonafide or spoofs [1]. However, more recent approaches are based upon deep Convolutional Neural Networks (CNN) and have been shown to significantly outperform previous approaches. The current state-of-the-art CNN based method [3] utilized multiple local patches of varying resolutions centered and aligned around fingerprint minutiae to train two-class MobileNet-v1 [8] and Inception v3 [9] networks.

We posit that another important limitation of many state-of-the-art FPAD algorithms that needs to be further addressed is that of their computational complexity. In particular, the recent CNN based FPAD algorithms are memory and processor intensive algorithms (e.g., the Inception v3 architecture utilized by several state-of-the-art approaches [3] is comprised of 27M parameters and requires 104 MB of memory) that will cause challenges on resource constrained devices such as smartphones. Even "lighter weight" CNN models such as MobileNet (used in [3]) will also add computational complexity to a resource constrained device, particularly when utilized in line with a current fingerprint recognition pipeline (first perform spoof detection, then perform matching). Therefore, in this work, we aim to alleviate some of the computational burden of the FPAD module in a fingerprint recognition system via a joint model which performs both fingerprint matching and FPAD.

4.2.2 Fingerprint Matching

A plethora of work has been done in the area of fingerprint matching. In our work, we are primarily interested in deep-learning-based solutions since they (i) have shown to outperform traditional solutions for fingerprint spoof detection [3] and on some databases, even for fingerprint matching [7, 10] and (ii) can be easily modified for simultaneous matching and spoof detection via a shared architecture and modified loss function.

One of the most well-known hand-crafted approaches for fingerprint recognition is the Minutiae Cylinder Code (MCC) [2]. More recently, CNNs have been used to extract fixed-length representations of fingerprints [7] and also for learning local minutiae descriptors [10]. In particular, the authors in [7] proposed a network called DeepPrint which demonstrated high levels of matching accuracy on several benchmark datasets. The limitation of DeepPrint's fixed-length representation is that it may fail when the probe and enrollment fingerprint are comprised of significantly non-overlapping portions of the fingerprint. In [10] a CNN was used to extract local minutiae descriptors to boost latent fingerprint matching accuracy. Given the high accuracy and open source code provided by [10], we incorporate their approach into our joint model.

4.3 Motivation

As an initial experiment to check whether there exists a relationship between fingerprint matching and spoof detection, we trained two separate models for spoof detection and matching, and visualized their GradCam [11] outputs, as shown in Fig. 4.3. We observed that initial layers for both the models extracted very similar features.

To further verify this observation, we trained a spoof detection network on top of the features extracted by DeepPrint [7]. Specifically, we used the Inception v3

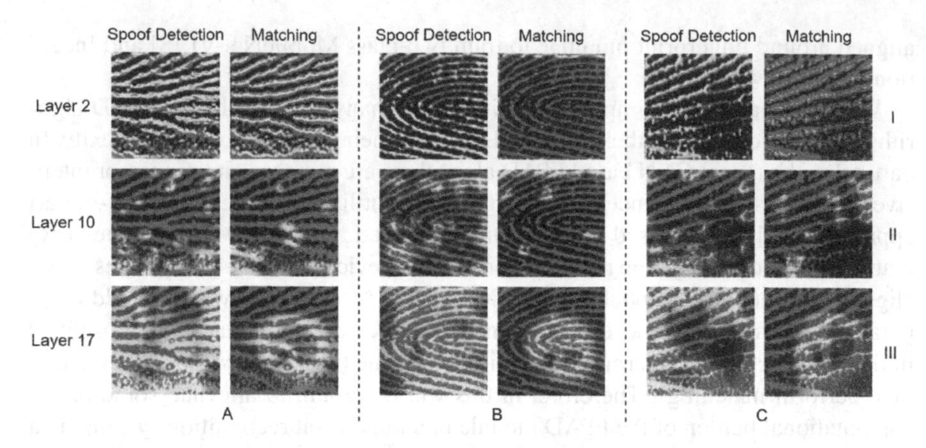

Fig. 4.3 GradCam [11] outputs on three sample fingerprint patches showing the areas-of-focus for the spoof detection and matching models. We can see that the initial layers for both the models extract similar features in the form of ridge information in the second layer and minutiae information in the tenth layer, and hence can be combined together. However, the deeper layers extract very different features, and hence require separate specialized networks for each task

Table 4.1 Custom spoof detection head

Layer	Kernel/Stride	Output size
conv (groups = 288)	$3 \times 3/1$	$36 \times 36 \times 288$
conv	$1 \times 1/1$	$36 \times 36 \times 512$
conv (groups = 512)	$3 \times 3/2$	$17 \times 17 \times 512$
conv	$1 \times 1/1$	$17 \times 17 \times 1024$
conv (groups = 1024)	$3 \times 3/2$	$8 \times 8 \times 1024$
Pool	$8 \times 8/8$	$1 \times 1 \times 1024$
Flatten		1024
Linear		2

[9] network as the student model[2] of DeepPrint by training it to extract embeddings from fingerprints as close as possible to the embeddings extracted by DeepPrint for the same input images obtained from the NIST SD-300 and NIST SD-302 datasets. Next, we use the intermediate feature maps of size 35×35 extracted from the *Mixed_5d* layer of our student model and pass them as inputs to a shallow spoof detection network. This shallow network is as described in Table 4.1. We observe that this spoof detection model is able to classify these intermediate identity-related feature maps as live or spoof with high accuracy (Table 4.2). It is important to note that our shallow spoof detection model is trained on features maps of size 35×35 which are much smaller than the original student model input size (of the fingerprint image) of 448×448.

[2] We have used a student model since the data used in the original paper is not publicly available and the authors of [7] agreed to share the weights of their model to use it as a teacher network.

Table 4.2 FPAD on LivDet 15 [12] using feature maps from a DeepPrint [7] student model (matcher) as input

Sensor	Cross match	Digital persona	Green Bit	HiScan	Avg.
ACE (%)	7.43	13.48	6.87	10.2	**9.49**

After establishing (experimentally) that common features can be used for both fingerprint matching and spoof detection, we focus our efforts on a single, joint model that is capable of performing both these tasks while maintaining the accuracy of published stand-alone approaches.

4.4 Methodology

To train a single model capable of both fingerprint spoof detection and fingerprint matching, we build a multi-branch CNN called DualHeadMobileNet (Fig. 4.4). The input to our model is a 96 × 96 minutiae-centered patch (resized to 224 × 224 according to the model input size). One branch of the CNN performs the task of spoof detection (outputting "spoofness" scores between 0 and 1 for the patch), while the other branch extracts a local minutiae descriptor of dimension 64 from the patch

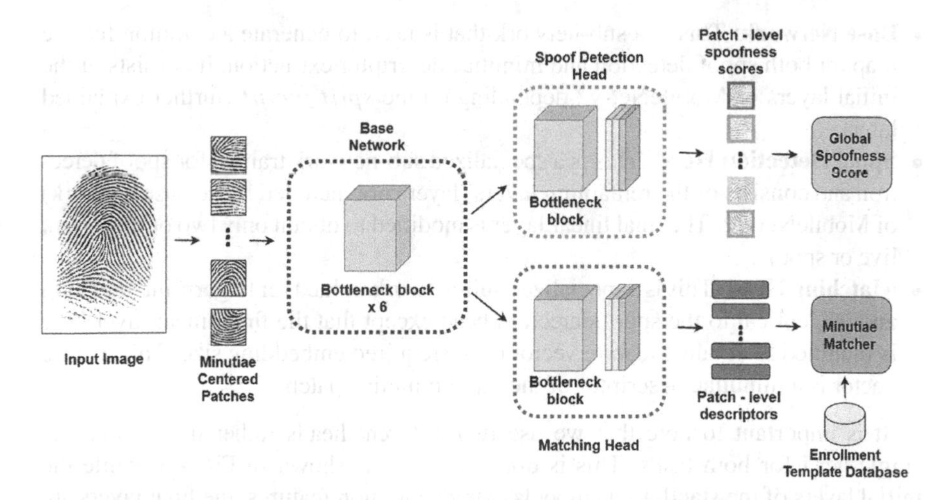

Fig. 4.4 Proposed architecture with $split\_point = 1$. Minutiae-centered patches of size 96 × 96 are extracted from the input image and fed to the base network which extracts a common feature map. This feature map is then fed to the spoof detection and matching heads to obtain patch-level spoofness scores and minutiae descriptors. The patch-level scores are averaged to obtain an image level spoofness score and the patch-level descriptors from two different fingerprints are fed to the matching algorithm of [10] to obtain a similarity score

(for the purpose of matching). Both branches share a common stem network. This parameter sharing is what enables us to reduce the time and memory constraints of our fingerprint recognition system. Rather than training two networks for two tasks, we share a number of parameters in the stem (given our hypothesis that these two tasks are related and can share a number of common parameters). To obtain a final spoof detection decision, we average the "spoofness" scores of all the minutiae-centered patches extracted from a given fingerprint. To conduct matching, we aggregate all of the minutiae descriptors extracted from all the minutiae-centered patches in a given fingerprint image and subsequently feed them to the minutiae matcher open sourced in [10].

4.4.1 DualHeadMobileNet (DHM)

Given the success of [3] in using MobileNet-v1 for spoof detection, we have used MobileNet-v2 [13] as the starting point for our DualHeadMobileNet. We have chosen MobileNet-v2 over MobileNet-v1 [8] because it (i) obtains higher classification accuracy on benchmark datasets (ii) uses 2 times fewer operations and (iii) needs 30 percent fewer parameters as compared to MobileNet-v1. Since we are highly cognizant of computational complexity, less parameters are a significant motivating factor in our selection.

We modify the MobileNet-v2 architecture for our experiments in the following manner:

- **Base Network**: This is a sub-network that is used to generate a common feature map for both spoof detection and minutiae descriptor extraction. It consists of the initial layers of MobileNet-v2 depending on the $split\_point$ (further explained later).
- **Spoof Detection Head**: This is a specialized sub-network trained for spoof detection and consists of the remaining layers (layers not included in the base network) of MobileNet-v2. The final linear layer is modified to obtain only two outputs (i.e., live or spoof).
- **Matching Head**: This is a specialized sub-network trained for fingerprint matching and is identical to the spoof detection head except that the final linear layer here is modified to obtain a feature vector of the required embedding size. This feature vector is a minutiae descriptor of the input minutiae patch.

It is important to note that we use two different heads rather than using one single head for both tasks. This is done because, as shown in Fig. 4.3, while the initial layers of the stand-alone models extract common features, the final layers are very specialized and extract only task-specific features. To indicate the number of bottleneck blocks in the base network and the separate head networks, we use the variable $split\_point$ where a network with $split\_point = x$ means that x bottleneck blocks are used in each head network, and 7-x bottleneck blocks are present in the base network. The complete model architecture is shown in Fig. 4.4.

Please refer to Sect. 4.6.4 for experiments using other networks as backbone architectures.

4.4.2 *Joint Training*

The two heads are trained separately to optimize the loss for their specific tasks while the common base network is trained to optimize a weighted sum of these individual losses, as shown in Eq. 4.1 where w_{sd} and w_m are the weights for the spoof detection and matching loss.

$$\mathcal{L}_j = w_m \mathcal{L}_m + w_{sd} \mathcal{L}_{sd} \tag{4.1}$$

The spoof detection head is trained via a cross-entropy loss ($\mathcal{L}_{sd}$) based on the ground truth of the input minutiae patch (i.e., if the input patch is a live or spoof). Meanwhile, the matching head outputs descriptors are regressed to ground truth minutiae descriptors using an L2-norm ($\mathcal{L}_m$). These ground truth minutiae descriptors are extracted from the minutiae patch using the algorithm in [10], i.e., the model in [10] is a teacher model which our DHM seeks to mimic as a student. Due to the non-availability of a large and patch-wise labeled public dataset, we use a student-teacher framework which eliminates the need for identity labels during the training procedure.

4.5 Experiments and Results

After training our DHM, we conduct experiments to demonstrate the capability of our joint model to accurately perform both spoof detection and matching. In doing so, we demonstrate that spoof detection and fingerprint matching are indeed correlated tasks. Please refer to Appendix 4.8 for implementation and hyperparameter details.

4.5.1 *Datasets*

For spoof detection, we report on the LivDet 2015 [12] and LivDet 2017 [14] datasets. For fingerprint matching, we report on the FVC 2000 [15], FVC 2002 [16], FVC 2004 [17] and FVC 2006 [18] datasets following the official protocols. Each sensor of the FVC 2002 and 2004 datasets contains 800 fingerprint images (100 fingers $\times$ 8 impressions), leading to 2800 genuine ($= \frac{8\times7}{2} \times 100$) and 4950 imposter ($= \frac{100\times99}{2}$) comparisons, while each sensor within the FVC 2006 dataset contains 1680 images (140 fingers $\times$ 12 impressions) leading to 9240 ($= \frac{12\times11}{2} \times 140$) genuine and 9730 ($= \frac{140\times139}{2}$) imposter comparisons (Fig. 4.5).

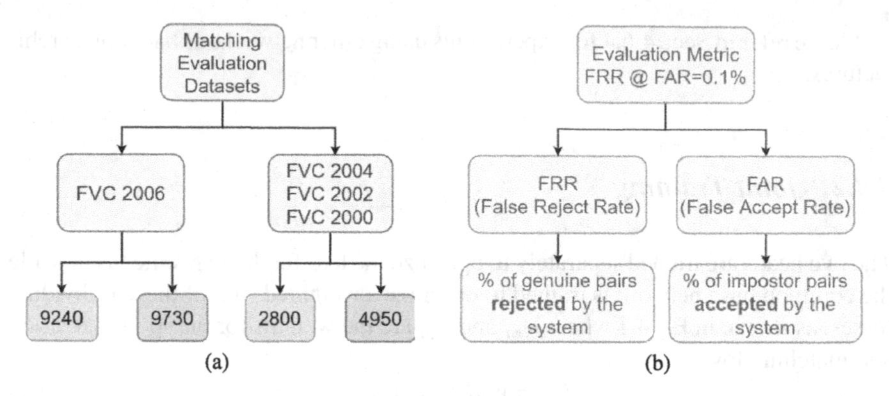

Fig. 4.5 **a** The green boxes represent the number of genuine pairs and the red boxes represent the number of imposter pairs in the datasets. **b** Description of the metric used to compare matching performances

The Orcanthus sensor of the LivDet 2017 dataset and non-optical sensors of the FVC datasets have been excluded for evaluation since we want to ensure that the sensor used to acquire our training images (e.g., an optical sensor from LiveDet training partition) uses similar sensing technology as the sensor used to acquire our testing images (e.g., an optical sensor from FVC). Therefore, we utilize only those subsets of data from LiveDet and FVC which both uses optical sensing. A future improvement to this work could be to make our model invariant to the differences in sensor characteristics observed across different sensing technologies.

4.5.2 Comparison with State-of-the-art Methods

4.5.2.1 Spoof Detection

We use our implementation of Fingerprint Spoof Buster [3] (current SOTA) as a FPAD baseline. Since the authors did not provide official code or models, we train our own models in order to present a fair comparison both in terms of accuracy and system constraints (refer to Sect. 4.5.3). For the reasons mentioned in Sect. 4.4.1, we use MobileNet-v2 (unless specified otherwise) instead of the MobileNet-v1 network. Also, since our goal is to develop a fast and memory-efficient system, we only train a single model using patches of size 96 × 96, as opposed to an ensemble of models trained on different patch sizes (as done by the authors in [3]). This baseline (referred to as *m*-FSB) achieves an average classification error of 1.48% on the LivDet 2015 dataset [12] compared to 0.98% reported in [3] (using an ensemble of models) and even outperforms the original model on two of the four sensors present in the dataset.

Table 4.3, shows the spoof detection results of our re-implementation of the baseline [3] as well as our joint model. The proposed joint model achieves comparable

Table 4.3 Spoof detection classification errors†

Dataset	Sensor	BPCER		APCER		ACER	
		m-FSB	DHM	*m*-FSB	DHM	*m*-FSB	DHM
LivDet 2015	Cross match	0.40	**0.20**	0.35	0.35	0.38	**0.28**
	Digital persona	2.90	**2.50**	**3.13**	4.07	**3.01**	3.28
	Green bit	0.40	0.40	0.60	**0.47**	0.50	**0.44**
	Hi scan	0.80	0.80	3.27	**2.74**	2.04	**1.77**
LivDet 2017	Digital persona	3.25	**3.07**	3.77	**3.72**	3.51	**3.40**
	Green bit	**3.06**	3.83	2.20	**2.00**	**2.63**	2.92

† *m*-FSB refers to the Fingerprint Spoof Buster [3] baseline and DHM refers to the proposed method.

results to the baseline stand-alone spoof detector and even outperforms it on 4 out of the 6 sensors of the LivDet 2015 [12] and LivDet 2017 [14] datasets. The joint model also outperforms the best performing algorithm of the LivDet 2017 Fingerprint Liveness Detection Competition (as per the results reported in [14]) on the Digital Persona sensor (ACE 3.4 vs. 4.29%) and achieves similar performance on the GreenBit sensor (ACE 2.92 vs. 2.86%).

4.5.2.2 Matching

Let the matching algorithm output similarity scores between two input fingerprints. Therefore, the impostor pairs should have low scores and the genuine pairs should have high scores. We pick a threshold score such that all scores lower than that are predicted as impostors and all scores greater than that are predicted as genuine. For the metric this threshold score is picked so that x% of impostor scores are predicted as False Accepts. Then we report the False Reject rate at this threshold. Lower the value, better is the performance.

We compare our results with the following baselines:

- **MCC** [2]: We use VeriFinger SDK v11 (commercial SDK) for minutiae detection and the official SDK provided by the authors for matching.
- **DeepPrint** [7]: We use the weights provided by the authors for feature extraction and follow the official protocol for calculating the matching scores.
- **LatentAFIS** [10]: Since we use this method as a teacher for training our matching branch, we also compare our results with this method. We use the weights and the matching algorithm open-sourced[3] by the authors to obtain the matching score.

[3] https://github.com/prip-lab/MSU-LatentAFIS.

Table 4.4 Comparison of the matching performance (FRR % @ FAR = 0.1%) on FVC datasets

Method	Train dataset	2006	2004		2002		2000	
		DB2A	DB1A	DB2A	DB1A	DB2A	DB1A	DB3A
VeriFinger v11	–	0.00	2.86	3.01	0.11	0.07	0.11	1.04
DeepPrint [7]	–	0.32	**2.43**	5.93	7.61	10.32	4.57	6.50
MCC [2]	–	0.03	7.64	<u>5.6</u>	1.57	0.71	1.86	2.43
LatentAfis [10] ‡	–	**0.00**	5.64	7.17	<u>0.82</u>	**0.46**	1.25	**1.79**
DHM ‡	CrossMatch	**0.00**	5.89	6.99	**0.75**	<u>0.50</u>	<u>1.04</u>	**1.79**
	DigitalPersona	**0.00**	6.36	6.81	1.14	0.61	1.14	2.36
	GreenBit	0.02	7.25	**5.34**	<u>0.82</u>	<u>0.50</u>	1.11	2.14
	HiScan	**0.00**	5.57	6.46	<u>0.82</u>	<u>0.50</u>	<u>1.04</u>	2.21
	DigitalPersona	<u>0.01</u>	6.32	7.03	0.93	0.54	1.32	<u>1.86</u>
	GreenBit	<u>0.01</u>	<u>5.25</u>	7.57	0.89	**0.46**	**0.96**	2.18

‡ This refers to the DualHeadMobileNet network trained using joint training as described in Sect. 4.4.2 with $split\_point = 0$, $w_{sd} = 1$ and $w_m = 10$. The top 4 rows for DHM correspond to the sensors in the LivDet 2015 dataset while the bottom 2 rows correspond to sensors in the LivDet 2017 dataset.
Best and second best results for the published matchers are in bold and underline respectively

We also provide the matching results of a commercial fingerprint matcher (VeriFinger v11 SDK) for reference. While our joint model does not outperform VeriFinger on its own, we note that (i) our joint model does outperform all existing baselines taken from the open literature in many testing scenarios and (ii) the existing published baselines are also inferior to Verifinger in most testing scenarios.

To best the performance of Verifinger, two problems would need to be solved: (1) We would need more discriminative minutiae descriptors than that which LatentAFIS currently provides as our ground truth for DHM. Indeed LatentAFIS is somewhat of an upperbound for DHM in terms of matching since LatentAFIS is the teacher of DHM. (2) We would have to further improve the minutiae matching algorithm of [10]. While both of these tasks constitute interesting research, they fall outside of the scope of this study which is focused on demonstrating the correlation of FPAD and matching and showing that the performance of existing systems for each task can be ***maintained*** when combining both tasks into a single network - not on obtaining SOTA minutiae matching performance. By maintaining the matching accuracy of LatentAFIS [10] with DHM with simultaneous FPAD, we have met these primary objectives.

It is important to note that the same joint models have been used for spoof detection (Table 4.3) and matching (Table 4.4), while the baselines are stand-alone models for their respective task. For example, the results corresponding to the fifth row in Table 4.4 and the first column (of the DHM) in Table 4.3 refer to the same model (weights) trained on the images of the CrossMatch sensor of the LivDet 15 [12] dataset.

Please refer to Appendix 4.6.3 for more experimental results.

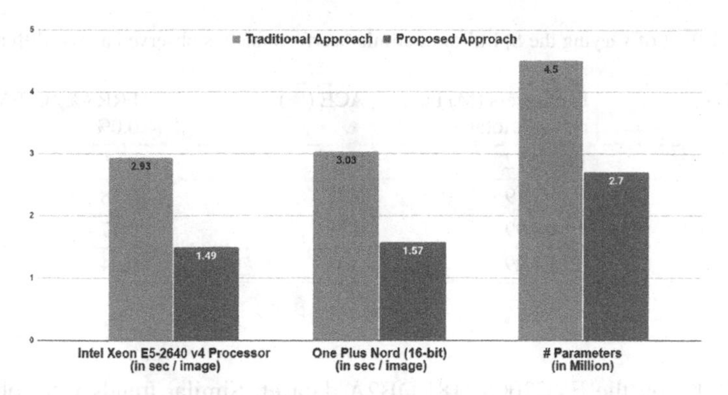

Fig. 4.6 Comparison of resource requirements during inference

4.5.3 Time and Memory

A fingerprint recognition system comprised of a "front-end" spoof detector such as that of m-FSB and a fingerprint matcher such as that in [10] requires 2.93 s of inference time per image[4], 1.46 s for spoof detection and 1.47 s for descriptor extraction for matching (assuming avg. number of minutiae of 49 as per the LivDet 15 dataset and batched processing) on an Intel Xeon E5-2640 v4 processor. In comparison, the **proposed approach consists of only a single network for both of the tasks and takes only 1.49 s, providing a speed up of 49.15%**. The number of parameters and the memory requirements of the proposed approach are 2.7 M and 10.38 MB compared to 4.5 M and 17.29 MB of the traditional approach (Fig. 4.6).

Since this reduction in space and time consumption is most useful in resource constrained environments, we also benchmark the inference times of the traditional method and our proposed joint method on a OnePlus Nord mobile phone. While the traditional pipeline takes 62.18 ms (per patch since mobile libraries do not support batch processing), the proposed approach takes only 32.03 ms for feature extraction, resulting in a 48.49% reduction in the feature extraction time. In this case, we first quantize the models (both baseline and joint) from 32-bits to 8-bits since quantization has shown to provide approximately 80% speed up without a significant drop in accuracy [19].

4.6 Ablation Study

All the ablation studies were done using the LivDet 2015 [12] CrossMatch dataset (training split) for training. In Sects. 4.6.1 and 4.6.2 we report the Spoof Detection ACE on LivDet 2015 [12] CrossMatch testing dataset and Matching FRR (%) @

[4] We ignore the time required for the minutiae matching since it is common to both the traditional and proposed pipeline and takes only $\sim$1 ms.

Table 4.5 Effect of varying the Split Point: similar performance is observed across different split points

Split point	Parameters (M) base network/total	ACE (%)	FRR (%) @ FAR = 0.0%
0	1.81/2.72	0.28	0.06
1	1.34/3.19	0.41	0.18
2	0.54/3.99	0.38	0.17
3	0.24/4.29	0.41	0.04

FAR = 0.0% on the FVC2006 [18] DB2A dataset. Similar trends were observed with other datasets.

4.6.1 Effect of Varying the Split Point

We vary the *split_point* of the DHM to further examine the correlation between fingerprint matching and spoof detection. A model with a higher *split_point* value indicates that the base network is shallower and hence the two specialized heads are more independent of each other. As shown in Table 4.5, we notice that there is very little improvement for both spoof detection and matching even when we allow deeper specialized heads (i.e., increase the *split_point*, and consequently the number of model parameters). This indicates that even when some bottleneck layers are allowed to train independently for the two tasks, they extract similar features. In a conventional fingerprint recognition pipeline, these redundant features would waste time and space, however our joint model eliminates these wasteful parameters by sharing them across both tasks (Fig. 4.7).

4.6.2 Effect of Suppression

In order to better understand the degree of dependence between these two tasks, we try to train models to perform only one task (e.g., matching) while suppressing any information which helps the other task (spoof detection, in this case). A decrease in matching performance due to suppression of spoof detection or vice versa would indicate a strong dependence between the two tasks. In this case, the two model heads are trained similarly as in Joint Training (Sect. 4.4.2), the only difference being in the merging of the two losses at the split point. In this case, the gradient flowing into the base network from the suppressed branch is first negated and then added to the gradient flowing from the other branch, similar to the technique proposed in [20] (refer diagram 4.8).

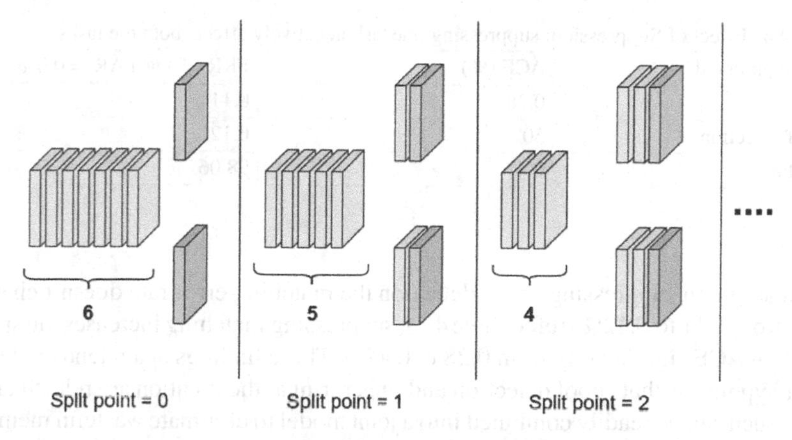

Fig. 4.7 Varying split point: Here the yellow blocks belong to the Base Network, green blocks belong to the Spoof Detection Head and the red blocks belong to the Matching Head

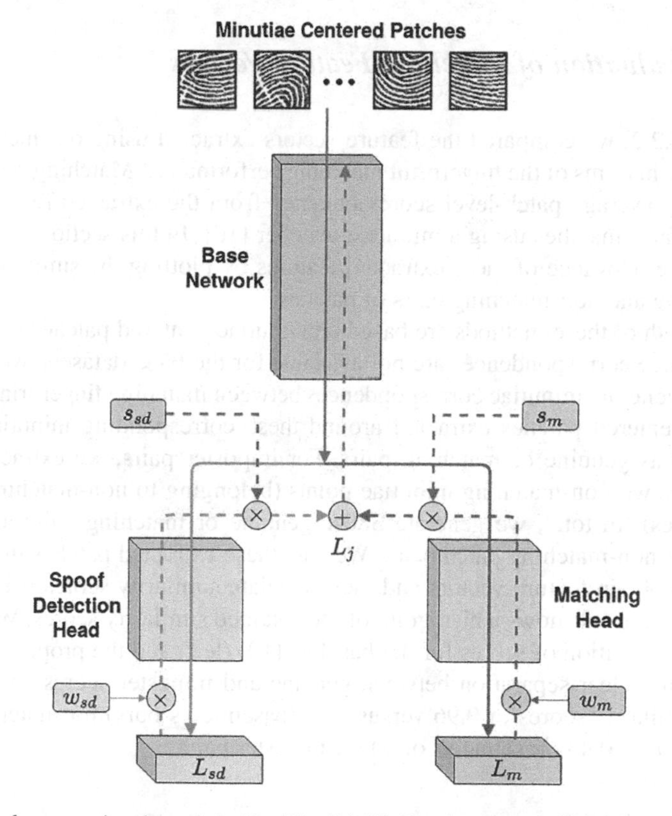

Fig. 4.8 Backpropagation through the DualHeadMobileNet: As shown in Fig. 4.8, the spoof detection and matching heads learn only from their individual weighted gradients (w_{sd} for spoof detection and w_m for matching), while the base network learns from their sum to serve as a common feature extractor. s_{sd} and s_m are set to -1 if spoof detection or matching (respectively) are suppressed, otherwise both are set to 1

Table 4.6 Effect of Suppression: suppressing one task negatively affects both the tasks

Loss suppressed	ACE (%)	FRR (%) @ FAR = 0.0%
None	0.28	0.11
Spoof detection	50	0.12
Matching	0.47	98.06

Although on suppressing spoof detection the matching error rate doesn't change much from 0.11 to 0.12% (refer Table 4.6), suppressing matching increases the spoof detection ACE significantly from 0.28 to 0.47%. These findings again lend evidence to our hypothesis that spoof detection and fingerprint authentication are related tasks and as such can be readily combined into a joint model to eliminate wasteful memory and computational time.

4.6.3 Evaluation of Matching Feature Vectors

In Sect. 4.5.2.2, we compared the feature vectors extracted using our method with the baseline in terms of the fingerprint matching performance. Matching scores were obtained by average patch-level scores obtained from the extracted features, after removing false matches using a minutiae matcher [10]. In this section, we directly compare the relevance of these extracted features by plotting the similarity scores for matching and non-matching pairs of patches.

Since both of these methods are based on minutiae-centered patches and ground truth minutiae correspondences are not available for the FVC datasets, we use VeriFinger to generate minutiae correspondences between matching fingerprint images. Minutiae-centered patches extracted around these corresponding minutiae points are treated as genuine or matching pairs. For imposter pairs, we extract patches around any two non-matching minutiae points (belonging to non-matching fingerprint images). In total, we generate 380K genuine or matching pairs and 380 K imposter or non-matching patch pairs. We pass these extracted patches through the network to obtain feature vectors and then calculate similarity scores using cosine distance. Figure 4.9 shows a histogram of the obtained similarity scores. We can see a similar distribution of scores for the baseline [10] (left) and the proposed method (right) with a clear separation between genuine and imposter scores. We observe average similarity scores of 0.96 versus 0.96 (baseline vs ours) for matching pairs and 0.41 versus 0.43 (baseline vs ours) for imposter pairs.

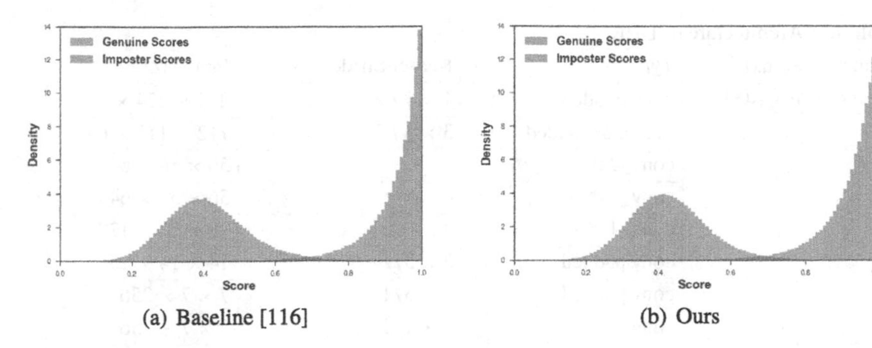

(a) Baseline [116]

(b) Ours

Fig. 4.9 Histogram plot of the obtained matching scores. We can see a similar distribution of scores for both cases with a clear separation between genuine and imposter scores

4.6.4 Robustness to Network Architecture

In DHM we have used MobileNet-v2 [13] as a starting point for our architecture. In order to examine the robustness of the proposed methodology to the underlying architecture, we also experiment with the popular ResNet-18 [21] and Inception v3 [22] networks and develop DualHeadResNet (DHR) and DualHeadInception (DHI) architectures analogous to DualHeadMobileNet. The architectures of DualHeadResNet (DHR) and DualHeadInception (DHI) are as follows.

4.6.5 DHR

We create the DHR network by splitting the ResNet-18 network [21] at the $conv\_5x$ block and reducing the number of channels from 512 to 256 for that block (since our aim is to obtain a compact model for both the tasks) as shown in Table 4.7. This new model consists of approximately 6.4M parameters (4M out of which are part of the base network and are hence common for both spoof detection and matching) in comparison to 10.4 M of a baseline system consisting of two separate ResNet-18 networks.

4.6.6 DHI

Similarly, we create the DHI network by splitting the Inception v3 network [22] after the *Mixed_7b* block as shown in Table 4.8. This new model consists of approximately 28M parameters (15.2M out of which are part of the base network and are hence common for both spoof detection and matching) in comparison to 43.7M of a baseline

Table 4.7 Architecture of DHR

Branch (Params)	Type	Kernel/Stride	Input size
Base Network (4M)	conv padded	$7 \times 7 / 2$	$224 \times 224 \times 3$
	maxpool padded	$3 \times 3 / 2$	$112 \times 112 \times 64$
	conv_2x*	–	$56 \times 56 \times 64$
	conv_3x*	–	$56 \times 56 \times 64$
	conv_4x*	–	$28 \times 28 \times 128$
	conv padded	$3 \times 3/2$	$14 \times 14 \times 256$
	conv padded	$3 \times 3/1$	$7 \times 7 \times 256$
	conv	$1 \times 1/2$	$7 \times 7 \times 256$
Spoof Detection Head (1.18M)	conv padded	$3 \times 3/1$	$4 \times 4 \times 256$
	conv padded	$3 \times 3/1$	$4 \times 4 \times 256$
	avg pool	4×4	$4 \times 4 \times 256$
	linear [256×2]	–	256
Matching Head (1.19M)	conv padded	$3 \times 3/1$	$4 \times 4 \times 256$
	conv padded	$3 \times 3/1$	$4 \times 4 \times 256$
	avg pool	4×4	$4 \times 4 \times 256$
	linear [256×64]	–	256

*ResNet blocks as defined in Table 1 of [21]

system consisting of two separate Inception v3 networks, while maintaining the accuracy for both the tasks.

As shown in Table 4.9, changing the underlying architecture has little effect on both the spoof detection and matching performance.

4.7 Failure Cases

We perform a qualitative analysis to understand the cases where our system fails. In particular, we focus on the fingerprint matching performance on the FVC 2004 DB1A dataset because our system (and also the baseline matching algorithm [10] on top of which we build our system) performs considerably worse than the best fingerprint matcher (DeepPrint [7]) for this particular database.

4.7.1 False Rejects

Figure 4.10a shows a subset of mated-minutiae pairs (obtained using VeriFinger and manually verified) across two different acquisitions of the same finger which

Table 4.8 Architecture of DHI

Branch (Params)	Type	Kernel/Stride	Input size
Base Network (15.2M)	conv	3 × 3 / 3	448 × 448 × 3
	conv	3 × 3 / 1	149 × 149 × 32
	conv padded	3 × 3 / 1	147 × 147 × 32
	pool	3 × 3 / 2	147 × 147 × 64
	conv	1 × 1 / 1	73 × 73 × 64
	conv	3 × 3 / 1	73 × 73 × 80
	pool	3 × 3 / 2	71 × 71 × 192
	3 × Inception A*	–	35 × 35 × 192
	5 × Inception B*	–	35 × 35 × 288
	2 × Inception C*	–	17 × 17 × 768
Spoof Detection Head (6M)	Inception C*	-	8 × 8 × 2048
	pool	8 × 8	8 × 8 × 2048
	linear [2048 × 2]	–	1 × 1 × 2048
	softmax	–	2
Matching Head (6.2M)	Inception C*	–	8 × 8 × 2048
	pool	8 × 8	8 × 8 × 2048
	linear [2048 × 64]	–	2048

*Inception A, Inception B and Inception C refer to the three types of Inception blocks defined in [9]

Table 4.9 Performance comparison of the DHR and DHI networks: proposed method is robust to the backbone architecture

Matching Performance (FRR (%) @ FAR = 0.1%)

Method	FVC 2006 DB2A	FVC 2004 DB1A	FVC 2004 DB2A
DHM	0.00	5.89	6.99
DHR	0.00	5.54	5.81
DHI	0.02	5.43	6.46

Spoof Detection Perf. (LivDet 15 CrossMatch sensor)

Method	ACE (%)	$\mathbf{E}_{fake}$ @ $\mathbf{E}_{live} = 0.1\%$
ResNet-18 [21]	0.5	0.9
DHR	0.48	0.69
Inception v3 [9]	0.44	0.55
DHI	0.28	0.76

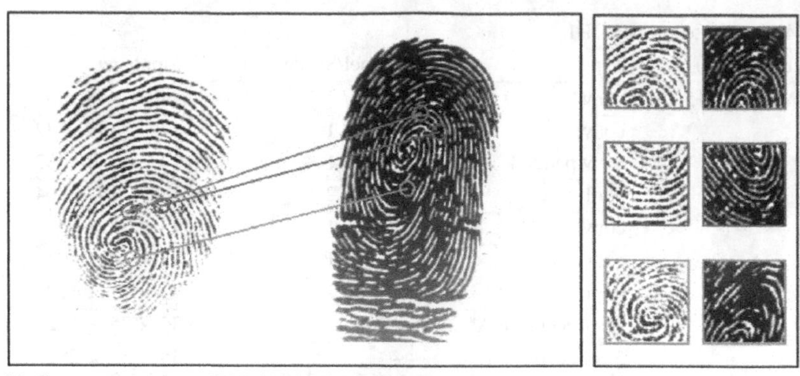

(a) Genuine pair from FVC 2004 DB1A falsely rejected by the DHM. Matching minutiae pairs missed by the DHM have been marked.

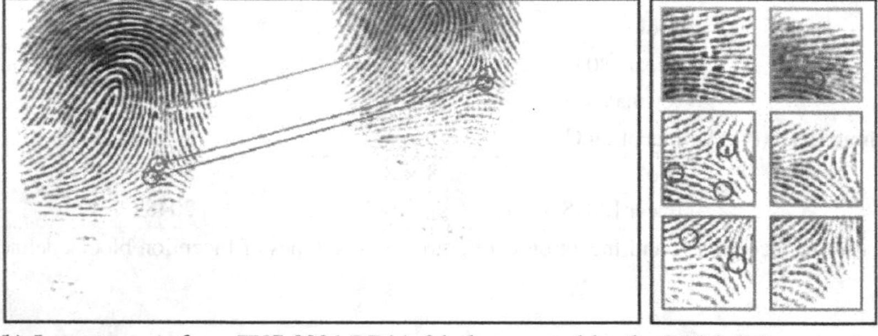

(b) Imposter pair from FVC 2004 DB1A falsely accepted by the DHM. False minutiae pair correspondences have been marked on the left and minutiae points differentiating the falsely corresponding minutiae patches have been marked in blue (right).

Fig. 4.10 Failure Analysis: In both Fig. 4.10a, b three minutiae pairs and their corresponding patches have been highlighted with different colors for ease of viewing

were rejected by DHM. Due to large distortion and difference in moisture, patches extracted around the same minutiae point (different acquisitions) are visually dissimilar and hence produce descriptors with low similarity. Global fingerprint matching techniques [7] which rely on the global ridge-flow information instead of local minutiae patches are more robust to local deformations and hence perform better in such cases.

4.7.2 False Accepts

Fig. 4.10b shows acquisitions from two different fingers that are accepted as a genuine pair by our matcher. Although both the global fingerprint images and local minutiae patches look very similar and have some similar non-matching minutiae patches, there also are imposter minutiae patches (marked in non-blue) with spurious minutiae (marked in blue circles) that should be possible a network to differentiate. We believe that the network learns the general ridge-flow structure instead of specific differences like missing minutiae points and is hence unable to differentiate between the two patches. These cases we believe can be addressed by incorporating attention maps or minutiae-heat maps (as done by [7]) to guide the network to focus on more discriminative features in the minutiae descriptor.

4.8 Implementation Details

Input images are first processed (as done by [23]) to crop out the region of interest from the background, followed by extraction of minutiae-centered and oriented patches (using VeriFinger SDK) of size 96 × 96 and their descriptors using [10]. We only used patches and descriptors of a single-type (minutiae-centered, size 96 × 96) instead of an ensemble of descriptors since (i) we did not observe much boost in the matching performance on using multiple descriptors and (ii) using an ensemble slows down the system considerably which goes against our motivation. For training, 20% of the data for each sensor was used for validation for the ReduceLROnPlateau scheduler (initial learning rate of 10^{-3} with patience 10 until 10^{-8}) and Adam optimizer. Weights for the spoof detection and matching loss, i.e., w_{sd} and w_m were set to 1 and 10 respectively according to the order of the losses. For inference, we have used the minutiae matching algorithm [10] to aggregate and compare the descriptors extracted by our network to obtain similarity scores between two fingerprints. All experiments have been performed using PyTorch framework and two Nvidia 2080-Ti GPUs.

4.9 Conclusion

Existing fingerprint recognition pipelines consist of FPAD followed by a matching module. These two tasks are often treated independently using separate algorithms. Our experimental results indicate that these tasks are indeed related. In practice, this enables us to train a single joint model capable of performing FPAD and authentication at levels comparable to published stand-alone models while reducing the memory and time of the fingerprint recognition systems by 50 and 40%. We have also shown that our algorithm is applicable to patch-based fingerprint recognition systems as well as full image recognition systems. In our ongoing research, we are investigating ways to further reduce the memory and computational complexity of fingerprint recognition systems, without sacrificing system accuracy. This will have tremendous benefits for fingerprint recognition systems running on resource constrained devices and communication channels.

References

1. Marasco E, Ross A (2014) A survey on antispoofing schemes for fingerprint recognition systems. ACM Comput Surv 47(2). https://doi.org/10.1145/2617756
2. Cappelli R, Ferrara M, Maltoni D (2010) Minutia cylinder-code: a new representation and matching technique for fingerprint recognition. IEEE Trans Pattern Anal Mach Intell 32(12):2128–2141
3. Chugh T, Cao K, Jain AK (2018) Fingerprint spoof buster: use of minutiae-centered patches. IEEE Trans Info For Sec 13(9):2190–2202
4. Marcialis GL, Roli F, Tidu A (2010) Analysis of fingerprint pores for vitality detection. In: 2010 20th international conference on pattern recognition, pp 1289–1292
5. Jain AK, Chen Y, Demirkus M (2007) Pores and ridges: high-resolution fingerprint matching using level 3 features. IEEE Pattern Anal Mach Intell 29(1):15–27
6. Ying X, Li X, Chuah MC (2018) Liveface: a multi-task CNN for fast face-authentication, pp 955–960. https://doi.org/10.1109/ICMLA.2018.00155
7. Engelsma JJ, Cao K, Jain AK (2019) Learning a fixed-length fingerprint representation. In: IEEE on pattern analysis and machine intelligence, p 1
8. Howard AG, Zhu M, Chen B, Kalenichenko D, Wang W, Weyand T, Andreetto M, Adam H (2017) Mobilenets: efficient convolutional neural networks for mobile vision applications. arXiv:1704.04861
9. Szegedy C, Vanhoucke V, Ioffe S, Shlens J, Wojna Z (2016) Rethinking the inception architecture for computer vision. In: 2016 IEEE conference on computer vision and pattern recognition (CVPR), pp 2818–2826
10. Cao K, Nguyen DL, Tymoszek C, Jain AK (2020) End-to-end latent fingerprint search. Trans Info For Sec 15:880–894. https://doi.org/10.1109/TIFS.2019.2930487
11. Selvaraju RR, Cogswell M, Das A, Vedantam R, Parikh D, Batra D (2019) Grad-cam: visual explanations from deep networks via gradient-based localization. Int J Comput Vis 128(2), 336–359. https://doi.org/10.1007/s11263-019-01228-7
12. Mura V, Ghiani L, Marcialis GL, Roli F, Yambay DA, Schuckers S (2015) Livdet 2015 fingerprint liveness detection competition 2015. In: IEEE 7th International conferences on biometrics theory, applications and systems (BTAS), pp 1–6
13. Sandler M, Howard A, Zhu M, Zhmoginov A, Chen L (2018) Mobilenetv2: inverted residuals and linear bottlenecks. In: 2018 IEEE/CVF conference on computer vision and pattern recognition, pp 4510–4520
14. Mura V, Orrù G, Casula R, Sibiriu A, Loi G, Tuveri P, Ghiani L, Marcialis G (2018) Livdet 2017 fingerprint liveness detection competition 2017
15. Maio D, Maltoni D, Cappelli R, Wayman JL, Jain AK (2002) Fvc 2000: fingerprint verification competition. IEEE Pattern Anal Mach Intell. https://doi.org/10.1109/34.990140
16. Maio D, Maltoni D, Cappelli R, Wayman JL, Jain AK (2002) Fvc2002: second fingerprint verification competition. In: Object recognition supported by user interaction for service robots. https://doi.org/10.1109/ICPR.2002.1048144
17. Maio D, Maltoni D, Cappelli R, Wayman JL, Jain AK (2004) Fvc 2004: third fingerprint verification competition. In: Zhang D, Jain AK (eds) Biometric authentication. Springer, Berlin, Heidelberg, pp 1–7
18. Cappelli RM, Ferrara AF, Maltoni D (2007) Fingerprint verification competition 2006. Biomet Technol Today 15(7):7–9 (2007). https://doi.org/10.1016/S0969-4765(07)70140-6. http://www.sciencedirect.com/science/article/pii/S0969476507701406
19. Chugh T, Jain AK (2018) Fingerprint presentation attack detection: generalization and efficiency
20. Ganin Y, Lempitsky V (2015) Unsupervised domain adaptation by backpropagation. In: Proceedings of the 32nd international conference on machine learning, ICML'15, vol 37, pp 1180–1189. JMLR.org
21. He K, Zhang X, Ren S, Sun J (2016) Deep residual learning for image recognition. In: 2016 IEEE conference on computer vision and pattern recognition (CVPR), pp 770–778

22. Szegedy C, Vanhoucke V, Ioffe S, Shlens J, Wojna Z (2016) Rethinking the inception architecture for computer vision. https://doi.org/10.1109/CVPR.2016.308
23. Gajawada R, Popli A, Chugh T, Namboodiri A, Jain AK (2019) Universal material translator: towards spoof fingerprint generalization. In: 2019 international conferences on biometrics (ICB), pp 1–8

Part II
Iris Biometrics

Chapter 5
Introduction to Presentation Attack Detection in Iris Biometrics and Recent Advances

Aythami Morales, Julian Fierrez, Javier Galbally, and Marta Gomez-Barrero

Abstract Iris recognition technology has attracted an increasing interest in the last decades in which we have witnessed a migration from research laboratories to real-world applications. The deployment of this technology raises questions about the main vulnerabilities and security threats related to these systems. Among these threats, presentation attacks stand out as some of the most relevant and studied. Presentation attacks can be defined as the presentation of human characteristics or artifacts directly to the capture device of a biometric system trying to interfere with its normal operation. In the case of the iris, these attacks include the use of real irises as well as artifacts with different levels of sophistication such as photographs or videos. This chapter introduces iris Presentation Attack Detection (PAD) methods that have been developed to reduce the risk posed by presentation attacks. First, we summarize the most popular types of attacks including the main challenges to address. Second, we present a taxonomy of PAD methods as a brief introduction to this very active research area. Finally, we discuss the integration of these methods into iris recognition systems according to the most important scenarios of practical application.

A. Morales (✉) · J. Fierrez (✉)
School of Engineering, Universidad Autonoma de Madrid, Madrid, Spain
e-mail: aythami.morales@uam.es

J. Fierrez
e-mail: julian.fierrez@uam.es

J. Galbally (✉)
eu-LISA / European Union Agency for the Operational Management of Large-Scale IT Systems in the Area of Freedom, Security and Justice, Tallinn, Estonia
e-mail: javier.galbally@eulisa.europa.eu

M. Gomez-Barrero (✉)
Hochschule Ansbach, Ansbach, Germany
e-mail: marta.gomez-barrero@hs-ansbach.de

S. Marcel et al. (eds.), *Handbook of Biometric Anti-Spoofing*, Advances in Computer Vision and Pattern Recognition, https://doi.org/10.1007/978-981-19-5288-3_5

5.1 Introduction

The iris is one of the most popular biometric characteristics within biometric recognition technologies. Since the earliest Daugman publications proposing the iris as a reliable method to identify individuals [1] to most recent approaches based on latest machine learning and computer vision techniques [2–4], iris recognition has evolved improving performance, ease of use, and security. Such advances have attracted the interest of researchers and companies boosting the number of products, publications, and applications. The first iris recognition capture devices were developed to work as stand-alone systems [5]. However, today iris recognition technology is included as an authentication service in some of the most important operating systems (e.g., Android and Microsoft Windows) and devices (e.g., laptop or desktop computers and smartphones). One-seventh of the world population (1.14 billion people) has been enrolled in the Aadhaar India national biometric ID program [6] and iris is one of the three biometric characteristics (in addition to fingerprint and face) employed for authentication in that program. The main advantages of iris as a means for personal authentication can be summarized as follows:

- The iris is generated during prenatal gestation and presents highly random patterns. Such patterns are composed of complex and interrelated shapes and colors. The highly discriminant characteristics of the iris make it possible that recognition algorithms reach performances comparable to the most accurate biometric characteristics [3].
- The genetic prevalence of iris is limited and therefore irises from people with shared genes are different. Both irises of a person are considered as different instances, which do not match each other.
- The iris is an internal organ of the eye that is externally visible. It can be acquired at a distance and the advances in capture devices allow to easily integrate iris recognition into portable devices [7, 8].

The fast deployment of iris recognition technologies in real applications has increased the concerns about its security. The applications of iris biometrics include a variety of different scenarios and security levels (e.g., banking, smartphone user authentication, and governmental ID programs). Among all threats associated to biometric systems, the resilience against presentation attacks (PAs) emerges as one of the most active research areas in the recent iris biometrics literature. The security of commercial iris systems has been questioned and put to test by users and researchers. For instance, in 2017, the Chaos Computer Club reported their successful attack to the Samsung Galaxy S8 iris scanner using a simple photograph and a contact lens [9]. In the context of biometric systems, PAs are defined as the presentation of human characteristics or artifacts directly to the capture device of a biometric system trying to interfere its normal operation [10, 11]. This definition includes spoofing attacks, evasion attacks, and the so-called zero-effort attacks. Most of the literature on iris PAD methods is focused on the detection of spoofing attacks. The term liveness detection is also employed in the literature to propose systems capable of classifying between bona fide samples and presentation attack instruments (PAIs) or artifacts used to attack biometric systems. Depending on the motivations of the attacker, we can distinguish two types of attacks:

- Impostor: the attacker tries to impersonate the identity of other subjects by using his own iris (e.g., zero-effort attacks) or a PAI mimicking the iris of the spoofed identity (e.g., photo, video, or synthetic iris). This type of attack requires a certain level of knowledge about the iris of the impersonated subject and the characteristics of the iris sensor in order to increase the success of the attack (see Sect. 5.2).
- Identity concealer: the attacker tries to evade recognition. Examples, in this case, include the enrollment of subjects with fake irises (e.g., synthetically generated) or modified irises (e.g., textured contact lens). These examples represent a way to masquerade real identities.

The first PAD approaches proposed in the literature were just theoretical exercises based on potential vulnerabilities [12]. In recent years, the number of publications focused on this topic has increased exponentially. Some of the PAD methods discussed in the recent literature have been inspired by methods proposed for other biometric characteristics such as face [13–15]. However, the iris has various particularities which can be exploited for PAD, such as the dynamic, fast, and involuntary responses of the pupil and the heterogeneous characteristics of the eye's tissue. The eye reacts according to the amount and nature of the light received. Another large group of PAD methods exploits these dynamic responses and involuntary signals produced by the eye.

The performance of iris PAD approaches can be measured according to different metrics. During the last years, the metrics proposed by the ISO/IEC 30107-3:2017 [10] have been adopted by the most important competitions [16]:

- Attack Presentation Classification Error Rate (APCER): measured as the percentage of attacks (usually from the same PAI) incorrectly classified as bonafide samples.[1]
- Bona fide Presentation Classification Error Rate (BPCER): defined as the percentage of bona fide samples classified as attacks.

In case of multiple PAIs or generation methods, the ISO/IEC 30107-3:2017 recommends to add the maximum value of APCER during the evaluation of different PAD approaches.

This chapter starts by presenting a description of the most important types of attacks from zero-effort attacks to the most sophisticated synthetic eyes. We then introduce iris PAD methods and their main challenges. The PAD methods are organized according to the nature of the features employed dividing them into three main groups: hardware-based, software-based, and challenge–response approaches.

---

[1] A bona fide presentation is defined in ISO/IEC 30107-3:2017 [10] as the interaction of the biometric capture subject and the biometric data capture subsystem in the fashion intended by the policy of the biometric system.

The rest of the chapter is organized as follows: Sect. 5.2 presents the main vulnerabilities of iris recognition systems with special attention to different types of PAs. Section 5.3 summarizes the PADmethods, while Sect. 5.4 presents the integration of PAD approaches with Iris Recognition Systems (IRSs). Finally, conclusions are presented in Sect. 5.5.

5.2 Vulnerabilities in Iris Biometrics

As already mentioned in the introduction, like any other biometric recognition technology, iris recognition is vulnerable to attacks. Figure 5.1 includes a typical block diagram of an IRS and its vulnerable points. The vulnerabilities depend on the characteristics of each module and cover communication protocols, data storage, or resilience against artifact presentations, among others. Several subsystems and not just one will define the security of an IRS:

- Sensor (V1): CCD/CMOS are the most popular sensors including visible and near-infrared images. The iris pattern is usually captured in the form of an image or video. The most important vulnerability is related to the presentation of PAIs (e.g., photos, videos, and synthetic eyes) that mimic the characteristics of real irises.
- Feature Extraction and Comparison modules (V2–V3): these software modules comprise the algorithms in charge of pre-processing, segmentation, generation of templates, and comparison. Attacks to these modules include the alteration of algorithms to carry out illegitimate operations (e.g., modified templates and altered comparison scores).
- Database (V5): the database is composed of structured data associated to the subject information, devices, and iris templates. Any alteration of this information can

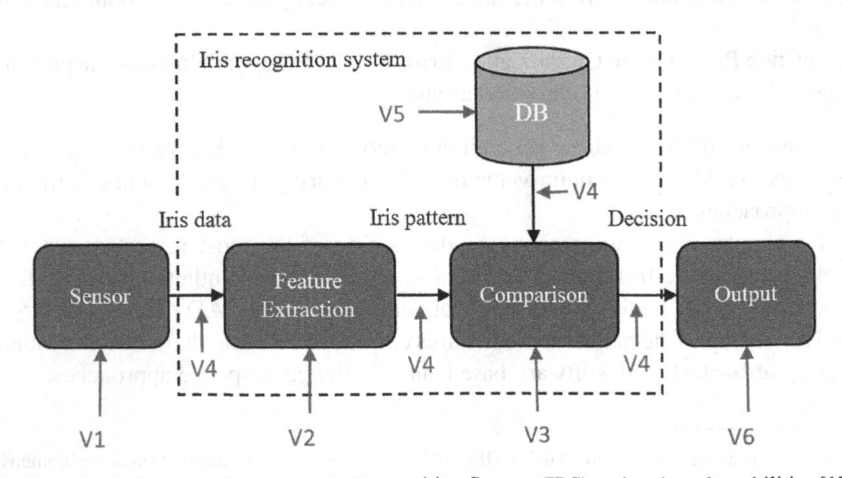

Fig. 5.1 Block diagram of a typical Iris Recognition System (IRS) and main vulnerabilities [17]

affect the final response of the system. The security level of the database storage differs depending on the applications. The use of encrypted templates is crucial to ensure the unlinkability between systems and attacks based on weak links.
- Communication channel and actuators (V4 and V6): include internal (e.g., communication between software modules) and external communications (e.g., communication with mechanical actuators or cloud services). The most important vulnerabilities rely on alterations of the information sent and received by the different modules of the IRS.

In this work, we will focus on PAs to the sensor (V1 vulnerabilities). Key properties of these attacks are their high attack success ratio (if the system is not properly protected) and the low amount of information about the system needed to perform the attack. Other vulnerabilities not covered by this work include attacks to the database (V5), to the software modules (V2–V3), the communication channels (V4), or actuators at the output (V6). This second group of vulnerabilities requires access to the system and countermeasures to these attacks are more related to general system security protocols. These attacks are beyond the scope of this chapter but should not be underestimated.

Regarding the nature of the PAI employed to spoof the system, the most popular PAs can be divided into the following categories:

- Zero-effort attacks: the attack is performed using the iris of the attacker that tries to take advantage of the False Match Rate (FMR) of the system.
- Photo and video attacks: the attack is performed displaying a printed photo, digital image, or video from the bona fide iris directly to the sensor of the IRS.
- Contact lens attacks: the attack is performed using iris patterns printed on contact lenses.
- Prosthetic eye attacks: the attack is performed using synthetic eyes generated to mimic the characteristics of real ones.
- Cadaver eye attacks: the attack is performed using eyes from postmortem subjects.

In the next subsections, each of these categories of attacks is discussed in more detail, paying special attention to three important features that define the risk posed by each of the threats: (1) information needed to perform the attack, (2) difficulty to generate the Presentation Attack Instrument (PAI), and (3) expected impact of the attack.

5.2.1 *Zero-effort Attacks*

In this attack, the impostor does not use any artifact or information about the identity under attack. The iris pattern from the impostor does not match the legitimate pattern and the success of the attack is exclusively related to the False Match Rate (FMR) of the system [18, 19]. Systems with high FMR will be more vulnerable to this type of attack. Note that the FMR is inversely related to the False Non-Match Rate (FNMR) and both are defined by the operational point of the system. An operation point setup to obtain a low FMR can produce an increment of the FNMR and therefore a higher number of false negatives (legitimate subjects are rejected).

- Information needed to perform the attack: no information needed.
- Generation of the PAIs: no need to generate a fake iris. The system is attacked using the real iris of the attacker.
- Expected impact of the attack: Most iris recognition systems present very low False Match Rates. The success rate of these attacks can be considered low.

5.2.2 *Photo and Video Attacks*

Photo attacks are probably the PAs against IRS most studied in the literature [14, 20–24]. They simply consist of presenting to the sensor an image of the attacked iris, either printed on a sheet of paper or displayed on a digital screen. These attacks are among the most popular ones due to three main factors.

First, with the advent of digital photography and social image sharing (e.g., Flickr, Facebook, Picasaweb, and others), headshots of attacked clients from which the iris can be extracted are becoming increasingly easy to obtain. Even though the face or the voice are characteristics more exposed to this new threat, iris patterns can also be obtained from high-resolution face images (e.g., 200 dpi resolution).

Second, it is relatively easy to print high-quality iris photographs using commercial cameras (up to 12 Megapixel sensors in most of the nowadays smartphones) and ink printers (1200 dpi in most of the commercial ink printers). Alternatively, most mobile devices (smartphones and tablets) are equipped with high-resolution screens capable of reproducing very natural images and videos in the visible spectrum (see Fig. 5.2).

Third, the latest advances in machine learning methods to generate synthetic images open new ways to attack IRS. Recent works demonstrated that it is possible to mimic the pattern of real images using Generative Adversarial Networks [25, 26]. The synthetic iris can be used to conduct PAs or to train PAD methods.

As a more sophisticated version of photo attacks, the literature has also largely analyzed the so-called "video attacks", that consist of replaying on a digital screen a video of the attacked iris [27–30]. These attacks are able to mimic not only the static patterns of the iris but also the dynamic information of the eye that could potentially be exploited to detect static photo attacks (e.g., blinking, dilation/contraction of the pupil). As a limitation compared to the simpler photo attacks, it is more difficult to obtain a video than an image of a given iris.

- Information needed to perform the attack: Image or video of the iris of the subject to be impersonated.
- Generation of the PAIs: It is relatively easy to obtain high-resolution face photographs from social media and internet profiles. Other options include capturing

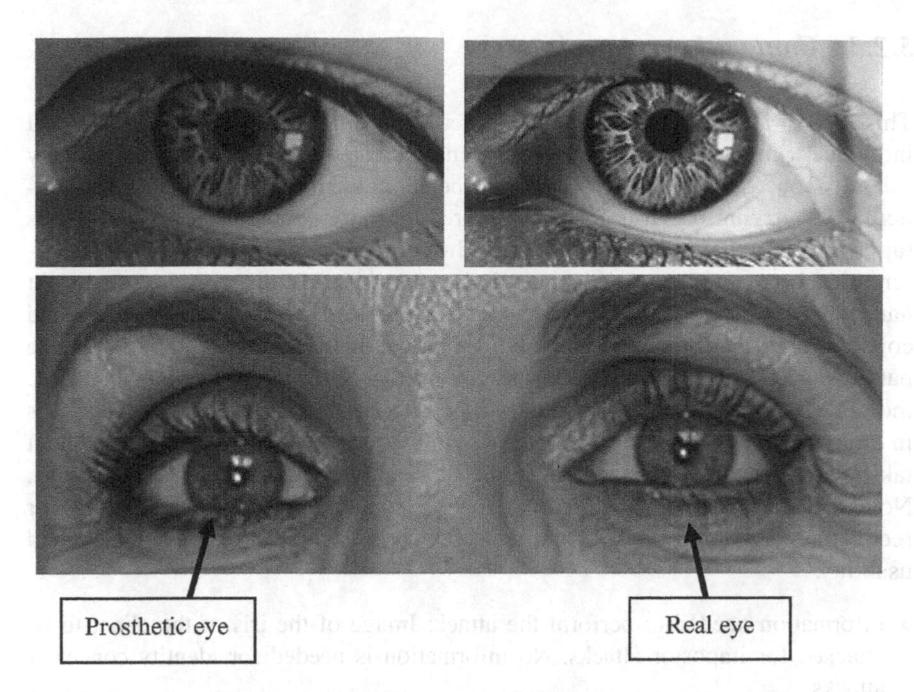

Fig. 5.2 Examples of Presentation Attack Instruments (PAIs): printed photo (top-left), screen photo (top-right) and prosthetic eye (bottom) adapted from Soper Brothers and Associates [http://www.soperbrothers.com]

it using a concealed camera. Once a photo is obtained, if it is of sufficient quality, the iris region can be printed and then presented to the iris acquisition sensor. A screen could also be used for presenting the photograph to the sensor. Another way to obtain the iris would be to steal the raw iris image acquired by an existing iris recognition system in which the subject being spoofed was already enrolled. In the case of video attacks, the availability of high-resolution videos of the iris is rather low in comparison with photographs. Nonetheless, it is relatively easy to capture a video in public spaces and the usage of long-range sensors open allows video capturing at a distance with high-resolution qualities.

- Expected impact of the attack: The literature offers a large number of approaches with good detection rates of printed photo attacks [14, 20–23, 31]. However, most of these methods exploit the lack of realism/quality of printed images in comparison with bona fide samples. The superior quality of new screens capable of reproducing digital images and video attacks with high definition represents a difficult challenge for PAD approaches based on visible spectrum imaging. To overcome this potential threat, new commercial scanners based on near-infrared cameras or even on 3D sensors have recently been proposed.

5.2.3 Contact Lens Attacks

This type of attack uses contact lenses created to mimic the pattern of a different individual (impostor attack) or contact lenses created to masquerade the identity (identity concealer attack). The case of a potential identity concealer attack is particularly worrying because nowadays more and more people wear contact lenses (approximately 125 million people worldwide wear contact lenses). We can differentiate between transparent contact lenses (used in general to correct some sight handicaps like myopia) and textured contact lenses (also known as printed). Textured contact lenses change the original iris information by the superposition of synthetic patterns. Although these contact lenses are mostly related to cosmetic applications, the same technology can be potentially used to print iris patterns from real subjects in order to carry out impostor attacks. If an individual is enrolled in the IRS without taking off the textured contact lenses, the IRS can be compromised at a later stage. Note that asking to remove transparent/corrective contact lenses before enrolment or recognition is a non-desirable solution as it clearly decreases the user comfort and usability.

- Information needed to perform the attack: Image of the iris of the client to be attacked for impostor attacks. No information is needed for identity concealer attacks.
- Generation of the PAIs: In comparison with photo or video attacks, the generation of textured contact lenses requires a more sophisticated method based on optometrist devices and protocols.The characteristics of the devices employed to manufacture the contact lenses are critical to define the performance of PAD methods. For example, the results reported in [16] suggested that the proposed methods presented a lower performance against unknown contact lens brands.
- Expected impact of the attack: These types of attacks represent a great challenge for either automatic PAD systems or visual inspection by humans. It has been reported by several researchers that it is actually possible to spoof iris recognition systems with well-made contact lenses [27, 30, 32–35].

5.2.4 Synthetic Eye Attacks

This type of attack is the most sophisticated. Prosthetic eyes have been used since the beginning of twentieth century to reduce the esthetic impact related to the absence of eyes (e.g., blindness, amputations, etc.). Current technologies for prosthetic manufacturing allow mimicking the most important attributes of the eye with very realistic results. The similarity goes beyond the visual appearance including manufacturing materials with similar physical properties (e.g., elasticity and density). The number of studies including attacks to iris biometric systems using synthetic eyes is still low [36].

- Information needed to perform the attack: Image of the eye of the client to be attacked.
- Generation of the PAIs: this is probably the most sophisticated attack method as it involves the generation of both 2D images and 3D structures. Manually made in the past, 3D printers and their application to the prosthetic field have revolutionized the generation of synthetic body parts.
- Expected impact of the attack: Although the number of studies is low, the detection of prosthetic eyes represents a big challenge. The detection of these attacks by techniques based on image features is difficult. On the other hand, PAD methods based on dynamic features can be useful to detect the unnatural dynamics of synthetic eyes.

5.2.5 *Cadaver Eye Attacks*

Postmortem biometric analysis is very common in forensic sciences. However, during the last years, this field of study attracted the interest of other research communities. As a result of this increasing interest, researchers have evaluated the potential of attacks based on eyes from postmortem subjects [37, 38]. This type of attack has been included in recent iris PAD competitions demonstrating its challenge [16].

- Information needed to perform the attack: no information needed.
- Generation of the PAIs: this type of attack is particularly difficult to perform as the attacker needs to have access to the cadaver of the person to be impersonated.
- Expected impact of the attack: Although the number of studies is low, the detection accuracy of this attack increases with respect to the time gap to the postmortem time horizon. As a living tissue, the pattern on the iris degrades over time once the subject is dead.

Table 5.1 summarizes the literature on iris PAs including the most popular public databases available for research purposes.

5.3 Presentation Attack Detection Approaches

These methods are also known in the literature as liveness detection, anti-spoofing, or artifact detection among others. The term PAD was adopted in the ISO/IEC 30107-1:2016 [10] and it is now largely accepted by the research community.

The different PAD methods can be categorized according to several characteristics. Some authors propose a taxonomy of PAD methods based on the nature of both methods and attacks: passive or active methods employed to detect static or dynamic attacks [40]. Passive methods include those capable of extracting features from samples obtained by traditional iris recognition systems (e.g., image from the iris sensor).

Table 5.1 Literature on presentation attack detection methods. Summary of the literature concerning iris Presentation Attack Detection (PAD) methods depending on the Presentation Attack Instrument (PAI)

Ref.	PAI	PAD	Database
[13]	Photo	Quality measures	Public
[39]	Photo	Wavelet measures	Proprietary
[14]	Photo	Deep features	Public
[40]	Video	Pupil dynamics	Proprietary
[41]	Video	Oculomotor Plant Char.	Proprietary
[42]	Video	Pupillary reflex	Proprietary
[43]	Contact Lens	Photometric features	Public
[44]	Contact Lens	2D & 3D features	Public
[35]	Contact lens	LBP features	Proprietary
[45]	Various	Hyperspectral features	Proprietary
[46]	Various	Pupillary reflex	Proprietary
[47]	Various	Various	Public
[48]	Various	Deep features	Public
[49]	Various	Deep features	Public
[26]	Various	Deep features	Public
[50]	Various	Deep features	Public

Active methods modify the recognition system in order to obtain features for the PAD method (e.g., dynamic illumination and challenge–response). Static attacks refer to those based on individual samples (e.g., image), while dynamic attacks include PAIs capable of changing with time (e.g., video or lens attacks).

In this chapter, we introduce the most popular PAD methods according to the nature of the features used to detect the forged iris: hardware-based, software-based, and challenge–response. The challenge–response and most of the hardware methods can be considered active approaches, as they need additional sensors or collaboration from the subject. On the other hand, most of the software methods employ passive approaches in which PAD features are directly obtained from the biometric sample acquired by the iris sensor. Fig. 5.3 presents a taxonomy of the iris PAD methods introduced in this chapter.

5.3.1 Hardware-Based Approaches

Also known as sensor-based approaches in the literature. These methods employ specific sensors (in addition to the standard iris sensor) to measure the biological and physical characteristics of the eye. These characteristics include physical properties

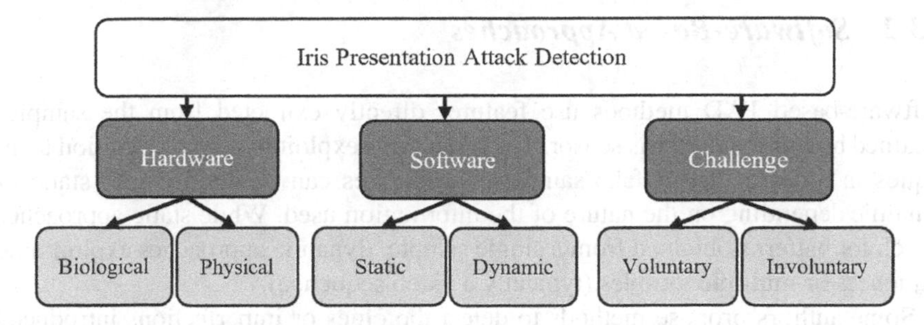

Fig. 5.3 Taxonomy of iris Presentation Attack Detection methods

of the eye (e.g., light absorption of the different eye layers or electrical conductivity), and biological properties (e.g., density of the eye tissues, melanin, or blood vessel structures in the eye). The most popular hardware-based PAD approaches that can be found in the literature belong to one of three groups:

- Multispectral imaging [51–54]: The eye includes complex anatomical structures enclosed in three layers. These layers are made of organic tissue with different spectrographic properties. The idea underlying these methods is to use the spectroscopic print of the eye tissues for PAD. Non-living tissue (e.g., paper, crystal from the screens, or synthetic materials including contact lenses) will present reflectance characteristics different from those obtained from a real eye. These approaches exploit the use of illumination with different wavelengths that vary according to the method proposed and the physical or biological characteristic being measured (e.g., hemoglobin presents an absorption peak in near-infrared bands).
- 3D imaging [22, 55]: The curvature and 3D nature of the eye has been exploited by researchers to develop PAD methods. The 3D profile of the iris is captured in [55] by using two near-infrared light sources and a simple 2D sensor. The idea underlying the method is to detect the shadows on real irises produced by non-uniform illumination provided from two different directions. Light-Field Cameras (LFCs) are used in [22] to acquire multiple depth images and detect the lack of volumetric profiles of photo attacks.
- Electrooculography [56]: The electric standing potential between the cornea and retina can be measured and the resulting signal is known as electrooculogram. This potential can be used as a liveness indicator but the acquisition of these signals is invasive and includes the placement of at least two electrodes in the eye region. Advances in non-intrusive new methods to acquire the electrooculogram can boost the interest in these approaches.

5.3.2 Software-Based Approaches

Software-based PAD methods use features directly extracted from the samples obtained by the standard iris sensor. These methods exploit pattern recognition techniques in order to detect fake samples. Techniques can be divided into static or dynamic depending on the nature of the information used. While static approaches search for patterns obtained from a single sample, dynamic approaches exploit time sequences or multiple samples (typically a video sequence).

Some authors propose methods to detect the clues or imperfections introduced by printing devices used during the manufacturing of PAIs (e.g., printing process for photo attacks). These imperfections can be detected by Fourier image decomposition [20, 21, 39], Wavelet analysis [27], or Laplacian transform [28]. All these methods employ features obtained from the frequency domain in order to detect artificial patterns in fake PAIs. Other authors have explored iris quality measures for PAD. The quality of biometric samples has a direct impact on the performance of biometric systems. The literature includes several approaches to measure the quality of image-based biometric samples. The application of quality measures as PAD features for iris biometrics has been studied in [13, 57]. These techniques exploit iris and image quality in order to detect photo attacks from real irises.

Advances in image processing techniques have also allowed to develop new PAD methods based on the analysis of features obtained at pixel level. These approaches include features obtained from gray level values [58], edges [34], or color [59]. The idea underlying these methods is that the texture of manufacturing materials shows different patterns due to the non-living properties of materials (e.g., density and viscosity). In this line, the method proposed in [59] analyzes image features obtained from near-infrared and visible spectrums. Local descriptors have been also used for iris PAD: Local Binary Patterns—LBP [35, 60–62], Binary Statistical Image Features—BSIF [15, 60], Scale-Invariant Feature Transform—SIFT [30, 60, 63], fusion of 2D and 3D features [44], and Local Phase Quantization—LPQ [60].

Finally, in [14], researchers evaluated for the first time the performance of deep learning techniques for iris photo attack detection with encouraging results. Then the same group of researchers studied how to use those networks to detect more challenging attacks in [64]. Since those initial works were based on deep learning, over the past few years, many works have continued to work in that direction, proposing novel PAD methods based on deep architectures including attention learning [48, 65], adversarial learning [26, 49], and several approaches based on popular convolutional features [66].

5.3.3 Challenge–Response Approaches

These methods analyze voluntary and involuntary responses of the human eye. The involuntary responses are part of the processes associated to the neuromotor activities of the eye, while the voluntary behavior are responses to specific challenges. Both voluntary and involuntary responses can be driven by external stimuli produced by the PAD system (e.g., changes in the intensity of the light, blink instructions, gaze tracking during dedicated challenges, etc.). The eye reacts to such external stimuli and these reactions can be used as a proof of life to detect attacks based on photos or videos. In addition, there are eye reactions inherent to a living body that can be measured in terms of signals (e.g., permanent oscillation of the eye pupil called hippus, involuntary eye movements called microsaccades, etc.). These reactions can be considered as involuntary responses non-controlled by the subject. The occurrence of these signals is used as a proof of life.

The pupil reactions in the presence of uniform light or lighting events were early proposed in [32] for PAD applications and more deeply studied in [40]. As mentioned above, the hippus are permanent oscillations of the pupil that are visible even with uniform illumination. These oscillations range from 0.3 to 0.7Hz and decline with age. The PAD methods based on hippus have been explored to detect photo attacks and prosthetic eye attacks [21, 67]. However, the difficulties to perform a reliable detection reduce the performance of these methods. Based on similar principles related to eye dynamics, the use of oculomotor plant models to serve as PAD methods was evaluated in [41].

The reflection of the light in the lens and cornea produces a well-known involuntary effect named Purkinje reflections. This effect is a reflection of the eye to external illumination. At least four Purkinje reflections are usually visible. The reflections change depending on the light source and these changes can be used for liveness detection [46, 52]. Simple photo and video attacks can be detected by these PAD methods by simply varying the illumination conditions. However, their performance against contact lens or synthetic eye attacks is not clear due to the natural reflections on real pupils (contact lens or photo attacks with pupil holes) or sophisticated fabrication methods (synthetic eyes).

5.4 Integration with Iris Recognition Systems

PAD approaches should be integrated into the iris recognition systems granting a correct and normal workflow. There are two basic integration schemes [68]:

- Parallel integration: the outputs of the IRS and PAD systems are combined before the decision module. The combination method depends on the nature of the output to be combined (e.g., score-level or decision-level fusion) [69].
- Serial integration: the sample is first analyzed by the PAD system. In case of a legitimate subject, the IRS processes the sample. Otherwise, the detection of an attack will avoid unnecessary recognition and the sample will be directly discarded.

The software-based PAD methods are usually included as modules (serial or parallel) in the feature extraction algorithms. A potential problem associated to the

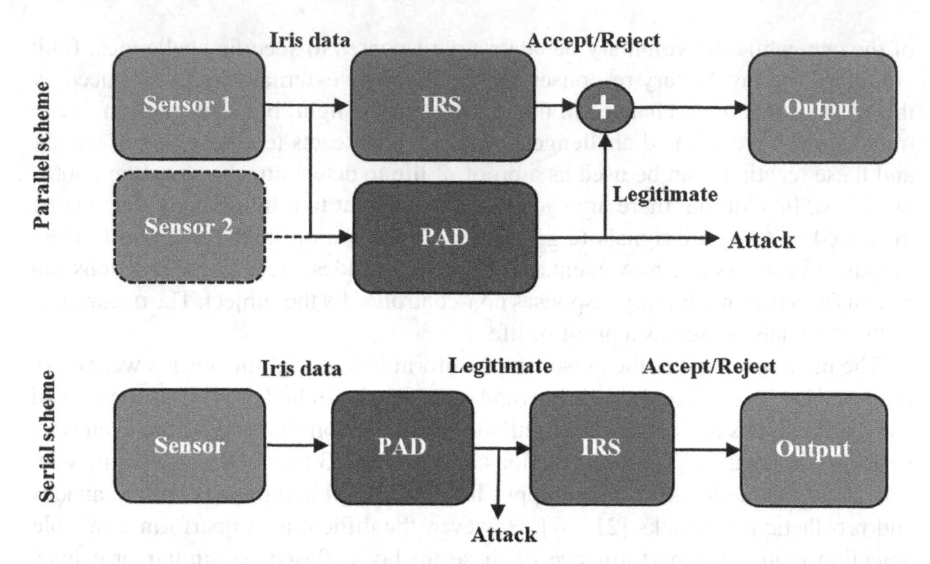

Fig. 5.4 Integration of Presentation Attack Detection (PAD) with Iris Recognition Systems (IRS) in parallel (top) and serial (bottom) schemes

inclusion of PAD software is a delay in the recognition time. However, most PAD approaches based on software methods report a low computational complexity that mitigates this concern. The automatic detection of contact lenses plays an important role in software-based approaches. The effects of wearing contact lenses can be critical in case of textured lenses. In [70], the authors reported that textured lenses can cause the FNMR to exceed 90%. The detection of contact lenses represents the first step in IRS and specific algorithms have been developed and integrated as a preprocessing module [14, 70, 71]. The final goal of these algorithms is to detect and filter the images to remove the synthetic pattern.

Hardware-based PAD approaches are usually integrated before the iris sensor (serial) or as an independent parallel module (see Fig. 5.4). In addition to the execution time concerns, hardware-based approaches increase the complexity of the system and the authentication process. Therefore, the main aspects to be analyzed during the integration of those approaches come from the necessity of dedicated sensors and its specific restrictions related to size, time, and cost. These are barriers that difficult the integration of hardware-based approaches into mobile devices (e.g., smartphones).

The main drawback of challenge-response approaches is the increased level of collaboration needed from the subject (either for serial or parallel schemes). This collaboration usually introduces delays in the recognition process and some subjects can perceive it as an unfriendly process.

5.5 Conclusions

Iris recognition systems have been improved over the last decade achieving better performance [16], more convenient acquisition at a distance [8], and full integration with mobile devices [7]. However, the robustness against attacks is still a challenge for the research community and industrial applications [18]. Researchers have shown the vulnerability of iris recognition systems and there is a consensus about the necessity of finding new methods to improve the security of iris biometrics. Among the different types of attacks, presentation attacks represent a key concern because of their simplicity and high attack success rates. The acquisition at a distance achieved by recent advances on new sensors and the public exposure of face, and therefore iris, make it relatively easy to obtain iris patterns and use them for malicious purposes. The literature on PAD methods is large including a broad variety of methods, databases, and protocols. Over the next years, it will be desirable to unify the research community into common benchmarks and protocols. Even if the current technology shows high detection rates for the simplest attacks (e.g., zero-effort and photo attacks), there are still challenges associated to the most sophisticated attacks such as those using textured contact lenses and synthetic eyes.

Future work in this area of iris PAD may exploit: (1) related research being conducted to recognize the periocular region [72], (2) other information from the face when the iris image or the biometric system as a whole includes a partial or a full face [73], and (3) related research in characterizing natural vs artificially generated faces (DeepFakes) using recent deep learning methods [74]. All these complementary fields of research can provide valuable information for improved PAD when the iris images or videos are accompanied by additional information from the face surrounding the iris.

Acknowledgements This work was mostly done (2nd Edition of the book) in the context of the TABULA RASA and BEAT projects funded under the 7th Framework Programme of EU. The 3rd Edition update has been made in the context of EU H2020 projects PRIMA and TRESPASS-ETN. This work was also partially supported by the Spanish project BIBECA (RTI2018-101248-B-I00 MINECO/FEDER) and by the DFG-ANR RESPECT Project (406880674).

References

1. Daugman J (1993) High confidence visual recognition of persons by a test of statistical independence. IEEE Trans Pattern Anal Mach Intell 15:1148–1161
2. Alonso-Fernandez F, Farrugia RA, Bigun J, Fierrez J, Gonzalez-Sosa E (2019) A survey of super-resolution in iris biometrics with evaluation of dictionary-learning. IEEE Access 7(1):6519–6544. https://doi.org/10.1109/ACCESS.2018.2889395
3. Burge MJ, Bowyer KW (eds) (2013) Handbook of iris recognition. Springer
4. Galbally J, Gomez-Barrero M (2017) Iris and periocular biometric recognition. In: Rathgeb C, Busch C (eds) Presentation attack detection in iris recognition. IET Digital Library, pp 235 –263

5. Flom L, Safir A (1987) Iris recognition system. US Patent US4641349 A
6. Abraham R, Bennett ES, Sen N, Shah NB (2017) State of AADHAAR report 2016–17. Technical report, IDinsight
7. Alonso-Fernandez F, Farrugia RA, Fierrez J, Bigun J (2019) Super-resolution for selfie biometrics: introduction and application to face and iris. Springer, pp 105–128. https://doi.org/10.1007/978-3-030-26972-2_5
8. Nguyen K, Fookes C, Jillela R, Sridharan S, Ross A (2017) Long range iris recognition: a survey. Pattern Recognit 72:123–143
9. Chaos Computer Club Berlin (2017) Chaos computer clubs breaks iris recognition system of the Samsung Galaxy S8. https://www.ccc.de/en/updates/2017/iriden
10. ISO/IEC CD 30107-1 (2016) information technology - biometrics - presentation attack detection - part 1: framework
11. Galbally J, Fierrez J, Ortega-Garcia J (2007) Vulnerabilities in biometric systems: attacks and recent advances in liveness detection. In: Proceedings of Spanish workshop on biometrics, (SWB)
12. Daugman J (1999) Biometrics. Personal identification in a networked society, Recognizing persons by their iris patterns. Kluwer Academic Publishers, pp 103–121
13. Galbally J, Marcel S, Fierrez J (2014) Image quality assessment for fake biometric detection: application to iris, fingerprint and face recognition. IEEE Trans Image Process 23:710–724
14. Menotti D, Chiachia G, Pinto A, Schwartz WR, Pedrini H, Falcao AX, Rocha A (2015) Deep representations for iris, face, and fingerprint spoofing detection. IEEE Trans Inf Forensics Secur 10:864–878
15. Raghavendra R, Busch C (2014) Presentation attack detection algorithm for face and iris biometrics. In: Proceedings of IEEE European signal processing conference (EUSIPCO), pp 1387–1391
16. Das P, Mcfiratht J, Fang Z, Boyd A, Jang G, Mohammadi A, Purnapatra S, Yambay D, Marcel S, Trokielewicz M, et al (2020) Iris liveness detection competition (LivDet-Iris)-the 2020 edition. In: 2020 IEEE international joint conference on biometrics (IJCB), pp 1–9
17. Ratha NK, Connell JH, Bolle RM (2001) Enhancing security and privacy in biometrics-based authentication systems. IBM Syst J 40(3):614–634
18. Hadid A, Evans N, Marcel S, Fierrez J (2015) Biometrics systems under spoofing attack. IEEE Signal Process Mag 32:20–30
19. Johnson P, Lazarick R, Marasco E, Newton E, Ross A, Schuckers S (2012) Biometric liveness detection: framework and metrics. In: Proceedings of NIST international biometric performance conference (IBPC)
20. Czajka A (2013) Database of iris printouts and its application: development of liveness detection method for iris recognition. In: Proceedings of international conference on methods and models in automation and robotics (MMAR), pp 28–33
21. Pacut A, Czajka A (2006) Aliveness detection for iris biometrics. In: Proceedings of IEEE international carnahan conference on security technology (ICCST), pp 122–129
22. Raghavendra R, Busch C (2014) Presentation attack detection on visible spectrum iris recognition by exploring inherent characteristics of light field camera. In: Proceedings of IEEE international joint conference on biometrics (IJCB)
23. Ruiz-Albacete V, Tome-Gonzalez P, Alonso-Fernandez F, Galbally J, Fierrez J, Ortega-Garcia J (2008) Direct attacks using fake images in iris verification. In: Proceedings of COST 2101 workshop on biometrics and identity management (BioID), LNCS-5372. Springer, pp 181–190
24. Thalheim L, Krissler J (2002) Body check: biometric access protection devices and their programs put to the test. ct magazine pp 114–121
25. Galbally J, Ross A, Gomez-Barrero M, Fierrez J, Ortega-Garcia J (2013) Iris image reconstruction from binary templates: an efficient probabilistic approach based on genetic algorithms. Comput Vis Image Underst 117(10):1512–1525. https://doi.org/10.1016/j.cviu.2013.06.003
26. Yadav S, Chen C, Ross A (2019) Synthesizing iris images using RaSGAN with application in presentation attack detection. In: Proceedings of the IEEE/CVF conference on computer vision and pattern recognition workshops

27. He X, Lu Y, Shi P (2009) A new fake iris detection method. In: Proceedings of IAPR/IEEE international conference on biometrics (ICB), LNCS-5558. Springer, pp 1132–1139
28. Raja KB, Raghavendra R, Busch C (2015) Presentation attack detection using laplacian decomposed frequency response for visible spectrum and near-infra-red iris systems. In: Proceedings of IEEE international conference on biometrics: theory and applications (BTAS)
29. Raja KB, Raghavendra R, Busch C (2015) Video presentation attack detection in visible spectrum iris recognition using magnified phase information. IEEE Trans Inf Forensics Secur 10:2048–2056
30. Zhang H, Sun Z, Tan T, Wang J (2011) Learning hierarchical visual codebook for iris liveness detection. In: Proceedings of IEEE international joint conference on biometrics (IJCB)
31. Yambay D, Doyle JS, Boyer KW, Czajka A, Schuckers S (2014) Livdet-iris 2013 - iris liveness detection competition 2013. In: Proceedings of IEEE international joint conference on biometrics (IJCB)
32. Daugman J (2004) Iris recognition and anti-spoofing countermeasures. In: Proceedings of international biometrics conference (IBC)
33. von Seelen UC (2005) Countermeasures against iris spoofing with contact lenses. In: Proceedings biometrics consortium conference (BCC)
34. Wei Z, Qiu X, Sun Z, Tan T (2008) Counterfeit iris detection based on texture analysis. In: Proceedings of IAPR international conference on pattern recognition (ICPR)
35. Zhang H, Sun Z, Tan T (2010) Contact lense detection based on weighted LBP. In: Proceedings of IEEE international conference on pattern recognition (ICPR), pp 4279–4282
36. Lefohn A, Budge B, Shirley P, Caruso R, Reinhard E (2003) An ocularist's approach to human iris synthesis. IEEE Trans Comput Graph Appl 23:70–75
37. Trokielewicz M, Czajka A, Maciejewicz P (2018) Presentation attack detection for cadaver iris. In: 2018 IEEE 9th international conference on biometrics theory, applications and systems (BTAS), pp 1–10
38. Trokielewicz M, Czajka A, Maciejewicz P (2020) Post-mortem iris recognition with deep-learning-based image segmentation. Image Vis Comput 94:103,866
39. He X, Lu Y, Shi P (2008) A fake iris detection method based on FFT and quality assessment. In: Proceedings of IEEE Chinese conference on pattern recognition (CCPR)
40. Czajka A (2015) Pupil dynamics for iris liveness detection. IEEE Trans Inf Forensics Secur 10:726–735
41. Komogortsev O, Karpov A (2013) Liveness detection via oculomotor plant characteristics: attack of mechanical replicas. In: Proceedings of international conference of biometrics (ICB)
42. Kanematsu M, Takano H, Nakamura K (2007) Highly reliable liveness detection method for iris recognition. In: Proceedings of SICE annual conference, international conference on instrumentation, control and information technology (ICICIT), pp 361–364
43. Czajka A, Fang Z, Bowyer K (2019) Iris presentation attack detection based on photometric stereo features. In: 2019 IEEE winter conference on applications of computer vision (WACV), pp 877–885
44. Fang Z, Czajka A, Bowyer KW (2020) Robust iris presentation attack detection fusing 2d and 3d information. IEEE Trans Inf Forensics Secur 16:510–520
45. Chen R, Lin X, Ding T (2012) Liveness detection for iris recognition using multispectral images. Pattern Recogn Lett 33:1513–1519
46. Lee EC, Yo YJ, Park KR (2008) Fake iris detection method using Purkinje images based on gaze position. Opt Eng 47:067,204
47. Yambay D, Becker B, Kohli N, Yadav D, Czajka A, Bowyer KW, Schuckers S, Singh R, Vatsa M, Noore A, Gragnaniello D, Sansone C, Verdoliva L, He L, Ru Y, Li H, Liu N, Sun Z, Tan T (2017) Livdet iris 2017 – iris liveness detection competition 2017. In: Proceedings of IEEE international joint conference on biometrics (IJCB), pp 1–6
48. Fang M, Damer N, Boutros F, Kirchbuchner F, Kuijper A (2021) Iris presentation attack detection by attention-based and deep pixel-wise binary supervision network. In: 2021 IEEE international joint conference on biometrics (IJCB), pp 1–8

49. Yadav S, Ross A (2021) Cit-gan: cyclic image translation generative adversarial network with application in iris presentation attack detection. In: Proceedings of the IEEE/CVF winter conference on applications of computer vision, pp 2412–2421
50. Sharma R, Ross A (2020) D-netpad: An explainable and interpretable iris presentation attack detector. In: 2020 IEEE international joint conference on biometrics (IJCB), pp 1–10
51. He Y, Hou Y, Li Y, Wang Y (2010) Liveness iris detection method based on the eye's optical features. In: Proceedings of SPIE optics and photonics for counterterrorism and crime fighting VI, p 78380R
52. Lee EC, Park KR, Kim J (2006) Fake iris detection by using Purkinje image. In: Proceedings of IAPR international conference on biometrics (ICB), pp 397–403
53. Lee SJ, Park KR, Lee YJ, Bae K, Kim J (2007) Multifeature-based fake iris detection method. Opt Eng 46:127,204
54. Park JH, Kang MG (2005) Iris recognition against counterfeit attack using gradient based fusion of multi-spectral images. In: Proceedings of international workshop on biometric recognition systems (IWBRS), LNCS-3781. Springer, pp 150–156
55. Lee EC, Park KR (2010) Fake iris detection based on 3D structure of the iris pattern. Int J Imaging Syst Technol 20:162–166
56. Krupiński R, Mazurek P (2012) Estimation of electrooculography and blinking signals based on filter banks. In: Proceedings of the 2012 international conference on computer vision and graphics, pp 156–163
57. Galbally J, Ortiz-Lopez J, Fierrez J, Ortega-Garcia J (2012) Iris liveness detection based on quality related features. In: Proceedings of IAPR international conference on biometrics (ICB), pp 271–276
58. He X, An S, Shi P (2007) Statistical texture analysis-based approach for fake iris detection using support vector machines. In: Proceedings of IAPR international conference on biometrics (ICB), LNCS-4642. Springer, pp 540–546
59. Alonso-Fernandez F, Bigun J (2014) Fake iris detection: a comparison between near-infrared and visible images. In: Proceedings IEEE International conference on signal-image technology and internet-based systems (SITIS), pp 546–553
60. Gragnaniello, D., Poggi, G., Sansone, C., Verdoliva, L.: An investigation of local descriptors for biometric spoofing detection. IEEE Trans. on Information Forensics and Security **10**, 849–863 (2015)
61. Gupta P, Behera S, Singh MVV (2014) On iris spoofing using print attack. In: IEEE international conference on pattern recognition (ICPR)
62. He Z, Sun Z, Tan T, Wei Z (2009) Efficient iris spoof detection via boosted local binary patterns. In: Proceedings IEEE international conference on biometrics (ICB)
63. Sun Z, Zhang H, Tan T, Wang J (2014) Iris image classification based on hierarchical visual codebook. IEEE Trans Pattern Anal Mach Intell 36:1120–1133
64. Silva P, Luz E, Baeta R, Pedrini H, Falcao AX, Menotti D (2015) An approach to iris contact lens detection based on deep image representations. In: Proceedings of conference on graphics, patterns and images (SIBGRAPI)
65. Chen C, Ross A (2021) An explainable attention-guided iris presentation attack detector. In: WACV (Workshops), pp 97–106
66. El-Din YS, Moustafa MN, Mahdi H (2020) Deep convolutional neural networks for face and iris presentation attack detection: survey and case study. IET Biometrics 9(5):179–193
67. Park KR (2006) Robust fake iris detection. In: Proceedings of articulated motion and deformable objects (AMDO), LNCS-4069. Springer, pp 10–18
68. Fierrez J, Morales A, Vera-Rodriguez R, Camacho D (2018) Multiple classifiers in biometrics. part 1: fundamentals and review. Inf Fusion 44:57–64
69. Biggio B, Fumera G, Marcialis G, Roli F (2017) Statistical meta-analysis of presentation attacks for secure multibiometric systems. IEEE Trans Pattern Anal Mach Intell 39(3):561–575
70. Bowyer KW, Doyle JS (2014) Cosmetic contact lenses and iris recognition spoofing. IEEE Comput 47:96–98

71. Yadav D, Kohli N, Doyle JS, Singh R, Vatsa M, Bowyer KW (2014) Unraveling the effect of textured contact lenses on iris recognition. IEEE Trans Inf Forensics Secur 9:851–862
72. Alonso-Fernandez F, Raja KB, Raghavendra R, Busch C, Bigun J, Vera-Rodriguez R, Fierrez J (2019) Cross-sensor periocular biometrics for partial face recognition in a global pandemic: comparative benchmark and novel multialgorithmic approach. arXiv:1902.08123
73. Gonzalez-Sosa E, Fierrez J, Vera-Rodriguez R, Alonso-Fernandez F (2018) Facial soft biometrics for recognition in the wild: recent works, annotation and COTS evaluation. IEEE Trans Inf Forensics Secur 13(8):2001–2014
74. Tolosana R, Vera-Rodriguez R, Fierrez J, Morales A, Ortega-Garcia J (2020) Deepfakes and beyond: a survey of face manipulation and fake detection. Inf Fusion 64:131–148. https://doi.org/10.1016/j.inffus.2020.06.014

Chapter 6
Pupil Size Measurement and Application to Iris Presentation Attack Detection

Adam Czajka, Benedict Becker, and Alan Johnson

Abstract This chapter presents a comprehensive study on the application of stimulated pupillary light reflex to presentation attack detection (PAD) that can be used in iris recognition systems. A pupil, when stimulated by visible light in a predefined manner, may offer sophisticated dynamic liveness features that cannot be acquired from dead eyes or other static objects such as printed contact lenses, paper printouts, or prosthetic eyes. Modeling of pupil dynamics requires a few seconds of observation under varying light conditions that can be supplied by a visible light source in addition to the existing near infrared illuminants used in iris image acquisition. The central element of the presented approach is an accurate modeling and classification of pupil dynamics that makes mimicking an actual eye reaction difficult. This chapter discusses new data-driven models of pupil dynamics based on recurrent neural networks and compares their PAD performance to solutions based on the parametric Clynes-Kohn model and various classification techniques. Experiments with 166 distinct eyes of 84 subjects show that the best data-driven solution, one based on long short-term memory, was able to correctly recognize 99.97% of attack presentations and 98.62% of normal pupil reactions. In the approach using the Clynes-Kohn parametric model of pupil dynamics, we were able to perfectly recognize abnormalities and correctly recognize 99.97% of normal pupil reactions on the same dataset with the same evaluation protocol as the data-driven approach. This means that the data-driven solutions favorably compare to the parametric approaches, which require model identification in exchange for a slightly better performance. We also show that

Benedict Becker—contributions: Sect. 4—"Data-driven Models of Pupil Dynamics" with appropriate parts of Sect. 5—"Results".
Alan Johnson—contributions: Sect. 6—"Open-Hardware And Open-Source Pupil Size Measurement Device".

A. Czajka (✉) · B. Becker · A. Johnson
University of Notre Dame, 384 Fitzpatrick Hall, 46556 Notre Dame, USA
e-mail: aczajka@nd.edu

B. Becker
e-mail: bbecker5@nd.edu

A. Johnson
e-mail: ajohns42@nd.edu

S. Marcel et al. (eds.), *Handbook of Biometric Anti-Spoofing*, Advances in Computer Vision and Pattern Recognition, https://doi.org/10.1007/978-981-19-5288-3_6

observation times may be as short as 3 s when using the parametric model, and as short as 2 s when applying the recurrent neural network without substantial loss in accuracy. Compared to the last edition, this chapter offers also an open-source/open-hardware Raspberry Pi Zero-based device, with Python implementation, for wireless pupil size measurement. It's a client-server system, in which Pi is responsible for appropriate camera stream acquisition and transferring data wirelessly to a host computer, which then performs data processing and pupil size estimation using algorithms based on convolutional neural networks. The proposed system can be used in various research efforts related to pupil size measurement, including PAD, where acquisition needs to be done without constraints related to the subject's position. Along with this chapter we offer: (a) all-time series representing pupil dynamics for 166 distinct eyes used in this study, (b) weights of the trained recurrent neural network offering the best performance, (c) source codes of the reference PAD implementation based on Clynes-Kohn parametric model, (d) all PAD scores that allows the reproduction of the plots presented in this chapter, and (e) all codes necessary to build a Raspberry Pi-based pupillometer.

6.1 Introduction

Presentation attack detection (PAD) is a key aspect of a biometric system's security. PAD refers to an automated detection of presentations to the biometric sensor that has the goal to interfere with an intended operation of the biometric system [1]. Presentation attacks may be realized in various ways, and using various *presentation attack instruments (PAI)*, such as presentation of fake objects or cadavers, non-conformant presentation, or even coerced use of biometric characteristics. This chapter focuses on the detection of liveness features of an iris, i.e.changes in pupil dilation under varying light conditions that indicate the authenticity of the eye. This liveness test can prove useful in both iris and ocular recognition systems.

Iris PAD has a significant representation in scientific literature. Most of the methods are based on static properties of the eye and iris to detect iris paper printouts and cosmetic contact lenses. Examples of such methods include the use of image quality metrics (Galbaly et al. [2], Wei et al. [3]) and texture descriptors such as Local Binary Patterns (Doyle et al. [4], Ojala et al. [5]), Local Phase Quantization (Ojansivu and Heikkilä [6]), Binary Gabor Pattern (Zhang et al. [7]), Hierarchical Visual Codebook (Sun et al. [8]), Histogram of Oriented Gradients (Dalal and Triggs [9]) and Binarized Statistical Image Features (Doyle and Bowyer [10], Raghavendra and Busch [11]). The recent advent of deep learning, especially convolutional neural networks, has caused a dynamic increase in solutions based on neural networks [12]. However, the results of the last LivDet-Iris 2020 competition [13] suggest that these static artifacts are still challenging when the algorithms are tested on data and attack types unknown to developers: the winning algorithms were able to achieve an average BCPER less than 1%, but the APCER for that method was as high as almost 60%!

The application of dynamic features of the eye is less popular. An interesting solution proposed by Raja et al. [14] was based on Eulerian Video Magnification that detects micro phase deformations of iris region. Also, Komogortsev et al. proposed an application of eyeball dynamics to detect mechanical eye replicas [15]. Pupil dynamics are rarely used in biometric presentation attack detection. Most existing approaches are based on methods proposed by Pacut and Czajka [16–18], which deploy a parametric Clynes-Kohn model of the pupil reaction. Readers interested in the most recent summary of the iris PAD topic are encouraged to look at the surveys by Czajka, Boyd et al. [19, 20].

To our knowledge, there are no publicly available databases of iris videos acquired under visible light stimuli. We have thus collected a dataset of pupil reactions measured for 166 distinct eyes when stimulating the eye with positive (a sudden increase of lightness) and negative (a sudden decrease of lightness) stimuli. Bona fide samples are represented by time series presenting pupil reaction to either positive, negative, or both stimuli. Attack samples are represented by time series acquired when the eye was not stimulated by the light. In many attack samples, one may still observe spontaneous pupil oscillation and noise caused by blinking, eye off-axis movements, or imprecise iris segmentation. These are good complications making the classification task more challenging. The measurements of pupil size (in a time series) are made available along with this chapter to interested research groups.

In this chapter, we present and compare two approaches to classify the acquired time series. The state-of-the-art approach is based on the parametric Clynes-Kohn model of pupil dynamics and various classifiers such as support vector machines, logistic regression, bagged trees, and *k* nearest neighbors. Solutions employing the parametric model require model identification for each presentation attack detection, which incorporates minimization procedures that often results in nondeterministic behavior. However, once trained this approach presents good generalization capabilities. The second approach presented in this chapter is based on data-driven models realized by four variants of recurrent neural networks, including long short-term memory. These solutions present slightly worse results than those based on the parametric model, but they deliver more deterministic outputs. It is also sometimes convenient to use an end-to-end classification algorithm, hence data-driven models may serve as an interesting PAD alternative to algorithms employing parametric models.

As the last contribution, we offer an open-source/open-hardware device for pupil size measurement (pupillometer). This is the client-server system based on Raspberry Pi Zero-, which transfers data wirelessly to a host computer, which then performs data processing and pupil size estimation.

In Sect. 6.2 we provide technical details of the employed dataset. Section 6.3 briefly summarizes the application of Clynes-Kohn model of pupil dynamics. In Sect. 6.4 we explain how the recurrent neural networks were applied to build data-driven models of pupil reflex. Results are presented in Sect. 6.5. Section 6.6 presents the offered pupillometer device, and in Sect. 6.7 we summarize the PAD approach, itemizing some its limitations.

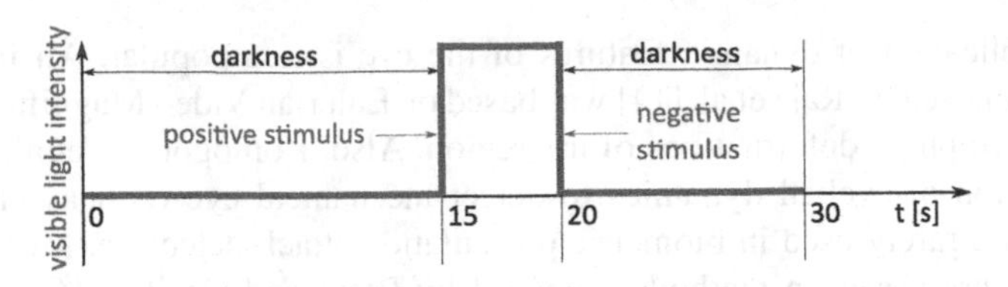

Fig. 6.1 Visible light intensity profile in the data collection for a single attempt

6.2 Database

6.2.1 Acquisition

The acquisition scenario followed the one applied in [18]. All images were captured in near infrared light ($\lambda = 850$ nm) and the recommendations for iris image quality given in ISO/IEC 19794-6 and ISO/IEC 29794-6 standards were easily met. The sensor acquired 25 images per second. Volunteers presented their eyes in a shaded box in which the sensor and four visible light-emitting diodes were placed to guarantee a stable acquisition environment throughout the experiment. The sensor was placed approximately 20 cm from subjects' eyes. This distance was based off of the optical characteristics of the lenses used in the study and camera resolution, to end up with iris images meeting the resolution requirements recommended by ISO/IEC 19794-6 (at least 120 pixels per iris diameter). A single acquisition attempt lasted 30 s: spontaneous pupil movement was recorded during the first 15 s, then the pupil reaction to a **positive light stimulus** (dark→light) was recorded during the next 5 seconds, and during the last 10 s a pupil reaction to a **negative light stimulus** (light→dark) was measured. Figure 6.1 presents a visible light profile applied in each attempt.

84 subjects presented their left and right eyes, except two subjects who presented only a single eye. Thus the database is comprised of measurements from 166 distinct eyes, with up to 6 attempts per eye. All attempts for a given subject were organized on the same day, except for four persons (8 eyes) who had their pupil dynamics measured 12 years before and then repeated (using the same hardware and software setup). The total number of attempts for all subjects is 435, and in each attempt we collected 750 iris images (30 s $\times$ 25 frames per second).

6.2.2 Estimation of Pupil Size

All samples were segmented independently using Open-Source IRIS [21] which implements a Viterbi algorithm to find a set of points on the iris boundaries. Least squares curve-fitting was then used to find two circles approximating the inner and outer iris boundaries. We tried to keep the acquisition setup identical during all

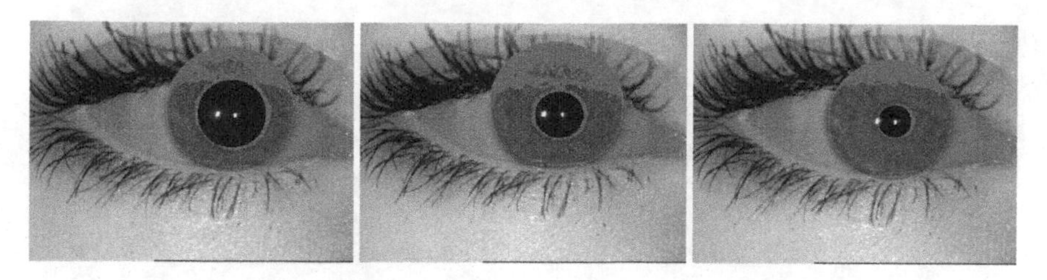

Fig. 6.2 Example iris images and segmentation results 0s (**left**), 1s (**middle**) and 5s (**right**) after positive light stimuli. Circular approximations of iris boundaries are shown in green, and irregular occlusions are shown in red

attempts. However, small adjustments to the camera before an attempt may have introduced different optical magnifications for each participant. Also, iris size may slightly differ across the population of subjects. Hence, despite a stable distance between a subject and a camera, we normalized the pupil radius by the iris radius to get the *pupil size* ranging from 0.2 to 0.8 in our experiments. Instead of using the iris radius calculated in each frame, we used its averaged value calculated for all frames considered in a given experiment. This approach nicely compensates for possible changes in the absolute iris size, and does not introduce additional noise related to fluctuations in outer boundary detection.

6.2.3 *Noise and Missing Data*

Figure 6.2 presents example iris images acquired immediately after positive light stimulus (increase of the light intensity at $t = 0$) along with the segmentation results. Correct segmentation delivers a valid pupil size for each video frame. However, there are at least two types of processing errors that introduce a natural noise into the data:

- *missing points:* the segmentation method may not be able to provide an estimate of pupil and/or iris position; this happens typically when the eye is closed completely (Fig. 6.3 left)
- *inaccurate segmentation:* caused mainly by low-quality data, due to factors such as blinking, off-axis gaze, or significant occlusion (Fig. 6.3 middle), or by occasional segmentation algorithm errors (Fig. 6.3 right).

These noisy measurements can be observed in each plot presented in Fig. 6.4 as points departing from the expected pupil size. Hence, we applied median filtering within a one-second window to smooth out most of the segmentation errors and get *de-noised pupil size* (cf. black dots in Fig. 6.4). De-noised pupil size was used in all parametric model experiments presented in this chapter. For data-driven models, we simply filled the gaps by taking the previous value of the pupil size as a predictor of a missing point.

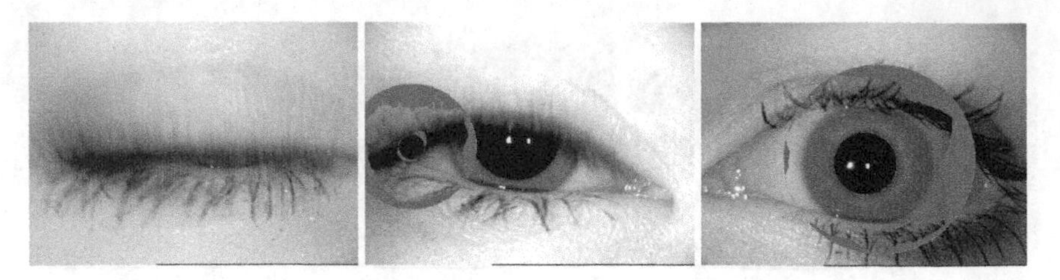

Fig. 6.3 Closed eyes prevent the segmentation from providing meaningful data (**left**). Significant eyelid coverage results in inaccurate segmentation (**middle**). Occasionally, the segmentation may also fail for a good-quality sample (**right**). All these cases introduce a natural noise into pupil dynamics time series. As in Fig. 6.2, circular approximations of iris boundaries are shown in green, and irregular occlusions are shown in red

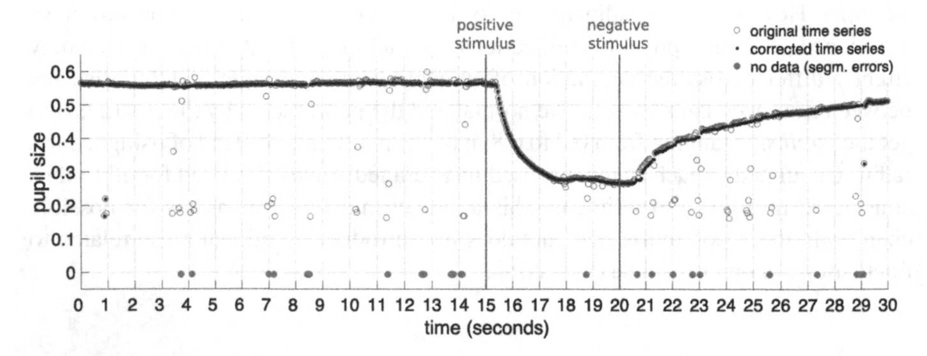

Fig. 6.4 Original (blue circles) and de-noised (black dots) pupil size depending on the stimulus. Original data points departing from the expected pupil size, including missing points marked as red dots, correspond to blinks and iris segmentation errors

6.2.4 Division of Data and Recognition Scenarios

To train the classifiers, both bona fide and attack examples were generated for each time series. The first 15 s and the last 5 s of each time series represent no pupil reaction, and thus were used to generate attack presentations. The time series starting from the 15th s and ending at the 25th s represent authentic reactions to light stimuli, and thus were used to generate bona fide presentations. We consider three scenarios of pupil stimulation:

- **s1:** presentation of only positive stimulus; in this case we observe the eye up to 5 s after the visible light is switched on,
- **s2:** presentation of only negative stimulus; in this case we observe the eye up to 5 s after the visible light is switched off, and
- **s3:** presentation of both stimuli sequentially; in this case we observe the eye for 10 s after the visible light is switched on and sequentially switched off.

Since the pupil reaction is different for each type of stimulus, we evaluate separate data-driven models. Also, when only one stimulus is used (positive or negative), the observation can be shorter than 5 seconds. Shorter observations are also investigated in this chapter.

6.3 Parametric Model of Pupil Dynamics

The parametric-model-based method for pupil dynamics recognition follows past works by Czajka [18] and uses the Clynes and Kohn pupil reaction model [22]. For completeness, we briefly characterize this approach in this subsection.

The Clynes and Kohn model, illustrated as a two-channel transfer function of a complex argument s in Fig. 6.5, accounts for asymmetry in pupil reaction y depending on the polarity of the stimulus x. Positive stimuli (darkness → lightness) engage two channels of the model, while for negative stimuli (lightness → darkness) the upper channel is cut by a nonlinear component, and only lower channel is used to predict the pupil size. That is, for positive stimuli each reaction is represented by a point in seven-dimensional space: three time constants (T_1, T_2 and T_3), two lag elements (τ_1 and τ_2) and two gains (K_r and K_i). For negative stimuli, we end up with three-dimensional feature space corresponding to lower channel parameters: K_i, T_3 and τ_2.

One can easily find the model response $y(t; \phi)$ in time domain by calculating the inverse Laplace transform, given a model shown in Fig. 6.5. Assuming that the light stimuli occurs at $t = 0$, the upper channel response

$$y_{\text{upper}}(t; \phi_1) = \begin{cases} -\frac{K_r}{T_1^2}(t - \tau_1)e^{-\frac{t-\tau_1}{T_1}} & \text{if } T_1 = T_2 \\ \frac{K_r}{T_2 - T_1}\left(e^{-\frac{t-\tau_1}{T_1}} - e^{-\frac{t-\tau_1}{T_2}}\right) & \text{otherwise} \end{cases} \tag{6.1}$$

where

$$\phi_1 = [K_r, T_1, T_2, \tau_1].$$

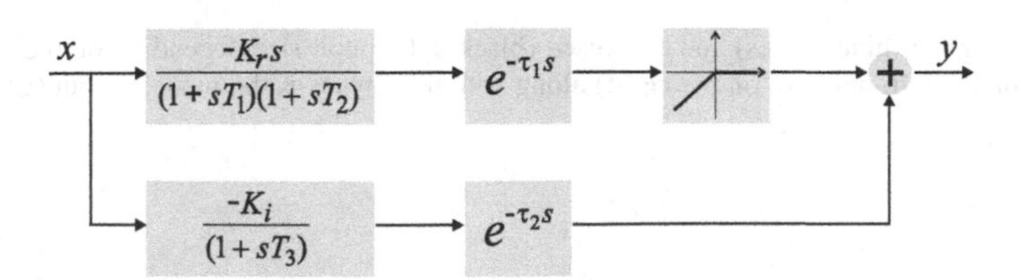

Fig. 6.5 Pupil dynamics model deployed in this work and derived from an original proposal of Kohn and Clynes [22]. Graph adapted from [17]

The lower channel response

$$y_{\text{lower}}(t; \phi_2) = -K_i\left(1 - e^{-\frac{t-\tau_2}{T_3}}\right), \tag{6.2}$$

where

$$\phi_2 = [K_i, T_3, \tau_2],$$

and the model output $y(t; \phi)$ is simply a sum of both responses

$$y(t; \phi) = y_{\text{upper}}(t; \phi_1) + y_{\text{lower}}(t; \phi_2), \tag{6.3}$$

where

$$\phi = [\phi_1, \phi_2] = [K_r, T_1, T_2, \tau_1, K_i, T_3, \tau_2].$$

In case of a negative light stimuli, the model output is based on the lower channel response since $y_{\text{upper}}(t; \phi_1) = 0$. The vector ϕ constitutes the *liveness features*, and is found as a result of the optimization process used to solve the model fitting problem. Figures 6.6 and 6.7 present model outputs obtained for example time series and the three recognition scenarios listed in Sect. 6.2.4. Figure 6.8 illustrates a correct model behavior in case of no stimuli.

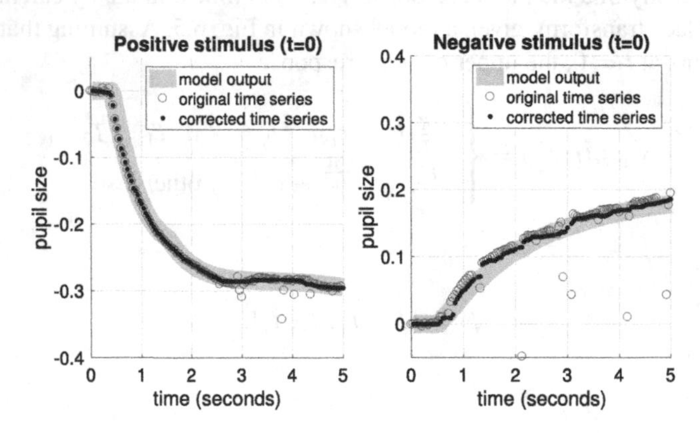

Fig. 6.6 Original (blue circles) and de-noised (black dots) pupil size depending on the stimulus (positive on the **left**, negative on the **right**) along with the Clynes-Kohn model output (gray thick line)

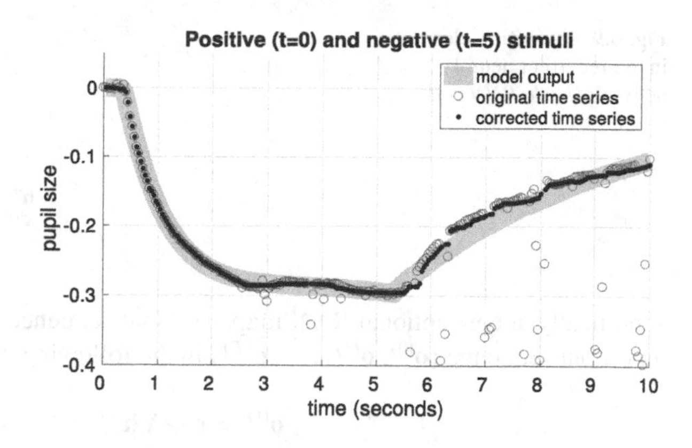

Fig. 6.7 Same as in Fig. 6.6, except that the results for a long reaction (10 seconds) after both stimuli (positive, then negative) are illustrated

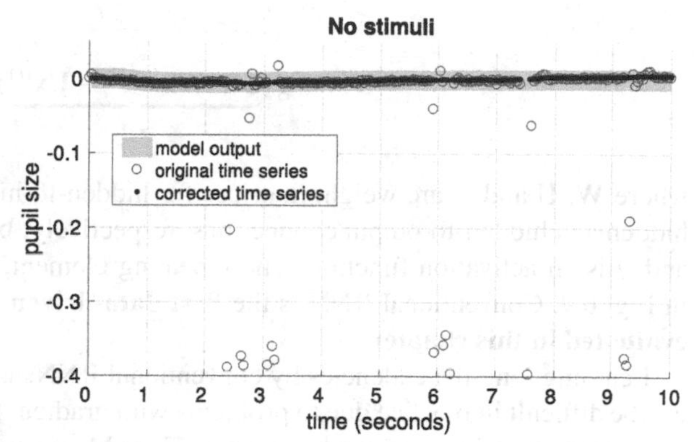

Fig. 6.8 Same as in Fig. 6.6, except that the pupil actual size and model output are shown when the eye is not stimulated by the light

6.4 Data-Driven Models of Pupil Dynamics

6.4.1 Variants of Recurrent Neural Networks

Recurrent neural networks (RNN) belong to a family of networks used for processing sequential data. RNNs can scale to much longer sequences than networks having no sequence-based specialization, such as convolutional neural networks (CNN). Regardless of the sequence length, the learned RNN-based model always has the same input size. These basic properties make the RNN a well-suited candidate to model the time series that represent pupil dynamics.

Graphs with cycles are used to model and visualize recurrent networks, and the cycles introduce dynamics by allowing the predictions made at the current time step to be influenced by past predictions. Parameter sharing in an RNN differs from the parameter sharing applied in a CNN: instead of having the same convolution kernel at each time step (CNN), the same transition function (from hidden state $h^{(t-1)}$ to hidden state $h^{(t)}$) with the same parameters is applied at every time step t. More

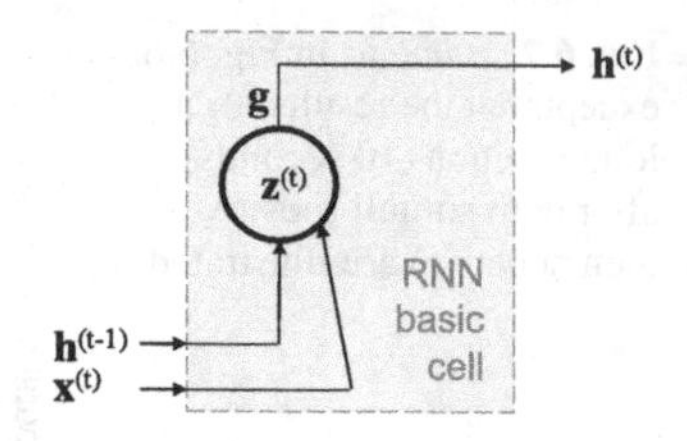

Fig. 6.9 Repeating element in the recurrent neural network (*basic RNN cell*)

specifically, a conventional RNN maps an input sequence $\left(\mathbf{x}^{(0)}, \mathbf{x}^{(1)}, \ldots, \mathbf{x}^{(T)}\right)$ into an output sequence $\left(\mathbf{o}^{(0)}, \mathbf{o}^{(1)}, \ldots, \mathbf{o}^{(T)}\right)$ in the following way:

$$\mathbf{o}^{(t)} = \mathbf{c} + \mathbf{V}\mathbf{h}^{(t)} \tag{6.4}$$

$$\mathbf{h}^{(t)} = g(\underbrace{\mathbf{b} + \mathbf{W}\mathbf{h}^{(t-1)} + \mathbf{U}\mathbf{x}^{(t)}}_{\mathbf{z}^{(t)}}) \tag{6.5}$$

where **W**, **U** and **V** are weight matrices for hidden-to-hidden (recurrent), input-to-hidden and hidden-to-output connections, respectively; **b** and **c** denote bias vectors; and g is an activation function. The repeating element, *basic RNN cell*, is shown in Fig. 6.9. Conventional RNN is the **first data-driven model of pupil dynamics evaluated in this chapter**.

Learning long-dependencies by conventional RNNs is theoretically possible, but can be difficult in practice due to problems with gradient flow [23]. Among different variants proposed to date, the Long Short-Term Memory (LSTM) [24] was designed to avoid the long-term dependency problem and allows the error derivatives to flow unimpeded. The LSTM repeating cell is composed of four nonlinear layers interacting in a unique way (in contrast to a conventional RNN, which uses just a single nonlinear layer in the repeating cell), Fig. 6.10. The LSTM hidden state $\mathbf{h}^{(t)}$ in time moment t can be expressed as:

$$\mathbf{h}^{(t)} = \gamma^{(t)} \circ g\left(\mathbf{m}^{(t)}\right), \tag{6.6}$$

where

$$\gamma^{(t)} = \sigma\Big(\underbrace{\mathbf{W}_{h\gamma}\mathbf{h}^{(t-1)} + \mathbf{U}_{x\gamma}\mathbf{x}^{(t)} + \mathbf{W}_{m\gamma}\mathbf{m}^{(t)} + \mathbf{b}_{\gamma}}_{\mathbf{z}_{\gamma}^{(t)}}\Big)$$

is the *output gate*, $\circ$ denotes the Hadamard product, and

$$\mathbf{m}^{(t)} = \alpha^{(t)} \circ \mathbf{m}^{(t-1)} + \beta^{(t)} \circ g\left(\mathbf{x}^{(t)}, \mathbf{h}^{(t-1)}\right)$$

is the LSTM cell state (the Constant Error Carousel, or CEC). $\alpha^{(t)}$ and $\beta^{(t)}$ are the so-called *forget gate* and *input gate*, namely,

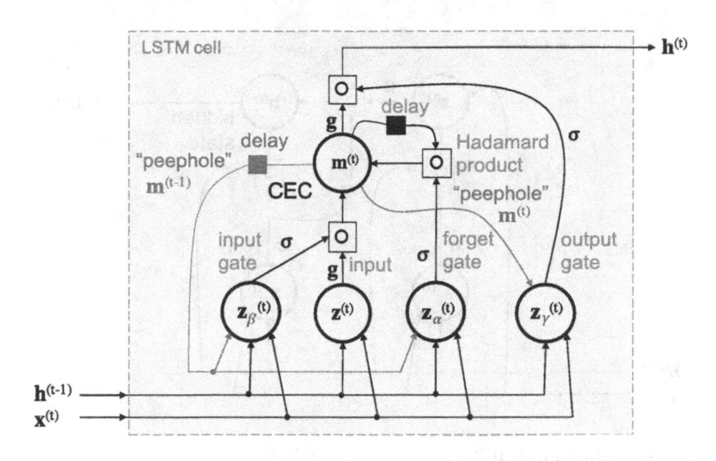

Fig. 6.10 Long Short-Term Memory cell (with peepholes)

$$\alpha^{(t)} = \sigma\big(\underbrace{\mathbf{W}_{h\alpha}\mathbf{h}^{(t-1)} + \mathbf{U}_{x\alpha}\mathbf{x}^{(t)} + \mathbf{W}_{m\alpha}\mathbf{m}^{(t-1)} + \mathbf{b}_\alpha}_{\mathbf{z}_\alpha^{(t)}}\big)$$

$$\beta^{(t)} = \sigma\big(\underbrace{\mathbf{W}_{h\beta}\mathbf{h}^{(t-1)} + \mathbf{U}_{x\beta}\mathbf{x}^{(t)} + \mathbf{W}_{m\beta}\mathbf{m}^{(t-1)} + \mathbf{b}_\beta}_{\mathbf{z}_\beta^{(t)}}\big),$$

where σ is a sigmoid activation function. The closer the gate's output is to 1, the more information the model will retain and the closer the output is to 0, the more the model will forget. Additional $\mathbf{W}_{m\alpha}$, $\mathbf{W}_{m\beta}$ and $\mathbf{W}_{m\gamma}$ matrices (compared to a conventional RNN) represent "peephole" connections in the gates to the CEC of the same LSTM. These connections were proposed by Gers and Schmidhuber [25] to overcome a problem of closed gates that can prevent the CEC from getting useful information from the past. LSTMs (with and without peepholes) are the **next two data-driven models of pupil dynamics evaluated in this work.**

Greff et al. [26] evaluated various types of recurrent networks, including variants of the LSTM, and found that coupling input and forget gates may end up with a similar or better performance as a regular LSTM with fewer parameters. The combination of the forget and input gates of the LSTM was proposed by Cho et al. [27] and subsequently named the Gated Recurrent Unit (GRU). The GRU merges the cell state and hidden state, that is

$$\mathbf{h}^{(t)} = \alpha^{(t)} \circ \mathbf{h}^{(t-1)} + (1 - \alpha^{(t)}) \circ g\big(\mathbf{x}^{(t)}, \gamma^{(t)} \circ \mathbf{h}^{(t-1)}\big), \tag{6.7}$$

where

$$\alpha^{(t)} = \sigma\big(\underbrace{\mathbf{W}_{h\alpha}\mathbf{h}^{(t-1)} + \mathbf{U}_{x\alpha}\mathbf{x}^{(t)} + \mathbf{b}_\alpha}_{\mathbf{z}_\alpha^{(t)}}\big)$$

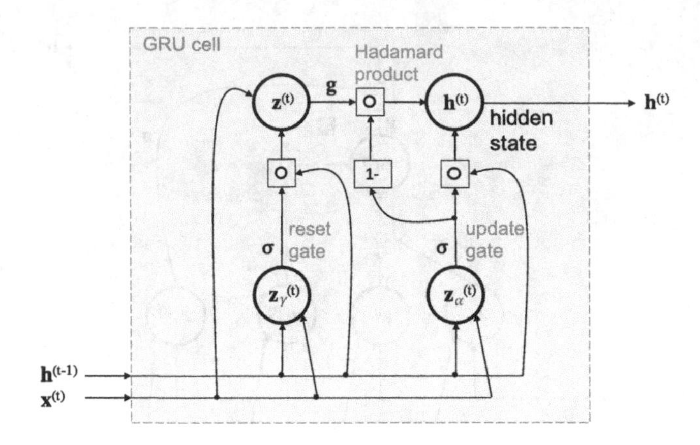

Fig. 6.11 Gated recurrent unit cell

is the *update gate*, and

$$\gamma^{(t)} = \sigma\big(\underbrace{\mathbf{W}_{h\gamma}\mathbf{h}^{(t-1)} + \mathbf{U}_{x\gamma}\mathbf{x}^{(t)} + \mathbf{b}_{\gamma}}_{\mathbf{z}_{\gamma}^{(t)}}\big)$$

is the *reset gate*. GRU uses no peephole connections and no output activation function. A repeating GRU cell in shown in Fig. 6.11. GRU will be **the fourth data-driven model of pupil dynamics deployed in this work.**

6.4.2 Implementation and Hyperparameters

In all data-driven models, a neural network with two layers was used: a type of recurrent neural network composed of 24 hidden neurons, and a perceptron layer composed of two neurons with a softmax output classifying samples as bona fide or attack classes. For training, a static learning rate of 0.0001 was used and the networks were trained using a batch size of 16 for a total of 200,000 iterations, or almost 87 epochs. In all cases, the optimizer used for training was RMSProp proposed by Hinton [28], the loss function used was categorical cross entropy and the network's weight initializer was "Xavier" initialization [29].

Each network took in time series data of the form 5×1 s intervals. The period of one second was chosen so that each time window would have a significant amount of data. Thus, the network input $\mathbf{x}^{(t)}$ is comprised of 25 data points that correspond to 25 iris images acquired per second. Shorter time windows would not provide enough context for the network and longer time windows would defeat the purpose of using recurrent networks. The length of training sequences for both bona fide and attack examples was kept the same.

6.5 Results

The parametric Clynes and Kohn model transforms each time series into a multi-dimensional point in the model parameter space. A binary classifier makes a decision whether the presented time series corresponds to an authentic reaction of the pupil, or to a reaction that is odd or noisy. Several classifiers were tested, namely linear and nonlinear support vector machines (SVM), logistic regression, bagged trees, and k nearest neighbors (kNN). In turn, the data-driven model based on recurrent neural network makes the classification all the way from the estimated pupil size to the decision. In all experiments (both for parametric and data-driven models) the same leave-one-out cross-validation was applied. That is, to make validation subject-disjoint, all the data corresponding to a single subject was left for validation, while the remaining data was used to train the classifier. This train-validation split could thus be repeated 84 times (equal to the number of subjects). The presented results are averages over all 84 of these validations.

We follow ISO/IEC 30107-1:2016 and use the following PAD-specific error metrics:

- **Attack Presentation Classification Error Rate (APCER)**: proportion of *attack presentations* incorrectly classified as *bona fide (genuine) presentations*,
- **Bona Fide Presentation Classification Error Rate (BPCER)**: proportion of *bona fide (genuine) presentations* incorrectly classified as *presentation attacks*.

Table 6.1 presents the results obtained from both parametric and data-driven models and for all three scenarios of eye stimulation (only positive, only negative, and both stimuli). The best solution in the **positive stimuli scenario** was based on the **parametric model** (Clynes-Kohn + SVM). It recognized 99.77% of the normal pupil reactions (BPCER = 0.23%) and 99.54% of the noise (APCER = 0.46%). The **data-driven solution** based on LSTM with no peephole connections achieved similar accuracy recognizing 98.62% of the normal pupil reactions (BPCER = 1.38%) and 99.77% of the noisy time series (APCER = 0.23%).

The **negative stimulus** was harder to classify by a **data-driven model**, as the best accuracy, obtained with a conventional recurrent neural network, recognized 96.09% of normal pupil reactions (BPCER = 3.91%) and 98.74% of noisy time series (APCER = 1.26%). In turn, **parametric model** with bagged trees or kNN classification was perfect in recognizing spontaneous pupil oscillations (APCER = 0) and correctly classified 99.77% of bona fide pupil reactions (BPCER = 0.23%).

Increasing the observation time to 10 s and applying **both positive and negative stimuli** (Table 6.1, last column) do not result in a more accurate solution. Therefore, we have investigated the winning solutions applied to positive or negative stimuli for **shorter times**, starting from 2 s. One should expect to get both APCER and BPCER below 1% when the parametric model is applied to time series of approximately 3 s, Fig. 6.12. For data-driven models, we do not observe a substantial decrease in performance even when the eye is observed for approximately 2 s, Fig. 6.13. These results increase the potential for practical implementations.

Table 6.1 Error rates (in %) for all modeling and classification approaches considered in this paper. Best approaches (in terms of average of APCER and BPCER) for each combination of model type and stimulus type are shown in bold font

Model	Variant		Positive stimulus (5 s)		Negative stimulus (5 s)		Both stimuli (10 s)	
			APCER	BPCER	APCER	BPCER	APCER	BPCER
Parametric	SVM [18]	Linear	**0.46**	**0.23**	1.15	0.23	1.61	0.23
		Polynomial	**0.46**	**0.23**	0.46	0.00	1.15	0.92
		Radial basis	**0.46**	**0.23**	0.00	0.46	0.69	0.69
	Logistic regression		0.92	0.92	0.92	0.23	1.61	0.23
	Bagged trees		0.69	0.46	0.23	0.23	**0.23**	**0.92**
	kNN	k = 1	0.92	0.23	**0.00**	**0.23**	1.15	0.46
		k =10	0.92	0.23	0.69	0.00	1.38	0.23
Data-driven	Basic RNN		0.46	1.84	**1.26**	**3.91**	1.15	1.15
	GRU		0.69	2.99	2.30	6.67	2.07	0.69
	LSTM	No peepholes	**0.23**	**1.38**	1.38	3.91	1.61	1.38
		With peepholes	0.46	1.84	1.95	3.45	**0.69**	**1.38**

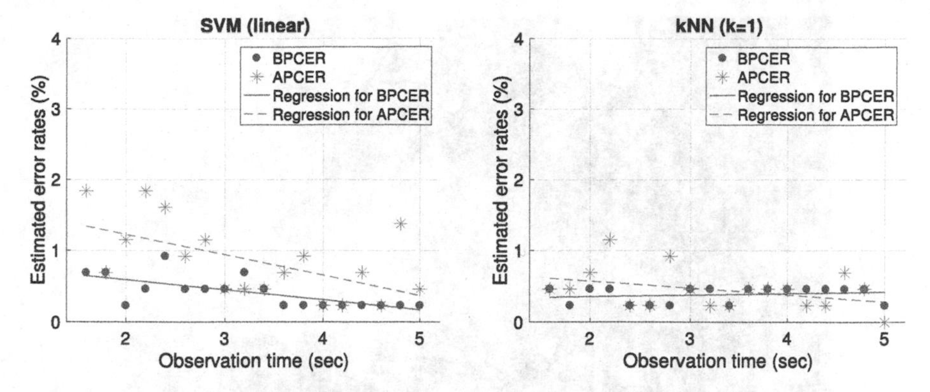

Fig. 6.12 The best configurations found for the entire positive and negative stimuli (5 seconds) analyzed for shorter horizons with a **parametric model**. Results for a positive stimulus (**left**) and a negative stimulus (**right**)

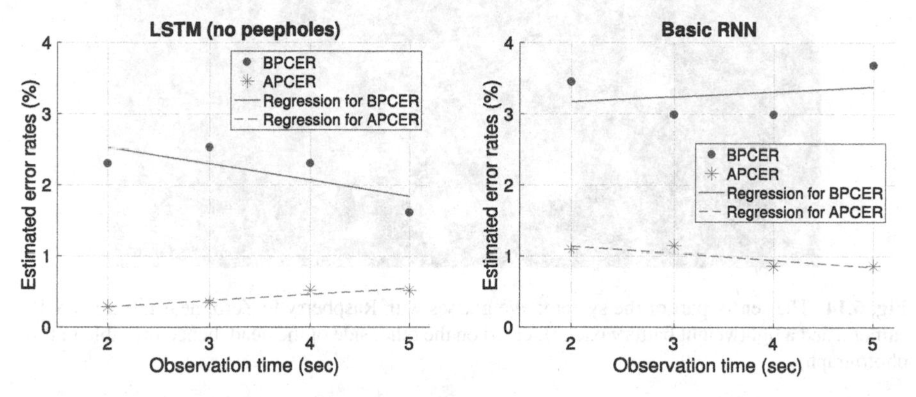

Fig. 6.13 Same as in Fig. 6.12, except that a **data-driven model** was used. Results for a positive stimulus (**left**) and a negative stimulus (**right**)

6.6 Open-Hardware and Open-Source Pupil Size Measurement Device

6.6.1 Brief Characteristics

The collection of quality videos of the eyes for the study described in previous sections of this chapter has been accomplished with a specially designed hardware, requiring each subject to precisely position their head in front of the camera and stay still for the entire 30 s acquisition. However, the pupillometry that is less constrained may have more value to biometrics. To our knowledge, there exists no open-source hardware and software for general pupillometry that is portable and affordable to anyone interested in collecting and analyzing videos of the eye. Thus, this Section offers an open source software and hardware system for general pupillometry, with

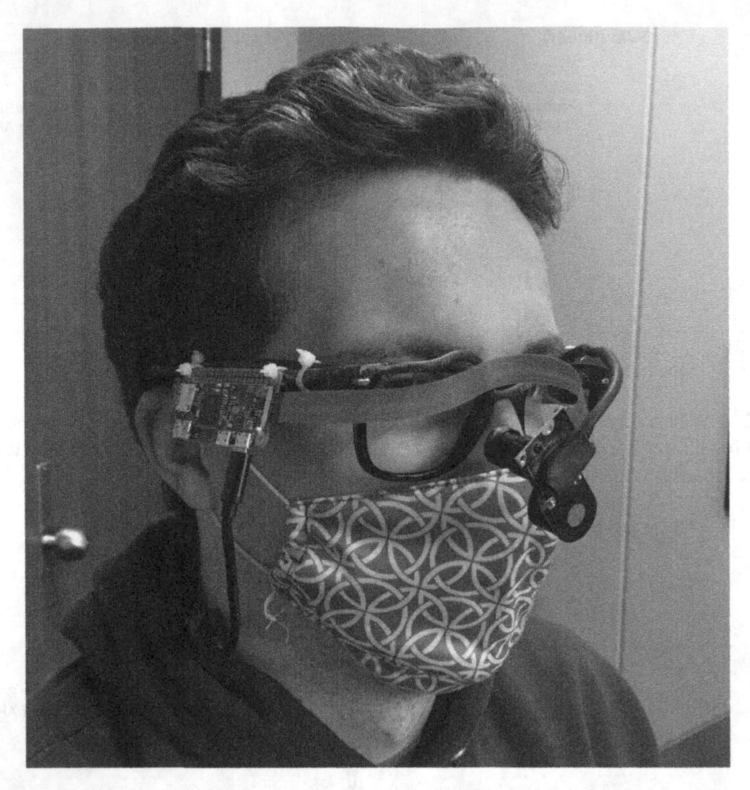

Fig. 6.14 The sensor part of the system: eye glasses with Raspberry Pi Zero, near infrared (NIR) camera, and a lightweight battery pack (located on the other side of the head, hence invisible in this photograph)

possible extensions into gaze detection and iris recognition. The hardware system consists of a Raspberry Pi Zero W and a compatible infrared camera attached to an ordinary pair of glasses, which can be worn comfortably by a subject whether they are still or actively moving or performing a task, as shown in Fig. 6.14. The novel protocol presented in this Section allows the Raspberry Pi Zero W to record video in real time and send the frames to a host computer with INET sockets. The host computer can interrupt the Pi if it detects poor-quality frames. The host computer then runs our algorithm to extract the pupil size.

Although other open source pupillometry hardware has been developed, the hardware proves impractical for an individual to recreate without substantial expertise or funding [30]. They tend to use expensive cameras and stationary systems with wires that prevent movement as well, making them less flexible for experiments. We are confident that the proposed project can be replicated by anyone with average computer skills and prove useful in various experimental applications in biometrics and psychology. There are several, hopefully appreciated novelties in this project:

- the connection between the glasses with Raspberry Pi and the host computer is wireless, so the subject is not constrained to any position relative to the system;
- the placement of a small camera makes the eye uncovered, so the individual can perform normal tasks when having their pupil size measured;
- it's an open source/open-hardware project, featuring affordable and easily replicable elements.

The complete software project can be accessed via GitHub,[1] as part of the reproducible materials for this chapter. This includes the client code run on the Pi, the host code run on the host computer, the neural network run on the host computer to obtain the proper coordinates for cropping, and general processing methods.

6.6.2 Related Works on Pupillometry

Pupillometry methods have been developed for at least the last 30 years. Accurate pupil measurements can be valuable for analysis of neuronal activity, preparation for surgery, and many more applications [30, 31]. Most methods, especially older ones, make use of eye image thresholding to get the pupil position. Nguyen et al. [32] present a simple method of pupillometry whereby the pixel intensity histogram is separated into three distinct peaks representing the pupil, iris, and noise, and the Purkinje image (PI). A region of interest (ROI) is also selected to reduce computation cost, and the pupil is simply segmented from the iris, and the PI is set to black. To recover from a blink, the last measured coordinates of the pupil are stored.

Once the segmented pupil is established, in order to measure its size Nowak et al. [33] used a method whereby they fit an ellipse over the segmented image. Their system measures pupil size in a binocular fashion with alternating lighting on each eye, lending itself more to clinical studies of pupil response to stimuli than general pupil measurement.

Iskander et al. [31] present a more complex algorithm for pupillometry, beginning with a detector based on quadruple axes symmetry indicator (QSI) to find the rough pupil center. From there, the coordinates of the image are converted into polar coordinates and the algorithm moves outwards from the pupil center to find the boundaries of the pupil, iris, and sclera on both the right and left sides of the eye. Like Nguyen et al., this algorithm is presented without any corresponding hardware, designed to run on any digital image.

The speed of accurate pupil measurement can vary greatly. Nasim et al. [34] present a method that necessitates the use of multiple cores or GPUs, but employs elliptical Hough Transform operations to detect the size of the pupil even in cases of occlusion by the eyelid. Alternatively, Miró et al. [35] propose a method that runs in real time and analyzes the vertical and horizontal areas of the ROI independently,

[1] https://github.com/aczajka/pupil-dynamics.

finding accurate values for the edge of the pupil. They also employ hardware that avoids the issue of PI's appearing on the pupil through the control of LED's on the device.

Open source pupillometry projects are less common, but De Souza et al. [30] describes an image acquisition system based around a FireWire camera. Their system involves the use of a goggles apparatus that covers both eyes, with the camera measuring the right eye. Continuous and event-driven modes are available for the device. As far as the algorithm, they select a small ROI around the pupil to minimize noise and computation cost, and they apply morphological operations to enhance the image of the segmented pupil. Although De Souza's system is similar to ours, he uses more complicated hardware and a pupillometry system consisting of googles, which may restrict possible experiments due to limiting the individual's vision.

6.6.3 Design

6.6.3.1 Hardware

The major hardware components of the project are the Raspberry Pi Zero W, the MakerFocus Raspberry Pi Zero Night Vision Camera, a lightweight battery pack, and a host computer. Neglecting the cost of a host computer, the cost of the setup was around $72 (see Table 6.2) when collecting all elements in the middle of 2021. The camera was equipped with an active near infrared (NIR) light, which improves the quality of the iris and pupil imaging, and provides the opportunity for more precise measurements and additional applications, such as eye tracking and iris recognition. It's noteworthy that any camera compatible with the Pi Zero with IR capabilities can be substituted for the MakerFocus product, although the software and methods were designed with the MakerFocus equipment. The distance between the camera and the eye for this device is approximately 5 cm. This distance was dictated by the camera lens angle of view and the goal of generating iris images meeting the resolution requirements recommended by ISO/IEC 19794-6.

Table 6.2 List of the main hardware components and their costs (mid 2021)

Component	Cost (USD)
Raspberry Pi Zero W	$23.95
Raspberry Pi Camera	$20.00
Micro SD Card	$5.00
Battery	$15.00
Glasses	$3.00
Ties, tape, wire	$5.00
Total	$71.95

6.6.3.2 Communication and Capture Protocol

Due to the relatively low processing power of the Raspberry Pi Zero W, we avoid full processing of the video on the device, opting to transfer pre-processed frames to a host computer in a client-server relationship. The communication protocol between the Pi Zero and host computer uses INET sockets, implemented with Python's socket module. In order to connect, the IP address of the host computer is input to the Raspberry Pi's program along with an open port, and the socket on the host computer is opened to a connection. In order to avoid the need to connect the Raspberry Pi to a desktop to start the program, we use SSH to interface with the Pi and boot headless in order to start the socket connection. Once the connection is made, an initial image is transferred from the Pi to the host computer along the socket.

Figure 6.15 presents the communication protocol of the proposed system. Upon receiving the initial full-frame image, the host computer runs an image segmentation based on the CCNet model [36] trained with images taken by the Raspberry Pi camera on a small dataset of 115 unique images. Standard data augmentation techniques (such as vertical and horizontal reflection, or Gaussian blurring) were applied to each training image to improve the performance of the network.

Blob detection is run on the output of the neural network to estimate the location of the eye and the coordinates are sent back to the Pi for appropriate cropping. The cropping is necessary so that the Pi is not sending back large image files, which would slow down the pipeline. The cropping reduces the size of the image to 240×360 centered around the pupil, which allows the system to operate at a respectable 5 frames per second. If the host computer determines the size of the pupil to be very low for $k = 5$ consecutive frames, it is assumed that the pupil is no longer within the cropped region, and the computer sends a signal to the Pi to restart the full-frame acquisition process.

6.6.3.3 Segmentation and Pupil Size Estimation

In order to demonstrate the hardware and software protocols, we implemented a system by which we measure the pupil size of the wearer over time, which allows us to see with reasonable accuracy the trends in pupil size over the course of acquisition. The segmentation algorithm we implement begins by running a Gaussian Mixture Model on a histogram of the pixel values, which allows to calculate a dynamic and noise-robust threshold for image binarization. As the pupil is a large dark region in the image, we select the darkest component as the pupil.

However, if we only select the darkest component then we are left with significant noise that can disrupt our radius measurement. Commonly, eyelashes and the border of the eyelid and sclera are included in this dark component. In order to eliminate these, we apply morphological operations and select the largest blob which will always be the pupil itself. Any other noisy components are discarded. Finally, the

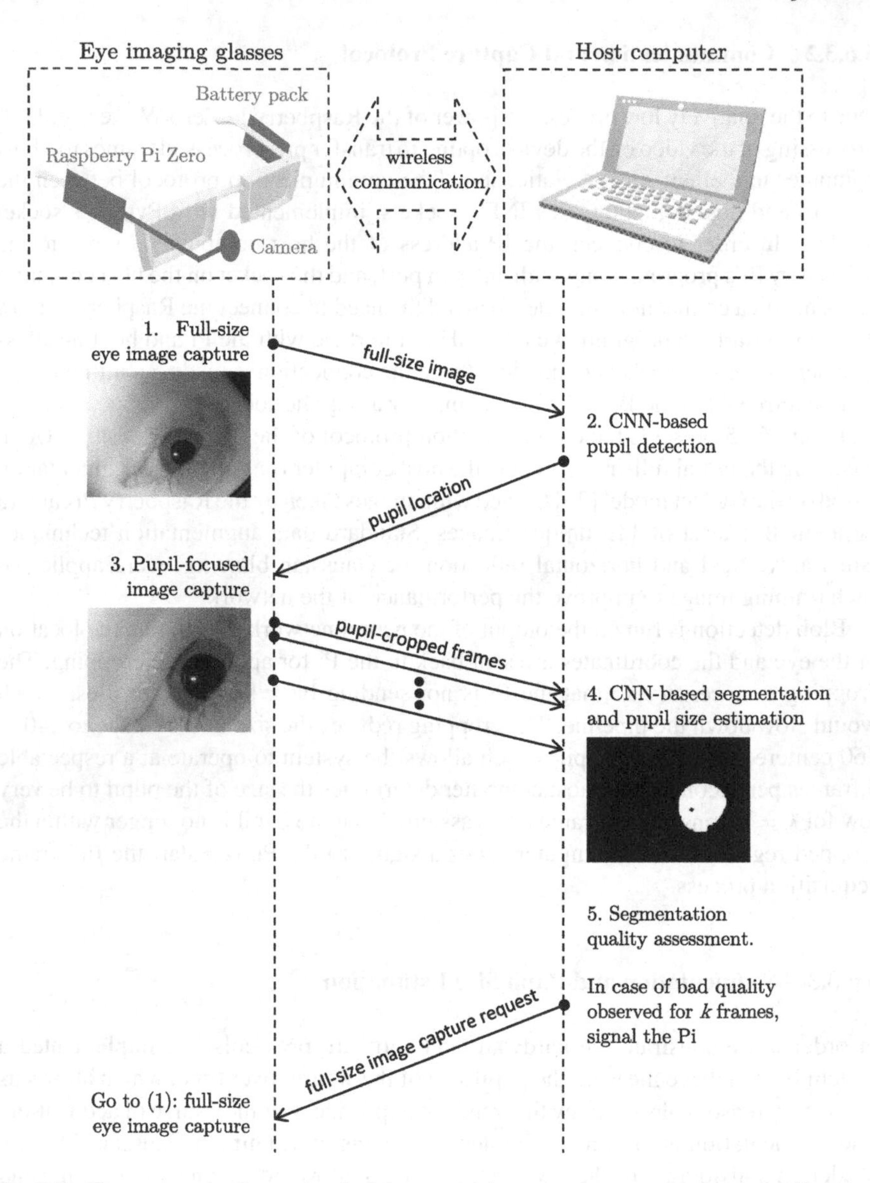

Fig. 6.15 The communication protocol of the proposed open-source open-hardware pupillometer

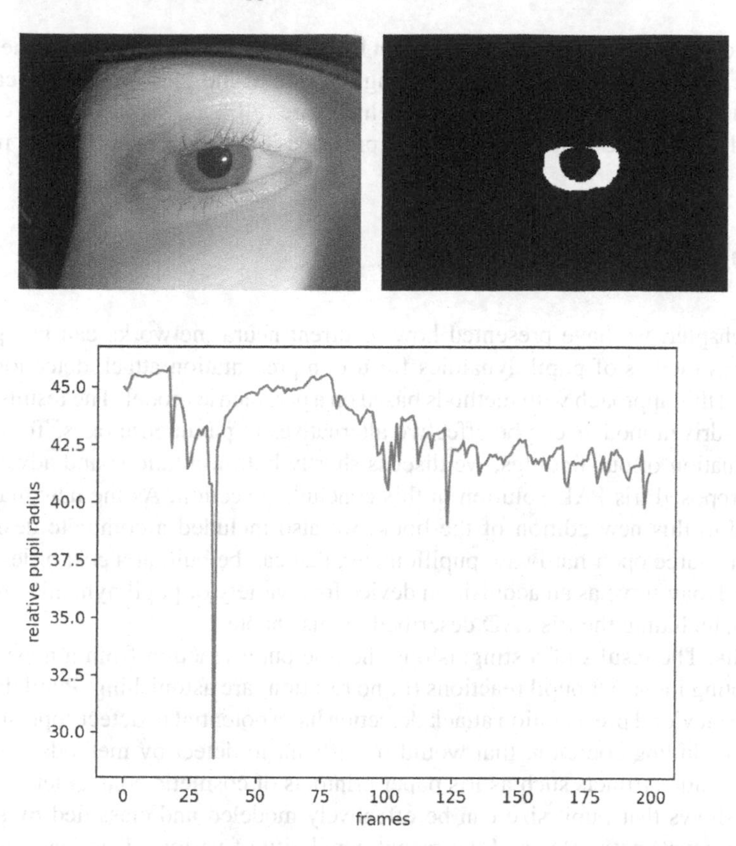

Fig. 6.16 Example full-frame sample captured by Pi-based pupillometer (left), it's segmentation done on a host computer (right), and an example pupil radius (measured in pixels) obtained from the proposed device over time (bottom). The negative peak corresponds to a blink. Note that this pupil estimation device employs the convolutional neural network for segmentation (offered with this Chapter), opposite to the OSIRIS software used previously in modeling and PAD method design

circular Hough transform is used to get the pupil size. Figure 6.16 shows an example cropped iris image, located pupil and the example pupil size signal, obtained from the proposed device.

6.6.4 Assembly Details

For readers interested in reproducing the device presented, only a few materials outside of the major hardware components are needed. The components are sturdy glasses, wire that holds its shape, electric tape, superglue, and zip ties. The Raspberry Pi can conveniently be attached to the side of the glasses through holes built into the frame by zip ties, while the battery can be attached on the other side in the same manner. The sturdy wire will be run along the side of the glasses, and the camera superglued to the end of it. Apart from hardware, all necessary software can be accessed through the virtual environment provided in the associated GitHub repo.[2]

6.7 Discussion

In this chapter, we have presented how recurrent neural networks can be applied to serve as models of pupil dynamics for use in presentation attack detection, and compared this approach with methods based on a parametric model. The results show that data-driven models can be effective alternatives to parametric ones. To make a fair evaluation of our findings, we discuss shortly both limitations and advantages of the proposed iris PAD solution in this concluding section. As the new material, prepared to this new edition of the book, we also included a complete design of the open-source open-hardware pupillometer, that can be built at a cost of less than $100, and may serve as an acquisition device for a variety of pupil dynamics-related research, including the iris PAD described in this chapter.

Merits. The results of distinguishing the true pupil reaction from a noisy input representing incorrect pupil reactions (or no reaction) are astonishing. Pupil dynamics as a behavioral presentation attack detection has a potential to detect sophisticated attacks, including coercion, that would be difficult to detect by methods designed to detect static artifacts such as iris paper printouts of cosmetic contact lenses. This chapter shows that pupil size can be effectively modeled and classified by simple recurrent neural networks, and the recent popularity of various deep learning software tools facilitates such implementations. The minimum observation time of 2–3 s required to achieve the best accuracy presented in this study makes this PAD close to practical implementations since a typical iris acquisition requires a few seconds even when only a still image is to be taken.

Limitations. It has been shown in literature that pupil reaction depends on many factors such as health conditions, emotional state, drug induction, and fatigue. It also depends on the level of ambient light. Therefore, the models developed in this work may not work correctly in adverse scenarios, unless we have knowledge or data that could be used to adapt our classifiers. Generating attack presentations for pupil dynamics is difficult, hence the next necessary step is the reformulation of this problem into one-class classification. Also, the application of negative or positive stimuli depends on the implementation environment. That is, it will be probably easier to stimulate the eye by increasing the level of light than evoke pupil dilation that requires a sudden decrease in the level of ambient light. An interesting observation is that BPCER (recognition of authentic eyes) for the data-driven solution is significantly higher (3–4%) than for the parametric-model-based approach (up to 1%). ACPER (recognition of no/odd reactions) remains similar in both approaches. This

[2] https://github.com/aczajka/pupil-dynamics.

may suggest that more flexible neural-network-based classification, when compared to a model specified by only 7 parameters, is more sensitive to subject-specific fluctuations of pupil size and presents lower generalization capabilities than methods based on a parametric model.

Acknowledgements The authors would like to thank Mr. Rafal Brize and Mr. Mateusz Trokielewicz, who collected the iris images in varying light conditions under the supervision of the first author. The application of Kohn and Clynes model was inspired by the research of Dr. Marcin Chochowski, who used parameters of this model as individual features in biometric recognition. This author, together with Prof. Pacut and Dr. Chochowski, has been granted a US patent No. 8,061,842 which partially covers the ideas related to parametric-model-based PAD and presented in this work.

References

1. ISO/IEC (2016) Information technology – biometric presentation attack detection – Part 1: Framework, 30107-3
2. Galbally J, Marcel S, Fierrez J (2014) Image quality assessment for fake biometric detection: application to iris, fingerprint, and face recognition. IEEE Trans Image Process (TIP) 23(2):710–724. https://doi.org/10.1109/TIP.2013.2292332
3. Wei Z, Qiu X, Sun Z, Tan T (2008) Counterfeit iris detection based on texture analysis. In: International conference on pattern recognition, pp 1–4. https://doi.org/10.1109/ICPR.2008.4761673
4. Doyle JS, Bowyer KW, Flynn PJ (2013) Variation in accuracy of textured contact lens detection based on sensor and lens pattern. In: IEEE international conference on biometrics: theory applications and systems (BTAS), pp 1–7. https://doi.org/10.1109/BTAS.2013.6712745
5. Ojala T, Pietikainen M, Maenpaa T (2002) Multiresolution gray-scale and rotation invariant texture classification with local binary patterns. IEEE Trans Pattern Anal Mach Intell (TPAMI) 24(7):971–987. https://doi.org/10.1109/TPAMI.2002.1017623
6. Ojansivu V, Heikkilä J (2008) Blur insensitive texture classification using local phase quantization. Springer, Berlin, Heidelberg, pp 236–243. https://doi.org/10.1007/978-3-540-69905-7_27
7. Zhang L, Zhou Z, Li H (2012) Binary gabor pattern: an efficient and robust descriptor for texture classification. In: IEEE international conference on image processing (ICIP), pp 81–84. https://doi.org/10.1109/ICIP.2012.6466800
8. Sun Z, Zhang H, Tan T, Wang J (2014) Iris image classification based on hierarchical visual codebook. IEEE Trans Pattern Anal Mach Intell (TPAMI) 36(6):1120–1133. https://doi.org/10.1109/TPAMI.2013.234
9. Dalal N, Triggs B (2005) Histograms of oriented gradients for human detection. In: IEEE international conference on computer vision and pattern recognition (CVPR), vol 1, pp 886–893. https://doi.org/10.1109/CVPR.2005.177
10. Doyle JS, Bowyer KW (2015) Robust detection of textured contact lenses in iris recognition using BSIF. IEEE Access 3:1672–1683. https://doi.org/10.1109/ACCESS.2015.2477470
11. Raghavendra R, Busch C (2015) Robust scheme for iris presentation attack detection using multiscale binarized statistical image features. IEEE Trans Inf Forensics Secur (TIFS) 10(4):703–715. https://doi.org/10.1109/TIFS.2015.2400393
12. Menotti D, Chiachia G, Pinto A, Schwartz W, Pedrini H, Falcao A, Rocha A (2015) Deep representations for iris, face, and fingerprint spoofing detection. IEEE Trans Inf Forensics Secur (TIFS) 10(4):864–879. https://doi.org/10.1109/TIFS.2015.2398817

13. Das P, McGrath J, Fang Z, Boyd A, Jang G, Mohammadi A, Purnapatra S, Yambay D, Marcel S, Trokielewicz M, Maciejewicz P, Bowyer K, Czajka A, Schuckers S, Tapia J, Gonzalez S, Fang M, Damer N, Boutros F, Kuijper A, Sharma R, Chen C, Ross A (2020) Iris liveness detection competition (LivDet-Iris) - The 2020 Edition. In: IEEE international joint conference on biometrics (IJCB), pp 1–9. https://doi.org/10.1109/IJCB48548.2020.9304941
14. Raja K, Raghavendra R, Busch C (2015) Video presentation attack detection in visible spectrum iris recognition using magnified phase information. IEEE Trans Inf Forensics Secur (TIFS) 10(10):2048–2056. https://doi.org/10.1109/TIFS.2015.2440188
15. Komogortsev OV, Karpov A, Holland CD (2015) Attack of mechanical replicas: liveness detection with eye movements. IEEE Trans Inf Forensics Secur (TIFS) 10(4):716–725. https://doi.org/10.1109/TIFS.2015.2405345
16. Pacut A, Czajka A (2006) Aliveness detection for iris biometrics. In: IEEE international carnahan conferences security technology (ICCST), pp 122–129. https://doi.org/10.1109/CCST.2006.313440
17. Czajka A, Pacut A, Chochowski M (2011) Method of eye aliveness testing and device for eye aliveness testing, United States Patent, US 8,061,842
18. Czajka A (2015) Pupil dynamics for iris liveness detection. IEEE Trans Inf Forensics Secur (TIFS) 10(4):726–735. https://doi.org/10.1109/TIFS.2015.2398815
19. Czajka A, Bowyer KW (2018) Presentation attack detection for iris recognition: An assessment of the state of the art. ACM Comput Surv (CSUR) 1(1):1–35. https://doi.org/10.1145/3232849. In press
20. Boyd A, Fang Z, Czajka A, Bowyer KW (2020) Iris presentation attack detection: where are we now? Pattern Recognit Lett 138:483–489. https://doi.org/10.1016/j.patrec.2020.08.018, https://www.sciencedirect.com/science/article/pii/S0167865520303226
21. Sutra G, Dorizzi B, Garcia-Salitcetti S, Othman N (2017) A biometric reference system for iris. OSIRIS version 4.1: http://svnext.it-sudparis.eu/svnview2-eph/ref_syst/iris_osiris_v4.1/. Accessed from 1 Aug 2017
22. Kohn M, Clynes M (2006) Color dynamics of the pupil. Ann N Y Acad Sci 156(2):931–950 (1969. Available online at Wiley Online Library
23. Bengio Y, Simard P, Frasconi P (1994) Learning long-term dependencies with gradient descent is difficult. IEEE Trans Neural Netw 5(2):157–166. https://doi.org/10.1109/72.279181
24. Hochreiter S, Schmidhuber J (1997) Long short-term memory. Neural Comput 9(8):1735–1780. https://doi.org/10.1162/neco.1997.9.8.1735
25. Gers FA, Schmidhuber J (2000) Recurrent nets that time and count. In: International joint conference on neural network (IJCNN), vol 3, pp 189–194. https://doi.org/10.1109/IJCNN.2000.861302
26. Greff K, Srivastava RK, Koutník J, Steunebrink BR, Schmidhuber J (2016) Lstm: a search space odyssey. IEEE Trans Neural Netw Learn Syst PP(99):1–11. https://doi.org/10.1109/TNNLS.2016.2582924
27. Cho K, van Merriënboer B, Bahdanau D, Bengio Y (2014) On the properties of neural machine translation: encoder–decoder approaches. In: Workshop on syntax, semantics and structure in statistical translation (SSST), pp 1–6
28. Hinton G, Srivastava N, Swersky K (2017) Neural networks for machine learning. Lecture 6a: overview of mini-batch gradient descent. http://www.cs.toronto.edu/~tijmen/csc321. Accessed from 28 April 2017
29. Glorot X, Bengio Y (2010) Understanding the difficulty of training deep feedforward neural networks. In: Teh YW, Titterington M (eds) international conference on artificial intelligence and statistics (AISTATS), Proceedings of machine learning research, vol 9. PMLR, Chia Laguna Resort, Sardinia, Italy, pp 249–256
30. de Souza JKS, da Silva Pinto MA, Vieira PG, Baron J, Tierra-Criollo CJ (2013) An open-source, firewire camera-based, labview-controlled image acquisition system for automated, dynamic pupillometry and blink detection. Comput Methods Prog Biomed 112(3):607–623. https://doi.org/10.1016/j.cmpb.2013.07.011, https://www.sciencedirect.com/science/article/pii/S0169260713002460

31. Iskander D, Collins M, Mioschek S, Trunk M (2004) Automatic pupillometry from digital images. IEEE Trans Biomed Eng 51(9):1619–1627. https://doi.org/10.1109/TBME.2004.827546
32. Nguyen AH, Stark LW (1993) Model control of image processing: pupillometry. Comput Med Imaging Graph 17(1):21–33. https://doi.org/10.1016/0895-6111(93)90071-T
33. Nowak W (2014) System and measurement method for binocular pupillometry to study pupil size variability. Biomedical Engineering OnLine
34. Nasim A, Maqsood A, Saeed T (2017) Multicore and GPU based pupillometry using parabolic and elliptic hough transform
35. Miro I, López-Gil N, Artal P (1999) Pupil meter and tracking system based in a fast image processing algorithm. Proc SPIE - Int Soc Opt Eng 3591:63–70. https://doi.org/10.1117/12.350615
36. Mishra S, Czajka A, Liang P, Chen DZ, Hu XS (2019) Cc-net: Image complexity guided network compression for biomedical image segmentation. In: The IEEE international symposium on biomedical imaging (ISBI). IEEE, Venice, Italy, pp 1–6. https://arxiv.org/abs/1901.01578

Chapter 7
Review of Iris Presentation Attack Detection Competitions

David Yambay, Priyanka Das, Aidan Boyd, Joseph McGrath, Zhaoyuan (Andy) Fang, Adam Czajka, Stephanie Schuckers, Kevin Bowyer, Mayank Vatsa, Richa Singh, Afzel Noore, Naman Kohli, Daksha Yadav, Mateusz Trokielewicz, Piotr Maciejewicz, Amir Mohammadi, and Sébastien Marcel

Abstract Biometric recognition systems have been shown to be susceptible to presentation attacks, the use of an artificial biometric in place of a live biometric sample from a genuine user. Presentation Attack Detection (PAD) is suggested as a solution to this vulnerability. The LivDet-Iris Liveness Detection Competition strives to showcase the state of the art in presentation attack detection by assessing the software-based iris Presentation Attack Detection (PAD) methods, as well as hardware-based

D. Yambay · P. Das · S. Schuckers (✉)
Clarkson University, 10 Clarkson Avenue Potsdam, Clarkson, NY 13699, USA
e-mail: sschucke@clarkson.edu

D. Yambay
e-mail: yambayda@clarkson.edu

P. Das
e-mail: prdas@clarkson.edu

A. Boyd · J. McGrath · Z. (Andy) Fang · A. Czajka (✉)
University of Notre Dame, 384 Fitzpatrick Hall of Engineering, Notre Dame, IN 46556, USA
e-mail: aczajka@nd.edu

A. Boyd
e-mail: aboyd3@nd.edu

J. McGrath
e-mail: jmcgrat3@alumni.nd.edu

K. Bowyer
Notre Dame University, Notre Dame, IN 46556, USA
e-mail: kwb@nd.edu

M. Vatsa (✉) · R. Singh (✉)
Department of CSE, IIT Jodhpur, Jodhpur 342037, RJ, India
e-mail: mvatsa@iitj.ac.in

R. Singh
e-mail: richa@iitj.ac.in

A. Noore
Texas A&M University-Kingsville, Kingsville, TX 78363, USA
e-mail: Afzel.Noore@tamuk.edu

S. Marcel et al. (eds.), *Handbook of Biometric Anti-Spoofing*, Advances in Computer Vision and Pattern Recognition, https://doi.org/10.1007/978-981-19-5288-3_7

iris PAD methods against multiple datasets of spoof and live fingerprint images. These competitions have been open to all institutions, industrial and academic, and competitors can enter as either anonymous or using the name of their institution. There have been four LivDet-Iris competitions organized to date: the series was launched in 2013, and the most recent competition was organized in 2020, with the other two happening in 2015 and 2017. This chapter briefly characterizes all four competitions, discusses the state of the art in iris PAD (from the independent evaluations point of view), and current needs to push the iris PAD reliability forward.

7.1 Introduction

Iris recognition is one of the most popular biometric modalities for large-scale governmental deployments [1, 2], commercial products [3] and financial applications [4] due to its several important properties: uniqueness, temporal stability, and performance. However, these systems have been shown to be vulnerable to biometric *presentation attacks*. Early presentation attack attempts employed patterned contact lenses and printed iris images, while more recent attack types include synthetically-generated irises [5, 6] and cadaver eyes [7].

Numerous competitions have been held in the past to address biometric matching performance, such as the Fingerprint Verification Competition held in 2000, 2002, 2004 and 2006 [8], Face Recognition Vendor Test (FRVT) [9], Iris Exchange (IREX) series [10], or the ICB Competition on Iris Recognition [11]. However, these early competitions did not consider presentation attacks, until the Fingerprint Liveness Detection Competition series started by the University of Cagliari and Clarkson University in 2009 to create a benchmark for measuring fingerprint presentation attack detection algorithms, similar to a matching performance. The 2009 Finger-

N. Kohli · D. Yadav
West Virginia University, Morgantown, WV 26506, USA
e-mail: nakohli@mix.wvu.edu

D. Yadav
e-mail: dayadav@mix.wvu.edu

M. Trokielewicz
Warsaw University of Technology, Warsaw, Poland
e-mail: mateusz.trokielewicz@pw.edu.pl

P. Maciejewicz
Medical University of Warsaw, Warsaw, Poland
e-mail: piotr.maciejewicz@wum.edu.pl

A. Mohammadi · S. Marcel
Idiap Research Institute, Rue Marconi 19, 1920 Martigny, Switzerland
e-mail: amir.mohammadi@idiap.ch

S. Marcel
e-mail: sebastien.marcel@idiap.ch

print Liveness Detection Competition pioneered the concept of liveness detection as part of a biometric modality in deterring spoof attacks. This expanded to LivDet-Iris series in 2013, focusing on iris biometrics [12], with Warsaw University of Technology, Poland (further Warsaw for brevity), and University of Notre Dame, USA (further Notre Dame for brevity), joining the team. With a core organizing consortium comprising three Universities (Clarkson, Warsaw, and Notre Dame), along with various partners throughout the years (West Virginia University, IIIT-Delhi, Medical University of Warsaw and Idiap), the LivDet-Iris has continued in 2015 [13], 2017 [14] and 2020 [15].

The goal of LivDet-Iris competition, since its conception, has been to allow researchers to test their own *presentation attack detection* (PAD) algorithms and systems on publicly available data sets, with results published to establish baselines for future research and to examine the state of the art as the PAD develops. Specifically, the most recent edition (2020) was focused on generalization capabilities of the iris PAD algorithms, partially measuring the algorithms' performance on unknown attack types. Each LivDet-Iris competition was open to two types of submissions: strictly software-based approaches to PAD (called *Part 1–Algorithms*), and software or hardware-based approaches in a fully-packaged device called *Part 2–Systems*. LivDet-Iris has not yet received hardware-based approaches, hence all editions to date conducted the evaluations of software-based approaches.

It is noteworthy that, in addition to fingerprint and iris PAD, also face recognition has observed dedicated PAD competitions, such as *Competition on Counter Measures to 2-D Facial Spoofing Attacks*, held in 2011 [16] and 2013 [17], or the *Face Anti-spoofing (Presentation Attack Detection) Challenges*, held at CVPR 2019, CVPR 2020 and ICCV 2021 [18]. This chapter, however, focuses solely on reviewing all the LivDet-Iris competitions organized to date, and their evolution over the years. The following sections will describe the methods used in testing for each of the LivDet-Iris competitions as well as descriptions of the datasets that have been generated from each competition. Also discussed are the trends across the different competitions that reflect advances in the art of presentation attacks as well as advances in the state of the art in presentation attack detection. Further, conclusions from LivDet-Iris competitions and their future are discussed.

7.2 Datasets

Each iteration of the competition features at least three different datasets of authentic and fake iris data. In 2013, 2015, and 2017 editions, there are two main presentation attack types: printed iris images and eyes wearing patterned contact lenses. Printed iris images offered by Warsaw University of Technology, were first used to impersonate another person in an example commercial iris recognition system (what made this dataset unique), while patterned contact lenses were used to obfuscate the attacker's natural iris pattern. The 2017 edition included also a hybrid type of attack, offered by the IIIT-Delhi, of eyes wearing cosmetic contact lenses, then printed on a paper.

The most recent 2020 edition added new attack types to a sequestered test corpus, never used in iris PAD competitions: cadaver samples (offered by Medical university of Warsaw and Warsaw University of Technology), artificial specimens, such as plastic eyes, and printed irises with transparent contact lenses on top to imitate the cornea. Competitors were able to use any publicly available data, including sets used in previous competitions, in their algorithm training. The paragraphs below discuss datasets used in each of the four LivDet-Iris competitions.

7.2.1 LivDet-Iris 2013 Data

The inaugural **LivDet-Iris 2013 dataset** consisted of data representing two attack types (textured contact lenses and iris images printed on a paper) from three iris sensors: DALSA (offered by Clarkson University), LG4000 (offered by University of Notre Dame), and IrisGuard AD100 (offered by Warsaw University of Technology). Numbers and types of images in each subset is shown in Table 7.1.

More specifically, textured contact lenses for the **LivDet-Iris 2013 Notre Dame subset** came from Johnson & Johnson [19], CIBA Vision [20], and Cooper Vision [21]. Contacts were purchased in varying colors from each manufacturer and some contacts are "toric" lenses, which means that they are designed to maintain a preferred orientation around the optical axis. The training and testing sets were split equally into three classes: (1) no contact lenses, (2) non-textured contact lenses, and (3) textured contact lenses. Sample images from the Notre Dame subset can be seen in Fig. 7.1.

The **LivDet-Iris 2013 Warsaw subset** was generated using printouts of iris images using a laser printer on matte paper and making a hole in place of the pupil to obtain a live eye specular reflection from the cornea when photographing spoofed irises. Two different printers were used to build the Warsaw subset: (a) HP LaserJet 1320 and (b) Lexmark c534dn. The HP LaserJet was used to print "low resolution" iris images of approximately 600 dpi, whereas the Lexmark was used to print "high resolution" images of 1200 dpi. Only high-resolution images were present in

Table 7.1 Numbers and types of samples included in the **LivDet-Iris 2013 datasets**. Attack Types: TCL = textured contact lenses; PP = paper printouts. Providers: Notre Dame = University of Notre Dame, USA; Warsaw = Warsaw University of Technology, Poland; Clarkson = Clarkson University, USA

Provider	Sensor	Attack types	Train set		Test set	
			Live	Spoof	Live	Spoof
Notre Dame	LG4000	TCL	2,000	1,000	800	400
Warsaw	IrisGuard AD100	PP	228	203	624	612
Clarkson	DALSA	TCL, PP	270	400	246	440

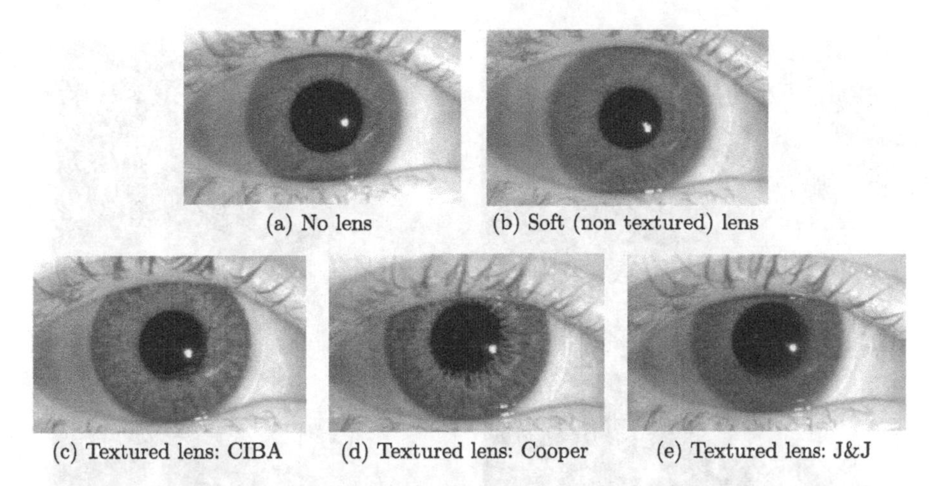

Fig. 7.1 Sample images from the **LivDet-Iris 2013 Notre Dame subset** showing the variety of cosmetic lens textures that are available from different manufacturers. All images are from the same subject eye

the training set, whereas both low and high resolution were used in the testing set. Sample images from the Warsaw subset can be seen in Fig. 7.2.

The **LivDet-Iris 2013 Clarkson subset** was created using patterned contact lenses. Images were collected using a near infra-red (NIR) video camera, DALSA. The camera is modified to capture in the NIR spectrum similar to commercial iris cameras. It captures a section of the face of each subject that includes both eyes. The eyes were then cropped out of the images to create the LivDet samples. A total of 64 eyes were used to create the live subset with varying illuminations and levels of blur. The training set contained 5 images per illumination, but the higher blur level was too strong and was removed giving 3 images per illumination in the testing set. The spoof set contained 6 subjects wearing 19 patterned contact lenses each. A list of patterned contacts used in the Clarkson subset is shown in Table 7.2, and sample images from that dataset can be seen in Fig. 7.3.

7.2.2 LivDet-Iris 2015 Data

Similarly to the 2013 edition, the attack types in the dataset for **LivDet-Iris 2015** edition included printed iris images and eyes wearing patterned contact lenses. Three datasets were delivered by Clarkson University (captured by DALSA and LG IrisAccess EOU2200 sensors) and Warsaw University of Technology (captured by IrisGuard AD100 sensor). The 2013 edition data served as an official training corpus, while new sequestered data served for testing of submitted algorithms. So the 2015 edition data is a superset of the 2013 edition data. Numbers and types of images in each subset is shown in Table 7.3.

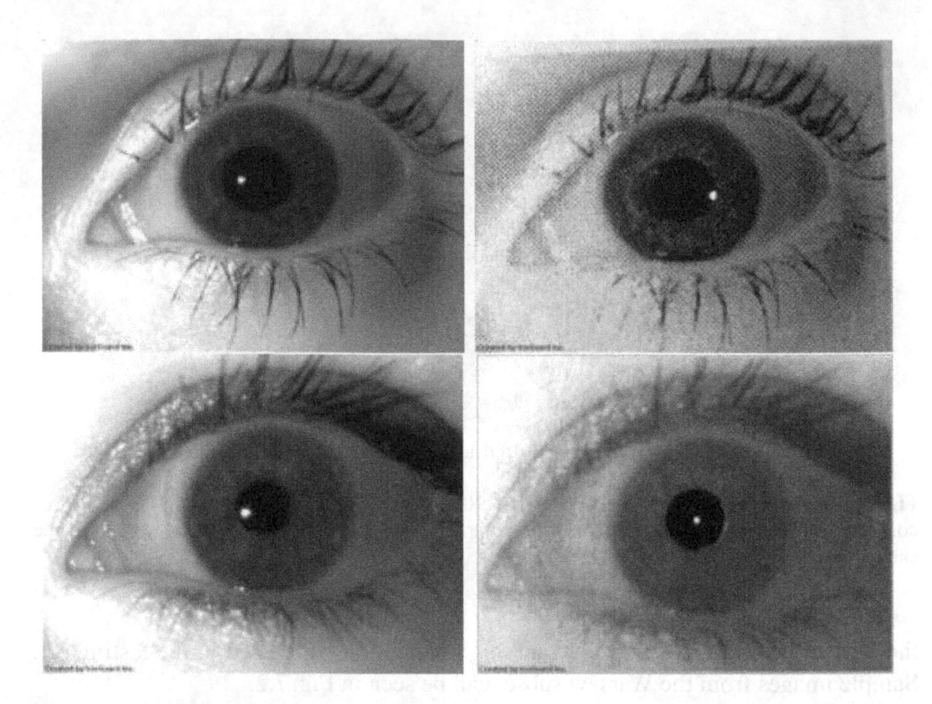

Fig. 7.2 Sample images from the **LivDet-Iris 2013 Warsaw subset**. Images of the authentic eyes are shown in the left column, and their fake counterparts are shown in the right column. Low and high-resolution printouts are presented in the upper and lower rows, respectively

More specifically, the **LivDet-Iris 2015 Warsaw subset** was created using the same Lexmark 534dn used previously in LivDet-Iris 2013 and only used 1200 dpi images. Printouts were completed using both color printing and black and white printing. In addition, holes were cut in place of the pupil in order to have a live user presented behind the printouts. This is to counter a camera that searches for specular reflection of a live cornea. Sample images for the Warsaw dataset can be seen in Fig. 7.4.

The **LivDet-Iris 2015 Clarkson subset** consisted images of both iris paper printouts and eyes wearing textured contact lenses. The printed images used a variety of configurations including 1200 dpi versus 2400 dpi, contrast adjustment versus raw images, pupil hole versus no pupil hole, and glossy paper versus matte paper. The printouts came from images collected from both the LG iris camera and from the DALSA camera. The LG camera is similar to the DALSA camera in that it captures video sequences, however the LG camera is a professional iris sensor. Sample images for both LG and DALSA can be seen in Figs. 7.5 and 7.6 respectively. A novel feature in the 2015 edition, compared to the 2013 edition, was that some selected combinations of contact lens brands and colors/patterns were excluded from the training set and used only during testing. The patterned contact lenses used by the Clarkson dataset is detailed in Table 7.4.

Table 7.2 Types of textured contact lenses used in the **LivDet-Iris 2013 Clarkson subset**

Number	Contact brand	Color
1	Freshlook Dimensions	Pacific Blue
2	Freshlook Dimensions	Sea Green
3	Freshlook Colorblends	Green
4	Freshlook Colorblends	Blue
5	Freshlook Colorblends	Brown
6	Freshlook Colors	Hazel
7	Freshlook Colors	Green
8	Freshlook Colors	Blue
9	Phantasee Natural	Turquoise
10	Phantasee Natural	Green
11	Phantasee Vivid	Green
12	Phantasee Vivid	Blue
13	Phantasee Diva	Black
14	Phantasee Diva	Brown
15	ColorVue BiggerEyes	Cool Blue
16	ColorVue BiggerEyes	Sweet Honey
17	ColorVue 3 Tone	Green
18	ColorVue Elegance	Aqua
19	ColorVue Elegance	Brown

Table 7.3 Numbers and types of samples included in the **LivDet-Iris 2015 datasets**. Attack Types: TCL = textured contact lenses; PP = paper printouts. Providers: Warsaw = Warsaw University of Technology, Poland; Clarkson = Clarkson University, USA

Provider	Sensor	Attack Types	Train set		Test set	
			Live	Spoof	Live	Spoof
Warsaw	IrisGuard AD100	PP	852	815	2,002	3,890
Clarkson	LG EOU2200	PP	400	846	378	900
Clarkson	LG EOU2200	TCL	400[a]	576	378[a]	576
Clarkson	DALSA	PP	700	846	378	900
Clarkson	DALSA	TCL	700(b	873	378[b]	558

a—same images as in Clarkson LG PP; b—same images as in Clarkson DALSA PP

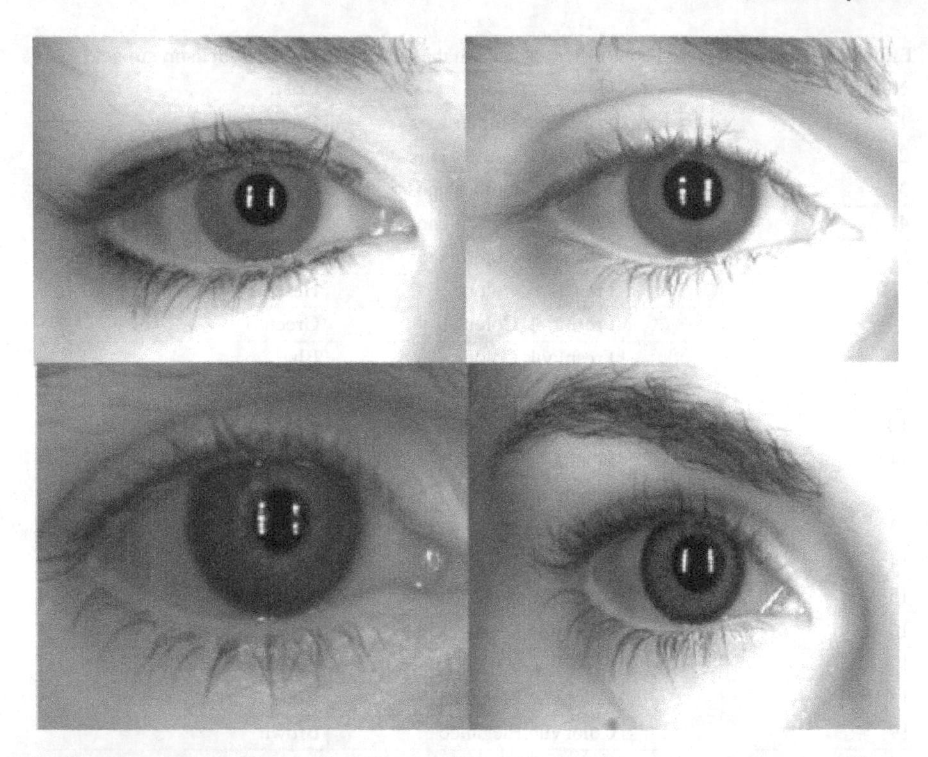

Fig. 7.3 Sample images from the **LivDet-Iris 2013 Clarkson subset**. Live iris images are shown on top, and spoofed irises are shown at the bottom

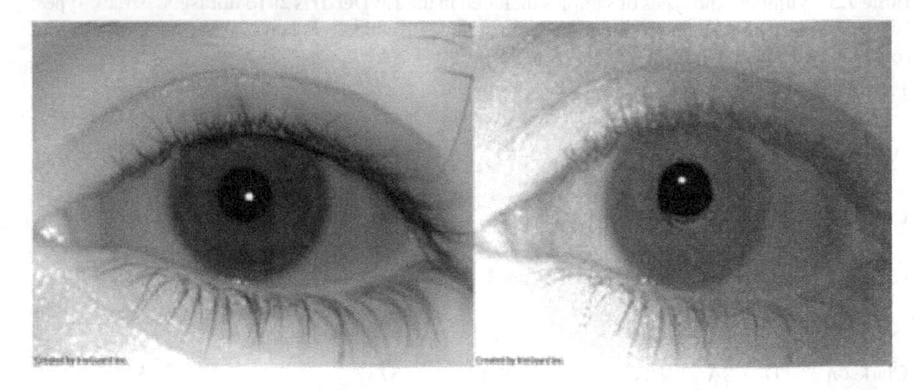

Fig. 7.4 Sample images from the **LivDet-Iris 2015 Warsaw subset**. The authentic iris is shown on the left, and a printouts is shown on the right

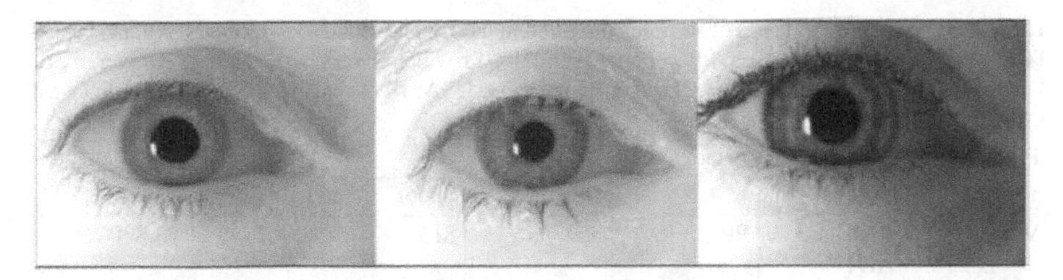

Fig. 7.5 Sample images from the **LivDet-Iris 2015 Clarkson LG subset**. Left to right: live, patterned contact lens, paper printout

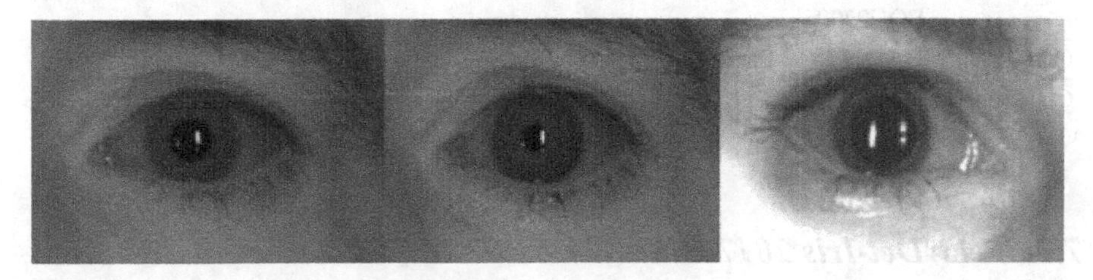

Fig. 7.6 Same as in Fig. 7.5, except that the DALSA camera was used to collect samples

Table 7.4 Types of textured contact lenses used in the **LivDet-Iris 2015 Clarkson DALSA and LG subsets**. Unknown (test set only) patterns are bolded

Number	Contact brand	Color
2	Expressions colors	Jade
3	Expressions colors	Blue
4	Expressions colors	Hazel
5	Air optix colors	Brown
6	Air optix colors	Green
7	Air optix colors	Blue
8	Freshlook colorblends	Brilliant Blue
9	Freshlook colorblends	Brown
10	Freshlook colorblends	Honey
11	Freshlook colorblends	Green
12	Freshlook colorblends	Sterling Gray
13	Freshlook one-day	Green
14	Freshlook one-day	Pure hazel
15	Freshlook one-day	Gray
16	Freshlook one-day	Blue
17	Air optix colors	Gray
18	Air optix colors	Honey
19	Expressions colors	Blue topaz
20	Expressions colors	Green

Table 7.5 Numbers and types of samples included in the **LivDet-Iris 2015 datasets**. Attack Types: TCL = textured contact lenses; PP = paper printouts. Providers: Warsaw = Warsaw University of Technology, Poland; Clarkson = Clarkson University, USA; IIITD-WVU: IIIT-Delhi, India/West Virginia University, USA

Provider	Sensor	Attack types	Train set		Test set	
			Live	Spoof	Live	Spoof
Warsaw	IrisGuard AD100	PP	852	815	2,002	3,890
Clarkson	LG EOU2200	PP	400	846	378	900
Clarkson	LG EOU2200	TCL	400[a]	576	378[a]	576
Clarkson	DALSA	PP	700	846	378	900
Clarkson	DALSA	TCL	700[b]	873	378[b]	558

[a]—same images as in Clarkson LG PP; [b]—same images as in Clarkson DALSA PP

7.2.3 *LivDet-Iris 2017 Data*

LivDet-Iris 2017 competition was joined by the IIIT-Delhi and West Virginia University, which collectively offered iris printouts and eyes with patterned contact lenses captured by a new (in LivDet-Iris series) sensor from IriTech (IriShield MK2120U). In addition, Clarkson University delivered images for the same two attack types collected by LG IrisAccess EOU2200, while University of Notre Dame and Warsaw University of Technology delivered samples for single attack types: textured contact lenses (captured by LG 4000 and IrisGuard AD100) and printouts (captured also with IrisGuard AD100), respectively. Numbers and types of images in each subset is shown in Table 7.5.

The training part of the **LivDet-Iris 2017 Warsaw subset** is composed of the entire 2015 edition of the Warsaw LivDet-Iris data. The test set is a new corpus collected with IrisGuard AD100 sensor (from different subjects, but the sensor used also to collect the training data), and comprising *known spoofs*, and with a proprietary iris capture device, built with the Aritech ARX-3M3C camera with SONY EX-View CCD sensor and Fujinon DV10X7.5A-SA2 lens. The latter subset was created to test algorithms on samples that differ from the training samples in some respect, hence it was called *unknown spoof* set. Examples from the Warsaw dataset are shown in Fig. 7.7.

The **LivDet-Iris 2017 Notre Dame subset** was built with samples taken from the earlier collection, NDCLD15 [22], acquired with LG 4000 and IrisGuard AD100 sensors. The textured contact lenses were manufactured by five different companies: Johnson & Johnson, Ciba, Cooper, UCL, and ClearLab. Samples from Ciba, UCL, and ClearLab were offered to the training and *known spoofs* sets, while samples from Cooper and Johnson & Johnson were used in the *unknown spoofs* test set. Examples from the Warsaw dataset are shown in Fig. 7.8.

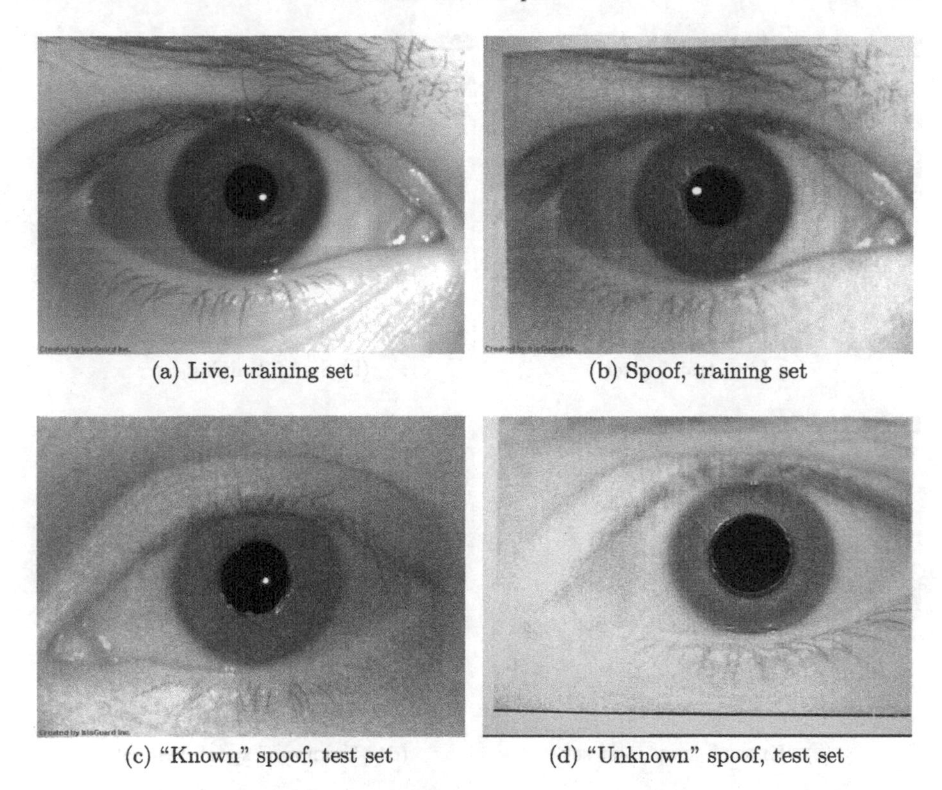

(a) Live, training set

(b) Spoof, training set

(c) "Known" spoof, test set

(d) "Unknown" spoof, test set

Fig. 7.7 Sample images from the **LivDet-Iris 2017 Warsaw subset**. Samples **a**, **b** and **c** are captured with IrisGuard AD100, while the image shown in **d** was captured by the proprietary sensor built with Aritech ARX-3M3C camera and Fujinon DV10X7.5A-SA2 lens

As in case of the Warsaw dataset, the **LivDet-Iris 2017 Clarkson subset** was a superset of their 2015 edition data. Clarkson offered samples representing live irises and two types of attacks: patterned contact lenses and printouts. The latter ones originated from both NIR iris scans (acquired by the LG IrisAccess EOU2200) as well as visible-light iris photographs (captured by iPhone 5 camera). Visible-light-sourced printouts and images of eyes wearing textured contact of selected brands (bolded in Table 7.6) were taken to the *unknown spoofs* test set.

Finally, the **LivDet-Iris 2017 IIITD-WVU subset** was an amalgamation of two, subject- and acquisition environment-disjoint databases offered by the IIIT-Delhi (training set) and by the West Virginia University (*unknown spoofs* test set). There was no *known spoofs* test set in the IIITD-WVU corpus. Two different printers, HP LaserJet Enterprise P3015 (in black and white mode) and Konica Minolta Bizhub C454E (in color mode), were used to create IIITD-WVU, and all samples were captured with the IriTech IriShield MK2120U sensor.

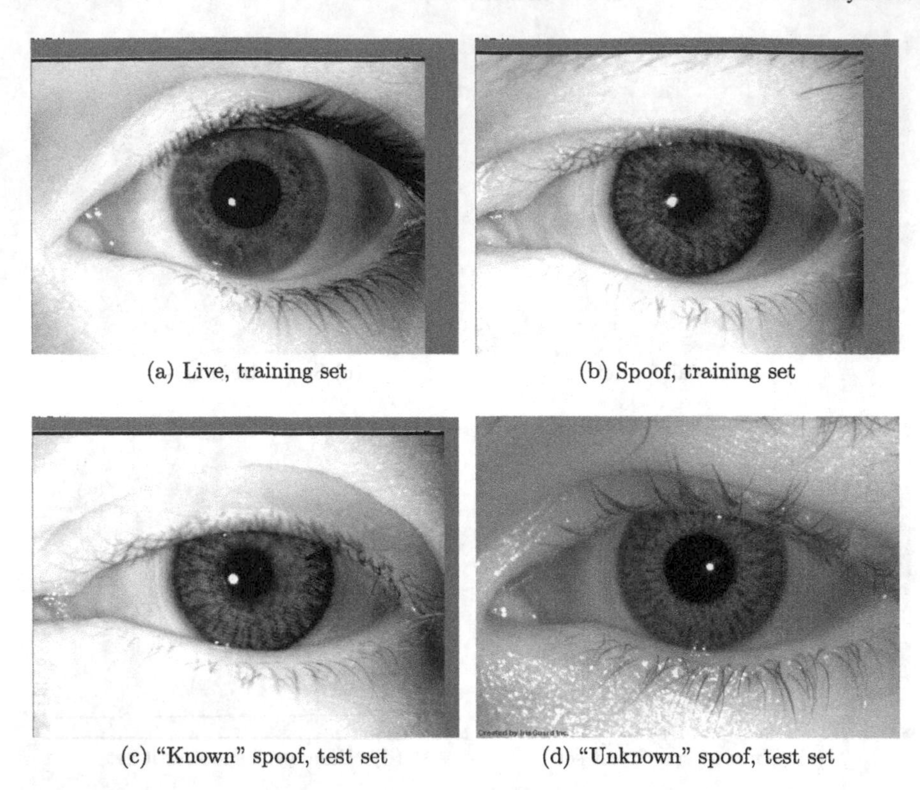

Fig. 7.8 Sample images from the **LivDet-Iris 2017 Notre Dame subset**. Samples **a**, **b** and **c** are captured with LG 4000 sensor, while the image shown in **d** was captured with the IrisGuard AD100 camera

7.2.4 LivDet-Iris 2020 Data

The last, 2020 edition of the LivDet-Iris competition series was unique with respect to datasets in several ways. First, there was no official training set suggested to be used by the competitors. In the era of deep learning and its huge appetite for data, the competitors had a complete freedom to use any data (publicly available as well as their in-house data) to train their algorithms. Second, in addition to paper printouts and eyes with textured contact lens (explored in all previous LivDet-Iris competitions), this last edition added never-used-before in public evaluations of iris PAD new attack types to the test set: cadaver irises, plastic eyes (with and without contact lens put on top), eyes displayed on a Kindle screen and iris printouts with contact lenses placed on them to imitate the cornea. Table 7.7 presents types and numbers of samples included into the test set.

More specifically, the **LivDet-Iris 2020 Warsaw subset** was this time composed solely of iris images captured after death, in the mortuary conditions and with the IriTech IriShield MK2120U. No pre-mortem samples were available. The images

Table 7.6 Types of textured contact lenses used in the **LivDet-Iris 2017 Clarkson LG EOU2200 subset**. Unknown (test set only) patterns are bolded

Number	Contact brand	Color
1	Acuvue define	Natural shimmer
2	Acuvue define	Natural shine
3	Acuvue define	Natural sparkle
4	Air optix	Green
5	Air optix	Brilliant Blue
6	Air optix	Brown
7	Air optix	Honey
8	Air optix	Hazel
9	Air optix	Sterling gray
10	Expressions colors	Aqua
11	Expressions colors	Hazel
12	Expressions colors	Brown
13	Expressions colors	Green
14	Expressions colors	Jade
15	Freshlook colorblends	Amethyst
16	Freshlook colorblends	Brown
17	Freshlook colorblends	Green
18	Freshlook colorblends	Turquoise
19	Freshlook colors	Blue
20	Freshlook colors	Green

originate from 42 deceased subjects, with the post-mortem intervals (PMI) ranging from 5h up to 369h. This subset is equivalent to the already published *Warsaw-BioBase-Post-Mortem-Iris v3.0* [23], which was not available before closing for LivDet-Iris 2020 submissions. So those who want to strictly follow the LivDet-Iris 2020 testing protocol, shall not use *Warsaw-BioBase-Post-Mortem-Iris v3.0* in training. Examples are shown in Fig. 7.9.

The **LivDet-Iris 2020 Notre Dame subset** was, as in previous editions, composed solely with images of eyes with textured contact lenses. In addition to brands previously used in LivDet-Iris—Johnson & Johnson and Ciba Vision—the Notre Dame team collected samples representing Bausch & Lomb contact lenses. Additionally, collected data had much more variability in the NIR illumination compared to previous editions of the team's LivDet-Iris datasets. Examples are shown in Fig. 7.10.

The **LivDet-Iris 2020 Clarkson subset** is the richest set in terms of newly proposed attack types. The Clarkson team offered never-used in LivDet-Iris images of irises displayed on Kindle device, as well as printouts and artificial (plastic) eyes with contact lenses placed on top to imitate a real cornea. Samples were captured with the Iris ID iCAM7000 sensor.

Table 7.7 Numbers and types of samples included into the **LivDet-Iris 2020 datasets**. Attack Types: TCL = textured contact lenses; PP = paper printouts; ED = electronic display; PPCL = prosthetic/printed eyes with contact lenses; CI = cadaver iris. Providers: Warsaw = Warsaw University of Technology, Poland; Notre Dame = University of Notre Dame; Clarkson = Clarkson University, USA

Provider	Sensor	Attack types	Test set	
			Live	Spoof
Warsaw	IriTech IriShield MK2120U	CI	0	1,094
Notre Dame	IrisGuard AD100, LG 4000	TCL	4,366	4,328
Clarkson	LG EOU2200	PP, TCL, ED, PPCL	965	1,679

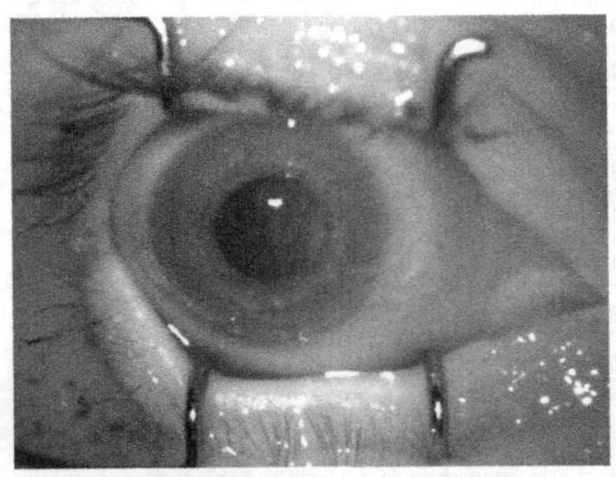

(a) Spoof (cadaver iris), PMI = 7h

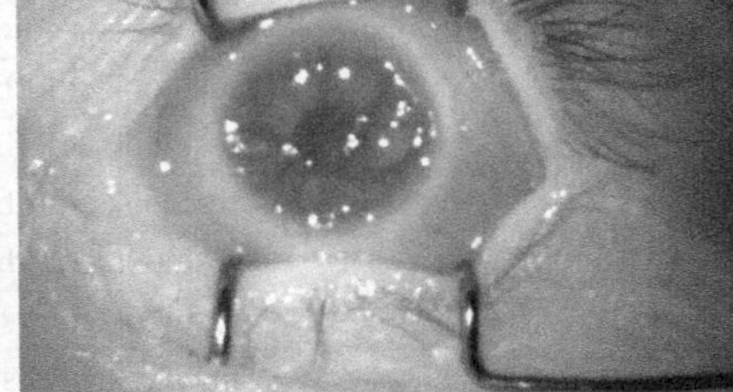

(b) Spoof (cadaver iris), PMI = 369h

Fig. 7.9 Sample images from the **LivDet-Iris 2020 Warsaw subset**. The Post-Mortem Interval (PMI) denotes the number of hours passed after demise until the moment when the picture was taken. Clear post-mortem decomposition deformations can be visible in a sample with a higher PMI. The metal retractors, used by a forensic technician, are rather clear indicators to algorithms that a given sample is a dead eye

A difference, compared to all previous editions of LivDet-Iris, is that competitors could run their algorithms multiple times on the test set (however, without an access to actual data) on the BEAT platform [24, 25]. This opportunity was optional, however, and none of the 2020 competitors decided to use BEAT. Hence, all LivDet-Iris 2020 submissions were evaluated locally by the organizers.

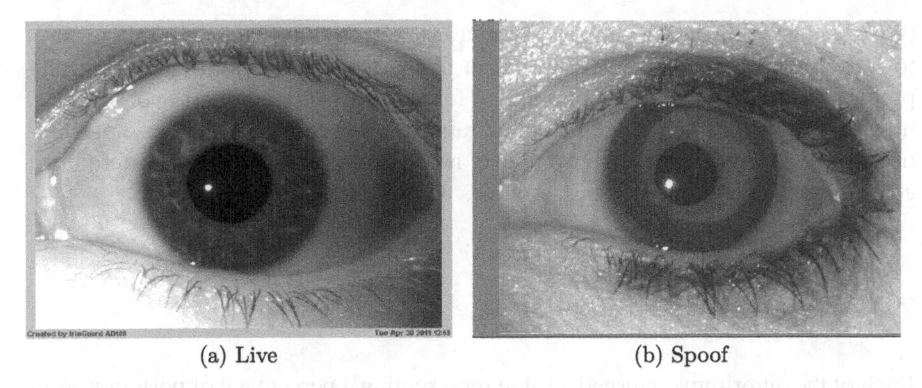

(a) Live (b) Spoof

Fig. 7.10 Sample images from the **LivDet-Iris 2020 Notre Dame subset**. Note an unusual appearance of the spoof sample: the textured contact lens covered the iris only partially, leaving a significant amount of authentic iris visible

7.3 Challenges

The LivDet-Iris 2013 and 2015 editions were organized in a similar, closed-set presentation attack detection regime. It means that the attack types and the main properties of collected samples were the same in the train and test sets. However, the 2017 and 2020 editions proposed more difficult challenges to participants.

In particular, in the 2017 edition, each dataset contained some form of unknown presentation attack information in the testing dataset, which was not always the case in previous iterations of the competition. The Clarkson dataset included two additional challenges beyond the unknown presentation attacks. The first of which is that included in the testing set were 3 patterned contact types that only obscure part of the iris: Acuvue Define is designed to accentuate the wearer's eyes and only cover a portion of the natural iris pattern. Also, the printed iris image testing set contained iris images that were captured by an iPhone 5 in the visible spectrum. The red channel was extracted and converted to grayscale and presented back to the sensor. Similarly, the Notre Dame dataset included data from contact lens brands that were not present in the training set. The Warsaw and IIITD-WVU testing datasets both had samples that were captured with a different camera than the training data, although resolution was kept consistent with the training sets. Additionally, the LivDet-Iris 2017 competition included a cross-sensor challenge. That is, the competitors had a chance to train their algorithms in a way to generalize well to various (unknown) sensors, or make their submissions sensor-specific. It was controlled by an execution flag provided as an argument. Consequently, the evaluation team was testing each submission with and without sensor specificity assumption, and both results were reported.

The 2020 edition is considered by the competition organizers as vastly more difficult to correctly classify the iris presentation attacks. This was the first edition organized in an open-set regime, in which the information about the attack types included into the test set was not fully revealed. The competitors knew the attack types, and were offered 5 samples per attack type (certainly, excluded from testing) to inform the participants about technical image formats and let them technically prepare their codes to a particular image format. This limited number of samples, however, made it difficult to use them in training, or forced applying few-shot learning strategies. Also, as mentioned above, no official training data was offered: the competitors were free to use any resources they deemed useful.

7.4 Performance Evaluation

Each of the algorithms returned a value representing a percentage of posterior probability of the live class (or a degree of "liveness") given the image, normalized in the range 0 to 100 (100 is the maximum degree of liveness, 0 means that the image is fake). The threshold value for determining liveness was set at 50. This threshold was known to all attendees before the competition, so they could curve their methods' scores at their discretion. The threshold was used to calculate Attack Presentation Classification Error Rate (APCER) and Bona Fide Presentation Classification Error Rate (BPCER) error estimators, where

- APCER is the rate of misclassified spoof images (spoof called live), and
- BPCER is the rate of misclassified live images (live called spoof).

Both APCER and BPCER are calculated for each dataset separately, as well as the average values across all datasets. To select a winner, the average of APCER and BPCER was calculated for each participant across datasets. The weight of importance between APCER to BPCER will change based on use case scenario. In particular, low BPCER is more important for low-security implementations such as unlocking phones, however low APCER is more important for high-security implementations. Due to this, APCER and BPCER are given equal weights in the entire LivDet-Iris competition series. Note that LivDet-Iris competitions before 2017, a different terminology was used to determine error rates: *FerrLive* was used in place of BPCER, and *FerrFake* was used in place of APCER. After having the ISO/IEC 30107-3 published in 2016, the LivDet-Iris series switched to APCER and BPCER to promote compliance with the ISO standardization efforts.

Error rate curves demonstrating a changing threshold are shown in the individual competition papers.

Processing time per image was measured but was not considered in selecting the winner. None of the algorithms submitted to LivDet-Iris series took unreasonably long time to make predictions. Processing time may be considered in future editions of the competition.

7.5 Summary of LivDet-Iris Results

In this section, we summarize the competition results for all four LivDet-Iris editions. Results show the improvement in accuracy for popular attack types (printouts and textured contact lens) across the three first competitions. However, more difficult quasi-open-set regime applied in the 2020 edition made the competition difficult, suggesting that the real-world iris PAD is far from being a solved problem. Certainly, more in-depth analysis of the results, as well as error rate curves, can be seen in the individual competition papers for each LivDet-Iris event.

7.5.1 Participants

The competition is open to all academic and industrial institutions. In three first editions (2013, 2015, and 2017), upon registration, each participant was required to sign a database release agreement detailing the proper usage of data made available through the competition. Participants were then given a database access letter with a user name and password to access the server to download the training data. There were no data license agreements applied in the 2020 edition, as there was no official training set offered to the competitors.

Participants had the ability to register as either an anonymous submission or by revealing their organizations. After results are tallied, each competitor was sent their organization's personal results and given the option to remain anonymous in the publication. Placement in the competition and results of other submissions were not provided to other competitors at that time. All competitors were always invited to co-author the competition paper. Some, but not all, competitors accepted this opportunity. Table 7.8 lists all competitors across all editions, respecting their final decisions to remain anonymous.

7.5.2 Trends in LivDet-Iris Across All Competitions

Unlike LivDet-Fingerprint, the number of competitors for LivDet-Iris has remained relatively constant from year to year. All LivDet-Iris competitions, except the 205 edition, were comprised of 3 competitors. LivDet-Iris 2015 had 4 total submissions.

The best results, in terms of performance, are detailed in Table 7.9. These values represent the best competitor algorithm taking the average error rates across all testing datasets. Average values are computed by taking the mean of the error rates with no adjustment for dataset size. The lack of adjustment for dataset size is due to the fact that each dataset is constructed with entirely different properties and has different attacks and quality levels. Also since each co-organizer creates its own dataset, the sizes of each dataset have the potential to be wildly different. Due to this, adjusting

Table 7.8 LivDet-Iris participants—all editions

Participant name	Country	Algorithm name	Edition	Winner?
University of Naples Federico II	Italy	Federico	2013	✓*
Universidad Autonoma de Madrid				
Biometric Recognition Group,	Spain	ATVS	2013	
Faculdade de Engenharia da Universidade do Porto	Portugal	Porto	2013	✓*
University of Naples Federico II	Italy	Federico	2015	✓
Anonymous 0	–	Anon0	2015	
Anonymous 1	–	Anon1	2015	
Anonymous 2	–	Anon2	2015	
Anonymous	–	Anon1	2017	✓
Universitá degli Studi di Napoli Federico II	Italy	UNINA	2017	
Chinese Academy of Sciences	China	CASIA	2017	
Universidad de Santiago de Chile/TOC Biometrics	Chile	USACH/TOC	2020	✓
Fraunhofer Institute	Germany	FraunhoferIGD	2020	
For Computer Graphics Research IGD				
Anonymous	–	Competitor-3	2020	

*Two competitors did not provide results on all datasets, and among them the best results were seen for Porto; Federico was the only competitor presenting results for all three datasets

Table 7.9 The best error rates (APCER and BPCER averaged over all competition datasets) achieved by the LivDet-Iris winners in all editions

Edition	Average BPCER (%)	Average APCER (%)	Winning algorithm
2013	28.56	5.72	Federico
2015	1.68	5.48	Federico
2017	3.36	14.71	Anon1
2020	0.46	59.10	USACH/TOC

error rates by dataset size when calculating the overall error rate would heavily skew the error rates toward an abnormally large dataset compared to other datasets, such as Warsaw from LivDet 2015.

So direct comparison among numbers in Tab. 7.9 is not possible. But we can use these numbers to roughly assess the level of difficulty of iris PAD, given the state of the development of both PAD methods (currently mostly deploying deep learning approaches) and the sophistication of attack types. Starting from LivDet-Iris 2013, the error rates of above 10% for BPCER were observed for all submissions, and the probability of falsely rejecting authentic eyes is much smaller for algorithms submitted to all next competitions. However, the opposite trend is observed for APCER across the 3 competitions, which started to be relatively small in the first two editions, larger in the 2017 edition, and sky-rocked to almost 60% in the last 2020 edition. Certainly the balance between BPCER and APCER can be controlled by the acceptance threshold. But it was one of the participants' tasks, who knew the fixed threshold used in testing, to set all hyperparameters of their submission, and curve their scores, if needed, to make their algorithms generalizing well to unknown spoofs present in the test sets. This is the main, and probably the most alarming observation after looking at all editions: the iris PAD is far from a problem we could consider as solved, and enlarging APCER may suggest low generalization capabilities of algorithms tested in the most fair and independent way. Attack type-specific analyses performed in the most recent LivDet-Iris 2020 competition have been included in the original LivDet-Iris 2020 paper [15].

7.6 Conclusions and Future of LivDet-Iris

Since the inception of the LivDet-Iris competition, it has been aimed to allow research centers and industry the ability to have an independent assessment of their presentation attack detection algorithms. While BPCER error rates have improved over time, it is evident that more improvements can be made in the area of generalization capabilities to unknown attack types, and unknown properties of known attack types along with cross-database settings. LivDet-Iris should be also appreciated for delivering important dataset benchmarks, which are (except for Warsaw subsets, withdrawn due to GDPR regulations) publicly available to other researchers. The 2020 test dataset is available via the BEAT platform.

LivDet-Iris will be continued, and the way forward is to move it completely to an online platform (such as BEAT) and allow participants to test their methods asynchronously, and several times a year. This will allow to (a) use test datasets that are not publicly available and are unknown to the research community (so no training would be possible with these samples), and (b) get the results instantaneously instead of biannually.

Acknowledgements The team at Clarkson University has been supported by the Center for Identification Technology Research and the National Science Foundation under Grant No. 1068055. Warsaw University of Technology team acknowledges the support of the NASK National Research Institute, Poland, for their support during the 2013, 2015 and 2017 editions.

References

1. Ghana T (2015) Somaliland introduce biometric voter verification. Biomet Technol Today 2015(10):3 –12. https://doi.org/10.1016/S0969-4765(15)30151-X
2. Unique Identification Authority of India: AADHAAR. https://uidai.gov.in. Accessed 29 Dec 2021
3. Samsung: How does the iris scanner work on Galaxy S9, Galaxy S9+, and Galaxy Note9. https://www.samsung.com/global/galaxy/what-is/iris-scanning/. Accessed 29 Dec 2021
4. PayEye: "the biggest pilot program for eye payments in the world". https://payeye.com/en/pilot-1-0/. Accessed 29 Dec 2021
5. Minaee S, Abdolrashidi A (2018) Iris-gan: learning to generate realistic iris images using convolutional gan
6. Yadav S, Chen C, Ross A (2019) Synthesizing iris images using rasgan with application in presentation attack detection. In: 2019 IEEE/CVF conference on computer vision and pattern recognition workshops (CVPRW), pp 2422–2430. https://doi.org/10.1109/CVPRW.2019.00297
7. Trokielewicz M, Czajka A, Maciejewicz P (2018) Presentation attack detection for cadaver iris. In: 2018 IEEE 9th international conference on biometrics theory, applications and systems (BTAS), pp 1–10 (2018). https://doi.org/10.1109/BTAS.2018.8698542
8. Cappelli R et al (2007) Fingerprint verification competition 2006. Biomet Technol Today 15(7):7–9. https://doi.org/10.1016/S0969-4765(07)70140-6
9. NIST: Face Recognition Vendor Test (FRVT) (2021). https://www.nist.gov/programs-projects/face-recognition-vendor-test-frvt
10. NIST: Iris Exchange (IREX) (2021). https://www.nist.gov/programs-projects/iris-exchange-irex-overview
11. Alonso-Fernandez F, Bigun J (2013) Halmstad university submission to the first icb competition on iris recognition (icir2013)
12. Yambay D et al (2014) Livdet-iris 2013-iris liveness detection competition 2013. https://doi.org/10.1109/BTAS.2014.6996283
13. Yambay D, Walczak B, Schuckers S, Czajka A (2017) LivDet-Iris 2015—Iris liveness detection. In: IEEE international conference on identity, security and behavior analysis (ISBA), pp 1–6
14. Yambay D, Becker B, Kohli N, Yadav D, Czajka A, Bowyer KW, Schuckers S, Singh R, Vatsa M, Noore A, Gragnaniello D, Sansone C, Verdoliva L, He L, Ru Y, Li H, Liu N, Sun Z, Tan T (2017) LivDet Iris 2017—iris liveness detection competition 2017. In: IEEE international joint conference on biometrics (IJCB), pp 1–6 (2017)

15. Das P, McGrath J, Fang Z, Boyd A, Jang G, Mohammadi A, Purnapatra S, Yambay D, Marcel S, Trokielewicz M, Maciejewicz P, Bowyer K, Czajka A, Schuckers S, Tapia J, Gonzalez S, Fang M, Damer N, Boutros F, Kuijper A, Sharma R, Chen C, Ross A (2020) Iris liveness detection competition (LivDet-Iris)—the 2020 edition. In: IEEE international joint conference on biometrics (IJCB), pp 1–9 (2020). https://doi.org/10.1109/IJCB48548.2020.9304941
16. Chakka MM, Anjos A, Marcel S, Tronci R, Muntoni D, Fadda G, Pili M, Sirena N, Murgia G, Ristori M, Roli F, Yan J, Yi D, Lei Z, Zhang Z, Li SZ, Schwartz WR, Rocha A, Pedrini H, Lorenzo-Navarro J, Castrillon-Santana M, Määttä J, Hadid A, Pietikainen M (2011) Competition on counter measures to 2-d facial spoofing attacks. In: 2011 international joint conference on biometrics (IJCB), pp 1–6. https://doi.org/10.1109/IJCB.2011.6117509
17. Chingovska I, Yang J, Lei Z, Yi D, Li SZ, Kahm O, Glaser C, Damer N, Kuijper A, Nouak A, Komulainen J, Pereira T, Gupta S, Khandelwal S, Bansal S, Rai A, Krishna T, Goyal D, Waris MA, Zhang H, Ahmad I, Kiranyaz S, Gabbouj M, Tronci R, Pili M, Sirena N, Roli F, Galbally J, Ficrrcz J, Pinto A, Pedrini H, Schwartz WS, Rocha A, Anjos A, Marcel S (2013) The 2nd competition on counter measures to 2d face spoofing attacks. In: 2013 international conference on biometrics (ICB), pp 1–6. https://doi.org/10.1109/ICB.2013.6613026
18. Face Anti-spoofing (Presentation Attack Detection) Challenges (2021). https://sites.google.com/qq.com/face-anti-spoofing
19. JohnsonJohnson (2013). http://www.acuvue.com/products-acuvue-2-colours
20. CibaVision (2013). http://www.freshlookcontacts.com
21. Vision C (2013) http://coopervision.com/contact-lenses/expressions-color-contacts
22. Doyle JS, Bowyer KW (2015) Robust detection of textured contact lenses in iris recognition using bsif. IEEE Access 3:1672–1683. https://doi.org/10.1109/ACCESS.2015.2477470
23. Trokielewicz M, Czajka A, Maciejewicz P (2020) Post-mortem iris recognition with deep-learning-based image segmentation. Image Vis Comput 94:103–866. https://doi.org/10.1016/j.imavis.2019.103866
24. Anjos A, El-Shafey L, Marcel S (2017) BEAT: an open-science web platform. In: International conference on machine learning (ICML). https://publications.idiap.ch/downloads/papers/2017/Anjos-ICML2017-2017.pdf
25. Anjos A, El-Shafey L, Marcel S (2017) BEAT: an open-source web-based open-science platform. https://arxiv.org/abs/1704.02319

Chapter 8
Intra and Cross-spectrum Iris Presentation Attack Detection in the NIR and Visible Domains

Meiling Fang, Fadi Boutros, and Naser Damer

Abstract Iris Presentation Attack Detection (PAD) is essential to secure iris recognition systems. Recent iris PAD solutions achieved good performance by leveraging deep learning techniques. However, most results were reported under intra-database scenarios, and it is unclear if such solutions can generalize well across databases and capture spectra. These PAD methods run the risk of overfitting because of the binary label supervision during the network training, which serves global information learning but weakens the capture of local discriminative features. This chapter presents a novel attention-based deep pixel-wise binary supervision (A-PBS) method. A-PBS utilizes pixel-wise supervision to capture the fine-grained pixel/patch-level cues and attention mechanism to guide the network to automatically find regions where most contribute to an accurate PAD decision. Extensive experiments are performed on six NIR and one visible-light iris databases to show the effectiveness and robustness of proposed A-PBS methods. We additionally conduct extensive experiments under intra-/cross-database and intra-/cross-spectrum for detailed analysis. The results of our experiments indicate the generalizability of the A-PBS iris PAD approach.

8.1 Introduction

Iris recognition systems are increasingly being deployed in many law enforcement and civil applications [1–3]. However, iris recognition systems are vulnerable to Presentation Attacks (PAs) [4, 5], performing to obfuscate the identity or impersonate

M. Fang (✉) · F. Boutros (✉) · N. Damer (✉)
Fraunhofer Institute for Computer Graphics Research IGD, Darmstadt, Germany
e-mail: meiling.fang@igd.fraunhofer.de

F. Boutros
e-mail: fadi.boutros@igd.fraunhofer.de

N. Damer
e-mail: naser.damer@igd.fraunhofer.de

S. Marcel et al. (eds.), *Handbook of Biometric Anti-Spoofing*, Advances in Computer Vision and Pattern Recognition, https://doi.org/10.1007/978-981-19-5288-3_8

a specific person. Such attacks can be performed by various methods ranging from printouts, video replay, or textured contact lenses, among others. As a result, the Presentation Attack Detection (PAD) domain has been developing solutions to mitigate security concerns in recognition systems.

Recent iris PAD works [6–10, 46] are competing to boost the performance using Convolution Neural Network (CNN) to facilitate discriminative feature learning. Even though the CNN-based algorithms achieved good results under intra-database setups, their generalizability across unseen attacks, databases, and spectra is still understudied. The results reported on the LivDet-Iris competitions verified the challenging nature of cross-PA and cross-database PAD. The LivDet-Iris is an international competition series launched in 2013 to assess the current state-of-the-art in the iris PAD field. The two most recent edition took place in 2017 [4] and 2020 [5]. The results reported in the LivDet-Iris 2017 [4] databases pointed out that there are still advancements to be made in the detection of iris PAs, especially under cross-PA, cross-sensor, or cross-database scenarios. Subsequently, LivDet-Iris 2020 [5] reported a significant performance degradation on novel PAs, showing that the iris PAD is still a challenging task. Iris PAD in the NIR domain has so far shown good performances and indicated the generalizability challenges under cross-database scenarios [9, 11, 46]. Nonetheless, studies addressing PAD algorithms in the visible spectrum are relatively limited [12–14]. Given that the integration of iris recognition in smart devices is on the rise [15–17], the study of iris PAD under the visible spectrum is essential. Furthermore, knowing that most iris PAD solutions are developed and trained for images captured in the NIR domain, an investigation of cross-spectrum iris PAD performance is much needed. To our knowledge, there is no existing work on that investigated the PAD performance under a cross-spectrum scenario. As a result, in this work, we further address the visible-light-based iris PAD and the cross-spectrum PAD scenario.

Most of the recent iris PAD solutions trained models by binary supervision (more details in Sect. 8.2), i.e., networks were only informed that an iris image is bona fide or attack, which may lead to overfitting. Besides, the limited binary information may be inefficient in locating the regions that contribute most to making an accurate decision. To target these issues, an Attention-based Pixel-wise Binary Supervision (A-PBS) network (See Fig. 8.1) is proposed. In this chapter, we adopt the A-PBS solution to perform extensive experiments under intra- and cross-spectrum scenarios. The main contributions of the chapter include (1) We present the A-PBS solution that successfully aims to capture subtle and fine-grained local features in attack iris samples with the help of spatially positional supervision and attention mechanism. (2) We perform extensive experiments on NIR-based LivDet-Iris 2017 databases, three publicly available NIR-based databases, and one visible-spectrum-based iris PAD database. The experimental results indicated that the A-PBS solution outperforms state-of-the-art PAD methods in most experimental settings, including cross-PA, cross-sensor, and cross-databases. (3) We additionally analyze the cross-spectrum performance of the presented PAD solutions. To our best knowledge, this is the first work in which the cross-spectrum iris PAD performance is investigated.

8.2 Related Works

CNN-based iris PAD: In recent years, many works [6–8, 10, 18, 46] leveraged deep learning techniques and showed great progress in iris PAD performance. Kuehlkamp et al. [46] proposed to combine multiple CNNs with the hand-crafted features. Nevertheless, training 61 CNNs requires high computational resources and can be considered as an over-tailored solution. Yadav et al. [18] employed the fusion of hand-crafted features with CNN features and achieved good results. Unlike fusing the hand-crafted and CNN-based features, Fang et al. [6] presented a multi-layer deep features fusion approach (MLF) by considering the characteristics of networks that different convolution layers encode the different levels of information. Apart from such fusion methods, a deep learning-based framework named Micro Stripe Analyses (MSA) [7, 9] was introduced to capture the artifacts around the iris/sclera boundary and showed a good performance on textured lens attacks. Yadav et al. [19] presented DensePAD method to detect PAs by utilizing DenseNet architecture [20]. Their experiments demonstrated the efficacy of DenseNet in the iris PAD task. Furthermore, Sharma and Ross [8] exploited the architectural benefits of DenseNet [20] to propose an iris PA detector (D-NetPAD) evaluated on a proprietary database and the LivDet-Iris 2017 databases. With the help of their private additional data, the fine-tuned D-NetPAD achieved good results on LivDet-Iris 2017 databases, however, scratch D-NetPAD failed in the case of cross-database scenarios. These works inspired us to use DenseNet [20] as the backbone for our A-PBS network architectures. Recently, Chen et al. [21] proposed an attention-guided iris PAD method to refine the feature maps of DenseNet [20]. However, this method utilized conventional sample binary supervision and did not report cross-database and cross-spectrum experiments to prove the generalizability of the additional attention module.

Limitations: Based on the recent iris PAD literature, it can be concluded that deep-learning-based methods boost the performance but still have the risk of overfitting under cross-PA, cross-database, and cross-spectrum scenarios. Some recent methods proposed the fusion of multiple PAD systems or features to improve the generalizability [6, 18, 46], which makes it challenging for deployment. One of the major reasons causing overfitting is the lack of availability of a sufficient amount of variant iris data for training networks. Another possible reason might be binary supervision. While the binary classification model provides useful global information, its ability to capture subtle differences in attacking iris samples may be weakened, and thus the deep features might be less discriminative. This possible cause motivates us to exploit binary masks to supervise the training of our PAD model, because a binary mask label may help to supervise the information at each spatial location. However, PBS may also lead to another issue, as the model misses the exploration of important regions due to the 'equal' focus on each pixel/patch. To overcome some of these difficulties, we propose the A-PBS architecture to force the network to find regions that should be emphasized or suppressed for a more accurate iris PAD decision. The detailed introduction of PBS and A-PBS can be found in Sect. 8.3. In addition to the challenges across database scenarios, another issue is that there is no existing research dedicated to exploring the generalizability of PAD methods across spectral scenarios. PAD research in the NIR domain [6–8, 18, 46] has attracted much attention, while few studies [12–14] investigated PAD performance in the visible spectrum. Furthermore, the generalizability of PAD methods under the cross-spectrum is unclear. Suppose a model trained on NIR data can be well generalized to visible-light data. In that case, it requires only low effort to transfer such solution to low-cost mobile devices [15–17], which simplifies its application in the real world. Therefore, this chapter explores the proposed PAD performance in cross-spectrum experimental settings.

8.3 Methodology

This section starts by introducing the DenseNet [20], which is used as a preliminary backbone architecture. Then, the Pixel-wise Binary Supervision (PBS) and Attention-based PBS (A-PBS) methods are described. We presented this approach initially in [22], however, we extend it here by investigating its advantages on different attack types, iris images captured in the visible spectrum, and cross-spectrum deployments. Figure 8.1 depicts an overview of our different methods. The first gray block (a) presents the basic DenseNet architecture with binary supervision, the second gray block (b) introduces the binary and PBS, and the third block (c) is the PBS with the fused multi-scale spatial attention mechanism (A-PBS).

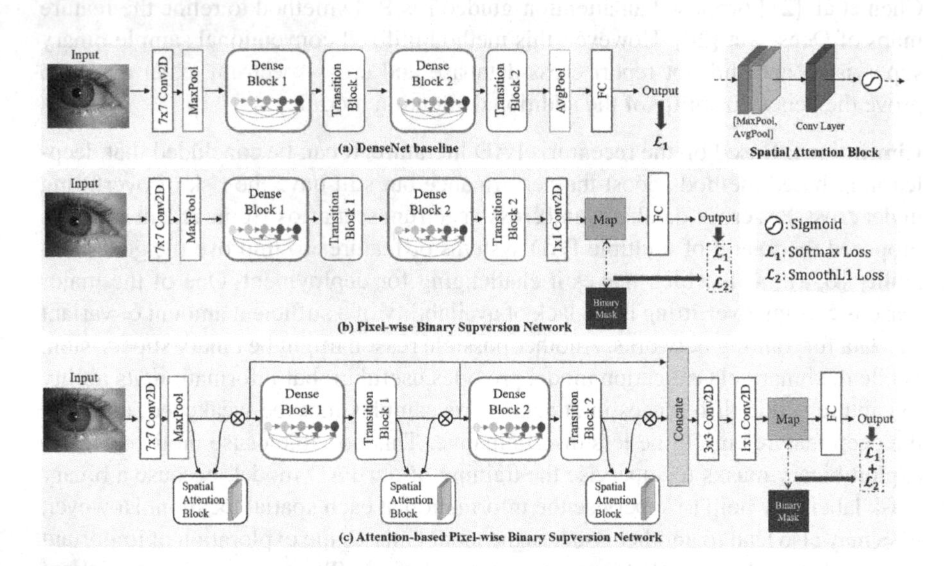

Fig. 8.1 An overview of **a** baseline DenseNet, **b** proposed PBS, and **c** proposed A-PBS networks

8.3.1 Baseline: DenseNet

DenseNet [20] presented direct connection between any two layers with the same feature-map size in a feed-forward fashion. The reasons inspiring our selection of DensetNet are: (1) DenseNets naturally integrate the properties of identity mappings and deep supervision following a simple connectivity rule. (2) DenseNet has already demonstrated its superiority in iris PAD [5, 8, 19]. Figure 8.1a illustrates that we reuse two dense and transition blocks of pre-trained DenseNet121. An average pooling layer and a fully-connected (FC) classification layer are sequentially appended, following the second transition block, to generate the final prediction to determine whether the iris image is bona fide or attack. PBS and A-PBS networks are extended on this basic architecture later.

8.3.2 Pixel-Wise Binary Supervision Network (PBS)

By reviewing the recent iris PAD literature [6–8, 46], it can be found that CNN-based methods outperformed hand-crafted feature-based methods. In typical CNN-based iris PAD solutions, networks are designed such that feeding pre-processed iris image as input to learn discriminative features between bona fide and artifacts. To that end, a FC layer is generally introduced to output a prediction score supervised by binary label (bona fide or attack). Recent face PAD works have shown that auxiliary supervision [23–25] achieved significant improvement in detection performance. Binary label supervised classification learns semantic features by capture global information but may lead to overfitting. Moreover, such embedded 'globally' features might lose the local detailed information in spatial position. These drawbacks give us the insight that adding pixel-wise binary along with binary supervision might improve the PAD performance. First, such supervision approach can be seen as a combination of patch-based and vanilla CNN-based methods. To be specific, each pixel-wise score in output feature map is considered as the score generated from the patches in an iris image. Second, the binary mask supervision would be provided for the deep embedding features in each spatial position. Figure 8.1b illustrates the network details that an intermediate feature map is produced before the final binary classification layer. The output from the *Transition Block 2* is 384 channels with the map size of 14×14. A 1×1 convolution layer is added to produce the intermediate map. In the end, an FC layer is utilized to generate a prediction score.

8.3.3 Attention-Based PBS Network (A-PBS)

The architecture of PBS is designed coarsely (simply utilizing the intermediate feature map) based on the DenseNet [20], which might be sub-optimal for iris PAD task. To enhance that, and inspired by Convolutional Block Attention Mechanism (CBAM) [26] and MLF [6], we propose an A-PBS method with multi-scale feature fusion (as shown in Fig. 8.1c).

Even though PBS boosts iris PAD performance under intra-database/-spectrum, it shows imperfect invariation under more complicated cross-PA, cross-database, and cross-spectrum scenarios (See results in Tables 8.5, 8.11, and 8.10). As a result, it is worth finding the important regions to focus on, although it contradicts learning *more* discriminative features. In contrast, the attention mechanism aims to automatically learn *essential* discriminate features from inputs that are relevant to PA detection. Woo et al. [26] presented an attention module consisting of the channel and spatial distinctive sub-modules, which possessed consistent improvements in various classification and detection tasks across different network architectures. Nonetheless, only spatial attention module is employed in our case due to the following reasons. The first reason is that the Squeeze-and-Excitation (SE)-based channel attention module focuses only on the inter-channel relationship by using dedicated global feature descriptors. Such channel attention module may lead to a loss of information (e.g., class-deterministic pixels) and may result in further performance degradation when the domain is shifted, e.g., different sensors and changing illumination. Second, a benefit of the spatial attention module is that the inter-spatial relationship of features is utilized. Specifically, it focuses on *'where'* is an informative region, which is more proper for producing intermediate feature maps for supervision. Furthermore, based on the fact that the network embeds different layers of information at different levels of abstraction, the MLF [6] approach confirmed that the fusing deep feature from multiple layers is beneficial to enhance the robustness of the networks in the iris PAD task. Nevertheless, we propose to fuse feature maps generated from different levels directly within the network instead of fusing features extracted from a trained model in MLF [6], because finding the best combination of network layers to fuse is a challenging task and difficult to generalize well, especially when targeting different network architectures.

Figure 8.1 illustrates that three spatial attention modules are appended after *MaxPool*, *Transition Block 1*, and *Transition Block 2*, respectively. The feature learned from the *MaxPool* or two *Transition Blocks* can be considered as low-, middle-, and high-level features and denoted as

$$\mathcal{F}_{level} \in \mathbb{R}^{C \times H \times W}, \quad level \in \{low, mid, high\}. \tag{8.1}$$

Then, the generated attention maps $\mathcal{A}_{level} \in \mathbb{R}^{H \times W}$ encoding where to emphasize or suppress are used to refine $\mathcal{F}_{level}$. The refined feature $\mathcal{F}'_{level}$ can be formulated as $\mathcal{F}'_{level} = \mathcal{F}_{level} \otimes \mathcal{A}_{level}$ where $\otimes$ is matrix multiplication. Finally, such three different level refined features are concatenated together and then fed into a 1×1 convolution layer to produce the pixel-wise feature map for supervision. It should be noticed that the size of convolutional kernel in three spatial attention modules is different. As mentioned earlier, the deeper the network layer, the more complex and abstract the extracted features. Therefore, we should use smaller convolutional kernels for deeper features to locate useful region. The kernel sizes of low-, middle-, and high-level layers are thus set to 7, 5, and 3, respectively. The experiments have been demonstrated later in Sect. 8.4 and showed that in most experimental setups, the A-PBS solution exhibited superior performance and generalizability in comparison to the PBS and DenseNet approaches.

8.3.4 Loss Function

In the training phase, Binary Cross Entropy (BCE) loss is used for final binary supervision. For the sake of robust PBS needed in iris PAD, Smooth L1 (SmoothL1) loss is utilized to help the network reduce its sensitivity to outliers in the feature map. The equations for SmoothL1 are shown below:

$$\mathcal{L}_{SmoothL1} = \frac{1}{n}\sum z \text{ , } where \;\; z = \begin{cases} \frac{1}{2} \cdot (y-x)^2, & if \;\; |y-x| < 1 \\ |y-x| - \frac{1}{2}, & otherwise \end{cases} \tag{8.2}$$

n is the amount number of pixels in the output map (14×14 in our case). The equation of BCE is:

$$\mathcal{L}_{BCE} = -[y \cdot \log p + (1-y) \cdot \log(1-p)] \, , \tag{8.3}$$

where y in both loss equations presents the ground truth label. x in SmoothL1 loss presents to the value in feature map, while p in BCE loss is predicted probability. The overall loss $\mathcal{L}_{overall}$ is formulated as $\mathcal{L}_{overall} = \lambda \cdot \mathcal{L}_{SmoothL1} + (1-\lambda) \cdot \mathcal{L}_{BCE}$. In our experiments, the λ is set to 0.2.

8.3.5 Implementation Details

In the training phase, we performed class balancing by under-sampling the majority class for the databases, whose distribution of bona fides and attacks are imbalanced in the training set. Data augmentation was performed during training using random horizontal flips with a probability of 0.5. The model weight of DenseNet, PBS, and A-PBS models ws first initialized by the base architecture DenseNet121 trained on the ImageNet dataset and then fine-tuned by iris PAD data, by considering the limited amount of iris data. The Adam optimizer was used for training with a initial learning rate of $1e^{-4}$ and a weight decay of $1e^{-6}$. To further avoid overfitting, the model was trained with the maximum 20 epochs and the learning rate halved every 6 epochs. The batch size is 64. In the testing stage, the binary output was used as a final prediction score. The proposed method was implemented using the Pytorch.

8.4 Experimental Evaluation

8.4.1 Databases

The DenseNet, PBS, and A-PBS were evaluated on multiple databases: three NIR-based databases comprising of textured contact lens attacks captured by different sensors [27–29], and three databases (Clarkson, Notre Dame and IIITD-WVU) from the LivDet-Iris 2017 competition [4] (also NIR-based). The Warsaw database in the LivDet-Iris 2017 is no longer publicly available due to General Data Protection Regulation (GDPR) issues. For the experiments on NDCLD13, NDCLD15, IIIT-CLI databases, five-fold cross-validation was performed due to no pre-defined training and testing sets. For the experiments in competition databases, we followed the defined data partition and experimental setting [4]. In addition to above NIR-based iris databases, we also perform experiments on another publicly available database where images were captured under the visible spectrum, named Presentation Attack Video Iris Database (PAVID) [12]. Subjects in each fold or defined partition are dis-joint. The image samples can be found in Fig. 8.2, and the summary of the used databases is listed in Tab 8.1.

NDCLD-2013: The NDCLD-2013 database comprises of 5100 NIR images and is conceptually divided into two sets based on capture sensors: (1) LG4000 including 4200 images captured by IrisAccess LG4000 camera, (2) AD100 consisting of 900 images captured by risGuard AD100 camera. Both the training and the test set are divided equally into no lens (bona fide), soft lens (bona fide), and textured lens (attack) classes.

NDCLD-2015: The 7300 images in the NDCLD-2015 [27] were captured by two sensors, IrisGuard AD100 and IrisAccess LG4000 under MIR illumination and controlled environments. The NDCLD15 contains iris images wearing no lenses, soft lenses, textured lenses.

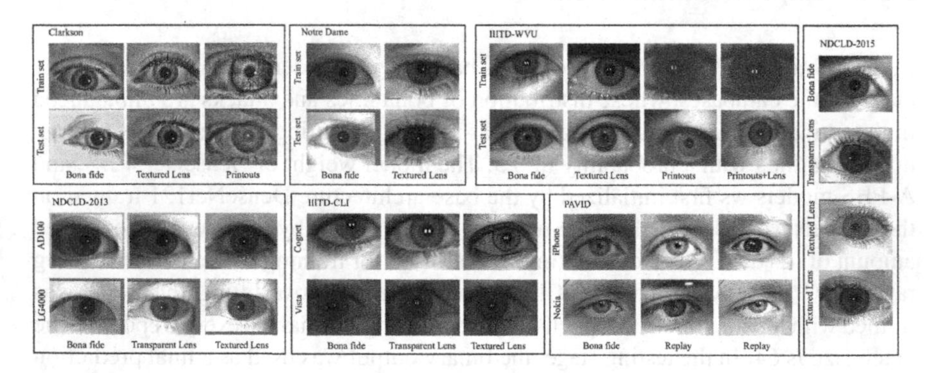

Fig. 8.2 Iris image samples from the used databases. It should be noted that transparent lens is classified as bona fide in our case. Only PAVID database was captured under the visible spectrum

Table 8.1 Characteristics of the used databases. All databases have the training and test sets based on their own experimental setting in the related papers. The Warsaw database in Iris-LivDet-2017 competition are no longer publicly available

Database		Spectrum	# Training	# Testing	Type of iris images
NDCLD-2015 [27]		NIR	6,000	1,300	BF, soft and textured lens
NDCLD-2013 [28]	LG4000	NIR	3,000	1,200	BF, soft and textured lens
	AD100	NIR	600	300	BF, soft and textured lens
IIIT-D CLI [29, 30]	Cognet	NIR	1,723	1,785	BF, soft and textured lens
	Vista	NIR	1,523	1,553	BF, soft and textured lens
LivDet-Iris 2017 [4]	Clarkson (cross-PAD)	NIR	4937	3158	BF, textured lens, printouts
	Notre Dame (cross-PA)	NIR	1,200	3,600	BF, textured lenses
	IIITD-WVU (cross-DB)	NIR	6,250	4,209	BF, textured lenses, printouts, lens printouts
PAVID [12]		VIS	180 [a]	612 [a]	BF, replay

BF: bona fide, VIS: visible light, NIR: near-infrared light
[a] The format of data is video, others are images

IIIT-D CLI: IIIT-D CLI database contains 6570 iris images of 101 subjects with left and right eyes. For each identity, three types of images were captured: (1) no lens, (2) soft lens, and (3) textured lens. Iris images are divided into two sets based on captured sensors: (1) Cogent dual iris sensor and (2) VistaFA2E single iris sensor.

LivDet-Iris 2017 Database: Though the new edition LivDet-Iris competition was held in 2020, we still evaluate the algorithms in databases provided by LivDet-Iris 2017 for several reasons: (1) No official training data was announced in the LivDet-Iris 2020 because the organizers encouraged the participants to use all available data (both publicly and proprietary) to enhance the effectiveness and robustness. (2) The test data is not publicly available. Consequently, to make a fair comparison with state-of-the-art algorithms on equivalent data, we use LivDet-Iris 2017 databases to restrict the evaluation factors to the algorithm itself rather than the data. (3) The LivDet-Iris 2017 competition databases are still valuable due to the challenging cross-PA and cross-database scenario settings. The Clarkson and Notre Dame database are designed for cross-PA scenarios, while the IIIT-WVU database is designed for a cross-database evaluation due to the different sensors and acquisition environments. The Clarkson testing set includes additional unknown visible-light image printouts and unknown textured lenses (unknown pattern). Moreover, Notre Dame focused on the unknown textured lenses. However, the Warsaw database is no longer publicly available.

Presentation Attack Video Iris Database (PAVID) [12]: PAVID is the video iris database collected using smartphones (Nokia Lumia 1020 and iPhone 5S) in the visible spectrum. PAVID contains 304 bona fide videos and 608 replay attack videos across 76 subjects. Moreover, PAVID was divided into three sets in the official protocol: training set including 180 videos, development set including 120 videos, and testing set including 608 videos. The development set defined in [12] was used only for determining the filter kernel of the Laplacian pyramid in [12], not for computing the decision threshold. Therefore, we omit the development set in our experiments.

8.4.2 Evaluation Metrics

The following metrics are used to measure the PAD algorithm performance: (1) Attack Presentation Classification Error Rate (APCER), the proportion of attack images incorrectly classified as bona fide samples, (2) Bona fide Presentation Classification Error Rate (BPCER), the proportion of bona fide images incorrectly classified as attack samples, (3) Half Total Error Rate (HTER), the average of APCER and BPCER. The APCER and BPCER follow the standard definition presented in the ISO/IEC 30107-3 [31] and are adopted in most PAD literature including in LivDet-Iris 2017. The threshold for determining the APCER and BPCER is 0.5 as defined in the LivDet-Iris 2017 protocol. In addition, for further comparison with the state-of-the-art iris PAD algorithms on IIITD-CLI [29, 30] database, we also report the Correct Classification Accuracy (CCR). CCR is the ratio between the total number of correctly classified images and the number of all classified presentations. Furthermore, to enable the direct comparison with [8], we evaluate the performance of our presented DenseNet, PBS, and A-PBS methods in terms of True Detection Rate (TDR) at a false detection rate of 0.2%, as [8] claims that this threshold is normally used to demonstrate the PAD performance in practice. TDR is 1 -APCER, and false detection rate is defined to be the same as BPCER, we, therefore, use BPCER. An Equal Error Rate (EER) locating at the intersection of APCER and BPCER is also reported under cross-database and cross-spectrum settings (results as shown in Tables 8.5, 8.10, and 8.11). The metrics beyond APCER and BPCER are presented to enable a direct comparison with reported results in state-of-the-arts.

8.5 Intra-Spectrum and Cross-Database Evaluation Results

This section presents the evaluation results on different databases and comparison to state-of-the-art algorithms. The comparison to state-of-the-arts depends mainly on the reported results in the literature, as most algorithms are not publicly available or their technical description is insufficient to ensure error-free re-implementation. Therefore, we aim to report the widest range of metrics used in other works to enable an extensive comparison. First, the results from different aspects/metrics on LivDet-Iris 2017 database are reported in Table 8.2, which compare our solution with state-of-the-art PAD methods, Table 8.3 that reports the results in terms of TDR at low BPCER, and Table 8.4 that investigates the performance on different PAs. Then, we demonstrate the experiments under cross-database scenarios by using the three databases in LivDet-Iris 2017 competition to verify the generalizability of our A-PBS solution. Furthermore, the results on NDCLD-2013/NDCLD-2015 and IIITD-CLI databases are presented in Tables 8.6, 8.7 and 8.8, respectively. We further perform the experiment on the PAVID database in visible spectrum (results in Table 8.9). In this section, we also provide explainability analyses using attention map visualizations for further visual reasoning of the presented solution.

8.5.1 Iris PAD in the NIR Spectrum

8.5.1.1 Results on the LivDet-Iris 2017 Databases

Table 8.2 summarizes the results in terms of APCER, BPCER, and HTER on the LivDet-Iris 2017 databases. We evaluate the algorithms on databases provided by LivDet-Iris 2017. The evaluation and comparison on LivDet-Iris 2020 are not included due to (1) no officially offered training data, (2) not publicly available test data. Moreover, LivDet-Iris 2017 databases are still considered as a challenging task, because the experimental protocols are designed for complicated cross-PA and cross-database scenarios. In this chapter, we aim to focus on the impact of the algorithm itself on PAD performance rather than the diversity of data. Consequently, to make a fair comparison with state-of-the-art algorithms on equivalent data, we compare to the Scratch version of the D-NetPAD results [8], because Pre-trained and Fine-tuned D-NetPAD used additional data (including part of Notre Dame test data) for training. This was not an issue with the other compared state-of-the-art methods.

It can be observed in Table 8.2 that A-PBS architecture achieves significantly improved performance in comparison to DenseNet and also slightly lower HTER values than the PBS model in all cases. For instance, the HTER value on Notre Dame is decreased from 8.14% by DenseNet and 4.97% by PBS to 3.94% by A-PBS. Although the slightly worse results on Notre Dame might be caused by the insufficient data in the training set, our PBS and A-PBS methods show significant superiority on the most challenging IIITD-WVU database. Moreover, Fig. 8.3 illustrates the

Table 8.2 Iris PAD performance of our presented DenseNet, PBS, and A-PBS solutions, and existing state-of-the-art algorithms on LivDet-Iris 2017 databases in terms of APCER (%), BPCER (%) and HTER (%) that determined by a threshold of 0.5. The *Winner* in the first column refers to the winner of each competition database. Bold numbers indicate the two lowest HTERs

Database	Metric	Winner [4]	SpoofNet [32]	Meta-Fusion [46]	D-NetPAD [8]	MLF [6]	MSA [7, 9]	DenseNet	PBS	A-PBS
Clarkson	APCER	13.39	33.00	18.66	5.78	–	–	10.64	8.97	6.16
	BPCER	0.81	0.00	0.24	0.94	–	–	0.00	0.00	0.81
	HTER	7.10	16.50	9.45	**3.36**	–	–	5.32	4.48	**3.48**
Notre Dame	APCER	7.78	18.05	4.61	10.38	2.71	12.28	16.00	8.89	7.88
	BPCER	0.28	0.94	1.94	3.32	1.89	0.17	0.28	1.06	0.00
	HTER	4.03	9.50	**3.28**	6.81	**2.31**	6.23	8.14	4.97	3.94
IIITD-WVU	APCER	29.40	0.34	12.32	36.41	5.39	2.31	2.88	5.76	8.86
	BPCER	3.99	36.89	17.52	10.12	24.79	19.94	17.95	8.26	4.13
	HTER	16.70	18.62	14.92	23.27	15.09	11.13	10.41	**7.01**	**6.50**

Table 8.3 Iris PAD performance reported in terms of TDR (%) at 0.2% BPCER on the LivDet-Iris 2017 databases. K indicates known test subset and U is unknown subset. The highest TDR is in bold

Database		TDR (%) @ 0.2% BPCER			
		D-NetPAD [8]	DenseNet	PBS	A-PBS
Clarkson		92.05	92.89	**94.02**	92.35
Notre dame	K	**100.00**	99.68	99.78	99.78
	U	66.55	58.33	76.89	**90.00**
IIITD-WVU		29.30	58.97	69.32	**72.00**

Table 8.4 Iris PAD performance reported based on each presentation attack on the LivDet-Iris-2017 database in terms of BPCER (%) and APCER (%). The Notre Dame database is omitted because it comprises only texture contact lens attack and the results are the same as in Table 8.2. It can be observed that textured contact lens attack is more challenging than printouts in most cases

Database	Clarkson			IIITD-WVU			
# Images	# 1485	# 908	# 765	# 704	# 1404	# 701	# 1402
Metric	BPCER	APCER (PR)	APCER (CL)	BPCER	APCER (PR)	APCER (CL)	APCER (PR-CL)
DenseNet	0.00	0.66	22.48	17.95	3.06	8.27	0.00
PBS	0.00	0.44	19.08	8.26	11.68	5.42	0.00
A-PBS	0.81	1.32	10.59	4.13	11.68	17.97	0.86

PR: printouts, CL: textured contact lens, PR-CL: printed textured contact lens

PAD score distribution of the bona fide and PAs for further analysis. The PAD score distribution generated by A-PBS shows an evident better separation between bona fide (green) and PAs (blue). In addition to reporting the results determined by a threshold of 0.5, we also measure the performance of DenseNet, PBS, and A-PBS in terms of its TDR at 0.2% BPCER (to follow state-of-the-art trends [8]) in Table 8.3. It is worth noting that our A-PBS method achieves the highest TDR value (90.00%) on unknown test set in Notre Dame, while the second-highest TDR is 76.89% achieved by PBS.

Furthermore, we explore the PAD performance based on each presentation attack in LivDet-Iris 2017 database [4]. Because the Notre Dame database contains only textured contact lenses, we report the results on Clarkson and IIITD-WVU databases in Table 8.4. The results show that textured contact lens attacks obtain higher APCER values than printouts attack in most cases, e.g., the APCER value on textured lens attack is 10.59% and on printouts is 1.52% both achieved by A-PBS solution. Hence, we conclude that contact lens is more challenging than printouts in most cases.

In addition to intra-dataset evaluation, we further evaluate the generalizability of our models under cross-database scenario, e.g., the model trained on Notre Dame is tested on Clarkson and IIITD-WVU. As shown in Table 8.5, the A-PBS model outperforms DenseNet and PBS in most cases, which verifying that additional spatial attention modules can reduce the overfitting of the PBS model and capture fine-

Table 8.5 Iris PAD performance measured under cross-database scenarios and reported in terms of EER (%), HTER (%), APCER (%), and BPCER (%). APCER and BPCER are determined by a threshold of 0.5. The lowest error rate is in bold

Train dataset	Notre Dame							
Test dataset	Clarkson				IIITD-WVU			
Metric	EER	HTER	APCER	BPCER	EER	HTER	APCER	BPCER
DenseNet	30.43	32.01	51.29	**12.73**	7.84	**8.81**	**5.93**	11.69
PBS	48.36	47.28	28.15	66.4	15.52	14.54	22.24	6.83
APBS	**20.55**	**23.24**	**14.76**	31.72	**6.99**	8.95	15.34	**2.56**
Train dataset	Clarkson							
Test dataset	Notre Dame				IIITD-WVU			
Metric	EER	HTER	APCER	BPCER	EER	HTER	APCER	BPCER
DenseNet	22.33	31.11	62.22	0.00	26.78	42.40	84.80	0.00
PBS	28.61	32.42	64.83	0.00	25.78	42.48	84.97	0.00
APBS	**21.33**	**23.08**	**46.16**	0.00	**24.47**	**34.17**	**68.34**	0.00
Train dataset	IIITD-WVU							
Test dataset	Notre Dame				Clarkson			
Metric	EER	HTER	APCER	BPCER	EER	HTER	APCER	BPCER
DenseNet	18.28	19.78	36.56	3.00	22.64	48.55	**0.00**	97.10
PBS	**12.39**	**16.86**	**33.33**	0.39	37.24	47.17	**0.00**	94.34
APBS	15.11	27.61	54.72	**0.33**	**21.58**	**21.95**	20.80	**32.10**

grained features. Furthermore, the DenseNet and A-PBS models trained on Notre Dame even exceed the prior state-of-the-arts when testing on the IIIT-WVU database (8.81% HTER by DenseNet and 8.95% by A-PBS, while the best prior state-of-the-art achieved 11.13% (see Table 8.2)). It should be noted that the APCER values on Notre Dame are significant higher by using models either trained on Clarkson or IIITD-WVU. Because Notre Dame training dataset contains only textured lens attacks while Clarkson and IIIT-WVU testing datasets comprise of both textured lens and printouts attacks, which makes this evaluation scenario partially considers unknown PAs. In such an unknown-PAs situation, our A-PBS method achieved significantly improved results. In general, the cross-database scenario is still a challenging problem since many D-EER values are above 20% (Table 8.5).

8.5.1.2 Results on the NDCLD-2013/2015 Database

Table 8.6 compares the iris PAD performance of our models with five state-of-the-art methods on NDCLD-2015 and two different subsets in the NDCLD-2013 database. It can be seen from Table 8.6 that our A-PBS model outperformed all methods on all databases, revealing the excellent effectiveness of a combination of PBS and attention module on textured contact lens attacks. In addition to comparison with

Table 8.6 Iris PAD performance of our proposed methods and existing state-of-the-arts on NDCLD-2013/-2015 databases with a threshold of 0.5. The best performance (in terms of lowest HTER) is in bold

Database	Metric	Presentation Attack Detection Algorithm (%)							
		LBP[35]	WLBP [36]	DESIST [37]	MHVF [18]	MSA [7, 9]	DenseNet	PBS	A-PBS
NDCLD-2015 [27]	ACPER	6.15	50.58	29.81	1.92	0.18	1.58	1.09	0.08
	BPCER	38.70	4.41	9.22	0.39	0.00	0.14	0.00	0.06
	HTER	22.43	27.50	19.52	1.16	0.09	0.86	0.54	**0.07**
NDCLD13 (LG4000) [28]	APCER	0.00	2.00	0.50	0.00	0.00	0.20	0.00	0.00
	BPCER	0.38	1.00	0.50	0.00	0.00	0.28	0.03	0.00
	HTER	0.19	1.50	0.50	**0.00**	**0.00**	0.24	0.02	**0.00**
NDCLD13 (AD100) [28]	APCER	0.00	9.00	2.00	1.00	1.00	0.00	0.00	0.00
	BPCER	11.50	14.00	1.50	0.00	0.00	0.00	0.00	0.00
	HTER	5.75	11.50	1.75	0.50	0.50	**0.00**	**0.00**	**0.00**

Table 8.7 Iris PAD performance reported in terms of TDR (%) at 0.2% BPCER on NDCLD-2013 and NDCLD-2015 databases. The best performance is in bold

Database	TDR (%) @ 0.2% BPCER		
	DenseNet	PBS	A-PBS
NDCLD-2015	99.45	99.84	**99.96**
NDCLD13 (LG4000)	99.75	**100.00**	**100.00**
NDCLD13 (AD100)	100.00	100.00	100.00
IIITD-CLI (Cognet)	99.02	**99.59**	99.57
IIITD-CLI (Vista)	100.00	100.00	100.00

state-of-the-art algorithms, we also report the TDR (%) at 0.2% BPCER in Table 8.7. It can be found that despite all three models produce similarly good results, A-PBS obtains slightly better performance than DenseNet and PBS. The near-perfect results on NDCLD-2013/-2015 databases hint at the obsolescence and limitations of the current iris PAD databases and call for the need for more diversity in iris PAD data.

8.5.1.3 Results on the IIITD-CLI Database

Since most of the existing works reported the results using CCR metric on IIITD-CLI database [29, 30], we also strictly follow its experimental protocol where we show the experimental results in Table 8.8. In addition to CCR, the TDR at 0.2% BPCER

Table 8.8 Iris PAD performance in terms of CCR (%) on IIITD-CLI database. The best performance is in bold

PAD algorithms	Cogent	Vista
Textural features [38]	55.53	87.06
WLBP [36]	65.40	66.91
LBP+SVM [35]	77.46	76.01
LBP+PHOG+SVM [39]	75.80	74.45
mLBP [29]	80.87	93.91
ResNet18 [40]	85.15	80.97
VGG [41]	90.40	94.82
MVANet [42]	94.90	95.11
DenseNet	99.37	**100.00**
PBS	99.62	**100.00**
A-PBS	**99.70**	**100.00**

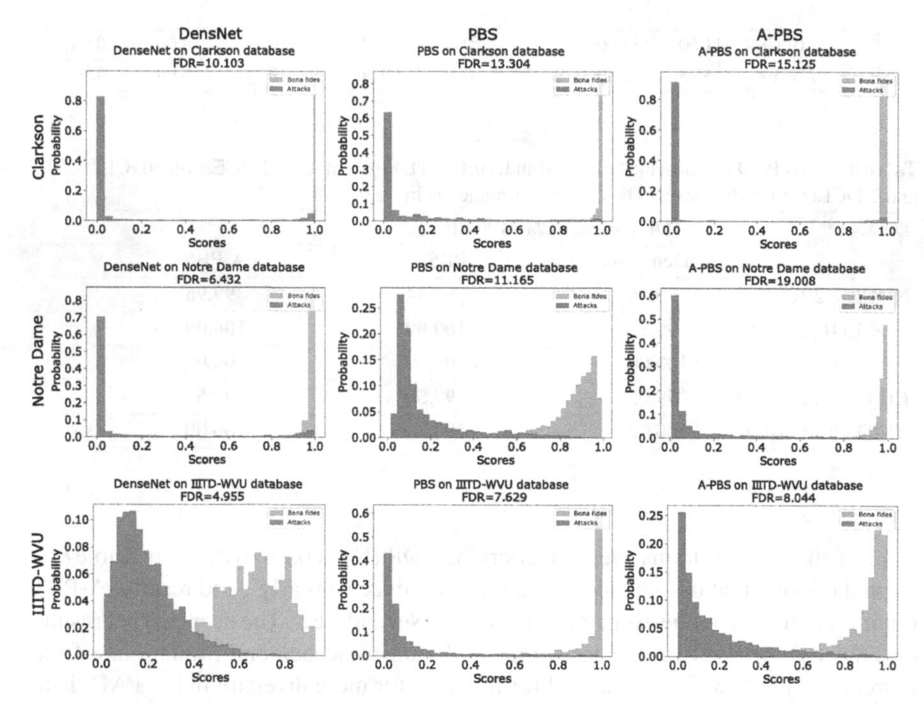

Fig. 8.3 PAD score distribution of bona fide (green) and PAs (blue) on the LivDet-Iris 2017 databases. The histogram from top to bottom are results on Clarkson, Notre Dame and IIITD-WVU databases, and the histograms from left to right are produced by DenseNet, PBS, and A-PBS, respectively. The larger separability (measured by Fisher Discriminant Ratio (FDR) [33, 34]) and smaller overlap indicate higher classification performance. It can be observed that our proposed A-PBS method achieved the highest FDR value on all three databases

is reported in Table 8.7. The experiments are performed on Cognet and Vista sensor subsets, respectively. As shown in Table 8.6, our PBS and A-PBS solutions outperform all hand-crafted and CNN-based methods by a large margin (99.79% on Cognet subset and 100.00% on Vista subset). The near-perfect classification performance obtained by DenseNet, PBS, and A-PBS reveals that despite the significant PAD improvements achieved by deep learning models, there is an urgent need for large-scale iris PAD databases to be built for future research and generalizability analysis.

8.5.2 Iris PAD in the Visible Spectrum

In addition to results on NIR databases, we also report results on the visible light-based PAVID database in Table 8.9. The experiments were demonstrated following the defined protocols in [12]. For example, the Nokia–iPhone setup refers to the training and testing data as bona fide videos captured using the Nokia phone and the attack videos captured by iPhone. Moreover, we provide the results under a grand test setup, where bona fide and attack data include videos captured by Nokia and iPhone. The models trained under grand-test setup will be used for cross-spectrum experiments later. It can be observed in Table 8.9 that deep-learning-based methods, including our A-PBS, outperform all the previously reported results on the PAVID database, which are hand-crafted feature-based PAD solutions. The DenseNet, PBS, and A-PBS methods obtain the best performance with all error rates of 0.00%.

8.6 Cross-Spectrum Evaluation Results

Most studies [7, 9, 18, 46] have presented PAD algorithms and verified their performance on NIR-based database. However, the performance of visible-light iris PAD has been understudied, especially under the cross-spectrum scenario. Therefore, we used the visible-light-based PAVID [12] and the NIR-based LivDet-Iris 2017 [4] databases to explore the effect of PAD performance across different spectra. The first scenario is the VIS-NIR where the models trained under the PAVID grand-test setup (visible spectrum) were evaluated on the test subsets of the NIR databses (Clarkson, Notre Dame, and IIIT-WVU), respectively. The evaluation results are presented in Table 8.10 and the bold numbers indicate the best performance (lowest error rates). It can be seen that our PBS and A-PBS outperform the trained from scratch DenseNet. However, all PAD methods do not generalize well on the Notre Dame database. One possible reason is that Notre Dame comprises only challenging textured lens attacks and no print/reply attacks. The PAVID database, used for training here, only includes reply attacks. One must note that both reply and print attacks involve the recapture of an artificially presented iris sample, unlike lens attacks. This recapture process can introduce artifacts identifiable by the PAD algorithms. Table 8.11 presents the

Table 8.9 Iris PAD performance of our proposed methods and established solutions on PAVID database with a threshold of 0.5. The results are reported based on APCER (%), BPCER (%), and HTER (%). Nokia—iPhone refers that the bona fide video is captured by Nokia while the replayed attack video is captured by iphone, and vice versa. Grand Test refers that both, bona fide and reply, videos are captured by Nokia and iphone. The best performance (the lowest HTER value) is in bold

Video	Metric	IQM-SVM [12, 43]	LBP-SVM [12, 44]	BSIF-SVM [12, 13]	STFT [12]	DenseNet	PBS	A-PBS
Nokia—iPhone	APCER	4.50	4.51	10.81	4.46	0.00	0.00	0.00
	BPCER	76.92	3.84	2.56	1.28	0.00	0.00	0.00
	HTER	40.71	4.18	6.68	2.87	**0.00**	**0.00**	**0.00**
Nokia—Nokia	APCER	3.57	2.67	0.89	2.68	0.00	0.00	0.00
	BPCER	57.31	4.87	6.09	1.21	0.00	0.00	0.00
	HTER	30.44	3.77	3.49	1.95	**0.00**	**0.00**	**0.00**
iPhone—iPhone	APCER	11.60	0.89	9.82	1.78	0.00	0.00	0.00
	BPCER	57.31	4.87	6.09	1.21	0.00	0.00	0.00
	HTER	34.45	2.88	7.96	1.49	**0.00**	**0.00**	**0.00**
iPhone—Nokia	APCER	10.71	3.54	8.92	0.00	0.00	0.00	0.00
	BPCER	76.92	3.84	2.56	1.28	0.00	0.00	0.00
	HTER	43.81	3.69	5.74	0.64	**0.00**	**0.00**	**0.00**
Grand-test	APCER	–	–	–	–	0.00	0.00	0.00
	BPCER	–	–	–	–	0.00	0.00	0.00
	HTER	–	–	–	–	**0.00**	**0.00**	**0.00**

results tested on the PAVID databases by using respective models trained on LivDet-Iris 2017 databases (the case of NIR-VIS). Similar observations can be found in Table 8.11 that the model trained on Notre Dame cannot generalize on the PAVID database, e.g., the lowest EER and HTER values are 52.34 and 50.63% obtained by our A-PBS solution. In contrast to the results on Notre Dame, the model trained on Clarkson and IIITD-WVU generalizes much better on the visible-light database. The lowest EER and HTER values are 1.34% achieved by A-PBS and 5.97% achieved by PBS methods, while DenseNet and A-PBS obtained similar error rates on IIITD-WVU. Moreover, we illustrate the PAD score distribution with the fisher discriminant ratio [33, 34], which measures the separability, for further analysis. Figures 8.4 and 8.5 present the results of case NIR-VIS and VIR-NIR, respectively. The PAD score distributions of the NIR-VIS case in Fig. 8.4 show that models trained on Notre Dame perform worse than those trained on Clarkson and IIIT-WVU (bona fide and attack scores almost completely overlap). Moreover, the model trained on PAVID also obtained the largest overlapping and the smallest FDR value in Fig. 8.5. One

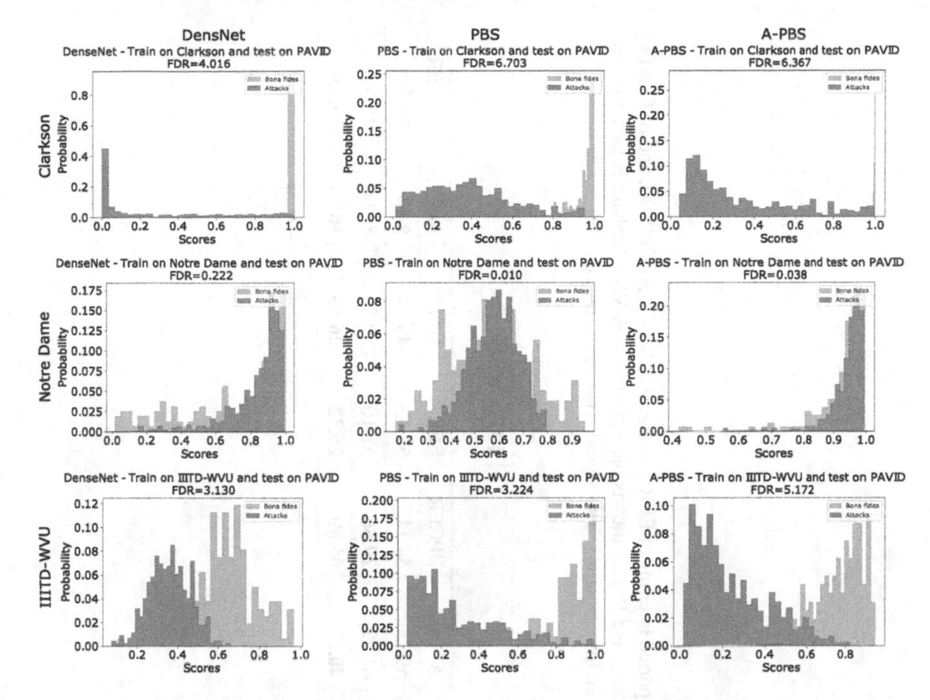

Fig. 8.4 PAD score distribution of bona fide (green) and PAs (blue) under cross-spectrum scenario (NIR-VIS). The models trained on the training subset of Clarkson (top), Notre Dame (middle), and IIITD-WVU (bottom) databases are used to evaluate on the test subset of PAVID database, and the histograms from left to right are produced by DenseNet, PBS, and A-PBS, respectively. The larger separability (measured by Fisher Discriminant Ratio (FDR) [33, 34]) and smaller overlap indicate higher classification performance

possible reason is the insufficient training data in Notre Dame (1,200 training data). However, the main reason might relate to the type of attacks and the lack of the recapturing process in the lens attacks, as mentioned earlier. This is also verified by the quantitative results in Table 8.10 and 8.11 (the APCER values are between 40.33% and 100.00%).

8.7 Visualization and Explainability

PBS is expected to learn more discriminative features by supervising each pixel/patch in comparison with binary supervised DenseNet. Subsequently, the A-PBS model, an extended model of PBS, is hypothesized to automatically locate the important regions that carry the features most useful for making an accurate iris PAD decision. To further verify and explain these assumptions, Score-Weighted Class Activation Mapping (Score-CAM) [45] is used to generate the visualizations for randomly

Table 8.10 Iris PAD performance measured under cross-spectrum scenarios and reported in terms of EER (%) and HTER (%), APCER(%), and BPCER(%). APCER and BPCER are determined by a threshold of 0.5. The training subset of the grand test on the visible-light-based PAVID database is used to train a model, and the testing subset of each database in the LivDet-Iris 2017 database is used for evaluation. The lowest error rate is in bold

Train database	PAVID											
Test database	Clarkson				Notre dame				IIITD-WVU			
Metric	EER	HTER	APCER	BPCER	EER	HTER	APCER	BPCER	EER	HTER	APCER	BPCER
DenseNet	37.78	36.69	**45.97**	27.41	56.39	56.69	59.28	54.11	54.43	49.94	**9.40**	49.94
PBS	**30.43**	37.12	66.23	**8.01**	55.67	55.39	81.22	**29.56**	51.10	50.59	82.66	18.52
A-PBS	33.41	**33.57**	46.20	20.94	**53.11**	**53.83**	**40.33**	65.89	**26.32**	**26.13**	36.30	**15.95**

Table 8.11 Iris PAD performance measured under cross-spectrum scenarios and reported in terms of EER (%), HTER (%), APCER(%), and BPCER(%). APCER and BPCER are determined by a threshold of 0.5. The training subset of the NIR-based Clarkson, Notre Dame, and IIITD-WVU is used to train a model, and the testing subset of the grand test on the PAVID database is used for evaluation. The lowest error rate is in bold

Train database	Clarkson				Notre Dame				IIITD-WVU			
Test database	PAVID											
Metric	EER	HTER	APCER	BPCER	EER	HTER	APCER	BPCER	EER	HTER	APCER	BPCER
DenseNet	6.04	13.53	23.94	3.13	57.49	61.40	95.30	27.50	**8.28**	8.07	**7.38**	8.75
PBS	4.47	**5.97**	**10.07**	1.88	56.38	57.94	**76.51**	39.38	13.43	14.15	17.67	10.63
A-PBS	**1.34**	12.98	25.95	**0.00**	**52.35**	**50.63**	100.00	**1.25**	8.63	**8.05**	11.63	**5.62**

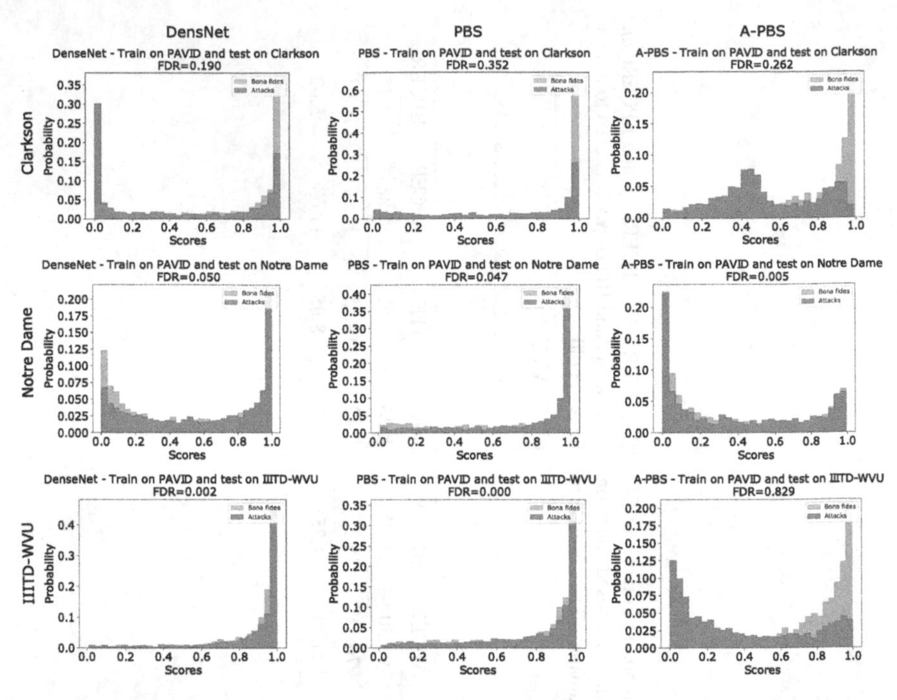

Fig. 8.5 PAD score distribution of bona fide (green) and PAs (blue) under cross-spectrum scenario (VIS-NIR). The model trained on PAVID database is used to test on the test subset of Clarkson, Notre Dame, and IIITD-WVU, respectively. The histogram top to bottom are test results on Clarkson, Notre Dame, and IIITD-WVU databases, and the histograms from left to right are produced by DenseNet, PBS, and A-PBS, respectively. The larger separability (measured by Fisher Discriminant Ratio (FDR) [33, 34]) and smaller overlap indicate higher classification performance

chosen bona fide and attack iris images (these images belong to the same identity) under intra-database and cross-spectrum scenarios as shown in Figs. 8.6 and 8.7.

Figure 8.6 illustrates the Score-CAM results on the NIR samples in the test subset of IIITD-WVU. We adopted models trained on the training subset of IIITD-WVU (NIR) and models trained on the training subset of PAVID (visible-light) to generate Score-CAMs, respectively. As shown in Fig. 8.6, it is clear that PBS and A-PBS models pay more attention to the iris region than DenseNet in both intra-database and cross-spectrum cases. The DenseNet model seems to lose some information due to binary supervision. Similar observations can be found in Fig. 8.7, where the NIR and visible-light models were tested on the visible images in the test subset of the PAVID database. In the visible intra-database case, DenseNet gained more attention on the eye region of visible-light images than of NIR images in Fig. 8.6. Moreover, in the cross-spectrum case in Fig. 8.7, the use of the attention module (A-PBS) has enabled the model to keep focusing on the iris area, while DenseNet and PBS lost some attention, especially on the attack samples. In general, the observations in Figs. 8.6 and 8.7 are consistent with the quantitative results in Tables 8.11 and 8.10

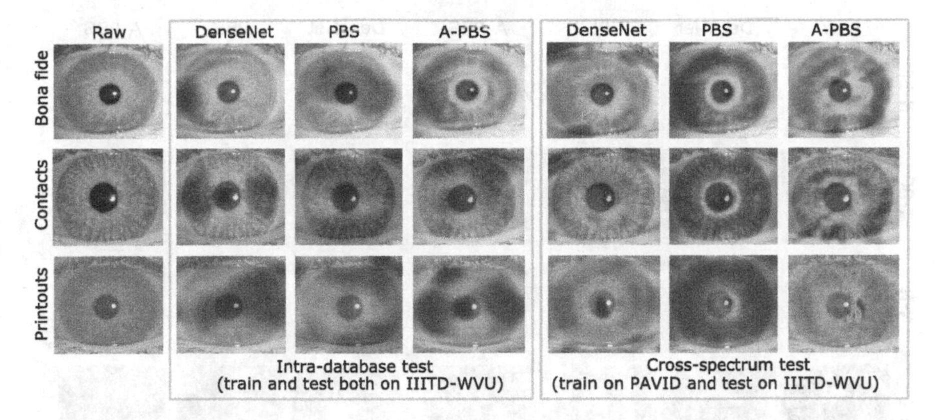

Fig. 8.6 Score-CAM visualizations for bona fide and attack samples in the IIITD-WVU test set under intra-database and cross-spectrum (VIS-NIR) scenarios. The darker the color of the region, the higher the attention on this area. The column from left to right refers to the raw samples, maps produced by DenseNet, PBS, and A-PBS model under two cases, respectively. The row from top to bottom refers to bona fide samples, textured contact lens, and printouts attack. PBS and A-PBS models pay more attention on iris region than DenseNet in both cases

that the training on visible-light and test on NIR data (VIS-NIR) is more challenging than the training on NIR and test on visible-light data (NIR-VIS) in our case. It might be caused by: (1) The perceived image quality of visible data in the PAVID database is relatively lower than NIR images (see samples in Fig. 8.2). (2) Some of the video frames in the PAVID database have an eye-blinking process, and thus some iris information (regions) will be hidden by eyelids and eyelashes. (3) While the used visible data (PAVID) contains only recaptured attacks (reply attacks), the NIR data contain both recaptured attacks (print attacks) and lens attacks, which makes it more difficult for a PAD trained on the visible images to perform properly on NIR attacks in our experiments.

8.8 Conclusion

This chapter focuses on the iris PAD performance in the NIR and visible domain, including challenging cross-database and cross-spectrum cases. The experiments were conducted using the novel attention-based pixel-wise binary supervision (A-PBS) method for iris PAD. A-PBS solution aimed to capture the fine-grained pixel/patch-level cues and utilize regions that contribute the most to an accurate PAD decision by utilizing an attention mechanism. The extensive experiments were performed on six publicly available iris PAD databases in the NIR spectrum (including LivDet-Iris 2017 competition databases) and one database in the visible spectrum. By observing intra-database and intra-spectrum experimental results, we concluded that (1) the results reported on respective attack types indicated that textured contact

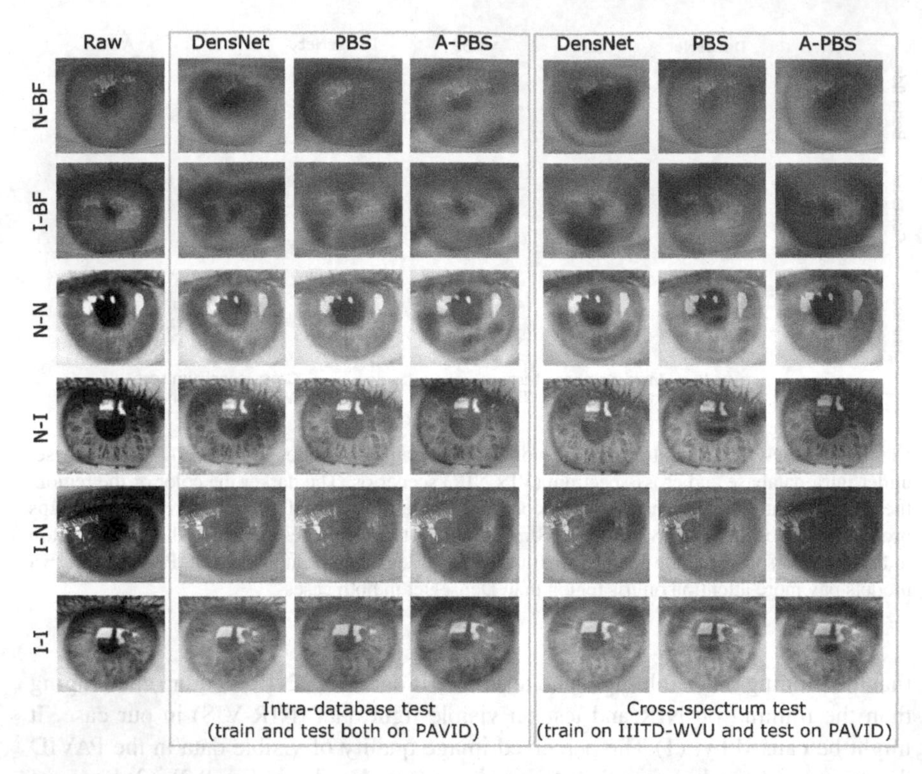

Fig. 8.7 Score-CAM visualizations for bona fide and attack samples in the PAVID test set under intra-database and cross-spectrum (NIR-VIS) scenarios. The darker the color of the region, the higher the attention on this area. The column from left to right refers to the raw samples, maps produced by DenseNet, PBS, and A-PBS model in two cases, respectively. N and I refer to Nokia and iPhone, respectively. The first two rows refer to the bona fide sample captured by Nokia and iPhone. The following four rows present the replay attack cases: Nokia-Nokia, Nokia-iPhone, iPhone-Nokia, and iPhone-iPhone. It is clear that the A-PBS model with an additional attention module is able to preserve relatively more attention on the iris region, especially for attack samples

lens attack is more challenging to detect correctly than printouts attack. (2) Cross-PA and cross-database are still challenging (EER values are over 20% in most cases). Furthermore, to our knowledge, this chapter is the first work to perform and analyze experiments under the cross-spectrum scenario. The experimental results showed that models trained on the visible spectrum do not generalize well on NIR data. It might also be caused by the limited visible data and its attack mechanism. In general, the A-PBS solution presents a superior PAD performance and high generalizability in the NIR and visible captured images, cross-database experiments, as well as cross-spectrum PAD deployments. The A-PBS also showed to focus the attention of the PAD models towards the iris region when compared to more traditional solutions.

8.9 Glossary

Presentation Attack (PA) is a presentation to the biometric capture subsystem with the goal of interfering with the operation of the biometric system.

Presentation Attack Detection (PAD) is a technique that automatically determines if the presentation is attack.

A-PBS: Attention-based Pixel-wise Binary Supervision (A-PBS) is our proposed attention-based pixel-wise supervision iris PAD method that utilizes pixel-wise supervision to capture the fine-grained patch-lebel cues and attention mechanism to guide the network to automatically find regions where most contribute to an accurate PAD decision.

Multi-Layer feature Fusion (MLF) is an iris PAD method that fuses the features extracted from last several convolutional layers. Such extracted information will be fused on two levels: feature-level and score-level.

Micro Stripe Analyses (MSA) is a solution for iris PAD, which extend the normalized iris texture and then sample such area as several individual srtipes. Finally, the majority vote will be used to make final decision.

Convolutional Block Attention Mechanism (CBAM) is an effective attention module for feed-forward networks. It can sequentially infer attention maps along channel and spatial channels.

Binary Cross Entropy (BCE) is a loss function that is used in binary classification tasks. BCE compares each of the predicted probabilities to actual class output, which can be either 0 or 1.

Attack Presentation Classification Error Rate (APCER) is the proportion of attack presentation incorrectly classified as bona fide presentations in a specific scenario.

Bona fide Presentation Classification Error Rate (BPCER) is the proportion of bona fide presentations incorrectly classified as attack presentations in a specific scenario.

Equal Error Rate (EER) is the value when the APCER and BPCER are equal.

Half Total Error Rate (HTER) is the average of APCER and BPCER.

Fisher Discriminant Ratio (FDR) is used to measure separability of classes, in our case, bona fide and attack scores. The higher the separability, the higher is the PAD performance.

Acknowledgements This research work has been funded by the German Federal Ministry of Education and Research and the Hessen State Ministry for Higher Education, Research and the Arts within their joint support of the National Research Center for Applied Cybersecurity ATHENE.

References

1. Jain AK, Nandakumar K, Ross A (2016) 50 years of biometric research: accomplishments, challenges, and opportunities. Pattern Recognit Lett 79:80–105. https://doi.org/10.1016/j.patrec.2015.12.013

2. Boutros F, Damer N, Raja KB, Ramachandra R, Kirchbuchner F, Kuijper A (2020) Iris and periocular biometrics for head mounted displays: segmentation, recognition, and synthetic data generation. Image Vis Comput 104:104,007. https://doi.org/10.1016/j.imavis.2020.104007
3. Boutros F, Damer N, Raja KB, Ramachandra R, Kirchbuchner F, Kuijper A (2020) On benchmarking iris recognition within a head-mounted display for AR/VR applications. In: 2020 IEEE international joint conference on biometrics, IJCB 2020, Houston, TX, USA, September 28—October 1, 2020. IEEE, pp 1–10. https://doi.org/10.1109/IJCB48548.2020.9304919
4. Yambay D, Becker B, Kohli N, Yadav D, Czajka A, Bowyer KW, Schuckers S, Singh R, Vatsa M, Noore A, Gragnaniello D, Sansone C, Verdoliva L, He L, Ru Y, Li H, Liu N, Sun Z, Tan T (2017) Livdet iris 2017—iris liveness detection competition 2017. In: 2017 IEEE international joint conference on biometrics, IJCB 2017, Denver, CO, USA, October 1–4, 2017. IEEE, pp 733–741. https://doi.org/10.1109/BTAS.2017.8272763
5. Das P, McGrath J, Fang Z, Boyd A, Jang G, Mohammadi A, Purnapatra S, Yambay D, Marcel S, Trokielewicz M, Maciejewicz P, Bowyer KW, Czajka A, Schuckers S, Tapia JE, Gonzalez S, Fang M, Damer N, Boutros F, Kuijper A, Sharma R, Chen C, Ross A (2020) Iris liveness detection competition (livdet-iris)—the 2020 edition. In: 2020 IEEE international joint conference on biometrics, IJCB 2020, Houston, TX, USA, September 28—October 1, 2020. IEEE, pp 1–9. https://doi.org/10.1109/IJCB48548.2020.9304941
6. Fang M, Damer N, Boutros F, Kirchbuchner F, Kuijper A (2020) Deep learning multi-layer fusion for an accurate iris presentation attack detection. In: IEEE 23rd international conference on information fusion, FUSION 2020, Rustenburg, South Africa, July 6–9, 2020. IEEE, pp 1–8. https://doi.org/10.23919/FUSION45008.2020.9190424
7. Fang M, Damer N, Kirchbuchner F, Kuijper A (2020) Micro stripes analyses for iris presentation attack detection. In: 2020 IEEE international joint conference on biometrics, IJCB 2020, Houston, TX, USA, September 28—October 1, 2020. IEEE, pp 1–10. https://doi.org/10.1109/IJCB48548.2020.9304886
8. Sharma R, Ross A (2020) D-NetPAD: an explainable and interpretable iris presentation attack detector. In: 2020 IEEE international joint conference on biometrics, IJCB 2020, Houston, TX, USA, September 28–October 1, 2020. IEEE, pp 1–10. https://doi.org/10.1109/IJCB48548.2020.9304880
9. Fang M, Damer N, Boutros F, Kirchbuchner F, Kuijper A (2021) Cross-database and cross-attack iris presentation attack detection using micro stripes analyses. Image Vis Comput 105:104,057. https://doi.org/10.1016/j.imavis.2020.104057
10. Fang M, Damer N, Kirchbuchner F, Kuijper A (2020) Demographic bias in presentation attack detection of iris recognition systems. In: 28th European signal processing conference, EUSIPCO 2020, Amsterdam, Netherlands, January 18–21, 2021. IEEE, pp 835–839. https://doi.org/10.23919/Eusipco47968.2020.9287321
11. Fang M, Damer N, Boutros F, Kirchbuchner F, Kuijper A (2022) The overlapping effect and fusion protocols of data augmentation techniques in iris PAD. Mach Vis Appl 33(1):8. https://doi.org/10.1007/s00138-021-01256-9
12. Raja KB, Raghavendra R, Busch C (2015) Presentation attack detection using Laplacian decomposed frequency response for visible spectrum and near-infra-red iris systems. In: IEEE 7th international conference on biometrics theory, applications and systems, BTAS 2015, Arlington, VA, USA, September 8–11, 2015. IEEE, pp 1–8. https://doi.org/10.1109/BTAS.2015.7358790
13. Raghavendra R, Busch C (2015) Robust scheme for iris presentation attack detection using multiscale binarized statistical image features. IEEE Trans Inf Forensics Secur 10(4):703–715. https://doi.org/10.1109/TIFS.2015.2400393
14. Yadav D, Kohli N, Vatsa M, Singh R, Noore A (2017) Unconstrained visible spectrum iris with textured contact lens variations: database and benchmarking. In: 2017 IEEE international joint conference on biometrics, IJCB 2017, Denver, CO, USA, October 1–4, 2017. IEEE, pp 574–580. https://doi.org/10.1109/BTAS.2017.8272744
15. SAMSUNG ELECTRONICS CO L How does the iris scanner work on galaxy s9, galaxy s9+, and galaxy note9? https://www.samsung.com/global/galaxy/what-is/iris-scanning/. Accessed 19 Apr 2021

16. Kentish P Is paying with your iris the future of transactions? polish start-up payeye certainly thinks so. https://emerging-europe.com/business/is_paying_with_your_iris_the_future_of_transactions_polish_start_up_payeye_certainly_thinks_so/. Accessed 07 July 2020
17. Raja KB, Raghavendra R, Busch C (2015) Iris imaging in visible spectrum using white LED. In: IEEE 7th international conference on biometrics theory, applications and systems, BTAS 2015, Arlington, VA, USA, September 8–11, 2015. IEEE, pp 1–8. https://doi.org/10.1109/BTAS.2015.7358769
18. Yadav D, Kohli N, Agarwal A, Vatsa M, Singh R, Noore A (2018) Fusion of handcrafted and deep learning features for large-scale multiple iris presentation attack detection. In: 2018 IEEE conference on computer vision and pattern recognition workshops, CVPR workshops 2018, Salt Lake City, UT, USA, June 18–22, 2018. Computer Vision Foundation/IEEE Computer Society, pp 572–579. https://doi.org/10.1109/CVPRW.2018.00099. http://openaccess.thecvf.com/content_cvpr_2018_workshops/w11/html/Yadav_Fusion_of_Handcrafted_CVPR_2018_paper.html
19. Yadav D, Kohli N, Vatsa M, Singh R, Noore A (2019) Detecting textured contact lens in uncontrolled environment using densepad. In: IEEE conference on computer vision and pattern recognition workshops, CVPR workshops 2019, Long Beach, CA, USA, June 16–20, 2019. Computer Vision Foundation/IEEE, pp 2336–2344. https://doi.org/10.1109/CVPRW.2019.00287. http://openaccess.thecvf.com/content_CVPRW_2019/html/Biometrics/Yadav_Detecting_Textured_Contact_Lens_in_Uncontrolled_Environment_Using_DensePAD_CVPRW_2019_paper.html
20. Huang G, Liu Z, van der Maaten L, Weinberger KQ (2017) Densely connected convolutional networks. In: 2017 IEEE conference on computer vision and pattern recognition, CVPR 2017, Honolulu, HI, USA, July 21–26, 2017. IEEE Computer Society, pp 2261–2269. https://doi.org/10.1109/CVPR.2017.243
21. Chen C, Ross A (2021) An explainable attention-guided iris presentation attack detector. In: IEEE winter conference on applications of computer vision workshops, WACV workshops 2021, Waikola, HI, USA, January 5–9, 2021. IEEE, pp 97–106. https://doi.org/10.1109/WACVW52041.2021.00015
22. Fang M, Damer N, Boutros F, Kirchbuchner F, Kuijper A (2021) Iris presentation attack detection by attention-based and deep pixel-wise binary supervision network. In: International IEEE joint conference on biometrics, IJCB 2021, Shenzhen, China, August 4–7, 2021. IEEE, pp 1–8. https://doi.org/10.1109/IJCB52358.2021.9484343
23. Liu Y, Jourabloo A, Liu X (2018) Learning deep models for face anti-spoofing: binary or auxiliary supervision. In: 2018 IEEE conference on computer vision and pattern recognition, CVPR 2018, Salt Lake City, UT, USA, June 18–22, 2018. Computer Vision Foundation/IEEE Computer Society, pp 389–398. https://doi.org/10.1109/CVPR.2018.00048. http://openaccess.thecvf.com/content_cvpr_2018/html/Liu_Learning_Deep_Models_CVPR_2018_paper.html
24. George A, Marcel S (2019) Deep pixel-wise binary supervision for face presentation attack detection. In: 2019 international conference on biometrics, ICB 2019, Crete, Greece, June 4–7, 2019. IEEE, pp 1–8. https://doi.org/10.1109/ICB45273.2019.8987370
25. Fang M, Damer N, Kirchbuchner F, Kuijper A (2022) Learnable multi-level frequency decomposition and hierarchical attention mechanism for generalized face presentation attack detection. In: 2022 IEEE winter conference on applications of computer vision, WACV 2022, Hawaii, USA, Jan 04–08, 2022. IEEE Computer Society, pp 3722–3731
26. Woo S, Park J, Lee J, Kweon IS (2018) CBAM: convolutional block attention module. In: Ferrari V, Hebert M, Sminchisescu C, Weiss Y (eds) Computer vision—ECCV 2018—15th European conference, Munich, Germany, September 8–14, 2018, proceedings, Part VII. Lecture Notes in Computer Science, vol 11211. Springer, pp 3–19. https://doi.org/10.1007/978-3-030-01234-2_1
27. Jr JSD, Bowyer KW (2015) Robust detection of textured contact lenses in iris recognition using BSIF. IEEE Access 3:1672–1683. https://doi.org/10.1109/ACCESS.2015.2477470

28. Jr JSD, Bowyer KW, Flynn PJ (2013) Variation in accuracy of textured contact lens detection based on sensor and lens pattern. In: IEEE sixth international conference on biometrics: theory, applications and systems, BTAS 2013, Arlington, VA, USA, September 29—October 2, 2013. IEEE, pp 1–7. https://doi.org/10.1109/BTAS.2013.6712745
29. Yadav D, Kohli N, Jr JSD, Singh R, Vatsa M, Bowyer KW (2014) Unraveling the effect of textured contact lenses on iris recognition. IEEE Trans Inf Forensics Secur 9(5):851–862. https://doi.org/10.1109/TIFS.2014.2313025
30. Kohli N, Yadav D, Vatsa M, Singh R (2013) Revisiting iris recognition with color cosmetic contact lenses. In: Fiérrez J, Kumar A, Vatsa M, Veldhuis RNJ, Ortega-Garcia J (eds) International conference on biometrics, ICB 2013, 4–7 June, 2013, Madrid, Spain. IEEE, pp 1–7. https://doi.org/10.1109/ICB.2013.6613021
31. International Organization for Standardization: ISO/IEC DIS 30107-3:2016: Information technology—biometric presentation attack detection—p 3: Testing and reporting (2017)
32. Kimura GY, Lucio DR, Jr ASB, Menotti D (2020) CNN hyperparameter tuning applied to iris liveness detection, pp 428–434. https://doi.org/10.5220/0008983904280434
33. Lorena AC, de Leon Ferreira de Carvalho ACP (2010) Building binary-tree-based multiclass classifiers using separability measures. Neurocomputing 73(16–18):2837–2845. https://doi.org/10.1016/j.neucom.2010.03.027
34. Damer N, Opel A, Nouak A (2014) Biometric source weighting in multi-biometric fusion: towards a generalized and robust solution. In: 22nd European signal processing conference, EUSIPCO 2014, Lisbon, Portugal, September 1–5, 2014. IEEE, pp 1382–1386. https://ieeexplore.ieee.org/document/6952496/
35. Gupta P, Behera S, Vatsa M, Singh R (2014) On iris spoofing using print attack. In: 22nd international conference on pattern recognition, ICPR 2014, Stockholm, Sweden, August 24–28, 2014. IEEE Computer Society, pp 1681–1686. https://doi.org/10.1109/ICPR.2014.296
36. Zhang H, Sun Z, Tan T (2010) Contact lens detection based on weighted LBP. In: 20th international conference on pattern recognition, ICPR 2010, Istanbul, Turkey, 23–26 August 2010. IEEE Computer Society, pp 4279–4282. https://doi.org/10.1109/ICPR.2010.1040
37. Kohli N, Yadav D, Vatsa M, Singh R, Noore A (2016) Detecting medley of iris spoofing attacks using DESIST. In: 8th IEEE international conference on biometrics theory, applications and systems, BTAS 2016, Niagara Falls, NY, USA, September 6–9, 2016. IEEE, pp 1–6. https://doi.org/10.1109/BTAS.2016.7791168
38. Wei Z, Qiu X, Sun Z, Tan T (2008) Counterfeit iris detection based on texture analysis. In: 19th international conference on pattern recognition (ICPR 2008), December 8–11, 2008, Tampa, Florida, USA. IEEE Computer Society, pp 1–4. https://doi.org/10.1109/ICPR.2008.4761673
39. Bosch A, Zisserman A, Muñoz X (2007) Representing shape with a spatial pyramid kernel. In: Sebe N, Worring M (eds) Proceedings of the 6th ACM international conference on image and video retrieval, CIVR 2007, Amsterdam, The Netherlands, July 9–11, 2007. ACM, pp 401–408. https://doi.org/10.1145/1282280.1282340
40. He K, Zhang X, Ren S, Sun J (2016) Deep residual learning for image recognition. In: 2016 IEEE conference on computer vision and pattern recognition, CVPR 2016, Las Vegas, NV, USA, June 27–30, 2016. IEEE Computer Society, pp. 770–778. https://doi.org/10.1109/CVPR.2016.90
41. Simonyan K, Zisserman A (2015) Very deep convolutional networks for large-scale image recognition. In: Bengio Y, LeCun Y (eds) 3rd international conference on learning representations, ICLR 2015, San Diego, CA, USA, May 7–9, 2015, conference track proceedings. http://arxiv.org/abs/1409.1556
42. Gupta M, Singh V, Agarwal A, Vatsa M, Singh R (2020) Generalized iris presentation attack detection algorithm under cross-database settings. In: 25th international conference on pattern recognition, ICPR 2020, virtual event/Milan, Italy, January 10–15, 2021. IEEE, pp 5318–5325. https://doi.org/10.1109/ICPR48806.2021.9412700
43. Galbally J, Marcel S, Fiérrez J (2014) Image quality assessment for fake biometric detection: application to iris, fingerprint, and face recognition. IEEE Trans Image Process 23(2):710–724. https://doi.org/10.1109/TIP.2013.2292332

44. Määttä J, Hadid A, Pietikäinen M (2021) Face spoofing detection from single images using micro-texture analysis. In: 2011 IEEE international joint conference on biometrics, IJCB 2011, Washington, DC, USA, October 11–13, 2011. IEEE Computer Society, pp 1–7. https://doi.org/10.1109/IJCB.2011.6117510
45. Wang H, Wang Z, Du M, Yang F, Zhang Z, Ding S, Mardziel P, Hu X (2020) Score-cam: score-weighted visual explanations for convolutional neural networks. In: 2020 IEEE/CVF conference on computer vision and pattern recognition, CVPR workshops 2020, Seattle, WA, USA, June 14–19, 2020. Computer Vision Foundation/IEEE, pp 111–119. https://doi.org/10.1109/CVPRW50498.2020.00020. https://openaccess.thecvf.com/content_CVPRW_2020/html/w1/Wang_Score-CAM_Score-Weighted_Visual_Explanations_for_Convolutional_Neural_Networks_CVPRW_2020_paper.html
46. Kuehlkamp A, da Silva Pinto A, Rocha A, Bowyer KW, Czajka A (2019) Ensemble of multi-view learning classifiers for cross-domain iris presentation attack detection. IEEE Trans Inf Forensics Secur 14(6):1419–1431. https://doi.org/10.1109/TIFS.2018.2878542

Part III
Face Biometrics

Chapter 9
Introduction to Presentation Attack Detection in Face Biometrics and Recent Advances

Javier Hernandez-Ortega, Julian Fierrez, Aythami Morales, and Javier Galbally

Abstract The main scope of this chapter is to serve as an introduction to face presentation attack detection, including key resources and advances in the field in the last few years. The next pages present the different presentation attacks that a face recognition system can confront, in which an attacker presents to the sensor, mainly a camera, a Presentation Attack Instrument (PAI), that is generally a photograph, a video, or a mask, with the target to impersonate a genuine user or to hide the actual identity of the attacker via obfuscation. First, we make an introduction of the current status of face recognition, its level of deployment, and its challenges. In addition, we present the vulnerabilities and the possible attacks that a face recognition system may be exposed to, showing that way the high importance of presentation attack detection methods. We review different types of presentation attack methods, from simpler to more complex ones, and in which cases they could be effective. Then, we summarize the most popular presentation attack detection methods to deal with these attacks. Finally, we introduce public datasets used by the research community for exploring vulnerabilities of face biometrics to presentation attacks and developing effective countermeasures against known PAIs.

J. Hernandez-Ortega (✉) · J. Fierrez (✉) · A. Morales (✉)
School of Engineering, Universidad Autonoma de Madrid, Madrid, Spain
e-mail: javier.hernandezo@uam.es

J. Fierrez
e-mail: julian.fierrez@uam.es

A. Morales
e-mail: aythami.morales@uam.es

J. Galbally (✉)
eu-LISA / European Union Agency for the Operational Management of Large-Scale IT Systems in the Area of Freedom, Security and Justice, Tallinn, Estonia
e-mail: javier.galbally@eulisa.europa.eu

S. Marcel et al. (eds.), *Handbook of Biometric Anti-Spoofing*, Advances in Computer Vision and Pattern Recognition, https://doi.org/10.1007/978-981-19-5288-3_9

9.1 Introduction

Over the last decades, there have been numerous technological advances that helped to bring new possibilities to people in the form of new devices and services. Some years ago, it would have been almost impossible to imagine having in the market devices like current smartphones and laptops, at affordable prices, that allow a high percentage of the population to have their own piece of top-level technology at home, a privilege that historically has been restricted to big companies and research groups.

Thanks to this quick advance in technology, specially in computer science and electronics, it has been possible to broadly deploy biometric systems for the first time. Nowadays, they are present in a high number of scenarios like border access control [1], surveillance [2], smartphone authentication [3], forensics [4], and on-line services like e-commerce and e-learning [5].

Among all the existing biometric characteristics, face recognition is currently one of the most extended. Face has been studied as a mean of recognition since the 60s, acquiring special relevance in the 90s following the evolution of computer vision [6]. Some interesting properties of the interaction of human faces with biometric systems are acquisition at a distance, non-intrusive, and the highly discriminant features of the face to perform identity recognition.

At present, face is one of the biometric characteristics with the highest economical and social impact due to several reasons:

- Face is one of the most largely deployed biometric modes at world level in terms of market quota [7]. Each day more and more manufacturers are including Face Recognition Systems (FRSs) in their products, like Apple with its Face ID technology. The public sector is also starting to use face recognition for a wide range of purposes like demographic analysis, identification, and access control [8].
- Face is adopted in most identification documents such as the ICAO-compliant biometric passport [9] or national ID cards [10].

Given their high level of deployment, attacks having a FRS as their target are not restricted anymore to theoretical scenarios, becoming a real threat. There are all kinds of applications and sensitive information that can be menaced by attackers. Providing each face recognition application with an appropriate level of security, as it is being done with other biometric characteristics, like iris or fingerprint, should be a top priority.

Historically, the main focus of research in face recognition has been given to the improvement of the performance at verification and identification tasks, that means, distinguishing better between subjects using the available information of their faces. To achieve that goal, a FRS should be able to optimize the differences between the facial features of each user and also the similarities among samples of the same user [11, 12]. Within the variability factors that can affect the performance of face recognition systems there are occlusions, low-resolution, different viewpoints, lighting, etc. Improving the performance of recognition systems in the presence of these variability factors is currently an active and challenging area in face recognition research [13–15].

Contrary to the optimization of their performance, the security vulnerabilities of face recognition systems have been much less studied in the past, and only over the recent few years some attention has been given to detecting different types of attacks [16–18].

Presentation Attacks (PA) can be defined as the presentation of human characteristics or artifacts directly to the input sensor of a biometric system, trying to interfere its normal operation. This category of attacks is highly present in real-world applications of biometrics since the attackers do not need to have access to the internal modules of the recognition system. For example, presenting a high-quality printed face of a legitimate user to a camera can be enough to compromise a face recognition system if it does not implement proper countermeasures against these artifacts.

The target of face Presentation Attack Detection (PAD) systems is the automated determination of presentation attacks. Face PAD methods aim to distinguish between a legitimate face and a Presentation Attack Instrument (PAI) that tries to mimic bona fide biometric traits. For example, a subset of PAD methods, referred to as liveness detection, involves measurement and analysis of anatomical characteristics or of involuntary and voluntary reactions, in order to determine if a biometric sample is being captured from a living subject actually present at the point of capture. By applying liveness detection, a face recognition system can become resilient against many presentation attacks like the printed photo attacks mentioned previously.

The rest of this chapter is organized as follows: Sect. 9.2 overviews the main vulnerabilities of face recognition systems, making a description of several presentation attack approaches. Section 9.3 introduces presentation attack detection techniques. Section 9.4 presents some available public databases for research and evaluation of face presentation attack detection. Section 9.5 discusses different architectures and applications of face PAD. Finally, concluding remarks and future lines of work in face PAD are drawn in Sect. 9.6.

9.2 Vulnerabilities in Face Biometrics

In the present chapter we concentrate on Presentation Attacks, i.e., attacks against the sensor of a FRS [19] (see point V1 in Fig. 9.1). Some relevant properties of these attacks are that they require low information about the attacked system and that they present a high success probability when the FRS is not properly protected.

On the contrary, indirect attacks (points V2-V7 in Fig. 9.1) are those attacks to the inner modules of the FRS, i.e., the preprocessing module, the feature extractor, the classifier, or the enrolling database. A detailed definition of indirect attacks to face systems can be found in [20]. Indirect attacks can be prevented by improving certain points of the FRS [21], like the communication channels, the equipment and infrastructure involved and the perimeter security. The techniques needed for improving those modules are more related to "classical" cybersecurity than to biometric tech-

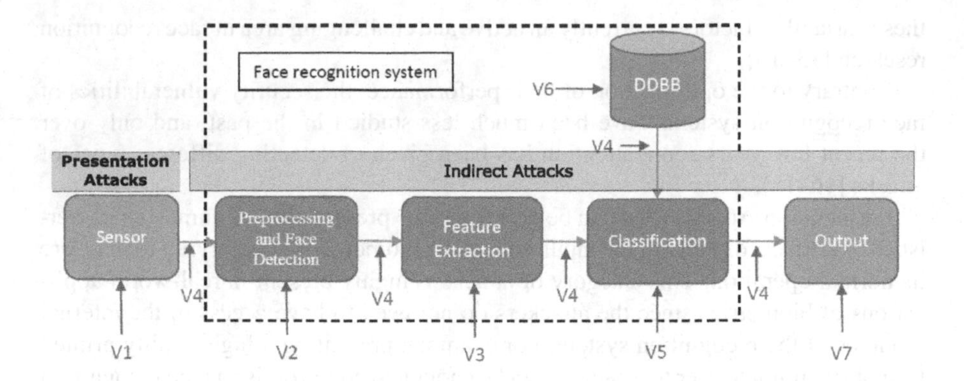

Fig. 9.1 Scheme of a generic biometric system. In this type of system, there exist several modules and points that can be the target of an attack (Vulnerabilities V1 to V7). Presentation attacks are performed at sensor level (V1), without the need of having access to the inner modules of the system. Indirect attacks (V2 to V7) can be performed at the databases, the matcher, the communication channels, etc.; in this type of attack the attacker needs access to the inner modules of the system and in most cases also specific information about their functioning

niques. These attacks and their countermeasures are beyond the scope of this book but should not be underestimated.

Presentation attacks are a purely biometric vulnerability that is not shared with other IT security solutions and that needs specific countermeasures. In these attacks, intruders use some type of artifact, typically artificial (e.g., a face photo, a mask, a synthetic fingerprint, or a printed iris image), or try to mimic the aspect of genuine users (e.g., gait, signature, or facial expression [22]) to present it to the acquisition scanner and fraudulently access the biometric system.

A high amount of biometric data are exposed, (e.g., photographs and videos on social media sites) showing the face, eyes, voice, and behavior of people. Presentation attackers are aware of this reality and take advantage of those sources of information to try to circumvent face recognition systems [23]. This is one of the well-known drawbacks of biometrics: "biometric characteristics are not secrets" [24]. In this context, it is worth noting that the factors that make face an interesting characteristic for person recognition, that is, images that can be taken at a distance and in a non-intrusive way, make it also specially vulnerable to attackers who want to use biometric information in an illicit manner.

In addition to being fairly easy to obtain a face image of the legitimate users, face recognition systems are known to respond weakly to presentation attacks, for example using one of these three categories of attacks:

1. Using a photograph of the user to be impersonated [25].
2. Using a video of the user to be impersonated [26, 27].
3. Building and using a 3D model of the attacked face, for example, an hyper-realistic mask [28].

The success probability of an attack may vary considerably depending on the characteristics of the FRS, for example, if it uses visible light or works in another range of the electromagnetic spectrum (e.g., infra-red lighting), if it has one or several sensors (e.g., 3D sensors, thermal sensors), the resolution, the lighting, and also depending on the characteristics of the PAI: quality of the texture, the appearance, the resolution of the presentation device, the type of support used to present the fake, or the background conditions.

Without implementing presentation attack detection measures most of the state-of-the-art facial biometric systems are vulnerable to simple attacks that a regular person would detect easily. This is the case, for example, of trying to impersonate a subject using a photograph of his face. Therefore, in order to design a secure FRS in a real scenario, for instance for replacing password-based authentication, Presentation Attack Detection (PAD) techniques should be a top priority from the initial planning of the system.

Given the discussion above, it could be stated that face recognition systems without PAD techniques are at clear risk, so a question often rises: What technique(s) should be adopted to secure them? The fact is that counterfeiting this type of threat is not a straightforward problem, as new specific countermeasures need to be developed and adopted whenever a new attack appears.

With the scope of encouraging and boosting the research in presentation attack detection techniques in face biometrics, there are numerous and very diverse initiatives in the form of dedicated tracks, sessions and workshops in biometric-specific and general signal processing conferences [29, 30]; organization of competitions [31–33]; and acquisition of benchmark datasets [27, 28, 34–36] that have resulted in the proposal of new presentation attack detection methods [19, 37]; standards in the area [38, 39]; and patented PAD mechanisms for face recognition systems [40, 41].

9.2.1 Presentation Attack Methods

Typically, a FRS can be spoofed by presenting to the sensor (e.g., a camera) a photograph, a video, or a 3D mask of a targeted person (see Fig. 9.2). There are other possibilities in order to circumvent a FRS, such as using makeup [42, 43] or plastic surgery. However, using photographs and videos is the most common type of attacks due to the high exposition of face (e.g., social media, video-surveillance), and the low cost of high-resolution digital cameras, printers, or digital screens.

Regarding the attack types, a general classification can be done by taking into account the nature and the level of complexity of the PAI used to attack: photo-based, video-based, and mask-based (as can be seen in Fig. 9.2). It must be remarked that this is only a classification of the most common types of attacks, but there are some complex attacks that may not fall specifically into in any of these categories, or that may belong to several categories at the same time. This is the case of DeepFake methods, that are usually defined as techniques able to create fake videos by swapping

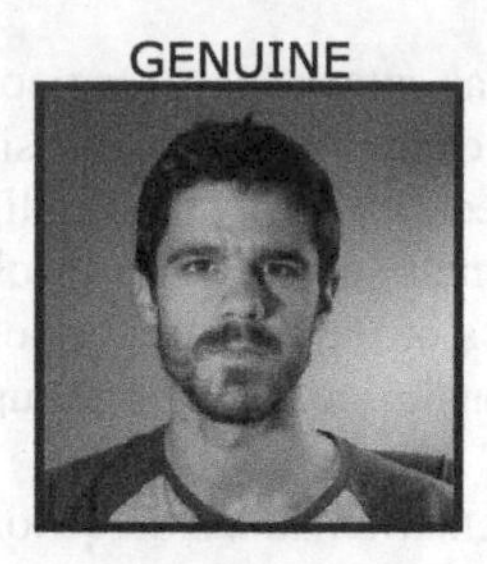

Fig. 9.2 Examples of face presentation attacks: The upper image shows an example of a genuine user, and below it there are some examples of presentation attacks, depending on the PAI shown to the sensor: a photo, a video, a 3D mask, DeepFakes, make-up, surgery, and others

the face of a person with the face of another person, and that could be classified into photo attacks, video attacks, or even mask attacks. In this chapter, we have classified those more complex and special attacks in a category named "Other attacks".

9.2.1.1 Photo Attacks

A photo attack consists in displaying a photograph of the attacked identity to the sensor of the face recognition system [44, 45] (see example in Fig. 9.2).

Photo attacks are the most critical type of attack to be protected from due to several factors. On the one hand, printing color images from the face of the genuine user is really cheap and easy to do. These are usually called print attacks in the literature [46]. Alternatively, the photos can be displayed in the high-resolution screen of a device (e.g., a smartphone, a tablet or a laptop [26, 34, 44]). On the other hand, it is also easy to obtain samples of genuine faces thanks to the recent growth of social media sites like Facebook, Twitter, and Instagram [23]. Additionally, with the price and size reduction experimented by digital cameras in recent years, it is also possible to obtain high-quality photos of a legitimate user simply by using a hidden camera.

Among the photo attack techniques, there are also more complex approaches like photographic masks. This technique consists in printing a photograph of the subject's face and then making holes for the eyes and the mouth [34]. This is a good way to avoid presentation attack detection techniques based on blink detection and in eyes and mouth movement tracking [47].

Even if these attacks may seem too simple to work in a real scenario, some studies indicate that many state-of-the-art systems are vulnerable to them [48–50]. Due to their simplicity, implementing effective countermeasures that perform well against them should be a must for any facial recognition system.

- Information needed to perform the attack: image of the face of the subject to be impersonated.
- Generation and acquisition of the PAIs: there are plenty of options to obtain high-quality face images of the users to be impersonated, e.g., social networks, internet profiles, and hidden cameras. Then, those photographs can be printed or displayed on a screen in order to present them to the sensor of the FRS.
- Expected impact of the attack: most basic face recognition systems are vulnerable to this type of attack if specific countermeasures are not implemented. However, the literature offers a large number of approaches with good detection rates of printed photo attacks [44, 51].

9.2.1.2 Video Attacks

Similar to the case of photo attacks, video acquisition of people intended to be impersonated is also becoming increasingly easier with the growth of public video sharing sites and social networks, or even using a hidden camera. Another reason to use this type of attack is that it increases the probability of success by introducing liveness appearance to the displayed fake biometric sample [52, 53].

Once a video of the legitimate user is obtained, one attacker could play it in any device that reproduces video (smartphone, tablet, laptop, etc.) and then present it to the sensor/camera [54], (see Fig. 9.2). This type of attack is often referred to in the literature as replay attacks, a more sophisticated version of the simple photo attacks.

Replay attacks are more difficult to detect, compared to photo attacks, as not only the face texture and shape is emulated but also its dynamics, like eye-blinking, mouth and/or facial movements [26]. Due to their higher sophistication, it is reasonable to assume that systems that are vulnerable to photo attacks will perform even worse with respect to video attacks, and also that being resilient against photo attacks does not mean to be equally strong against video attacks [34]. Therefore, specific countermeasures need to be developed and implemented, e.g., an authentication protocol based on challenge-response [47].

- Information needed to perform the attack: video of the face of the subject to be impersonated.
- Generation and acquisition of the PAIs: similar to the case of photo attacks, obtaining face videos of the users to be impersonated is relatively easy thanks to the growth of video sharing platforms (YouTube, Twitch) and social networks (Face-

book, Instagram), and also using hidden cameras. The videos are then displayed on a screen in order to present them to the sensor of the FRS.
- Expected impact of the attack: like in the case of photo attacks most face recognition systems are inherently vulnerable to these attacks, and countermeasures based on challenge-response or in the appearance of the faces are normally implemented. With these countermeasures, classic video attacks have a low success rate.

9.2.1.3 Mask Attacks

In this type of attack, the PAI is a 3D mask of the user's face. The attacker builds a 3D reconstruction of the face and presents it to the sensor/camera. Mask attacks require more skills to be well executed than the previous attacks, and also access to extra information in order to construct a realistic mask of the genuine user [55, 56].

There are different types of masks depending on the complexity of the manufacturing process and the amount of data that is required. Some examples ordered from simpler to more complex are

- The simplest method is to print a 2D photograph of the user's face and then stick it to a deformable structure. Examples of this type of structures could be a t-shirt or a plastic bag. Finally, the attacker can put the bag on his face and present it to the biometric sensor [34]. This attack can mimic some deformable patterns of the human face, allowing to spoof some low-level 3D face recognition systems.
- Image reconstruction techniques can generate 3D models from 2 or more pictures of the genuine user's face, e.g., one frontal photo and a profile photo. Using these photographs, the attacker could be able to extrapolate a 3D reconstruction of the real face[1] (see Fig. 9.2). This method is unlikely to spoof top-level 3D face recognition systems, but it can be an easy and cheap option to spoof a high number of standard systems.
- A more sophisticated method consists in making directly a 3D capture of a genuine user's face [28, 56, 57] (see Fig. 9.3). This method entails a higher level of difficulty than the previous ones since a 3D acquisition can be done only with dedicated equipment and it is complex to obtain without the cooperation of the end-user. However, this is becoming more feasible and easier each day with the new generation of affordable 3D acquisition sensors [58].

When using any of the two last methods, the attacker would be able to build a 3D mask with the model he has computed. Even though the price of 3D printing devices is decreasing, 3D printers with sufficient quality and definition are still expensive. See references [56, 57] for examples of 3D-printed masks. There are some companies where such 3D face models may be obtained for a reasonable price.[2]

---

[1] https://3dthis.com, https://www.reallusion.com/character-creator/headshot.

[2] http://real-f.jp, https://shapify.me, and http://www.sculpteo.com.

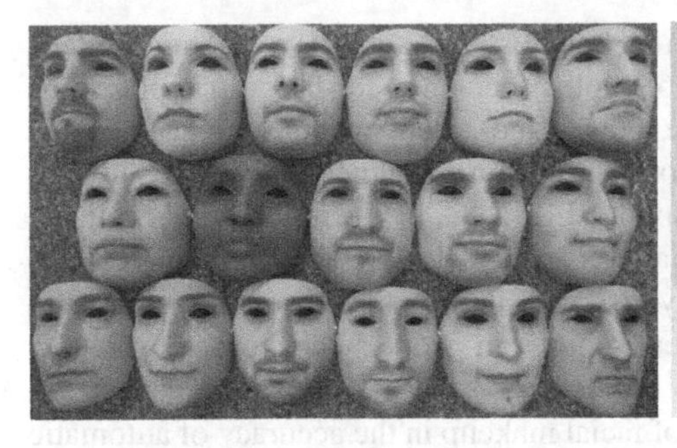
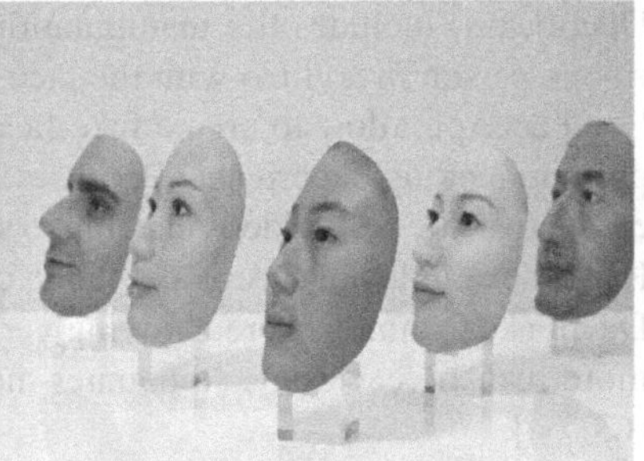

Fig. 9.3 **Examples of 3D masks.** (Left) The 17 hard-resin facial masks used to create the 3DMAD dataset, from [28]. (Right) Face Masks built by Real-F reproducing even the eyeballs and finest skin details

This type of attack may be more likely to succeed due to the high realism of the spoofs. As the complete structure of the face is imitated, it becomes difficult to find effective countermeasures. For example, the use of depth information becomes inefficient against this particular threat.

These attacks are far less common than the previous two categories because of the difficulties mentioned above to generate the spoofs. Despite the technical complexity, mask attacks have started to be systematically studied thanks to the acquisition of the first specific databases which include masks of different materials and sizes [28, 55–57, 59, 60].

- Information needed to perform the attack: 2D masks can be created using only one face image of the user to be impersonated. However, 3D realistic masks usually need 3 or more face images acquired from different angles.
- Generation and acquisition of the PAIs: compared to photo and video attacks it is more difficult to generate realistic 3D masks since the attacker needs images of high quality captured from different and complementary angles. Using the photographs, face masks can be ordered to third companies for a reasonable price.
- Expected impact of the attack: these attacks are more challenging than photo and video attacks because of the higher realism of the PAIs. However, they are less common due to the difficulty in generating the masks.

9.2.1.4 Other Attacks

There are other possibilities in order to circumvent a face recognition system, such as using DeepFake techniques, facial makeup, or modifications via plastic surgery.

Together with the recent availability of large-scale face databases, the progress of deep learning methods like Generative Adversarial Networks (GANs) [61] has led to the emergence of new realistic face manipulation techniques that can be used to spoof face recognition systems. The term Identity Swap (commonly known as

DeepFakes) includes the manipulation techniques consisting in replacing the face of one person in a video with the face of another person. User-friendly applications like FaceApp allow to create fake face images and videos without the need of any previous coding experience. Public information from social networks can be used to create realistic fake videos capable to spoof a FRS, e.g., by means of a replay attack. Recent examples of DeepFake video databases are Celeb-DF [27] and DFDC [62]. DeepFake techniques are evolving very fast, with their outputs becoming more and more realistic each day, so counter-measuring them is a very challenging problem [63, 64].

Works like [42] studied the impact of facial makeup in the accuracy of automatic face recognition systems. They focused their work on determining if facial makeup can affect the matching accuracy of a FRS. They acquired two different databases of females with and without makeup and they tested the accuracy of several face recognition systems when using those images. They concluded that the accuracy of the FRS is hugely impacted by the presence of facial makeup.

Nowadays, the technology advancements and the social acceptance have led to a higher presence of plastic surgery among the population. Therefore, the authors of [65, 66] focused their research on determining the level of affectation of face recognition accuracy when using images with presence of plastic surgery that modifies facial appearance. In [65] they reported a high reduction of face recognition accuracy (around a 30%) of several matching algorithms when comparing images with and without plastic surgery modifications.

- Information needed to perform the attack: in the case of DeepFake attacks, PAIs can be created only with a few photographs of the targeted subject. Makeup and surgery also need information about the face of the user to be impersonated.
- Generation and acquisition of the PAIs: similar to the case of the previous types of attacks, obtaining face images and videos of the users to be impersonated can be easy thanks to video sharing platforms and social networks. Then, the DeepFake videos can be displayed on a screen to present them to the camera of the FRS. Makeup-based attacks need of certain skills to achieve a high-quality makeup. Attacks based on surgery modifications are much harder to achieve since they need of highly qualified professionals and of some recovery time after the modifications.
- Expected impact of the attack: DeepFake attacks are more difficult to detect than other attacks and how to prevent them is a very active research area nowadays. Surgery attacks are also very difficult to detect since they are actually faces and not synthetic fakes. Makeup attacks, on the other hand, can be detected more easily using PAD techniques based on texture or color, like in the case of photo and video attacks.

9.3 Presentation Attack Detection

Face recognition systems are designed to differentiate between genuine users, not to determine if the biometric sample presented to the sensor is legitimate or a fake. A presentation attack detection method is usually accepted to be any technique that is able to automatically distinguish between legitimate biometric characteristics presented to the sensor and artificially produced PAIs.

Presentation attack detection can be done in four different ways [16]: (*i*) with dedicated hardware to detect an evidence of liveness, which is not always possible to deploy, (*ii*) with a challenge-response method where a presentation attack can be detected by requesting the user to interact with the system in a specific way, (*iii*) employing recognition algorithms intrinsically resilient against attacks, and (*iv*) with software that uses already available sensors to detect any pattern characteristic of live traits. Figure 9.4 shows how PAD methods are organized according to this proposed taxonomy.

- **Hardware-based:** PAD approaches based on dedicated hardware usually take benefit of special sensors like Near Infrared (NIR) cameras [37], thermal sensors [67], Light Field Cameras (LFC) [68], multi-spectral sensors [69], and 3D cameras [70]. Using the unique properties of the different types of dedicated hardware, the biological and physical characteristics of legitimate faces can be measured and distinguished from PAIs more easily, e.g., processing temperature information with thermal cameras [71] and estimating the 3D volume of the artifacts thanks to 3D acquisition sensors [70]. However, these approaches are not popular even though they tend to achieve high presentation detection rates, because in most systems the required hardware is expensive and not broadly available.
- **Challenge-Response:** These PAD methods present a challenge to the user, i.e., completing a predefined task, or expose them to a stimulus in order to record their voluntary or involuntary response. Then that response is analyzed to decide if the access attempt comes from a legitimate user or from an attacker. An example of this approach can be found in [72], where the authors studied the involuntary eye response of users exposed to a visual stimulus. Other examples are [47] where the authors requested the users to make specific movements with their eyes and [73] where the users were indicated to say some predefined words. Challenge-response

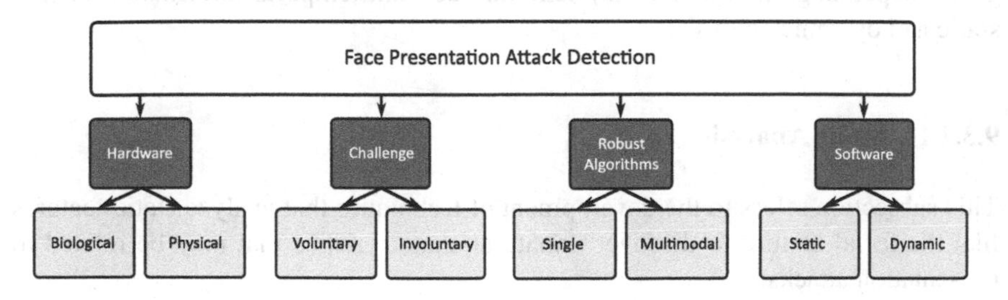

Fig. 9.4 Taxonomy of face presentation attack detection methods

methods can be effective against many presentation attacks but they usually require more time and cooperation from users' side, something that is not always possible or desirable.

- **Robust algorithms:** Existing face recognition systems can be designed or trained to learn how to distinguish between legitimate faces and PAIs, making them inherently robust to some types of presentation attacks. However, developing face recognition algorithms intrinsically robust against presentation attacks is not straightforward and the most common approach consists in relying on multimodal biometrics to increment the security level thanks to the information coming from the other biometric modalities.

The limitations of these three types of PAD methods, together with the easiness of deploying software-based PAD methods, have made most of the literature on face Presentation Attack Methods (PAD) to be focused on running software-based PAD algorithms over already deployed hardware. This is why in the next part of the chapter we focus on describing software-based PAD and the different categories that compose this type of approach.

9.3.1 *Software-Based Face PAD*

Software-based PAD methods are convenient in most of the cases since they allow to upgrade the countermeasures in existing systems without needing new pieces of hardware and permitting authentication to be done in real time without extra user interaction. Table 9.1 shows a selection of relevant PAD works based on software techniques, including information about the type of images they use, the databases in which they are evaluated, and the types of features they analyze. The table also illustrates the current dominance of deep learning among PAD methods, since like in many other research areas, during the last years most state-of-the-art face PAD works have changed from methods based on hand-crafted features to deep learning approaches based on architectures like Convolutional Neural Networks and Generative Adversarial Networks.

Regardless of whether they belong to one category (hand-crafted) or the other (deep learning), software-based PAD methods can be divided into two main categories depending on whether they take into account temporal information or not: static and dynamic analysis.

9.3.1.1 Static Analysis

This subsection refers to the development of techniques that analyze static features like the facial texture to discover unnatural characteristics that may be related to presentation attacks.

Table 9.1 Selection of relevant works in software-based face PAD

Method	Year	Type of images	Database used	Type of features
[74]	2009	Visible and IR photo	Private	Color (reflectance)—Hand Crafted
[46]	2011	RGB video	PRINT-ATTACK	Face-background motion—Hand Crafted
[26]	2012	RGB video	REPLAY-ATTACK	Texture based—Hand Crafted
[75]	2013	RGB photo and video	NUAA PI, PRINT-ATTACK and CASIA FAS	Texture based—Hand Crafted
[51]	2013	RGB photo and video	PRINT-ATTACK and REPLAY ATTACK	Texture based—Hand Crafted
[44]	2013	RGB video	PHOTO-ATTACK	Motion correlation analysis—Hand Crafted
[76]	2014	RGB video	REPLAY-ATTACK	Image Quality based—Hand Crafted
[77]	2015	RGB video	Private	Color (challenge reflections)—Hand Crafted
[78]	2016	RGB video	3DMAD and private	rPPG (color based)—Hand Crafted
[79]	2017	RGB video	OULU-NPU	Texture based—Hand Crafted
[37]	2018	RGB and NIR video	3DMAD and private	rPPG (color based)—Hand Crafted
[35]	2019	RGB, Depth and NIR video	WMCA	Fine-tuned face recog. features—Deep Learning
[64]	2020	RGB video	Celeb-DF v2 and DFDC	rPPG (color based)—Deep Learning
[80]	2021	RGB, Depth, Thermal, and NIR video	WMCA, MLFP, and SiW-M	Spoof-specific info.—Deep Learning
[81]	2021	RGB video	3DMAD and HKBU-MARsV2	rPPG (color based)—Deep Learning

The key idea of the texture-based approach is to learn and detect the structure of facial micro-textures that characterize real faces but not fake ones. Micro-texture analysis has been effectively used in detecting photo attacks from single face images: extraction of texture descriptions such as Local Binary Patterns (LBP) [26] or Gray-Level Co-occurrence Matrices (GLCM) followed by a learning stage to perform discrimination between textures.

For example, the recapturing process by a potential attacker, the printing of an image to create a spoof, usually introduces quality degradation in the sample, making it possible to distinguish between a genuine access attempt and an attack, by analyzing their textures [76].

The major drawback of texture-based presentation attack detection is that high-resolution images are required in order to extract the fine details from the faces that are needed for discriminating genuine faces from presentation attacks. These countermeasures will not work properly with bad illumination conditions that make the captured images to have bad quality in general.

Most of the time, the differences between genuine faces and artificial materials can be seen in images acquired in the visual spectrum with or without a preprocessing stage. However, sometimes a translation to a more proper feature space [82], or working with images from outside the visible spectrum [83] is needed in order to distinguish between real faces and spoof-attack images.

Additionally to the texture, there are other properties of the human face and skin that can be exploited to differentiate between real and fake samples. Some of these properties are: absorption, reflection, scattering, and refraction [74].

This type of approaches may be useful to detect photo attacks, video attacks, and also mask attacks, since all kinds of spoofs may present texture or optical properties different than real faces.

In recent years, to improve the accuracy of traditional static face PAD methods the researchers have been focused on applying the power of deep learning to face PAD mainly using transfer-learning from face recognition. This technique makes possible to adapt facial features learned for face recognition to face presentation attack detection without the need of a huge amount of labeled data. This is the case of [35] where the authors transfer facial features learned for face recognition and used them for detecting presentation attacks. Finally, to increase the generalization ability of their method to unseen attacks, they fused the decisions of different models trained with distinct types of attacks.

However, even though deep learning methods have shown to be really accurate when evaluated in intra-database scenarios, their performance usually drops significantly when they are tested under inter-database scenarios. Deep learning models are capable of learning directly from data, but they are normally overfitted to the training databases, causing poor generalization of the resulting models when facing data from other sources. To avoid this, the most recent works in the literature face this problem by implementing domain generalization techniques. For example, the authors of [80] introduced a novel loss function to force their model to learn a compact embedding for genuine faces while being far from the embeddings of the different presentation attacks.

9.3.1.2 Dynamic Analysis

These techniques have the target of distinguishing presentation attacks from genuine access attempts based on the analysis of motion. The analysis may consist in detecting any physiological sign of life, for example: pulse [84], eye-blinking [85], facial expression changes [22], or mouth movements. This objective is achieved using knowledge of the human anatomy and physiology.

As stated in Sect. 9.2, photo attacks are not able to reproduce all signs of life because of their static nature. However, video attacks and mask attacks can emulate blinking, mouth movements, etc. Related to these types of presentation attacks, it can be assumed that the movement of the PAIs, differs from the movement of real human faces which are complex nonrigid 3D objects with deformations.

One simple approximation to this type of countermeasures consists in trying to find correlations between the movement of the face and the movement of the background with respect to the camera [44, 54]. If the fake face presented contains also a piece of fake background, the correlation between the movement of both regions should be high. This could be the case of a replay attack, in which the face is shown on the screen of some device. This correlation in the movements allows to evaluate the degree of synchronization within the scene during a defined period of time. If there is no movement, as in the case of a fixed support attack, or too much movement, as in a hand-based attack, the input data is likely to come from a presentation attack. Genuine authentication will usually have uncorrelated movement between the face and the background, since user's head generally moves independently from the background.

Another example of this type of countermeasure is [53] where the authors propose a method for detecting face presentation attacks based on properties of the scenario and the facial surfaces such as albedo, depth, and reflectance.

A high number of the software-based PAD techniques are based on liveness detection without needing any special help of the user. These presentation attack detection techniques aim to detect some physiological signs of life such as eye-blinking [75, 85, 86], facial expression changes [22], and mouth movements.

Other works like [51] provide more evidence of liveness using Eulerian video magnification [87] applying it to enhance small changes in face regions, that often go unnoticed. Some changes that are amplified thanks to this technique are, for example, small color and motion changes on the face caused by the human blood flow, by finding peaks in the frequency domain that correspond to the human heartbeat rate. Works like [37, 64, 81] use remote photoplethysmography for liveness detection, and more specifically 3D mask PAD, without relying on the appearance features of the spoof like the texture, shape, etc.

As mentioned above, motion analysis approaches usually require some level of motion between different head parts or between the head and the background. Sometimes this can be achieved through user cooperation [77]. Therefore, some of these techniques can only be used in scenarios without time requirements as they may need time for analyzing a piece of video and/or for recording the user's response to a command. Due to the nature of these approaches, some videos and well-performed mask attacks may deceive the countermeasures.

9.4 Face Presentation Attacks Databases

In this section, we overview some publicly available databases for research in face PAD. The information contained in these datasets can be used for the development and evaluation of new face PAD techniques against presentation attacks.

As it has been mentioned in the past sections, with the recent spread of biometric applications, the threat of presentation attacks has grown, and the biometric community is starting to acquire large and comprehensive databases to make recognition systems more resilient against presentation attacks.

International competitions have played a key role to promote the development of PAD measures. These competitions include the recent LivDet-Face 2021 [36], the CelebA-Spoof Challenge 2020 [27], the ChaLearn Face Antispoofing Attack Detection Challenge 2019 [88], the Multi-modal Face Anti-spoofing (Presentation Attack Detection) Challenge 2019 [89], the Competition on Generalized Face Presentation Attack Detection in Mobile Authentication Scenarios 2017 [33], and the 2011 and 2013 2D Face Anti-Spoofing contests [31, 32].

Despite the increasing interest of the community in studying the vulnerabilities of face recognition systems, the availability of PAD databases is still scarce. The acquisition of new datasets is highly difficult because of two main reasons:

- Technical aspects: the acquisition of presentation attack data offers additional challenges to the usual difficulties encountered in the acquisition of standard biometric databases [90] in order to correctly capture similar fake data than the present in real attacks (e.g., generation of multiple types of PAIs).
- Legal aspects: as in the face recognition field in general, data protection limits the distribution or sharing of biometric databases among research groups. These legal restrictions have forced most laboratories or companies working in the field of presentation attacks to acquire their own datasets usually small and limited.

In the area of face recognition PAD, we can find the following public databases (ordered chronologically following their publication date):

- The NUAA Photo Imposter Database (NUAA PI DB) [25] was one of the first efforts to generate a large public face PAD dataset. It contains images of real-access attempts and print attacks of 15 users. The images contain frontal faces with a neutral expression captured using a webcam. Users were also told to avoid eye blinks. The attacks are performed using printed photographs on photographic paper. Examples from this database can be seen in Fig. 9.5. The NUAA PI DB is property of the Nanjing University of Aeronautics and Astronautics, and it can be obtained at http://parnec.nuaa.edu.cn/_upload/tpl/02/db/731/template731/pages/xtan/NUAAImposterDB_download.html.
- The PRINT-ATTACK DB [46] represents another step in the evolution of face PAD databases, both in terms of the size (50 different users were captured) and of the types of data acquired (it contains video sequences instead of still images). It only considers the case of photo attacks. It consists of 200 videos of real accesses and

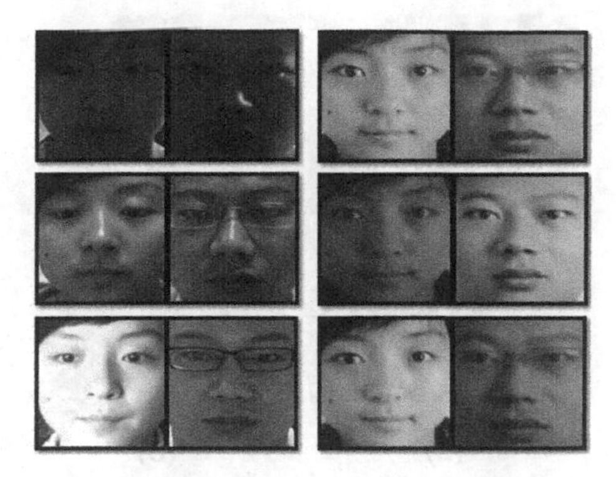

Fig. 9.5 Samples from the NUAA Photo Imposter Database [25]. Samples from two different users are shown. Each row corresponds to one different session. In each row, the left pair are from a live human and the right pair from a photo fake. Images have been taken from [25]

200 videos of print attack attempts from 50 different users. Videos were recorded under two different backgrounds and illumination conditions. Attacks were carried out with hard copies of high-resolution photographs of the 50 users, printed on plain A4 paper. The PRINT-ATTACK DB is property of the Idiap Research Institute, and it can be obtained at https://www.idiap.ch/en/dataset/printattack.

- The REPLAY-ATTACK database citeChingovskaspsBIOSIGsps2012 is an extension of the PRINT-ATTACK database. It contains short videos of both real-access and presentation attack attempts of 50 different subjects. The attack attempts present in the database are 1300 photo and video attacks using mobile phones and tablets under different lighting conditions. The attack attempts are also distinguished depending on how the attack device is held: hand-based and fixed-support. Examples from this database can be seen in Fig. 9.6. It can be obtained at https://www.idiap.ch/en/dataset/replayattack/.
- The 3D MASK-ATTACK DB (3DMAD) [28], as its name indicates, contains information related to mask attacks. As described above, all previous databases contain attacks performed with 2D spoofing artifacts (i.e., photo or video) that are very rarely effective against systems capturing 3D face data. It contains access attempts of 17 different users. The attacks were performed with real-size 3D masks manufactured by ThatsMyFace.com.[3] For each access attempt a video was captured using the Microsoft Kinect for Xbox 360, that provides RGB data and also depth information. That allows to evaluate both 2D and 3D PAD techniques, and also their fusion [57]. Example masks from this database can be seen in Fig. 9.3. 3DMAD is property of the Idiap Research Institute, and it can be obtained at https://www.idiap.ch/dataset/3dmad.
- The OULU-NPU DB [79] contains information of presentation attacks acquired with mobile devices. Nowadays mobile authentication is one of the most relevant scenarios due to the wide spread of the use of smartphones. However, in most datasets the images are acquired in constrained conditions. This type of data may present motion, blur, and changing illumination conditions, backgrounds and head poses. The database consists of 5, 940 videos of 55 subjects recorded in three dis-

[3] http://www.thatsmyface.com/.

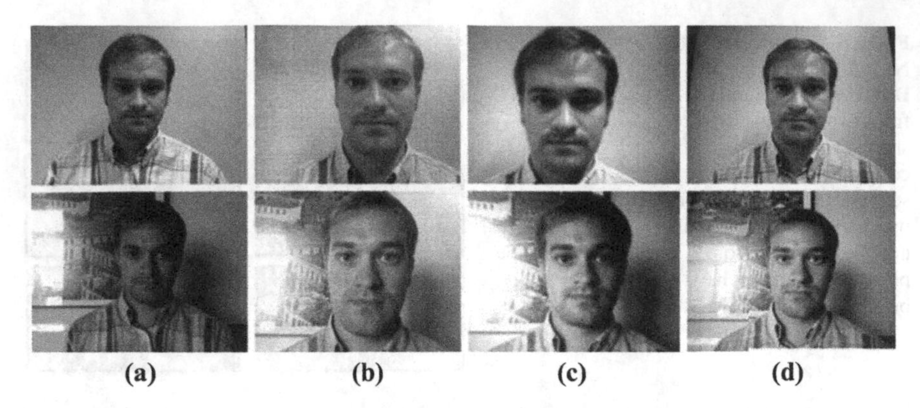

Fig. 9.6 Examples of real and fake samples from the REPLAY-ATTACK DB [26]. The images come from videos acquired in two illumination and background scenarios (controlled and adverse). The first row belongs to the controlled scenario while the second row represents the adverse conditions. **a** Shows real samples, **b** shows samples of a printed photo attack, **c** corresponds to a LCD photo attack, and **d** to a high-definition photo attack

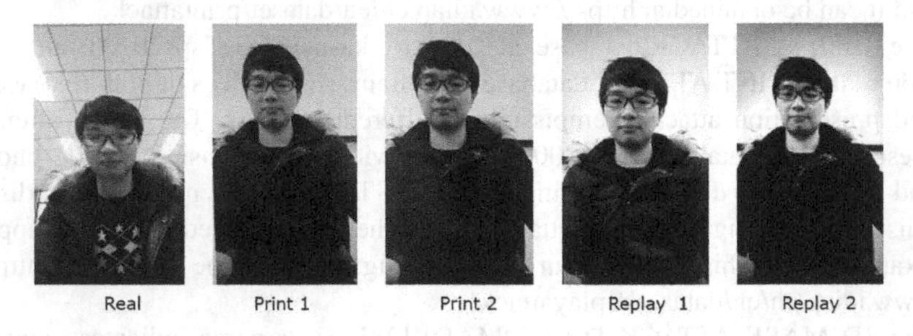

Fig. 9.7 Examples of bona-fide and attack samples from OULU-NPU DB [79]. The images come from videos acquired with mobile devices. The figure shows a legitimate sample, two examples of print attacks, and other two examples of replay attacks. Image extracted from https://sites.google.com/site/oulunpudatabase

tinct illumination conditions, with 6 different smartphone models. The resolution of all videos is 1920 × 1080 and it comprehends print and video-replay attacks. The OULU-NPU DB is property of the University of Oulu and it has been used in the IJCB 2017 Competition on Generalized Face Presentation Attack Detection [33]. Examples from this database can be seen in Fig. 9.7 and it can be obtained at https://sites.google.com/site/oulunpudatabase/.

- The Custom Silicone Mask Attack Dataset (CSMAD) [56] (collected at the Idiap Research Institute) contains 3D mask presentation attacks. It is comprised of face data from 14 subjects, from which 6 subjects have been selected to construct realistic silicone masks (made by Nimba Creations Ltd.). The database contains 87 bona-fide videos and 159 attack videos captured under four different lighting conditions. CSMAD is composed of RGB, Depth, NIR, and thermal videos of

10 seconds of duration. The database can be obtained at https://www.idiap.ch/en/dataset/csmad.

- The Spoof in the Wild (SiW) Database [91] provides bona-fide and presentation attacks from 165 different subjects. For each of those subjects, the database contains 8 live and up to 20 spoof videos recorded at 30 fps and 1920×1080 pixels of resolution, making a total of 4,478 video sequences. The database was acquired in order to have several types of distances to faces, poses, illumination conditions, and facial expressions. The collection of PAIs comprehends paper print photo attacks and video-replay attacks.
- The Wide Multi-Channel Presentation Attack Database (WMCA) [35] consists of video recordings of 10 seconds for both bona-fide attempts and presentation attacks. The database is composed of RGB, Depth, NIR, and thermal videos of 72 different subjects, with 7 sessions for each subject. WCMA contains 2D and 3D presentation attacks including print, replay, and paper and silicone mask attacks. The total number of videos in the database is 1,679 (1,332 are attacks).
- The CelebA-Spoof Database [27] is a large-scale face anti-spoofing dataset with 625,537 images from 10,177 different subjects. The database includes labels for 43 attributes related to the face, the illumination conditions, the environment, and the spoof types. The spoof images were captured on 2 different environments and under 4 illumination conditions using 10 different acquisition devices. CelebA-Spoof contains 10 different attack types, e.g., print attacks, replay attacks, and 3D masks. The database can be obtained at https://github.com/Davidzhangyuanhan/CelebA-Spoof.
- The High-Quality Wide Multi-Channel Attack Database (HQ-WMCA) [92] is an extension of the WMCA database. However, they differ in several important aspects like frame rate and resolution and the use of a new sensor for the ShortWave InfraRed spectrum (SWIR) during the acquisition. Additionally, it contains a wider range of attacks than the previous database, incorporating obfuscation attacks, where the attacker tries to hide its identity. In the database, there are 555 bona-fide presentations from 51 participants and the remaining 2,349 are presentation attacks. RGB, NIR, SWIR, thermal, and depth information was acquired, with each recording containing data in 14 different bands, including 4 NIR and 7 SWIR wavelengths. The PAIs used can be grouped into ten different categories ranging from glasses to flexible masks (including makeup).
- The LiveDet Database [36] is a dataset built as a combination of data from two of the organizers of the Face Liveness Detection Competition (LivDet-Face 2021), i.e., Clarkson University (CU) and Idiap Research Institute. The final database contains data from 48 subjects, with a total of 724 images and 814 videos (of 6 seconds) acquired using 5 different sensors including reflex cameras and mobile devices. 8 different presentation artifacts were used for the images and 9 for the videos, comprehending paper photographs, photographs shown from digital screens (e.g., a laptop), paper masks, 3D silicone masks, and video replays. Additional details of LiveDet can be found at https://face2021.livdet.org/.

Finally, we include the description of two of the most challenging DeepFake databases up to date [93, 94], i.e., Celeb-DF v2 and DFDC. The videos in these databases contain DeepFakes with a large range of variations in face sizes, illumination, environments, pose variations, etc. DeepFake videos can be used, for example, to create a video with the face of a user and use it for a replay attack against a FRS.

- Celeb-DF v2 [27] is a database that consists of 590 legitimate videos extracted from YouTube, corresponding to celebrities of different gender, age, and ethnic group. Regarding fake videos, a total of 5,639 videos were created swapping faces using DeepFake technology. The average length of the face videos is around 13 seconds (at 30 fps).
- DFDC Preview Database [95] is one of the latest public databases, released by Facebook in collaboration with other institutions like Amazon, Microsoft, and the MIT. The DFDC Preview dataset consists of 1,131 legitimate videos from 66 different actors, ensuring realistic variability in gender, skin tone, and age. A total of 4,119 videos were created using two different DeepFake generation methods by swapping subjects with similar appearances.

In Table 9.2, we show a comparison of the most relevant features of all the databases described in this section.

9.5 Integration with Face Recognition Systems

In order to create a face recognition system resistant to presentation attacks, the proper PAD techniques have to be selected. After that, the integration of the PAD countermeasures with the FRS can be done at different levels, namely, score-level or decision-level fusion [96, 97].

The first possibility consists in using score level fusion as shown in Fig. 9.8. This is a popular approach due to its simplicity and the good results given in fusion of multimodal biometric systems [98–100]. In this case, the biometric data enter at the same time to both the face recognition system and the PAD system, and each one computes its own scores. Then the scores from each system are combined into a new final score that is used to determine if the sample comes from a genuine user or not. The main advantage of this approach is its speed, as both modules, i.e., the PAD and face recognition modules, perform their operations at the same time. This fact can be exploited in systems with good parallel computation specifications, such as those with multicore/multithread processors.

Another common way to combine PAD and face recognition systems is a serial scheme, as in Fig. 9.9, in which the PAD system makes its decision first, and only if the samples are determined to come from a living person, then they are processed by the face recognition system. Thanks to this decision-level fusion, the FRS will search for the identity that corresponds to the biometric sample knowing previously that the sample does not come from a presentation attack. Different from the parallel

Table 9.2 Features of the main public databases for research in face PAD. Comparison of the most relevant features of each of the databases described in this chapter

Database	Users # (real/fakes)	Samples # (real/fakes)	Attack types	Support	Attack illumination
NUAA PI [25]	15/15	5,105/7,509	Photo	Held	Uncont.
REPLAY-ATTACK [26, 44, 46]	50/50	200/1,000	Photo and Replay	Held and Fixed	Cont. and Uncont.
3DMAD [28]	17/17	170/85	Mask	Held	Cont.
OULU-NPU [79]	55/55	1,980/3,960	Photo and Replay	Mobile	Uncont.
CSMAD [56]	14/6	87/159	Photo and Replay	Held and Fixed	Cont.
SiW [91]	165/165	1,320/3,158	Photo and Replay	Held	Uncont.
WMCA [35]	72/72	347/1,332	Photo, Replay, and Mask	Held	Uncont.
CelebA-Spoof [27]	10,177/10,177	202,599/422,938	Photo, Replay, and Mask	Held	Uncont.
HQ-WMCA [92]	51/51	555/2,349	Photo, Replay, Mask, Makeup, others.	Held	Uncont.
LiveDet [36]	48/48	125/689	Photo, Replay, and Mask	Held	Uncont.
CelebDF-v2 [27]	59/59	590/5,639	DeepFakes	–	Uncont.
DFDC Preview [95]	66/66	1,131/4,119	DeepFakes	–	Uncont.

Containing also PHOTO-ATTACK DB and PRINT-ATTACK DB

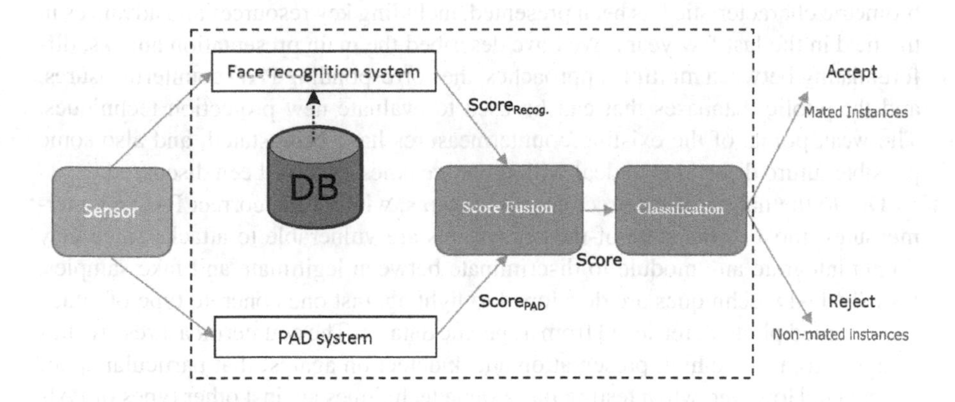

Fig. 9.8 Scheme of a parallel score-level fusion between a PAD and a face recognition system. In this type of scheme, the input biometric data is sent at the same time to both the face recognition system and the PAD system, and each one generates a independent score, then the two scores are fused to take one unique decision

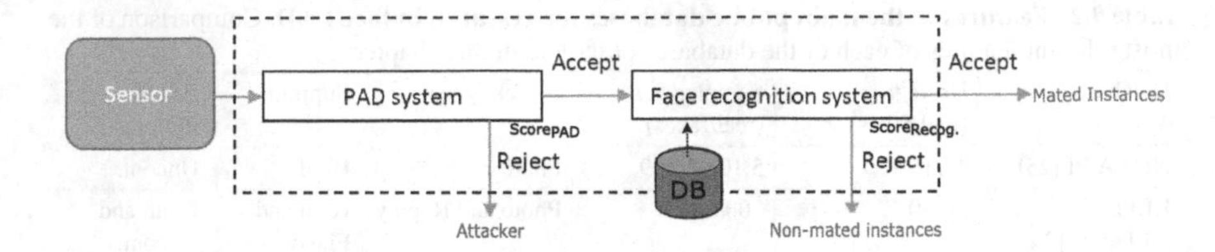

Fig. 9.9 Scheme of a serial fusion between a PAD and a face recognition system. In this type of scheme the PAD system makes its decision first, and only if the samples are determined to come from a living person, then they are processed by the face recognition system

approach, in the serial scheme the average time for an access attempt will be longer due to the consecutive delays of the PAD and the face recognition modules. However, this approach avoids extra work to the face recognition system in the case of a PAD attack, since it should be detected in an early stage.

9.6 Conclusion and Look Ahead on Face PAD

Face recognition systems are increasingly being deployed in a diversity of scenarios and applications. Due to this widespread use, they have to withstand a high variety of attacks. Among all these threats, one with high impact are presentation attacks.

In this chapter, an introduction of the strengths and vulnerabilities of face as a biometric characteristic has been presented, including key resources and advances in the field in the last few years. We have described the main presentation attacks, differentiating between multiple approaches, the corresponding PAD countermeasures, and the public databases that can be used to evaluate new protection techniques. The weak points of the existing countermeasures have been stated, and also some possible future directions to deal with those weaknesses have been discussed.

Due to the nature of face recognition systems, without the correct PAD countermeasures, most of the state-of-the-art systems are vulnerable to attacks since they do not integrate any module to discriminate between legitimate and fake samples. Usually, PAD techniques are developed to fight against one concrete type of attack (e.g., printed photos), retrieved from a specific dataset. The countermeasures are thus designed to achieve high presentation attack detection against that particular spoof technique. However, when testing these same techniques against other types of PAIs (e.g., video-replay), usually the system is unable to efficiently detect them. There is one important lesson to be learned from this fact: there is not a superior PAD technique that outperforms all the others in all conditions; so knowing which technique to use against each type of attack is a key element. It would be interesting to use different countermeasures that have proved to be effective against particular types of PAIs, in order to develop fusion schemes that combine their results, achieving that way a high performance against a variety of presentation attacks data [16, 98, 99]. This problem becomes more relevant in the case of DeepFakes, a term that includes those methods capable of generating images and videos with very realistic face spoofs using deep learning methods and few input data. When dealing with DeepFakes, cross-data generalization is one of the main open problems at present. The majority of the most accurate DeepFake detection solutions nowadays are highly overfitted to the techniques present in their training databases, therefore, their detection accuracy is usually poor when facing fakes created using other techniques unseen during training.

In addition, as technology progresses constantly, new hardware devices and software techniques continue to appear. From the detection point of view, it is also important to keep track of this quick technological progress in order to use it to develop more efficient presentation attack detection techniques. For example, using the power of deep learning and some of its associated techniques like transfer-learning has shown to improve the accuracy of PAD methods in the recent years [35]. Additionally, focusing the research on the biological nature of biometric characteristics (e.g., thermogram, blood flow, etc.) should be considered [64, 78], as the standard techniques based on texture and movement seem to be inefficient against some PAIs.

Additionally, it is of the utmost importance to collect new databases with new scenarios in order to develop more effective PAD methods. Otherwise, it will be difficult to grant an acceptable level of security of face recognition systems. However, it is especially challenging to recreate realistic attacking conditions in a laboratory evaluation. Under controlled conditions, systems are tested against a restricted number of typical PAIs. These restrictions make it unfeasible to collect a database with all the different fake spoofs that may be found in the real world.

To conclude this introductory chapter, it could be said that even though a great amount of work has been done to fight face presentation attacks, there are still big challenges to be addressed in this topic, due to the evolving nature of the attacks, and the critical applications in which these systems are deployed in the real world.

Acknowledgements This work was mostly done (2nd Edition of the book) in the context of the TABULA RASA and BEAT projects funded under the 7th Framework Programme of EU. The 3rd Edition update has been made in the context of EU H2020 projects PRIMA and TRESPASS-ETN. This work was also partially supported by the Spanish project BIBECA (RTI2018-101248-B-I00 MINECO/FEDER).

References

1. Galbally J, Ferrara P, Haraksim R, Psyllos A, Beslay L (2019) Study on face identification technology for its implementation in the Schengen information system. Joint Research Centre, Ispra, Italy, Rep. JRC-34751
2. Tome P, Fierrez J, Vera-Rodriguez R, Nixon MS (2014) Soft biometrics and their application in person recognition at a distance. IEEE Trans Inf Forensics Secur 9(3):464–475

3. Alonso-Fernandez F, Farrugia RA, Fierrez J, Bigun J (2019) Super-resolution for selfie biometrics: introduction and application to face and iris. In: Selfie biometrics. Springer, pp 105–128
4. Tome P, Vera-Rodriguez R, Fierrez J, Ortega-Garcia J (2015) Facial soft biometric features for forensic face recognition. Forensic Sci Int 257:271–284
5. Hernandez-Ortega J, Daza R, Morales A, Fierrez J, Ortega-Garcia J (2020) edBB: biometrics and behavior for assessing remote education. In: AAAI workshop on artificial intelligence for education (AI4EDU)
6. Turk MA, Pentland AP (1991) Face recognition using eigenfaces. In: Computer society conference on computer vision and pattern recognition (CVPR), pp 586–591
7. International biometric group and others: biometrics market and industry report 2009-2014 (2007)
8. Richardson R (2021). https://www.gmfus.org/news/facial-recognition-public-sector-policy-landscape
9. Gipp B, Beel J, Rössling I (2007) ePassport: the world's new electronic passport. A Report about the ePassport's Benefits, Risks and it's Security. CreateSpace
10. Garcia C (2004) Utilización de la firma electrónica en la Administración española iv: Identidad y firma digital. El DNI electrónico, Administración electrónica y procedimiento administrativo
11. Jain AK, Li SZ (2011) Handbook of face recognition. Springer
12. Tistarelli M, Champod C (2017) Handbook of biometrics for forensic science. Springer
13. Zhao J, Cheng Y, Xu Y, Xiong L, Li J, Zhao F, Jayashree K, Pranata S, Shen S, Xing J et al (2018) Towards pose invariant face recognition in the wild. In: IEEE conference on computer vision and pattern recognition (CVPR), pp 2207–2216
14. Li P, Prieto L, Mery D, Flynn PJ (2019) On low-resolution face recognition in the wild: comparisons and new techniques. IEEE Trans Inf Forensics Secur 14(8):2000–2012
15. Gonzalez-Sosa E, Fierrez J, Vera-Rodriguez R, Alonso-Fernandez F (2018) Facial soft biometrics for recognition in the wild: recent works, annotation and COTS evaluation. IEEE Trans Inf Forensics Secur 13(8):2001–2014
16. Hadid A, Evans N, Marcel S, Fierrez J (2015) Biometrics systems under spoofing attack: an evaluation methodology and lessons learned. IEEE Signal Process Mag 32(5):20–30
17. Li L, Correia PL, Hadid A (2018) Face recognition under spoofing attacks: countermeasures and research directions. IET Biom 7(1):3–14
18. Galbally J, Fierrez J, Ortega-Garcia J (2007) Vulnerabilities in biometric systems: attacks and recent advances in liveness detection. In: Proceedings of Spanish workshop on biometrics, (SWB)
19. Galbally J, Marcel S, Fierrez J (2014) Biometric antispoofing methods: a survey in face recognition. IEEE Access 2:1530–1552
20. Gomez-Barrero M, Galbally J, Fierrez J, Ortega-Garcia J (2013) Multimodal biometric fusion: a study on vulnerabilities to indirect attacks. In: Iberoamerican congress on pattern recognition. Springer, pp 358–365
21. Martinez-Diaz M, Fierrez J, Galbally J, Ortega-Garcia J (2011) An evaluation of indirect attacks and countermeasures in fingerprint verification systems. Pattern Recognit Lett 32:1643–1651
22. Pena A, Serna I, Morales A, Fierrez J, Lapedriza A (2021) Facial expressions as a vulnerability in face recognition. In: IEEE international conference on image processing (ICIP), pp 2988–2992
23. Newman LH (2016). https://www.wired.com/2016/08/hackers-trick-facial-recognition-logins-photos-facebook-thanks-zuck/
24. Goodin D (2008) Get your german interior minister's fingerprint here. The Register 30
25. Tan X, Li Y, Liu J, Jiang L (2010) Face liveness detection from a single image with sparse low rank bilinear discriminative model. Computer Vision–ECCV, pp 504–517
26. Chingovska I, Anjos A, Marcel S (2012) On the effectiveness of local binary patterns in face anti-spoofing. In: IEEE BIOSIG

27. Li Y, Yang X, Sun P, Qi H, Lyu S (2020) Celeb-DF: a large-scale challenging dataset for DeepFake forensics. In: IEEE/CVF conference on computer vision and pattern recognition (CVPR)
28. Erdogmus N, Marcel S (2014) Spoofing face recognition with 3D masks. IEEE Trans Inf Forensics Secur 9(7):1084–1097
29. Proceedings IEEE International Conference on Acoustics Speech Signal Process (ICASSP) (2017)
30. Proceedings of IEEE/IAPR international joint conference on biometrics (IJCB) (2017)
31. Chakka MM, Anjos A, Marcel S, Tronci R, Muntoni D, Fadda G, Pili M, Sirena N, Murgia G, Ristori M, Roli F, Yan J, Yi D, Lei Z, Zhang Z, Li SZ, Schwartz WR, Rocha A, Pedrini H, Lorenzo-Navarro J, Castrillón-Santana M, Määttä J, Hadid A, Pietikäinen M (2011) Competition on counter measures to 2-D facial spoofing attacks. In: IEEE international joint conference on biometrics (IJCB)
32. Chingovska I, Yang J, Lei Z, Yi D, Li SZ, Kahm O, Glaser C, Damer N, Kuijper A, Nouak A et al (2013) The 2nd competition on counter measures to 2D face spoofing attacks. In: International conference on biometrics (ICB)
33. Boulkenafet Z, Komulainen J, Akhtar Z, Benlamoudi A, Samai D, Bekhouche S, Ouafi A, Dornaika F, Taleb-Ahmed A, Qin L et al (2017) A competition on generalized software-based face presentation attack detection in mobile scenarios. In: International joint conference on biometrics (IJCB), pp 688–696
34. Zhang Z, Yan J, Liu S, Lei Z, Yi D, Li SZ (2012) A face antispoofing database with diverse attacks. In: International conference on biometrics (ICB), pp 26–31
35. George A, Mostaani Z, Geissenbuhler D, Nikisins O, Anjos A, Marcel S (2019) Biometric face presentation attack detection with multi-channel convolutional neural network. IEEE Trans Inf Forensics Secur 15:42–55
36. Purnapatra S, Smalt N, Bahmani K, Das P, Yambay D, Mohammadi A, George A, Bourlai T, Marcel S, Schuckers S et al (2021) Face liveness detection competition (LivDet-Face)-2021. In: IEEE international joint conference on biometrics (IJCB)
37. Hernandez-Ortega J, Fierrez J, Morales A, Tome P (2018) Time analysis of pulse-based face anti-spoofing in visible and NIR. In: IEEE CVPR computer society workshop on biometrics
38. ISO: information technology security techniques security evaluation of biometrics, ISO/IEC Standard ISO/IEC 19792:2009, 2009. International Organization for Standardization (2009). https://www.iso.org/standard/51521.html
39. ISO: Information technology – Biometric presentation attack detection – Part 1: Framework. International Organization for Standardization (2016). https://www.iso.org/standard/53227.html
40. Kim J, Choi H, Lee W (2011) Spoof detection method for touchless fingerprint acquisition apparatus. Korea Pat 1(054):314
41. Raguin DH (2020) System for presentation attack detection in an iris or face scanner. US Patent 10,817,722
42. Dantcheva A, Chen C, Ross A (2012) Can facial cosmetics affect the matching accuracy of face recognition systems? In: IEEE international conference on biometrics: theory, applications and systems (BTAS). IEEE, pp 391–398
43. Chen C, Dantcheva A, Swearingen T, Ross A (2017) Spoofing faces using makeup: an investigative study. In: IEEE international conference on identity, security and behavior analysis (ISBA)
44. Anjos A, Chakka MM, Marcel S (2013) Motion-based counter-measures to photo attacks in face recognition. IET Biom 3(3):147–158
45. Peng F, Qin L, Long M (2018) Face presentation attack detection using guided scale texture. Multimed Tools Appl 77(7):8883–8909
46. Anjos A, Marcel S (2011) Counter-measures to photo attacks in face recognition: a public database and a baseline. In: International joint conference on biometrics (IJCB)
47. Shen M, Wei Y, Liao Z, Zhu L (2021) IriTrack: face presentation attack detection using iris tracking. In: ACM on Interactive, Mobile, Wearable and Ubiquitous Technologies 5(2)

48. Nguyen D, Bui Q (2009) Your face is NOT your password. BlackHat DC
49. Scherhag U, Raghavendra R, Raja KB, Gomez-Barrero M, Rathgeb C, Busch C (2017) On the vulnerability of face recognition systems towards morphed face attacks. In: IEEE international workshop on biometrics and forensics (IWBF)
50. Ramachandra R, Venkatesh S, Raja KB, Bhattacharjee S, Wasnik P, Marcel S, Busch C (2019) Custom silicone face masks: Vulnerability of commercial face recognition systems & presentation attack detection. In: IEEE international workshop on biometrics and forensics (IWBF)
51. Bharadwaj S, Dhamecha TI, Vatsa M, Singh R (2013) Computationally efficient face spoofing detection with motion magnification. In: Proceedings of the ieee conference on computer vision and pattern recognition workshops, pp 105–110
52. da Silva Pinto A, Pedrini H, Schwartz W, Rocha A (2012) Video-based face spoofing detection through visual rhythm analysis. In: SIBGRAPI conference on graphics, patterns and images, pp 221–228
53. Pinto A, Goldenstein S, Ferreira A, Carvalho T, Pedrini H, Rocha A (2020) Leveraging shape, reflectance and albedo from shading for face presentation attack detection. IEEE Trans Inf Forensics Secur 15:3347–3358
54. Kim Y, Yoo JH, Choi K (2011) A motion and similarity-based fake detection method for biometric face recognition systems. IEEE Trans Consum Electron 57(2):756–762
55. Liu S, Yang B, Yuen PC, Zhao G (2016) A 3D mask face anti-spoofing database with real world variations. In: Proceedings of the ieee conference on computer vision and pattern recognition workshops, pp 100–106
56. Bhattacharjee S, Mohammadi A, Marcel S (2018) Spoofing deep face recognition with custom silicone masks. In: IEEE international conference on biometrics theory, applications and systems (BTAS)
57. Galbally J, Satta R (2016) Three-dimensional and two-and-a-half-dimensional face recognition spoofing using three-dimensional printed models. IET Biom 5(2):83–91
58. Intel (2021). https://www.intelrealsense.com
59. Kose N, Dugelay JL (2013) On the vulnerability of face recognition systems to spoofing mask attacks. In: (ICASSP)international conference on acoustics, speech and signal processing. IEEE, pp 2357–2361
60. Liu S, Lan X, Yuen P (2020) Temporal similarity analysis of remote photoplethysmography for fast 3d mask face presentation attack detection. In: IEEE/CVF winter conference on applications of computer vision, pp 2608–2616
61. Goodfellow I, Pouget-Abadie J, Mirza M, Xu B, Warde-Farley D, Ozair S, Courville A, Bengio Y (2014) Generative adversarial nets. Adv Neural Inf Process Syst 27
62. Dolhansky B, Bitton J, Pflaum B, Lu J, Howes R, Wang M, Canton Ferrer C (2020) The deepfake detection challenge dataset. arXiv:2006.07397
63. Tolosana R, Vera-Rodriguez R, Fierrez J, Morales A, Ortega-Garcia J (2020) Deepfakes and beyond: a survey of face manipulation and fake detection. Inf Fusion 64:131–148
64. Hernandez-Ortega J, Tolosana R, Fierrez J, Morales A (2021) DeepFakesON-Phys: DeepFakes detection based on heart rate estimation. In: AAAI conference on artificial intelligence workshops
65. Singh R, Vatsa M, Bhatt HS, Bharadwaj S, Noore A, Nooreyezdan SS (2010) Plastic surgery: a new dimension to face recognition. IEEE Trans Inf Forensics Secur 5(3):441–448
66. Aggarwal G, Biswas S, Flynn PJ, Bowyer KW (2012)(2012) A sparse representation approach to face matching across plastic surgery. In: IEEE workshop on the applications of computer vision (WACV), pp 113–119
67. Bhattacharjee S, Mohammadi A, Marcel S (2018) Spoofing deep face recognition with custom silicone masks. In: International conference on biometrics theory, applications and systems (BTAS). IEEE
68. Raghavendra R, Raja KB, Busch C (2015) Presentation attack detection for face recognition using light field camera. IEEE Trans Image Process 24(3):1060–1075

69. Yi D, Lei Z, Zhang Z, Li SZ (2014) Face anti-spoofing: multi-spectral approach. In: Handbook of biometric anti-spoofing. Springer, pp 83–102
70. Lagorio A, Tistarelli M, Cadoni M, Fookes C, Sridharan S (2013) Liveness detection based on 3D face shape analysis. In: International workshop on biometrics and forensics (IWBF). IEEE
71. Sun L, Huang W, Wu M (2011) TIR/VIS correlation for liveness detection in face recognition. In: International conference on computer analysis of images and patterns. Springer, pp 114–121
72. Sluganovic I, Roeschlin M, Rasmussen KB, Martinovic I (2016) Using reflexive eye movements for fast challenge-response authentication. In: ACM SIGSAC conference on computer and communications security, pp 1056–1067
73. Chou CL (2021) Presentation attack detection based on score level fusion and challenge-response technique. J Supercomput 77(5):4681–4697
74. Kim Y, Na J, Yoon S, Yi J (2009) Masked fake face detection using radiance measurements. JOSA A 26(4):760–766
75. Yang J, Lei Z, Liao S, Li SZ (2013) Face liveness detection with component dependent descriptor. In: International conference on biometrics (ICB)
76. Galbally J, Marcel S, Fierrez J (2014) Image quality assessment for fake biometric detection: application to iris, fingerprint, and face recognition. IEEE Trans Image Process 23(2):710–724
77. Smith DF, Wiliem A, Lovell BC (2015) Face recognition on consumer devices: reflections on replay attacks. IEEE Trans Inf Forensics Secur 10(4):736–745
78. Li X, Komulainen J, Zhao G, Yuen PC, Pietikäinen M (2016) Generalized face anti-spoofing by detecting pulse from face videos. In: IEEE international conference on pattern recognition (ICPR), pp 4244–4249
79. Boulkenafet Z, Komulainen J, Li L, Feng X, Hadid A (2017) OULU-NPU: a mobile face presentation attack database with real-world variations. In: IEEE international conference on automatic face gesture recognition, pp 612–618
80. George A, Marcel S (2020) Learning one class representations for face presentation attack detection using multi-channel convolutional neural networks. IEEE Trans Inf Forensics Secur 16:361–375
81. Yu Z, Li X, Wang P, Zhao G (2021) TransRPPG: remote photoplethysmography transformer for 3d mask face presentation attack detection. IEEE Signal Process Lett
82. Zhang D, Ding D, Li J, Liu Q (2015) PCA based extracting feature using fast fourier transform for facial expression recognition. In: Transactions on engineering technologies, pp 413–424
83. Gonzalez-Sosa E, Vera-Rodriguez R, Fierrez J, Patel V (2017) Exploring body shape from mmW images for person recognition. IEEE Trans Inf Forensics Secur 12(9):2078–2089
84. Hernandez-Ortega J, Fierrez J, Morales A, Diaz D (2020) A comparative evaluation of heart rate estimation methods using face videos. In: IEEE conference on computers, software, and applications (COMPSAC)
85. Daza R, Morales A, Fierrez J, Tolosana R (2020) mEBAL: a multimodal database for eye blink detection and attention level estimation. In: ACM international conference on multimodal interaction (ICMI)
86. Pan G, Wu Z, Sun L (2008) Liveness detection for face recognition. In: Recent advances in face recognition. InTech
87. Wu HY, Rubinstein M, Shih E, Guttag J, Durand F, Freeman W (2012) Eulerian video magnification for revealing subtle changes in the world. ACM Trans Graph 31(4)
88. Liu A, Wan J, Escalera S, Jair Escalante H, Tan Z, Yuan Q, Wang K, Lin C, Guo G, Guyon I et al (2019) Multi-modal face anti-spoofing attack detection challenge at CVPR2019. In: IEEE/CVF conference on computer vision and pattern recognition workshops (CVPRw)
89. Zhang S, Wang X, Liu A, Zhao C, Wan J, Escalera S, Shi H, Wang Z, Li SZ (2019) A dataset and benchmark for large-scale multi-modal face anti-spoofing. In: IEEE/CVF conference on computer vision and pattern recognition (CVPR), pp 919–928
90. Ortega-Garcia J, Fierrez J, Alonso-Fernandez F, Galbally J, Freire MR, Gonzalez-Rodriguez J, Garcia-Mateo C, Alba-Castro JL, Gonzalez-Agulla E, Otero-Muras E et al (2010) The

multiscenario multienvironment biosecure multimodal database (BMDB). IEEE Trans Pattern Anal Mach Intell 32(6):1097–1111
91. Liu Y, Jourabloo A, Liu X (2018) Learning deep models for face anti-spoofing: Binary or auxiliary supervision. In: IEEE/CVF conference on computer vision and pattern recognition (CVPR), pp 389–398
92. Heusch G, George A, Geissbühler D, Mostaani Z, Marcel S (2020) Deep models and shortwave infrared information to detect face presentation attacks. IEEE Trans Biom Behav Identity Sci 2(4):399–409
93. Neves JC, Tolosana R, Vera-Rodriguez R, Lopes V, Proenca H, Fierrez J (2020) GANprintR: improved fakes and evaluation of the state of the art in face manipulation detection. IEEE J Sel Top Signal Process 14(5):1038–1048
94. Tolosana R, Romero-Tapiador S, Fierrez J, Vera-Rodriguez R (2021) DeepFakes evolution: analysis of facial regions and fake detection performance. In: IAPR international conference on pattern recognition workshops (ICPRw)
95. Dolhansky B, Howes R, Pflaum B, Baram N, Ferrer CC (2019) The DeepFake detection challenge (DFDC) preview dataset. arXiv:1910.08854
96. Chingovska I, Anjos A, Marcel S (2013) Anti-spoofing in action: joint operation with a verification system. In: Proceedings of the ieee conference on computer vision and pattern recognition workshops, pp 98–104
97. Fierrez J (2006) Adapted fusion schemes for multimodal biometric authentication. PhD Thesis, Universidad Politecnica de Madrid
98. de Freitas Pereira T, Anjos A, De Martino JM, Marcel S (2013) Can face anti-spoofing countermeasures work in a real world scenario? In: International conference on biometrics (ICB)
99. Fierrez J, Morales A, Vera-Rodriguez R, Camacho D (2018) Multiple classifiers in biometrics. Part 1: fundamentals and review. Inf Fusion 44:57–64
100. Ross AA, Nandakumar K, Jain AK (2006) Handbook of multibiometrics. Springer Science & Business Media

Chapter 10
Recent Progress on Face Presentation Attack Detection of 3D Mask Attack

Si-Qi Liu and Pong C. Yuen

Abstract With the development of 3D reconstruction and 3D printing technologies, customizing super-real 3D facial mask at an affordable cost becomes feasible, which brings a big challenge to face presentation attack detection (PAD). With the boosting of face recognition application in wide scenarios, there is an urgent need to solve the 3D facial mask attack problem. Since this appearance and material of 3D masks could vary in a much larger range compared with traditional 2D attacks, 2D face PAD solutions introduced in previous chapters may not work any longer. In order to attract more attention to 3D mask face PAD, this book chapter summarizes the progress in the past few years, as well as publicly available datasets. Finally, open problems and possible future directions are discussed.

10.1 Background and Motivations

Face presentation attack, where a fake face of an authorized user is present to cheat or obfuscate the face recognition system, becomes one of the greatest challenges in practice. With the increase of the face recognition applications in wide scenarios, this security concern has been attracting increasing attentions [1, 2]. The conventional face image or video attacks can be easily conducted through prints or screens of mobile devices. Face images or videos can also be easily collected from Internet with the popularity of social networks. Numerous efforts have been devoted to conventional face Presentation Attack Detection (PAD) [2–19] and encouraging progress has been achieved. Recent research indicates that 3D mask face presentation attack (PA) can be conducted at an affordable price with the off-the-shelf 3D printing and open-source 3D reconstruction technologies. Thatsmyface 3D mask shown in Fig. 10.1a can be made with a frontal face image with a few attributed key points. Making a super-real 3D facial mask such as the REAL-f shown in Fig. 10.1b is also

S.-Q. Liu · P. C. Yuen (✉)
Department of Computer Science, Hong Kong Baptist University, Kowloon, Hong Kong
e-mail: pcyuen@comp.hkbu.edu.hk

S.-Q. Liu
e-mail: siqiliu@comp.hkbu.edu.hk

S. Marcel et al. (eds.), *Handbook of Biometric Anti-Spoofing*, Advances in Computer Vision and Pattern Recognition, https://doi.org/10.1007/978-981-19-5288-3_10

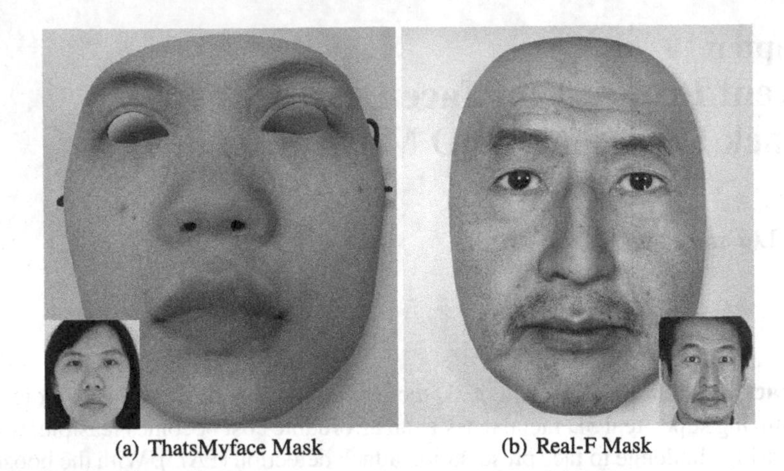

Fig. 10.1 High-resolution sample images of Thatsmyface mask and REAL-f mask (The figure of the genuine subjects of REAL-f mask is from http://real-f.jp/)

feasible although it requires more detailed facial texture and 3D facial structure. Compared with 2D prints or screen display attacks, 3D masks not only can keep the facial appearance with detailed skin texture but also preserve the vivid 3D human facial structures. In addition, decorative masks and transparent masks [19] can also be used to obfuscate the system at very low costs. The traditional 2D face PAD approaches, especially for those data-driven methods [17–21], may not work in this case.

Existing 3D mask PAD can be mainly categorized into appearance-based approach, motion-based approach, and remote-Photoplethysmography-based approach. Similar to conventional image and video attacks, 3D masks may contain some defects due to the low 3D printing quality. Therefore, using appearance cues such as the texture and color becomes a possible solution and impressive results have been achieved [22, 23]. However, due to the data-drive nature of these methods, they may not be able to adapt to the large appearance variations of different types of masks. Applying domain adaptation technique shows promising improvements on traditional 2D PAD [20], while very few works address their effectiveness on 3D mask attack [23]. The motion-based approaches could be effective in detecting rigid 3D masks which block subtle facial movements of genuine faces [24]. However, the flexible silicone mask [25] and transparent mask [19] are able to preserve such subtle movement of the facial skin, which make the motion-based approaches less reliable.

As such, an intrinsic liveness cue that is independent of the appearance and motion pattern variations of different masks is necessary. Recent studies found that the heartbeat signal can be extracted through normal RGB camera based on the color variation of the facial skin. When applying it to the 3D mask PAD, since 3D mask can block the light transmission, a periodic heartbeat signal can only be detected on a genuine face but not on a masked face [26]. This remote photoplethysmography (rPPG)-based approach exhibits outstanding generalizability to different types of masks. Also, due to the fragileness of facial rPPG signals, efforts keep being put on improving the robustness under different recording environments.

Organization of this chapter is summarized as follows. Background and motivations of 3D mask PAD are given above. Next, we introduce the publicly available 3D mask datasets and evaluation protocols in Sect. 10.2. Then we discuss existing 3D mask PAD methods in Sect. 10.3 based on the liveness cues they rely on. In Sect. 10.4, typical methods of different liveness cues are evaluated. Finally, open challenges of 3D mask face PAD are discussed in Sect. 10.5.

The previous version of this chapter [27] has been published in second edition of the handbook [28]. In this version, we summarize the newly created datasets and 3D mask PAD methods with the updated experiments and discussion.

10.2 Publicly Available Datasets and Experiments Evaluation Protocol

10.2.1 Datasets

Existing 3D mask attack datasets have covered different types of masks under various lighting conditions with different camera settings. Early datasets such as **3D M**ask **A**ttack **D**ataset (3DMAD) [29] focus on off-the-shelf customized 3D masks, i.e., the paper cut masks and 3D printed masks from Thatsmyface (shown in Fig. 10.2). Although these masks are in relatively low quality and can be detected easily through appearance defects, it has been proved that they can successfully spoof many exiting popular face recognition systems [22]. The Inter Session Variability (ISV) modeling method [30] achicves around a Spoofing False Acceptance Rate (SFAR) of around 30%. After that, the high-quality real-F mask is introduced in **H**ong **K**ong **B**aptist **U**niversity 3D **M**ask **A**ttack with **R**eal-World Variations (HKBU-MARs) dataset [26], where the appearance difference can hardly be identified. As shown in Fig. 10.3, REAL-f using the 3D scan and "Three-Dimension Photo Form (3DPF)" technique to model the 3D structure and print the facial texture can achieve higher appearance quality and 3D modeling accuracy than the Thatsmyface mask [26]. By observing the detailed textures such as the hair, wrinkle, or even eyes' vessels, without the context information, even a human can hardly identify whether it is a genuine face or not. As such, there is an urgent need to address the 3Dmask face PAD. High-quality plaster resin masks are also adopted in CASIA-SURF 3Dmask [18] and **Hi**gh-**Fi**delity (HiFi) Mask dataset [23]. To mimic the facial motion of genuine face, flexible silicone gel mask is employed in **S**ilicone **M**ask **A**ttack **D**ataset (SMAD) [31], ERPA dataset [32], **C**ustom **S**ilicone **M**ask **A**ttack **D**ataset (CSMAD) [25], **W**ide **M**ulti-**C**hannel presentation **A**ttack (WMCA) dataset [19], and **M**ulti-spectral **L**atex Mask-based Video **F**ace **P**resentation Attack (MLFP) dataset [33]. These customized resin or silicone masks are at high costs (2000–4000 USD per mask) so the number

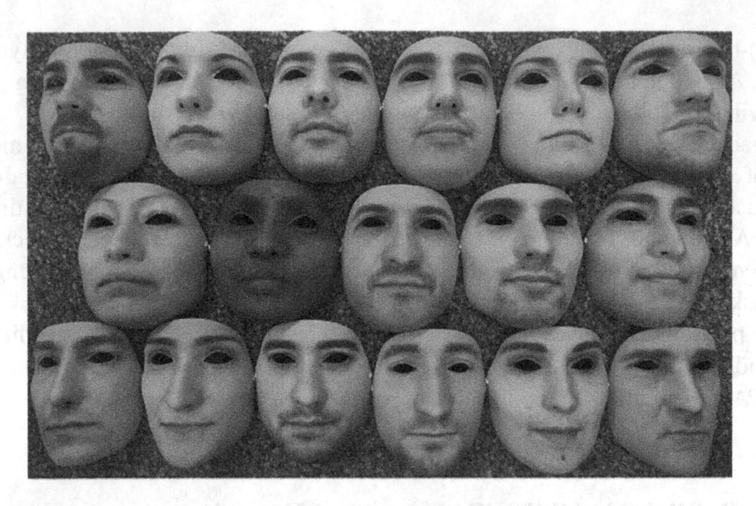

Fig. 10.2 The 17 customized Thatsmyface masks used in 3DMAD dataset [22]

of mask for each dataset is relatively small. Transparent 3D masks with vivid local specular reflection are employed in recently released datasets such as SiW-M [34] and HiFi Mask datasets [23].

Lighting conditions and camera variation are the other two major aspects that affect detection performance. Early datasets focus on one camera setting under controlled lighting conditions. Since the development of HKBU-MARs V2 [35], various lighting conditions with different camera settings are considered in 3D mask attack datasets. The web camera, mobile device camera, mirrorless camera, and industrial camera are used to record videos under six common lighting conditions (shown in Fig. 10.4.) in HKBU-MARs V2. Infrared camera and thermal camera are evaluated in ERPA dataset [32] where infrared thermal image can effectively differentiate mask from genuine face. WMCA dataset covers day light, office light, and LED lamps light and videos are recorded by mainstream off-the-shelf thermal cameras [19]. The newly released datasets tend to use mobile devices for recording. Popular smartphone cameras from different brands are adopted CASIA-SURF 3Dmask dataset [18] and HiFiMask dataset [23] including iPhone, Samsung, Huawei, Xiaomi, and Vivo. These datasets contain a relatively larger number of subjects and complex light settings including norm light, dark light, backlight, front light, side light, sunlight, and flashlight with a periodic variation. The detailed comparison of existing 3D mask attack datasets are summarized in Table 10.1. Typical datasets are discussed in details in the following subsections.

10.2.1.1 3DMAD

The **3D M**ask **A**ttack **D**ataset (3DMAD) [29] contains a total of 255 videos of 17 subjects with their customized Thatsmyface masks as shown in Fig. 10.2. Eye

Table 10.1 Comparison of existing 3D mask attack datasets

Dataset	Year	Skin tone	#Subject	#Mask	Mask type	Scenes	Light. condition	Camera	#Live videos	#Masked videos	#Total videos
3DMAD [22]	2013	Asian, African	17	17	Paper, ThatsMyFace	Office	Studio light	Kinect	170	85	255
3DFS-DB [36]	2016	Asian	26	26	Plastic	Office	Studio light	Kinect, Carmine-1.09	260	260	520
BRSU [37]	2016	Asian, White, African.	137	6	Silicone, Plastic Resin, Latex	Disguise, Counter-feiting	Studio light	SWIR, Color	0	141	141
HKBU-MARs V1+ [38]	2018	Asian	12	12	ThatsMyFace, REAL-f	Office	Room light	Logitech C920	120	60	180
HKBU-MARs V2 [35]	2016	Asian	12	12	ThatsMyFace, REAL-f	Office	Room light, Low light, Bright light, Warm light, Side light, Up-side light	Logitech C920, Indust.-Cam, EOS-M3, Nexus 5, iPhone 6, Samsung S7, Sony TabletS	504	504	1008
SMAD [31]	2017	–	–	–	Silicone	–	Varying lighting	Varying cam.	65	65	130
MLFP [33]	2017	White, African	10	7	Latex, Paper	Indoor, Outdoor	Day light	Visible-RGB, Near infrared, Thermal	150	1200	1350
ERPA [32]	2017	White, African	5	6	Resin, Silicone	Indoor	Room light	Xenics-Gobi, Thermal-Cam. Intel Realsense SR300	–	–	86
CSMAD [25]	2018	Asian, White.	14	6	Silicone	Indoor	ceiling light, Studio light, Side light	Intel RealSense SR300, Compact-Pro	88	160	248
WMCA [19]	2019	Asian, White, African.	72	*	Plastic, Silicone, Paper	Indoor	Office light, LED lamps, day light	Intel RealSense SR300, Seek-Thermal, Compact-Pro.	347	1332	1679
HQ-WMCA [39]	2020	White-	51	*	Plastic, Silicone, Paper	Indoor	Office light, halogen lamps, LED lamps, day light	Basler acA1921-150uc, acA1920-150um, Xenics Bobcat-640-GigE, Gobi-640-GigE, Intel Realsense D415	555	2349	2904
CASIA-SURF 3DMask [18]	2020	Asian	48	48	Plaster	Indoor, Outdoor	Normal light, Back light, Front light, Side light, Sun light, Shadow	Apple, Huawei, Samsung	288	864	1152
HiFiMask [23]	2021	Asian, White, African.	75	75	Transprnt., Plaster, Resin	White, Green, Tricolor, Sunshine, Shadow, Motion	Normal light, Dim light, Bright light, Back light, Side light, Top light	iPhone11, iPhoneX, MI10, P40, S20, Vivo, HJIM	13650	40950	54600

* Denotes the information is not mentioned in the related paper

regions and nose holes are cut out for a better wearing experience. Thatsmyface mask is proved to be good enough to spoof facial recognition system in [29]. Videos are recorded by Kinect under studio lighting condition and contain color and depth data of size 640×480 at 30 fps. Details of 3DMAD can be found at https://www.idiap.ch/dataset/3dmad.

10.2.1.2 HKBU-MARs

The **H**ong **K**ong **B**aptist **U**niversity 3D **M**ask **A**ttack with **R**eal-World Variations (HKBU-MARs) dataset [35] contains variations of mask type, camera setting, and lighting condition so it can be used to evaluate the generalization ability of face PAD systems across different application scenarios in practice. The HKBU-MARs V2 contains 12 subjects with 12 masks where 6 masks are from Thatsmyface and the other 6 are from REAL-f. Samples of two types of masks are shown in Fig. 10.3. Considering the various application environment of face recognition system, 7 different cameras from the stationary and mobile devices are involved. For stationary applications, webcam (Logitech C920), entry-level mirrorless camera (Canon EOS M3), and industrial camera are used to represent different types of face acquisition systems. Mobile device includes mainstream off-the-shelf smartphones: iPhone 6, Samsung S7, Nexus 5, and Sony Tablet S. HKBU-MARs V2 also considers 6 lighting conditions, i.e., low light, room light, bright light, warm light, side light, and upside light (shown in Fig. 10.4) to cover typical face recognition scenes. Detailed information can be found at http://rds.comp.hkbu.edu.hk/mars/. HKBU-MARs V2 [26] and its preliminary version—HKBU-MARsV1+—have been publicly available at http://rds.comp.hkbu.edu.hk/mars/. The preliminary HKBU-MARsV1+ contains 12 masks

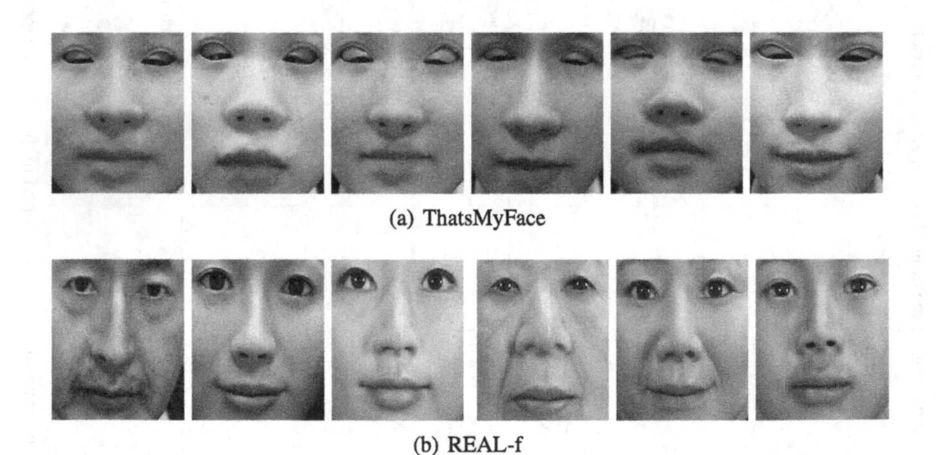

(a) ThatsMyFace

(b) REAL-f

Fig. 10.3 Sample mask images from the HKBU-MARs dataset. **a** ThatsMyFace mask and **b** REAL-f mask

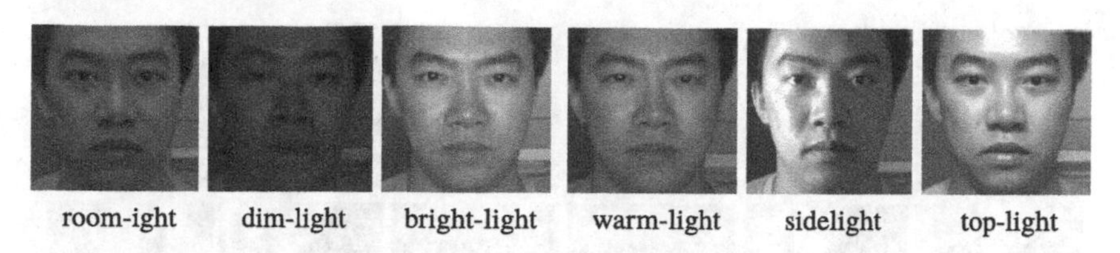

Fig. 10.4 Different lighting conditions in HKBU-MARs dataset

including 6 Thatsmyface masks and 8 REAL-f masks. It is recorded through a Logitech C920 webcam at a resolution of 1280 $\times$ 720 under natural lighting conditions.

10.2.1.3 SMAD

Silicone masks (such as the one used in Hollywood movie *Mission Impossible*) with soft surface can retain both the appearance and facial motions. The **S**ilicone **M**ask **A**ttack **D**atabase (SMAD) [31] collects 65 real and 65 attack videos from online resources under unconstrained scenarios. Genuine videos are from audition, interview, or hosting shows under different illumination, background, and camera settings. Decorations such as hair or mustache are included for silicone mask attack videos. Different from manually created datasets, the time length of videos in SMAD varies from 3 to 15 s. The silicone masks from online sources are not customized according to the real faces. Video URLs of SMAD can be found at http://www.iab-rubric.org/resources.html.

10.2.1.4 MLFP

The appearance of 3D mask can be super real with the advancement of 3D printing techniques. Soft silicone mask is able to preserve facial expression. The appearance-based or motion-based solution built on visible spectrum may not be robust. Multi-spectral can be the future direction with the reduction of costs of related sensors. **M**ulti-spectral **L**atex Mask-based Video **F**ace **P**resentation Attack (MLFP) dataset is built to improve the development of multi-spectral-based face PAD method [33]. There are 10 subjects with 7 latex and 3 paper masks and in total 1350 videos in MLFP. Masks are also not customized according to genuine faces. MLFP contains three different spectrums: visible, near infrared, and thermal. Environmental variations in MLFP cover indoor and outdoor scenes with fixed and random backgrounds. Detailed information can be found at http://iab-rubric.org/resources/mlfp.html.

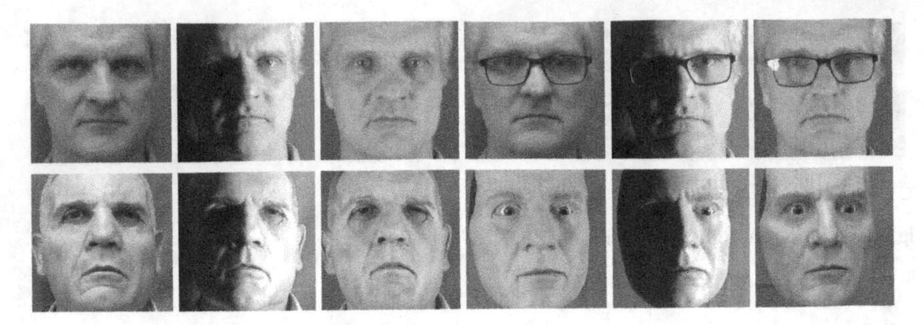

Fig. 10.5 Sample images from CSMAD [25] including room ceiling light, side light, bright studio light; glasses decoration; mask put on real person; and mask put on a fixed stand

10.2.1.5 CSMAD

The **C**ustom **S**ilicone **M**ask **At**tack **D**ataset (CSMAD) [25] contains 14 subjects and 6 customized high-quality silicone masks. To increase the variability of the dataset, mask attacks in CSMAD are presented in different ways: wore by different attackers or put on a stationary shelf. Eyeglass is also employed for some of the genuine faces. Four common lighting conditions are involved, i.e., room ceiling light, studio halogen lamplight, and severe side light from left and right. To ease the preprocessing steps when developing 3D mask PAD methods, all samples are recorded with a green uniform background. For camera device, RealSense SR300 are used to record the RGB, depth, and infrared video data and the Compact Pro is used to obtain the thermal information. Each video is of approximately 10 s with around 300 frames. Sample images of CSMAD are shown in Fig. 10.5. Detailed information can be found at https://www.idiap.ch/en/dataset/csmad.

10.2.1.6 WMCA

The **W**ide **M**ulti-**C**hannel presentation **A**ttack (WMCA) dataset [19] contains 72 subjects with three types of 3D masks including rigid customized realist mask, off-the-shelf decorative plastic mask, transparent plastic mask, customized flexible mask, and customized paper cut mask. To simulate daily life situations, some bona fide subjects are asked to wear glasses with black rim. 1941 short videos (10 s) are recorded by four types of cameras with several channels including color, depth, infrared, and thermal. Intel RealSense SR300 is used to capture the color, depth, and near-infrared information. Seek Thermal Compact PRO and Xenics Gobi-640-GigE are adopted to record thermal information. uEye SE USE2.0 is used to record high-quality infrared video. The first three cameras are synchronized when recording and the last one (uEye SE USE2.0) works after that with the LED NIR light on. All cameras are attached together on an optical mounting post and videos are aligned precisely through a checkerboard with different thermal characteristics. The RGB and NIR videos are

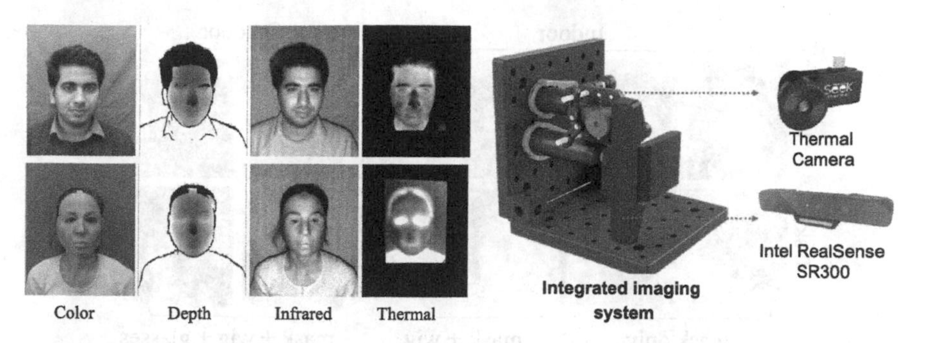

Fig. 10.6 Samples of color, depth, infrared, and thermal image captured by the integrated imaging system in WMCA [19]

Fig. 10.7 Samples of customized realist rigid mask, off-the-shelf decorative plastic mask, transparent plastic mask, customized flexible mask, and customized paper cut mask in WMCA [19]. The right subfigure shows different lighting conditions and background of the six sessions

in full HD and HD mode and the resolution of thermal video is 320×240. WMCA contains 7 sessions with an interval of 5 months with different environmental conditions. Three common lighting conditions are covered, i.e., ceiling office light, side light, and day light. Sample images of different types of masks, lighting conditions, and modalities are shown in Figs. 10.6 and 10.7.

Except for 3D masks, WMCA also consists disguise glasses with fake eyes and paper glasses, fake mannequin heads, wigs of men and women, and makeup. Considering the development of remote photoplethysmography (rPPG)-based study, Blood-Volume Pulse (BVP) signal and respiration signal are also recorded, using the device SA9308M and SA9311M from Thought Technologies, respectively. Detailed information of WMCA can be found at https://www.idiap.ch/en/dataset/wmca. The WMCA has been extended into High-Quality WMCA (HQ-WMCA) with better data quality and additional shortwave infrared (SWIR) sensor. Detailed information of HQ-WMCA can be found at https://www.idiap.ch/en/dataset/hq-wmca.

10.2.1.7 CASIA-SURF 3DMask

The CASIA-SURF 3DMask dataset contains 48 subjects (21 males and 27 females, age 23–62) with their customized high-quality 3D printed plastic masks. To simulate

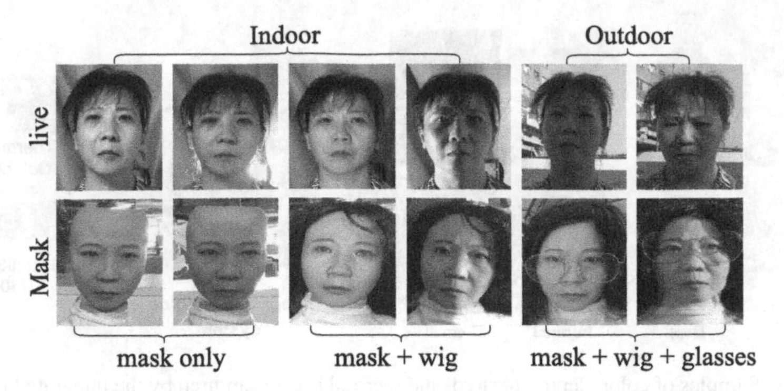

Fig. 10.8 Samples of six lighting conditions from indoor and outdoor scene and mask with decorations in CASIA-SURF 3DMask dataset [18]

daily-life scenarios, videos are recorded by latest smartphones from three popular brands, i.e., Apple, Huawei, and Samsung with frame rate 30 fps and full HD resolution. Six common lighting conditions are covered including normal light, back light, front light, side light, outdoor sunlight, and in shadow. In addition, two realistic decorations, i.e., wig and glasses, are applied when conducting a 3D mask attack. Sample images of different light settings and masks are shown in Fig. 10.8. In sum, CASIA-SURF 3DMask dataset includes 288 genuine face videos from six lighting conditions and 864 masked videos from six lighting conditions and three appearances (with and without decoration).

10.2.1.8 HiFiMask

The **Hi**gh-**Fi**delity Mask dataset (HiFiMask [23]) contains 75 subjects and their customized masks where each one-third of them are Asian, Caucasian, and African. HiFiMask employs three types of masks including high-fidelity plaster masks, resin masks, and transparent masks. Different from previous 3D mask attack datasets, six complex scenes are covered including white light, green light, periodic flashlight, outdoor sunlight, outdoor shadow, and head motion. Specifically, to challenge the rPPG-based PAD approach, the frequency of periodic flashlight is at $[0.7, 4]$ Hz which matches the heartbeat frequency range of a healthy person. Under each scene, six lighting conditions are manipulated using additional light sources including natural light, dim light, bright light, back light, side light, and top light. For recording devices, seven smartphones from mainstream brands (Apple, Xiaomi, Huawei, and Samsung) are involved to ensure the high resolution and good image quality of videos. In the data acquisition process, subjects are uniformly dressed and sit in front of the camera with small head movements. Masks are worn on a real person so it also contains the head motion. To simulate the real application scenarios, random decorations are applied such as wigs, glasses, sunglasses, hats of difference, and wigs. An irrelevant

Fig. 10.9 Samples of HiFiMask dataset [23] with six imaging devices (first row), different decorations on mask attacks (second row), six lighting conditions (third row), and six application scenarios (fourth row)

pedestrian is asked to walk around in the background. The typical variations covered in HiFiMask dataset are shown in Fig. 10.9.

10.2.2 Evaluation Protocols

Evaluation protocols of 3D mask face PAD mainly cover the intra-dataset, cross-dataset, and intra/cross-variation scenarios to analyze the discriminability and generalizability. The intra-dataset evaluation is conducted on one dataset by separating the subjects into non-overlapping sets for training, development (validation), and testing. The cross-variation and cross-dataset evaluations are designed to simulate the scenarios where the variations of training set are limited and different from those of testing set.

10.2.2.1 Intra-Dataset Testing

Intra-dataset evaluation follows the standard rule where a dataset is partitioned into different sets for training and testing. The development set is used to tune the parameters (e.g., threshold of required false accept rate) for the testing set in real application scenarios. Leave-one-out cross-validation (LOOCV) in [22] is one of the typical implementations where in each iteration one subject is selected for testing and the rests are divided for training and development. For 3DMAD with 17 subjects, in each iteration, after selecting one subject's data as the testing set, for the remaining 16 subjects, the first 8 subjects are chosen for training and the remaining 8 subjects are for development. LOOCV is updated by random partition to avoid the effect caused by the order of subjects [26]. Recently, considering the time consumption of deep-learning-based methods, LOOCV with multiple rounds becomes not applicable. Fixed partitions of training and testing set tend to be the dominant selection, such as the Protocol 1 of HiFiMask dataset [23].

10.2.2.2 Cross-Dataset Testing

Cross-dataset evaluation protocol is designed to simulate the practical scenarios where the training data are different from the testing samples. Under cross-dataset evaluation, model is trained on one dataset and test on the others. Due to the limited number of subjects of 3D mask attack datasets, the result based on one partition may be biased. A practical choice is to select part of the data from one dataset for training and summarize the final results by conducting several rounds of experiments. For instance, for cross-dataset evaluation between HKBU-MARsV1 and 3DMAD, five subjects are randomly selected from one dataset as the training set and all subjects of 3DMAD are involved for testing [26]. For deep-learning-based methods with a large parameter scale, the entire source datasets could be used for training. For instance, in [23], the model is trained on the entire HiFiMask dataset and tested on HKBU-MARs and CASIA-SURF 3DMask.

10.2.2.3 Intra-variation Testing

For the datasets that contain multiple types of variations, e.g., mask type, imaging device, and lighting condition, the intra-variation testing protocol is designed to evaluate the robustness of a 3D mask face PAD method when encountering variation(s) that has been seen during training stage. For specific types of variation(s) (fixed mask type, camera, and lighting condition), LOOCV or multi-fold fixed partitions can also be conducted for comparison, such as the grandtest protocol of WMCA dataset [19]. Although the intra-variation testing may not match the practical scenarios, it is useful to evaluate the discriminability of a 3D mask PAD method.

10.2.2.4 Cross-Variation Testing

The cross-variation evaluate protocol aims to test the generalization ability across different types of variation which simulate the common scenarios that the training data is not able to cover all variations in testing. The leave-one-variation-out cross-validation (LOVO) proposed for HKBU-MARs dataset is one of the typical implementations [35]. Different from LOOCV, in each iteration, one condition of a specific type of variation is selected for training and the rest are used for testing. For example, for the LOVO on camera-type variations, the samples captured by one camera are selected for training and samples collected by the rest cameras are selected for testing. When evaluating one type of variation, the other variations such as mask and lighting conditions are fixed. For HKBU-MARs, the LOVO of cameras under two mask types and six lighting conditions involves a total of 2×6 sets of LOOCV evaluations [35]. Similarly, to evaluate the robustness when encountering unseen variation, the unseen attack protocol is also used in WMCA dataset [19] while the partition of training and testing set for each attack is fixed, considering the data requirement of deep-learning-based methods. For models with larger parameter scales, more than one type of variation can be used for training. For instance, protocol 2-"unseen" of HiFiMask dataset using one type of mask for testing and rests for training [23].

10.3 Methods

In this section, we introduce and analyze recently proposed methods that related to 3D mask PAD methods mentioned in previous version of this chapter [27].

10.3.1 Appearance-Based Approach

As the appearance of a face in printed photos or videos is different from the real face, several texture-based methods have been used for face PAD and achieve encouraging results [7, 11, 15, 40]. The 3D mask also contains the quality defect that results in the appearance difference from a genuine face, due to the imperfection precision problems of 3D printing technology. For example, the skin texture and detailed facial structures in masks as shown in Fig. 10.1a have perceivable differences compared to those in real faces.

10.3.1.1 Conventional Approach

The Multi-scale LBP (MS-LBP) [7] is an intuitive solution and has been evaluated in [22]. From a normalized face image, MS-LBP extracts LBP features with different radius (scale) to mine detailed information. As reported in [22], MS-LBP achieves 99.4% Area Under Curve (AUC), 5% Equal Error Rate (EER) on the Morpho dataset.[1] Color Texture Analysis (CTA) method [41] which applies LBP on different color spaces can also achieve similar performance. Although results are promising with the above methods, recent studies indicate their poor generalizability under cross-

[1] http://www.morpho.com.

dataset scenarios [16, 42]. It is reported that the MS-LBP is less effective (22.6% EER and 86.8% AUC) on HKBU-MARsV1 due to the super-real mask—REAL-f [26].

Since the texture differences between 3D masks and real faces mainly located on fine-grained details, analyzing the textures in the frequency domain can be effective. RDWT-Haralick [43] adopts redundant discrete wavelet transform (RDWT) and Haralick descriptors [44] to extract liveness features at different scales. The input image is divided into 3×3 blocks and then the Haralick descriptors are extracted from the four sub-bands of the RDWT results and the original image. For video input, features are extracted from multiple frames and concatenated. Principal component analysis (PCA) is adopted for dimension reduction and the final result is obtained through Support Vector Machine (SVM) with linear kernel. It is reported that the RDWT-Haralick feature can achieve 100% accuracy and 0% HTER on 3DMAD dataset [43].

Instead of mining the texture details, 3D geometric appearance also contains the perceivable differences between genuine faces and 3D masks. Tang et al. investigate dissimilarities of micro-shape of the 3D face mesh (acquired through a 3D scanner) via the principal curvature measures [45]-based 3D geometric attribute [46]. A histogram of principle curvature measures (HOC) is designed to represent 3D face based on a shape-description-oriented 3D facial description scheme. As reported in [46], HOC can reach 92% true acceptance rate (TAR) when FAR equals to 0.01 on Morpho dataset.

10.3.1.2 Deep-Learning-Based Approach

With the popularity of deep learning, it is also intuitive to design deep networks to extract liveness features for face PAD [47, 48]. Different from knowledge-driven solutions, liveness features can be directly learned through standard CNN. In [47], an interactive scheme where the network structure and convolution filter are optimized simultaneously is designed. 100% detection accuracy and 0% HTER on 3DMAD dataset are achieved. Due to the large searching space of the complex design, the network needs to be trained by well-organized large-scale data. In addition, due to the intrinsic data-driven nature [42], the over-fitting problem of the deep-learning-based methods in cross-dataset scenarios remains unsolved.

Newly proposed deep-learning-based methods mainly target on 2D face presentation attacks which may also be applied for 3D mask PAD. To grub more details from a limited number of training data, auxiliary tasks, i.e., depth map regression, rPPG regression [17], are added during learning the feature encoder. A pre-obtained depth map is used to supervise the depth regression of a genuine face. For mask faces, the depth map is set to zeros. Low-level and high-level features from different layers are aggregated and the final decision is made by summarizing the predicted value of all pixels [18]. The depth map can be regarded as a pixel-level supervision that forces the network to learn the fine-grained details from small local patches. Dense supervision is further investigated in [49] and turned out that similar performance can be achieved by simple pixel-wise binary labels. The proposed pixel-wise binary supervision (DeepPixBiS) network in [49] contains a pretrained DenseNet [50] backbone concatenated by a 2D convolutional transition block which normalizes and down samples the multi-dimensional feature maps into binary feature map. Binary Cross-Entropy (BCE) loss is applied for pixel-wise supervision and the final outputs.

Similar to the CTA work, different color spaces are also considered in deep-learning-based methods. Jia et al. construct a two-stream network to extract fine-grained information based on factorized bilinear coding to RGB and YCbCr color spaces [51]. Deep Tree Network (DTN) aims to improve discriminability by encouraging the awareness of different types of spoofing attacks (3D mask types). Specifically, semantic attributes of pre-defined attacks are learned and the seen and unseen attack types are adaptively categorized in an unsupervised manner. The network architecture largely affects the overall performance, Yu et al. propose NAS-FAS to search for an optimized network based on the newly designed central difference convolutional (CDC) kernel [18]. Inspired by the LBP feature, CDC kernel aims to encode the difference between central and adjacent neighbors in a learnable way. Experiments in [21] show that DTN and NAS-FAS can achieve better generalizability on HKBU-MARs, compared with depth regression auxiliary method [17], while for unconstrained HiFiMask the performance is still unsatisfying.

The handcrafted features are difficult to be robust across multiple kinds of application scenarios. On the other hand, the deep-network-based features require significantly large and representative training datasets. Ishan et al. propose to use deep dictionary [52] via greedy learning algorithm (DDGL) for PAD [31]. It is reported that DDGL can achieve impressive performance on both the photo, video, hard mask, and silicone mask attacks. On the other hand, its generalizability under cross-dataset scenarios is less promising since the DDGL is also based on the feature learning framework and the performance, to some extent, still depends on the training data. Applying domain adaptation framework into face PAD has attracted increasing attention. However, most of them focus on traditional 2D presentation attacks [20] and very few of them addresses the 3D mask attack. Liu et al. propose a context-aware learning framework (CCL) which compares image pairs under varies context, such as the genuine and masked face of the same subject. To alleviate the effect of hard positive pairs, outlier pairs are removed with the proposed guided dropout module. Here the various contexts can be regarded as different domains with different lighting conditions, imaging devices, disguises, or mask types. CCL can be applied to existing backbone and networks such as the ResNet50, CDCN [18], and auxiliary depth regression [17] method. Promising improvements are achieved on HiFiMask and WMCA [19] datasets for both intra-variation (seen) and cross-variation (unseen) evaluation protocols.

10.3.1.3 Multi-modality Approach

Recent studies indicate that the 3D mask material can be well identified using invisible light spectrums, such as near-infrared, thermal [19, 25], and other modalities such as depth image or 3D data gathered through structured light. The multi-spectrum analysis [13, 19] on them can effectively aggregate liveness information from different frequencies. Specifically, the thermal image clearly reflects the cut-out region (i.e., eyes region) of masked face. Also, 3D mask partially blocks the heat and removes the detailed heat distribution of a genuine face (as shown in Fig. 10.6-Thermal). The multi-channel CNN (MC-CNN) proposed in [19] concatenates CNN features from different modalities as the input of the fully connected layer classifier, which achieves promising results on WMCA dataset. It also performs well when facing unseen mask types. Under grandtest protocol MC-CNN reaches 0.3% ACER on the testing sets. However, these multi-spectrum or multi-modality approaches require additional sensors to collect extra data, which is not applicable for existing widely deployed RGB-based face recognition systems. With the reduction of manufacturing costs of these imaging devices, the multi-modality solution will play an important role in the future.

10.3.2 Motion-Based Approach

Genuine faces always contain unconscious subtle facial motion during the access of a face recognition system. For rigid 3D masks such as the Thatsmyface and REAL-f mask, the facial motion or facial expression could be blocked. As such, the motion-based methods on 2D face PAD can be applied to detect 3D mask attacks. Dynamic textures [11] or Histograms of Oriented Optical Flow (HOOF) [53] are proven to be effective in detecting the rigid 3D mask. Multi-feature Videolet encodes the texture with motion information [54] of both facial and surrounding regions. Specifically, multi-LBP is adopted to extract the texture feature and HOOF from different time slots of the input video is adopted to extract the motion feature. The multi-feature Videolet can reach 0% EER on the 3DMAD and also perform well on detecting the conventional 2D presentation attacks. Shao et al. exploit the dynamic information from texture details in lower convolutional layers [24]. Given an input face video, fine-grained textures in feature channels of each frame are first extracted through a pretrained VGG [55]. Then the dynamic information of textures is estimated via optical flow [56]. To exploit the most discriminative dynamic information, a channel-discriminability constraint is learned by minimizing intra-class variance and maximizing inter-class variance. This deep dynamic feature achieves good performance on 3DMAD and HKBU-MARsV1 dataset under intra- and cross-dataset evaluations.

10.3.3 Remote-Photoplethysmography-Based Approach

Different from the aforementioned conventional solutions, the rationale of the rPPG-based approach is based on contact Photoplethysmography (PPG), a biomedical monitoring technique that measures the volumetric of an organ optically. For heart-beat signal measurement, a pulse oximeter is attached to skin to measure the changes

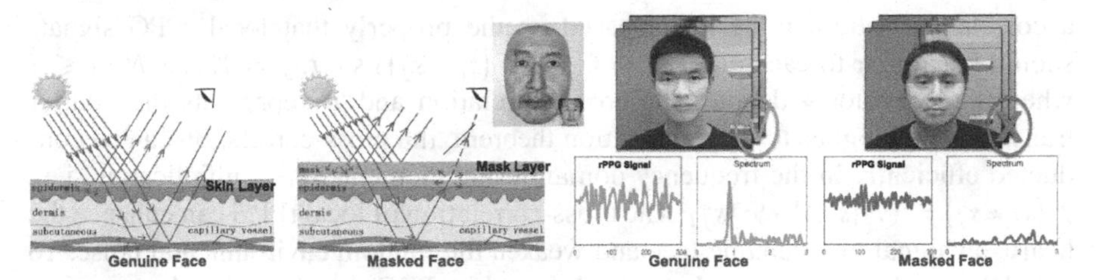

Fig. 10.10 Principle of using rPPG for 3D mask face PAD [26]

in light absorption of tissue when the heart pumps blood in a cardiac cycle [57]. The heartbeat can also be measured remotely through remote photoplethysmography (rPPG). Different from PPG that requires skin contact, rPPG can be measured remotely through a normal RGB camera (e.g., webcam and camera on mobile phone) under ambient illumination.

It has been proved that rPPG is a robust liveness cue for 3D mask face PAD [26, 58]. As shown in Fig. 10.10, the ambient light penetrates the skin layer and illuminates the subcutaneous layer. During cardiac cycles, the blood volume changes and results in a periodic color variation, i.e., heartbeat signal. The signal transmits back through the skin layer and can be observed by a normal RGB camera. While for a masked face, the light is blocked by the mask layer and the source heartbeat signal can also hardly go through the mask layer again. Consequently, a strong periodic heartbeat signal can be obtained from a genuine face and a noisy signal is expected for a masked face.

10.3.3.1 rPPG-Based 3D Mask Presentation Attack Detection

An intuitive solution is to measure the amplitude of global rPPG signals from genuine and masked faces [58]. Given an input face video, the Viola-Jones face detector [59] is used to find the bounding box for landmark detection. A customized region of interest is then defined to extract three raw pulse signals from the RGB channels. Temporal filters are applied to remove frequencies that are not relevant for pulse analysis. Then the liveness feature is extracted by analyzing the signal amplitude in frequency domain. Noise ratio of the three channels is extracted from the power density curves. SVM is used for classification. This global rPPG (GrPPG) method achieves 4.71% EER on 3DMAD and 1.58% EER on two REAL-f masks [58] and most of the error cases fall on false rejection. This is because the amplitude of heartbeat signal is weak and fragile due the factors like darker skin tone and facial motion.

To enhance the robustness, Liu et al. exploit a local rPPG solution for 3D mask face PAD [26]. The input face is divided into N local regions based on the facial landmarks. CHROM [60] is adopted as the rPPG extractor, which is robust to small facial motion and allows flexible skin ROI selection. Given local rPPG signals $[s_1, s_2, \dots, s_N]^\mathsf{T}$, a correlation pattern is designed based on the properly that local rPPG signals should be similar to each other: $\boldsymbol{x} = Conc(\mathcal{F}\{\boldsymbol{s}_i \star \boldsymbol{s}_j\})$ s.t. $i, j = 1, \ldots, N,\ i \leqslant j$ where the operator $\star$ denotes the cross-correlation and $\mathcal{F}$ represents the Fourier transform. Analogous to the convolution theorem, the cross-correlation can be conducted efficiently in the frequency domain using element-wise multiplication, i.e., $\mathcal{F}\{\boldsymbol{s}_i \star \boldsymbol{s}_j\} = (\mathcal{F}\{\boldsymbol{s}_i\})^* \cdot \mathcal{F}\{\boldsymbol{s}_j\}$. The cross-correlation of local rPPG can enhance the heartbeat signal of a genuine face and weaken the random environmental noise. To avoid the local facial regions that provide unstable rPPG signals, a confidence map is learned by analyzing the principal component of multiple local rPPG signals. Finally, the local rPPG correlation feature is fed into an SVM where the metric is adjusted by the learned confidence map.

The cross-correlation of local rPPG signals can boost the robustness to random environmental noise. While in real-world scenarios, periodic noise may be dominant and introduced false acceptance errors, such as the camera motion or flashlight. To handle it, the rPPG correspondence feature (CFrPPG) is proposed [38]. A heartbeat verification template from multiple local rPPG signals is then used to construct the liveness feature by measuring the correspondence between the learned heartbeat template and the local rPPG spectrums. Background rPPG signals are regarded as a reference of global noise and used to regularize the template learning. PPGSec [61] also employs the background rPPG pattern in the final liveness feature, combined with the raw rPPG spectrums of facial rPPG signals. CFrPPG is further extended into multi-channel CFrPPG (MCCFrPPG) by considering the temporal heartbeat variation during the observation [62]. Each channel contains the liveness information of a smaller time slot so the heartbeat variation can be encoded to achieve better robustness. rPPG-based solution can also be boosted on top of deep learning technique. TransrPPG extracts the local rPPG signals map and uses an input of the popular fully attentional vision transformer network [63] which achieves consistently good results on 3DMAD and HKBU-MARs V1+ dataset.

Although the abovementioned rPPG-based methods exhibit encouraging performance, they require at least 10 s observation time, which may not fit the application scenarios like smartphone unlock or e-payment. To shorten the observation time while keeping the performance, TSrPPG is proposed by analyzing the similarity of local facial rPPG signals in the time domain [64]. TsrPPG consists of three types of temporal similarity features of facial and background local rPPG signals based on rPPG shape and phase properties. The final decision is obtained through score level fusion and SVM classifier. Evaluation in [64] shows that TSrPPG improves the performance with large gap given 1 s observation time.

10.4 Experiments

Experiments on HKBU-MARsV1+ and 3DMAD are conducted to evaluate the performance of the appearance-based, motion-based, and rPPG-based methods. In particular, MS-LBP [22], CTA [40], VGG16 finetuned on ImageNet, and FaceBagNet-

RGB (FBNet-RGB) [65] are evaluated. Recently proposed network and learning strategy can also be applied to 3D mask PAD so we also test the popular CDCN [18] and DeepPixBiS [49] which achieve encouraging performance on 2D face PAD. Considering the limited number of samples of 3D mask attack datasets, we manually pretrain the backbone of CDCN on ImageNet by concatenating two blocks to extract high-level semantic features which expand the number of channels into 512. Two FC layers are used as classifier. Following the released code of [18], we adopt commonly used data augmentations including random erasing, horizontal flip, and cut out to increase the variation. For DeepPixBiS, we follow the original settings in [49] and directly use pretrained DenseNet161 as the backbone. Following the released code of [49], we adopt random rotation and horizontal flip augmentation.

For rPPG-based methods, global rPPG solution (GrPPG) [58], local rPPG solution (LrPPG) [26], CFrPPG [38], and MCCFrPPG [62] are implemented and evaluated under both intra- and cross-dataset testing protocols. For the appearance-based methods, only the first frame of the input video is used. HOOF [54], RDWT-Haralick [43], and convMotion [24] are compared on 3DMAD and HKBU-MARsV1 (8 subjects version). It is also noted that only the HOOF part of the videoLet [54] method is adopted to compare the motion cue with other methods. The final results are obtained through MATLAB SVM with RBF kernel for MS-LBP, HOOF, RDWT-Haralick, and LrPPG and linear kernel for GrPPG.

To evaluate the performance, Half Total Error Rate (HTER) [2, 22], AUC, EER, and Bona fide Presentation Classification Error Rate (BPCER) when Attack Presentation Classification Error Rate (APCER) equals 0.1 and 0.01 are used as the evaluation criteria. For the intra-dataset test, HTER is evaluated on the development set and testing set, which are named as HTER_dev and HTER_test, respectively. A ROC curve with BPCER and APCER is used for qualitative comparisons.

10.4.1 Intra-Dataset Evaluation

LOOCV on 3DMAD and HKBU-MARsV1+ datasets are conducted to evaluate the discriminability. For the 3DMAD dataset, 8 subjects are randomly selected for training and the remaining 8 are used for development. For the HKBU-MARsV1+ dataset, 5 subjects are randomly selected as training set and the remaining 6 are used as development set. Due to the randomness in partition, 20 rounds of experiments are conducted. For CDCN and DeepPixBis, 10 rounds of experiments are conducted due to the their high computation costs.

Table 10.2 demonstrates the effectiveness of the texture-based method on detecting low-quality 3D mask while the performance on the HKBU-MARsV1+ indicates the limitation when facing hyper-real masks. The DeepPixBis, MS-LBP, and convMotion that fuses deep learning with motion liveness cues achieve the best performance on 3DMAD. Similar performance is achieved by the fine-tuned CNN, FBNet-RGB, CDCN, and CTA, because of the appearance quality defect of Thatsmyface masks. We also note the consistent performance of CDCN and DeepPixBis on the

Table 10.2 Comparison of results under intra-dataset protocol on the 3DMAD dataset

	HTER_dev (%)	HTER_test (%)	EER (%)	AUC	BPCER@ APCER=0.1	BPCER@ APCER=0.01
MS-LBP [22]	0.15 ± 0.6	1.56 ± 5.5	0.67	100.0	**0.00**	0.42
CTA [40]	2.78 ± 3.6	4.40 ± 9.7	4.24	99.3	1.60	11.8
VGG16	1.58 ± 1.6	1.93 ± 3.4	2.18	99.7	0.27	5.00
FBNet-RGB [65]	3.91 ± 2.4	5.66 ± 9.7	5.54	98.6	2.21	19.9
DeepPixBis [49]	**0.00 ± 0.0**	0.86 ± 2.9	**0.00**	**100.0**	0.00	**0.00**
CDCN [18]	6.55 ± 6.0	4.20 ± 7.1	8.34	96.7	6.62	59.6
HOOF [54]	32.9 ± 6.5	33.8 ± 20.7	34.5	71.9	62.4	85.2
RDWT-Haralick [43]	7.88 ± 5.4	10.0 ± 16.2	7.43	94.0	2.78	89.1
convMotion [24]	0.09 ± 0.1	**0.84 ± 0.6**	0.56	100.0	0.00	0.30
GrPPG [58]	13.5 ± 4.3	13.3 ± 13.3	14.4	92.2	16.4	36.0
PPGSec [61]	15.2 ± 4.4	15.9 ± 14.6	16.4	90.7	20.5	36.7
LrPPG [26]	7.12 ± 4.0	7.60 ± 13.0	8.03	96.2	7.36	14.8
CFrPPG [38]	5.95 ± 3.3	6.82 ± 12.1	7.44	96.8	6.51	13.6
MCCFrPPG [62]	4.42 ± 2.3	5.60 ± 8.8	5.54	98.5	4.27	11.0
TransrPPG† [63]	–	–	2.38	98.77	–	13.73

† Results are from paper [63]

two datasets, which demonstrate the effectiveness of pixel-level supervision. HOOF achieves consistent performance on the 3DMAD and HKBU-MARsV1 (Table 10.3) while the precisions are below the average.

Note that the early rPPG-based methods (GrPPG and PPGSec) perform better on 3DMAD than on HKBU-MARsV1+ due to the difference in compression algorithm. The default H.264 compression in Logitech C920 may remove the subtle skin color variation that cannot be observed by human eyes and therefore reduce the signal-to-noise ratio (SNR) of rPPG signal. Specifically, LrPPG achieves better results due to the robustness of the cross-correlation model and confidence map. In Fig. 10.11, we find that the major error classifications fall on the false rejection due to the fragileness of facial rPPG signals. The CFrPPG improves the performance by identifying the liveness information among noise rPPG signals. MCCFrPPG further enhances the robustness by considering the heartbeat variation during observation. We can also note the higher performance on HKBU-MARsV1+. This may be due to the facial resolution of videos from 3DMAD (around 80×80) are smaller than the videos from HKBU-MARsV1+ (around 200×200). rPPG signal quality also depends on the number of pixels of the region of interest and local rPPG extracted from smaller local facial regions increases this effect. We also report the results of TransrPPG which performs better on 3DMAD. Video compression could be the reason since TransrPPG directly learns the feature from the temporal rPPG signal.

10.4.2 Cross-Dataset Evaluation

To evaluate the generalization ability across different datasets, we conduct cross-dataset experiments where the training and test samples are from two different datasets. In particular, for train on 3DMAD and test on HKBU-MARsV1+ (3DMAD→HKBU-MARsV1+), we randomly select eight subjects from 3DMAD for training and use the remaining nine subjects from 3DMAD for development. For train on HKBU-MARsV1+ and test on 3DMAD (HKBU-MARsV1+→3DMAD), we randomly select six subjects from HKBU-MARsV1+ as the training set and the remaining six subjects as the development set. We also conduct 20 rounds of experiments.

As shown in Table 10.4 and Fig. 10.12, the rPPG-based methods achieve close results as the intra-dataset testing, which shows their encouraging robustness. The main reason behind the rPPG-based method's success is that the rPPG signals for different people under different environments are consistent, so the features of genuine and fake faces can be well separated. On the other hand, the performance of MS-LBP and HOOF drops. Specifically, the appearance-based methods, MS-LBP and RDWT-Haralick, and the data-driven TransrPPG achieve better results for HKBU-MARsV1+→3DMAD, since the HKBU-MARsV1+ contains two types of masks while the 3DMAD contains one (the classifier can generalize better when it is trained with larger data variance). In contrast, convMotion perform worse for HKBU-MARsV1+→3DMAD since the model learned from high facial resolution samples

Table 10.3 Comparison of results under intra-dataset protocol on the HKBU-MARsV1+ dataset

	HTER_dev (%)	HTER_test (%)	EER (%)	AUC	BPCER@ APCER=0.1	BPCER@ APCER=0.01
MS-LBP [22]	20.5 ± 8.9	24.0 ± 25.6	22.5	85.8	48.6	95.1
CTA [40]	22.4 ± 10.4	23.4 ± 20.5	23.0	82.3	53.7	89.2
VGG16	16.7 ± 6.7	18.8 ± 15.0	18.0	90.2	31.0	73.8
FBNet-RGB [65]	35.0 ± 11.3	36.1 ± 26.0	36.2	67.3	78.3	96.5
DeepPixBis [49]	8.33 ± 11.7	7.22 ± 11.3	11.6	94.1	15.8	95.0
CDCN [18]	4.97 ± 7.5	4.83 ± 7.6	8.70	96.0	7.77	66.2
HOOF* [54]	25.4 ± 12.8	23.8 ± 23.9	28.9	77.9	75.0	95.6
RDWT-Haralick* [43]	16.3 ± 6.8	10.3 ± 16.3	18.1	87.3	31.3	81.7
convMotion* [24]	**2.31 ± 1.7**	6.36 ± 3.5	8.47	98.0	6.72	27.0
GrPPG [58]	15.4 ± 6.7	15.4 ± 20.4	16.5	89.5	18.4	33.4
PPGSec [61]	14.2 ± 5.8	15.6 ± 16.4	16.8	91.0	20.7	35.4
LrPPG [26]	8.43 ± 2.9	8.67 ± 8.8	9.07	97.0	8.51	38.9
CFrPPG [38]	3.24 ± 2.0	4.10 ± 4.9	3.95	99.3	1.00	18.7
MCCFrPPG [62]	2.85 ± 1.8	**3.38 ± 4.8**	**3.35**	**99.7**	**0.43**	**7.08**
TransrPPG† [63]	–	–	8.47	96.82	–	29.79

* Denotes that results are based on HKBU-MARsV1
† Results are from paper [63]

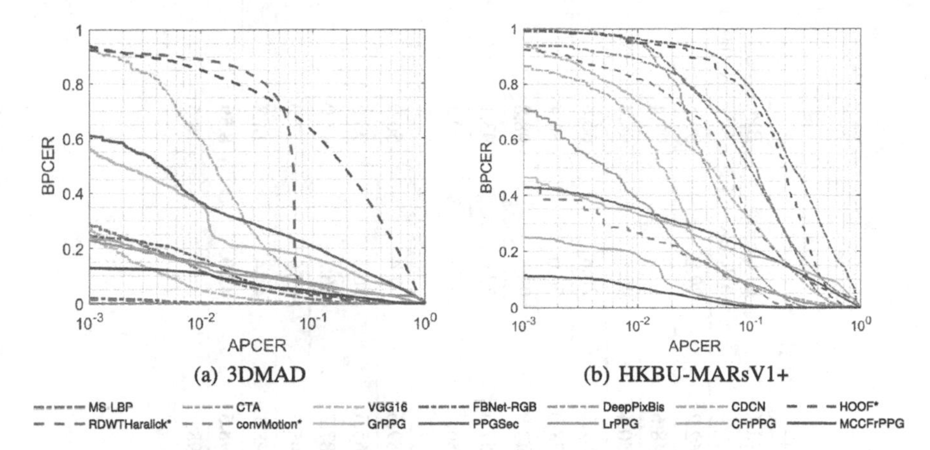

Fig. 10.11 Average ROC curves of two evaluations under intra-dataset protocol

may not be able to adapt to low-resolution samples in 3DMAD. It is also noted that the RDWT-Haralick feature achieves better performance than MS-LBP as it analyzes the texture differences from different scales with redundant discrete wavelet transform [43]. The HOOF fails on the cross-dataset as for different dataset, the motion patterns based on optical flow may vary with different recording settings. We can also note the huge performance drop of CDCN. The central difference encoding in convolution kernel may help in investigating the fine-grained structure information. However, it still suffers from the overfitting problem when the dataset variation is small. The most effective fine-grained cues learned on one type of mask may not be able to generalize to others. Figure 10.13 visualizes typical cases of predicted depth map of CDCN under intra-dataset evaluation on 3DMAD. The inconsistent predictions of the same subject in Fig. 10.13c indicate that the model can hardly learn generalized cues. Figure 10.13a also demonstrates the sensitivity of expression variation where the mouth motion causes the wrong prediction.

10.5 Discussion and Open Challenges

3D mask has been proven to be effective to spoof the face recognition system [22, 23] due to the advancement of 3D printing and 3D reconstruction techniques. 3D mask face PAD has drawn increasing attention with the boosting numbers of related publications. In this chapter, we revealed the progress of 3D mask PAD, summarized the datasets with different evaluation protocols, and analyzed the pros and cons of existing approaches based on their liveness cues. Still, there are challenges that remain open.

Table 10.4 Cross-dataset evaluation results between 3DMAD and HKBU-MARsV1+

	3DMAD→HKBU-MARsV1+				HKBUMARsV1+→3DMAD			
	HTER↓	AUC↑	BPCER@ APCER=0.1↓	BPCER@ APCER=0.01↓	HTER↓	AUC↑	BPCER@ APCER=0.1↓	BPCER@ APCER=0.01↓
MS-LBP [22]	36.8 ± 2.9	60.7	87.5	97.0	41.3 ± 14.0	62.2	89.2	99.5
CTA [40]	71.8 ± 2.1	45.9	96.8	99.3	55.7 ± 8.7	48.6	89.9	97.4
VGG16	49.4 ± 1.7	72.1	86.4	99.1	62.5 ± 7.4	50.4	90.5	98.7
FBNet-RGB [65]	34.0 ± 1.4	73.6	65.7	97.8	12.3 ± 10.6	89.6	26.8	66.8
DeepPixBiS [49]	29.1 ± 3.9	74.3	81.5	97.3	17.0 ± 9.1	78.0	56.0	86.4
CDCN [18]	17.7 ± 2.7	45.5	94.7	99.3	22.0 ± 8.8	43.8	93.4	98.8
HOOF* [54]	51.8 ± 12.0	47.3	84.1	92.1	42.4 ± 4.1	57.6	77.6	93.1
RDWT-Haralick* [43]	23.5 ± 4.7	68.3	71.6	98.1	13.8 ± 7.5	86.7	37.1	90.3
convMotion* [24]	6.06 ± 2.1	96.2	13.8	33.9	9.88 ± 1.3	89.6	27.4	56.0
GrPPG [58]	35.9 ± 4.5	67.2	75.8	97.6	36.5 ± 6.8	66.5	86.3	98.6
PPGSec [61]	14.4 ± 1.4	91.8	16.9	25.8	19.1 ± 2.3	87.2	26.2	45.0
LrPPG [26]	4.46 ± 0.9	98.9	1.33	31.2	8.46 ± 0.3	95.3	8.79	17.0
CFrPPG [38]	4.23 ± 0.3	99.0	2.83	19.9	4.81 ± 0.4	98.1	4.44	14.3
MCCFrPPG [62]	**3.46 ± 0.6**	**99.6**	**0.25**	**7.63**	**4.78 ± 0.8**	**98.5**	**4.00**	**8.59**
TransrPPG† [63]	–	91.3	–	47.6	–	98.3	–	18.5

* Denotes that results are based on HKBU-MARsV1
† Results are from paper [63]

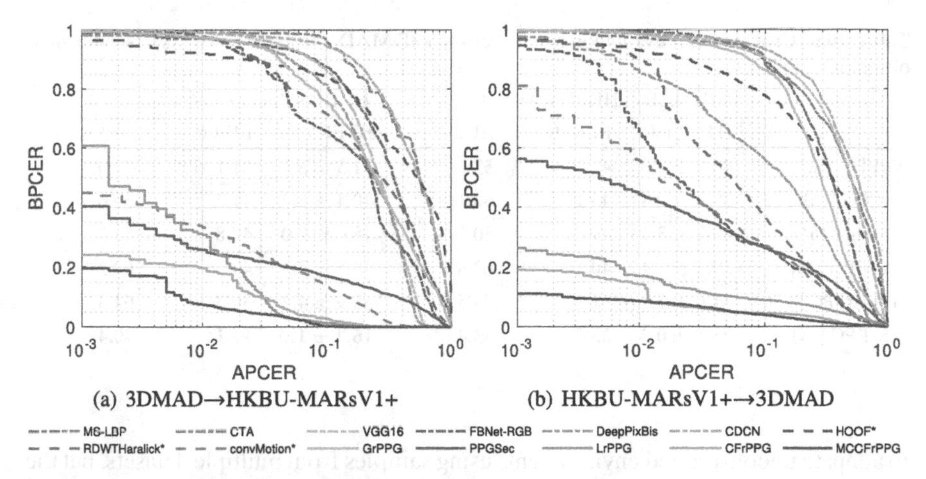

Fig. 10.12 Average ROC curves of under cross-dataset protocol

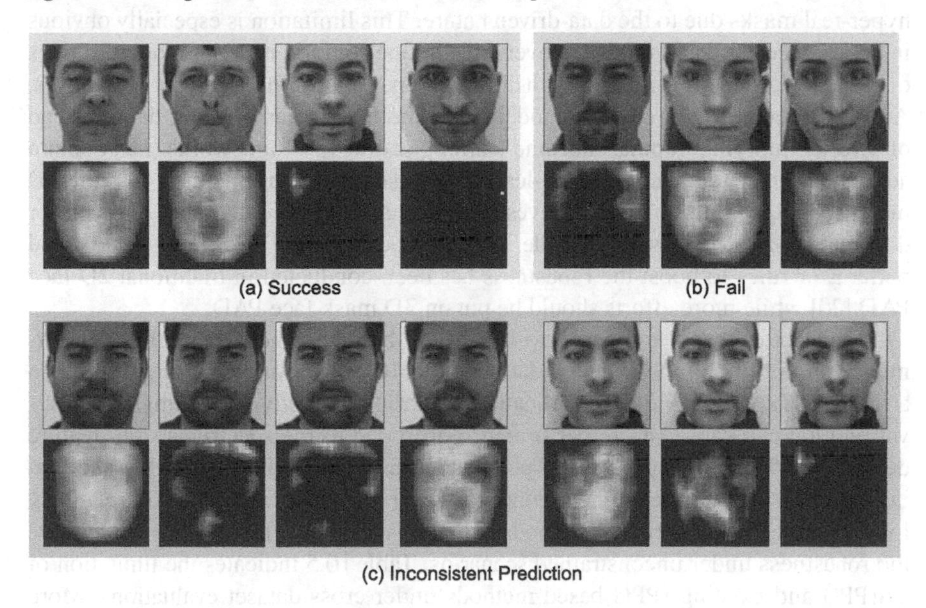

Fig. 10.13 Visualization of depth map predicted by CDCN [18] under intra-dataset evaluation on 3DMAD

The scale and variations of 3D mask datasets increase largely in terms of the number of subjects, types of masks, lighting conditions, imaging devices, and decorations, which enable the evaluation of generalization capabilities in real-world scenarios. With the popularity of mobile applications, the built-in camera of mobile devices and unconstrained daily lighting conditions becomes the mainstream in the recently built datasets. However, although these improvements have been made, the number of hyper-real masks is still limited due to the expensive costs. A model can be trained

Table 10.5 Cross-dataset evaluation results between 3DMAD and HKBU-MARsV1+ with short observation time (1 s)

	3DMAD→HKBUMARsV1+			HKBUMARsV1+→3DMAD		
	HTER (%)	EER (%)	AUC (%)	HTER (%)	EER (%)	AUC (%)
GrPPG [58]	46.8 ± 3.0	47.5	53.6	31.5 ± 3.8	31.1	70.0
LrPPG [26]	39.2 ± 0.8	43.1	60.1	40.4 ± 2.7	41.7	60.6
PPGSec [61]	49.7 ± 3.1	49.2	50.7	48.0 ± 2.0	47.8	52.6
CFrPPG [38]	39.2 ± 1.4	40.1	63.6	40.1 ± 2.3	40.6	62.3
TransrPPG [63]	31.6 ± 1.9	31.7	73.5	38.9 ± 4.2	39.3	64.3
TSrPPG [64]	**23.5 ± 0.5**	**23.5**	**83.4**	**16.1 ± 1.0**	**17.1**	**90.4**

to adapt to unconstrained environments using samples from multiple datasets, but the feature learned from 3D masks with specific quality defects may still fail on detecting hyper-real masks due to the data-driven nature. This limitation is especially obvious in appearance-based methods. To overcome it, one solution is to put more resources on building larger scale datasets with different types of customized hyper-real masks. Another direction is to develop model-based methods given existing data. Instead of directly applying popular machine learning settings or frameworks (i.e., domain adaptation or generalization, meta-learning, noise-label learning, etc.) into the 3D mask PAD application, we may investigate domain knowledge that is intrinsic for different types of masks and encode into the model design. Analyzing generalized patterns or rules to boost the robustness has been conducted in traditional 2D face PAD [20] while more efforts should be put on 3D mask face PAD.

On the other hand, although existing rPPG-based methods reach high performance on public 3D mask attack datasets, the 10 s observation time will result in a bad user experience in real application scenarios. Directly shortening the observation time can cause severe performance drop due to the spectrum-based feature design. TSrPPG [64] highlights this point and improves the robustness when shortening the observation time to one second. However, the performance is still relatively low compared with previous methods. There also exists a large space to improve the robustness under unconstrained scenarios. Table 10.5 indicates the limitation of TSrPPG and existing rPPG-based methods under cross-dataset evaluations. More challenging settings with unconstrained lighting conditions remain unsolved. The handcrafted TSrPPG may not be optimal for different scenarios. A possible future direction is to learn more robust temporal liveness features using deep learning techniques with the domain knowledge of the handcrafted design. Recent research also points out the sensitivity of rPPG estimation when facing compressed videos [66]. Built-in video compression is widely adopted in off-the-shelf imaging devices such as the web camera, laptop camera, and smartphone front camera. Under dim lighting conditions, video compression will further degrade the video quality and contaminate the rPPG signal. Improving the robustness to video compression is another direction of the rPPG-based approach.

Acknowledgements This project is partially supported by Hong Kong RGC General Research Fund HKBU 12200820.

References

1. Galbally J, Marcel S, Fierrez J (2014) Biometric antispoofing methods: a survey in face recognition. IEEE Access 2:1530–1552
2. Hadid A, Evans N, Marcel S, Fierrez J (2015) Biometrics systems under spoofing attack: an evaluation methodology and lessons learned. IEEE Signal Process Mag 32(5):20–30
3. Rattani A, Poh N, Ross A (2012) Analysis of user-specific score characteristics for spoof biometric attacks. In: CVPRW
4. Evans NW, Kinnunen T, Yamagishi J (2013) Spoofing and countermeasures for automatic speaker verification. In: Interspeech, pp 925–929
5. Pavlidis I, Symosek P (2000) The imaging issue in an automatic face/disguise detection system. In: Computer vision beyond the visible spectrum: methods and applications
6. Tan X, Li Y, Liu J, Jiang L (2010) Face liveness detection from a single image with sparse low rank bilinear discriminative model. Computer Vision–ECCV, pp 504–517
7. Määttä J, Hadid A, Pietikäinen M (2011) Face spoofing detection from single images using micro-texture analysis. In: IJCB
8. Anjos A, Marcel S (2011) Counter-measures to photo attacks in face recognition: a public database and a baseline. In: International joint conference on biometrics (IJCB)
9. Zhang Z, Yan J, Liu S, Lei Z, Yi D, Li SZ (2012) A face antispoofing database with diverse attacks. In: International conference on biometrics (ICB), pp 26–31
10. Pan G, Sun L, Wu Z, Lao S (2007) Eyeblink-based anti-spoofing in face recognition from a generic webcamera. In: ICCV
11. de Freitas Pereira T, Komulainen J, Anjos A, De Martino JM, Hadid A, Pietikäinen M, Marcel S (2014) Face liveness detection using dynamic texture. EURASIP J Image Video Process 1:1–15
12. Kose N, Dugelay JL (2014) Mask spoofing in face recognition and countermeasures. Image Vis Comput 32(10):779–789
13. Yi D, Lei Z, Zhang Z, Li SZ (2014) Face anti-spoofing: multi-spectral approach. In: Handbook of biometric anti-spoofing. Springer, pp 83–102
14. Galbally J, Marcel S, Fierrez J (2014) Image quality assessment for fake biometric detection: application to iris, fingerprint, and face recognition. IEEE Trans Image Process 23(2):710–724
15. Kose N, Dugelay JL (2013) Shape and texture based countermeasure to protect face recognition systems against mask attacks. In: CVPRW
16. Wen D, Han H, Jain AK (2015) Face spoof detection with image distortion analysis. IEEE Trans Inf Forensics Secur 10(4):746–761
17. Liu Y, Jourabloo A, Liu X (2018) Learning deep models for face anti-spoofing: binary or auxiliary supervision. In: CVPR, pp 389–398
18. Yu Z, Wan J, Qin Y, Li X, Li SZ, Zhao G (2020) Nas-fas: static-dynamic central difference network search for face anti-spoofing. arXiv:2011.02062
19. George A, Mostaani Z, Geissenbuhler D, Nikisins O, Anjos A, Marcel S (2019) Biometric face presentation attack detection with multi-channel convolutional neural network. IEEE Trans Inf Forensics Secur 15:42–55
20. Wang G, Han H, Shan S, Chen X (2021) Unsupervised adversarial domain adaptation for cross-domain face presentation attack detection. IEEE Trans Inf Forensics Secur 16:56–69
21. Yu Z, Qin Y, Li X, Zhao C, Lei Z, Zhao G (2021) Deep learning for face anti-spoofing: a survey. arXiv:2106.14948

22. Erdogmus N, Marcel S (2014) Spoofing face recognition with 3D masks. IEEE Trans Inf Forensics Secur 9(7):1084–1097
23. Liu A, Zhao C, Yu Z, Wan J, Su A, Liu X, Tan Z, Escalera S, Xing J, Liang Y et al (2021) Contrastive context-aware learning for 3d high-fidelity mask face presentation attack detection. IEEE Trans Inf Forensics Secur 17:2497–2507
24. Rui Shao XL, Yuen PC (2017) Deep convolutional dynamic texture learning with adaptive channel-discriminability for 3d mask face anti-spoofing. In: IJCB
25. Bhattacharjee S, Mohammadi A, Marcel S (2018) Spoofing deep face recognition with custom silicone masks. In: International conference on biometrics theory, applications and systems (BTAS). IEEE
26. Liu S, Yuen PC, Zhang S, Zhao G (2016) 3d mask face anti-spoofing with remote photoplethysmography. In: ECCV
27. Liu SQ, Yuen PC, Li X, Zhao G (2019) Recent progress on face presentation attack detection of 3d mask attacks. Handbook of biometric anti-spoofing, pp 229–246
28. Marcel S, Nixon MS, Fierrez J, Evans N (2019) Handbook of biometric anti-spoofing: Presentation attack detection. Springer
29. Erdogmus N, Marcel S (2013) Spoofing in 2d face recognition with 3d masks and anti-spoofing with kinect. In: BTAS
30. Wallace R, McLaren M, McCool C, Marcel S (2011) Inter-session variability modelling and joint factor analysis for face authentication. In: IJCB
31. Manjani I, Tariyal S, Vatsa M, Singh R, Majumdar A (2017) Detecting silicone mask based presentation attack via deep dictionary learning. TIFS
32. Bhattacharjee S, Marcel S (2017) What you can't see can help you-extended-range imaging for 3d-mask presentation attack detection. In: International conference of the biometrics special interest group (BIOSIG). IEEE, pp 1–7
33. Agarwal A, Yadav D, Kohli N, Singh R, Vatsa M, Noore A (2017) Face presentation attack with latex masks in multispectral videos. SMAD 13:130
34. Liu Y, Stehouwer J, Jourabloo A, Liu X (2019) Deep tree learning for zero-shot face anti-spoofing. In: CVPR
35. Liu S, Yang B, Yuen PC, Zhao G (2016) A 3D mask face anti-spoofing database with real world variations. In: Proceedings of the IEEE conference on computer vision and pattern recognition workshops, pp 100–106
36. Galbally J, Satta R (2016) Three-dimensional and two-and-a-half-dimensional face recognition spoofing using three-dimensional printed models. IET Biom 5(2):83–91
37. Steiner H, Kolb A, Jung N (2016) Reliable face anti-spoofing using multispectral swir imaging. In: 2016 international conference on biometrics (ICB). IEEE, pp 1–8
38. Liu SQ, Lan X, Yuen PC (2018) Remote photoplethysmography correspondence feature for 3d mask face presentation attack detection. In: ECCV, pp 558–573
39. Heusch G, George A, Geissbühler D, Mostaani Z, Marcel S (2020) Deep models and shortwave infrared information to detect face presentation attacks. IEEE Trans Biom Behav Identity Sci
40. Boulkenafet Z, Komulainen J, Hadid A (2016) Face spoofing detection using colour texture analysis. IEEE Trans Inf Forensics Secur 11(8):1818–1830
41. Boulkenafet Z, Komulainen J, Hadid A (2015) Face anti-spoofing based on color texture analysis. In: ICIP
42. de Freitas Pereira T, Anjos A, De Martino JM, Marcel S (2013) Can face anti-spoofing countermeasures work in a real world scenario? In: International conference on biometrics (ICB)
43. Agarwal A, Singh R, Vatsa M (2016) Face anti-spoofing using haralick features. In: BTAS
44. Haralick RM, Shanmugam K et al (1973) Textural features for image classification. IEEE Trans Syst Man Cybern 1(6):610–621
45. Cohen-Steiner D, Morvan JM (2003) Restricted delaunay triangulations and normal cycle. In: Proceedings of the nineteenth annual symposium on Computational geometry, pp 312–321
46. Tang Y, Chen L (2017) 3d facial geometric attributes based anti-spoofing approach against mask attacks. In: FG

47. Menotti D, Chiachia G, Pinto A, Robson Schwartz W, Pedrini H, Xavier Falcao A, Rocha A (2015) Deep representations for iris, face, and fingerprint spoofing detection. IEEE Trans Inf Forensics Secur 10(4):864–879
48. Yang J, Lei Z, Li SZ (2014) Learn convolutional neural network for face anti-spoofing. arXiv:1408.5601
49. George A, Marcel S (2019) Deep pixel-wise binary supervision for face presentation attack detection. In: 2019 international conference on biometrics (ICB). IEEE, pp 1–8
50. Huang G, Liu Z, Van Der Maaten L, Weinberger KQ (2017) Densely connected convolutional networks. In: Proceedings of the IEEE conference on computer vision and pattern recognition, pp 4700–4708
51. Jia S, Li X, Hu C, Guo G, Xu Z (2020) 3d face anti-spoofing with factorized bilinear coding. IEEE Trans Circuits Syst Video Technol
52. Tariyal S, Majumdar A, Singh R, Vatsa M (2016) Deep dictionary learning. IEEE Access 4:10096–10109
53. Chaudhry R, Ravichandran A, Hager G, Vidal R (2009) Histograms of oriented optical flow and binet-cauchy kernels on nonlinear dynamical systems for the recognition of human actions. In: CVPR
54. Siddiqui TA, Bharadwaj S, Dhamecha TI, Agarwal A, Vatsa M, Singh R, Ratha N (2016) Face anti-spoofing with multifeature videolet aggregation. In: ICPR
55. Simonyan K, Zisserman A (2014) Very deep convolutional networks for large-scale image recognition. arXiv:1409.1556
56. Barron JL, Fleet DJ, Beauchemin SS (1994) Performance of optical flow techniques. Int J Comput Vis
57. Shelley K, Shelley S (2001) Pulse oximeter waveform: photoelectric plethysmography. In: Lake C, Hines R, Blitt C (eds) Clinical monitoring. WB Saunders Company, pp 420–428
58. Li X, Komulainen J, Zhao G, Yuen PC, Pietikäinen M (2016) Generalized face anti-spoofing by detecting pulse from face videos. In: IEEE international conference on pattern recognition (ICPR), pp 4244–4249
59. Viola P, Jones M (2001) Rapid object detection using a boosted cascade of simple features. In: CVPR
60. de Haan G, Jeanne V (2013) Robust pulse rate from chrominance-based rppg. IEEE Trans Biomed Eng 60(10):2878–2886
61. Nowara EM, Sabharwal A, Veeraraghavan A (2017) Ppgsecure: biometric presentation attack detection using photopletysmograms. In: FG
62. Liu SQ, Lan X, Yuen PC (2021) Multi-channel remote photoplethysmography correspondence feature for 3d mask face presentation attack detection. IEEE Trans Inf Forensics Secur 16:2683–2696
63. Yu Z, Li X, Wang P, Zhao G (2021) TransRPPG: remote photoplethysmography transformer for 3d mask face presentation attack detection. IEEE Signal Process Lett
64. Liu S, Lan X, Yuen P (2020) Temporal similarity analysis of remote photoplethysmography for fast 3d mask face presentation attack detection. In: IEEE/CVF winter conference on applications of computer vision, pp 2608–2616
65. Shen T, Huang Y, Tong Z (2019) Facebagnet: bag-of-local-features model for multi-modal face anti-spoofing. In: IEEE/CVF conference on computer vision and pattern recognition workshops (CVRPw)
66. McDuff DJ, Blackford EB, Estepp JR (2017) The impact of video compression on remote cardiac pulse measurement using imaging photoplethysmography. In: FG

Chapter 11
Robust Face Presentation Attack Detection with Multi-channel Neural Networks

Anjith George and Sébastien Marcel

Abstract Vulnerability against presentation attacks remains a challenging issue limiting the reliable use of face recognition systems. Though several methods have been proposed in the literature for the detection of presentation attacks, majority of these methods fail in generalizing to unseen attacks and environments. Since the quality of attack instruments keeps getting better, the difference between bona fide and attack samples is diminishing making it harder to distinguish them using the visible spectrum alone. In this context, multi-channel presentation attack detection methods have been proposed as a solution to secure face recognition systems. Even with multiple channels, special care needs to be taken to ensure that the model generalizes well in challenging scenarios. In this chapter, we present three different strategies to use multi-channel information for presentation attack detection. Specifically, we present different architecture choices for fusion, along with ad hoc loss functions as opposed to standard classification objective. We conduct an extensive set of experiments in the HQ-WMCA dataset, which contains a wide variety of attacks and sensing channels together with challenging unseen attack evaluation protocols. We make the protocol, source codes, and data publicly available to enable further extensions of the work.

11.1 Introduction

While face recognition technology has become a ubiquitous method for biometric authentication, the vulnerability to presentation attacks (PA) (also known as "spoofing attacks") is a major concern when used in secure scenarios [1, 2]. These attacks can be either *impersonation* or *obfuscation* attacks. Impersonation attacks attempt to gain access by masquerading as someone else and obfuscation attacks attempt to evade face recognition systems. While many methods have been suggested in the literature to address this problem, most of these methods fail in generalizing

---

A. George (✉) · S. Marcel (✉)
Idiap Research Institute, Martigny, Switzerland
e-mail: anjith.george@idiap.ch

S. Marcel
e-mail: sebastien.marcel@idiap.ch

S. Marcel et al. (eds.), *Handbook of Biometric Anti-Spoofing*, Advances in Computer Vision and Pattern Recognition, https://doi.org/10.1007/978-981-19-5288-3_11

to unseen attacks [3, 4]. Another challenge is poor generalization across different acquisition settings, such as sensors and lighting. In a practical scenario, it is not possible to anticipate all the types of attacks at the time of training a presentation attack detection (PAD) model. Moreover, a PAD system is expected to detect new types of sophisticated attacks. It is therefore important to have unseen attack robustness in PAD models.

Majority of the literature deals with the detection of these attacks with RGB cameras. Over the years, many feature-based methods have been proposed using color, texture, motion, liveliness cues, histogram features [5], local binary pattern [6, 7], and motion patterns [8] for performing PAD. Recently, several Convolutional Neural Network (CNN)-based methods have also been proposed including 3DCNN [9], part-based models [10], and so on. Some works have shown that using auxiliary information in the form of binary or depth supervision improves performance [11, 12]. In depth supervision, the model is trained to regress the depth map of the face as an auxiliary supervision. However, most of these methods have been designed specifically for 2D attacks and the performance of these methods against challenging 3D and partial attacks is poor [13]. Moreover, these methods suffer from poor unseen attack robustness.

The performance of RGB-only models deteriorates with sophisticated attacks such as 3D masks and partial attacks. Due to the limitations of visible spectrum alone, several multi-channel methods have been proposed in literature such as [2, 14–25] for face PAD. Essentially, it becomes more difficult to fool a multi-channel PAD system as it captures complementary information from different channels. Deceiving different channels at the same time requires considerable effort. Multi-channel methods have proven to be effective, but this comes at the expense of customized and expensive hardware. This could make these systems difficult to deploy widely, even if they are robust. Nevertheless, well-known commercial systems like Apple's Face ID [26] demonstrate the robustness of multi-channel PAD. A variety of channels are available for PAD, e.g., RGB, depth, thermal, near-infrared (NIR) spectra [20], short-wave infrared (SWIR) spectra [15, 21], ultraviolet [27], light field imagery [28], hyper-spectral imaging [29], etc.

Even when using multiple channels, the models tend to overfit to attacks seen in the training set. While the models could perform perfectly in attacks seen in the training set, degradation in performance is often observed when confronted with unseen attacks in real-world scenarios. This is a common phenomenon with most of the machine learning algorithms, and this problem is aggravated in case of a limited amount of training data. The models, in the lack of strong priors, could overfit to the statistical biases of specific datasets it was trained on and could fail in generalizing to unseen samples. Multi-channel methods also suffer from an increased possibility of overfitting as they increase the number of parameters due to the extra channels.

In this work, we present three different strategies to fuse multi-channel information for presentation attack detection. We consider early, late, and hybrid fusion approaches and evaluate their performance in a multi-channel setting. The joint representation helps in identifying important discriminative information in detection of the attacks.

The main contributions of this work are listed below:

- We present different strategies for the fusion of multi-channel information for presentation attack detection.
- An extensive set of experiments in the HQ-WMCA database, which contains a wide variety of attacks, using both seen and unseen attack protocols.

Additionally, the source code and protocols to reproduce the results are available publicly.[1]

11.2 Related Works

Majority of the literature in face PAD is focused on the detection of 2D attacks and uses feature-based methods [5, 6, 8, 30, 31] or CNN-based methods. Recently, CNN-based methods have been more successful as compared to feature-based methods [11, 12, 32, 33]. These methods usually leverage the quality degradation during "recapture" and are often useful only for the detection of attacks like 2D prints and replays. Sophisticated attacks like 3D masks [34] are more harder to detect using RGB information alone and pose serious threat to the reliability of face recognition systems [4].

11.2.1 RGB Only Approaches (Feature Based and CNNs)

11.2.1.1 Feature-Based Approaches for Face PAD

For PAD using visible spectrum images, several methods such as detecting motion patterns [8], color texture, and histogram-based methods in different color spaces, and variants of Local Binary Patterns (LBP) in grayscale [5] and color images [6, 7] have shown good performance. Image-quality-based feature [35] is one of the successful methods available in prevailing literature. Methods identifying moiré patterns [36], and image distortion analysis [37], use the alteration of the images due to the replay artifacts. Most of these methods treat PAD as a binary classification problem which may not generalize well for unseen attacks [38].

Chingovska et al. [39] studied the amount of client-specific information present in features used for PAD. They used this information to build client-specific PAD methods. Their method showed a 50% relative improvement and better performance in unseen attack scenarios.

Arashloo et al. [40] proposed a new evaluation scheme for unseen attacks. Authors have tested several combinations of binary classifiers and one-class classifiers. The performance of one-class classifiers was better than binary classifiers in the unseen

[1] https://gitlab.idiap.ch/bob/bob.paper.cross_modal_focal_loss_cvpr2021.

attack scenario. A variant of binarized statistical image features (BSIF), BSIF-TOP was found successful in both one-class and two-class scenarios. However, in cross-dataset evaluations, image quality features were more useful. Nikisins et al. [38] proposed a similar one-class classification framework using one-class Gaussian Mixture Models (GMM). In the feature extraction stage, they used a combination of Image Quality Measures (IQM). The experimental part involved an aggregated database consisting of REPLAY-ATTACK [7], REPLAY-MOBILE [1], and MSU-MFSD [37] datasets. A good review of related works on face PAD in color channel and available databases can be found in [41].

Heusch and Marcel [42] recently proposed a method for using features derived from remote photoplethysmography (rPPG). They used the long-term spectral statistics (LTSS) of pulse signals obtained from available methods for rPPG extraction. The LTSS features were combined with support vector machines (SVM) for PA detection. Their approach obtained better performance than state-of-the-art methods using rPPG in four publicly available databases.

11.2.1.2 CNN-Based Approaches for Face PAD

Recently, several authors have reported good performance in PAD using convolutional neural networks (CNN). Gan et al. [9] proposed a 3DCNN-based approach, which utilized the spatial and temporal features of the video. The proposed approach achieved good results in the case of 2D attacks, prints, and videos. Yang et al. [43] proposed a deep CNN architecture for PAD. A preprocessing stage including face detection and face landmark detection is used before feeding the images to the CNN. Once the CNN is trained, the feature representation obtained from CNN is used to train an SVM classifier and used for the final PAD task. Boulkenafet et al. [44] summarized the performance of the competition on mobile face PAD. The objective was to evaluate the performance of the algorithms under real-world conditions such as unseen sensors, different illumination, and presentation attack instruments. In most of the cases, texture features extracted from color channels performed the best. Li et al. [45] proposed a 3DCNN architecture, which utilizes both the spatial and temporal nature of videos. The network was first trained after data augmentation with a cross-entropy loss, and then with a specially designed generalization loss, which acts as a regularization factor. The Maximum Mean Discrepancy (MMD) distance among different domains is minimized to improve the generalization property.

There are several works involving various auxiliary information in the CNN training process, mostly focusing on the detection of 2D attacks. Authors use either 2D or 3D CNNs. The main problem of CNN-based approaches mentioned above is the lack of training data, which is usually required to train a network from scratch. One broadly used solution is fine-tuning, rather than a complete training, of the networks trained for face recognition or image classification tasks. Another issue is the poor generalization in cross-database and unseen attack tests. To circumvent these issues, some researchers have proposed methods to train a CNN using auxiliary tasks, which is shown to improve generalization properties. These approaches are discussed below.

Liu et al. [32] presented a novel method for PAD with auxiliary supervision. Instead of training a network end-to-end directly for the PAD task, they used the CNN-RNN model to estimate the depth with pixel-wise supervision and estimate remote photoplethysmography (rPPG) with sequence-wise supervision. The estimated rPPG and depth were used for the PAD task. The addition of the auxiliary task improved the generalization capability.

Atoum et al. [11] proposed a two-stream CNN for 2D presentation attack detection by combining a patch-based model and holistic depth maps. For the patch-based model, an end-to-end CNN was trained. In the depth estimation, a fully convolutional network was trained using the entire face image. The generated depth map was converted to a feature vector by finding the mean values in the $N \times N$ grid. The final PAD score was obtained by fusing the scores from the patch and depth CNNs.

Shao et al. [33] proposed a deep-convolutional-network-based architecture for 3D mask PAD. They tried to capture the subtle differences in facial dynamics using CNN. Feature maps obtained from the convolutional layer of a pre-trained VGG network were used to extract features in each channel. The optical flow was estimated using the motion constraint equation in each channel. Further, the dynamic texture was learned using the data from different channels. The proposed approach achieved an AUC (area under curve) score of 99.99% in the 3DMAD dataset.

George et al. [12] presented an approach for detection of presentation attacks using a training strategy leveraging both binary and pixel-wise binary loss. The method achieved superior intra- as well as cross-database performance when fine-tuned from pre-trained DenseNet blocks, showing the effectiveness of the proposed loss function.

In [46], George and Marcel have shown that fine-tuning vision transformer models work well in both intra- as well as cross-database settings. However, the computational complexity of these models makes it harder to deploy these models in edge devices.

11.2.1.3 One-Class Classifier-Based Approaches

Most of these methods handle the PAD problem as binary classification, which results in classifiers overfitting to the known attacks resulting in poor generalization to unseen attacks. We focus the further discussion on the detection of unseen attacks. However, methods working for unseen attacks must perform accurately for known attacks as well. One naive solution for such a task is one-class classifiers (OCC). OCC provides a straightforward way of handling the unseen attack scenario by modeling the distribution of the *bona fide* class alone.

Arashloo et al. [40] and Nikisins et al. [38] have shown the effectiveness of one-class methods against unseen attacks. Even though these methods performed better than binary classifiers in an unseen attack scenario, the performance in known attack protocols was inferior to that of binary classifiers. Xiong et al. [47] proposed unseen PAD methods using auto-encoders and one-class classifiers with texture features extracted from images. However, the performance of the methods compared to recent CNN-based methods is very poor. CNN-based methods outperform most of the feature-based baselines for PAD task. Hence, there is a clear need for one-class classifiers or anomaly detectors in the CNN framework. One of the drawbacks of one-class model is that they do not use the information provided by the known attacks. An anomaly detector framework that utilizes the information from the known attacks could be more efficient.

Perera and Patel [48] presented an approach for one-class transfer learning in which labeled data from an unrelated task is used for feature learning. They used two loss functions, namely, descriptive loss and compactness loss to learn the representations. The data from the class of interest is used to calculate the compactness loss whereas an external multi-class dataset is used to compute the descriptive loss. Accuracy of the learned model in classification using another database is used as the descriptive loss. However, in the face PAD problem, this approach would be challenging since the *bona fide* and attack classes appear very similar.

Fatemifar et al. [49] proposed an approach to ensemble multiple one-class classifiers for improving the generalization of PAD. They introduced a class-specific normalization scheme for the one-class scores before fusion. Seven regions, three one-class classifiers, and representations from three CNNs were used in the pool of classifiers. Though their method achieved better performance as compared to client independent thresholds, the performance is inferior to CNN-based state-of-the-art methods. Specifically, many CNN-based approaches have achieved 0% Half Total Error Rate (HTER) in Replay-Attack and Replay-Mobile datasets. Moreover, the challenging unseen attack scenario is not evaluated in this work.

Pérez-Cabo et al. [50] proposed a PAD formulation from an anomaly detection perspective. A deep metric learning model is proposed, where a triplet focal loss is used as a regularization for "metric-softmax", which forces the network to learn discriminative features. The features learned in such a way are used together with an SVM with Radial Basis Function (RBF) kernel for classification. They have performed several experiments on an aggregated RGB-only dataset showing the improvement made by their proposed approach. However, the analysis is mostly limited to RGB-only models and 2D attacks. Challenging 3D and partial attacks are not considered in this work. Specifically, the effectiveness in challenging unknown attacks (2D vs. 3D) is not evaluated.

Recently, Liu et al. [51] proposed an approach for the detection of unknown spoof attacks as Zero-Shot Face Anti-spoofing (ZSFA). They proposed a Deep Tree Network (DTN) which partitions the attack samples into semantic sub-groups in an unsupervised manner. Each tree node in their network consists of a Convolutional Residual Unit (CRU) and a Tree Routing Unit (TRU). The objective is to route the unknown attacks to the most proper leaf node for correctly classifying them. They have considered a wide variety of attacks in their approach and their approach achieved superior performance compared to the considered baselines.

Jaiswal et al. [52] proposed an end-to-end deep learning model for PAD that used unsupervised adversarial invariance. In their method, the discriminative information and nuisance factors are disentangled in an adversarial setting. They showed that by retaining only discriminative information, the PAD performance improved for the same base architecture. Mehta et al. [53] trained an Alexnet model with a combination of cross-entropy and focal losses. They extracted the features from Alexnet and trained a two-class SVM for the PAD task. However, results in challenging datasets such as OULU and SiW were not reported.

Recently, Joshua and Jain [54] utilized multiple Generative Adversarial Networks (GAN) for spoof detection in fingerprints. Their method essentially consisted of training a Deep Convolutional GAN (DCGAN) [55] using only the *bona fide* samples. At the end of the training, the generator is discarded, and the discriminator is used as the PAD classifier. They combined the results from different GANs operating on different features. However, this approach may not work well for face images as the recaptured images look very similar to the *bona fide* samples.

11.2.2 Multi-channel Methods

In general, most of the visible spectrum-based PAD methods try to detect the subtle differences in image quality when it is recaptured. With the advances in sensor and printer technology, the quality of the generated PA instruments improves over time. The high fidelity of PAs might make it difficult to recognize the subtle differences between bona fide and PAs. For 3D attacks, the problem is even more severe. As the technology to make detailed masks is available, it becomes very hard to distinguish between *bona fide* and presentation attacks by just using visible spectrum imaging. Many researchers have suggested using multi-spectral and extended range imaging to solve this issue [14, 15].

Raghavendra et al. [14] presented an approach using multiple spectral bands for face PAD. The main idea is to use complementary information from different bands. To combine multiple bands they observed a wavelet-based feature level fusion and a score fusion methodology. They experimented with detecting print attacks prepared using different kinds of printers. They obtained better performance with score-level fusion as compared to the feature fusion strategy.

Erdogmus and Marcel [2] evaluated the performance of several face PAD approaches against 3D masks using the 3DMAD dataset. This work demonstrated that 3D masks could fool PAD systems easily. They achieved HTER of 0.95% and 1.27% using simple LBP features extracted from color and depth images captured with Kinect.

Steiner et al. [15] presented an approach using multi-spectral SWIR imaging for face PAD. They considered four wavelengths—935 nm, 1060 nm, 1300 nm, and 1550 nm. In their approach, they trained an SVM for classifying each pixel as a skin pixel or not. They defined a Region Of Interest (ROI) where the skin is likely to be present, and skin classification results in the ROI are used for classifying PAs. The approach obtained 99.28% accuracy in per pixel skin classification.

Dhamecha et al. [16] proposed an approach for PAD by combining the visible and thermal image patches for spoofing detection. They classified each patch as either *bona fide* or attack and used the *bona fide* patches for subsequent face recognition pipeline.

In [18], Bhattacharjee et al. showed that it is possible to spoof commercial face recognition systems with custom silicone masks. They also proposed to use the mean temperature of the face region for PAD.

Bhattacharjee et al. [19] presented a preliminary study of using multi-channel information for PAD. In addition to visible spectrum images, they considered thermal, near-infrared, and depth channels. They showed that detecting rigid masks and 2D attacks is simple in thermal and depth channels, respectively. Most of the attacks can be detected with a similar approach with combinations of different channels, where the features and combinations of channels to use are found using a learning-based approach.

Wang et al. [56] proposed multimodal face presentation attack detection with a ResNet-based network using both spatial and channel attentions. Specifically, the approach was tailored for the *CASIA-SURF* [57] database which contained RGB, near-infrared, and depth channels. The proposed model is a multi-branch model where the individual channels and fused data are used as inputs. Each input channel has its own feature extraction module and the features extracted are concatenated in a late fusion strategy. Followed by more layers to learn a discriminative representation for PAD. The network training is supervised by both center loss and softmax loss. One key point is the use of spatial and channel attention to fully utilize complementary information from different channels. Though the proposed approach achieved good results in the *CASIA-SURF* database, the challenging problem of unseen attack detection is not addressed.

Parkin et al. [58] proposed a multi-channel face PAD network based on ResNet. Essentially, their method consists of different ResNet blocks for each channel followed by fusion. Squeeze and excitation modules (SE) are used before fusing the channels, followed by remaining residual blocks. Further, they add aggregation blocks at multiple levels to leverage inter-channel correlations. Their approach achieved state-of-the-art results in *CASIA-SURF* [57] database. However, the final model presented is a combination of 24 neural networks trained with different attack-specific folds, pre-trained models, and random seeds, which would increase the computation greatly.

11.2.3 Open Challenges in PAD

In general, presentation attack detection in a real-world scenario is still challenging. Most of the PAD methods available in prevailing literature try to solve the problem for a limited number of presentation attack instruments. Though some success has been achieved in addressing 2D presentation attacks, the performance of the algorithms in realistic 3D masks and other kinds of attacks is poor.

As the quality of attack instruments evolves, it becomes increasingly difficult to discriminate between *bona fide* and PAs in the visible spectrum alone. In addition, more sophisticated attacks, like 3D silicone masks, make PAD in visual spectra challenging. These issues motivate the use of multiple channels, making PAD systems harder to bypass.

We argue that the accuracy of the PAD methods can get better with a multi-channel acquisition system. Multi-channel acquisition from consumer-grade devices can improve performance significantly. Hybrid methods, combining both extended hardware and software could help in achieving good PAD performance in real-world scenarios. We extend the idea of a hybrid PAD framework and develop a multi-channel framework for presentation attack detection.

11.3 PAD Approach

We present three different strategies to fuse multi-channel information for the presentation attack detection task. Different stages of the PAD framework are described in this section.

11.3.1 Preprocessing

The PAD pipeline acts on the cropped facial images. For the RGB image, the preprocessing stage consists of face detection and landmark localization using the MTCNN [59] framework, followed by alignment. The detected face is aligned by making the eye centers horizontal followed by resizing them to a resolution of 224×224. For the non-RGB images, a normalization method using the median absolute deviation (MAD) [60] is used to normalize the face image to an 8-bit range. The raw images from RGB and other channels are already spatially registered so that the same transformation can be used to align the face in the non-RGB channels.

11.3.2 Network Architectures for Multi-channel PAD

From the prevailing literature, it has been observed that multi-channel methods are robust against a wide range of attacks [20–23]. Broadly, there are four different strategies to fuse the information from multiple channels, they are (1) early fusion, meaning the channels are stacked at the input level (for example, MC-PixBiS [21]). The second strategy is late fusion, meaning the representations from different networks are combined at a later stage similar to feature fusion (for example, MCCNN [20]). A third strategy is a hybrid approach where information from multiple levels is combined as in [58] or [61]. A fourth strategy is score-level fusion where individual networks are trained separately for different channels and score-level fusion is performed on the scalar scores from each channel. However, the score fusion performs poorly compared to other methods since it does not use cross-channel relations efficiently. The details of the fusion strategies used are presented in the following subsection.

11.3.2.1 Late Fusion: Multi-channel CNN (MCCNN-OCCL-GMM)

This architecture uses a late fusion strategy for combining multiple channels for the face PAD problem. The main idea in the Multi-Channel CNN (MC-CNN) is to use the joint representation from multiple modalities for PAD, using transfer learning from a pre-trained face recognition network [20]. The underlying hypothesis is that the joint representation in the face space could contain discriminative information for PAD. This network consists of three parts: low- and high-level convolutional/pooling layers, and fully connected layers, as shown in Fig. 11.1. As noted in [62], high-level features in deep convolutional neural networks trained in the visual spectrum are domain independent, i.e., they do not depend on a specific modality. Consequently, they can be used to encode face images collected from different image-sensing domains. The parameters of this CNN can then be split into higher level layers (shared among the different channels) and lower level layers (known as Domain-Specific Units). By concatenating the representation from different channels and using fully connected layers, a decision boundary for the appearance of bona fide and attack presentations can be learned via back-propagation. During training, low-level layers are adapted separately for different modalities, while shared higher level layers remain unaltered. In the last part of the network, embeddings extracted from all modalities are concatenated, and two fully connected layers are added. The first fully connected layer has ten nodes and the second one has one node. Sigmoidal activation functions are used in each fully connected layer, as in the original implementation [20]. These layers, added on top of the concatenated representations, are tuned exclusively for the PAD task using the binary cross-entropy as the loss function.

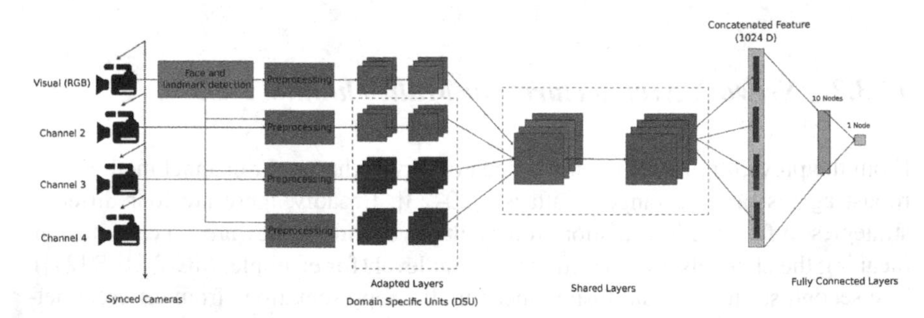

Fig. 11.1 Block diagram of the MC-CNN network. The gray color blocks in the CNN part represent layers which are not re-trained, and other colored blocks represent re-trained/adapted layers. Note that the original approach from [20] is depicted here: it takes grayscale, infrared, depth, and thermal data as input. The channels used can be changed depending of the available channels

The MC-CNN approach hence introduces a novel solution for multimodal PAD problems, leveraging a pre-trained network for face recognition when a limited amount of data is available for training PAD systems. Note that this architecture can be easily extended for an arbitrary number of input channels.

Later, in [23], this work was extended to a one-class implementation utilizing a newly proposed one-class contrastive loss (OCCL) and Gaussian mixture model. Essentially, the new loss function forces the network to learn a compact embedding for the bona fide channel, making sure that attacks are far from the bona fide attacks. This network learned is used as a fixed feature extractor and used together with a one-class Gaussian mixture model to perform the final classification. This approach yielded better results in unseen attacks.

11.3.2.2 Early Fusion: Multi-channel Pixel-Wise Binary Supervision (MC-PixBiS)

This architecture showcases the use of early fusion for a multi-channel PAD system. The Multi-Channel Deep Pixel-wise Binary Supervision network (MC-PixBiS) is a multi-channel extension of a recently published work on face PAD using legacy RGB sensors [12]. The main idea in [12] is to use pixel-wise supervision as an auxiliary supervision. The pixel-wise supervision forces the network to learn shared representations, and it acts as a patch-wise method (see Fig. 11.2). To extend this network for a multimodal scenario, the method proposed in [63] was used, i.e., averaging the filters in the first layer and replicating the weights for different modalities.

The general block diagram of the framework is shown in Fig. 11.2 and is based on DenseNet [64]. The first part of the network contains eight layers, and each layer consists of two dense blocks and two transition blocks. The dense blocks consist of dense connections between every layer with the same feature map size, and the transition blocks normalize and down sample the feature maps. The output from the eighth layer is a map of size 14×14 with 384 features. A 1×1 convolution layer is added along with sigmoid activation to produce the binary feature map. Further, a

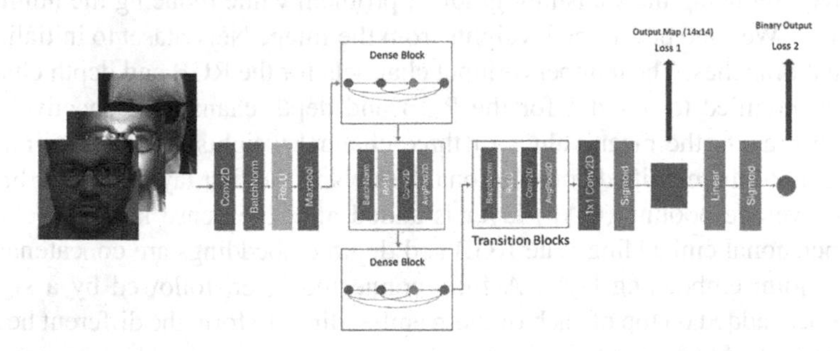

Fig. 11.2 MC-PixBiS architecture with pixel-wise supervision. Input channels are stacked before being passed to a series of dense blocks

fully connected layer with sigmoid activation is added to produce the binary output. A combination of losses is used as the objective function to minimize

$$\mathcal{L} = \lambda \mathcal{L}_{pix} + (1 - \lambda)\mathcal{L}_{bin} \tag{11.1}$$

where $\mathcal{L}_{pix}$ is the binary cross-entropy loss applied to each element of the 14×14 binary output map and $\mathcal{L}_{bin}$ is the binary cross-entropy loss on the network's binary output. A λ value of 0.5 was used in our implementation. Even though both losses are used in training, in the evaluation phase, only the pixel-wise map is used: the mean value of the generated map is used as a score reflecting the probability of bona fide presentation.

11.3.2.3 Hybrid (Multi-head): Cross-Modal Focal Loss (RGBD-MH(CMFL))

The architecture presented here shows a hybrid approach to presentation attack detection [61]. A multi-head architecture that follows a hybrid fusion strategy is detailed here. The architecture of the network is shown in Fig. 11.4. Essentially, the architecture consists of a two-stream network with separate branches for the component channels. The embeddings from the two channels are combined to form the third branch. Fully connected layers are added to each of these branches to form the final classification head. These three heads are jointly supervised by a loss function which forces the network to learn discriminative information from individual channels as well as the joint representation, reducing the possibility of overfitting. The multi-head structure also makes it possible to perform scoring even when a channel is missing at test time, meaning that we can do scoring with RGB branch alone (just using the score from the RGB head) even if the network was trained on a combination of two channels.

The individual branches are comprised of the first eight blocks (following the DeepPixBiS architecture [12]) from DenseNet architecture (densenet161) proposed by Huang et al. [65]. In the DenseNet architecture, each layer is connected to every other layer, reducing the vanishing gradient problem while reducing the number of parameters. We used pre-trained weights from the Image Net dataset to initialize the individual branches. The number of input channels for the RGB and depth channels has been modified to 3 and 1 for the RGB and depth channels, respectively. For the depth branch, the mean values of three-channel weights are used to initialize the weights of the modified convolutional kernels in the first layer. In each branch, a global average pooling (GAP) layer is added after the dense layers to obtain a 384-dimensional embedding. The RGB and depth embeddings are concatenated to form the joint embedding layer. A fully connected layer, followed by a sigmoid activation, is added on top of each of these embeddings to form the different heads in the framework. At training time, each of these heads is supervised by a separate loss function. At test time, the score from the RGB-D branch is used as the PAD score.

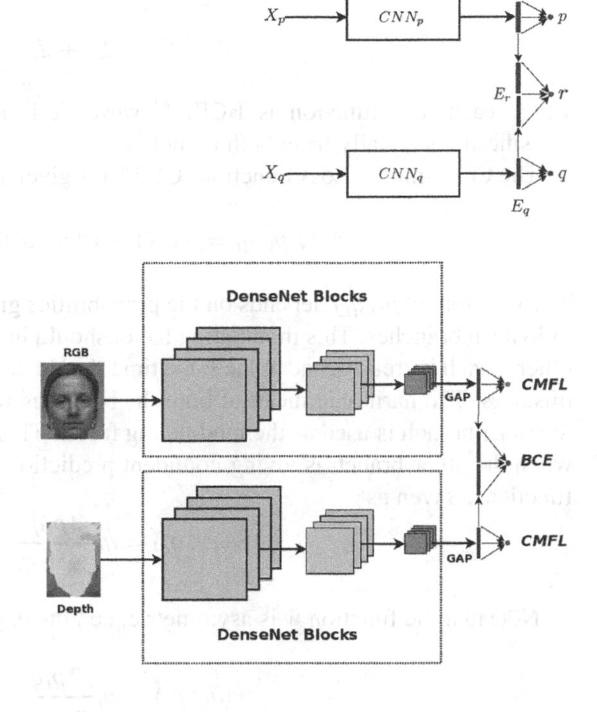

Fig. 11.3 Diagram of the two-stream multi-head model, showing the embeddings and probabilities from individual and joint branches. This can be extended to multiple heads as well

Fig. 11.4 The proposed framework for PAD. A two-stream multi-head architecture is used following a late fusion strategy. Heads corresponding to individual channels are supervised by the proposed cross-modal focal loss (CMFL), while the joint model is supervised by binary cross-entropy (BCE)

A cross-modal focal loss (CMFL) to supervise the individual channels is also proposed in this work [61]. The core idea is that, when one of the channels can correctly classify a sample with high confidence, then the loss contribution of the sample in the other branch can be reduced. If a channel can correctly classify a sample confidently, then we don't want the other branch to penalize the model more. CMFL forces each branch to learn robust representations for individual channels, which can then be utilized with the joint branch, effectively acting as an auxiliary loss function.

The idea of relaxing the loss contribution of samples correctly classified is similar to the Focal Loss [66] used in object detection problems. In Focal Loss, a modulating factor is used to reduce the loss contributed by samples that are correctly classified with high confidence. We use this idea by modulating the loss factoring in the confidence of the sample in the current and the alternate branch.

Consider the two-stream multi-branch multi-head model in Fig. 11.3. X_p and X_q denote the image inputs from different modalities, and E_p, E_q, and E_r denote the corresponding embeddings for the individual and joint representations. In each branch, after the embedding layer, a fully connected layer (followed by a sigmoid layer) is present which provides classification probability. The variables p, q, and r denote these probabilities (Fig. 11.4).

The naive way to train a model is to use BCE loss on all three branches as

$$\mathcal{L} = \mathcal{L}_p + \mathcal{L}_q + \mathcal{L}_r \tag{11.2}$$

where each loss function is BCE. However, this approach penalizes all miss-classifications equally from both branches.

The Cross-modal Loss Function (CMFL) is given as follows:

$$CMFLp_{\mathrm{t}}, q_{\mathrm{t}} = -\alpha_{\mathrm{t}}(1 - w(p_t, q_t))^{\gamma} \log(p_{\mathrm{t}}) \tag{11.3}$$

The function $w(p_t, q_t)$ depends on the probabilities given by the channels from two individual branches. This modulating factor should increase as the probability of the other branch increases, and at the same time should be able to prevent very confident mistakes. The harmonic mean of both the branches weighted by the probability of the other branch is used as the modulating factor. This reduces the loss contribution when the other branch is giving confident predictions. And the expression for this function is given as

$$w(p_t, q_t) = q_t \frac{2p_t q_t}{p_t + q_t} \tag{11.4}$$

Note that the function w is asymmetric, i.e., the expression for $w(q_t, p_t)$ is

$$w(q_t, p_t) = p_t \frac{2p_t q_t}{p_t + q_t} \tag{11.5}$$

meaning the weight function depends on the probability of the other branch. Now we use the proposed loss function as auxiliary supervision, and the overall loss function to minimize is given as

$$\mathcal{L} = (1 - \lambda)\mathcal{L}_{\mathrm{CE}(r_{\mathrm{t}})} + \lambda(\mathcal{L}_{CMFLp_{\mathrm{t}},q_{\mathrm{t}}} + \mathcal{L}_{CMFLq_{\mathrm{t}},p_{\mathrm{t}}}) \tag{11.6}$$

The value of λ was non-optimally as 0.5 for the study. When the probability of the other branch is zero, then the loss is equivalent to standard cross-entropy. The loss contribution is reduced when the other branch can correctly classify the sample, i.e., when an attack example is misclassified by network CNN_p, the network CNN_p is penalized unless model CNN_q can classify the attack sample with high confidence. As the $w(p, q) \rightarrow 1$, the modulating factor goes to zero, meaning if one channel can classify it perfectly, then the other branch is less penalized. Also, the focusing parameter γ can be adapted to change the behavior of the loss curve. We used an empirically obtained value of $\gamma = 3$ in all our experiments.

11.4 Experiments

We have used the *HQ-WMCA* dataset for the experiments, which contains a wide variety of 2D, 3D, and partial attacks, collected from different channels such as color, thermal, infrared, depth, and short-wave infrared.

11.4.1 Dataset: HQ-WMCA

The High-Quality Wide Multi-Channel Attack (*HQ-WMCA*) [21, 67] dataset consists of 2904 short multi-channel video recordings of both bona fide and presentation attacks. This database again consists of a wide variety of attacks including both obfuscation and impersonation attacks. Specifically, the attacks considered are print, replay, rigid mask, paper mask, flexible mask, mannequin, glasses, makeup, tattoo, and wig (Fig. 11.5). The database consists of recordings from 51 different subjects, with several channels including color, depth, thermal, infrared (spectra), and short-wave infrared (spectra). In this work, we consider the RGB channel captured with Basler acA1921-150uc camera and depth image captured with Intel RealSense D415.

11.4.2 Protocols

We use the grand_test as well as the leave-one-out (LOO) attack protocols distributed with the *HQ-WMCA* dataset. Specifically, in the LOO protocols, one attack is left out in the train and development set and the evaluation set consists of bona fide and the attack which was left out in the train and development set. This constitutes the unseen attack protocols or zero-shot attack protocols. The performance of the PAD methods in these protocols gives a more realistic estimate of their robustness against unseen attacks in real-world scenarios. In addition, we performed experiments with known attack protocols to evaluate the performance in a known attack scenario.

Fig. 11.5 Attacks present in *HQ-WMCA* dataset: **a** Print, **b** Replay, **c** Rigid mask, **d** Paper mask, **e** Flexible mask, **f** Mannequin, **g** Glasses, **h** Makeup, **i** Tattoo, and **j** Wig. Image taken from [21]

11.4.3 Metricsx

For the evaluation of the algorithms, we have used the ISO/IEC 30107-3 metrics [68], Attack Presentation Classification Error Rate (APCER), and Bona fide Presentation Classification Error Rate (BPCER) along with the Average Classification Error Rate (ACER) in the *eval* set. We compute the threshold in the *dev* set for a BPCER value of 1%, and this threshold is applied in the *eval* set to compute the reported metrics.

$$ACER = \frac{APCER + BPCER}{2}. \tag{11.7}$$

11.4.4 Implementation Details

We performed data augmentation during the training phase with random horizontal flips with a probability of 0.5. The combined loss function is minimized with Adam Optimizer [69]. A learning rate of 1×10^{-4} was used with a weight decay parameter of 1×10^{-5}. We used a mini-batch size of 64, and the network was trained for 25 epochs on a GPU grid. The architecture and the training framework were implemented using the PyTorch [70] library.

11.4.5 Baselines

For a fair comparison with state of the art, we have implemented three different multi-channel PAD approaches from literature as described in Sect. 11.3.2 for the RGB-D channels. Besides, we also introduce a multi-head architecture supervised with BCE alone, as another baseline for comparison. The baselines implemented are listed below

RGB-DeepPixBiS: This is an RGB-only CNN-based system [12], trained using both binary and pixel-wise binary loss function. This model is used as a baseline for comparing with multi-channel models.

MC-PixBiS: This is a CNN-based system [12], extended to multi-channel scenario as described in [21] trained using both binary and pixel-wise binary loss function. This model uses RGB and depth channels stacked together at the input level.

MCCNN-OCCL-GMM: This model is the multi-channel CNN system proposed to learn one-class model using the one-class contrastive loss (OCCL) and Gaussian mixture model as reported in [23]. The model was adapted to accept RGB-D channels as the input.

MC-ResNetDLAS: This is the reimplementation of the architecture from [58], which won the first prize in the "CASIA-SURF" challenge, extending it to RGB-D channels, based on the open-source implementation [71]. We used the initialization from the best pre-trained model as suggested in [58] followed by retraining in the current protocols using RGB-D channels.

RGBD-MH-BCE: This uses the multi-head architecture shown in Fig. 11.4, where all the branches are supervised by binary cross-entropy (BCE). In essence, this is equivalent to setting the value of $\gamma = 0$, in the expression for the cross-modal loss function. This is shown as a baseline to showcase the improvement by the new multi-head architecture alone and to contrast with the performance change with the new loss function.

RGBD-MH-CMFL: This uses the multi-head architecture shown in Fig. 11.4, where individual branches are supervised by CMFL loss and joint branch is supervised by BCE loss.

11.4.6 Experiments and Results

Results in HQ-WMCA dataset: Table 11.1 shows the performance of different methods in the grandtest protocol, which evaluates the performance of the methods in a known attack scenario, meaning all attacks are distributed equally across train, development, and test set. From the ACER values, it can be seen that the multi-head architecture performs the best followed by MC-PixBiS architecture. The RGB alone model (RGB-DeepPixBiS) also performs reasonably well in this protocol. The corresponding ROC plots for the evaluation set are shown in Fig. 11.6. It can be seen from the ROC that the RGB-only model outperforms the multi-head model in the evaluation set opposite to the results in Table 11.1. In ACER evaluations, the threshold used is selected from the development set. The ROC plots only depict the performance in the evaluation set without considering the threshold selected from the development set causing the discrepancy. ACER reported shows more realistic performance estimates in real-world scenarios as the thresholds are fixed in advance according to specific performance criteria.

Table 11.1 Performance of the multi-channel systems in the grandtest protocol of HQ-WMCA dataset. The values reported are obtained with a threshold computed for BPCER 1% in *dev* set

	APCER	BPCER	ACER
RGB-DeepPixBiS	9.2	0.0	4.6
MC-PixBiS	9.7	0.0	4.8
MCCNN-OCCL-GMM	7.9	11.4	9.7
MC-ResNetDLAS	8.0	6.4	7.2
RGBD-MH-BCE	4.0	2.0	3.0
RGBD-MH-CMFL	6.6	0.1	3.3

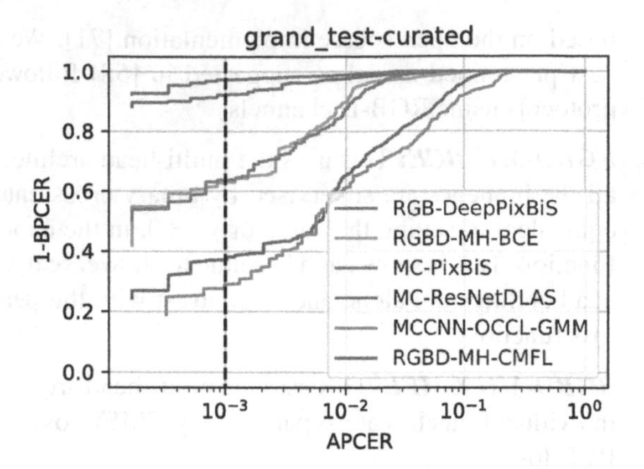

Fig. 11.6 ROC plot for evaluation set in the grandtest protocol in HQ-WMCA dataset

The *HQ-WMCA* dataset consists of challenging attacks, specifically, there are different types of partial attacks such as *Glasses* which occupy only a part of the face. These attacks are much harder to detect when they are not seen in the training set, as they appear very similar to bona fide samples. The analysis we performed is similar to a worst-case analysis since it specifically focuses on the unseen attack robustness. The experimental results in the LOO protocols of *HQ-WMCA* are tabulated in Table 11.2. Overall, *MCCNN-OCCL-GMM* and *MC-ResNetDLAS* do not perform well in the LOO protocols of *HQ-WMCA* database. In addition, the *MC-PixBiS* method also performs poorly in the unseen attack protocols of the *HQ-WMCA* dataset. This could be due to the challenging nature of the attacks in the database. The RGB-only method *RGB-DeepPixBiS* performs reasonably well overall too. It can be seen that the multi-head architecture, *RGBD-MH-BCE*, already improves the results as compared to all the baselines with an average ACER of 13.3 ± 16.5. With the addition of the *CMFL* loss, the ACER further improves to $11.6 \pm 14.8\%$. The results indicate that the proposed architecture already improves the performance in challenging attacks, and the proposed loss further improves the results achieving state-of-the-art results in the *HQ-WMCA* dataset.

From Table 11.2, it can be seen that the effectiveness of fusion strategies is different for unseen PA scenarios. While most of the multi-channel methods struggle to achieve good performance in detecting unseen *Flexiblemasks*, the *RGB-DeepPixBiS* achieves much better performance in this case. A similar trend can be seen in the case of *Makeup* attack as well. This could be due to the lack of additional information provided by the depth channel in these cases. The depth information in the case of these attacks is very similar to that of bona fide samples. However, multi-channel method provides a significant boost in performance in detecting *Rigidmask*, *Tattoo*, and *Replay* attacks. Attacks like *Papermask* and *Mannequins* are easier to detect in most of the cases due to the distinct appearance compared to bona fide samples. The multi-head architecture improves the performance compared to other baselines in most of the sub-protocols. The ROC plots for the eval set for the corresponding protocols are shown in Fig. 11.7.

Table 11.2 Performance of the multi-channel systems method in **unseen** protocols of HQ-WMCA dataset. The values reported are obtained with a threshold computed for BPCER 1% in *dev* set

	Flexiblemask	Glasses	Makeup	Mannequin	Papermask	Rigidmask	Tattoo	Replay	Mean±Std
RGB-DeepPixBiS [12]	5.8	49.3	23.8	0.0	0.0	25.9	13.6	6.0	15.5± 15.8
MC-PixBiS [12]	29.9	49.9	29.4	0.1	0.0	32.5	5.7	9.6	19.6±17.1
MCCNN-OCCL-GMM [23]	14.2	32.7	22.0	1.5	7.1	33.7	4.2	36.6	19.0±13.2
MC-ResNetDLAS [58]	23.5	50.0	33.8	1.0	2.6	31.0	5.7	15.5	20.3±16.2
RGBD-MH-BCE	16.7	38.1	43.3	0.4	1.3	3.0	2.0	2.3	13.3±16.5
RGBD-MH-CMFL [61]	14.8	37.4	34.9	0.0	0.4	2.4	2.4	1.0	**11.6±14.8**

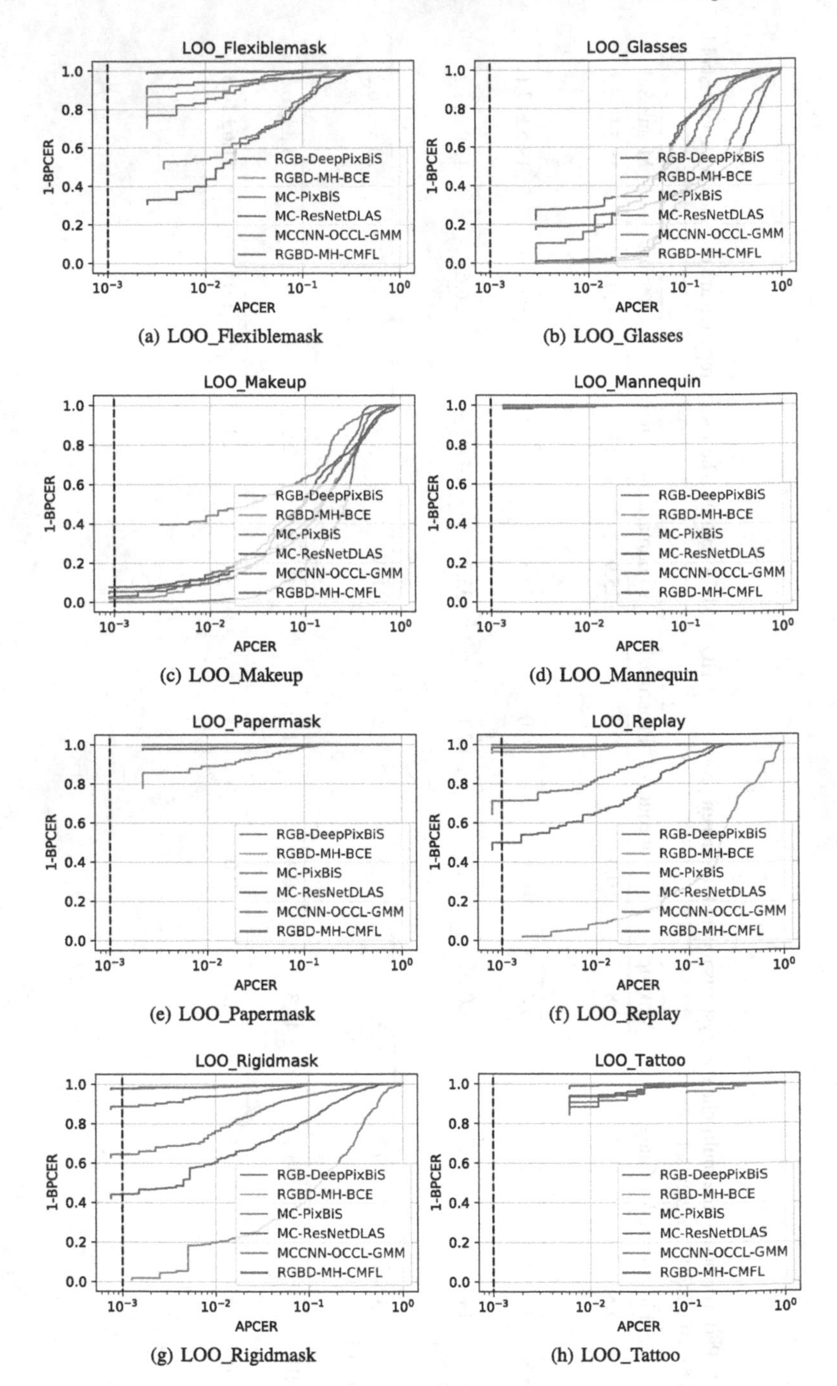

(a) LOO_Flexiblemask (b) LOO_Glasses (c) LOO_Makeup (d) LOO_Mannequin (e) LOO_Papermask (f) LOO_Replay (g) LOO_Rigidmask (h) LOO_Tattoo

Fig. 11.7 ROC plots for the evaluation set in **unseen attack** protocols of HQ-WMCA dataset

Table 11.3 Ablation study using only one channel at deployment time

	RGB	Depth
RGBD-MH-BCE	15.4±16.1	34.2±11.6
RGBD-MH-CMFL	**12.0±13.9**	**30.6±17.5**

Table 11.4 Computational and parameter complexity comparison

Model	Compute	Parameters
MCCNN-OCCL-GMM [23]	14.51 GMac	50.3 M
MC-PixBiS [12]	4.7 GMac	3.2M
MC-ResNetDLAS [58]	15.34 GMac	69.29 M
RGBD-MH(CMFL) [61]	9.16 GMac	6.39 M

Performance with missing channels: We evaluate the performance of the multi-head models when evaluated with only a single channel at test time. Consider a scenario where the model was trained with RGB and depth, and at the test time, only one of the channels is available. We compare with the mean performance in the *HQ-WMCA* dataset, with RGB and depth alone at test time. The results are shown in Table 11.3. For the baseline *RGBD-MH-BCE*, using RGB alone at test time the error rate is 15.4 ± 16.1, whereas for the proposed approach it improves to 12.0 ± 13.9. The performance improves for the depth channel as well.

From Table 11.3, it can be seen that the performance improves, as compared to using BCE even when using a single channel at the time of deployment. This shows that the performance of the system improves when the loss contributions of samples that are not possible to classify by that modality are reduced. Forcing the individual networks to learn a decision boundary leads to overfitting resulting in poor generalization.

11.4.7 Computational Complexity

Here we compare the complexity of the models in terms of parameters and compute required (for RGB and Depth channels). The comparison is shown in Table 11.4. It can be seen that the parameter and compute for late fusion (MCCNN-OCCL-GMM) is quite high. A lot of additional parameters are added for each channel before fusion which increases the total number of parameters. The MC-ResNetDLAS also suffers from a high number of parameters and compute. The early fusion method, MC-PixBiS, with the truncated DenseNet architecture saves compute and parameters a lot compared to others. Thanks to fusing the channels at the input level, the parameter increase is just for the first convolutional filter keeping rest of the operations the same. This makes it easy to add more channels as the rest of the network remains the same except for the first convolutional filter. Lastly, the RGBD-MH(CMFL) is composed of the PixBiS model for each of the channels, and hence roughly double the number of parameters and compute compared to the PixBiS model.

11.4.8 Discussions

From the results, it was observed that the late fusion method MCCNN-OCCL-GMM performed poorly compared to other methods. Also, this strategy increases the number of parameters with the increase in the number of channels. The MC-PixBiS model, on the other hand, does not increase the number of parameters with the increase in the number of channels. Each additional channel only changes the parameters in the first convolutional filter, which is negligible compared to the total number of parameters. In short, the early fusion method is more scalable to an increasing number of channels as the computational complexity is very less. However, this method cannot handle a missing channel scenario in a real-world application.

From the performance evaluations, it was seen that the RGBD-MH(CMFL) architecture achieves the best performance. Usage of multiple heads forces the network to learn multiple redundant features from individual channels as well as from the joint representation. This effectively acts as a regularization mechanism for the network, preventing overfitting to seen attacks. Further, one limitation of the multi-head representation is that the network is forced to leave discriminative features from all the channels, which may not be trivial. The CMFL loss proposed effectively addresses this issue by dynamically modulating the loss contribution from individual channels. Comparing the computational complexity, this model is relatively simpler compared to the late fusion model. Nevertheless, this model is more complex compared to the early fusion approach with nearly double the number of parameters, however, with the addition of parameters and the formulation it can be seen that the architecture itself improves the robustness, and again with the use of CMFL loss, the performance further improves indicating a good performance complexity tradeoff. The cross-modal focal loss function modulates the loss contribution of samples based on the confidence of individual channels. The framework can be trivially extended to multiple channels and different classification problems where information from one channel alone is inadequate for classification. This loss forces the network to learn complementary, discriminative, and robust representations for the component channels. The structure of the framework makes it possible to train models using all the available channels and to deploy with a subset of channels. One limitation of the framework is that the addition of more channels requires more branches which increases the parameters linearly with the number of channels. While we have selected RGB and depth channels for this study, mainly due to the availability of off-the-shelf devices consisting of these channels, it is trivial to extend this study to other combinations of channels as well, for instance, RGB-Infrared and RGB-Thermal.

11.5 Conclusions

In this chapter, we have presented different approaches using multi-channel information for presentation attack detection. All the approaches have their merits and limitations, however, we have conducted an extensive analysis of the unseen attack robustness as a worst-case performance evaluation. As multi-channel methods are required for safety-critical applications, robustness against unseen attacks is an essential requirement. In the evaluations, we have noted that the performance is much better for the multi-head architecture, thanks to the CMFL loss. The CMFL loss forces the network to learn complementary, discriminative, and robust representations for the component channels. This work can be straightforwardly extended to other combinations of channels and architectures as well. The *HQ-WMCA* database and the source code and protocols will be made available to download for research purposes. This will certainly foster further research in multi-channel face presentation attack detection in the future.

Acknowledgements Part of this research is based upon work supported by the Office of the Director of National Intelligence (ODNI), Intelligence Advanced Research Projects Activity (IARPA), via IARPA R&D Contract No. 2017-17020200005. The views and conclusions contained herein are those of the authors and should not be interpreted as necessarily representing the official policies or endorsements, either expressed or implied, of the ODNI, IARPA, or the U.S. Government. The U.S. Government is authorized to reproduce and distribute reprints for Governmental purposes notwithstanding any copyright annotation thereon.

References

1. Costa-Pazo A, Bhattacharjee S, Vazquez-Fernandez E, Marcel S (2016) The replay-mobile face presentation-attack database. In: 2016 international conference of the biometrics special interest group (BIOSIG). IEEE, pp 1–7
2. Erdogmus N, Marcel S (2014) Spoofing face recognition with 3D masks. IEEE Trans Inf Forensics Secur 9(7):1084–1097
3. de Freitas Pereira T, Anjos A, De Martino JM, Marcel S (2013) Can face anti-spoofing countermeasures work in a real world scenario? In: International conference on biometrics (ICB)
4. Purnapatra S, Smalt N, Bahmani K, Das P, Yambay D, Mohammadi A, George A, Bourlai T, Marcel S, Schuckers S et al (2021) Face liveness detection competition (livdet-face)-2021. In: 2021 IEEE international joint conference on biometrics (IJCB). IEEE, pp 1–10
5. Boulkenafet Z, Komulainen J, Hadid A (2015) Face anti-spoofing based on color texture analysis. In: ICIP
6. Määttä J, Hadid A, Pietikäinen M (2011) Face spoofing detection from single images using micro-texture analysis. In: IJCB
7. Chingovska I, Anjos A, Marcel S (2012) On the effectiveness of local binary patterns in face anti-spoofing. In: Proceedings of the 11th international conference of the biometrics special interest group, EPFL-CONF-192369
8. Anjos A, Marcel S (2011) Counter-measures to photo attacks in face recognition: a public database and a baseline. In: International joint conference on biometrics (IJCB)

9. Gan J, Li S, Zhai Y, Liu C (2017) 3d convolutional neural network based on face anti-spoofing. In: 2017 2nd international conference on multimedia and image processing (ICMIP). IEEE, pp 1–5
10. Li L, Xia Z, Li L, Jiang X, Feng X, Roli F (2017) Face anti-spoofing via hybrid convolutional neural network. In: 2017 international conference on the frontiers and advances in data science (FADS). IEEE, pp 120–124
11. Atoum Y, Liu Y, Jourabloo A, Liu X (2017) Face anti-spoofing using patch and depth-based cnns. In: 2017 IEEE international joint conference on biometrics (IJCB). IEEE, pp 319–328
12. George A, Marcel S (2019) Deep pixel-wise binary supervision for face presentation attack detection. In: 2019 international conference on biometrics (ICB). IEEE, pp 1–8
13. Liu Y, Stehouwer J, Jourabloo A, Liu X (2019) Deep tree learning for zero-shot face anti-spoofing. In: CVPR
14. Raghavendra R, Raja KB, Venkatesh S, Busch C (2017) Extended multispectral face presentation attack detection: an approach based on fusing information from individual spectral bands. In: 2017 20th international conference on information fusion (Fusion). IEEE, pp 1–6
15. Steiner H, Kolb A, Jung N (2016) Reliable face anti-spoofing using multispectral swir imaging. In: 2016 international conference on biometrics (ICB). IEEE, pp 1–8
16. Dhamecha TI, Nigam A, Singh R, Vatsa M (2013) Disguise detection and face recognition in visible and thermal spectrums. In: 2013 international conference on biometrics (ICB). IEEE, pp 1–8
17. Agarwal A, Yadav D, Kohli N, Singh R, Vatsa M, Noore A (2017) Face presentation attack with latex masks in multispectral videos. SMAD 13:130
18. Bhattacharjee S, Mohammadi A, Marcel S (2018) Spoofing deep face recognition with custom silicone masks. In: 2018 IEEE 9th international conference on biometrics theory, applications and systems (BTAS). IEEE, pp 1–7
19. Bhattacharjee S, Marcel S (2017) What you can't see can help you-extended-range imaging for 3d-mask presentation attack detection. In: International conference of the biometrics special interest group (BIOSIG). IEEE, pp 1–7
20. George A, Mostaani Z, Geissenbuhler D, Nikisins O, Anjos A, Marcel S (2019) Biometric face presentation attack detection with multi-channel convolutional neural network. In: IEEE transactions on information forensics and security, pp 1. https://doi.org/10.1109/TIFS.2019.2916652
21. Heusch G, George A, Geissbühler D, Mostaani Z, Marcel S (2020) Deep models and shortwave infrared information to detect face presentation attacks. IEEE Trans Biom Behav Identity Sci 2(4):399–409
22. George A, Marcel S (2020) Can your face detector do anti-spoofing? face presentation attack detection with a multi-channel face detector. In: Idiap Research Report, Idiap-RR-12-2020
23. George A, Marcel S (2020) Learning one class representations for face presentation attack detection using multi-channel convolutional neural networks. In: IEEE transactions on information forensics and security, pp 1
24. George A, Marcel S (2021) Multi-channel face presentation attack detection using deep learning. Springer International Publishing, Cham, pp 269–304. https://doi.org/10.1007/978-3-030-74697-1_13
25. George A, Geissbuhler D, Marcel S (2022) A comprehensive evaluation on multi-channel biometric face presentation attack detection. arXiv:2202.10286
26. About face id advanced technology (2021). https://support.apple.com/en-gb/HT208108
27. Siegmund D, Kerckhoff F, Magdaleno JY, Jansen N, Kirchbuchner F, Kuijper A (2020) Face presentation attack detection in ultraviolet spectrum via local and global features. In: 2020 international conference of the biometrics special interest group (BIOSIG). IEEE, pp 1–5
28. Raghavendra R, Raja KB, Busch C (2015) Presentation attack detection for face recognition using light field camera. IEEE Trans Image Process 24(3):1060–1075
29. Kaichi T, Ozasa Y (2021) A hyperspectral approach for unsupervised spoof detection with intra-sample distribution. In: 2021 IEEE international conference on image processing (ICIP). IEEE, pp 839–843

30. Ramachandra R, Busch C (2017) Presentation attack detection methods for face recognition systems: a comprehensive survey. ACM Comput Surv (CSUR) 50(1):8
31. Heusch G, Marcel S (2019) Remote blood pulse analysis for face presentation attack detection. In: Handbook of biometric anti-spoofing. Springer, pp 267–289
32. Liu Y, Jourabloo A, Liu X (2018) Learning deep models for face anti-spoofing: binary or auxiliary supervision. In: Proceedings of the ieee conference on computer vision and pattern recognition, pp 389–398
33. Rui Shao XL, Yuen PC (2017) Deep convolutional dynamic texture learning with adaptive channel-discriminability for 3d mask face anti-spoofing. In: IJCB
34. Bhattacharjee S, Mohammadi A, Marcel S (2018) Spoofing deep face recognition with custom silicone masks. In: International conference on biometrics theory, applications and systems (BTAS). IEEE
35. Galbally J, Marcel S, Fierrez J (2014) Image quality assessment for fake biometric detection: application to iris, fingerprint, and face recognition. IEEE Trans Image Process 23(2):710–724
36. Patel K, Han H, Jain AK, Ott G (2015) Live face video vs. spoof face video: Use of moiré patterns to detect replay video attacks. In: 2015 international conference on biometrics (ICB). IEEE, pp 98–105
37. Wen D, Han H, Jain AK (2015) Face spoof detection with image distortion analysis. IEEE Trans Inf Forensics Secur 10(4):746–761
38. Nikisins O, Mohammadi A, Anjos A, Marcel S (2018) On effectiveness of anomaly detection approaches against unseen presentation attacks in face anti-spoofing. In: The 11th IAPR international conference on biometrics (ICB 2018), EPFL-CONF-233583
39. Chingovska I, Dos Anjos AR (2015) On the use of client identity information for face anti-spoofing. IEEE Trans Info Forensics Secur 10(4):787–796
40. Arashloo SR, Kittler J, Christmas W (2017) An anomaly detection approach to face spoofing detection: a new formulation and evaluation protocol. IEEE Access 5:13868–13882
41. Sepas-Moghaddam A, Pereira F, Correia PL (2018) Light field-based face presentation attack detection: reviewing, benchmarking and one step further. IEEE Trans Inf Forensics Secur 13(7):1696–1709
42. Heusch G, Marcel S (2018) Pulse-based features for face presentation attack detection
43. Yang J, Lei Z, Li SZ (2014) Learn convolutional neural network for face anti-spoofing. arXiv:1408.5601
44. Boulkenafet Z, Komulainen J, Akhtar Z, Benlamoudi A, Samai D, Bekhouche S, Ouafi A, Dornaika F, Taleb-Ahmed A, Qin L et al (2017) A competition on generalized software-based face presentation attack detection in mobile scenarios. In: International joint conference on biometrics (IJCB), pp 688–696
45. Li H, He P, Wang S, Rocha A, Jiang X, Kot AC (2018) Learning generalized deep feature representation for face anti-spoofing. IEEE Trans Inf Forensics Secur 13(10):2639–2652
46. George A, Marcel S (2021) On the effectiveness of vision transformers for zero-shot face anti-spoofing. In: 2021 IEEE international joint conference on biometrics (IJCB). IEEE, pp 1–8
47. Xiong F, AbdAlmageed W (2018) Unknown presentation attack detection with face rgb images. In: 2018 IEEE 9th international conference on biometrics theory, applications and systems (BTAS). IEEE, pp 1–9
48. Perera P, Patel VM (2019) Learning deep features for one-class classification. IEEE Trans Image Process 28(11):5450–5463
49. Fatemifar S, Awais M, Arashloo SR, Kittler J (2019) Combining multiple one-class classifiers for anomaly based face spoofing attack detection. In: International conference on biometrics (ICB)
50. Pérez-Cabo D, Jiménez-Cabello D, Costa-Pazo A, López-Sastre R.J (2019) Deep anomaly detection for generalized face anti-spoofing. In: Proceedings of the ieee conference on computer vision and pattern recognition workshops, pp 0
51. Liu Y, Stehouwer J, Jourabloo A, Liu X (2019) Deep tree learning for zero-shot face anti-spoofing. In: The IEEE conference on computer vision and pattern recognition (CVPR)

52. Jaiswal A, Xia S, Masi I, AbdAlmageed W (2019) Ropad: Robust presentation attack detection through unsupervised adversarial invariance. arXiv:1903.03691
53. Mehta S, Uberoi A, Agarwal A, Vatsa M, Singh R Crafting a panoptic face presentation attack detector
54. Engelsma JJ, Jain AK (2019) Generalizing fingerprint spoof detector: learning a one-class classifier. arXiv:1901.03918
55. Radford A, Metz L, Chintala S (2015) Unsupervised representation learning with deep convolutional generative adversarial networks. arXiv:1511.06434
56. Wang G, Lan C, Han H, Shan S, Chen X (2019) Multi-modal face presentation attack detection via spatial and channel attentions. In: PIEEE/CVF conference on computer vision and pattern recognition workshops (CVRPw)
57. Zhang S, Wang X, Liu A, Zhao C, Wan J, Escalera S, Shi H, Wang Z, Li SZ (2018) Casia-surf: a dataset and benchmark for large-scale multi-modal face anti-spoofing. arXiv:1812.00408
58. Parkin A, Grinchuk O (2019) Recognizing multi-modal face spoofing with face recognition networks. In: IEEE/CVF conference on computer vision and pattern recognition workshops (CVRPw)
59. Zhang K, Zhang Z, Li Z, Qiao Y (2016) Joint face detection and alignment using multitask cascaded convolutional networks. IEEE Signal Process Lett 23(10):1499–1503
60. Nikisins O, George A, Marcel S (2019) Domain adaptation in multi-channel autoencoder based features for robust face anti-spoofing. In: 2019 international conference on biometrics (ICB). IEEE, pp 1–8
61. George A, Marcel S (2021) Cross modal focal loss for RGBD face anti-spoofing. In: IEEE/CVF conference on computer vision and pattern recognition (CVPR)
62. Pereira TdF, Anjos A, Marcel S (2019) Heterogeneous face recognition using domain specific units. IEEE Trans Inf Forensics Secur (TIFS) 14(7):1803–1816
63. Wang L, Xiong Y, Wang Z, Qiao Y, Lin D, Tang X, Van Gool L (2016) Temporal segment networks: towards good practices for deep action recognition. In: European conference on computer vision (ECCV). Springer, pp 20–36
64. Huang G, Liu Z, Van Der Maaten L, Weinberger KQ (2017) Densely connected convolutional networks. In: IEEE conference on computer vision and pattern recognition (CVPR), vol 1, p 3
65. Huang G, Liu Z, Van Der Maaten L, Weinberger KQ (2017) Densely connected convolutional networks. In: Proceedings of the IEEE conference on computer vision and pattern recognition, pp 4700–4708
66. Lin TY, Goyal P, Girshick R, He, K, Dollár P (2018) Focal loss for dense object detection
67. Mostaani ZA, Heusch G, Geissenbuhler D, Marcel S (2020) The high-quality wide multi-channel attack (hq-wmca) database
68. Information technology –International Organization for Standardization (2016) Standard, international organization for standardization
69. Kingma DP, Ba J (2015) Adam: a method for stochastic optimization. In: ICLR (Poster)
70. Paszke A, Gross S, Chintala S, Chanan G, Yang E, DeVito Z, Lin Z, Desmaison A, Antiga L, Lerer A (2017) Automatic differentiation in pytorch. In: NIPS-W
71. ResNetDLAS. https://github.com/AlexanderParkin/ChaLearn_liveness_challenge/ (2021). [Online; accessed 1-Feb-2021]

Chapter 12
Review of Face Presentation Attack Detection Competitions

Zitong Yu, Jukka Komulainen, Xiaobai Li, and Guoying Zhao

Abstract Face presentation attack detection (PAD) has received increasing attention ever since the vulnerabilities to spoofing have been widely recognized. The state of the art in unimodal and multi-modal face anti-spoofing has been assessed in eight international competitions organized in conjunction with major biometrics and computer vision conferences in 2011, 2013, 2017, 2019, 2020 and 2021, each introducing new challenges to the research community. In this chapter, we present the design and results of the five latest competitions from 2019 until 2021. The first two challenges aimed at evaluating the effectiveness of face PAD in multi-modal setup introducing near-infrared (NIR) and depth modalities in addition to colour camera data, while the latest three competitions focused on evaluating domain and attack type generalization abilities of face PAD algorithms operating on conventional colour images and videos. We also discuss the lessons learnt from the competitions and future challenges in the field in general.

12.1 Introduction

Presentation attacks (PAs) [1], commonly referred to also as spoofing, pose a serious security issue to biometric systems. Automatic face recognition (AFR) systems, in particular, are easy to be deceived, e.g. using images of the targeted person published on the web or captured from distance. Many works, such as [2], have concluded that face recognition systems are vulnerable to sensor-level attacks launched with

Z. Yu (✉) · J. Komulainen · X. Li (✉) · G. Zhao (✉)
Center for Machine Vision and Signal Analysis, University of Oulu, Oulu, Finland
e-mail: zitong.yu@oulu.fi

X. Li
e-mail: xiaobai.li@oulu.fi

G. Zhao
e-mail: guoying.zhao@oulu.fi

J. Komulainen (✉)
Visidon Ltd, Oulu, Finland
e-mail: jukka.komulainen@oulu.fi

S. Marcel et al. (eds.), *Handbook of Biometric Anti-Spoofing*, Advances in Computer Vision and Pattern Recognition, https://doi.org/10.1007/978-981-19-5288-3_12

different presentation attack instruments (PAI), such as prints, displays and wearable 3D masks. The vulnerability to PAs is one of the main reasons for the lack of public confidence in AFR systems, especially in high-security applications, such as mobile payment services, which has created a necessity for robust solutions to counter spoofing.

One possible solution is to include a dedicated face presentation attack detection (PAD) component into AFR systems. Face PAD, commonly referred to also as face anti-spoofing (FAS) or liveness detection, aims at automatically differentiating whether the presented face biometric sample originates from a bona fide subject or an artefact. Based on the used imaging modalities, face PAD schemes can be broadly categorized into unimodal and multi-modal methods. Unimodal face PAD systems usually exploit efficient visual features extracted from conventional colour (RGB) camera data for binary classification (i.e. bona fide vs. attack), thus they can be easily deployed in most practical AFR scenarios but with limited accuracy. On the contrary, multi-modal methods [3] introduce some additional imaging modalities (e.g. depth, near-infrared (NIR) or thermal infrared sensor data) that can capture specific intrinsic differences between the bona fide and attack samples but with extra hardware costs. For example, the depth maps obtained from 2D printout and display face artefacts using 3D sensors usually have flat and close-to-zero distributions in facial regions.

Despite the recent progress in deep learning-based face anti-spoofing methods [4, 5] with powerful representation capacity, it is difficult to tell what are the best or most promising feature learning-based approaches for generalized face PAD. Along with the development in manufacturing technologies, it has become even cheaper for an attacker to exploit known vulnerabilities of AFR systems with different kinds of face artefacts, such as a realistic 3D mask made of plaster. Simulating AFR scenarios with various attacks, environmental conditions, acquisition devices and subjects is extremely time-consuming and expensive, but the domain shifts caused by such covariates have a significant impact on the face PAD performance. These issues have been already explored with several public unimodal datasets, such as [6–10]. However, these benchmarks have been yet rather small-scale in terms of number of subjects, samples and AFR scenarios and, consequently, the corresponding evaluation protocols have been too limited (e.g. in terms of unknown PAs and acquisition conditions in the test set). Moreover, there have been no public benchmarks and protocols for evaluating multi-modal face PAD schemes until recently.

Competitions play a key role in advancing the research on face PAD and provide valuable insights for the entire face recognition community. It is important to organize collective evaluations regularly in order to assess, or ascertain, the current state of the art and gain insight on the robustness of different approaches using a common platform. Also, new, more challenging public datasets are often collected and introduced within such collective efforts to the research community for future development and benchmarking use. The quality of PAIs keeps improving as technology (e.g. 2D/3D printers and displays) gets cheaper and better, which is another reason why benchmark datasets need to be updated regularly. Open contests are likely to inspire researchers and engineers beyond the field to participate, and their outside-the-box thinking may lead to new ideas on the problem of face PAD and novel countermeasures.

In the context of face PAD, altogether eight international competitions [11–18] have been organized in conjunction with major biometrics and computer vision conferences in 2011, 2013, 2017, 2019, 2020 and 2021, each introducing new challenges to the research community. In this chapter, we focus on analysing the design and results of the five latest competitions [11, 15–18] from 2019 until 2021, while an extensive review of the first three competitions [12–14] can be found in [19]. The key features of the five most recent face PAD competitions are summarized in Table 12.1.

The multi-modal face anti-spoofing challenge organized in 2019 (referred to as CVPR2019 challenge) [11] provided an initial assessment of multi-modal countermeasures to various kinds of print attacks by introducing a precisely defined test protocol for evaluating the performance of the face PAD solutions with three modalities (i.e. colour, depth and NIR). In 2020, the cross-ethnicity face anti-spoofing challenge (referred to as CVPR2020 challenge) [15] extended the previous CVPR2019 challenge with several new factors (e.g. unseen ethnicities and 3D mask attacks), and having separate competition tracks for unimodal (colour) and multi-modal (colour, depth and NIR) data. While the datasets used in these first two contests contained a limited number of subjects and samples, the CelebA-Spoof Challenge in 2020 (referred to as ECCV2020 challenge) [18] provided an assessment on the performance of face PAD methods on the largest publicly available unimodal (colour) benchmark dataset. The LivDet Face 2021 liveness detection competition (referred to as IJCB2021 LivDet-Face) [17] focused on the challenging domain generalization issues and unseen attacks, with separate competition tracks for methods using colour image and video data. In order to bridge the gap between competitions (i.e. laboratory conditions) and real-world application scenarios, the test data was practically concealed as only few samples of attacks were provided. Finally, to further evaluate the performance of face PAD approaches under challenging 3D mask attacks, the 3D high-fidelity mask attack detection challenge in 2021 (referred to as ICCV2021

Table 12.1 Summary of the recent five face PAD competitions organized from 2019 until 2021

Competition	Modality	Highlight	Limitation
CVPR2019 challenge [11]	RGB, depth, NIR	First multi-modal face PAD challenge	With only print attacks
CVPR2020 challenge [15]	RGB, depth, NIR	Cross-ethnicity and cross-PAI testing	Testing with only print and mask PAs
ECCV2020 challenge [18]	RGB	Largest dataset with rich (43) attributes	Limited domain shift in testing set
IJCB2021 LivDet-Face [17]	RGB	Unseen testing set with rich (9) PAIs	No training set provided
ICCV2021 challenge [16]	RGB	Largest 3D mask dataset and open-set protocol	Limited (only three) mask types

challenge) [16] was conducted on the largest publicly available 3D mask dataset with a novel open-set evaluation protocol.

The remainder of the chapter is organized as follows. First, we will recapitulate the organization, solutions as well as results of the five most recent face PAD competitions in Sect. 12.2. Then, in Sect. 12.3, we will discuss the lessons learnt from the competitions and future challenges in the field of face PAD in general. Finally, Sect. 12.4 summarizes the chapter and presents conclusions drawn from the competitions discussed here.

12.2 Review of Recent Face PAD Competitions

We begin our review by first introducing two multi-modal face PAD competitions, namely, CVPR2019 and CVPR2020 challenges, in Sects. 12.2.1 and 12.2.2, respectively, where the latter one included also a competition track for unimodal (colour) data. Then, three latest unimodal (colour) face PAD competitions, namely, ECCV2020, IJCB2021 LivDet-Face and ICCV2021 challenges are reviewed in remaining Sects. 12.2.3, 12.2.4 and 12.2.5, respectively.

12.2.1 Multi-modal Face Anti-spoofing Attack Detection Challenge (CVPR2019)

The first three face PAD competitions [19] organized in conjunction with International Joint Conference on Biometrics (IJCB) 2011, International Conference on Biometrics (ICB) 2013 and IJCB2017 focused on photo (i.e. both printed and digital) and video-replay attack detection relying on small-scale datasets (i.e. PRINT-ATTACK [13], REPLAY-ATTACK [14], OULU-NPU [12]) for training, tuning and testing. To be more specific, these datasets have an insufficient number of subjects (<60) and data samples (<6,000 videos) compared with databases used in the field of image classification, e.g. ImageNet [20] and face recognition, e.g. CASIA-WebFace [21], which severely limits the development and testing of data-driven deep model-based approaches for generalized face PAD. Also, due to the lack of variation in the face PAD datasets, the deep models have been suffering from overfitting and learning database-specific information instead of generalized feature representations capturing the disparities in the inherent fidelity characteristics between bona fide samples and different kinds of face artefacts. Another missing feature in previous face PAD competitions has been the availability of multi-modal facial information in addition to conventional visible light colour (RGB) data. This kind of extended range imaging information might be very helpful for developing more robust face PAD methods for practical real-world AFR applications. In order to address the limitations of previous competitions, the Chalearn multi-modal face anti-spoofing attack detection challenge[1] [11] was held in conjunction with the Conference on Computer Vision and Pattern Recognition (CVPR) in 2019. The competition was based on a newly collected large-scale multi-modal face anti-spoofing dataset, namely, CASIA-SURF [22, 23], which consists of 1,000 subjects and 21,000 video clips in three modalities (colour, depth and NIR). The goal of this competition was to push the research progress in AFR applications, where plenty of data and multiple modalities can be considered to be available.

The CVPR2019 challenge was run in the CodaLab[2] platform and consisted of two phases: development phase (December 22, 2018–March 6, 2019) and final phase (March 6, 2019–March 10, 2019). More than 300 academic and industrial institutions worldwide participated in this challenge, and finally 13 teams entered into the final stage. A summary with the names and affiliations of these teams is presented in Table 12.2. Compared with the previous competitions [12–14], the majority of the final participants (10 out of 13) of this competition came from the industry, which indicates the increased need for reliable liveness detection products in daily life applications. Furthermore, one highlight of the CVPR2019 challenge is that the three top-performing teams (VisionLabs[3], ReadSense[4], and Feather[5]) released their source code in GitHub and summarized their approaches in the related CVPR workshop papers [24–27], enhancing the fairness, transparency and reproducibility of the solutions so that they can be easily facilitated by the face recognition community.

12.2.1.1 Dataset

The CASIA-SURF dataset [22, 23] was the largest face PAD database in terms of number of subjects and videos at the time of the CVPR2019 challenge. Each sample of the dataset was associated with three modalities (colour, depth and NIR) captured using an Intel RealSense SR300 camera. Samples from each subject consist of one live video clip and one video clip of each six different attack presentations. A total number of 1,000 subjects and 21,000 videos were captured to build the dataset. Representative samples of bona fide and attack samples across the three modalities are illustrated in Fig. 12.1.

The CASIA-SURF dataset considers six different kinds of print attacks, where a person is holding:

- **Attack 1:** A flat face photo from which the eye regions are cut.
- **Attack 2:** A curved face photo from which the eye regions are cut.
- **Attack 3:** A flat face photo from which the eye and nose regions are cut.
- **Attack 4:** A curved face photo from which the eye and nose regions are cut.

---

[1] https://sites.google.com/qq.com/face-anti-spoofing/welcome/challengecvpr2019.

[2] https://competitions.codalab.org/competitions/20853.

[3] https://github.com/AlexanderParkin/ChaLearn_liveness_challenge.

[4] https://github.com/SeuTao/CVPR19-Face-Anti-spoofing.

[5] https://github.com/SoftwareGift/FeatherNets_Face-Anti-spoofing-Attack-Detection-Challenge-CVPR2019.

Table 12.2 Teams and affiliations listed in the final ranking of the CVPR2019 challenge [11]

Ranking	Team Name	Affiliation
1	VisionLabs	VisionLabs
2	ReadSense	ReadSense
3	Feather	Intel
4	Hahahaha	Megvii
5	MAC-adv-group	Xiamen University
6	ZKBH	Biomhope
7	VisionMiracle	VisonMarcle
8	GradiantResearch	Gradiant
9	Vipl-bpoic	ICT, CAS
10	Massyhnu	Hunan University
11	AI4all	BUPT
12	Guillaume	Idiap Research Institute
Invited team	Vivi	Baidu

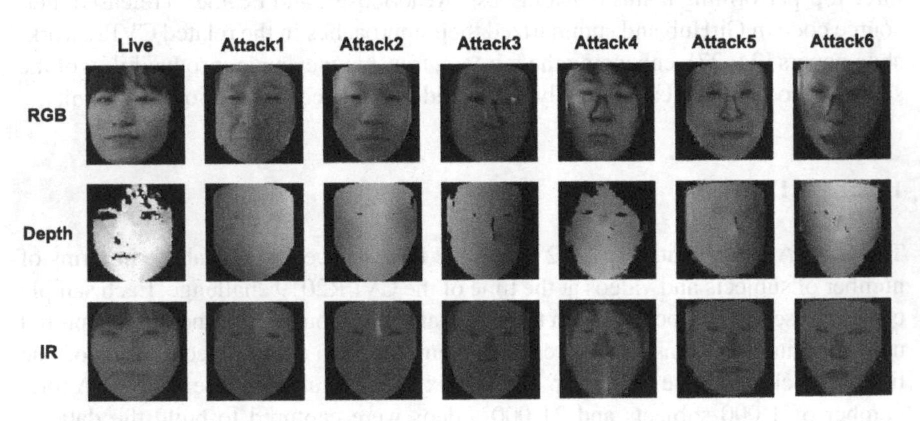

Fig. 12.1 Samples of a live face and six kinds of print attacks from the CASIA-SURF dataset [22, 23]

- **Attack 5:** A flat face photo from which eyes, nose and mouth regions are cut.
- **Attack 6:** A curved face photo from which the eye, nose and mouth regions have been cut.

The samples in the CASIA-SURF dataset were pre-processed for the competition as follows: (1) the dataset is split in three subject-disjoint partitions: train, validation and test sets, with 300, 100 and 600 subjects, respectively, when the corresponding number of videos is 6,300 (2,100 per modality), 2,100 (700 per modality) and 12,600 (4,200 per modality), (2) only every tenth frame from each video was selected to reduce the size of the competition dataset, which resulted in 148K, 48K and 295K frames for the three subsets, respectively, and (3) to mitigate the effect of pre-processing methods (e.g. face detection and alignment) and limit the problem of face PAD to the actual facial information, the background information was masked out pixel-wise from original data, thus only pre-cropped aligned facial images were provided for each modality.

12.2.1.2 Evaluation Protocol and Metrics

In order to focus on the generalization to unknown attacks, the organizers provided only a part of the CASIA-SURF dataset for training, i.e. for each subject only a subset of the PA types was available. Hence, the participants were given about 30K frames for training and 9.6K frames for validation. Note that the attacks in the test set differ from the attacks in the training set, therefore a successful model should avoid intra-attack overfitting, which was a common issue in the earlier face PAD competitions. The challenge comprised development and final stages. The detailed protocols are described as follows.

Protocol in development phase: *(22 December 2018–6 March 2019)*. During the development phase, the participants had access to the labelled training and unlabelled validation samples. Training data included the bona fide samples and three kinds of PAs (4, 5, 6), whereas the validation data consisted of bona fide samples and three other types of PAs (1, 2, 3). The participants were able to submit predictions on the validation partition and receive immediate response via the leaderboard using the CodaLab platform. As it can be observed from Fig. 12.1, the attacks (4, 5, 6) in the validation set differ in appearance (partial cuts in eyes, nose and mouth regions) from attacks (1, 2, 3), which made the task of face PAD challenging.

Protocol in final phase: *(6 March 2019–10 March 2019)*. During the final phase, the labels for the validation subset were made available to the participants, so that they could leverage the additional labelled data for tuning to alleviate the domain gap between different attack types. The participants had to make predictions on the unlabelled test partition and upload their solutions to the CodaLab platform. The considered test set was formed from bona fide samples and three kinds of PAs (1, 2, 3). The final ranking of the participants was determined based on the performance of the submitted systems on the test set. To be eligible for prizes, the top solutions had to release their source code under a licence of their choice and provide a fact sheet describing their solution. All codes would be rerun and verified by the organizing team after the final submission phase. For the sake of reproducibility and fairness, the final ranking of the teams was based on the verified results.

Evaluation metrics: The recently standardized ISO/IEC 30107-3[6] [1] metrics, including Attack Presentation Classification Error Rate (APCER), Bona Fide Presentation Classification Error Rate (BPCER) and Average Classification Error Rate (ACER), were adopted as part of the used evaluation criteria. They can be formulated as

[6] https://www.iso.org/standard/67381.html.

$$APCER = FP/(FP + TN), \quad (12.1)$$

$$BPCER = FN/(FN + TP), \quad (12.2)$$

$$ACER = (APCER + BPCER)/2, \quad (12.3)$$

where TP, FP, TN and FN correspond to true positive, false positive, true negative and false negative, respectively. APCER and BPCER are used to measure the error rates of attack or bona fide samples, respectively. Similar to the common metrics in AFR systems, the Receiver Operating Characteristic (ROC) curve was also considered for examining a suitable operating point trade-off in the False Positive Rate (FPR) and True Positive Rate (TPR) regarding the requirements of real-world biometric applications. Finally, the operating point of TPR@FPR=10^{-4} was selected as the leading evaluation measure for the CVPR2019 challenge, while the ACER was used as an additional evaluation criterion.

12.2.1.3 Results and Discussion

In this subsection, we summarize all the face PAD solutions reaching the final stage in terms of method keywords, backbone models, pre-training data, modalities, fusion schemes and loss functions. Finally, the overall results are analysed and discussed.

Summary of the participating solutions: Table 12.3 summarizes the face PAD solutions of the 13 participating teams and the baseline method. Different from the previous three competitions (i.e. IJCB2011 [13], ICB2013 [14] and IJCB2017 [12]), none of the final teams used traditional face PAD methods, such as hand-crafted image quality/texture descriptors [44], and liveness cues like eye blinking, facial expression changes and mouth movements [45]. Instead, all the submitted face PAD solutions relied on data-driven model-based feature extractors, such as ResNet [28] and VGG16 [41]. Furthermore, most of the approaches were multi-modal, combining two or three modalities, while only three teams (Hahaha, VisionMiracle and AI4all) relied on unimodal (depth-based) PAD solution. It can be seen from the last column in Table 12.3 that several kinds of multi-modal fusion strategies (e.g. input-level data fusion, feature-level Squeeze-and-Excitation (SE) [28] fusion and score-level fusion) were used. Regarding the use of pre-training data, two teams (VisionLabs and GradiantResearch) leveraged pre-trained models from related face analysis tasks (e.g. face recognition models on CASIA-WebFace [21] and face PAD models on GRAD-GPAD [38]) to mitigate the issues with overfitting. It is worth to note that all three top-performing solutions (VisionLabs, ReadSense and Feather) adopted ensemble strategy to aggregate the predictions from multiple variant models.

Result analysis: The results and ROC curves of the participating teams on the test data are shown in Table 12.4 and Fig. 12.2, respectively. It can be observed that the winning team (VisionLabs) achieved TPR = 99.8739%@FPR=10^{-4}, and the FN = 27 and FP = 3 on the test set. In fact, different application scenarios have different requirements for each indicator. For example, to meet the higher security needs of

Table 12.3 Summary of the face PAD methods for all participating teams and baseline method [11]. 'SE' denotes Squeeze-and-Excitation [28] and 'BCE' denotes binary cross-entropy

Team	Method	Backbone model	Pre-training data	Modalities	Fusion scheme and loss function
VisionLabs [24]	Fine-tuning Ensembling	ResNet34 [28] Resnet50 [28]	CASIA-WebFace [21] AFAD-Lite [29] MSCeleb1M [30] Asian dataset [31]	RGB Depth NIR	SE Fusion Score fusion BCE loss
ReadSense [25]	Bag-of-local features Ensembling	SE-ResNeXt [32]	No	RGB Depth NIR	SE Fusion Score fusion BCE loss
Feather [27]	Ensembling	Fishnet [33] MobileNetV2 [34]	No	Depth NIR	Score fusion BCE loss
Hahahaha	Depth only	ResNeXt [32]	Imagenet [20]	Depth	BCE loss
MAC-adv-group	Feature fusion	ResNet34	No	RGB Depth NIR	Feature fusion BCE loss
ZKBH	Regression model	ResNet18	No	RGB Depth NIR	Data fusion Regression loss
VisionMiracle	Modified Shufflenet-V2 Depth only	Shufflenet-V2 [35]	No	Depth	BCE loss
Baseline [22, 23]	Feature fusion	ResNet18	No	RGB Depth NIR	SE fusion BCE loss
GradiantResearch	Deep metric learning	Inception [36]	VGGFace2 [37] GRAD-GPAD [38]	RGB Depth NIR	Logistic regression ensemble
Vipl-bpoic [26]	Attention mechanism [39]	ResNet18	No	RGB Depth NIR	Data fusion Centre loss [40] BCE loss
Massyhnu	Ensembling	9 softmax classifiers	No	RGB Depth NIR	Colour information fusion BCE loss
AI4all	Depth only	VGG16 [41]	No	Depth	BCE loss
Guillaume	Multi-Channel CNN No RGB	LightCNN [42]	Yes	Depth NIR	Data fusion BCE loss
Vivi	Dense cross-modal-attention model	DenseNet [43]	Yes	RGB Depth NIR	Feature fusion Score fusion BCE loss

Table 12.4 Results and rankings of the stage teams [11] at the final stage. The best results are bolded. (∗ denotes Vivi that is affiliated with the sponsor and did not participate in the final ranking)

Team Name	FP	FN	APCER (%)	BPCER (%)	ACER (%)	TPR (%)@FPR = 10e-4
VisionLabs	**3**	27	**0.0074**	0.1546	0.0810	**99.8739**
ReadSense	77	**1**	0.1912	**0.0057**	0.0985	99.8052
Feather	48	53	0.1192	0.1392	0.1292	98.1441
Hahahaha	55	214	0.1366	1.2257	0.6812	93.1550
MAC-adv-group	825	30	2.0495	0.1718	1.1107	89.5579
ZKBH	396	35	0.9838	0.2004	0.5921	87.6618
VisionMiracle	119	83	0.2956	0.4754	0.3855	87.2094
GradiantResearch	787	250	1.9551	1.4320	1.6873	63.5493
Baseline	1542	177	3.8308	1.0138	2.4223	56.8381
Vipl-bpoic	1580	985	3.9252	5.6421	4.7836	39.5520
Massyhnu	219	621	0.5440	3.5571	2.0505	29.2990
AI4all	273	100	0.6782	0.5728	0.6255	25.0601
Guillaume	5252	1869	13.0477	10.7056	11.8767	0.1595
Vivi*	7	15	0.0173	0.0859	**0.0516**	99.8282

an access control system, the FP is required to be as small as possible. With respect to this criterion, VisionLabs performed very well as only three attack samples were misclassified as bona fide. On the contrary, a small FN value is crucial from usability point of view, where the team ReadSense achieved the best result (FN=1) due to the effectiveness of local patch inputs. In overall, the first eight teams were performing better than the baseline method [22, 23] in terms of FP and TPR@FPR $= 10^{-4}$, indicating the valuable outputs and insightful solutions of this challenge.

As shown in Table 12.4, the results of the three top-performing teams on the test set were clearly superior compared with the other teams. By combining Table 12.3 with Table 12.4, we can conclude that ensemble learning performed more robustly compared to single model-based solutions under the same conditions. The ROC curves of all the participating teams are illustrated in Fig. 12.2. It can be seen that three teams (i.e. VisionLabs, ReadSense and Vivi) were significantly better than other teams on the test set. For instance, the TPR@FPR=10^{-4} values of these three teams are relatively close to each other and superior compared to the other teams. The characteristics of the three modalities are different, as the colour data is rich in details, the depth data is sensitive to the distance, while the NIR data captures better the face-specific skin reflectance properties, for instance. Therefore, the teams Vivi and Vipl-bpoic introduced attention mechanisms into the face PAD task, enforcing the models to focus on different informative regions among three modalities. Similarly, the team Feather used a cascaded architecture with two subnetworks, where two modalities of the CASIA-SURF dataset (i.e. depth and NIR data) were examined subsequently by each network. Some teams considered also facial landmarks (i.e. Hahahaha) and

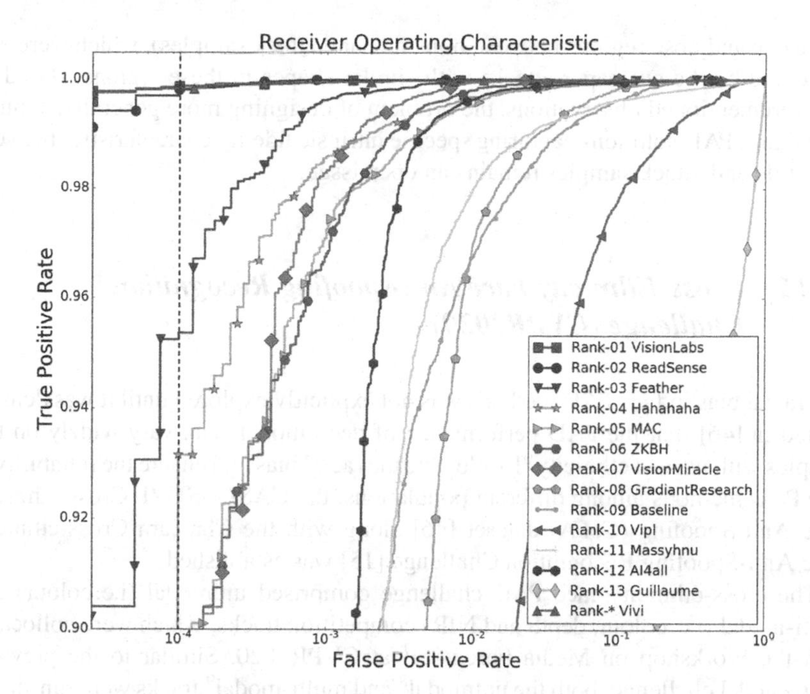

Fig. 12.2 ROC curves of the teams on test set at the final stage [11]

colour space conversions (i.e. MAC-adv-group and Massyhnu) for PAD. Instead of conventional binary classification-based PAD model, the team ZKBH constructed a regression model to supervise the model to learn local cues in the eye regions. In order to generalize better to unseen attacks, the team GradiantResearch reformulated face PAD as an anomaly detection problem using deep metric learning.

Discussion: Although most of the proposed solutions achieved superior performance compared with the provided multi-modal SE fusion-based baseline, there were still some limitations in the CVPR2019 challenge. Originally, the main research question was to explore new efficient multi-modal fusion schemes for combining the colour, depth and NIR modalities. However, no novel or otherwise insightful multi-modal fusion strategies were proposed in the end. Most of the teams applied very simple data-level and score-level fusion in a greedy search manner, which is likely to fail when evaluating a method on an unknown multi-modal dataset. Furthermore, the two top-performing teams adopted the feature-level SE fusion strategy directly from the baseline method, which again was not very fruitful from the multi-modal challenge point of view. Many of the top-performing solutions exploited an ensemble of multiple models to boost the performance. However, while pushing the efficiency, model ensembling also increases the complexity of the whole solution, which is not practical in real-world conditions, especially considering the limitations of mobile and embedded platforms. Finally, some solutions were considerably inspired by the human-observed priors in the CASIA-SURF dataset (e.g. the apparent discrepancy in the eye and nose regions between bona fide and attack samples), which were easily fooled by the cut paper attacks with similar shapes in these regions. Based on the aforementioned observations, the problem of designing more generalized multi-modal face PAD solutions capturing specific intrinsic fidelity characteristics between bona fide and attack samples remains an open issue.

12.2.2 *Cross-Ethnicity Face Anti-spoofing Recognition Challenge (CVPR2020)*

The racial bias in face PAD methods was not explicitly explored until it was demonstrated in [46] that the PAD performance of deep models can vary widely on test samples with unseen ethnicity. To alleviate the racial bias and ensure the reliability of face PAD methods among different populations, the CASIA-SURF Cross-ethnicity Face Anti-Spoofing (CeFA) dataset [46] along with the Chalearn Cross-ethnicity Face Anti-Spoofing Recognition Challenge [15] was established.

The cross-ethnicity face PAD challenge comprised unimodal (i.e. colour) and multi-modal (i.e. colour, depth and NIR) competition tracks, which were collocated with the Workshop on Media Forensics[7] at CVPR2020. Similar to the previous multi-modal challenge, both the unimodal[8] and multi-modal[9] tracks were run simultaneously using the CodaLab platform. The competition attracted 340 teams in the development stage, with 11 and eight teams finally entering the actual evaluation stage for the unimodal and multi-modal face PAD tracks, respectively. A summary of the names and affiliations of teams that entered the final stage as well as their final rankings are shown in Tables 12.5 and 12.6 for the unimodal and multi-modal tracks, respectively. From the tables, it can be seen that most participants came from industrial institutions, indicating the increasing need for reliable and robust PAD systems in practical AFR applications. Interestingly, the team VisionLabs was not only the winner of the unimodal track of the CVPR2020 challenge, but also the winner of the earlier multi-modal CVPR2019 challenge [11]. In addition, the team BOBO from University of Oulu (the authors' team) proposed novel Central Difference Convolution (CDC) [47, 48] and contrastive depth loss (CDL) [49] methods for feature learning, achieving the first and second place in multi-modal and unimodal tracks, respectively.

[7] https://sites.google.com/view/wmediaforensics2020/home.
[8] https://competitions.codalab.org/competitions/22151.
[9] https://competitions.codalab.org/competitions/22036.

Table 12.5 Names, affiliations and rankings of the participating systems in the unimodal track [15]

Ranking	Team name	Affiliation
1	VisionLabs	VisionLabs
2	BOBO	Zitong Yu, University of Oulu
3	Harvest	Jiachen Xue, Horizon
4	ZhangTT	Zhang Tengteng, CMB
5	Newland_tianyan	Xinying Wang, Newland Inc.
6	Dopamine	Wenwei Zhang, huya
7	IecLab	Jin Yang, HUST
8	Chuanghwa Telecom Lab	Li-Ren Hou, Chunghwa Telecom
9	Wgqtmac	Guoqing Wang, ICT
10	Hulking	Yang, Qing, Intel
11	Dqiu	Qiudi

Table 12.6 Names, affiliations and rankings of the participating systems in the multi-modal track [15]

Ranking	Team name	Affiliation
1	BOBO	Zitong Yu, University of Oulu
2	Super	Zhihua Huang, USTC
3	Hulking	Qing Yang, Intel
4	Newland_tianyan	Zebin Huang, Newland Inc.
5	ZhangTT	Tengteng Zhang, CMB
6	Harvest	Yuxi Feng, Horizon
7	Qyxqyx	Yunxiao Qin, NWPU
8	Skjack	Sun Ke, XMU

12.2.2.1 Dataset

The CASIA-SURF CeFA [46] was the largest face anti-spoofing dataset at the time of the CVPR2020 competition, covering three ethnicities (i.e. Africa, East Asia and Central Asia), three modalities (i.e. colour, depth and NIR), 1, 607 subjects and four different PA types (i.e. prints, video-replays, 3D print and silica gel masks). The multi-modal videos were captured using an Intel RealSense SR300 camera with resolution of 1280×720 pixels for each video frame at 30 frames per second. The data was pre-processed in similar way as in the previous CVPR2019 multi-modal challenge [11] (see Sect. 12.2.1.1). The CASIA-SURF CeFA was the first public dataset designed for exploring also the racial bias of face PAD methods. Some samples of the CASIA-SURF CeFA dataset are shown in Fig. 12.3.

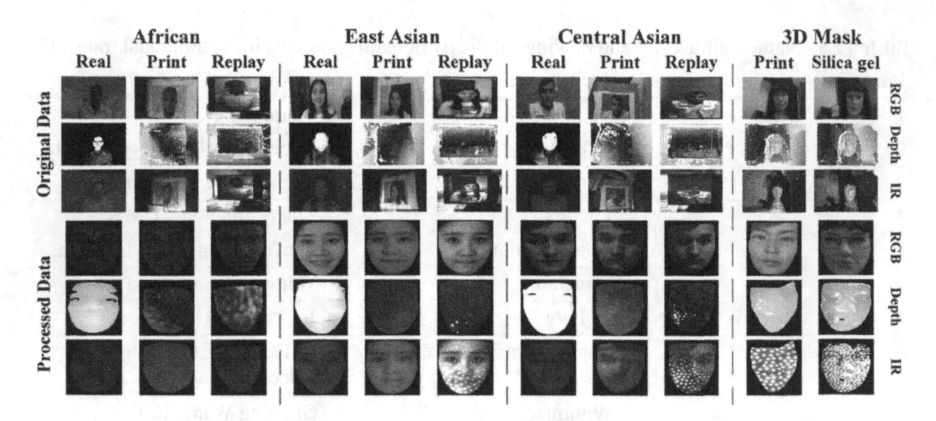

Fig. 12.3 Samples from the CASIA-SURF CeFA dataset [46], consisting of 1, 607 subjects, three different ethnicities (i.e. Africa, East Asia, and Central Asia), four PA types (i.e. print, video-replay, 3D print and silica gel mask) and three modalities (i.e. colour, depth and NIR)

The main motivation of CASIA-SURF CeFA dataset is to serve as a benchmark that allows evaluating the generalization of PAD methods across different ethnicities, PAIs and modalities under varying scenarios using four specific protocols:

- **Protocol 1:** Cross-ethnicity generalization of PAD methods is evaluated by using one ethnicity for training and validation, while the two remaining ones are used as unseen ethnicities for testing.
- **Protocol 2:** Cross-PAI generalization of PAD methods is evaluated by using print or video-replay attack for training and validation, while the remaining three attacks are used as unknown PA types for testing.
- **Protocol 3:** Cross-modality generalization of PAD methods is evaluated by using one modality for training and validation, while the two remaining ones are used as unknown modalities for testing.
- **Protocol 4:** Cross-ethnicity and cross-PAI generalization of PAD methods is evaluated simultaneously by combining the first two protocols, i.e. using one ethnicity and PAI for training and validation, while the remaining two ethnicities as well as three PAIs are used for testing.

The most challenging Protocol 4 was adopted for ranking the methods in both unimodal and multi-modal tracks of the competition. As shown in Table 12.7, this protocol consists of three subsets: training, validation and test sets, containing 200, 100 and 200 subjects for each ethnicity, respectively. Note that the remaining 107 subjects correspond to the 3D masks attacks. Since there are three ethnicities in CASIA-SURF CeFA, in total three sub-protocols (i.e. 4_1, 4_2 and 4_3 in Table 12.7) were adopted in the CVPR2020 challenge. In addition to the racial variation, the unknown PAIs introduced in the test sets made the competition even more challenging.

Table 12.7 Protocols and statistics of the Protocol 4 of the CASIA-SURF CeFA [46] dataset, where 'A', 'C' and 'E' denote Africa, Central Asia and East Asia, respectively

Subset	Subjects (one ethnicity)	Ethnicity			PAIs	# images (RGB)		
		4_1	4_2	4_3		4_1	4_2	4_3
Train	1–200	A	C	E	Video-Replay	33,713	34,367	33,152
Valid	201–300	A	C	E	Video-Replay	17,008	17,693	17,109
Test	301–500	C&E	A&E	A&C	Print+Mask	105,457	102,207	103,420

12.2.2.2 Evaluation Protocol and Metrics

The challenge comprised development and final stages. The detailed protocols are described as follows.

Protocol in development phase: *(13 December 2019–1 March 2020)*. During the development phase, the participants had access to the labelled training set and unlabelled validation set. Since the Protocol 4 of the CASIA-SURF CeFA dataset used in this competition comprised three sub-protocols (see Table 12.7), the participants first needed to train a model for each sub-protocol and then predict the scores for each corresponding validation set. Finally, the participants had to merge the predicted scores of the three sub-protocols and submit the resulting final scores to the CodaLab platform, where an immediate response was seen in the public leaderboard.

Protocol in final phase: *(1 March 2020–10 March 2020)*. During the final phase, the labelled validation set and the unlabelled testing set were released. The participants could first utilize the labels of the validation set for model selection to improve the generalization on the test data. All results of the three sub-protocols were made publicly available online in terms of APCER, BPCER and ACER. Like with the OULU-NPU dataset [6] used in the IJCB2017 competition [12], the mean and variance of evaluated metrics across the three different sub-protocols were calculated and included in the final results.

Note that in order to fairly compare the performance of different submitted systems, the use of external training datasets or pre-trained models was explicitly prohibited in the CVPR2020 challenge. All participants were encouraged to release their source codes under feasible licences and to provide a fact sheet describing their solution. All codes would be rerun and verified by the organizing team after the final submission phase. For the sake of reproducibility and fairness, the final ranking of the teams was based on the verified results.

Evaluation metrics: Similar to the previous CVPR2019 competition [11], also in the CVPR2020 challenge, the standardized ISO/IEC 30107-3 [1] metrics (i.e. APCER, BPCER and ACER) were considered as the main evaluation criteria (see Sect. 12.2.1.2). Also, ROC curves were included for visualization purposes and additional result analysis. The final rankings were based on the ACER metric on the test set because it has been widely used for evaluating the performance of face PAD systems in the literature and majority of the previous face PAD competitions.

The ACER threshold was determined by calculating the Equal Error Rate (EER) operating point on the validation set.

12.2.2.3 Results and Discussion

In this section, we first summarize the methods and report the results of the unimodal track, and then analyse the solutions as well as the results of the multi-modal track. Finally, we provide general discussion on the proposed algorithms and competition.

Solutions of the unimodal (colour) track: Table 12.8 summarizes the face PAD solutions of the teams participated in the unimodal track. The source codes of ten teams, including VisionLabs[10], BOBO[11], Harvest[12], ZhangTT[13], Newland-tianyan[14], Dopamine[15], IecLab[16], Chungwa-Telecom[17], Wgqtmac[18], and Hulking[19], were made publicly available. It was not surprising that every team adopted end-to-end learning-based approaches due to the strong representation capacity of modern deep models. Regarding the model inputs, most of the teams used the provided facial colour images directly, while the winning team VisionLabs considered two kinds of pre-processing methods for dynamic inputs (i.e. optical flow [50] and rank pooling [51] images). As for the backbone networks, only the team Dopamine adopted spatio-temporal 3D convolutional neural network (CNN) model, while the others relied on 2D CNNs (mostly ResNet). Most of the solutions treated face PAD as a binary classification problem via simple binary cross-entropy (BCE) loss, but few teams (i.e. BOBO, Harvest and ZhangTT) considered pixel-wise depth loss, temporal continuous L1 regression loss and multi-class softmax cross-entropy (CE) loss, respectively. It is interesting to see from the last two columns of Table 12.8 that most of the solutions leveraged dynamic cues but did not adopt complex model ensemble strategy.

Results of the unimodal (colour) track: The final results of the 11 participated teams are shown in Table 12.9. The final ranking was based on the mean ACER computed over the three sub-protocols. The EER thresholds from the validation set are also reported in Table 12.9. The threshold values for the best performing algorithms were either extremely large (e.g. more than 0.9 for BOBO) or small (e.g. 0.01 for Harvest). On the contrary, VisionLabs's algorithm was more stable with threshold values around 0.5 and achieved the highest accuracy in detecting the PA samples (APCER = 2.72%), while Wgqtmac's algorithm obtained the best results in terms

[10] https://github.com/AlexanderParkin/CASIA-SURF_CeFA.

[11] https://github.com/ZitongYu/CDCN/tree/master/FAS_challenge_CVPRW2020.

[12] https://github.com/yueyechen/cvpr20.

[13] https://github.com/ZhangTT-race/CVPR2020-SingleModal.

[14] https://github.com/XinyingWang55/RGB-Face-antispoofing-Recognition.

[15] https://github.com/xinedison/huya_face.

[16] https://github.com/1relia/CVPR2020-FaceAntiSpoofing.

[17] https://drive.google.com/open?id=1ouL1X69KlQEUl72iKHl0-_UvztlW8f_l.

[18] https://github.com/wgqtmac/cvprw2020.git.

[19] https://github.com/muyiguangda/cvprw-face-project.

Table 12.8 Summary of the top-performing solutions in the unimodal track of the CVPR2020 challenge, where 'S/D' indicates Static/Dynamic, 'OF' and 'RP' for optical flow [50] and rank pooling [51], 'BCE', 'CDC' and 'CDL' binary cross-entropy, central difference convolution and contrastive depth loss, respectively

Team name	Method (keywords)	Input	Backbone	Loss function	S/D	Ensemble
VisionLabs	Creating artificial modalities	OF+RP	SimpleNet [52]	BCE loss	D	No
BOBO	CDC, CDL, Attention	RGB	CDCN [48]	Depth loss	S	Yes
Harvest	Motion-aware labels	RGB	ResNet101	L1 loss	D	Yes
ZhangTT	Quality tensor	Grayscale	ResNet [28]	4-class CE loss	D	No
Newland-tianyan	Subtracted neighbourhood mean	RGB	5-layer network	BCE loss	D	No
Dopamine	Multi-task learning	RGB	ResNet100	BCE + face ID loss	S	No
IecLab	Authenticity+expression features	RGB	3DResNet [53]	BCE loss	D	No
Chunghwa Telecom	Bag-of-local features	RGB	MIMAMO-Net [54]	BCE loss	S	Yes
Wgqtmac	Warmup strategy	RGB	ResNet18	BCE loss	S	No
Hulking	Frame vote module	RGB patch	PipeNet [51]	BCE loss	D	No
Dqiu	–	RGB	ResNet50	BCE loss	S	No
Baseline	Hybrid feature fusion	RGB+RP	SD-Net [46]	BCE loss	D	No

Table 12.9 The results of the unimodal track of the CVPR2020 challenge [46]. Avg±Std denotes the mean and variance computed across the three sub-protocols. The best results are shown in bold

Team Name	Threshold	FP	FN	APCER(%)	BPCER(%)	ACER(%)	Rank
VisionLabs	0.34±0.48	2±2	21±9	**0.11±0.11**	5.33±2.37	**2.72±1.21**	1
BOBO	0.97±0.02	129±67	10±2	7.18±3.74	2.50±0.50	4.84±1.79	2
Harvest	0.01±0.00	85±47	55±10	4.74±2.62	13.83±2.55	9.28±2.28	3
ZhangTT	0.9	97±37	75±31	5.40±2.10	18.91±7.88	12.16±2.89	4
Newland-tianyan	0.67±0.11	282± ±239	44±62	15.66±13.33	11.16±15.67	13.41±3.77	5
Dopamine	0.07±0.11	442±168	10±12	24.59±9.37	2.50±3.12	13.54±3.95	6
IecLab	0.40±0.07	597±103	24±2	33.16±5.76	6.08±0.72	19.62±2.59	7
Chunghwa Telecom	0.86±0.06	444±93	76±34	24.66±5.16	19.00±8.69	21.83±1.82	8
Wgqtmac	0.80±0.22	928±310	2±3	51.57±17.24	**0.66±0.94**	26.12±8.15	9
Hulking	0.76±0.08	810±199	78±53	45.00±11.07	19.50±13.27	32.25±3.18	10
Dqiu	1.00±0.00	849±407	116±48	47.16±22.62	29.00±12.13	38.08±15.57	11
Baseline	1.00±0.00	1182±300	30±25	65.66±16.70	7.58±6.29	36.62±5.76	

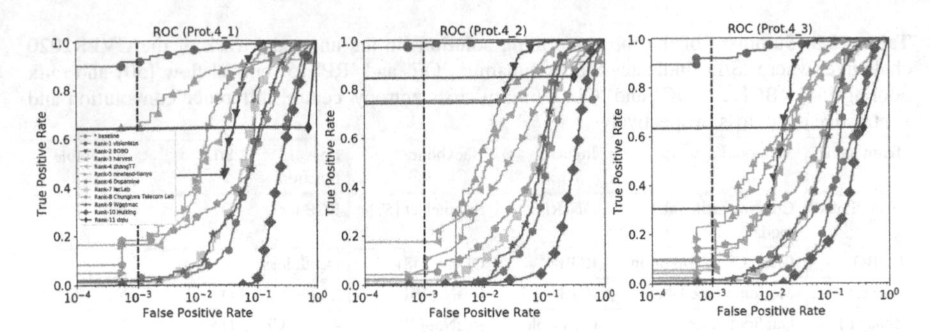

Fig. 12.4 The ROC curves for the 12 teams participating in the unimodal track of the CVPR2020 challenge [46]. From left to right are the ROC curves for protocol 4_1, 4_2 and 4_3, respectively

of BPCER (0.66%). In overall, the first ten teams were performing better than the baseline method [46] in terms of ACER. The top three teams obtained excellent ACER values below 10%, and the team VisionLabs achieved the first place with a clear margin.

The ROC curves of the three sub-protocols are given in Fig. 12.4 to further analyse the trade-off between APCER and BPCER, i.e. tuning the operating point according to the requirements of a given real-world application. The results of the winning team VisionLabs (blue curve) on all three sub-protocols are clearly superior compared to others, indicating the benefits of optical flow-based motion clues and rank pooling images in improving the generalization performance. However, the TPR value of the remaining teams decreases rapidly as the FPR reduces (e.g. TPR@FPR $= 10^{-3}$ values for these teams are almost zero). In addition, although the ACER of the team Harvest was worse than that of the team BOBO, its TPR@FPR $= 10^{-3}$ was significantly better than that of BOBO. It was mainly because the values of FP and FN samples for the team Harvest were relatively close to each other (see Table 12.9).

Solutions of the multi-modal track: Table 12.10 summarizes the face PAD solutions of the teams that participated in the multi-modal track. The source codes of seven teams, including BOBO, Super[20], Hulking[21], Newland-tianyan[22], ZhangTT[23], Qyxqyx[24] and Skjack[25], were made publicly available. Most of the teams exploited all the three modalities (colour, depth and NIR) for feature- and score-level fusion, except for the teams ZhangTT and Harvest who considered only depth and NIR modalities. There were no teams using data-level fusion strategy. As for the architectures and loss functions, teams BOBO and Qyxqyx adopted MM-CDCN [55] and DepthNet [8] with pixel-wise supervision, while the other teams relied on ResNet

[20] https://github.com/hzh8311/challenge2020_face_anti_spoofing.

[21] https://github.com/ZhangTT-race/CVPR2020-SingleModal.

[22] https://github.com/Huangzebin99/CVPR-2020.

[23] https://github.com/ZhangTT-race/CVPR2020-MultiModal.

[24] https://github.com/qyxqyx/FAS_Chalearn_challenge.

[25] https://github.com/skJack/challange.git.

Table 12.10 Summary of the top-ranked solutions of the multi-modal track in the CVPR2020 challenge

Team name	Modality	Fusion	Backbone	Loss function	S/D	Ensemble
BOBO	RGB, Depth, NIR	Feature & score level	MM-CDCN [55]	Depth loss	S	Yes
Super	RGB, Depth, NIR	SE fusion in feature level	ResNet34/50	BCE loss	S	Yes
Hulking	RGB, Depth, NIR	Feature level	PipeNet [56]	BCE loss	D	No
Newland-tianyan	Grayscale, Depth, NIR	Score level	Resnet9	BCE loss	S	No
ZhangTT	Depth, NIR	Feature level	ID-Net	BCE loss	S	Yes
Harvest	NIR	No fusion	–	Triplet loss	S	No
Qyxqyx	RGB, Depth, NIR	Score level	DepthNet [8]	BCE+BinaryMap loss	S	Yes
Skjack	RGB, Depth, NIR	Feature level	Resnet9	BCE loss	S	No
Baseline	RGB, Depth, NIR, RP	Feature level	PSMM-Net [46]	BCE loss	D	No

Table 12.11 The results of the multi-modal track in the CVPR2020 challenge [46]. Avg±Std indicates the mean and variance across the three folds and the best results are shown in bold

Team name	Threshold	FP	FN	APCER (%)	BPCER (%)	ACER (%)	Rank
BOBO	0.95±0.02	19±11	4±2	1.05±0.62	**1.00±0.66**	**1.02±0.59**	1
Super	1.0±0.00	11.33±7.76	11±6	0.62±0.43	2.75±1.50	1.68±0.54	2
Hulking	0.98±0.02	58±35	4±4	3.25±1.98	1.16±1.12	2.21±1.26	3
Newland-tianyan	1.00±0.00	4±4	17±12	**0.24±0.25**	4.33±3.12	2.28±1.66	4
ZhangTT	0.87±0.07	56±51	17±17	3.11±2.87	4.41±4.25	3.76±2.02	5
Harvest	0.92±0.04	104±84	13±12	5.77±4.69	3.33±3.21	4.55±3.82	6
Qyxqyx	0.95±0.05	92±142	26±23	5.12±7.93	6.66±5.86	5.89±4.04	7
Skjack	0.00±0.00	1012±447	47±45	56.24±24.85	11.75±11.37	33.99±7.08	8
Baseline	0.39±0.52	872±463	62±43	48.46±25.75	15.58±10.86	32.02±7.56	

with BCE loss. In contrast to the unimodal track, the use of static cues and ensemble models was popular in the multi-modal track.

Results of the multi-modal track: The results of the eight teams participating in the final stage are shown in Table 12.11. The team BOBO team achieved the best performance in terms of BPCER = 1.00% and ACER = 1.02%, and the team Super ranked second with a minor margin ACER = 1.68%. It is worth noting that the team Newland-tianyan achieved the best results in terms of APCER (0.24%). Similar to the unimodal track, most of the participating teams had relatively large EER thresholds calculated on the validation set, especially the teams Super and Newland-tianyan with threshold values of 1.0, indicating that the samples would be easily classified as

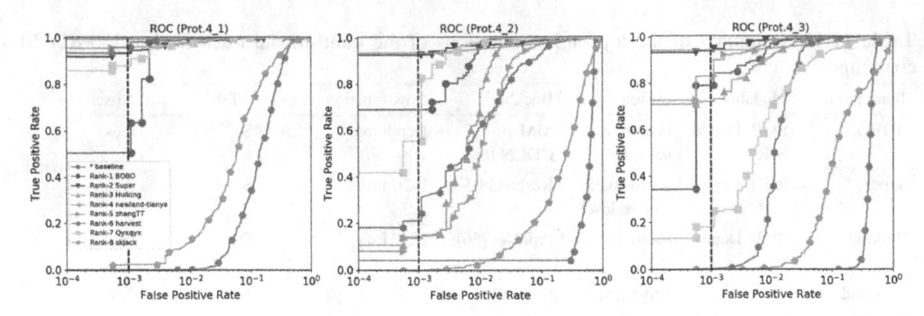

Fig. 12.5 The ROC curves of nine teams in the multi-modal track [46]. From left to right are the ROC curves on protocol 4_1, 4_2 and 4_3, respectively

anomalies. In addition, it can be seen that the ACER values of the four top teams were 1.02%, 1.68%, 2.21% and 2.28%, all outperforming the best performance (2.72%) reported in the unimodal track. This suggests that the additional modalities are indeed useful in improving robustness of face PAD under the challenging cross-ethnicity and cross-PAI conditions.

The ROC curves of the solutions in multi-modal track are shown in Fig. 12.5. From Table 12.11 and Fig. 12.5, we can find that even though the ACER values of the top two algorithms were relatively close, the stability of the team Super (brown curve) is better than that of the team BOBO (blue curve). For instance, the TPR@FPR $= 10^{-3}$ values for Super and Newland-tianyan were better than that of BOBO on all three sub-protocols. In other words, compared with the BCE loss-based solution, the depth-wise supervision in team BOBO's solution might cause larger bias between metrics ACER and TPR@FPR $= 10^{-3}$.

Discussion: From Tables 12.9 and 12.11 of the competition results, we can find that the EER thresholds computed on the validation set for both unimodal and the multi-modal track were generally high, indicating that the proposed algorithms might easily make over-confident or biased decisions. The reason behind this might be twofold: (1) the biased distributions of the CASIA-SURF CeFA dataset, e.g. as the environments for the attack samples were more diverse, while bona fide samples were usually recorded indoor and (2) the lack of generalization when the algorithm faces unknown PA types and ethnicities. Moreover, rethinking the evaluation metric for ranking the solutions is necessary. It can be seen from the both tracks that some solutions (e.g. the team BOBO) could achieve excellent ACERs but, on the other hand, unsatisfying ROC curves, especially at operating points with low FPR.

12.2.3 CelebA-Spoof Challenge on Face Anti-spoofing (ECCV2020)

Despite both the CVPR2019 [11] and CVPR2020 [15] competitions were successful in benchmarking the generalization of unimodal and multi-modal face PAD methods in challenging settings, the amount of data (number of images $< 150{,}000$ and subjects ≤ 1000) and domain diversity (only indoor conditions) of these two previous contests were still limited for evaluating the performance of FAS methods 'in the wild'. Recently, a large-scale face PAD dataset, namely, CelebA-Spoof [57], containing 625,537 face images of 10,177 subjects, was released. It is still the largest publicly available face PAD dataset in terms of the number of images and subjects. Leveraging the CelebA-Spoof dataset, the CelebA-Spoof Challenge on Face Anti-Spoofing was organized in conjunction with the European Conference on Computer Vision (ECCV) 2020 Workshop on Sensing, Understanding and Synthesizing Humans[26]. The goal of this competition was to boost the research on large-scale face anti-spoofing.

The ECCV2020 challenge was also hosted in the CodaLab platform[27]. After registering to the competition, each team was allowed to submit their models to the Amazon Web Services (AWS) and allocated with one 16 GB Tesla V100 GPU to perform online evaluation on the hidden test set. The encrypted prediction files, including the results for each data sample in the hidden test set, were sent to the teams via an automatically generated email after their requested online evaluation was finished. The teams were required to upload their encrypted prediction files to the CodaLab platform for ranking the algorithms.

The ECCV2020 challenge lasted for 9 weeks from 28 August 2020 to 31 October 2020. During the contest, the participants had access to the public CelebA-Spoof dataset and were restricted to use only the public CelebA-Spoof training dataset for building their models. The results of the challenge were announced on 10 February 2021. A total number of 134 participants registered for the competition, while 19 teams made valid submissions in the end. The details and results of top five teams are shown in Table 12.12. It is surprising to see that all top three teams were from industry, indicating even increasing attention on PAD for real-world AFR applications. It is worth noting that the top three teams achieved TPR=100%@FPR=$5*10^{-3}$, indicating the effectiveness of the solutions for large-scale face PAD on the CelebA-Spoof dataset.

12.2.3.1 Dataset

The ECCV2020 challenge employed the CelebA-Spoof dataset [57] for training and evaluation purposes. The CelebA-Spoof is a large-scale face PAD dataset that has 625,537 images corresponding to 10,177 subjects, including 43 rich attributes on face,

---

[26] https://sense-human.github.io/.

[27] https://competitions.codalab.org/competitions/26210.

Table 12.12 Final results of the top-5 teams in the ECCV2020 challenge

Ranking	Team	User	Affiliation	$TPR @FPR = 10^{-3}$	$TPR @FPR = 5 * 10^{-3}$	$TPR @FPR = 10^{-6}$
1	ZOLOZ	ZOLOZ	ZOLOZ	1.00000	1.00000	**1.00000**
2	MM	liujeff	Meituan	1.00000	1.00000	0.99991
3	AFO	winboyer	Meituan	1.00000	1.00000	0.99918
4	k_	k_	–	0.99973	0.99927	0.98026
5	SmartQ	SmartQ	–	0.99963	0.99872	0.96938

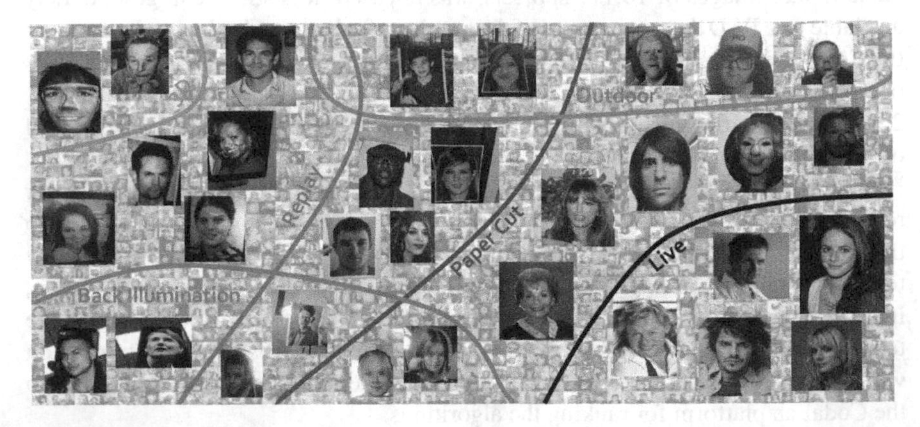

Fig. 12.6 Representative samples and attributes in the CelebA-Spoof dataset [57]

illumination, environment and PA types. Bona fide facial images were selected from the CelebA dataset [58] but they were also manually examined to find and remove possible 'attack' samples, including posters, advertisements and artistic drawings. The corresponding attack samples with four different PAIs (i.e. 2D print, cut paper, video-replay and 3D paper mask) were collected and annotated to form the CelebA-Spoof database. Among the 43 rich attributes, 40 attributes describe the bona fide images, including all facial components and accessories (e.g. skin, nose, eye, eyebrow, lip, hair, hat and eyeglasses), while the three remaining attributes describe the attack samples, including PAI, environments (e.g. indoor and outdoor) and illumination conditions (e.g. strong, weak, back, dark and normal). Some typical samples in the CelebA-Spoof database are shown in Fig. 12.6.

12.2.3.2 Evaluation Protocol and Metrics

The training, validation and test sets of the original CelebA-Spoof database were split into subject-disjoint folds with a ratio of 8 : 1 : 1. The compilation of the hidden test data devised for the ECCV2020 challenge was as the same as for the public test set.

Table 12.13 Summary of the top-ranked solutions [18] in the ECCV2020 challenge

Team	Input	Model	Ensemble strategy
ZOLOZ	Face	FOCUS, AENet, ResNet, Attack types, Noise Print	Heuristic voting strategy
MM	Face + Patches	CDCN++, LGSC, SE-ResNet50, EfficientNet-B7, SE-ResNeXt50	Weight-after-sorting
AFO	Patches	CDCN, CDC-DAN, SE-ResNeXt26, Lightweighted network	Weighted summation

All the teams participating in the competition were restricted to train their algorithms using only the training subset of the publicly available CelebA-Spoof dataset, thus the use of both proprietary and public external datasets was explicitly forbidden.

Unlike in the two previous contests (i.e. CVPR2019 [11] and CVPR2020 [15]), TPR@FPR-based evaluation criteria were adopted for the ECCV2020 challenge. The TPR@FPR $= 5^{-3}$ determined the final ranking but also TPR@FPR $= 10^{-3}$ and TPR@FPR $= 10^{-4}$ values were reported. In the case if the TPR@FPR $= 5^{-3}$ for two submitted algorithms were the same, the one with higher TPR@FPR $= 10^{-4}$ would rank better.

12.2.3.3 Results and Discussion

In this section, we first analyse the top three solutions as well as their results, and then we discuss the algorithms and challenge in general.

Analysis on the top three solutions: Table 12.13 summarizes the FAS solutions of the top three teams. It is not surprising that all the best performing solutions exploited ensembles of multiple deep models to achieve more robust performance. As the use of external training data in addition to the competition dataset was explicitly forbidden, the features of a single model can easily overfit and learn specific attack cues, while stacking the feature representations of multiple (deep) models can be more generalizing to alleviate this issue. To be more specific, the two best teams ZOLOZ and MM utilized more advanced ensembling strategies compared with the team AFO considering only straightforward weighted summation. The team ZOLOZ proposed a heuristic voting scheme at the score level to form robust combinations of different models, whereas the team MM proposed a novel 'weighting-after-sorting' strategy based on particle swarm optimization (PSO) [59] algorithm for their model ensembles. All three teams utilized at least five different deep model architectures of which some (e.g. FOCUS [18], AENet [57], Noise Print [60], CDCN [48], CDCN++ [48], LGSC [61], CDC-DAN [18]) were aiming at capturing fine-grained (pixel-wise) fidelity characteristics, while some others (e.g. ResNet [28], SE-ResNet [62], EfficientNet [63], ResNeXt [32]) focused on extracting semantic cues that could be complementary to improve face PAD performance. As for the model inputs, both whole facial images and local image patches were utilized by team MM, while the team ZOLOZ and AFO considered only the whole facial images and local image patches, respectively. It can be seen from Table 12.12 that all top three teams achieved excellent PAD performance, reaching TPR>0.999@FPR=10^{-6}, indicating the effectiveness of deep multi-model ensembles on the competition data.

Discussion: Although the aforementioned best solutions achieved very promising results on the CelebA-Spoof dataset, there were still some shortcomings with the ECCV2020 challenge. The hidden test set is rather similar compared with the training data because the CelebA-Spoof dataset was simply divided into subject-disjoint training, development and test folds, thus not explicitly taking into account specific known issues related to domain generalization or unknown PAs. Furthermore, there were no restrictions on the size or number of deep models, which was also disappointing from real-world deployment point of view. Finally, compared with previous two face PAD challenges (i.e. CVPR2019 [11] and CVPR2020 [46]), detailed ablation studies (e.g. impact of each sub-model prior stacking) were missing, as well as the source codes of the solutions were not made public available in the ECCV2020 challenge, thus limiting the transparency and, consequently, usefulness of the whole competition to the FAS community.

12.2.4 LivDet-Face 2021—Face Liveness Detection Competition (IJCB2021)

Recent literature surveys (e.g. [4, 5]) have concluded that both hand-crafted and deep features yield in satisfying classification performance in identifying known PAIs but often fail to detect unknown PAIs and more sophisticated face artefacts. Therefore, continuous efforts are necessary to update face anti-spoofing algorithms to detect rapidly evolving PAs. Although earlier the CVPR2020 cross-ethnicity face PAD challenge considered also a cross-PAI setting (i.e. training on the video-replay attacks and testing on the print and mask PAIs), the types and quality of the unknown PAIs were still limited from the generalized PAD point of view. To address this issue, the LivDet-Face 2021—Face Liveness Detection Competition[28] was organized in conjunction with IJCB2021.

The registration for the IJCB2021 LivDet-Face competition began on 15 February 2021 and ended on 25 April 2021, while the final submission deadline was 30 April 2021. The objective of the competition was to evaluate the performance of the state-of-the-art face PAD algorithms against traditional and novel PAIs. The competition had two separate tracks for image and video data, and the competitors were allowed to participate in both tracks. Different from all previous competitions, IJCB2021

[28] https://face2021.livdet.org/.

LivDet-Face contest did not provide any specific training dataset to the participants, thus the competitors were free to use any proprietary and/or publicly available data to train their algorithms, replicating more realistic and challenging practical AFR application scenarios.

Both academic and industrial organizations were welcome to participate in IJCB2021 LivDet-Face competition anonymously or non-anonymously. In total, 30 international teams registered to the competition, including ten submissions for the image track and six submissions for the video track. Finally, six submission could be successfully tested by the organizers for the image track and five submission for the video track. Unsuccessful tests were due to software issues, which were communicated with the participants.

12.2.4.1 Dataset

No official training dataset was shared by the organizers of the IJCB2021 LivDet-Face competition. Instead, participants were encouraged to use any data available to them (i.e. from both public and proprietary sources) to train and tune their algorithms. The organizers shared only few (no more than two) examples of the known PAIs to familiarize the competitors with the test dataset, while the remaining samples of the disclosed PAI types were considered as unknown to the competitors. The test dataset used in the IJCB2021 LivDet-Face competition was a combination of data from two of the organizing institutions: Clarkson University (CU) and Idiap Research Institute. The dataset consisted of 724 images (135 bona fide and 589 PAI samples) and 814 videos (125 bona fide and 689 PAI samples) for the image and video tracks, respectively. The data was collected from in total 48 live subjects using altogether five different sensors (digital single-lens reflex (DSLR) camera, iPhone X, Samsung Galaxy S9, Google Pixel and Basler aA1920-150uc). The length of the videos in the test dataset was up to six seconds. Eight PAIs for the image track and nine PAIs for the video track were included in the dataset (see Table 12.14 for a summary and Fig. 12.7 for typical examples for each PAI type).

Regarding the 2D PAIs, 100 low-quality (LQ) print paper, 100 high-quality (HQ) photo paper attacks, 100 static display (SD) and video-replay (VR) attacks on laptop screen, and 100 2D photo mask attacks were collected from 25 live subjects using four different sensors. Specifically, the 100 video-replay attacks were used only as an unknown PAI for the video category of the competition, thus were not introduced in the few validation samples that were shared with the competitors. In addition to the 2D attacks, three different qualities of 3D masks (low, medium and high) as well as silicon mask attacks were included in the test dataset for the both competition tracks. A total number of 24 images and 24 videos of low-quality (LQ) 3D masks were created corresponding to six live subjects. Also, 12 medium-quality (MQ) 3D mask and 12 high-quality (HQ) images/videos corresponding to three live subjects were included in the test dataset. The HQ 3D masks were kept as an unknown PAIs for the competitors in the test dataset. In total, 141 image/video samples of wearable 3D silicon masks were collected using five different sensors.

Table 12.14 Summary of the test dataset used in the IJCB2021 LivDet-Face [17]

Class	Types of PAIs	Images	Videos	Sensors
Live	–	135	125	DSLR, iPhone X, Samsung Galaxy S9, Google Pixel
PAI	Laptop Display (DL)	100	100	DSLR, iPhone X, Samsung Galaxy S9, Google Pixel
PAI	Photo Mask (PM)	100	100	DSLR, iPhone X, Samsung Galaxy S9, Google Pixel
PAI	Low-Quality Paper Display	100	100	DSLR, iPhone X, Samsung Galaxy S9, Google Pixel
PAI	High-Quality Paper Display	100	100	DSLR, iPhone X, Samsung Galaxy S9, Google Pixel
PAI	Low-Quality 3D Mask	24	24	DSLR, iPhone X, Samsung Galaxy S9, Google Pixel
PAI	Medium-Quality 3D Mask	12	12	DSLR, iPhone X, Samsung Galaxy S9, Google Pixel
PAI	High-Quality 3D Mask	12	12	DSLR, iPhone X, Samsung Galaxy S9, Google Pixel
PAI	Silicon Mask	141	141	DSLR, iPhone X, Samsung Galaxy S9, Google Pixel, Basler acA1920-150uc
PAI	Video Display (VD)	–	100	DSLR, iPhone X, Samsung Galaxy S9, Google Pixel

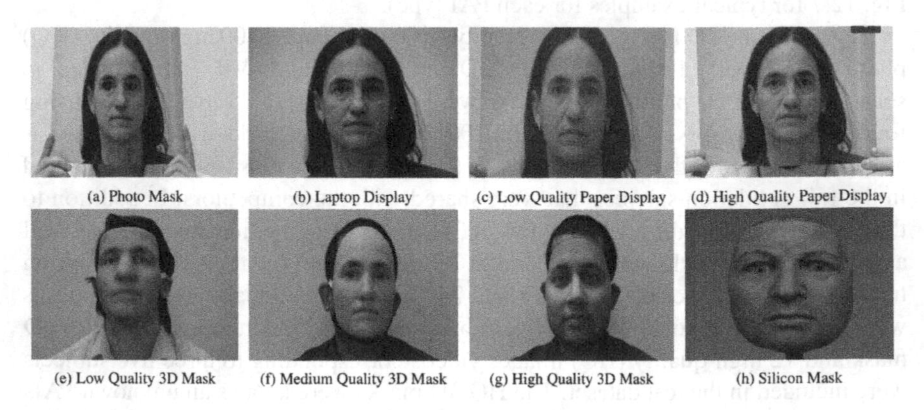

Fig. 12.7 Samples of each PA type present in the IJCB2021 LivDet-Face test set [17]

12.2.4.2 Evaluation Protocol and Metrics

During the IJCB2021 LivDet-Face competition, at least two samples of the majority of the considered PAIs (except the high-quality 3D masks and video-replay attacks) were shared with the competitors of both image and video tracks as a small validation set to fine-tune their algorithms. The performance of an algorithm for each sample was determined by an output ('liveness') score ranging between 0 and 100 with a threshold of 50, while the score of 1000 indicates undetected samples. The test samples with scores less than 50 were classified as PA, whereas the scores of 50 and above were classified as bona fide. Most of the competitors normalized their output scores at their end and provided a score of 0, 100 or 1000 (if undetected) based on their classification output. If the submitted algorithms provided a score of 1000 for the PAs, the result was considered as a correct decision as the algorithm was able to reject an attack, thus is not included in attack presentation classification errors. A score of 1000 for bona fide samples were considered as incorrect, thus was accumulated in bona fide classification errors.

Following the recommendations of the ISO/IEC 30107-3 [1] standard, APCER, BPCER and ACER (see, Sect. 12.2.1.2) were used as the evaluation metrics like in the CVPR2020 challenge. Since all algorithms were required to deliver normalized liveness scores in the range of 0–100, $t = 50$ was used as the decision threshold to calculate the APCER and BPCER. The final ranking of teams was based on the ACER calculated over all the test samples.

12.2.4.3 Results and Discussion

In this section, we first analyse the solutions of both tracks. Then, the result analysis of the image and video tracks is presented. Finally, we discuss the algorithms and the challenge in general.

Analysis on the solutions: Table 12.15 summarizes the FAS solutions of the six teams. The teams SiMiT Lab and NTNU Gjøvik utilized several public datasets for training, including the 2D PAI (e.g. Replay Mobile [64], SiW [8], Oulu-NPU [6], SWAN [65], CASIA-FASD [10]) and 3D mask attack (e.g. 3DMAD [66] and NTNU-Silicon Mask [67]) data. The team Fraunhofer IGD improved the generalization of their algorithm by using the 50 attacks and bona fide samples from the Real Mask Attack Database (CRMA) [68] as unknown development data to tune the decision threshold.

Like in the previous three challenges discussed already in this chapter, most of the solutions in IJCB2021 LivDet-Face competition used ensembles of multiple deep models to achieve more robust performance. The teams Fraunhofer IGD, SiMiT Lab, FaceMe, little tiger and NTNU Gjøvik stacked 12, 2, 3, 5 and 6 models in their submitted systems to make the final PAD decision. Three kinds of models were considered in these solutions: (1) hand-crafted models with digital signal processing, (2) pixel-wise supervised models (e.g. DeepPixBis [69], DepthNet [8] and CDCN [47]) and (3) BCE supervised models (e.g. ResNet [28], ResNeXt [32], VGG [41], Incep-

Table 12.15 Summary of the solutions of in IJCB2021 LivDet-Face [17]

Team	Training/validation	Model	Ensemble strategy
Fraunhofer IGD	CRMA for validation	12 models (DeepPixBis, ResNeXt, etc.)	FDR weights
SiMiT Lab	Replay Mobile, SiW, Oulu-NPU, 3DMAD for training	DeepPixBis, EfficientNet-B7	Mean score fusion
CLFM	–	CDCN	–
FaceMe	–	3 models (DepthNet, Digital signal processor, etc.)	–
little tiger	Glint360k for pre-training	5 models (ResNet50, ResNext26, CDCN, etc.)	–
NTNU Gjøvik	SWAN, CASIA-FASD, NTNU-Silicon Mask for training	6 models (Resnet18, Resnet50, InceptionV3, 9 VGG1, VGG16, Alexnet) with 2 linear SVM	Majority voting

tion [70] and Alexnet [71]). The team Fraunhofer IGD adopted Fisher discriminant ratio (FDR) weights [72] for 12 models to get a combined face PAD decision, while the team NTNU Gjøvik considered a simple majority voting strategy to make decision from six models, and the team SiMiT Lab fused the scores from two models with shuffled patch-wise supervision [73]. Some competitors (e.g. teams CLFM and FaceMe) were lacking these kinds of details in their method descriptions, thus making it impossible to draw conclusions on some factors in our result analysis, including data fusion approaches and training/tuning data.

Results of the image track: Table 12.16 summarizes the results of the image track. The team Fraunhofer IGD was the winner, obtaining the lowest ACER = 16.47%, followed by the team CLFM with a narrow margin in ACER = 18.71%. The winning team Fraunhofer IGD achieved the lowest BPCER = 15.33% among the six competitors. The six competitors achieved highly varying performances across the different PAI types. The algorithm submitted by the team Fraunhofer IGD detected all the low-quality paper display attacks, and the algorithm by CLFM successfully detected all the low-quality 3D mask samples. Team CLFM's algorithm also performed the best with APCER = 10% for high-quality photo paper display samples but achieved an unsatisfying BPCER = 24.08%. The team FaceMe, who achieved third place in the image track, achieved the best APCER = 3% for laptop display samples and second best BPCER = 16.06%. The team SiMiT Lab successfully detected all the medium-quality 3D mask samples with APCER = 0%, which was best among all the competitors, but achieved also the worst BPCER = 51.09%.

Table 12.16 Results of the image track in the IJCB2021 LivDet-Face competition. The APCER is, respectively, calculated for each type of PAI, and then averaged for final ACER calculation

Team	Paper		Replay	2D Mask	3D Mask				BPCER(%)	ACER(%)	Ranking
	LQ	HQ			LQ	MQ	HQ	Silicon			
Fraunhofer IGD	**0**	24	45	14.7	4.17	8.33	**14.29**	**16.31**	**15.33**	**16.47**	1
CLFM	6.06	**10**	8	**5.88**	**0**	16.67	21.43	34.75	24.08	18.71	2
FaceMe	22.22	11	**3**	11.76	66.67	66.66	50	57.45	16.06	20.72	3
little tiger	41.41	52	4	58.82	54.17	25	28.57	82.98	21.17	33.92	4
SiMiT Lab	7.07	18	43	15.68	16.66	**0**	42.85	80.85	51.09	42.05	5
Anonymous	78.78	86	77	89.21	87.5	83.33	100	98.58	16.79	49.35	6

By comparing the performance of the top two ranked solutions in the image track, it is obvious that the models performed better against low-quality PAIs than higher quality PAIs. The team Fraunhofer IGD obtained APCER = 0% for the low-quality paper display and APCER = 24% for high-quality paper display. Similarly, the team CLFM obtained APCER = 6.06% for low-quality paper display, and APCER = 10% for high-quality paper display. The same trend can be observed for the different quality of 3D face masks. The team Fraunhofer IGD obtained APCER = 4.17% for low-quality 3D masks compared with APCER = 8.33% for medium-quality 3D masks, APCER = 14.29% for high-quality 3D masks and APCER = 16.31% for high-quality silicon masks. Similarly, the team CLFM achieved APCER = 0% against low-quality 3D masks compared with APCER = 16.67% for medium-quality 3D masks, APCER = 21.43% for high-quality 3D masks and APCER = 34.75% for high-quality silicon masks.

Results of the video track: Table 12.17 summarizes the results of the five solutions in the video track of the IJCB2021 LivDet-Face competition. The team FaceMe was the winner with the ACER = 13.81% followed by the team Fraunhofer IGD with a narrow margin in ACER = 14.49%. The lowest BPCER = 4.76% was achieved by the team NTNU Gjøvik. The team CLFM performed well in detecting paper and video-replay attacks (with APCER $<$ 3.30% for all scenarios) but ranked third due to the bad BPCER performance (39.68%). The team NTNU Gjøvik-V2 performed well against 3D face mask attacks, achieving APCER = 0% for the three different types of 3D masks but also the worst BPCER (51.59%).

It can be observed that the top two solutions performed better against low-quality PAIs compared to higher quality PAIs. For example, the performance of the team Fraunhofer IGD against low-quality paper display was APCER = 1%, while APCER = 25.25% was obtained against the high-quality paper display attacks. The same trend can be observed for the different quality of 3D face masks as well. The performance of the team Fraunhofer IGD against low-quality 3D masks was APCER = 4%, which was obviously better than that of the medium-quality (APCER = 9.09%) and high-quality silicon masks (APCER = 34.75%). However, the performance of high-quality 3D masks was better than that of any other 3D mask category with APCER = 0%.

Discussion: Compared with the three earlier competitions [11, 15, 18] introduced in this chapter, a significant degradation in the overall performance can be observed. This can be due to several factors, such as (1) increased complexity in the test dataset with nine different PAI types; (2) introduction of three novel attack types with limited availability, or not covered at all, in the public datasets; (3) lack of specific competition training dataset, i.e. choice of training data up to the competitors; (4) domain shift between the training and test conditions in terms of environmental factors, sensors, quality of PAIs, and the introduction of unknown PAIs. The results of this competition indicate that generalized face PAD is still far away from a solved research problem.

Despite its important findings on the current state of face PAD 'in the wild', the IJCB2021 LivDet-Face competition still had some shortcomings. First, as there were no pre-defined training sets and, consequently, no restrictions on the diversity and scale of the private or public datasets, it is unfair to evaluate the performance of the

Table 12.17 Results of the video track in the IJCB2021 LivDet-Face competition

Team	Paper		Replay		2D Mask	3D Mask				BPCER(%)	ACER(%)	Ranking
	LQ	HQ	SD	VR		LQ	MQ	HQ	Silicon			
FaceMe	8	10.10	18	16	6.93	40	45.45	38.46	9.22	14.29	**13.81**	1
Fraunhofer IGD	**1**	25.25	29	9	1	4	9.09	**0**	12.77	16.67	14.49	2
CLFM	4	**4.04**	**8**	**1**	**0**	**0**	27.27	7.69	**1.42**	39.68	21.49	3
NTNU Gjøvik-V1	50	59.60	83	75	18.81	36	18.18	46.15	21.28	**4.76**	26.51	4
NTNU Gjøvik-V2	5	9.09	32	20	1	**0**	**0**	**0**	33.33	51.59	34.05	5

different approaches. It is simply impossible to explore the hidden reasons behind the differences in performance and to tell if an algorithm is actually better than another one, or is it in fact a matter of the amount and quality of training data. Second, no ablation studies or open-source code for most of the solutions were provided, which again limits the transparency and, consequently, usefulness of the results and findings to the FAS community. Finally, as the complexity of the proposed systems was not limited, mainly huge ensembles of deep models were adopted by the participants, while none of the teams proposed interesting novel efficient face PAD approaches.

12.2.5 3D High-Fidelity Mask Face Presentation Attack Detection Challenge (ICCV2021)

As seen in two recent competitions (i.e. CVPR2020 [15] and IJCB2021 LivDet-Face [17]), face PAD performance of state-of-the-art methods drops significantly under unknown 3D mask attacks. However, the previous competition datasets contained in general only a limited number of samples and types of 3D facial masks, thus there is a large gap in between the existing benchmarks and 3D mask attack detection in real-world conditions. To alleviate the threats posed by 3D mask attacks and to improve the reliability of face PAD methods under emerging types of 3D mask attacks in various scenarios, the 3D High-Fidelity Mask Face Presentation Attack Detection Challenge [16] was organized in conjunction with the International Conference on Computer Vision (ICCV) in 2021 using the very recently constructed CASIA-SURF High-Fidelity Mask (HiFiMask) dataset [74].

The ICCV2021 challenge was conducted using the CodaLab platform[29], attracting 195 teams from all over the world. A summary with the names and affiliations of teams that entered the final stage of the contest is shown in Table 12.18. Again, the majority of the final participants came from industrial institutions, and all the six best performing teams represented companies. This indicates clearly that mask attack detection is no longer limited to academic research but also a crucial problem in real-world AFR applications. The results of the top three teams were far better than the baseline results [74], thus greatly improving the performance of 3D high-fidelity mask attack detection.

12.2.5.1 Dataset

HiFiMask [74] is currently the largest 3D face mask PAD dataset, consisting of 54, 600 videos corresponding to 75 subjects with three skin tones, including 25 subjects in yellow, white and black, respectively. The database contains three high-fidelity masks for each identity, which are made of transparent, plaster and resin materials, respectively. During the acquisition process, six complex scenes were

[29] https://competitions.codalab.org/competitions/30910.

Table 12.18 Names, affiliations and final ranking of the teams participating in the ICCV2021 challenge

Ranking	Team name	Leader name	Affiliation
1	VisionLabs	Oleg Grinchuk	Visionlabs.ai
2	WeOnlyLookOnce	Ke-Yue Zhang	Tencent Youtu Lab
3	CLFM	Samuel Huang	FaceMe
4	oldiron666	Zezheng Wang	Kuaishou Technology
5	Reconova AI-LAB	Mingmu Chen	Reconova Technology
6	Inspire	Jiang Hao	Bytedance Ltd.
7	Piercing Eyes	Hyokong	National University of Singapore
8	msxf_cvas	Liang Gao, MaShang	Consumer Finance Co.,Ltd
9	VIC_FACE	Cheng Zhen	Meituan
10	DXM-DI-AI-CV-TEAM	Weitai Hu	Du Xiaoman Financial
11	fscr	Artem Petrov	Peter the Great St. Petersburg Polytechnic University
12	VIPAI	Yao Xiao	Zhejiang University
13	reconova-ZJU	Zhishan Li	Zhejiang University
14	sama_cmb	Yifan Chen	Chinese Merchants Bank(CMB)
15	Super	Yu He	Technische Universität München
16	ReadFace	Zhijun Tong	ReadFace
17	LsyL6	Dongxiao Li	Zhejiang University
18	HighC	Minzhe Huang	Akuvox (Xiamen) Networks Co., Ltd.

considered for recording the videos (i.e. white light, green light, periodic three-colour light, outdoor sunshine, outdoor shadow and motion blur). For each scene, there are six videos captured under different lighting conditions (i.e. normal, dim, bright, back, side and top) to explore the impact of directional lighting. Periodic lighting within [0.7, 4] Hz in the first three scenarios (see the first three columns of the last row in Fig. 12.8 for example) tries to mimic the natural human pulse variations to fool remote photoplethysmography (rPPG)-based mask detection technology (e.g. [75]). Finally, seven mainstream imaging devices (i.e. iPhone11, iPhone X, MI10, P40, S20, Vivo and HJIM) were utilized for recording the videos in order to ensure high resolution and imaging quality corresponding to modern mobile devices. The original videos were not provided due to huge amount of data. In order to decrease the size of the dataset, the organizers sampled every tenth frame of each video and applied a fast face detector [76] to remove most of the background information from the sampled video frames, thus the final competition data consisted of coarsely pre-cropped facial

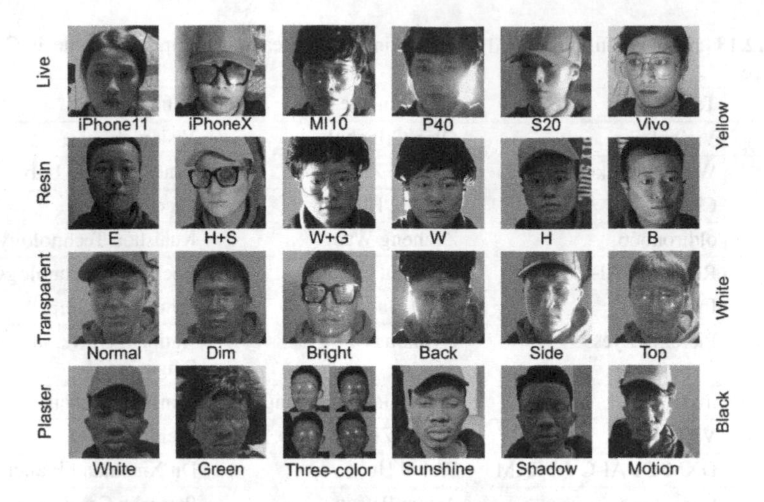

Fig. 12.8 Samples from the HiFiMask dataset [74]. The first row shows six kinds of imaging sensors. The second row shows six kinds of appendages, of which E, H, S, W, G and B are the abbreviations for 'empty', 'hat', 'sunglasses', 'wig', 'glasses' and 'messy background', respectively. The third row shows six kinds of illumination conditions, and the fourth row represents six deployment scenarios

images. Some typical samples of pre-processed video frames in the HiFiMask dataset are presented in Fig. 12.8.

In order to increase the difficulty of the competition and meet the real-world deployment requirements, an 'open-set' test protocol was utilized to comprehensively evaluate the discriminative and generalization power of face PAD algorithms. In other words, the training and development sets contained only subsets of common mask types and operating scenarios, while there were more general mask types and scenarios in the test set. Thus, the distribution of test set was more complicated compared to the training and development sets in terms of mask types, scenes, lighting and imaging devices. Such 'open-set' protocol considers explicitly both 'seen' and 'unseen' domains as well as mask types for evaluation, which is also more valuable from real-world face PAD deployment point of view.

As shown in Table 12.19, every skin tone, part of mask types, such as transparent and resin materials (1, 3); part of scenes, such as white light, outdoor sunshine and motion blur (1, 4, 6); part of lighting conditions, such as normal, bright, back and top (1, 3, 4, 6) and part of imaging devices, such as iPhone 11, iPhone X, MI10, P40 (1, 2, 3, 4) are included in the training and development subsets. All skin tones, mask types, scenes, lighting conditions and imaging devices are present in the test subset. For clarity, the organization of the dataset and quantity of videos for each sub-protocol of the challenge are shown in Table 12.19.

Table 12.19 Statistical information for the protocols used in the ICCV2021 challenge. Note that 1, 2 and 3 in the third column mean 'transparent', 'plaster' and 'resin' masks, respectively. The numbers in the fourth, fifth and sixth columns are explained in Sect. 12.2.5.1

Subset	Subject	Mask type	Scene	Light	Sensor	# Live num.	# Mask num.	# All num.
Train	45	1&3	1&4&6	1&3&4&6	1&2&3&4	1,610	2,105	3,715
Dev	6	1&3	1&4&6	1&3&4&6	1&2&3&4	210	320	536
Test	24	1~3	1~6	1~6	1~7	4,335	13,027	17,362

12.2.5.2 Evaluation Protocol and Metrics

The challenge comprised two stages as follows:

Development phase: *(19 April 2021–10 June 2021)*. During the development phase, the participants had access to the labelled training data and unlabelled development data. The samples in the training set were labelled with the bona fide, two types of masks (1, 3), three types of scenes (1, 4, 6), four kinds of lighting conditions (1, 2, 4, 6) and four imaging sensors (1, 2, 3, 4). The labels of the validation data were not provided to the participants in development phase. Instead, participants could build their models on the labelled training data and then submit their predictions on the development data and receive immediate feedback via the competition leaderboard in the CodaLab platform.

Final phase: *(10 June 2021–20 June 2021)*. During the final phase, the labels for the development set were also made available to the participants and the unlabelled test set was also released. The competitors had to make their predictions on the test samples and upload their solutions to the challenge platform. The organizers did then rerun the best performing algorithms, and the final ranking of the participants was obtained based from the verified results on the test data. To further facilitate the outcome and findings of the competition to the face PAD community, the best performing teams were encouraged to make their source code publicly available and provide fact sheets describing their solutions.

Evaluation metrics: Similar to the previous two competitions (i.e. CVPR2020 [15] and IJCB2021 LivDet-Face [17]), the ISO/IEC 30107-3 [1] standardized metrics APCER, BPCER and ACER were selected as the evaluation criteria, and the ACER was the leading evaluation criterion for ranking the submitted systems in the ICCV2021 challenge. The ACER threshold on the test set was determined based on the EER operating point on the development data.

12.2.5.3 Results and Discussion

In this section, we first summarize solutions in the 3D Mask face PAD challenge. Then, we provide our result analysis. Finally, the algorithms and the challenge are discussed in general.

Table 12.20 Summary of the solutions in the ICCV2021 challenge [16]. 'DSFD' and 'DBO' denote dual shot face detector [77] and deep bilateral operator [78], respectively

Team name	Pre-processing	Backbone	Branch	Loss
VisionLabs	DSFD detector	EfficientNet-B0	6	BCE
WeOnlyLookOnce	DSFD detector	ResNet12	2	3-class CE with label smoothing
CLFM	Crop mouth region	CDCN++	1	BCE
Oldiron666	Whole face	Resnet6 (SimSiam)	1	BCE+MSE+Contrastive loss
Reconova-AI-Lab	RetinaFace detector	ResNet50+YoLoV3-FPN	3	BCE+Focal loss
inspire	RetinaFace detector	SE-ResNeXt101	1	BCE+MSE+Contrastive loss
Piercing Eye	Face detection	CDCN	2	Depth regression+BCE
msxf cvas	Face detection+alignment	ResNet34	1	4-class CE loss
VIC FACE	Face detection	CDCN with DBO	1	Depth regression
DXM-DI-AI-CV-TEAM	–	DepthNet	1	BCE with meta learning

Analysis of the solutions: Table 12.20 summarizes the solutions of the most teams participating in the ICCV2021 challenge. The source code of the winning team VisionLabs was released[30] while the detailed descriptions as well as the ablation studies of the top three ranked solutions can be found in [79–81]. Different from all the previous competitions, the ICCV2021 challenge accepted only the results obtained with single (deep) model-based systems, thus ensemble strategy with multiple models was explicitly prohibited. As a result, several solutions (e.g. the teams VisionLabs, WeOnlyLookOnce, Reconova-AI-Lab and Piercing Eye) developed multi-branch architectures, which aimed at capturing more diverse PAD-specific feature representations. Also, team VIC FACE proposed a novel architecture in its solution, which integrated the deep bilateral operator in the original CDCN [82] in order to learn more intrinsic features via aggregating multi-level bilateral macro- and micro-texture information. Most of the teams adopted DSFD [77] or RetinaFace [83] detector for pre-processing to localize more fine-grained facial region and to filter out partial low-quality face mask attacks. The teams considered face PAD mainly as a binary classification task using BCE loss or as a depth regression problem, but two teams (i.e. WeOnlyLookOnce and msxf cvas) also forced the models to learn more mask type-aware features via fine-grained multiple class CE loss.

Result analysis: The results and ranking of the top 18 teams are shown in Table 12.21. The ACER performance of the top three teams was relatively close (within $<$ 3.3%). One major reason behind this is that the original pre-processed images in the HiFiMask dataset correspond to very coarsely cropped facial regions, including also outliers (i.e. non-facial samples), due to the limited accuracy of the

[30] https://github.com/AlexanderParkin/chalearn_3d_hifi.

Table 12.21 The final results and team rankings of the ICCV2021 challenge [16]. The best results are shown in bold

Team name	FP	FN	APCER (%)	BPCER (%)	ACER (%)	Rank
VisionLabs	492	**101**	3.777	**2.330**	**3.053**	1
WeOnlyLookOnce	**242**	193	**1.858**	4.452	3.155	2
CLFM	483	118	3.708	2.722	3.215	3
oldiron666	644	115	4.944	2.653	3.798	4
Reconova-AI-LAB	277	276	2.126	6.367	4.247	5
Inspire	760	176	5.834	4.060	4.947	6
Piercing Eyes	887	143	6.809	3.299	5.054	7
msxf_cvas	752	232	5.773	5.352	5.562	8
VIC_FACE	1152	104	8.843	2.399	5.621	9
DXM-DI-AI-CV-TEAM	1100	181	8.444	4.175	6.310	10
fscr	794	326	6.095	7.520	6.808	11
VIPAI	1038	268	7.968	6.182	7.075	12
reconova-ZJU	1330	183	10.210	4.221	7.216	13
sama_cmb	1549	188	11.891	4.337	8.114	14
Super	780	454	5.988	10.473	8.230	15
ReadFace	1556	202	11.944	4.660	8.302	16
LsyL6	2031	138	15.591	3.183	9.387	17
HighC	1656	340	12.712	7.843	18	

used efficient face detector [76]. Thus, the additional pre-processing with DSFD and mouth region cropping was very beneficial in removing the outlier samples affecting negatively the training of deep models and focusing on the details in the actual facial information discriminating attacks from bona fide samples. The team VisionLabs achieved the best BPCER = 2.33% with only 101 FN samples, while WeOnly-LookOnce had the lowest APCER = 1.858% with 242 FP samples, indicating that the features from multiple facial region-based branches and vanilla/CDC branches benefited the spoof cue representations. Moreover, the ACER performance of all the teams was evenly distributed ranging from 3% to 10%, which not only indicates the rationality and selectivity of the 3D mask face PAD challenge but also demonstrated the value of the HiFiMask dataset for future face PAD studies.

Discussion: Some general observations on 3D mask face attack detection can be concluded based on findings of the ICCV2021 competition: (1) the accurate facial localization and subregion information was very beneficial, focusing on the actual discriminative local facial details and avoiding learning and extraction of irrelevant features for face PAD and (2) multi-branch-based feature learning was a widely used framework by the participating teams, which benefits from the shared low-level feature learning, capturing diverse separated multi-branch features for generalized 3D mask description. However, there were still some shortcomings in the ICCV2021 challenge. The fidelity and the collection of the 3D mask types were still limited. For instance, the face mask in the fourth row and second column of Fig. 12.8 has a very artificial appearance, while also more 3D attack types, such as paper and silicone masks, and wax faces should be considered. Furthermore, only every tenth frame for each video was provided. Such a low frame rate makes it impossible to recover facial physiological signals and, consequently, to study rPPG-based 3D mask detection [84, 85], for instance.

12.3 Discussion

All the recent five competitions were successful in consolidating and benchmarking the current state of the art in face PAD. In the following, we provide general observations and further discussion on the lessons learnt, model architectures and potential future challenges.

12.3.1 General Observations

It is apparent that the used datasets and evaluation protocols, and also the recent advances in the state of the art, reflect the trends in face PAD schemes seen in the different contests. The algorithms proposed in the context of the first two multi-modal competitions (i.e. CVPR2019 [11] and CVPR2020 [15]) exploited the evident visual cues that we humans can observe in the multi-modal imaging data of the CASIA-SURF and CASIA-SURF CeFA datasets, including plain/structural discrepancies in the depth modality, material surface reflection differences in the NIR modality and natural human movement in the dynamic modality. Most of the solutions adopted exhausted model and hyperparameter selection strategies for each modality and ensembling the best combination of different modalities and models for robust PAD performance in the multi-modal setting, i.e. TPR = 99.8739%@FPR = 10e-4 on the CASIA-SURF and ACER = 1.02% on the CASIA-SURF CeFA datasets. In these two competitions, the performance gains relied mostly on powerful deep models and ensembling strategies, while the essence of multi-modal fusion was not truly explored. This gave, disappointingly, limited insight to the multi-modal face PAD community, thus it is necessary to rethink these kinds of multi-modal PAD settings, especially from the efficient fusion point of view, in the upcoming benchmarks.

In the latest two competitions (i.e. IJCB2021 LivDet-Face [17] and ICCV2021 [16]) on unimodal colour camera-based face PAD, generalized and intrinsic bona fide/attack cues and motion analysis were hardly used. The proposed RGB data-based features were overfitting in the training data, thus generalized poorly to the test sets. One reason to the unsatisfactory performance is that the test sets included more unseen high-fidelity 3D mask types (e.g. plaster masks in the ICCV2021 [16] and high-quality 3D masks in the IJCB2021 LivDet-Face [17] challenge). Although it was nice to see a diverse set of advanced deep learning-based systems and further improved versions of the provided baseline method, it was a bit disappointing that entirely novel generalized face PAD solutions, e.g. for zero-shot unseen attack (especially against 3D masks) detection, were not proposed. The best performances in the IJCB2021 LivDet-Face (ACER = 16.47%/13.81% for image/video tracks) and the ICCV2021 (ACER = 3.053%) competitions were still limited, considering the needs of real-world use cases. However, as seen in the CVPR2020 competition [15], the major issues with domain shift and unseen PAs can be at least partially alleviated by introducing depth and NIR sensors as these additional modalities provide more accurate 3D shape information for 2D PAI detection and material reflection information for discriminating realistic 3D masks among other face artefacts from genuine human skin.

Unlike with the first three face PAD contests (i.e. IJCB2011 [13], ICB2013 [14] and IJCB2017 [12]), it is interesting to observe that most of the participants came from the industry in the competitions organized from 2019. For example, majority of the final participants (10 out of 13) of the CVPR2019 challenge [11] and (8 out of 11) of the unimodal track of the CVPR2020 challenge [15] came from industrial institutions. Also, in the most recent ICCV2021 challenge [16], the majority (13 out of 18) of the final participants came from companies, and the six top-performing teams were all from industry, indicating the steadily growing interest and need for practical and reliable face PAD solutions in commercial AFR products.

12.3.2 Lessons Learnt

The competitions have given valuable lessons on designing databases and test protocols, and competitions in general. In the CVPR2019 [11] and ECCV2020 [18]) challenges, the best performances on the test data reached TPR = 99.8739%@FPR = 10e-4 and TPR = 100%@FPR = 10e-6, respectively. However, the problem of face PAD has not been solved as the error rates (ACER = 16.47%/13.81%) of the more recent IJCB2021 LivDet-Face competition reveal that the state of the art in face anti-spoofing still suffers from significant generalization issues in unknown operating conditions. Therefore, we should also rethink the design of evaluation protocols in competitions like the CVPR2019 [11] and ECCV2020 [18]) challenges. In the CVPR2019 challenge, the provided validation data had similar distribution with the test data, and the validation set was also allowed to be included in training the face PAD models. Furthermore, the training, validation and testing sets were split in subject-disjoint manner in the ECCV2020 challenge when the domain covariates (e.g. biometric sensors and lighting conditions) and PAIs are too similar. As a result, even using the off-the-shelf deep models with powerful representation learning capacity, the participants could easily train an ensemble of several overfitting models in these two competitions, performing well not only on the training and validation sets but also on the test data. It is necessary to mimic the requirements of real-world biometric applications and to design and capture even more challenging and larger scale datasets with unknown domains and unseen PA types in the test sets.

TPR@FPR was utilized in the CVPR2019 and ECCV2020 challenges as the ranking criteria, while ACER was adopted in the three remaining recent face PAD competitions. Despite being widely used in large-scale biometric evaluations (e.g. face recognition), TPR@FPR was first utilized in face PAD competitions due to emerging larger scale face PAD datasets (e.g. the CASIA-SURF [23] with 295,000 frame samples and CelebA-Spoof [57] with over 62,000 samples), which made the calculation of TPR@FPR = 1e-4 or even TPR@FPR = 1e-6 possible. In most of the competitions (4 out of 5), the results were reported using the mainstream metrics APCER, BPCER and ACER recommended in ISO/IEC 30107-3 standard [1]. However, the selection of score threshold value for computing the ACER is worth discussing from both contest and database design point of views. For the CVPR2019 [11], CVPR2020 [15] and ICCV2021 [16] challenges, the ACERs for the test sets were determined by the EER operating point on the validation sets, while in the IJCB2021 LivDet-Face [17]) evaluation, the fixed threshold of 50 (normalized output liveness scores from 0 to 100) was utilized directly, and it was up to the participant how the scores were normalized within the valid range of liveness scores. The latter evaluation method with fixed range of liveness scores and ACER threshold sounds intuitive from interpretability and real-world applications point of views, and adequate considering the lack of specific validation set in the IJCB2021 LivDet-Face competition. The former approach seems to be more reasonable but with some drawbacks. For instance, significant discrepancies in performance can be observed among the ranked solutions when comparing their results using the ROC-related TPR@FPR metrics and ACER. As an example, the second best solution in terms of ACER ranking in Table 12.9 performed poorly in terms of TPR@FPR = 1e-3 in Fig. 12.4. In practical biometric applications, the suitable operating point depends highly on the application context. However, when looking at a single ACER value, the misclassification rates between bona fide and attack classes are often ignored, even though they are a crucial piece of information. A method performing well in terms of ACER at the selected threshold might suffer from severely imbalanced APCER/BPCER ratio. For instance, one of the metrics might clamp to zero, while the other one can be relatively high. In this case, the ACER fails to point out that the PAD system is able to detect all attacks but rejects many bona fide samples (or vice versa), thus making it impossible to judge its performance trade-off in the APCER and BPCER, i.e. security and usability. Examples of such system behaviour can be seen in Table 12.9, where the teams VisionLabs and Wgqtmac show *0.11%/5.33%* and *51.57%/0.66%* APCER/BPCER ratios, respectively.

The CVPR2019 [11], CVPR2020 [15] and ICCV2021 [16] challenges can be considered to be more transparent and fair due to the richer amount of publicly available details about the participating teams and the best performing solutions, whereas the ECCV2020 [18] and IJCB2021 LivDet-Face [17] contests reported only limited information about the teams and evaluated solutions. In general, competitions should encourage the participants to provide authentic public details on the registered teams and open-source implementations, as well as detailed ablation analysis of the best performing solutions. The authentic team information is useful for avoiding malicious registration situations where teams consisting of similar members from the same institution could make more submission entries, thus less cost of trial and error. The open-source codes and detailed ablation analysis would benefit the reproducibility of the solutions and mitigate the possibility of cheating by using manually annotated competition test data. The best solution to prevent 'data peeking' would be to keep the evaluation set, including unseen test conditions, inaccessible during algorithm development phase and to conduct independent (third-party) evaluations, in which the organizers run the provided executables or source codes of the submitted systems on the competition data. Another option would be to hide some 'anchor' samples from the development set (with randomized file names) in the evaluation data and releasing the augmented test set once the development set scores have been submitted (fixed), as done in the BTAS 2016 Speaker Anti-spoofing Competition [86]. The scores of the anchor videos could be used for checking whether the scores for the development and test sets have been generated by the same system. At minimum, the organizers should be able to retrain and rerun the best performing models following the official competition protocols to check if the submitted solutions have been calibrated, or even trained, on the test set, and determine the final ranking of the teams based on the verified results.

The ICCV2021 challenge [16] provided also a fairer evaluation of the proposed algorithms per se by limiting the influence caused by differences in the amount of training data and number of ensembled (deep) models. The use of the same training data and a fixed number of models would be fairer for comparing different algorithms. Otherwise, the competitions just assess and ascertain how far the participants can push the face PAD performance with 'black-box' methods on the specific benchmark, while not gaining actual insight in the effectiveness of the different proposed algorithms under the same conditions. For instance, we observed that the performance gains with the best solutions of the CVPR2019 [11] and ECCV2020 [18] challenges were largely due to the use of large-scale pre-training data and huge ensemble models, respectively. Another option would be to conduct separate ablation studies on the competition test data by evaluating the solutions trained on the same training data in order to find out at least the impact of data, especially in the case of proprietary datasets.

The CASIA-SURF [22, 23] and the CASIA-SURF CeFA datasets [46] used in the CVPR2019 [11] and CVPR2020 [15] challenges, respectively, provide pre-cropped and aligned facial images for each modality. Furthermore, the background information has been pixel-wise masked out to mitigate the effect of different face detection and alignment methods and limit the problem of face PAD to the actual facial information instead of exploiting the domain-specific contextual cues. The findings of the ICCV2021 challenge [16] suggest, however, that the use of proper pre-processing (e.g. face or facial attribute localization) can, in fact, significantly improve the PAD performance. The best performing teams used additional pre-processing steps to focus on the actual discriminative facial details and specific subregion information and to avoid learning and extraction of irrelevant features for face PAD. Data pre-processing needs definitely further attention in future because it is an understudied subject in face PAD research and an important component in complete face PAD solutions used in real-world AFR applications. While artificially restricting the original facial images and videos into pre-cropped bounding boxes or pixel-wise masked regions mitigates the issues related to sharing and working with huge datasets, and exploiting dataset-specific contextual cues for better PAD performance, the heavily pre-processed data can also limit the novelty of proposed solutions and usability of the dataset for experimental analysis during the competitions and, more importantly, in future studies in the research field. Therefore, the research community needs to find means for providing large-scale face PAD datasets also with unprocessed facial images and videos in order to be able to evaluate complete face PAD solutions and conduct comprehensive ablation studies.

12.3.3 Summary on Model Architectures

Model architectures play a vital role in extracting PAD-specific feature representations, thus it is worth investigating how to select suitable deep models for unimodal and multi-modal face PAD in light of the competition results. Table 12.22 shows the performance of the methods based on a single model architecture (i.e. without ensemble models) in the CVPR2019 [11], CVPR2020 [15], IJCB2021 LivDet-Face [17] and ICCV2021 competitions [16]. In the multi-modal challenges (see the second and third columns in Table 12.22), the methods based on the DenseNet121 and CDCN backbone models achieved the best results. Compared with the ResNet fam-

Table 12.22 ACER (%) performance comparison of the single model architectures used in the CVPR2019 [11], CVPR2020 [15], IJCB2021 LivDet-Face [17] and ICCV2021 competitions [16]

Model	CVPR2019 challenge	CVPR2020 challenge		IJCB2021 LivDet-Face	ICCV2021 challenge
	Multi-modal	Multi-modal	Unimodal	Unimodal	Unimodal
VGG16 [41]	0.6255	–	–	–	–
ResNet18 [28]	2.4223	–	26.12	–	–
ResNet34 [28]	1.1107	1.68	–	–	5.562
ResNet101 [28]	–	–	9.28	–	–
ResNeXt101 [32]	0.6812	–	–	–	–
SE-ResNeXt101 [62]	0.0985	–	–	–	4.947
DenseNet121 [87]	**0.0516**	–	–	–	–
Inception [88]	1.6873	–	–	–	–
ShuffleNet-V2 [35]	0.3855	–	–	–	–
EfficientNet-B0 [63]	–	–	–	–	**3.053**
DepthNet [8]	–	5.89	–	–	6.31
CDCN [48]	–	**1.02**	**4.84**	**18.71**	5.054
CDCN++ [48]	–	–	–	–	**3.215**

ily (e.g. ResNet18, ResNet34 and ResNeXt101), the DenseNet121 extracts denser contextual semantic features, which benefits from the hidden feature representations between the different modalities. The generic backbones supervised by BCE loss focus on high-level semantic feature representations, while the CDCN with central difference convolution and multi-level fusion module is based on pixel-wise supervision, aiming at capturing more fine-grained disparities in the intrinsic fidelity characteristics between bona fide samples and face artefacts. In the unimodal challenges, the EfficientNet-B0 and CDCN performed the best in the IJCB2021 LivDet-Face and ICCV2021 challenges, respectively. It is interesting to notice that both of these two architectures have been discovered using automatic neural architecture search (NAS) [89], indicating the promising potential of NAS in searching optimal task-aware architectures also for face PAD. Despite the differences in hyperparameter settings, the architectures highlighted in bold in Table 12.22 can be recommended and treated as valuable prior knowledge in selecting models for the upcoming unimodal and multi-modal face PAD challenges.

In the ECCV2020 and IJCB2021 LivDet-Face competitions, most of the top teams considered ensemble models combining different architectures. For instance, the best performing solution in the ECCV2020 challenge (see Table 12.13) consists of five kinds of models (i.e. FOCUS, AENet, ResNet, Attack type, Noise Print) while 12 different models (e.g. DeepPixBis, ResNeXt, etc.) are combined in the winning solution of the IJCB2021 LivDet-Face competition (see Table 12.15). It can be concluded that ensemble strategy works the best in these kinds of competition settings, but the contribution of each model on the final performance remains unclear due to lack of proper ablation studies. However, one common feature among these ensemble solutions combining mixed architectures is that both generic, high-level backbones (e.g. ResNet [28] and ResNeXt [32]) with BCE loss and PAD-specific architectures (e.g. DeepPixBis [69] and AENet [57]) with pixel-wise supervision are used. Thus, the feature representations from these two different approaches seem to be compatible and complementary towards more generalized face PAD solutions.

12.3.4 Future Challenges

The test cases in the current competitions measuring the generalization across different covariates are still rather limited. Especially the domain diversity should be increased, as the samples in most (4 out of 5) of the competitions were recorded in indoor office locations with no more than three ethnicities. Regarding the PA species, recently introduced challenging partial face attacks (e.g. half 3D mask, makeup and tattoo) have not been yet considered in face PAD contests. Another one issue is the imbalanced long-tailed data distribution across different PAs, where some more challenging attack types (e.g. high-fidelity masks) have usually been represented only with a few samples due to their high manufacturing costs. A large-scale test set ‘in the wild’ with diverse domain conditions and PA types, as well as more balanced data distribution, will be eventually needed to achieve more realistic evaluation settings.

Most of the existing face PAD competitions have not considered the efficiency or costs of the proposed solutions, as no constraints on either model size and number of models have been given. As a result, the best performing algorithms have been usually ensembles of several deep models, which gives insight how robust PAD performance can be reached with the current state-of-the-art methods on the competition data. However, blindly pushing to maximum detection performance without any restrictions encourages the participants to exploit and combine off-the-shelf components instead of trying to invent truly novel and effective solutions that could be also deployed in real-world applications on mobile and embedded platforms with restricted resources. Although the organizers of the latest ICCV2021 challenge [16] explicitly informed that results obtained with fusion of deep models are rejected and the computational cost of a single model should be less than 100G FLOPs, there are still no evaluation metrics for measuring the trade-off between accuracy and efficiency.

Despite two multi-modal face PAD competitions have been already conducted, it is a bit disappointing that only few solutions pursued to introduce new kinds of advanced multi-modal fusion algorithms. Many major manufacturers have already included multi-modal camera systems into their products, including mobile devices and laptops, thus there is an urgent need to explore novel multi-modal fusion algorithms instead of just ensembling models with different modalities. Furthermore, not all the modalities are always available, as the selection of multi-modal data sources depends on the deployment scenario in question. Thus, it would be useful to investigate performance in settings where a model is trained on multiple modalities but evaluated on partial or arbitrary combinations of the modalities. Among emerging imaging technologies, depth and NIR cameras have been already considered in the two multi-modal competitions, thus it would be interesting to include also more advanced sensors, such as short-wave infrared (SWIR) [90]) or even hyperspectral imaging [91], in the upcoming collective evaluations.

Apart from conventional PAs, two kinds of physical adversarial attacks (i.e. recognition and PA aware) could be considered for generic face PAD. For example, special printed eyeglasses [92], hats [93] and stickers [94] synthesized by adversarial generators have been demonstrated to effectively circumvent deep-learning-based AFR systems when worn by an attacker. Besides recognition-aware adversarial attacks, adversarial print and replay attacks [95] with specific perturbation injected before physical broadcast have been developed to fool face PAD systems. Therefore, it can be expected to be necessary to consider a diverse set of physical adversarial attacks in future competitions. In addition to PAs, there are many vicious digital manipulation attacks (e.g. 'deepfakes' [96]) that can be applied against AFR systems. Despite the differences in generation techniques and visual quality, some of these attacks still have coherent properties and artefacts. In [97], a unified digital and physical face attack detection framework is proposed to learn joint representations for coherent attacks. Therefore, another interesting challenge to tackle in upcoming contests assessing the robustness of face biometric systems would be to simultaneously detect both digital and physical attacks.

12.4 Conclusions

Competitions play a vital role in consolidating the recent trends and assessing the state of the art in face PAD. This chapter introduced the design and results of the five latest international competitions on unimodal and multi-modal face PAD organized from 2019 until 2021. These contests have been important milestones in advancing the research on face PAD to the next level, as each competition has offered new challenges to the research community and resulted in novel countermeasures and new insight. The industrial participants have dominated most of these competitions, which indicates the strong need for robust anti-spoofing solutions in real-world applications. The first two face PAD competitions (i.e. CVPR2019 [11] and CVPR2020 [15]) provided initial assessments of the state of the art in multi-modal face PAD algorithms under mainstream PAs in diverse conditions, while the three most recent face PAD competitions (i.e. ECCV2020 [18], IJCB2021 LivDet-Face [17] and ICCV2021 [16]) benchmarked conventional colour (RGB) camera-based PAD algorithms on larger scale datasets, challenging unseen attacks, and high-quality 3D mask attacks, respectively. Although several solutions proposed in the first two multi-modal competitions achieved satisfying PAD performance, more comprehensive multi-modal datasets and evaluation protocols on generalized PAD are still needed, especially considering the situation in which one or more modalities are missing in the test phase. On the contrary, none of the systems proposed in the context of the latest two unimodal competitions managed to achieve satisfying PAD performance under unseen attacks detection in unknown operating conditions. Thus, more diverse, larger scale unimodal face PAD datasets are still needed to develop and evaluate more robust learning-based algorithms.

Acknowledgements This work was supported by Infotech Oulu and the Academy of Finland for Academy Professor project EmotionAI (grants 336116, 345122) and ICT 2023 project (grant 345948).

References

1. ISO (2016) ISO/IEC JTC 1/SC 37 Biometrics: Information technology biometric presentation attack detection part 1: Framework. In: https://www.iso.org/obp/ui/iso
2. Mohammadi A, Bhattacharjee S, Marcel S (2017) Deeply vulnerable: a study of the robustness of face recognition to presentation attacks. IET Biometrics
3. Wan J, Guo G, Escalera S, Escalante HJ, Li SZ (2020) Multi-modal face presentation attack detection. Synth Lect Comput Vis 9(1):1–88
4. Ming Z, Visani M, Luqman MM, Burie JC (2020) A survey on anti-spoofing methods for facial recognition with RGB cameras of generic consumer devices. J Imaging 6(12):139
5. Yu Z, Qin Y, Li X, Zhao C, Lei Z, Zhao G (2021) Deep learning for face anti-spoofing: a survey. arXiv:2106.14948
6. Boulkenafet Z, Komulainen J, Li L, Feng X, Hadid A (2017) Oulu-npu: a mobile face presentation attack database with real-world variations. In: 2017 12th IEEE international conference on automatic face & gesture recognition (FG 2017). IEEE, pp 612–618

7. Chingovska I, Anjos A, Marcel S (2012) On the effectiveness of local binary patterns in face anti-spoofing. In: International conference of the biometrics special interest group (BIOSIG)
8. Liu Y, Jourabloo A, Liu X (2018) Learning deep models for face anti-spoofing: Binary or auxiliary supervision. In: IEEE/CVF conference on computer vision and pattern recognition (CVPR), pp 389–398
9. Wen D, Han H, Jain AK (2015) Face spoof detection with image distortion analysis. IEEE Trans Inf Forensics Secur 10(4):746–761
10. Zhang Z, Yan J, Liu S, Lei Z, Yi D, Li SZ (2012) A face antispoofing database with diverse attacks. In: International conference on biometrics (ICB), pp 26–31
11. Liu A, Wan J, Escalera S, Escalante HJ, Tan Z, Yuan Q, Wang K, Lin C, Guo G, Guyon I, Li SZ (2019) Multi-modal face anti-spoofing attack detection challenge at CVPR2019. In: IEEE/CVF conference on computer vision and pattern recognition workshops (CVPRW), pp 1601–1610
12. Boulkenafet Z, Komulainen J, Akhtar Z, Benlamoudi A, Samai D, Bekhouche S, Ouafi A, Dornaika F, Taleb-Ahmed A, Qin L, et al (2017) A competition on generalized software-based face presentation attack detection in mobile scenarios. In: International joint conference on biometrics (IJCB), pp 688–696
13. Chakka MM, Anjos A, Marcel S, Tronci R, Muntoni D, Fadda G, Pili M, Sirena N, Murgia G, Ristori M, et al (2011) Competition on counter measures to 2-d facial spoofing attacks. In: International joint conference on biometrics (IJCB)
14. Chingovska I, Yang J, Lei Z, Yi D, Li SZ, Kahm O, Glaser C, Damer N, Kuijper A, Nouak A, et al (2013) The 2nd competition on counter measures to 2D face spoofing attacks. In: International conference on biometrics (ICB)
15. Liu A, Li X, Wan J, Liang Y, Escalera S, Escalante HJ, Madadi M, Jin Y, Wu Z, Yu X, Tan Z, Yuan Q, Yang R, Zhou B, Guo G, Li SZ (2021) Cross-ethnicity face anti-spoofing recognition challenge: a review. IET Biomet 10(1):24–43
16. Liu A, Zhao C, Yu Z, Su A, Liu X, Kong Z, Wan J, Escalera S, Escalante HJ, Lei Z, et al (2021) 3d high-fidelity mask face presentation attack detection challenge. In: IEEE/CVF international conference on computer vision (ICCV) workshops, pp 814–823
17. Purnapatra S, Smalt N, Bahmani K, Das P, Yambay D, Mohammadi A, George A, Bourlai T, Marcel S, Schuckers S, et al (2021) Face liveness detection competition (livdet-face)-2021. In: 2021 IEEE international joint conference on biometrics (IJCB). IEEE, pp 1–10
18. Zhang Y, Yin Z, Shao J, Liu Z, Yang S, Xiong Y, Xia W, Xu Y, Luo M, Liu J, et al (2021) Celeba-spoof challenge 2020 on face anti-spoofing: methods and results. arXiv:2102.12642
19. Komulainen J, Boulkenafet Z, Akhtar Z.: Review of face presentation attack detection competitions. In: Marcel S, Nixon MS, Fierrez J, Evans N (eds) Handbook of biometric anti-spoofing: presentation attack detection. Springer, pp 291–317
20. Deng J, Dong W, Socher R, Li LJ, Li K, Fei-Fei L (2009) ImageNet: a large-scale hierarchical image database. In: IEEE/CVF conference on computer vision and pattern recognition (CVPR), pp 248–255
21. Yi D, Lei Z, Liao S, Li SZ (2014) Learning face representation from scratch. arXiv:1411.7923
22. Zhang S, Liu A, Wan J, Liang Y, Guo G, Escalera S, Escalante HJ, Li SZ (2020) CASIA-SURF: a large-scale multi-modal benchmark for face anti-spoofing. IEEE Trans Biom Behav Identity Sci 2(2):182–193
23. Zhang S, Wang X, Liu A, Zhao C, Wan J, Escalera S, Shi H, Wang Z, Li SZ (2019) A dataset and benchmark for large-scale multi-modal face anti-spoofing. In: IEEE/CVF conference on computer vision and pattern recognition (CVPR), pp 919–928
24. Parkin A, Grinchuk O (2019) Recognizing multi-modal face spoofing with face recognition networks. In: IEEE/CVF conference on computer vision and pattern recognition workshops (CVRPW)
25. Shen T, Huang Y, Tong Z (2019) Facebagnet: bag-of-local-features model for multi-modal face anti-spoofing. In: IEEE/CVF conference on computer vision and pattern recognition workshops (CVRPW)
26. Wang G, Lan C, Han H, Shan S, Chen X (2019) Multi-modal face presentation attack detection via spatial and channel attentions. In: IEEE/CVF conference on computer vision and pattern recognition workshops (CVRPW)

27. Zhang P, Zou F, Wu Z, Dai N, Mark S, Fu M, Zhao J, Li K (2019) FeatherNets: convolutional neural networks as light as feather for face anti-spoofing. In: IEEE/CVF conference on computer vision and pattern recognition workshops (CVPRW)
28. He K, Zhang X, Ren S, Sun J (2016) Deep residual learning for image recognition. In: Proceedings of the IEEE conference on computer vision and pattern recognition, pp 770–778
29. Niu Z, Zhou M, Wang L, Gao X, Hua G (2016) Ordinal regression with multiple output CNN for age estimation. In: IEEE conference on computer vision and pattern recognition (CVPR), pp 4920–4928
30. Guo Y, Zhang L, Hu Y, He X, Gao J (2016) MS-Celeb-1M: challenge of recognizing one million celebrities in the real world. Electron Imaging **2016**(11)
31. Zhao J, Cheng Y, Xu Y, Xiong L, Li J, Zhao F, Jayashree K, Pranata S, Shen S, Xing J, et al (2018) Towards pose invariant face recognition in the wild. In: IEEE conference on computer vision and pattern recognition (CVPR), pp 2207–2216
32. Xie S, Girshick R, Dollár P, Tu Z, He K (2017) Aggregated residual transformations for deep neural networks. In: IEEE conference on computer vision and pattern recognition (CVPR), pp 1492–1500
33. Sun S, Pang J, Shi J, Yi S, Ouyang W (2018) FishNet: a versatile backbone for image, region, and pixel level prediction. In: Advances in neural information processing systems, pp 762–772
34. Sandler M, Howard A, Zhu M, Zhmoginov A, Chen LC (2018) MobileNetV2: inverted residuals and linear bottlenecks. In: IEEE conference on computer vision and pattern recognition (CVPR), pp 4510–4520
35. Ma N, Zhang X, Zheng HT, Sun J (2018) ShuffleNet v2: Practical guidelines for efficient CNN architecture design. In: European conference on computer vision (ECCV), pp 116–131
36. Szegedy C, Ioffe S, Vanhoucke V, Alemi AA (2017) Inception-v4, inception-ResNet and the impact of residual connections on learning. In: AAAI conference on artificial intelligence (AAAI)
37. Cao Q, Shen L, Xie W, Parkhi OM, Zisserman A (2018) VGGFace2: a dataset for recognising faces across pose and age. In: IEEE international conference on automatic face & gesture recognition (FG), pp 67–74
38. Costa-Pazo A, Jiménez-Cabello D, Vazquez-Fernandez E, Alba-Castro JL, López-Sastre RJ (2019) Generalized presentation attack detection: a face anti-spoofing evaluation proposal. In: International conference on biometrics (ICB)
39. Woo S, Park J, Lee JY, So Kweon I (2018) CBAM: convolutional block attention module. In: European conference on computer vision (ECCV), pp 3–19
40. Wen Y, Zhang K, Li Z, Qiao Y (2016) A discriminative feature learning approach for deep face recognition. In: European conference on computer vision. Springer, pp 499–515
41. Simonyan K, Zisserman A (2014) Very deep convolutional networks for large-scale image recognition. arXiv:1409.1556
42. Wu X, He R, Sun Z, Tan T (2018) A light cnn for deep face representation with noisy labels. IEEE Trans Inf Forensics Secur 13(11):2884–2896
43. Zhu Y, Newsam S (2017) Densenet for dense flow. In: IEEE international conference on image processing (ICIP), pp 790–794
44. Boulkenafet Z, Komulainen J, Hadid A (2016) Face spoofing detection using colour texture analysis. IEEE Trans Inf Forensics Secur 11(8):1818–1830
45. Pan G, Sun L, Wu Z, Lao S (2007) Eyeblink-based anti-spoofing in face recognition from a generic webcamera. In: International Conference on Computer Vision (ICCV)
46. Liu A, Tan Z, Wan J, Escalera S, Guo G, Li SZ (2021) CASIA-SURF CeFA: a benchmark for multi-modal cross-ethnicity face anti-spoofing. In: IEEE winter conference on applications of computer vision (WACV), pp 1179–1187
47. Yu Z, Wan J, Qin Y, Li X, Li SZ, Zhao G (2020) Nas-fas: static-dynamic central difference network search for face anti-spoofing. arXiv:2011.02062
48. Yu Z, Zhao C, Wang Z, Qin Y, Su Z, Li X, Zhou F, Zhao G (2020) Searching central difference convolutional networks for face anti-spoofing. In: IEEE/CVF Conference on computer vision and pattern recognition (CVPR)

49. Wang Z, Yu Z, Zhao C, Zhu X, Qin Y, Zhou Q, Zhou F, Lei Z (2020) Deep spatial gradient and temporal depth learning for face anti-spoofing. In: IEEE/CVF conference on computer vision and pattern recognition (CVPR)
50. Horn BK, Schunck BG (1981) Determining optical flow. Artif Intell 17(1–3):185–203
51. Fernando B, Gavves E, Oramas J, Ghodrati A, Tuytelaars T (2016) Rank pooling for action recognition. IEEE Trans Pattern Anal Mach Intell 39(4):773–787
52. Parkin A, Grinchuk O (2020) Creating artificial modalities to solve RGB liveness. arXiv:2006.16028
53. Hara K, Kataoka H, Satoh Y (2018) Can spatiotemporal 3d CNNs retrace the history of 2d CNNs and ImageNet? In: IEEE/CVF conference on computer vision and pattern recognition (CVPR), pp 6546–6555
54. Deng D, Chen Z, Zhou Y, Shi B (2020) MIMAMO Net: integrating micro-and macro-motion for video emotion recognition. In: AAAI conference on artificial intelligence (AAAI), vol 34, pp 2621–2628
55. Yu Z, Qin Y, Li X, Wang Z, Zhao C, Lei Z, Zhao G (2020) Multi-modal face anti-spoofing based on central difference networks. In: IEEE/CVF conference on computer vision and pattern recognition workshops (CVPRW), pp 650–651
56. Yang Q, Zhu X, Fwu JK, Ye Y, You G, Zhu Y (2020) PipeNet: selective modal pipeline of fusion network for multi-modal face anti-spoofing. In: IEEE/CVF conference on computer vision and pattern recognition workshops (CVPRW), pp 644–645
57. Zhang Y, Yin Z, Li Y, Yin G, Yan J, Shao J, Liu Z (2020) Celeba-spoof: large-scale face anti-spoofing dataset with rich annotations. In: European conference on computer vision (ECCV). Springer, pp 70–85
58. Liu Z, Luo P, Wang X, Tang X (2015) Deep learning face attributes in the wild. In: IEEE international conference on computer vision, pp 3730–3738
59. Kennedy J, Eberhart R (1995) Particle swarm optimization. In: IEEE international conference on neural networks (ICNN), vol 4, pp 1942–1948
60. Jourabloo A, Liu Y, Liu X (2018) Face de-spoofing: anti-spoofing via noise modeling. In: European Conference on Computer Vision (ECCV), pp 290–306
61. Feng H, Hong Z, Yue H, Chen Y, Wang K, Han J, Liu J, Ding E (2020) Learning generalized spoof cues for face anti-spoofing. arXiv:2005.03922
62. Hu J, Shen L, Sun G (2018) Squeeze-and-excitation networks. In: IEEE conference on computer vision and pattern recognition (CVPR), pp 7132–7141
63. Tan M, Le Q (2019) Efficientnet: rethinking model scaling for convolutional neural networks. In: International conference on machine learning (ICML), pp 6105–6114
64. Costa-Pazo A, Bhattacharjee S, Vazquez-Fernandez E, Marcel S (2016) The replay-mobile face presentation-attack database. In: International conference of the biometrics special interest group (BIOSIG)
65. Ramachandra R, Stokkenes M, Mohammadi A, Venkatesh S, Raja K, Wasnik P, Poiret E, Marcel S, Busch C (2019) Smartphone multi-modal biometric authentication: database and evaluation. arXiv:1912.02487
66. Erdogmus N, Marcel S (2014) Spoofing face recognition with 3D masks. IEEE Trans Inf Forensics Secur 9(7):1084–1097
67. Ramachandra R, Venkatesh S, Raja KB, Bhattacharjee S, Wasnik P, Marcel S, Busch C (2019) Custom silicone face masks: Vulnerability of commercial face recognition systems & presentation attack detection. In: IEEE international workshop on biometrics and forensics (IWBF)
68. Fang M, Damer N, Kirchbuchner F, Kuijper A (2021) Real masks and fake faces: on the masked face presentation attack detection. Pattern Recogn 123:108398. https://doi.org/10.1016/j.patcog.2021.108398
69. George A, Marcel S (2019) Deep pixel-wise binary supervision for face presentation attack detection. In: 2019 international conference on biometrics (ICB). IEEE, pp 1–8
70. Szegedy C, Vanhoucke V, Ioffe S, Shlens J, Wojna Z (2016) Rethinking the inception architecture for computer vision. In: IEEE conference on computer vision and pattern recognition (CVPR), pp 2818–2826

71. Krizhevsky A, Sutskever I, Hinton G (2012) ImageNet classification with deep convolutional neural networks. In: Neural information processing systems (NIPS)
72. Damer N, Opel A, Nouak A (2014) Biometric source weighting in multi-biometric fusion: towards a generalized and robust solution. In: European signal processing conference (EUSIPCO), pp 1382–1386
73. Kantarcı A, Dertli H, Ekenel HK (2021) Shuffled patch-wise supervision for presentation attack detection. In: International conference of the biometrics special interest group (BIOSIG)
74. Liu A, Zhao C, Yu Z, Wan J, Su A, Liu X, Tan Z, Escalera S, Xing J, Liang Y, et al (2021) Contrastive context-aware learning for 3d high-fidelity mask face presentation attack detection. arXiv:2104.06148
75. Li X, Komulainen J, Zhao G, Yuen PC, Pietikäinen M (2016) Generalized face anti-spoofing by detecting pulse from face videos. In: IEEE international conference on pattern recognition (ICPR), pp 4244–4249
76. Zhang S, Zhu X, Lei Z, Shi H, Wang X, Li SZ (2017) Faceboxes: a CPU real-time face detector with high accuracy. In: International joint conference on biometrics (IJCB)
77. Li J, Wang Y, Wang C, Tai Y, Qian J, Yang J, Wang C, Li J, Huang F (2019) DSFD: dual shot face detector. In: IEEE/CVF conference on computer vision and pattern recognition (CVPR), pp 5060–5069
78. Yu Z, Li X, Niu X, Shi J, Zhao G (2020) Face anti-spoofing with human material perception. In: European conference on computer vision (ECCV), pp 557–575
79. Chen S, Yao T, Zhang K, Chen Y, Sun K, Ding S, Li J, Huang F, Ji R (2021) A dual-stream framework for 3d mask face presentation attack detection. In: IEEE/CVF international conference on computer vision (ICCV), pp 834–841
80. Grinchuk O, Parkin A, Glazistova E (2021) 3d mask presentation attack detection via high resolution face parts. In: IEEE/CVF international conference on computer vision (ICCV), pp 846–853
81. Huang S, Cheng WH, Cheng R (2021) Single patch based 3d high-fidelity mask face anti-spoofing. In: IEEE/CVF international conference on computer vision (ICCV), pp 842–845
82. Yu Z, Qin Y, Zhao H, Li X, Zhao G (2021) Dual-cross central difference network for face anti-spoofing. In: International joint conference on artificial intelligence (IJCAI), pp 1281–1287
83. Deng J, Guo J, Ververas E, Kotsia I, Zafeiriou S (2020) RetinaFace: single-shot multi-level face localisation in the wild. In: IEEE/CVF conference on computer vision and pattern recognition (CVPR), pp 5203–5212
84. Liu SQ, Lan X, Yuen PC (2021) Multi-channel remote photoplethysmography correspondence feature for 3d mask face presentation attack detection. IEEE Trans Inf Forensics Secur 16:2683–2696
85. Yu Z, Li X, Wang P, Zhao G (2021) TransRPPG: remote photoplethysmography transformer for 3d mask face presentation attack detection. IEEE Signal Process Lett
86. Korshunov P, Marcel S, Muckenhirn H, Gonçalves AR, Mello AGS, Violato RPV, Simoes FO, Neto MU, de Assis Angeloni M, Stuchi JA, Dinkel H, Chen N, Qian Y, Paul D, Saha G, Sahidullah M (2016) Overview of BTAS 2016 speaker anti-spoofing competition. In: IEEE international conference on biometrics theory, applications and systems (BTAS)
87. Huang G, Liu Z, Van Der Maaten L, Weinberger KQ (2017) Densely connected convolutional networks. In: Proceedings of the IEEE conference on computer vision and pattern recognition, pp 4700–4708
88. Szegedy C, Liu W, Jia Y, Sermanet P, Reed S, Anguelov D, Erhan D, Vanhoucke V, Rabinovich A (2015) Going deeper with convolutions. In: IEEE conference on computer vision and pattern recognition (CVPR)
89. Elsken T, Metzen JH, Hutter F (2019) Neural architecture search: a survey. J Mach Learn Res 20(1):1997–2017
90. Heusch G, George A, Geissbühler D, Mostaani Z, Marcel S (2020) Deep models and shortwave infrared information to detect face presentation attacks. IEEE Trans Biom Behav Identity Sci 2(4):399–409

91. Kaichi T, Ozasa Y (2021) A hyperspectral approach for unsupervised spoof detection with intra-sample distribution. In: IEEE international conference on image processing (ICIP), pp 839–843
92. Sharif M, Bhagavatula S, Bauer L, Reiter MK (2019) A general framework for adversarial examples with objectives. ACM Trans Priv Secur (TOPS) 22(3):1–30
93. Komkov S, Petiushko A (2021) AdvHat: Real-world adversarial attack on arcface face id system. In: IEEE International Conference on Pattern Recognition (ICPR), pp 819–826
94. Guo Y, Wei X, Wang G, Zhang B (2021) Meaningful adversarial stickers for face recognition in physical world. arXiv:2104.06728
95. Zhang B, Tondi B, Barni M (2019) Attacking CNN-based anti-spoofing face authentication in the physical domain. arXiv:1910.00327
96. Rossler A, Cozzolino D, Verdoliva L, Riess C, Thies J, Nießner M (2019) FaceForensics++: Learning to detect manipulated facial images. In: IEEE/CVF international conference on computer vision (ICCV)
97. Deb D, Liu X, Jain AK (2021) Unified detection of digital and physical face attacks. arXiv:2104.02156

Part IV
Voice Biometrics

Chapter 13
Introduction to Voice Presentation Attack Detection and Recent Advances

Md Sahidullah, Héctor Delgado, Massimiliano Todisco, Andreas Nautsch, Xin Wang, Tomi Kinnunen, Nicholas Evans, Junichi Yamagishi, and Kong-Aik Lee

Abstract Over the past few years, significant progress has been made in the field of presentation attack detection (PAD) for automatic speaker recognition (ASV). This includes the development of new speech corpora, standard evaluation protocols and advancements in front-end feature extraction and back-end classifiers. The use of standard databases and evaluation protocols has enabled, for the first time, the meaningful benchmarking of different PAD solutions. This chapter summarises the progress, with a focus on studies completed in the last 3 years. The article presents a summary of findings and lessons learned from three ASVspoof challenges, the first community-led benchmarking efforts. These show that ASV PAD remains an

M. Sahidullah (✉)
Université de Lorraine, CNRS, Inria, LORIA, 54000 Nancy, France
e-mail: sahidullahmd@gmail.com

H. Delgado (✉)
Nuance Communications, Madrid, Spain
e-mail: hector.delgado@nuance.com

M. Todisco · A. Nautsch · N. Evans (✉)
Department of Digital Security, EURECOM (France), Biot, France
e-mail: evans@eurecom.fr

M. Todisco
e-mail: massimiliano.todisco@eurecom.fr

X. Wang
National Institute of Informatics (Japan), Tokyo, Japan
e-mail: wangxin@nii.ac.jp

T. Kinnunen
School of Computing, University of Eastern Finland (Finland), Joensuu, Finland
e-mail: tkinnu@cs.uef.fi

J. Yamagishi
National Institute of Informatics (Japan), Tokyo, Japan
e-mail: jyamagis@nii.ac.jp

K.-A. Lee
Institute for Infocomm Research, A*STAR (Singapore), Singapore, Singapore
e-mail: kongaik.lee@ieee.org

S. Marcel et al. (eds.), *Handbook of Biometric Anti-Spoofing*, Advances in Computer Vision and Pattern Recognition, https://doi.org/10.1007/978-981-19-5288-3_13

unsolved problem and further attention is required to develop generalised PAD solutions which have the potential to detect diverse and previously unseen spoofing attacks.

13.1 Introduction

Automatic Speaker Verification (ASV) technology aims to recognise individuals using samples of the human voice signal [1, 2]. Most ASV systems operate on estimates of the spectral characteristics of voice in order to recognise individual speakers. ASV technology has matured in recent years and now finds application in a growing variety of real-world authentication scenarios involving both *logical* and *physical* access. In logical access scenarios, ASV technology can be used for remote person authentication via the Internet or traditional telephony. In many cases, ASV serves as a convenient and efficient alternative to more conventional password-based solutions, one prevalent example being person authentication for Internet and mobile banking. Physical access scenarios include the use of ASV to protect personal or secure/sensitive facilities, such as domestic and office environments. With the growing, widespread adoption of smartphones and voice-enabled smart devices, such as intelligent personal assistants all equipped with at least one microphone, ASV technology stands to become even more ubiquitous in the future.

Despite its appeal, the now-well-recognised vulnerability to manipulation through presentation attacks (PAs), also known as spoofing, has dented confidence in ASV technology. As identified in ISO/IEC 30107-1 standard [3], the possible locations of presentation attack points in a typical ASV system are illustrated in Fig. 13.1. Two of the most vulnerable places in an ASV system are marked by 1 and 2, corresponding to physical access and logical access. This work is related to these two types of attacks.

Unfortunately, ASV is arguably more prone to PAs than other biometric systems based on traits or characteristics that are less easily acquired; samples of a given person's voice can be collected readily by fraudsters through face-to-face or telephone conversations and then replayed in order to manipulate an ASV system. Replay attacks are furthermore only one example of ASV PAs. More advanced voice conversion or speech synthesis algorithms can be used to generate particularly effective PAs using only modest amounts of voice data collected from a target person.

There are a number of ways to prevent PA problems. The first one is based on a text-prompted system which uses an utterance verification process [4]. The user needs to utter a specific text, prompted for authentication by the system which requires a text verification system. Secondly, as human can never reproduce an identical speech signal, some countermeasures use template matching or audio fingerprinting to verify whether the speech utterance was presented to the system earlier [5]. Thirdly, some work looks into statistical acoustic characterisation of authentic speech and speech created with presentation attack methods or spoofing techniques [6]. Our focus is on the last category, which is more convenient in a practical scenario for both text-

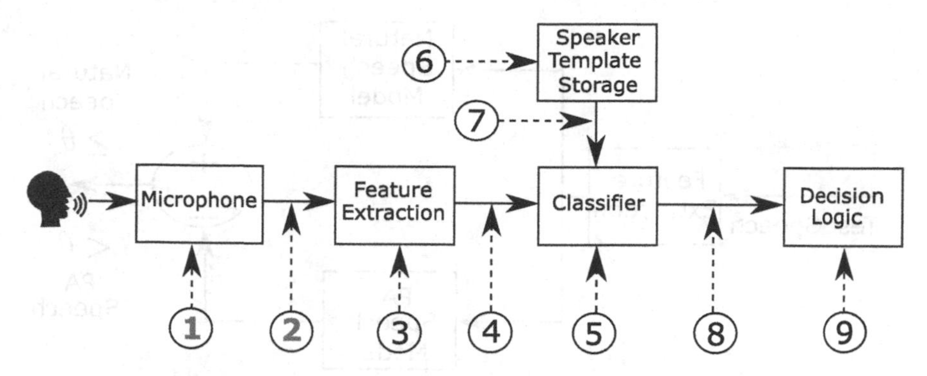

Fig. 13.1 Possible attack locations in a typical ASV system. 1: microphone point, 2: transmission point, 3: override feature extractor, 4: modify probe to features, 5: override classifier, 6: modify speaker database, 7: modify biometric reference, 8: modify score and 9: override decision

dependent and text-independent ASVs. In this case, given a speech signal, S, PA detection here, the determination of whether S is a natural or PA speech can be formulated as a hypothesis test:

- H_0: S is natural speech.
- H_1: S is created with PA methods.

A likelihood ratio test can be applied to decide between H_0 and H_1. Suppose that $\mathbf{X} = \{\mathbf{x}_1, \mathbf{x}_2, ..., \mathbf{x}_N\}$ are the acoustic feature vectors of N speech frames extracted from S, then the logarithmic likelihood ratio score is given by

$$\Lambda(\mathbf{X}) = \log p(\mathbf{X}|\lambda_{H_0}) - \log p(\mathbf{X}|\lambda_{H_1}) \qquad (13.1)$$

In (13.1), λ_{H_0} and λ_{H_1} are the acoustic models to characterise the hypotheses correspondingly for natural speech and PA speech. The parameters of these models are estimated using training data for natural and PA speech. A typical PAD system is shown in Fig. 13.2. A test speech can be accepted as natural or rejected as PA speech with help of a threshold, θ computed on some development data. If the score is greater than or equal to the threshold, it is accepted; otherwise, rejected. The performance of the PA system is assessed by computing the *Equal Error Rate* (EER) metric. This is the error rate for a specific value of a threshold where two error rates, i.e. the probability of a PA speech detected as being natural speech (known as false acceptance rate or FAR) and the probability of a natural speech being misclassified as a PA speech (known as false rejection rate or FRR), are equal. Sometimes *Half Total Error Rate* (HTER) is also computed [7]. This is the average of FAR and FRR which are computed using a decision threshold obtained with the help of the development data.

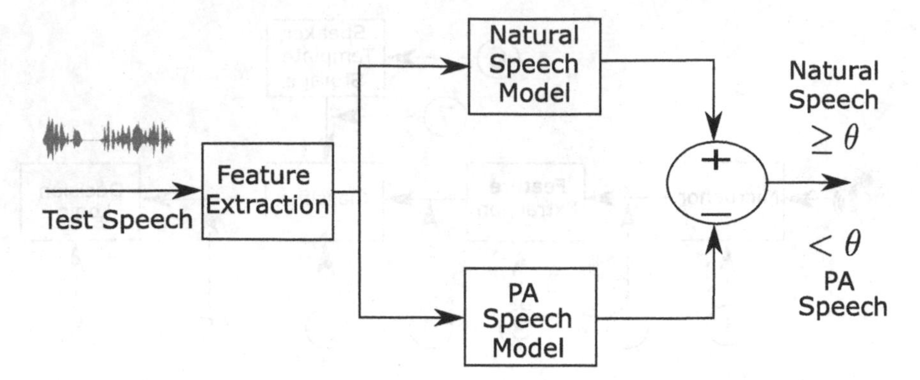

Fig. 13.2 Block diagram of a typical presentation attack detection system

Awareness and acceptance of the vulnerability to PAs have generated a growing interest in developing solutions to presentation attack detection (PAD), also referred to as spoofing countermeasures. These are typically dedicated auxiliary systems which function in tandem with ASV in order to detect and deflect PAs. The research in this direction has progressed rapidly in the last 3 years, due partly to the release of several public speech corpora and the organisation of PAD challenges for ASV. This article, a continuation of the Chap. [8] in the first edition of the Handbook for Biometrics [9] presents an up-to-date review of the different forms of voice presentation attacks, broadly classified in terms of impersonation, replay, speech synthesis and voice conversion. The primary focus is nonetheless on the progress in PAD. The chapter reviews the most recent work involving a variety of different features and classifiers. Most of the work covered in the chapter relates to that conducted using the two most popular and publicly available databases, which were used for the two ASVspoof challenges co-organised by the authors. The chapter concludes with a discussion of research challenges and future directions in PAD for ASV.

13.2 Basics of ASV Spoofing and Countermeasures

Spoofing or presentation attacks are performed on a biometric system at the sensor or acquisition level to bias score distributions toward those of genuine clients, thus provoking increases in the false acceptance rate (FAR). This section reviews four well-known ASV spoofing techniques and their respective countermeasures: impersonation, replay, speech synthesis and voice conversion. Here, we mostly review the work in the pre-ASVspoof period, as well as some very recent studies on presentation attacks.

13.2.1 *Impersonation*

In speech impersonation or mimicry attacks, an intruder speaker intentionally modifies his or her speech to sound like the target speaker. Impersonators are likely to copy lexical, prosodic and idiosyncratic behaviour of their target speakers presenting a potential point of vulnerability concerning speaker recognition systems.

13.2.1.1 Spoofing

There are several studies about the consequences of mimicry on ASV. Some studies concern attention to the voice modifications performed by professional impersonators. It has been reported that impersonators are often particularly able to adapt the fundamental frequency (F0) and occasionally also the formant frequencies towards those of the target speakers [10–12]. In studies, the focus has been on analysing the vulnerability of speaker verification systems in the presence of voice mimicry. The studies by Lau et al. [13, 14] suggest that if the target of impersonation is known in advance and his or her voice is 'similar' to the impersonator's voice (in the sense of automatic speaker recognition score), then the chance of spoofing an automatic recogniser is increased. In [15], the experiments indicated that professional impersonators are potentially better impostors than amateur or naive ones. Nevertheless, the voice impersonation was not able to spoof the ASV system. In [10], the authors attempted to quantify how much a speaker is able to approximate other speakers' voices by selecting a set of prosodic and voice source features. Their prosodic- and acoustic-based ASV results showed that two professional impersonators imitating known politicians increased the identification error rates.

More recently, a fundamentally different study was carried out by Panjwani et al. [16] using crowdsourcing to recruit both amateur and more professional impersonators. The results showed that impersonators succeed in increasing their average score, but not in exceeding the target speaker score. All of the above studies analysed the effects of speech impersonation either at the acoustic or speaker recognition score level, but none proposed any countermeasures against impersonation. In a recent study [17], the experiments aimed to evaluate the vulnerability of three modern speaker verification systems against impersonation attacks and to further compare these results to the performance of non-expert human listeners. It is observed that, on average, the mimicry attacks lead to increased error rates. The increase in error rates depends on the impersonator and the ASV system.

The main challenge, however, is that no large speech corpora of impersonated speech exists for the quantitative study of impersonation effects on the same scale as for other attacks, such as text-to-speech synthesis and voice conversion, where generation of simulated spoofing attacks as well as developing appropriate countermeasures is more convenient.

13.2.1.2 Countermeasures

While the threat of impersonation is not fully understood due to limited studies involving small datasets, it is perhaps not surprising that there is no prior work investigating countermeasures against impersonation. If the threat is proven to be genuine, then the design of appropriate countermeasures might be challenging. Unlike the spoofing attacks discussed below, all of which can be assumed to leave traces of the physical properties of the recording and playback devices, or signal processing artefacts from synthesis or conversion systems, impersonators are live human beings who produce entirely natural speech.

13.2.2 Replay

Replay attacks refer to the use of pre-recorded speech from a target speaker, which is then replayed through some playback device to feed the system microphone. These attacks require no specific expertise nor sophisticated equipment, thus they are easy to implement. Replay is a relatively low-technology attack within the grasp of any potential attacker even without specialised knowledge in speech processing. Several works in the earlier literature report significant increases in error rates when using replayed speech. Even if replay attacks may present a genuine risk to ASV systems, the use of a prompted phrase has the potential to mitigate the impact.

13.2.2.1 Spoofing

The study on the impact of a replay attack on ASV performance was very limited until recently before the release of AVspoof [18] and ASVspoof 2017 corpus. The earlier studies were conducted either on simulated or on real replay recording from far field.

The vulnerability of ASV systems to replay attacks was first investigated in a text-dependent scenario [19], where the concatenation of recorded digits was tested against a hidden Markov model (HMM)-based ASV system. Results showed an increase in the FAR from 1 to 89% for male speakers and from 5 to 100% for female speakers.

The work in [20] investigated text-independent ASV vulnerabilities through the replaying of far-field recorded speech in a mobile telephony scenario where signals were transmitted by analogue and digital telephone channels. Using a baseline ASV system based on *joint factor analysis* (JFA), the work showed an increase in the EER of 1% to almost 70% when impostor accesses were replaced by replayed spoof attacks.

A physical access scenario was considered in [21]. While the baseline performance of the Gaussian mixture model—universal background model (GMM-UBM) ASV system was not reported, experiments showed that replay attacks produced a FAR of 93%.

The work in [18] introduced audio-visual spoofing (AVspoof) database for replay attack detection where the replayed signals are collected and played back using different low-quality (phones and laptops) and high-quality (laptop with loud speakers) devices. The study reported that FARs for replayed speech was 77.4 and 69.4% for male and female, respectively, using a total variability system speaker recognition system. In this study, the EER for bona fide trials was 6.9 and 17.5% for those conditions. This study also includes a presentation attack where speech signals created with voice conversion and speech synthesis were used in playback attacks. In that case, higher FAR was observed, particularly when a high-quality device is used for playback.

13.2.2.2 Countermeasures

A countermeasure for replay attack detection in the case of text-dependent ASV was reported in [5]. The approach is based upon the comparison of new access samples with stored instances of past accesses. New accesses which are deemed too similar to previous access attempts are identified as replay attacks. A large number of different experiments, all relating to a telephony scenario, showed that the countermeasures succeeded in lowering the EER in most of the experiments performed. While some form of text-dependent or challenge-response countermeasure is usually used to prevent replay attacks, text-independent solutions have also been investigated. The same authors in [20] showed that it is possible to detect replay attacks by measuring the channel differences caused by far-field recording [22]. While they show spoof detection error rates of less than 10% it is feasible that today's state-of-the-art approaches to channel compensation will render some ASV systems still vulnerable.

Two different replay attack countermeasures are compared in [21]. Both are based on the detection of differences in channel characteristics expected between licit and spoofed access attempts. Replay attacks incur channel noise from both the recording device and the loudspeaker used for replay and thus the detection of channel effects beyond those introduced by the recording device of the ASV system thus serves as an indicator of replay. The performance of a baseline GMM-UBM system with an EER of 40% under spoofing attack falls to 29% with the first countermeasure and a more respectable EER of 10% with the second countermeasure.

In another study [23], a speech database of 175 subjects has been collected for different kinds of replay attacks. Other than the use of genuine voice samples for the legitimate speakers in playback, the voice samples recorded over the telephone channel were also used for unauthorised access. Further, a far-field microphone is used to collect the voice samples as eavesdropped (covert) recording. The authors proposed an algorithm motivated by music recognition system used for comparing recordings on the basis of the similarity of the local configuration of maxima pairs extracted from spectrograms of verified and reference recordings. The experimental results show the EER of playback attack detection to be as low as 1.0% on the collected data.

13.2.3 Speech Synthesis

Speech synthesis, commonly referred to as text-to-speech (TTS), is a technique for generating intelligible, natural-sounding artificial speech for any arbitrary text. Speech synthesis is used widely in various applications including in-car navigation systems, e-book readers, voice-over for the visually impaired and communication aids for the speech impaired. More recent applications include spoken dialogue systems, communicative robots, singing speech synthesisers and speech-to-speech translation systems.

Typical speech synthesis systems have two main components [24]: text analysis followed by speech waveform generation, which is sometimes referred to as the front end and back end, respectively. In the text analysis component, input text is converted into a linguistic specification consisting of elements such as phonemes. In the speech waveform generation component, speech waveforms are generated from the produced linguistic specification. There are emerging end-to-end frameworks that generate speech waveforms directly from text inputs without using any additional modules.

Many approaches have been investigated, but there have been major paradigm shifts every ten years. In the early 1970s, the speech waveform generation component used very low-dimensional acoustic parameters for each phoneme, such as formants, corresponding to vocal tract resonances with hand-crafted acoustic rules [25]. In the 1980s, the speech waveform generation component used a small database of phoneme units called *diphones* (the second half of one phoneme plus the first half of the following) and concatenated them according to the given phoneme sequence by applying signal processing, such as linear predictive (LP) analysis, to the units [26]. In the 1990s, larger speech databases were collected and used to select more appropriate speech units that matched both phonemes and other linguistic contexts such as lexical stress and pitch accent in order to generate high-quality natural-sounding synthetic speech with the appropriate prosody. This approach is generally referred to as *unit selection*, and is nowadays used in many speech synthesis systems [27–31].

In the late 2000s, several machine learning based data-driven approaches emerged. 'Statistical parametric speech synthesis' was one of the more popular machine learning approaches [32–35]. In this approach, several acoustic parameters are modelled using a time-series stochastic generative model, typically an HMM. HMMs represent not only the phoneme sequences but also various contexts of the linguistic specification. Acoustic parameters generated from HMMs and selected according to the linguistic specification are then used to drive a vocoder, a simplified speech production model in which speech is represented by vocal tract parameters and excitation parameters in order to generate a speech waveform. HMM-based speech synthesisers [36, 37] can also learn speech models from relatively small amounts of speaker-specific data by adapting background models derived from other speakers based on the standard model adaptation techniques drawn from speech recognition, i.e. maximum likelihood linear regression (MLLR) [38, 39].

In the 2010s, deep learning has significantly improved the performance of speech synthesis and led to a significant breakthrough. First, various types of deep neural networks are used to improve the prediction accuracy of the acoustic parameters [40, 41]. Investigated architectures include recurrent neural network [42–44], residual/highway network [45, 46], autoregressive network [47, 48] and generative adversarial networks (GAN) [49–51]. Furthermore, in the late 2010s, conventional waveform generation modules typically used signal processing and text analysis modules that used natural language processing were substituted by neural networks. This allows for neural networks capable of directly outputting the desired speech waveform samples from the desired text inputs. Successful architectures for direct waveform modelling include dilated convolutional autoregressive neural network, known as 'Wavenet' [52] and hierarchical recurrent neural network, called 'SampleRNN' [53]. Finally, we have also seen successful architectures that totally remove the hand-crafted linguistic features obtained through text analysis by relying on sequence-to-sequence systems. This system is called Tacotron [54]. As expected, the combination of these advanced models results in a very high-quality end-to-end TTS synthesis system [55, 56] and recent results reveal that the generated synthetic speech sounds as natural as human speech [56].

For more details and technical comparisons, please see the results of Blizzard Challenge, which annually compares the performance of speech synthesis systems built on the common database over decades [57, 58].

13.2.3.1 Spoofing

There is a considerable volume of research in the literature which has demonstrated the vulnerability of ASV to synthetic voices generated with a variety of approaches to speech synthesis. Experiments using formant, diphone and unit selection-based synthetic speech in addition to the simple cut-and-paste of speech waveforms have been reported [19, 20, 59].

ASV vulnerabilities to HMM-based synthetic speech were first demonstrated over a decade ago [60] using an HMM-based, text-prompted ASV system [61] and an HMM-based synthesiser where acoustic models were adapted to specific human speakers [62, 63]. The ASV system scored feature vectors against speaker and background models composed of concatenated phoneme models. When tested with human speech, the ASV system achieved a FAR of 0% and a false rejection rate (FRR) of 7%. When subjected to spoofing attacks with synthetic speech, the FAR increased to over 70%, however, this work involved only 20 speakers.

Larger scale experiments using the Wall Street Journal corpus containing in the order of 300 speakers and two different ASV systems (GMM-UBM and SVM using Gaussian supervectors) were reported in [64]. Using an HMM-based speech synthesiser, the FAR was shown to rise to 86 and 81% for the GMM-UBM and SVM systems respectively representing a genuine threat to ASV. Spoofing experiments using HMM-based synthetic speech against a forensics speaker verification tool *BATVOX* was also reported in [65] with similar findings. Therefore, the above speech synthesisers were chosen as one of the spoofing methods in the ASVspoof 2015 database.

Spoofing experiments using the above-advanced DNNs or using spoofing-specific strategies such as GAN have not yet been properly investigated. Only a relatively small-scale spoofing experiment against a speaker recognition system using Wavenet, SampleRNN and GAN is reported in [66].

13.2.3.2 Countermeasures

Only a small number of attempts to discriminate synthetic speech from natural speech had been investigated before the ASVspoof challenge started. Previous work has demonstrated the successful detection of synthetic speech based on prior knowledge of the acoustic differences of specific speech synthesisers, such as the dynamic ranges of spectral parameters at the utterance level [67] and variance of higher order parts of mel-cepstral coefficients [68].

There are some attempts which focus on acoustic differences between vocoders and natural speech. Since the human auditory system is known to be relatively insensitive to phase [69], vocoders are typically based on a minimum-phase vocal tract model. This simplification leads to differences in the phase spectra between human and synthetic speech, differences which can be utilised for discrimination [64, 70].

Based on the difficulty in reliable prosody modelling in both unit selection and statistical parametric speech synthesis, other approaches to synthetic speech detection use F0 statistics [71, 72]. F0 patterns generated for the statistical parametric speech synthesis approach tend to be over-smoothed and the unit selection approach frequently exhibits 'F0 jumps' at concatenation points of speech units.

After the ASVspoof challenges took place, various types of countermeasures that work for both speech synthesis and voice conversion have been proposed. Please read the next section for the details of the recently developed countermeasures.

13.2.4 Voice Conversion

Voice conversion, in short, VC VC: voice conversion, is a spoofing attack against automatic speaker verification using an attacker's natural voice which is converted towards that of the target. It aims to convert one speaker's voice towards that of another and is a sub-domain of voice transformation [73]. Unlike TTS, which requires text input, voice conversion operates directly on speech inputs. However, speech waveform generation modules such as vocoders may be the same as or similar to those for TTS.

A major application of VC is to personalise and create new voices for TTS synthesis systems and spoken dialogue systems. Other applications include speaking aid devices that generate more intelligible voice sounds to help people with speech disorders, movie dubbing, language learning and singing voice conversion. The field has also attracted increasing interest in the context of ASV vulnerabilities for almost two decades [74].

Most voice conversion approaches require a parallel corpus where source and target speakers read out identical utterances and adopt a training phase which typically requires frame- or phone-aligned audio pairs of the source and target utterances and estimates transformation functions that convert acoustic parameters of the source speaker to those of the target speaker. This is called 'parallel voice conversion'. Frame alignment is traditionally achieved using dynamic time warping (DTW) on the source–target training audio files. Phone alignment is traditionally achieved using *automatic speech recognition* (ASR) and phone-level forth alignment. The estimated conversion function is then applied to any new audio files uttered by the source speaker [75].

A large number of estimation methods for the transformation functions have been reported starting in the late 1980s. In the late 1980s and 90s, simple techniques employing vector quantisation (VQ) with codebooks [76] or segmental codebooks [77] of paired source–target frame vectors were proposed to represent the transformation functions. However, these VQ methods introduced frame-to-frame discontinuity problems.

In the late 1990s and 2000s, *joint density Gaussian mixture model* (JDGMM) based transformation methods [78, 79] were proposed and have since then been actively improved by many researchers [80, 81]. This method still remains popular even now. Although this method achieves smooth feature transformations using a locally linear transformation, this method also has several critical problems such as over-smoothing [82–84] and over-fitting [85, 86] which leads to muffled quality of speech and degraded speaker similarity.

Therefore, in early 2010, several alternative linear transformation methods were developed. Examples are partial least square (PLS) regression [85], tensor representation [87], a trajectory HMM [88], mixture of factor analysers [89], local linear transformation [82] or noisy channel models [90].

In parallel to the linear-based approaches, there have been studies on non-linear transformation functions such as support vector regression [91], kernel partial least square [92], conditional restricted Boltzmann machines [93], neural networks [94, 95], highway network [96] and RNN [97, 98]. Data-driven frequency warping techniques [99–101] have also been studied.

Recently, deep learning has changed the above standard procedures for voice conversion and we can see many different solutions now. For instance, variational autoencoder or sequence-to-sequence neural networks enable us to build VC systems without using frame level alignment [102, 103]. It has also been shown that a cycle-consistent adversarial network called 'CycleGAN' [104] is one possible solution for building VC systems without using a parallel corpus. Wavenet can also be used as a replacement for the purpose of generating speech waveforms from converted acoustic features [105].

The approaches to voice conversion considered above are usually applied to the transformation of spectral envelope features, though the conversion of prosodic features such as fundamental frequency [106–109] and duration [107, 110] has also been studied.

For more details and technical comparisons, please see the results of Voice Conversion Challenges that compare the performance of VC systems built on a common database [111, 112].

13.2.4.1 Spoofing

When applied to spoofing, the aim of voice conversion is to synthesise a new speech signal such that the extracted ASV features are close in some sense to the target speaker. Some of the first works relevant to text-independent ASV spoofing were reported in [113, 114]. The work in [113] showed that baseline EER increased from 16% to 26% thanks to a voice conversion system which also converted prosodic aspects not modeled in typical ASV systems. This work targeted the conversion of spectral-slope parameters and showed that the baseline EER of 10% increased to over 60% when all impostor test samples were replaced with converted voices. Moreover, signals subjected to voice conversion did not exhibit any perceivable artefacts indicative of manipulation.

The work in [115] investigated ASV vulnerabilities to voice conversion based on JDGMMs [78] which requires a parallel training corpus for both source and target speakers. Even if the converted speech could be easily detectable by human listeners, experiments involving five different ASV systems showed their universal susceptibility to spoofing. The FAR of the most robust, JFA system increased from 3% to over 17%. Instead of vocoder-based waveform generation, unit selection approaches can be applied directly to feature vectors coming from the target speaker to synthesise converted speech [116]. Since they use target speaker data directly, unit-selection approaches arguably pose a greater risk to ASV than statistical approaches [117]. In the ASVspoof 2015 challenge, we therefore had chosen these popular VC methods as spoofing methods.

Other work relevant to voice conversion includes attacks referred to as artificial signals. It was noted in [118] that certain short intervals of converted speech yield extremely high scores or likelihoods. Such intervals are not representative of intelligible speech but they are nonetheless effective in overcoming typical ASV systems which lack any form of speech quality assessment. The work in [118] showed that artificial signals optimised with a genetic algorithm provoke increases in the EER from 10% to almost 80% for a GMM-UBM system and from 5% to almost 65% for a factor analysis (FA) system.

13.2.4.2 Countermeasures

Here, we provide an overview of countermeasure methods developed for the VC attacks before the ASVspoof challenge began.

Some of the first works to detect converted voice draws on related work in synthetic speech detection [119]. In [70, 120], cosine phase and modified group delay function (MGDF)-based countermeasures were proposed. These are effective in detecting converted speech using vocoders based on the minimum phase. In VC, it is, however, possible to use natural phase information extracted from a source speaker [114]. In this case, they are unlikely to detect converted voice.

Two approaches to artificial signal detection are reported in [121]. Experimental work shows that supervector-based SVM classifiers are naturally robust to such attacks and that all the spoofing attacks they used could be detected by using an utterance-level variability feature, which detected the absence of the natural and dynamic variabilities characteristic of genuine speech. A related approach to detect converted voice is proposed in [122]. Probabilistic mappings between source and target speaker models are shown to typically yield converted speech with less short-term variability than genuine speech. Therefore, the thresholded, average pair-wise distance between consecutive feature vectors was used to detect converted voice with an EER of under 3%.

Due to the fact that the majority of VC techniques operate at the short-term frame level, more sophisticated long-term features such as temporal magnitude and phase modulation feature can also detect converted speech [123]. Another experiment reported in [124] showed that local binary pattern analysis of sequences of acoustic vectors can also be used for successfully detecting frame-wise JDGMM-based converted voice. However, it is unclear whether these features are effective in detecting recent VC systems that consider long-term dependency such as recurrent or autoregressive neural network models.

After the ASVspoof challenges took place, new countermeasures that work for both speech synthesis and voice conversion were proposed and evaluated. See the next section for a detailed review of the recently developed countermeasures.

13.3 Summary of the Spoofing Challenges

A number of independent studies confirm the vulnerability of ASV technology to spoofed voice created using voice conversion, speech synthesis and playback [6]. Early studies on speaker anti-spoofing were mostly conducted on in-house speech corpora created using a limited number of spoofing attacks. The development of countermeasures using only a small number of spoofing attacks may not offer the generalisation ability in the presence of different or unseen attacks. There was a lack of publicly available corpora and evaluation protocol to help with comparing the results obtained by different researchers.

The ASVspoof[1] initiative aims to overcome this bottleneck by making available standard speech corpora consisting of a large number of spoofing attacks, evaluation protocols and metrics to support a common evaluation and the benchmarking of different systems. The speech corpora were initially distributed by organising an evaluation challenge. In order to make the challenge simple and to maximise participation, the ASVspoof challenges so far involved only the detection of spoofed speech; in effect, to determine whether a speech sample is genuine or spoofed. A training set and development set consisting of several spoofing attacks were first shared with the challenge participants to help them develop and tune their anti-spoofing algorithm. Next, the evaluation set without any label indicating genuine or spoofed speech was distributed, and the organisers asked the participants to submit scores within a specific deadline. Participants were allowed to submit scores of multiple systems. One of these systems was designated as the primary submission. Spoofing detectors for all primary submissions were trained using only the training data in the challenge corpus. Finally, the organisers evaluated the scores for benchmarks and ranking. The evaluation keys were subsequently released to the challenge participants. The challenge results were discussed with the participants in a special session at INTERSPEECH conferences, which also involved sharing knowledge and receiving useful feedback. To promote further research and technological advancements, the datasets used in the challenge are made publicly available.

The ASVspoof challenges have been organised four times so far. The first was held in 2015 and the latest in 2021. A summary of the speech corpora used in the first two challenges are shown in Table 13.1. The third edition of the challenge has two separate tracks and its summary is shown in Table 13.2. In the first two challenges, EER metric was used to evaluate the performance of spoofing detector. The EER is computed by considering the scores of genuine files as positive scores and those of spoofed files as negative scores. A lower EER means more accurate spoofing countermeasures. In practice, the EER is estimated using a specific *receiver operating characteristics convex hull* (ROCCH) technique with an open-source implementation[2] originating from outside the ASVspoof consortium. While the EER is retained as the secondary metric, the 2019 edition of the challenge migrates to a new primary metric in the form of ASV-centric *tandem decision cost function* (t-DCF) [125] which is of the form:

$$\text{t-DCF}_{\text{norm}}^{\text{min}} = \min_{s} \left\{ \beta P_{\text{miss}}^{\text{cm}}(s) + P_{\text{fa}}^{\text{cm}}(s) \right\}, \tag{13.2}$$

where β depends on application parameters (priors, costs) *and* ASV performance (miss, false alarm and spoof miss rates), while $P_{\text{miss}}^{\text{cm}}(s)$ and $P_{\text{fa}}^{\text{cm}}(s)$ are the CM miss and false alarm rates at threshold s. The minimum in (13.2) is taken over all thresholds on given data (development or evaluation) with a known key, corresponding to oracle calibration. For further details about the newly introduced cost function, we refer to [125, 126].

---

[1] http://www.asvspoof.org/.

[2] https://sites.google.com/site/bosaristoolkit/.

Table 13.1 Summary of the datasets used in ASVspoof 2015 and 2017 challenge

	ASVspoof 2015 [127]	ASVspoof 2017 [128]
Theme	Detection of artificially generated speech	Detection of replay speech
Speech format	$F_s = 16$ kHz, 16 bit PCM	$F_s = 16$ kHz, 16 bit PCM
Natural speech	Recorded using high-quality microphone	Recorded using different smartphones
Spoofed speech	Created with seven VC and three SS methods	Collected 'in the wild' by crowdsourcing using different microphones and playback devices from diverse environments
Spoofing types in train/dev/eval	5/5/10	3/10/57
No of speakers in train/dev/eval	25/35/46	10/8/24
No of genuine speech files in train/dev/eval	3750/3497/9404	1508/760/1298
No of spoofed speech files in train/dev/eval	12625/49875/184000	1508/950/12008

Table 13.2 Summary of the datasets used in two tracks of ASVspoof 2019 challenge [130]

	Logical access	Physical acces
Theme	Detection of artificially generated speech	Detection of replay speech
Speech format	$F_s = 16$ kHz, 16 bit FLAC	$F_s = 16$ kHz, 16 bit FLAC
Natural speech	Recorded using high-quality microphone	Recorded in simulated room environment
Spoofed speech	Created with 19 VC and TTS methods	Created with simulated room environment
Spoofing types in train/dev/eval	6/6/13	9/9/9
No of speakers in train/dev/eval	20/20/67	20/20/67
No of genuine speech files in train/dev/eval	2580/2548/7355	5400/5400/18090
No of spoofed speech	22800/22296/63882	48600/24300/116640

In the following subsections, we briefly discuss the first three challenges. For more interested readers, [127] contains details of the 2015 edition while [128] and [129] discuss the results of the 2017 and 2019 edition, respectively.

13.3.1 *ASVspoof 2015*

The first ASVspoof challenge involved the detection of artificial speech created using a mixture of voice conversion and speech synthesis techniques [127]. The dataset was generated with ten different artificial speech generation algorithms. The ASVspoof 2015 was based upon a larger collection spoofing and anti-spoofing (SAS) corpus (v1.0) [131] that consists of both natural and artificial speech. Natural speech was recorded from 106 human speakers using a high-quality microphone and without significant channel or background noise effects. In a speaker disjoint manner, the full database was divided into three subsets called the training, development and evaluation set. Five of the attacks (S1–S5), named as *known attacks*, were used in the training and development set. The other five attacks, S6–S10, called *unknown attacks*, were used only in the evaluation set, along with the known attacks. Thus, this provides the possibility of assessing the generalisability of the spoofing detectors. The detailed evaluation plan is available in [132], describing the speech corpora and challenge rules.

Ten different spoofing attacks used in the ASVspoof 2015 are listed below:

- **S1**: a simplified frame selection (FS)-based voice conversion algorithm, in which the converted speech is generated by selecting target speech frames.
- **S2**: the simplest voice conversion algorithm which adjusts only the first mel-cepstral coefficient (C1) in order to shift the slope of the source spectrum to the target.
- **S3**: a speech synthesis algorithm implemented with the HMM-based speech synthesis system (HTS3) using speaker adaptation techniques and only 20 adaptation utterances.
- **S4**: the same algorithm as S3, but using 40 adaptation utterances.
- **S5**: a voice conversion algorithm implemented with the voice conversion toolkit and with the Festvox system[3].
- **S6**: a VC algorithm based on joint density Gaussian mixture models (GMMs) and maximum likelihood parameter generation considering global variance.
- **S7**: a VC algorithm similar to S6, but using line spectrum pair (LSP) rather than mel-cepstral coefficients for spectrum representation.
- **S8**: a tensor-based approach to VC, for which a Japanese dataset was used to construct the speaker space.
- **S9**: a VC algorithm which uses kernel-based partial least square (KPLS) to implement a non-linear transformation function.
- **S10**: an SS algorithm implemented with the open-source MARY text-to-speech system (MaryTTS)[4].

More details of how the SAS corpus was generated can be found in [131].

The organisers also confirmed the vulnerability to spoofing by conducting speaker verification experiments with this data and demonstrating considerable performance

---

[3] http://www.festvox.org/.

[4] http://mary.dfki.de/.

Table 13.3 Performance of top five systems in ASVspoof 2015 challenge (ranked according to the average % EER for all attacks) with respective features and classifiers

System	Avg. EER for			System
Identifier	Known	Unknown	All	Description
A [133]	0.408	2.013	1.211	*Features*: mel-frequency cepstral coefficients (MFCC), Cochlear filter cepstral coefficients plus instantaneous frequency (CFCCIF) *Classifier*: GMM
B [134]	0.008	3.922	1.965	*Features*: MFCC, MFPC, cosine-phase principal coefficients (CosPhasePCs). *Classifier*: Support vector machine (SVM) with i-vectors
C [135]	0.058	4.998	2.528	*Feature*: DNN-based with filterbank output and their deltas as input. *Classifier*: Mahalanobis distance on s-vectors
D [136]	0.003	5.231	2.617	*Features*: log magnitude spectrum (LMS), residual log magnitude spectrum (RLMS), group delay (GD), modified group delay (MGD), instantaneous frequency derivative (IF), baseband phase difference (BPD) and pitch synchronous phase (PSP). *Classifier*: Multilayer perceptron (MLP)
E [137]	0.041	5.347	2.694	*Features*: MFCC, product spectrum MFCC (PS-MFCC), MGD with and without energy, weighted linear prediction group delay cepstral coefficients (WLP-GDCCs) and MFCC cosine-normalised phase-based cepstral coefficients (MFCC-CNPCCs). *Classifier*: GMM

degradation in the presence of spoofing. With a state-of-the-art probabilistic linear discriminant analysis (PLDA)-based ASV system, it is shown that in the presence of spoofing, the average EER for ASV increases from 2.30% to 36.00% for male and 2.08% to 39.53% for female [127]. This motivates the development of the anti-spoofing algorithm.

For ASVspoof 2015, the challenge evaluation metric was the average EER. It is computed by calculating EERs for each attack and then taking average. The dataset was requested by 28 teams from 16 countries and 16 teams returned primary submissions by the deadline. A total of 27 additional submissions were also received. Anonymous results were subsequently returned to each team, who were then invited to submit their work to the ASVspoof special session for INTERSPEECH 2015.

Table 13.3 shows the performance of the top five systems in the ASVspoof 2015 challenge. The best-performing system [133] uses a combination of *mel-cesptral* and *cochlear filter cepstral coefficients plus instantaneous frequency* features with GMM back end. In most cases, the participants have used a fusion of multiple feature-based systems to get better recognition accuracy. Variants of cepstral features computed from the magnitude and phase of short-term speech are widely used for the detection of spoofing attacks. As a back end, GMM was found to outperform more advanced classifiers like i-vectors, possibly due to the use of short segments of high-quality speech not requiring treatment for channel compensation and background noise reduction. All the systems submitted in the challenge are reviewed in more detail [138].

13.3.2 ASVspoof 2017

The ASVspoof 2017 is the second automatic speaker verification anti-spoofing and countermeasures challenge. Unlike the 2015 edition that used very high-quality speech material, the 2017 edition aims to assess spoofing attack detection with 'out in the wild' conditions. It focuses exclusively on replay attacks. The corpus originates from the recent *text-dependent RedDots* corpus,[5] whose purpose was to collect speech data over mobile devices, in the form of smartphones and tablet computers, by volunteers from across the globe.

The replayed version of the original *RedDots* corpus was collected through a crowdsourcing exercise using various replay configurations consisting of varied devices, loudspeakers and recording devices, under a variety of different environments across four European countries within the EU Horizon 2020-funded OCTAVE project[6], (see [128]). Instead of covert recording, we made a 'short-cut' and took the digital copy of the target speakers' voice to create the playback versions. The collected corpus is divided into three subsets: training, development and evaluation. Details of each are presented in Table 13.1. All three subsets are disjoint in terms of speakers and data collection sites. The training and development subsets were collected at three different sites. The evaluation subset was collected at the same three sites and also included data from two new sites. Data from the same site include different recordings and replaying devices and from different acoustic environments. The evaluation subset contains data collected from 161 replay sessions in 62 unique replay configurations[7]. More details regarding replay configurations can be found in [128, 139].

The primary evaluation metric is 'pooled' EER. In contrast to the ASVspoof 2015 challenge, the EER is computed from scores pooled across all the trial segments rather than condition averaging. A baseline[8] system based on common GMM back-end classifier with constant Q cepstral coefficient (CQCC) [140, 141] features was provided to the participants. This configuration is chosen as baseline as it has shown the best recognition performance on ASVspoof 2015. The baseline is trained using

[5] https://sites.google.com/site/thereddotsproject/.

[6] https://www.octave-project.eu/.

[7] A **replay configuration** refers to a unique combination of room, replay device and recording device while a **session** refers to a set of source files, which share the same replay configuration.

[8] See *Appendix A.2. Software packages*.

either combined training and development data (B01) or training data (B02) alone. The baseline system does not involve any kind of optimisation or tuning with respect to [140]. The dataset was requested by 113 teams, of which 49 returned primary submissions by the deadline. The results of the challenge were disseminated at a special session consisting of two slots at INTERSPEECH 2017.

Most of the systems are based on standard spectral features, such as CQCCs, MFCCs, and *perceptual linear prediction* (PLP). As a back end, in addition to the classical GMM to model the replay and non-replay classes, it has also exploited the power of deep classifiers, such as *convolutional neural network* (CNN) or *recurrent neural network* (RNN). A fusion of multiple features and classifiers is also widely adopted by the participants. A summary of the top-10 primary systems is provided in Table 13.4. Results in terms of EER of the 49 primary systems and the baseline B01 and B02 are shown in Fig. 13.3.

13.3.3 ASVspoof 2019

Unlike the two previous challenges, the ASVspoof 2019 addresses both synthetic speech detection and playback attack detection tasks in two different scenarios named *logical access* (LA) and *physical access* (PA). The dataset for both scenarios is derived from the VCTK-based corpus which includes speech data captured from 107 speakers (46 males, 61 females) [147]. Both LA and PA databases are themselves divided into three partitions, namely, training, development and evaluation which comprise the audio data from 20 (8 male, 12 female), 10 (4 male, 6 female) and 48 (21 male, 27 female) speakers, respectively. The three partitions are disjoint in terms of speakers but the recording conditions for all source data are identical.

The ASVspoof 2019 challenge uses t-DCF and EER as performance evaluation metrics. The orgnaisers set the parameters for t-DCF computation and mentioned them in the evaluation plan [148]. The ASV system developed by the organisers relies on Kaldi x-vector recipe for VoxCeleb[9]. Similar to ASVspoof 2017, the ASV 2019 challenge has used two baseline spoofing countermeasures systems[10] based on two different acoustic frontends, namely CQCC and LFCC, and GMM classifier. Those two baselines are denoted as B01 and B02, respectively. Corresponding to each scenario, the challenge participants were required to submit two sets of scores: primary and single on both development and evaluation parts. The primary system may be a fusion of classifiers consisting of multiple subsystems. The single system scores, in turn, should correspond to exactly one of the subsystems in such fusion.

---

[9] https://github.com/kaldi-asr/kaldi/tree/master/egs/voxceleb/v2.

[10] https://www.asvspoof.org/asvspoof2019/ASVspoof_2019_baseline_CM_v1.zip.

Table 13.4 Summary of top ten primary submissions to ASVspoof 2017. Systems' IDs are the same received by participants in the evaluation. The column 'Training' refers to the part of data used for training: train (T) and/or development (D)

ID	Features	Post-proc.	Classifiers	Fusion	#Subs.	Training	Performances on eval subset (EER%)
S01 [142]	Log-power Spectrum, LPCC	MVN	CNN, GMM, TV, RNN	Score	3	T	6.73
S02 [143]	CQCC, MFCC, PLP	WMVN	GMM-UBM, TV-PLDA, GSV-SVM, GSV-GBDT, GSV-RF	Score	–	T	12.34
S03	MFCC, IMFCC, RFCC, LFCC, PLP, CQCC, SCMC, SSFC	–	GMM, FF-ANN	Score	18	T+D	14.03
S04	RFCC, MFCC, IMFCC, LFCC, SSFC, SCMC	–	GMM	Score	12	T+D	14.66
S05 [144]	Linear filterbank feature	MN	GMM, CT-DNN	Score	2	T	15.97
S06	CQCC, IMFCC, SCMC, Phrase one-hot encoding	MN	GMM	Score	4	T+D	17.62
S07	HPCC, CQCC	MVN	GMM, CNN, SVM	Score	2	T+D	18.14
S08 [145]	IFCC, CFCCIF, Prosody	–	GMM	Score	3	T	18.32
S09	SFFCC	No	GMM	None	1	T	20.57
S10 [146]	CQCC	–	ResNet	None	1	T	20.32

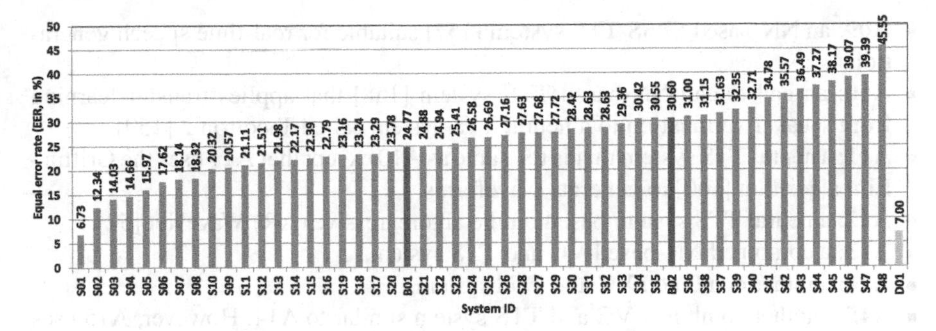

Fig. 13.3 Performance of the two baseline systems (B01 and B02) and the 49 primary systems (S01–S48 in addition to late submission D01) for the ASVspoof 2017 challenge. Results are in terms of the replay/non-replay EER (%)

13.3.3.1 LA Scenario

The LA database contains bona fide speech and spoofed speech data. The spoofed data are generated using 17 different TTS and VC systems. Data used for the training of TTS and VC systems also comes from the VCTK database but there is no overlap with the data contained in the 2019 database. Six of these systems are designated as *known attacks*, with the other 11 being designated as *unknown attacks*. The training and development sets contain known attacks only, whereas the evaluation set contains 2 known and 11 unknown spoofing attacks.

In total, 19 different attacks, namely A01–19, are created with those 17 different TTS and VC systems listed below. Attacks A01–06 are used in the training and development set, whereas the remaining A07–13 are included in the evaluation set.

- **A01**: a neural network (NN)-based TTS system. This system follows the standard NN-based statistical parametric speech synthesis (SPSS) framework [149] and uses a powerful neural waveform generator called WaveNet [52].
- **A02**: an NN-based TTS system similar to A01 except that the WORLD vocoder [150] rather than WaveNet is used to generate waveforms.
- **A03**: an NN-based TTS system similar to A02. It is built using recipes in an open-source TTS toolkit called Merlin [151].
- **A04**: a waveform concatenation TTS system based on the MaryTTS platform [152][11] using the voice building plugin (v5.4) [153].
- **A05**: an NN-based VC system that uses a variational autoencoder (VAE) [154] as the VC conversion model.
- **A06**: a transfer function-based VC system [155].
- **A07**: an NN-based TTS system with a generative adversarial network (GAN)-based post-filter.
- **A08**: an NN-based TTS system similar to A01, however, this employs a neural source-filter waveform model [156] instead of WaveNet.

[11] https://github.com/marytts/.

- **A09**: an NN-based SPSS TTS system [157] suitable for real-time speech generation.
- **A10**: an end-to-end NN-based TTS system [158] that applies transfer learning from speaker verification to a neural TTS system called Tacotron 2 [159].
- **A11**: a neural TTS system that is the same as A10 except that A11 uses the Griffin–Lim algorithm [160] to generate waveforms.
- **A12**: a neural TTS system based on the autoregressive (AR) WaveNet [52].
- **A13**: a combined NN-based VC and TTS system.
- **A14**: another combined VC and TTS system.
- **A15**: another combined VC and TTS system similar to A14. However, A15 uses speaker-dependent WaveNet vocoders rather than the STRAIGHT vocoder to generate waveforms.
- **A16**: a waveform concatenation TTS system that uses the same algorithm as A04 but with different training data.
- **A17**: an NN-based VC system that uses the same VAE-based framework as A05. However, rather than using the WORLD vocoder, **A17** uses a generalised direct waveform modification method [161, 162] for waveform generation.
- **A18**: a non-parallel VC system [163] inspired by the standard *i-vector* framework used in text-independent ASV.
- **A19**: a transfer function-based VC system using the same algorithm as A06 but with different training data.

In total, 48 teams submitted the scores for the LA scenario. Twenty-seven of them outperformed both the baselines. Results of 48 submissions along with the baselines are shown in Fig. 13.4. Most systems used different CNN-based architectures such as ResNet and LCNN. The details of the results are reported in [129].

13.3.3.2 PA Scenario

The PA dataset contains bona fide speech and replay audio data. Both types of data are generated according to a simulation [164–166] of their presentation to the microphone of an ASV system within a reverberant acoustic environment. Replayed speech is assumed first to be captured with a recording device before being replayed using a non-linear replay device. Training and development data is created according to 27 different acoustic and 9 different replay configurations resulting in 243 conditions. Acoustic configurations comprise an exhaustive combination of three categories of room sizes, three categories of reverberation and three categories of speaker/talker-to-ASV microphone distances. On the other hand, replay configurations comprise three categories of attacker-to-talker recording distances and three categories of loudspeaker quality. Evaluation data is generated in the same manner as training and development data, but with different, random acoustic and replay configurations. The details of the dataset are discussed in [130, 148, 167].

In the PA scenario, 50 teams submitted the scores and 32 of them outperformed both the baselines. Results of the 50 teams along with the baseline are shown in

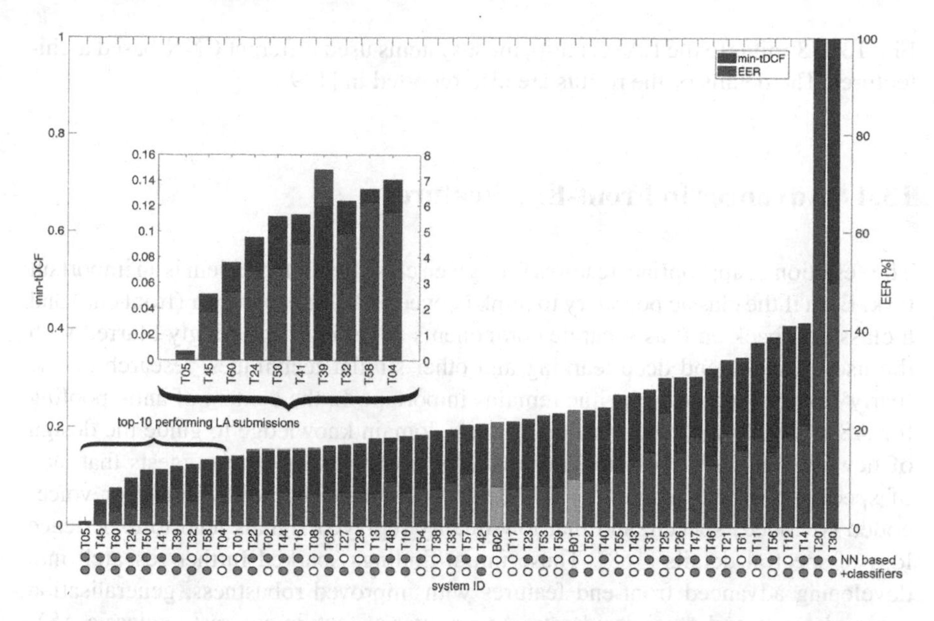

Fig. 13.4 Performance of the two baseline systems (B01 and B02) and the 48 primary systems for the LA scenario of the ASVspoof 2019 challenge. Results are in terms of the min-tDCF and EER (%)

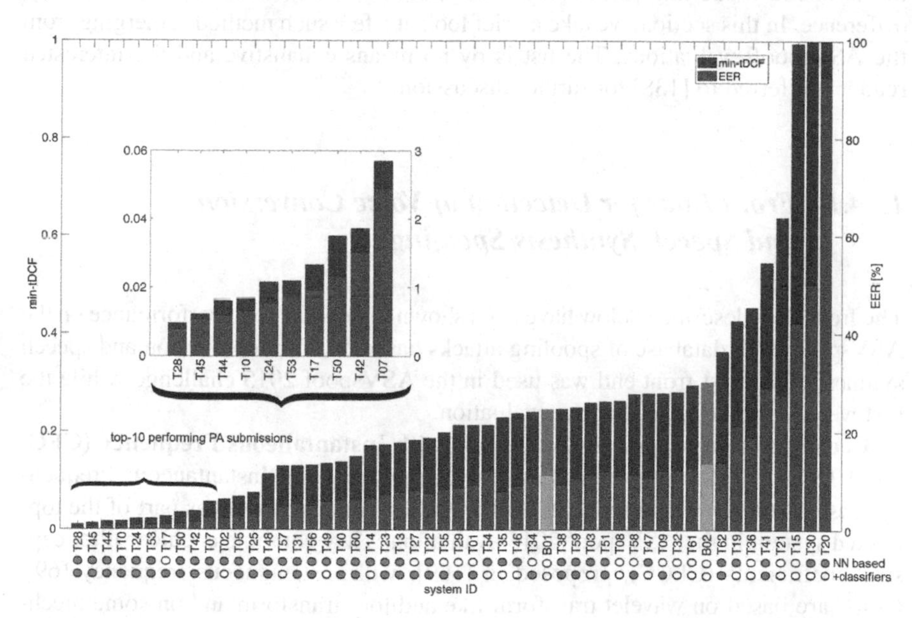

Fig. 13.5 Performance of the two baseline systems (B01 and B02) and the 50 primary systems for the PA scenario of the ASVspoof 2019 challenge. Results are in terms of the min-tDCF and EER (%)

Fig. 13.5. Similar to the LA scenario, most systems used different CNN-based architectures. The details of the results are also reported in [129].

13.4 Advances in Front-End Features

The selection of appropriate features for a given classification problem is an important task. Even if the classic boundary to think between a feature extractor (front end) and a classifier (back end) as separate components is getting increasingly blurred with the use of end-to-end deep learning and other similar techniques, research on the 'early' components in a pipeline remains important. In the context of anti-spoofing for ASV, this allows the utilisation of one's domain knowledge to guide the design of new discriminative features. For instance, earlier experience suggests that lack of spectral [70] and temporal [123] detail is characteristic of synthetic or voice-coded (vocoded) speech, and that low-quality replayed signals tend to experience loss of spectral details [168]. These initial findings sparked further research into developing advanced front-end features with improved robustness, generalisation across datasets and other desiderata. As a matter of fact, in contrast to classic ASV (without spoofing attacks) where the most significant advancements have been in the back-end modelling [2], in ASV anti-spoofing, the features seem to make the difference. In this section, we take a brief look at a few such methods emerging from the ASVspoof evaluations. The list is by no means exhaustive and the interested reader is referred to [138] for further discussion.

13.4.1 Front Ends for Detection of Voice Conversion and Speech Synthesis Spoofing

The front ends described below have been shown to provide good performance on the ASVspoof 2015 database of spoofing attacks based on voice conversion and speech synthesis. The first front end was used in the ASVspoof 2015 challenge, while the rest were proposed later after the evaluation.

Cochlear Filter Cepstral Coefficients with Instantaneous Frequency (CFCCIF). CFCCIF: Cochlear Filter Cepstral Coefficients with Instantaneous FrequencyThese features were introduced in [133] and successfully used as part of the top-ranked system in the ASVspoof 2015 evaluation. They combine cochlear filter cepstral coefficients (CFCC), proposed in [169], with instantaneous frequency [69]. CFCC are based on wavelet transform-like auditory transform and on some mechanisms of the cochlea of the human ear, such as hair cells and nerve spike density. To compute CFCC with instantaneous frequency (CFCCIF), the output of the nerve spike density envelope is multiplied by the instantaneous frequency, followed by the derivative operation and logarithm non-linearity. Finally, the Discrete Cosine

Fig. 13.6 Block diagram of CQCC feature extraction process

Transform (DCT) is applied to decorrelate the features and obtain a set of cepstral coefficients.

Linear Frequency Cepstral Coefficients (LFCC). LFCCsLFCC: Linear Frequency Cepstral Coefficients are very similar to the widely used mel-frequency cepstral coefficients (MFCCs) [170], though the filters are placed in equal sizes for linear scale. This front end is widely used in speaker recognition and has been shown to perform well in spoofing detection [171]. This technique performs a windowing on the signal, computes the magnitude spectrum using the short-time Fourier transform (STFT), followed by logarithm non-linearity and the application of a filterbank of linearly spaced N triangular filters to obtain a set of N log-density values. Finally, the DCT is applied to obtain a set of cepstral coefficients.

Constant Q Cepstral Coefficients (CQCC). This feature was proposed in [140, 141] for spoofing detection and it is based on the Constant Q Transform (CQT)CQT: Constant Q Transform [172]. The CQT is an alternative time–frequency analysis tool to the STFT that provides variable time and frequency resolution. It provides greater frequency resolution at lower frequencies but greater time resolution at higher frequencies. Figure 13.6 illustrates the CQCCCQCC: Constant Q Cepstral Coefficients extraction process. The CQT spectrum is obtained, followed by logarithm non-linearity and by a linearisation of the CQT geometric scale. Finally, cepstral coefficients are obtained through the DCT.

As an alternative to CQCC, infinite impulse response constant-Q transform cepstrum (ICQC) features [173] use the infinite impulse response—constant Q transform [174], an efficient constant Q transform based on the IIR filtering of the fast Fourier transform (FFT) spectrum. It delivers multiresolution time-frequency analysis in a linear scale spectrum which is ready to be coupled with traditional cepstral analysis. The IIR-CQT spectrum is followed by the logarithm and decorrelation, either through the DCT or principal component analysis.

Deep features for spoofing detection. All of the above three feature sets are hand-crafted and consist of a fixed sequence of standard digital signal processing operations. An alternative approach, seeing increased popularity across different machine learning problems, is to learn the feature extractor from a given data by using deep learning techniques [175, 176]. In speech-related applications, these features are widely employed for improving recognition accuracy [177–179]. The work in [180] uses deep neural network to generate bottleneck features for spoofing detection, that is, the activations of a hidden layer with a relatively small number of nodes compared to the size of other layers. The study in [181] investigates various features based on deep learning techniques. Different feed-forward DNNs are used to obtain frame-level deep features. Input acoustic features consisting of filterbank outputs with their first derivatives are used to train the network to discriminate between the natural and spoofed speech classes, and the output of hidden layers are taken as deep features which are then averaged to obtain an utterance-level descriptor. RNNs are also proposed to estimate utterance-level features from input sequences of acoustic features. In another recent work [182], the authors have investigated deep features based on filterbank trained with the natural and artificial speech data. A feed-forward neural network architecture called here as filterbank neural network (FBNN) is used here that includes a linear hidden layer, a sigmoid hidden layer and a softmax output layer. The number of nodes in the output is six; and of them, five are for the number of spoofed classes in the training set, and the remaining one is for natural speech. The filterbanks are learned using the stochastic gradient descent algorithm. The cepstral features extracted using these DNN-based features are shown to be better than the hand-crafted cepstral coefficients.

Scattering cepstral coefficients. This feature for spoofing detection was proposed in [183]. It relies upon *scattering spectral decomposition* [184, 185]. This transform is a hierarchical spectral decomposition of a signal based on wavelet filterbanks (constant Q filters), modulus operator and averaging. Each level of decomposition processes the input signal (either the input signal for the first level of decomposition or the output of a previous level of decomposition) through the wavelet filterbank and takes the absolute value of filter outputs, producing a scalogram. The scattering coefficients at a certain level are estimated by windowing the scalogram signals and computing the average value within these windows. A two-level scattering decomposition has been shown to be effective for spoofing detection [183]. The final feature vector is computed by taking the DCT of the vector obtained by concatenating the logarithms of the scattering coefficients from all levels and retaining the first a few coefficients. The 'interesting' thing about scattering transform is its stability to small signal deformation and more details of the temporal envelopes than MFCCs [183, 184].

Fundamental frequency variation features. The prosodic features are not as successful as cepstral features in detecting artificial speech on ASVspoof 2015, though some earlier results on PAs indicate that pitch contours are useful for such tasks [6]. In a recent work [186], the author use fundamental frequency variation (FFV) for this. The FFV captures pitch variation at the frame level and provides complementary information on cepstral features [187]. The combined system gives a very promising performance for both known and unknown conditions on ASVspoof evaluation data.

Phase-based features. The phase-based features are also successfully used in PAD systems for ASVspoof 2015. For example, relative phase shift (RPS) and modified group delay (MGD) based features are explored in [188]. The authors in [189] have investigated relative phase information (RPI) features. Though the performances on seen attacks are promising with these phase-based features, the performances noticeably degrade for unseen attacks, particularly for S10.

General observations regarding front ends for artificial speech detection. Beyond the feature extraction method used, there are two general findings common to any front end [133, 141, 171, 173]. The first refers to the use of dynamic coefficients.

The first and second derivatives of the static coefficients, also known as velocity and acceleration coefficients, respectively are found important to achieve good spoofing detection performance. In some cases, the use of only dynamic features is superior to the use of static plus dynamic coefficients [171]. This is not entirely surprising, since voice conversion and speech synthesis techniques may fail to model the dynamic properties of the speech signals, introducing artefacts that help the discrimination of spoofed signals. The second finding refers to the use of speech activity detection. In experiments with ASVspoof 2015 corpus, it appears that the silence regions also contain useful information for discriminating between natural and synthetic speech. Thus, retaining non-speech frames turns out to be a better choice for this corpus [171]. This is likely due to the fact that non-speech regions are usually replaced with noise during the voice conversion or speech synthesis operation. However, this could be a database-dependent observation, thus detailed investigations are required.

13.4.2 Front Ends for Replay Attack Detection

The following front ends have been proposed for the task of replay spoofing detection and evaluated in replayed speech databases such as the BTAS 2016 and ASVspoof 2017. Many standard front ends, such as MFCC, LFCC and PLP, have been combined to improve the performance of replay attack detection. Other front ends proposed for synthetic and converted speech detection (CFCCIF, CQCC) have been successfully used for the replay detection task. In general, and in opposition to the trend for synthetic and converted speech detection, the use of static coefficients has been shown to be crucial for achieving good performance. This may be explained by the nature of the replayed speech detection task, where detecting changes in the channel captured by static coefficients helps with the discrimination of natural and replayed speech. Two additional front ends are described next.

Inverted mel-frequency cepstral coefficients (IMFCC). This front end is relatively simple and similar to the standard MFCC. The only difference is that the filterbank follows an inverted mel scale, that is, it provides an increasing frequency resolution (narrower filters) when frequency increases, and a decreased frequency resolution (wider filters) for decreasing frequency, unlike the mel scale [190]. This front end was used as part of the top-ranked system of the Biometrics: Theory, Applications and Systems (BTAS) 2016 speaker anti-spoofing competition [7].

Features based on convolutional neural networks. In the recent ASVspoof 2017 challenge, the use of deep learning frameworks for feature learning was proven to be key in achieving good replay detection performance. In particular, convolutional neural networks have been successfully used to learn high-level utterance-level features which can later be classified with simple classifiers. As part of the top-ranked system [142] in the ASVspoof 2017 challenge, a light convolutional neural network architecture [191] is fed with truncated normalised FFT spectrograms (to force fixed data dimensions). The network consists of a set of convolutional layers, followed by a fully connected layer. The last layer contains two outputs with softmax activation corresponding to the two classes. All layers use the max-feature-map activation function [191], which acts as a feature selector and reduces the number of feature maps by half on each layer. The network is then trained to discriminate between the natural and spoofed speech classes. Once the network is trained, it is used to extract a high-level feature vector which is the output of the fully connected layer. All the test utterances are processed to obtain high-level representations, which are later classified with an external classifier.

Other hand-crafted features. Many other features have also been used for replayed speech detection in the context of the ASVspoof 2017 database. Even if the performances of single systems using such features are not always high, they are shown to be complementary when fused at the score level [192], similar to conventional ASV research outside of the spoofing detection. These features include MFCC, IMFCC, rectangular filter cepstral coefficients (RFCCs), PLP, CQCC, spectral centroid magnitude coefficients (SCMC), subband spectral flux coefficient (SSFC) and variable length Teager energy operator energy separation algorithm-instantaneous frequency cosine coefficients (VESA-IFCC). Though, of course, one usually then has to further train the fusion system, which makes the system more involved concerning practical applications.

13.5 Advances in Back-End Classifiers

In the natural versus spoof classification problem, two main families of approaches have been adopted, namely generative and discriminative. Generative approaches include those of GMM-based classifiers and i-vector representations combined with support vector machines (SVMs). As for discriminative approaches, deep learning-based techniques have become more popular. Finally, new deep learning end-to-end solutions are emerging. Such techniques perform the typical pipeline entirely through deep learning, from feature representation learning and extraction to the final classification. While including such approaches in the traditional classifiers category may not be the most precise, they are included in this classifiers section for simplicity.

13.5.1 Generative Approaches

Gaussian mixture model (GMM) classifiers. Considering two classes, namely natural and spoofed speech, one GMM can be learned for each class using appropriate training data. In the classification stage, an input utterance is processed to obtain its likelihood with respect to the natural and spoofed models. The resulting classification score is the log-likelihood ratio between the two competing hypotheses; in effect, those of the input utterances belonging to the natural and to the spoofed classes. A high score supports the former hypothesis, while a low score supports the latter. Finally, given a test utterance, classification can be performed by thresholding the obtained score. If the score is above the threshold, the test utterance is classified as natural, and otherwise, it is classified as spoof. Many proposed anti-spoofing systems use GMM classifiers [133, 140, 171, 173, 180, 183, 193].

I-vector. The state-of-the-art i-vector paradigm for speaker verification [194] has been explored for spoofing detection [195, 196]. Typically, an i-vector is extracted from an entire speech utterance and used as a low-dimensional, high-level feature which is later classified by means of a binary classifier, commonly cosine distance measure or support vector machine (SVM). Different amplitude- and phase-based front ends [134, 142] can be employed for the estimation of i-vectors. A recent work shows that data selection for i-vector extractor training (also known as **T** matrix) is an important factor for achieving completive recognition accuracy [197].

13.5.2 Discriminative Approaches

DNN classifiers. Deep learning-based classifiers have been explored for use in the task of natural and spoofed speech discrimination. In [180, 198], several front ends are evaluated with neural network classifier consisting of several hidden layers with sigmoid nodes and softmax output, which is used to calculate utterance posteriors. However, the implementation detail of the DNNs—such as the number of nodes, the cost function, the optimisation algorithm and the activation functions—is not precisely mentioned in those work and the lack of this very relevant information makes it difficult to reproduce the results.

In a recent work [199], a five-layer DNN spoofing detection system is investigated for ASVspoof 2015 which uses a novel scoring method, termed in the paper as *human log-likelihoods* (HLLs). Each of the hidden layers has 2048 nodes with a sigmoid activation function. The network has six softmax output layers. The DNN is implemented using a computational network toolkit[12] and trained with stochastic gradient descent methods with dynamics information of acoustic features, such as spectrum-based cepstral coefficients (SBCC) and CQCC as input. The cross-entropy function is selected as the cost function and the maximum training epoch is chosen as 120. The mini-batch size is set to 128. The proposed method shows considerable PAD detection performance. The author obtain an EER for S10 of 0.255% and average EER for all attacks of 0.045% when used with CQCC acoustic features. These are the best-reported performance in ASVspoof 2015 so far.

DNN-based end-to-end approaches. End-to-end systems aim to perform all the stages of a typical spoofing detection pipeline, from feature extraction to classification, by learning the network parameters involved in the process as a whole. The advantage of such approaches is that they do not explicitly require prior knowledge of the spoofing attacks as required for the development of acoustic features. Instead, the parameters are learned and optimised from the training data. In [200], a convolutional long short-term memory (LSTM) deep neural network (CLDNN) [201]

[12] https://github.com/Microsoft/CNTK.

is used as an end-to-end solution for spoofing detection. This model receives input in the form of a sequence of raw speech frames and outputs a likelihood for the whole sequence. The CLDNN performs time–frequency convolution through CNN to reduce spectral variance, long-term temporal modelling by using an LSTM, and classification using a DNN. Therefore, it is entirely an end-to-end solution which does not rely on any external feature representation. The works in [142, 202] propose other end-to-end solutions by combining convolutional and recurrent layers, where the first act as a feature extractor and the second models the long-term dependencies and acts as a classifier. Unlike the work in [200], the input data is the FFT spectrogram of the speech utterance and not the raw speech signal. In [203], the authors have investigated CNN-based end-to-end system for PAD where the raw speech is used to jointly learn the feature extractor and classifier. Score-level combination of this CNN system with standard long-term spectral statistics-based system shows considerable overall improvement.

13.6 Other PAD Approaches

While most of the studies in voice PAD detection research focus on algorithmic improvements for discriminating natural and artificial speech signals, some recent studies have explored utilising additional information collected using special additional hardware to protect ASV system from presentation attacks [204–207]. Since an intruder can easily collect voice samples for the target speakers using covert recording; the idea there is to detect and recognise supplementary information related to the speech production process. Moreover, by its nature, that supplementary information is difficult, if not impossible, to mimic using spoofing methods in a practical scenario. These PAD techniques have shown excellent recognition accuracy in the spoofed condition, at the cost of additional set-up in the data acquisition step.

The work presented in [205, 206] utilises the phenomenon of *pop noise*, which is a distortion in human breath when it reaches a microphone [208]. During natural speech production, the interactions between the airflow and the vocal cavities may result in a sort of plosive burst, commonly known as pop noise, which can be captured via a microphone. In the context of professional audio and music production, pop noise is unwanted and is eliminated during the recording or mastering process. In the context of ASV, however, it can help in the process of PAD. The basic principle is that a replay sound from a loudspeaker does not involve the turbulent airflow generating the pop noise as in natural speech. The authors in [205, 206] have developed a pop noise detector which eventually distinguishes natural speech from playback recording as well as synthetic speech generated using VC and SS methods. In experiments with 17 female speakers, a tandem detection system that combines both single- and double-channel pop noise detection gives the lowest ASV error rates in the PA condition.

The authors in [204] have introduced the use of a smartphone-based *magnetometer* to detect voice presentation attacks. The conventional loudspeakers, which are used for playback during access to the ASV systems, generate sound using acoustic transducer and generate a magnetic field. The idea, therefore, is to capture the use of loudspeaker by sensing the magnetic field which would be absent from human vocals. Experiments were conducted using playback from 25 different conventional loudspeakers, ranging from low end to high end placed at different distances from the smartphone that contains the ASV system. A speech corpus of five speakers was collected for the ASV experiments executed using an open-source ASV toolkit, SPEAR.[13] Experiments were conducted with other datasets, using a similarly limited number of speakers. The authors demonstrated that magnetic field-based detection can be reliable for the detection of playback within 6–8 cm from the smartphone. They further developed a mechanism to detect the size of the sound source to prevent the use of small speakers, such as earphones.

The authors in [209, 210] utilise certain acoustics concepts to prevent ASV systems from PAs. They first introduced a method [209] that estimates dynamic sound source position (articulation position within mouth) of some speech sounds using a small array using *microelectromechanical systems* (MEMS) microphones embedded in mobile devices and compare it with loudspeakers, which have a flat sound source. In particular, the idea is to capture the dynamics of *time-difference-of-arrival* (TDOA) in a sequence of speech sounds to the microphones of the smartphone. Such unique TDOA changes, which do not exist under replay conditions, are used for detecting replay attacks. The similarities between the TDOAs of test speech and user templates are measured using probability function under Gaussian assumption and correlation measure as well as their combinations. Experiments involving 12 speakers and three different types of smartphones demonstrate a low EER and high PAD accuracy. The proposed method is seen to remain robust despite the change in smartphones during the test and the displacements.

In [210], the same research group has used the idea of the *Doppler effect* to detect the replay attack. The idea here is to capture the *articulatory gestures* of the speakers when they speak a pass phrase. The smartphone acts as a Doppler radar and transmits a high-frequency tone at 20 kHz from the built-in speaker and senses the reflections using the microphone during the authentication process. The movement of the speaker's articulators during vocalisation creates a speaker-dependent Doppler frequency shift at around 20 kHz, which is stored along with the speech signal during the speaker-enrollment process. During a playback attack, the Doppler frequency shift will be different due to the lack of articulatory movements. Energy-based frequency features and frequency-based energy features are computed from a band of 19.8 kHz and 20.2 kHz. These features are used to discriminate between the natural and replayed voice, and the similarity scores are measured in terms of the Pearson correlation coefficient. Experiments are conducted with a dataset of 21 speakers and using three different smartphones. The data also includes test speech for replay attack with different loudspeakers and for impersonation attack with four different impersonators. The proposed system was demonstrated to be effective in achieving low EER for both types of attacks. Similar to [209], the proposed method indicated robustness to the phone placement.

---

[13] https://www.idiap.ch/software/bob/docs/bob/bob.bio.spear/stable/index.html.

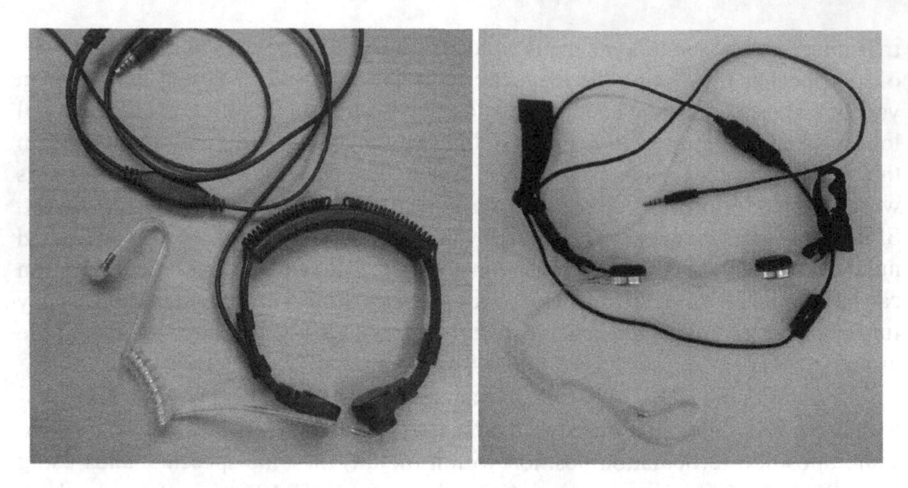

Fig. 13.7 Throat microphones used in [207] [Reprinted with permission from IEEEACM Transactions on (T-ASL) Audio, Speech, and Language Processing]

The work in [207] introduces the use of a specific non-acoustic sensor, *throat microphone* (TM), or laryngophone, to enhance the performance of the voice PAD system. An example of such microphones is shown in Fig. 13.7. The TM is used with a conventional acoustic microphone (AM) in a dual-channel framework for robust speaker recognition and PAD. Since this type of microphone is attached to the speaker's neck, it would be difficult for the attacker to obtain a covert recording of the target speaker's voice. Therefore, one possibility for the intruder is to use the stolen recording from an AM and try to record it back using a TM for accessing the ASV system. A speech corpus of 38 speakers was collected for the ASV experiments. The dual-channel set-up yielded considerable ASV for both licit and spoofed conditions. The performance is further improved when this ASV system is integrated with the dual channel-based PAD. The authors show zero FAR for replay imposters by decision fusion of ASV and PAD.

All of the above new PAD methods deviating from the 'mainstream' of PAD research in ASV are reported to be reliable and useful in specific application scenarios for identifying presentation attacks. The methods are also fundamentally different and difficult to compare in the same settings. Since the authors focus on the methodological aspects, experiments are mostly conducted on a dataset of a limited number of speakers. Extensive experiments with more subjects from diverse environmental conditions should be performed to assess their suitability for real-world deployment.

13.7 Future Directions of Anti-spoofing Research

The research in ASV anti-spoofing is becoming popular and well recognised in the speech processing and voice-biometric community. The state-of-the-art spoofing detector gives promising accuracy in the benchmarking of spoofing countermeasures. Further work is needed to address a number of specific issues regarding its practical use. A number of potential topics for consideration in further work are now discussed.

- **Noise, reverberation and channel effect.** Recent studies indicate that spoofing countermeasures offer little resistance to additive noise [211, 212], reverberation [213] and channel effect [214] even though their performances on 'clean' speech corpus are highly promising. The relative degradation of performance is actually much worse than the degradation of a typical ASV system under the similar mismatch condition. One reason could be that, at least until the ASVspoof 2017 evaluation, the methodology developed has been driven in clean, high-quality speech. In other words, the community might have developed its methods implicitly for laboratory testing. The commonly used speech enhancement algorithms also fail to reduce the mismatch due to environmental differences, though multi-condition training [212] and more advanced training methods [215] have been found useful. The study presented in [214] shows considerable degradation of PAD performance even in *matched* acoustic conditions. The feature settings used for the original corpus give lower accuracy when both training and test data are digitally processed with the telephone channel effect. These are probably because the spoofing artefacts themselves act as extrinsic variabilities which degrade the speech quality in some way. Since the task of spoofing detection is related to detecting those artefacts, the problem becomes more difficult in the presence of small external effects due to variation in environment and channel. These suggest further investigations need to be carried out for the development of robust spoofing countermeasures.
- **Generalisation of spoofing countermeasures**. The generalisation property of spoofing countermeasures for detecting new kinds of speech presentation attack is an important requirement for their application in the wild. The study explores that countermeasure methods trained with a class of spoofing attacks fail to generalise this for other classes of spoofing attack [192, 216]. For example, PAD systems trained with VC- and SS-based spoofed speech give a very poor performance for playback detection [217]. The results of the first two ASVspoof challenges also reveal that detecting the converted speech created with an 'unknown' method or the playback voice recording in a new replay session is difficult to detect. These clearly indicate the overfitting of PAD systems with available training data. Therefore, further investigation should be conducted to develop attack-independent universal spoofing detector. Other than the unknown attack issue, generalisation is also an important concern for cross-corpora evaluation of the PAD system [218]. This specific topic is discussed in Chap. 19 of this book.
- **Investigations with new spoofing methods**. The studies of converted spoof speech mostly focused on methods based on classical signal processing and machine

learning techniques. Recent advancements in VC and SS research with deep learning technology show significant improvements in creating high-quality synthetic speech [52]. The GAN [219] can be used to create (generator) spoofed voices with relevant feedback from the spoofing countermeasures (discriminator). Some preliminary studies demonstrate that the GAN-based approach can make speaker verification systems more vulnerable to presentation attacks [66, 220]. More detailed investigations should be conducted in this direction for the development of countermeasure technology to guard against this type of advanced attack.

- **Joint operations of PAD and ASV**. The ultimate goal of developing PAD system is to protect the recogniser, the ASV system from imposters with spoofed speech. So far, the majority of the studies focused on the evaluation of standalone countermeasures. The integration of these two systems is not a trivial number of reasons. First, standard linear output score fusion techniques, being extensively used to combine homogenous ASV system, are not appropriate since the ASV and its countermeasures are trained to solve two different tasks. Second, an imperfect PAD can increase the false alarm rate by rejecting genuine access trials [221]. Third, and more fundamentally, it is not obvious whether improvements in standalone spoofing countermeasures should improve the overall system as a whole: a nearly perfect PAD system with close to zero EER may fail to protect ASV system in practice if not properly calibrated [222]. In a recent work [223], the authors propose a modification in a GMM-UBM-based ASV system to make it suitable for both licit and spoofed conditions. The joint evaluation of PAD and ASV, as well as their combination techniques, certainly deserves further attention. Among other feedback received from the attendees of the ASVspoof 2017 special session organised during INTERSPEECH 2017, it was proposed that the authors of this chapter consider shifting the focus from standalone spoofing to more ASV-centric solutions in future. We tend to agree. In our recent work [224], we propose a new cost function for joint assessment of PAD and ASV systems. In another work [225], we propose a new fusion method for combining scores of countermeasures and recognisers. This work also explores speech features which can be used both for PAD and ASV.

13.8 Conclusion

This contribution provides an introduction to the different voice presentation attacks and their detection methods. It then reviews previous works with a focus on recent progress in assessing the performance of PAD systems. We have also briefly reviewed two recent ASVspoof challenges organised for the detection of voice PAs. This study includes a discussion of recently developed features and the classifiers which are predominantly used in ASVspoof evaluations. We further include an extensive survey on alternative PAD methods. Apart from the conventional voice-based systems that use statistical properties of natural and spoofed speech for their discrimination, these recently developed methods utilise a separate hardware for the acquisition of other signals such as pop noise, throat signal and extrasensory signals with smartphones for PAD. The current status of these non-mainstream approaches to PAD detection is somewhat similar to the status of the now more-or-less standard methods for artificial speech and replay PAD detection some three to four years ago: they are innovative and show promising results, but the pilot experiments have been carried out on relatively small and/or proprietary datasets, leaving an open question as to how scalable or generalisable these solutions are in practice. Nonetheless, in the long run, and noting especially the rapid development of speech synthesis technology, it is likely that the quality of artificial/synthetic speech will eventually be indistinguishable from that of natural human speech. Such future spoofing attacks therefore could not be detected using the current mainstream techniques that focus on spectral or temporal details of the speech signal but will require novel ideas that benefit from auxiliary information, rather than just the acoustic waveform.

In the past 3 years, the progress in voice PAD research has been accelerated by the development and free availability of speech corpus such as the ASVspoof series, SAS, BTAS 2016 and AVSpoof. The work discussed several open challenges which show that this problem requires further attention to improving robustness due to mismatch condition, generalisation to a new type of presentation attacks and so on. Results from joint evaluations with integrated ASV system are also an important requirement for practical applications of PAD research. We think, however, that this extensive review will be of interest not only to those involved in voice PAD research but also to voice-biometrics researchers in general.

Appendix A. Action Towards Reproducible Research

A.1: Speech Corpora

1. Spoofing and Anti-Spoofing (SAS) database v1.0: This database presents the first version of a speaker verification spoofing and anti-spoofing database, named SAS corpus [226]. The corpus includes nine spoofing techniques, two of which are speech synthesis, and seven are voice conversion.
   Download link: http://dx.doi.org/10.7488/ds/252
2. ASVspoof 2015 database: This database has been used in the first Automatic Speaker Verification Spoofing and Countermeasures Challenge (ASVspoof 2015). Genuine speech is collected from 106 speakers (45 male, 61 female) and with no significant channel or background noise effects. Spoofed speech is generated from the genuine data using a number of different spoofing algorithms. The full dataset is partitioned into three subsets, the first for training, the second for development and the third for evaluation.
   Download link: http://dx.doi.org/10.7488/ds/298

3. ASVspoof 2017 database: This database has been used in the Second Automatic Speaker Verification Spoofing and Countermeasures Challenge: ASVspoof 2017. This database makes extensive use of the recent text-dependent RedDots corpus, as well as a replayed version of the same data. It contains a large amount of speech data from 42 speakers collected from 179 replay sessions in 62 unique replay configurations.
   Download link: http://dx.doi.org/10.7488/ds/2313
4. ASVspoof 2019 database: This database has been used in the third Automatic Speaker Verification Spoofing and Countermeasures Challenge: ASVspoof 2019. This database has two independent subsets: logical access (LA) and physical (PA). Both of the subsets contain speech data from 107 speakers (46 male, 61 female) with no significant background noise effects as in ASVspoof 2015.
   Download link: https://datashare.ed.ac.uk/handle/10283/3336

A.2: Software Packages

1. Feature extraction techniques for anti-spoofing: This package contains the MATLAB implementation of different acoustic feature extraction schemes as evaluated in [171].
   Download link: http://cs.joensuu.fi/~sahid/codes/AntiSpoofing_Features.zip
2. Baseline spoofing detection package for ASVspoof 2017 corpus: This package contains the MATLAB implementations of two spoofing detectors employed as baseline in the official ASVspoof 2017 evaluation. They are based on linear-frequency cepstral coefficients (LFCCs) and constant Q cepstral coefficients (CQCC) [141] and Gaussian mixture model classifiers.
   Download link: http://audio.eurecom.fr/software/ASVspoof2017_baseline_countermeasures.zip
3. Baseline spoofing detection package for ASVspoof 2019: This package contains the MATLAB implementation of two official baseline spoofing detectors. Similar to ASVspoof 2017, they are based on CQCC and LFCC with GMM as classifier backend.
   Download link: https://www.asvspoof.org/asvspoof2019/ASVspoof_2019_baseline_CM_v1.zip
4. Software package for t-DCF metric computation: This package contains the implementations of the t-DCF metric. This also computes the EER.
   Download link: MATLAB: https://www.asvspoof.org/asvspoof2019/tDCF_matlab_v1.zip
   Python: https://www.asvspoof.org/asvspoof2019/tDCF_python_v1.zip

References

1. Kinnunen T, Li H (2010) An overview of text-independent speaker recognition: from features to supervectors. Speech Commun 52(1):12–40 https://doi.org/10.1016/j.specom.2009.08.009. www.sciencedirect.com/science/article/pii/S0167639309001289
2. Hansen J, Hasan T (2015) Speaker recognition by machines and humans: a tutorial review. IEEE Signal Process Mag 32(6):74–99
3. ISO/IEC 30107 (2016) Information technology—biometric presentation attack detection. International Organization for Standardization
4. Kinnunen T, Sahidullah M, Kukanov I, Delgado H, Todisco M, Sarkar A, Thomsen N, Hautamäki V, Evans N, Tan ZH (2016) Utterance verification for text-dependent speaker recognition: a comparative assessment using the reddots corpus. In: Proceedings of the interspeech, pp 430–434
5. Shang W, Stevenson M (2010) Score normalization in playback attack detection. In: Proceedings of the ICASSP. IEEE, pp 1678–1681
6. Wu Z, Evans N, Kinnunen T, Yamagishi J, Alegre F, Li H (2015) Spoofing and countermeasures for speaker verification: a survey. Speech Commun 66:130–153
7. Korshunov P, Marcel S, Muckenhirn H, Gonçalves A, Mello A, Violato R, Simoes F, Neto M, de Angeloni AM, Stuchi J, Dinkel H, Chen N, Qian Y, Paul D, Saha G, Sahidullah M (2016) Overview of BTAS 2016 speaker anti-spoofing competition. In: 2016 IEEE 8th international conference on biometrics theory, applications and systems (BTAS), pp 1–6
8. Evans N, Kinnunen T, Yamagishi J, Wu Z, Alegre F, DeLeon P (2014) Speaker recognition anti-spoofing. In: Marcel S, Li SZ, Nixon M (eds) Handbook of biometric anti-spoofing. Springer
9. Marcel S, Li SZ, Nixon M (eds) (2014) Handbook of biometric anti-spoofing: trusted biometrics under spoofing attacks. Springer
10. Farrús Cabeceran M, Wagner M, Erro D, Pericás H (2010) Automatic speaker recognition as a measurement of voice imitation and conversion. Int J Speech Lang Law 1(17):119–142
11. Perrot P, Aversano G, Chollet G (2007) Voice disguise and automatic detection: review and perspectives. Progress in nonlinear speech processing, pp 101–117
12. Zetterholm E (2007) Detection of speaker characteristics using voice imitation. In: Speaker classification II. Springer, pp 192–205
13. Lau Y, Wagner M, Tran D (2004) Vulnerability of speaker verification to voice mimicking. In: Proceedings of 2004 international symposium on intelligent multimedia, video and speech processing, 2004. IEEE, pp 145–148
14. Lau Y, Tran D, Wagner M (2005) Testing voice mimicry with the YOHO speaker verification corpus. In: International conference on knowledge-based and intelligent information and engineering systems. Springer, pp 15–21
15. Mariéthoz J, Bengio S (2005) Can a professional imitator fool a GMM-based speaker verification system? Technical report, IDIAP
16. Panjwani S, Prakash A (2014) Crowdsourcing attacks on biometric systems. In: Symposium on usable privacy and security (SOUPS 2014), pp 257–269
17. Hautamäki R, Kinnunen T, Hautamäki V, Laukkanen AM (2015) Automatic versus human speaker verification: the case of voice mimicry. Speech Commun 72:13–31
18. Ergunay S, Khoury E, Lazaridis A, Marcel S (2015) On the vulnerability of speaker verification to realistic voice spoofing. In: IEEE international conference on biometrics: theory, applications and systems, pp 1–8
19. Lindberg J, Blomberg M (1999) Vulnerability in speaker verification-a study of technical impostor techniques. In: Proceedings of the European conference on speech communication and technology, vol 3, pp 1211–1214
20. Villalba J, Lleida E (2010) Speaker verification performance degradation against spoofing and tampering attacks. In: FALA 10 workshop, pp 131–134

21. Wang ZF, Wei G, He QH (2011) Channel pattern noise based playback attack detection algorithm for speaker recognition. In: 2011 international conference on machine learning and cybernetics, vol 4, pp 1708–1713
22. Villalba J, Lleida E (2011) Preventing replay attacks on speaker verification systems. In: 2011 IEEE international Carnahan conference on security technology (ICCST). IEEE, pp 1–8
23. Gałka J, Grzywacz M, Samborski R (2015) Playback attack detection for text-dependent speaker verification over telephone channels. Speech Commun 67:143–153
24. Taylor P (2009) Text-to-Speech synthesis. Cambridge University Press
25. Klatt DH (1980) Software for a cascade/parallel formant synthesizer. J Acoust Soc Am 67:971–995
26. Moulines E, Charpentier F (1990) Pitch-synchronous waveform processing techniques for text-to-speech synthesis using diphones. Speech Commun 9:453–467
27. Hunt A, Black AW (1996) Unit selection in a concatenative speech synthesis system using a large speech database. In: Proceedings of the ICASSP, pp 373–376
28. Breen A, Jackson P (1998) A phonologically motivated method of selecting nonuniform units. In: Proceedings of the ICSLP, pp 2735–2738
29. Donovan RE, Eide EM (1998) The IBM trainable speech synthesis system. In: Proceedings of the ICSLP, pp 1703–1706
30. Beutnagel B, Conkie A, Schroeter J, Stylianou Y, Syrdal A (1999) The AT&T Next-Gen TTS system. In: Proceedings of the joint ASA, EAA and DAEA meeting, pp 15–19
31. Coorman G, Fackrell J, Rutten P, Coile B (2000) Segment selection in the L & H realspeak laboratory TTS system. In: Proceedings of the ICSLP, pp 395–398
32. Yoshimura T, Tokuda K, Masuko T, Kobayashi T, Kitamura T (1999) Simultaneous modeling of spectrum, pitch and duration in HMM-based speech synthesis. In: Proceedings of the Eurospeech, pp 2347–2350
33. Ling ZH, Wu YJ, Wang YP, Qin L, Wang RH (2006) USTC system for Blizzard Challenge 2006 an improved HMM-based speech synthesis method. In: Proceedings of the Blizzard challenge workshop
34. Black A (2006) CLUSTERGEN: a statistical parametric synthesizer using trajectory modeling. In: Proceedings of the Interspeech, pp 1762–1765
35. Zen H, Toda T, Nakamura M, Tokuda K (2007) Details of the Nitech HMM-based speech synthesis system for the Blizzard challenge 2005. IEICE Trans Inf Syst E90-D(1):325–333
36. Zen H, Tokuda K, Black AW (2009) Statistical parametric speech synthesis. Speech Commun 51(11):1039–1064
37. Yamagishi J, Kobayashi T, Nakano Y, Ogata K, Isogai J (2009) Analysis of speaker adaptation algorithms for HMM-based speech synthesis and a constrained SMAPLR adaptation algorithm. IEEE Trans Speech Audio Lang Process 17(1):66–83
38. Leggetter CJ, Woodland PC (1995) Maximum likelihood linear regression for speaker adaptation of continuous density hidden Markov models. Comput Speech Lang 9:171–185
39. Woodland PC (2001) Speaker adaptation for continuous density HMMs: a review. In: Proceedings of the ISCA workshop on adaptation methods for speech recognition, p 119
40. Ze H, Senior A, Schuster M (2013) Statistical parametric speech synthesis using deep neural networks. In: Proceedings of the ICASSP, pp 7962–7966
41. Ling ZH, Deng L, Yu D (2013) Modeling spectral envelopes using restricted boltzmann machines and deep belief networks for statistical parametric speech synthesis. IEEE Trans Audio Speech Lang Proc 21(10):2129–2139
42. Fan Y, Qian Y, Xie FL, Soong F (2014) TTS synthesis with bidirectional LSTM based recurrent neural networks. In: Proceedings of the interspeech, pp 1964–1968
43. Zen H, Sak H (2015) Unidirectional long short-term memory recurrent neural network with recurrent output layer for low-latency speech synthesis. In: Proceedings of the ICASSP, pp 4470–4474
44. Wu Z, King S (2016) Investigating gated recurrent networks for speech synthesis. In: Proceedings of the ICASSP, pp 5140–5144

45. Wang X, Takaki S, Yamagishi J (2016) Investigating very deep highway networks for parametric speech synthesis. In: 9th ISCA speech synthesis workshop, pp 166–171
46. Wang X, Takaki S, Yamagishi J (2018) Investigating very deep highway networks for parametric speech synthesis. Speech Commun 96:1–9
47. Wang X, Takaki S, Yamagishi J An autoregressive recurrent mixture density network for parametric speech synthesis. In: Proceedings of the ICASSP, pp 4895–4899
48. Wang X, Takaki S, Yamagishi J (2017) An RNN-based quantized F0 model with multi-tier feedback links for text-to-speech synthesis. In: Proceedings of the interspeech, pp 1059–1063
49. Saito Y, Takamichi S, Saruwatari H (2017) Training algorithm to deceive anti-spoofing verification for DNN-based speech synthesis. In: Proceedings of the ICASSP, pp 4900–4904
50. Saito Y, Takamichi S, Saruwatari H (2018) Statistical parametric speech synthesis incorporating generative adversarial networks. IEEE/ACM Trans Audio Speech Lang Proc 26(1):84–96
51. Kaneko T, Kameoka H, Hojo N, Ijima Y, Hiramatsu K, Kashino K (2017) Generative adversarial network-based postfilter for statistical parametric speech synthesis. In: Proceedings of the ICASSP, pp 4910–4914
52. Van Oord DA, Dieleman, S, Zen H, Simonyan K, Vinyals O, Graves A, Kalchbrenner N, Senior A, Kavukcuoglu K (2016) Wavenet: a generative model for raw audio. arXiv:1609.03499
53. Mehri S, Kumar K, Gulrajani I, Kumar R, Jain S, Sotelo J, Courville A, Bengio Y (2016) Samplernn: an unconditional end-to-end neural audio generation model. arXiv:1612.07837
54. Wang Y, Skerry-Ryan R, Stanton D, Wu Y, Weiss R, Jaitly N, Yang Z, Xiao Y, Chen Z, Bengio S, Le Q, Agiomyrgiannakis Y, Clark R, Saurous R (2017) Tacotron: towards end-to-end speech synthesis. In: Proceedings of the interspeech, pp 4006–4010
55. Gibiansky A, Arik S, Diamos G, Miller J, Peng K, Ping W, Raiman J, Zhou Y (2017) Deep voice 2: multi-speaker neural text-to-speech. In: Advances in neural information processing systems, pp 2966–2974
56. Shen J, Schuster M, Jaitly N, Skerry-Ryan R, Saurous R, Weiss R, Pang R, Agiomyrgiannakis Y, Wu Y, Zhang Y, Wang Y, Chen Z, Yang Z (2018) Natural tts synthesis by conditioning wavenet on mel spectrogram predictions. In: Proceedings of the ICASSP
57. King S (2014) Measuring a decade of progress in text-to-speech. Loquens 1(1):006
58. King S, Wihlborg L, Guo W (2017) The blizzard challenge 2017. In: Proceedings of the Blizzard Challenge Workshop, Stockholm, Sweden
59. Foomany F, Hirschfield A, Ingleby M (2009) Toward a dynamic framework for security evaluation of voice verification systems. In: IEEE Toronto international conference on science and technology for humanity (TIC-STH), 2009, pp 22–27
60. Masuko T, Hitotsumatsu T, Tokuda K, Kobayashi T (1999) On the security of HMM-based speaker verification systems against imposture using synthetic speech. In: Proceedings of the EUROSPEECH
61. Matsui T, Furui S (1995) Likelihood normalization for speaker verification using a phoneme- and speaker-independent model. Speech Commun 17(1–2):109–116
62. Masuko T, Tokuda K, Kobayashi T, Imai S (1996) Speech synthesis using HMMs with dynamic features. In: Proceedings of the ICASSP
63. Masuko T, Tokuda K, Kobayashi T, Imai S (1997) Voice characteristics conversion for HMM-based speech synthesis system. In: Proceedings of the ICASSP
64. De Leon PL, Pucher M, Yamagishi J, Hernaez I, Saratxaga I (2012) Evaluation of speaker verification security and detection of HMM-based synthetic speech. IEEE Trans Audio Speech Lang Process 20(8):2280–2290
65. Galou G (2011) Synthetic voice forgery in the forensic context: a short tutorial. In: Forensic speech and audio analysis working group (ENFSI-FSAAWG), pp 1–3
66. Cai W, Doshi A, Valle R (2018) Attacking speaker recognition with deep generative models. arXiv:1801.02384
67. Satoh T, Masuko T, Kobayashi T, Tokuda K (2001) A robust speaker verification system against imposture using an HMM-based speech synthesis system. In: Proceedings of the Eurospeech

68. Chen LW, Guo W, Dai LR (2010) Speaker verification against synthetic speech. In: 2010 7th international symposium on Chinese spoken language processing (ISCSLP), pp 309–312
69. Quatieri TF (2002) Discrete-Time speech signal processing: principles and practice. Prentice-Hall, Inc
70. Wu Z, Chng E, Li H (212) Detecting converted speech and natural speech for anti-spoofing attack in speaker recognition. In: Proceedings of the interspeech
71. Ogihara A, Unno H, Shiozakai A (2005) Discrimination method of synthetic speech using pitch frequency against synthetic speech falsification. IEICE Trans Fund Electron Commun Comput Sci 88(1):280–286
72. De Leon P, Stewart B, Yamagishi J (2012) Synthetic speech discrimination using pitch pattern statistics derived from image analysis. In: Proceedings of the interspeech 2012, Portland, Oregon, USA
73. Stylianou Y (2009) Voice transformation: a survey. In: Proceedings of the ICASSP, pp 3585–3588
74. Pellom B, Hansen J (1999) An experimental study of speaker verification sensitivity to computer voice-altered imposters. In: Proceedings of the ICASSP, vol 2, pp 837–840
75. Mohammadi S, Kain A (2017) An overview of voice conversion systems. Speech Commun 88:65–82
76. Abe M, Nakamura S, Shikano K, Kuwabara H (1988) Voice conversion through vector quantization. In: Proceedings of the ICASSP, pp 655–658
77. Arslan L (1999) Speaker transformation algorithm using segmental codebooks (STASC). Speech Commun 28(3):211–226
78. Kain A, Macon M (1998) Spectral voice conversion for text-to-speech synthesis. In: Proceedings of the ICASSP, vol 1, pp 285–288
79. Stylianou Y, Cappé O, Moulines E (1998) Continuous probabilistic transform for voice conversion. IEEE Trans Speech Audio Process 6(2):131–142
80. Toda T, Black A, Tokuda K (2007) Voice conversion based on maximum-likelihood estimation of spectral parameter trajectory. IEEE Trans Audio Speech Lang Process 15(8):2222–2235
81. Kobayashi K, Toda T, Neubig G, Sakti S, Nakamura S (2014) Statistical singing voice conversion with direct waveform modification based on the spectrum differential. In: Proceedings of the interspeech
82. Popa V, Silen H, Nurminen J, Gabbouj M (2012) Local linear transformation for voice conversion. In: Proceedings of the ICASSP. IEEE, pp 4517–4520
83. Chen Y, Chu M, Chang E, Liu J, Liu R (2003) Voice conversion with smoothed GMM and MAP adaptation. In: Proceedings of the EUROSPEECH, pp 2413–2416
84. Hwang HT, Tsao Y, Wang HM, Wang YR, Chen SH (2012) A study of mutual information for GMM-based spectral conversion. In: Proceedings of the interspeech
85. Helander E, Virtanen T, Nurminen J, Gabbouj M (2010) Voice conversion using partial least squares regression. IEEE Trans Audio Speech Lang Process 18(5):912–921
86. Pilkington N, Zen H, Gales M (2011) Gaussian process experts for voice conversion. In: Proceedings of the interspeech
87. Saito D, Yamamoto K, Minematsu N, Hirose K (2011) One-to-many voice conversion based on tensor representation of speaker space. In: Proceedings of the interspeech, pp 653–656
88. Zen H, Nankaku Y, Tokuda K (2011) Continuous stochastic feature mapping based on trajectory HMMs. IEEE Trans Audio Speech Lang Process 19(2):417–430
89. Wu Z, Kinnunen T, Chng E, Li H (2012) Mixture of factor analyzers using priors from non-parallel speech for voice conversion. IEEE Signal Process Lett 19(12)
90. Saito D, Watanabe S, Nakamura A, Minematsu N (2012) Statistical voice conversion based on noisy channel model. IEEE Trans Audio Speech Lang Process 20(6):1784–1794
91. Song P, Bao Y, Zhao L, Zou C (2011) Voice conversion using support vector regression. Electron Lett 47(18):1045–1046
92. Helander E, Silén H, Virtanen T, Gabbouj M (2012) Voice conversion using dynamic kernel partial least squares regression. IEEE Trans Audio Speech Lang Process 20(3):806–817

93. Wu Z, Chng E, Li H (2013) Conditional restricted boltzmann machine for voice conversion. In: The first IEEE China summit and international conference on signal and information processing (ChinaSIP). IEEE
94. Narendranath M, Murthy H, Rajendran S, Yegnanarayana B (1995) Transformation of formants for voice conversion using artificial neural networks. Speech Commun 16(2):207–216
95. Desai S, Raghavendra E, Yegnanarayana B, Black A, Prahallad K (2009) Voice conversion using artificial neural networks. In: Proceedings of the ICASSP. IEEE, pp 3893–3896
96. Saito Y, Takamichi S, Saruwatari H (2017) Voice conversion using input-to-output highway networks. IEICE Trans Inf Syst E10(08):1925–1928
97. Nakashika T, Takiguchi T, Ariki Y (2015) Voice conversion using RNN pre-trained by recurrent temporal restricted boltzmann machines. IEEE/ACM Trans Audio Speech Lang Process (TASLP) 23(3):580–587
98. Sun L, Kang S, Li K, Meng H (2015) Voice conversion using deep bidirectional long short-term memory based recurrent neural networks. In: Proceedings of the ICASSP, pp 4869–4873
99. Sundermann D, Ney H (2003) VTLN-based voice conversion. In: Proceedings of the 3rd IEEE international symposium on signal processing and information technology, 2003. ISSPIT 2003. IEEE, pp 556–559
100. Erro D, Moreno A, Bonafonte A (2010) Voice conversion based on weighted frequency warping. IEEE Trans Audio Speech Lang Process 18(5):922–931
101. Erro D, Navas E, Hernaez I (2013) Parametric voice conversion based on bilinear frequency warping plus amplitude scaling. IEEE Trans Audio Speech Lang Process 21(3):556–566
102. Hsu CC, Hwang HT, Wu YC, Tsao Y, Wang HM (2017) Voice conversion from unaligned corpora using variational autoencoding wasserstein generative adversarial networks. In: Proceedings of the interspeech 2017, pp 3364–3368
103. Miyoshi H, Saito Y, Takamichi S, Saruwatari H (2017) Voice conversion using sequence-to-sequence learning of context posterior probabilities. In: Proceedings of the interspeech 2017, pp 1268–1272
104. Fang F, Yamagishi J, Echizen I, Lorenzo-Trueba J (2018) High-quality nonparallel voice conversion based on cycle-consistent adversarial network. In: Proceedings of the ICASSP 2018
105. Kobayashi K, Hayashi T, Tamamori A, Toda T (2017) Statistical voice conversion with wavenet-based waveform generation. In: Proceedings of the interspeech, pp 1138–1142
106. Gillet B, King S (2003) Transforming F0 contours. In: Proceedings of the EUROSPEECH, pp 101–104
107. Wu CH, Hsia CC, Liu TH, Wang JF (2006) Voice conversion using duration-embedded bi-HMMs for expressive speech synthesis. IEEE Trans Audio Speech Lang Process 14(4):1109–1116
108. Helander E, Nurminen J (2007) A novel method for prosody prediction in voice conversion. In: Proceedings of the ICASSP, vol 4. IEEE, pp IV–509
109. Wu Z, Kinnunen T, Chng E, Li H (2010) Text-independent F0 transformation with non-parallel data for voice conversion. In: Proceedings of the interspeech
110. Lolive D, Barbot N, Boeffard O (2008) Pitch and duration transformation with non-parallel data. Speech Prosody 2008:111–114
111. Toda T, Chen LH, Saito D, Villavicencio F, Wester M, Wu Z, Yamagishi J (2016) The voice conversion challenge 2016. In: Proceedings of the interspeech, pp 1632–1636
112. Wester M, Wu Z, Yamagishi J (2016) Analysis of the voice conversion challenge 2016 evaluation results. In: Proceedings of the interspeech, pp 1637–1641
113. Perrot P, Aversano G, Blouet R, Charbit M, Chollet G (2005) Voice forgery using ALISP: indexation in a client memory. In: Proceedings of the ICASSP, vol 1. IEEE, pp 17–20
114. Matrouf D, Bonastre JF, Fredouille C (2006) Effect of speech transformation on impostor acceptance. In: Proceedings of the ICASSP, vol 1. IEEE, pp I–I
115. Kinnunen T, Wu Z, Lee K, Sedlak F, Chng E, Li H (2012) Vulnerability of speaker verification systems against voice conversion spoofing attacks: The case of telephone speech. In: Proceedings of the ICASSP. IEEE, pp 4401–4404

116. Sundermann D, Hoge H, Bonafonte A, Ney H, Black A, Narayanan S (2006) Text-independent voice conversion based on unit selection. In: Proceedings of the ICASSP, vol 1, pp I–I
117. Wu Z, Larcher A, Lee K, Chng E, Kinnunen T, Li H (2013) Vulnerability evaluation of speaker verification under voice conversion spoofing: the effect of text constraints. In: Proceedings of the interspeech, Lyon, France
118. Alegre F, Vipperla R, Evans N, Fauve B (2012) On the vulnerability of automatic speaker recognition to spoofing attacks with artificial signals. In: 2012 EURASIP conference on European conference on signal processing (EUSIPCO)
119. De Leon PL, Hernaez I, Saratxaga I, Pucher M, Yamagishi J (2011) Detection of synthetic speech for the problem of imposture. In: Proceedings of the ICASSP, Dallas, USA, pp 4844–4847
120. Wu Z, Kinnunen T, Chng E, Li H, Ambikairajah E (2012) A study on spoofing attack in state-of-the-art speaker verification: the telephone speech case. In: Proceedings of the Asia-Pacific signal information processing association annual summit and conference (APSIPA ASC), pp 1–5. IEEE
121. Alegre F, Vipperla R, Evans N (2012) Spoofing countermeasures for the protection of automatic speaker recognition systems against attacks with artificial signals. In: Proceedings of the interspeech
122. Alegre F, Amehraye A, Evans N (2013) Spoofing countermeasures to protect automatic speaker verification from voice conversion. In: Proceedings of the ICASSP
123. Wu Z, Xiao X, Chng E, Li H (2013) Synthetic speech detection using temporal modulation feature. In: Proceedings of the ICASSP
124. Alegre F, Vipperla R, Amehraye A, Evans N (2013) A new speaker verification spoofing countermeasure based on local binary patterns. In: Proceedings of the interspeech, Lyon, France
125. Kinnunen T, Lee K, Delgado H, Evans N, Todisco M, Sahidullah M, Yamagishi J, Reynolds DA (2018) t-DCF: a detection cost function for the tandem assessment of spoofing countermeasures and automatic speaker verification. In: Proceedings of the Odyssey, Les Sables d'Olonne, France
126. Kinnunen T, Delgado H, Evans N, Lee KA, Vestman V, Nautsch A, Todisco M, Wang X, Sahidullah M, Yamagishi J et al (2020) Tandem assessment of spoofing countermeasures and automatic speaker verification: fundamentals. IEEE/ACM Trans Audio Speech Lang Process 28:2195–2210
127. Wu Z, Kinnunen T, Evans N, Yamagishi J, Hanilçi C, Sahidullah M, Sizov A (2015) ASVspoof 2015: the first automatic speaker verification spoofing and countermeasures challenge. In: Proceedings of the interspeech
128. Kinnunen T, Sahidullah M, Delgado H, Todisco M, Evans N, Yamagishi J, Lee K (2017) The ASVspoof 2017 challenge: assessing the limits of replay spoofing attack detection. In: Interspeech
129. Nautsch A, Wang X, Evans N, Kinnunen TH, Vestman V, Todisco M, Delgado H, Sahidullah M, Yamagishi J, Lee KA (2021) Asvspoof 2019: spoofing countermeasures for the detection of synthesized, converted and replayed speech. IEEE Trans Biomet Behav Identity Sci 3(2):252–265
130. Wang X, Yamagishi J, Todisco M, Delgado H, Nautsch A, Evans N, Sahidullah M, Vestman V, Kinnunen T, Lee KA, Juvela L, Alku P, Peng YH, Hwang HT, Tsao Y, Wang HM, Maguer SL, Becker M, Henderson F, Clark R, Zhang Y, Wang Q, Jia Y, Onuma K, Mushika K, Kaneda T, Jiang Y, Liu LJ, Wu YC, Huang WC, Toda T, Tanaka K, Kameoka H, Steiner I, Matrouf D, Bonastre JF, Govender A, Ronanki S, Zhang JX, Ling ZH (2020) Asvspoof 2019: a large-scale public database of synthesized, converted and replayed speech. Comput Speech Lang 64:101–114
131. Wu Z, Khodabakhsh A, Demiroglu C, Yamagishi J, Saito D, Toda T, King S (2015) SAS: a speaker verification spoofing database containing diverse attacks. In: Proceedings of the IEEE international conferences on acoustics, speech, and signal processing (ICASSP)

132. Wu Z, Kinnunen T, Evans N, Yamagishi J (2014) ASVspoof 2015: automatic speaker verification spoofing and countermeasures challenge evaluation plan. http://www.spoofingchallenge.org/asvSpoof.pdf
133. Patel T, Patil H (2015) Combining evidences from mel cepstral, cochlear filter cepstral and instantaneous frequency features for detection of natural versus spoofed speech. In: Proceedings of the interspeech
134. Novoselov S, Kozlov A, Lavrentyeva G, Simonchik K, Shchemelinin V (2016) STC anti-spoofing systems for the ASVspoof 2015 challenge. In: Proceedings of the IEEE international conferences on acoustics, speech, and signal processing (ICASSP), pp 5475–5479
135. Chen N, Qian Y, Dinkel H, Chen B, Yu K (2015) Robust deep feature for spoofing detection-the SJTU system for ASVspoof 2015 challenge. In: Proceedings of the interspeech
136. Xiao X, Tian X, Du S, Xu H, Chng E, Li H (2015) Spoofing speech detection using high dimensional magnitude and phase features: the NTU approach for ASVspoof 2015 challenge. In: Proceedings of the Interspeech (2015)
137. Alam M, Kenny P, Bhattacharya G, Stafylakis T (2015) Development of CRIM system for the automatic speaker verification spoofing and countermeasures challenge 2015. In: Proceedings of the interspeech
138. Wu Z, Yamagishi J, Kinnunen T, Hanilçi C, Sahidullah M, Sizov A, Evans N, Todisco M, Delgado H (2017) Asvspoof: the automatic speaker verification spoofing and countermeasures challenge. IEEE J Sel Top Signal Process 11(4):588–604
139. Delgado H, Todisco M, Sahidullah M, Evans N, Kinnunen T, Lee K, Yamagishi J (2018) ASVspoof 2017 version 2.0: meta-data analysis and baseline enhancements. In: Proceedings of the Odyssey 2018 the speaker and language recognition workshop, pp 296–303
140. Todisco M, Delgado H, Evans N (2016) A new feature for automatic speaker verification anti-spoofing: constant Q cepstral coefficients. In: Proceedings of the Odyssey: the speaker and language recognition workshop, Bilbao, Spain, pp 283–290
141. Todisco M, Delgado H, Evans N (2017) Constant Q cepstral coefficients: a spoofing countermeasure for automatic speaker verification. Comput Speech Lang 45:516–535
142. Lavrentyeva G, Novoselov S, Malykh E, Kozlov A, Kudashev O, Shchemelinin V (2017) Audio replay attack detection with deep learning frameworks. In: Proceedings of the interspeech, pp 82–86
143. Ji Z, Li Z, Li P, An M, Gao S, Wu D, Zhao F (2017) Ensemble learning for countermeasure of audio replay spoofing attack in ASVspoof2017. In: Proceedings interspeech, pp 87–91
144. Li L, Chen Y, Wang D, Zheng T (2017) A study on replay attack and anti-spoofing for automatic speaker verification. In: Proceedings of the interspeech, pp 92–96
145. Patil H, Kamble M, Patel T, Soni M (2017) Novel variable length teager energy separation based instantaneous frequency features for replay detection. In: Proceedings of the interspeech, pp 12–16
146. Chen Z, Xie Z, Zhang W, Xu X (2017) ResNet and model fusion for automatic spoofing detection. In: Proceedings of the interspeech, pp 102–106
147. CSTR VCTK Corpus: English Multi-speaker Corpus for CSTR Voice Cloning Toolkit. https://doi.org/10.7488/ds/1994. Accessed 3 Sept 2019
148. Yamagishi J, Todisco M, Sahidullah M, Delgado H, Wang X, Evans N, Kinnunen T, Lee KA, Vestman V, Nautsch A (2019) ASVspoof 2019: Automatic speaker verification spoofing and countermeasures challenge evaluation plan. Technical report, ASVspoof Consortium
149. Zen H, Senior A, Schuster M (2013) Statistical parametric speech synthesis using deep neural networks. In: Proceedings of the ICASSP, pp 7962–7966
150. Morise M, Yokomori F, Ozawa K (2016) WORLD: a vocoder-based high-quality speech synthesis system for real-time applications. IEICE Trans Inf Syst 99(7):1877–1884
151. Wu Z, Watts O, King S (2016) Merlin: An open source neural network speech synthesis system. In: Speech synthesis workshop SSW 2016
152. Schröder M, Charfuelan M, Pammi S, Steiner I (2011) Open source voice creation toolkit for the MARY TTS platform. In: Proceedings of the interspeech, pp 3253–3256

153. Steiner I, Le Maguer S (2018) Creating new language and voice components for the updated MaryTTS text-to-speech synthesis platform. In: 11th language resources and evaluation conference (LREC), Miyazaki, Japan, pp 3171–3175
154. Hsu CC, Hwang HT, Wu YC, Tsao Y, Wang HM (2016) Voice conversion from non-parallel corpora using variational auto-encoder. In: 2016 Asia-Pacific signal and information processing association annual summit and conference (APSIPA). IEEE, pp 1–6
155. Matrouf D, Bonastre J, Fredouille C (2006) Effect of speech transformation on impostor acceptance. In: 2006 IEEE international conference on acoustics speech and signal processing proceedings, vol 1, pp I–I
156. Wang X, Takaki S, Yamagishi J (2019) Neural source-filter-based waveform model for statistical parametric speech synthesis. In: ICASSP 2019–2019 IEEE international conference on acoustics, speech and signal processing (ICASSP), pp 5916–5920
157. Zen H, Agiomyrgiannakis Y, Egberts N, Henderson F, Szczepaniak P (2016) Fast, compact, and high quality lstm-rnn based statistical parametric speech synthesizers for mobile devices. In: Proceedings of the interspeech, pp 2273–2277
158. Jia Y, Zhang Y, Weiss R, Wang Q, Shen J, Ren F, Nguyen P, Pang R, Moreno IL, Wu Y, et al (2018) Transfer learning from speaker verification to multispeaker text-to-speech synthesis. In: Advances in neural information processing systems, pp 4480–4490
159. Shen J, Pang R, Weiss RJ, Schuster M, Jaitly N, Yang Z, Chen Z, Zhang Y, Wang Y, Skerrv-Ryan R et al (2018) Natural tts synthesis by conditioning wavenet on mel spectrogram predictions. In: 2018 IEEE international conference on acoustics, speech and signal processing (ICASSP), pp 4779–4783. IEEE
160. Griffin DW, Lim JS (1984) Signal estimation from modified short-time Fourier transform. IEEE Trans Acoust Speech Signal Process 32(2):236–243
161. Huang WC, Wu YC, Kobayashi K, Peng YH, Hwang HT, Lumban Tobing P, Tsao Y, Wang HM, Toda T (2019) Generalization of spectrum differential based direct waveform modification for voice conversion. In: Proceedings of the SSW10
162. Kobayashi K, Toda T, Nakamura S (2018) Intra-gender statistical singing voice conversion with direct waveform modification using log-spectral differential. Speech Commun 99:211–220
163. Kinnunen T, Juvela L, Alku P, Yamagishi J (2017) Non-parallel voice conversion using i-vector PLDA: towards unifying speaker verification and transformation. In: 2017 IEEE international conference on acoustics, speech and signal processing (ICASSP), pp 5535–5539
164. Campbell DR, Palomäki KJ, Brown G (2005) A MATLAB simulation of "shoebox" room acoustics for use in research and teaching. Comput Inf Syst J 9(3). ISSN 1352-9404
165. Vincent E (2008) Roomsimove. http://homepages.loria.fr/evincent/software/Roomsimove_1.4.zip
166. Novak A, Lotton P, Simon L (2015) Synchronized swept-sine: theory, application, and implementation. J Audio Eng Soc 63(10):786–798. http://www.aes.org/e-lib/browse.cfm?elib=18042
167. Todisco M, Wang X, Vestman V, Sahidullah M, Delgado H, Nautsch A, Yamagishi J, Evans N, Kinnunen TH, Lee KA (2019) ASVspoof 2019: future horizons in spoofed and fake audio detection. In: Proceedings of the interspeech, pp 1008–1012
168. Wu Z, Gao S, Cling E, Li H (2014) A study on replay attack and anti-spoofing for text-dependent speaker verification. In: Proceedings of the Asia-Pacific signal information processing association annual summit and conference (APSIPA ASC), pp 1–5. IEEE
169. Li Q (2009) An auditory-based transform for audio signal processing. In: 2009 IEEE workshop on applications of signal processing to audio and acoustics. IEEE, pp 181–184
170. Davis S, Mermelstein P (1980) Comparison of parametric representations for monosyllabic word recognition in continuously spoken sentences. IEEE Trans Acoust Speech Signal Process 28(4):357–366
171. Sahidullah M, Kinnunen T, Hanilçi C (2015) A comparison of features for synthetic speech detection. In: Proceedings of the interspeech. ISCA, pp 2087–2091

172. Brown J (1991) Calculation of a constant Q spectral transform. J Acoust Soc Am 89(1):425–434
173. Alam M, Kenny P (2017) Spoofing detection employing infinite impulse response - constant Q transform-based feature representations. In: Proceedings of the European signal processing conference (EUSIPCO)
174. Cancela P, Rocamora M, López E (2009) An efficient multi-resolution spectral transform for music analysis. In: Proceedings of the international society for music information retrieval conference, pp 309–314
175. Bengio Y (2009) Learning deep architectures for AI. Found Trends Mach Learn 2(1):1–127
176. Goodfellow I, Bengio Y, Courville A, Bengio Y (2016) Deep learning. MIT Press, Cambridge
177. Tian Y, Cai M, He L, Liu J (2015) Investigation of bottleneck features and multilingual deep neural networks for speaker verification. In: Proceedings of the interspeech, pp 1151–1155
178. Richardson F, Reynolds D, Dehak N (2015) Deep neural network approaches to speaker and language recognition. IEEE Signal Process Lett 22(10):1671–1675
179. Hinton G, Deng L, Yu D, Dahl GE, Mohamed RA, Jaitly N, Senior A, Vanhoucke V, Nguyen P, Sainath TN, Kingsbury B (2012) Deep neural networks for acoustic modeling in speech recognition: the shared views of four research groups. IEEE Signal Process Mag 29(6):82–97
180. Alam M, Kenny P, Gupta V, Stafylakis T (2016) Spoofing detection on the ASVspoof2015 challenge corpus employing deep neural networks. In: Proceedings of the Odyssey: the speaker and language recognition workshop, Bilbao, Spain, pp 270–276
181. Qian Y, Chen N, Yu K (2016) Deep features for automatic spoofing detection. Speech Commun 85:43–52
182. Yu H, Tan ZH, Zhang Y, Ma Z, Guo J (2017) DNN filter bank cepstral coefficients for spoofing detection. IEEE Access 5:4779–4787
183. Sriskandaraja K, Sethu V, Ambikairajah E, Li H (2017) Front-end for antispoofing countermeasures in speaker verification: scattering spectral decomposition. IEEE J Sel Topi Signal Process 11(4):632–643. https://doi.org/10.1109/JSTSP.2016.2647202
184. Andén J, Mallat S (2014) Deep scattering spectrum. IEEE Trans Signal Process 62(16):4114–4128
185. Mallat S (2012) Group invariant scattering. Commun Pure Appl Math 65:1331–1398
186. Pal M, Paul D, Saha G (2018) Synthetic speech detection using fundamental frequency variation and spectral features. Comput Speech Lang 48:31–50
187. Laskowski K, Heldner M, Edlund J (2008) The fundamental frequency variation spectrum. In: Proceedings of fonetik, vol 2008, pp 29–32
188. Saratxaga I, Sanchez J, Wu Z, Hernaez I, Navas E (2016) Synthetic speech detection using phase information. Speech Commun 81:30–41
189. Wang L, Nakagawa S, Zhang Z, Yoshida Y, Kawakami Y (2017) Spoofing speech detection using modified relative phase information. IEEE J Sel Top Signal Process 11(4):660–670
190. Chakroborty S, Saha G (2009) Improved text-independent speaker identification using fused MFCC & IMFCC feature sets based on Gaussian filter. In J Signal Process 5(1):11–19
191. Wu X, He R, Sun Z, Tan T (2018) A light CNN for deep face representation with noisy labels. IEEE Trans Inf Forens Secur 13(11):2884–2896
192. Goncalves AR, Violato RPV, Korshunov P, Marcel S, Simoes FO (2017) On the generalization of fused systems in voice presentation attack detection. In: 2017 international conference of the biometrics special interest group (BIOSIG), pp 1–5. https://doi.org/10.23919/BIOSIG.2017.8053516
193. Paul D, Pal M, Saha G (2016) Novel speech features for improved detection of spoofing attacks. In: Proceedings of the annual IEEE India conference (INDICON)
194. Dehak N, Kenny P, Dehak R, Dumouchel P, Ouellet P (2011) Front-end factor analysis for speaker verification. IEEE Trans Audio Speech Lang Process 19(4):788–798
195. Khoury E, Kinnunen T, Sizov A, Wu Z, Marcel S (2014) Introducing i-vectors for joint anti-spoofing and speaker verification. In: Proceedings of the interspeech
196. Sizov A, Khoury E, Kinnunen T, Wu Z, Marcel S (2015) Joint speaker verification and antispoofing in the i-vector space. IEEE Trans Inf Forens Secur 10(4):821–832

197. Hanilçi C (2018) Data selection for i-vector based automatic speaker verification anti-spoofing. Digit Signal Process 72:171–180
198. Tian X, Wu Z, Xiao X, Chng E, Li H (2016) Spoofing detection from a feature representation perspective. In: Proceedings of the IEEE international conference on acoustics, speech, and signal processing (ICASSP), pp 2119–2123
199. Yu H, Tan ZH, Ma Z, Martin R, Guo J (2018) Spoofing detection in automatic speaker verification systems using DNN classifiers and dynamic acoustic features. IEEE Trans Neural Netw Learn Syst PP(99):1–12
200. Dinkel H, Chen N, Qian Y, Yu K (2017) End-to-end spoofing detection with raw waveform cldnns. In: Proceedings of the IEEE international conferences on acoust speech signal process (ICASSP), pp 4860–4864
201. Sainath T, Weiss R, Senior A, Wilson K, Vinyals O (2015) Learning the speech front-end with raw waveform CLDNNs. In: Proceedings of the interspeech
202. Zhang C, Yu C, Hansen JHL (2017) An investigation of deep-learning frameworks for speaker verification antispoofing. IEEE J Sel Top Signal Process 11(4):684–694
203. Muckenhirn H, Magimai-Doss M, Marcel S (2017) End-to-end convolutional neural network-based voice presentation attack detection. In: 2017 IEEE international joint conference on biometrics (IJCB), pp 335–341
204. Chen S, Ren K, Piao S, Wang C, Wang Q, Weng J, Su L, Mohaisen A (2017) You can hear but you cannot steal: defending against voice impersonation attacks on smartphones. In: 2017 IEEE 37th international conference on distributed computing systems (ICDCS), pp 183–195. IEEE
205. Shiota S, Villavicencio F, Yamagishi J, Ono N, Echizen I, Matsui T (2015) Voice liveness detection algorithms based on pop noise caused by human breath for automatic speaker verification. In: Proceedings of the interspeech
206. Shiota S, Villavicencio F, Yamagishi J, Ono N, Echizen I, Matsui T (2016) Voice liveness detection for speaker verification based on a tandem single/double-channel pop noise detector. In: Odyssey
207. Sahidullah M, Thomsen D, Hautamäki R, Kinnunen T, Tan ZH, Parts R, Pitkänen M (2018) Robust voice liveness detection and speaker verification using throat microphones. IEEE/ACM Trans Audio Speech Lang Process 26(1):44–56
208. Elko G, Meyer J, Backer S, Peissig J (2007) Electronic pop protection for microphones. In: 2007 IEEE workshop on applications of signal processing to audio and acoustics, pp 46–49. IEEE
209. Zhang L, Tan S, Yang J, Chen Y (2016) Voicelive: a phoneme localization based liveness detection for voice authentication on smartphones. In: Proceedings of the 2016 ACM SIGSAC conference on computer and communications security, pp 1080–1091. ACM
210. Zhang L, Tan S, Yang J (2017) Hearing your voice is not enough: an articulatory gesture based liveness detection for voice authentication. In: Proceedings of the 2017 ACM SIGSAC conference on computer and communications security. ACM, pp 57–71
211. Hanilçi C, Kinnunen T, Sahidullah M, Sizov A (2016) Spoofing detection goes noisy: an analysis of synthetic speech detection in the presence of additive noise. Speech Commun 85:83–97
212. Yu H, Sarkar A, Thomsen D, Tan ZH, Ma Z, Guo J (2016) Effect of multi-condition training and speech enhancement methods on spoofing detection. In: Proceedings of the international workshop on sensing, processing and learning for intelligent machines (SPLINE)
213. Tian X, Wu Z, Xiao X, Chng E, Li H (2016) An investigation of spoofing speech detection under additive noise and reverberant conditions. In: Proceedings of the interspeech
214. Delgado H, Todisco M, Evans N, Sahidullah M, Liu W, Alegre F, Kinnunen T, Fauve B (2017) Impact of bandwidth and channel variation on presentation attack detection for speaker verification. In: 2017 international conference of the biometrics special interest group (BIOSIG), pp 1–6
215. Qian Y, Chen N, Dinkel H, Wu Z (2017) Deep feature engineering for noise robust spoofing detection. IEEE/ACM Trans Audio Speech Lang Process 25(10):1942–1955

216. Korshunov P, Marcel S (2016) Cross-database evaluation of audio-based spoofing detection systems. In: Proceedings of the interspeech
217. Paul D, Sahidullah M, Saha G (2017) Generalization of spoofing countermeasures: a case study with ASVspoof2015 and BTAS 2016 corpora. In: Proceedings of the IEEE international conferences on acoustics, speech, and signal processing (ICASSP). IEEE, pp 2047–2051
218. Lorenzo-Trueba J, Fang F, Wang X, Echizen I, Yamagishi J, Kinnunen T (2018) Can we steal your vocal identity from the Internet?: initial investigation of cloning Obama's voice using GAN, WaveNet and low-quality found data. In: Proceedings of the Odyssey: the speaker and language recognition workshop
219. Goodfellow I, Pouget-Abadie J, Mirza M, Xu B, Warde-Farley D, Ozair S, Courville A, Bengio Y (2014) Generative adversarial nets. In: Advances in neural information processing systems, pp 2672–2680
220. Kreuk F, Adi Y, Cisse M, Keshet J (2018) Fooling end-to-end speaker verification by adversarial examples. arXiv:1801.03339
221. Sahidullah M, Delgado H, Todisco M, Yu H, Kinnunen T, Evans N, Tan ZH (2016) Integrated spoofing countermeasures and automatic speaker verification: an evaluation on ASVspoof 2015. In: Proceedings of the interspeech
222. Muckenhirn H, Korshunov P, Magimai-Doss M, Marcel S (2017) Long-term spectral statistics for voice presentation attack detection. IEEE/ACM Trans Audio Speech Lang Process 25(11):2098–2111
223. Sarkar A, Sahidullah M, Tan ZH, Kinnunen T (2017) Improving speaker verification performance in presence of spoofing attacks using out-of-domain spoofed data. In: Proceedings of the interspeech
224. Kinnunen T, Lee K, Delgado H, Evans N, Todisco M, Sahidullah M, Yamagishi J, Reynolds D (2018) t-DCF: a detection cost function for the tandem assessment of spoofing countermeasures and automatic speaker verification. In: Proceedings of the Odyssey: the speaker and language recognition workshop
225. Todisco M, Delgado H, Lee K, Sahidullah M, Evans N, Kinnunen T, Yamagishi J (2018) Integrated presentation attack detection and automatic speaker verification: common features and Gaussian back-end fusion. In: Proceedings of the interspeech
226. Wu Z, De Leon P, Demiroglu C, Khodabakhsh A, King S, Ling ZH, Saito D, Stewart B, Toda T, Wester M, Yamagishi Y (2016) Anti-spoofing for text-independent speaker verification: an initial database, comparison of countermeasures, and human performance. IEEE/ACM Trans Audio Speech Lang Process 24(4):768–783

Chapter 14
A One-class Model for Voice Replay Attack Detection

Xingliang Cheng, Lantian Li, Mingxing Xu, Dong Wang, and Thomas Fang Zheng

Abstract Replay attack poses a serious security concern for automatic speaker verification systems. Most of the existing replay detection methods cast the task to a binary classification problem. In this article, by analyzing distributions of genuine and replayed speech with a specifically designed database and summarizing the known artifacts in existing datasets, we show the potential shortcomings of the two-class approach in both discrimination and generalization, and discuss the advantage of the one-class approach. As a demonstration, we present our recent investigation on a novel one-class-based replay detection method, which models the discrepancy between the test speech and the enrollment speech by a Gaussian mixture model, thus casting the replay detection task to an out-of-distribution detection task.

14.1 Introduction

Automatic speaker verification (ASV) aims to identify a person by speech signals [1–4]. ASV has been regarded as a new generation biometric identification technique, due to its features of easy use, non-intrusive, non-touching, and low privacy leakage. Recently, ASV has found broad applications, in particular those requiring remote information access [5].

X. Cheng (✉) · L. Li (✉) · M. Xu (✉) · D. Wang (✉) · T. F. Zheng (✉)
Tsinghua University, Beijing, China
e-mail: chengxl16@mails.tsinghua.edu.cn

L. Li
e-mail: lilt@tsinghua.edu.cn

M. Xu
e-mail: xumx@tsinghua.edu.cn

D. Wang
e-mail: wangdong99@mails.tsinghua.edu.cn

T. F. Zheng
e-mail: fzheng@tsinghua.edu.cn

S. Marcel et al. (eds.), *Handbook of Biometric Anti-Spoofing*, Advances in Computer Vision and Pattern Recognition, https://doi.org/10.1007/978-981-19-5288-3_14

In the meantime, the security of ASV systems has attracted much attention. There are mainly four kinds of attacks for ASV systems [6]: speaker imitation, voice conversion, speech synthesis, and recording and replay (R&R). Recently, the adversarial attack has emerged to be a new risk and attracted much research effort [7, 8]. In this chapter, we focus on the replay attack: recording the voice of the target user and then conducting the authentication by replaying the recorded speech with a loudspeaker [9]. Replay attack poses a serious practical risk to ASV systems for several reasons. Firstly, it is very easy to conduct, just recording and replay; Secondly, ASV systems are highly vulnerable to replayed speech, which has been reported by several authors [9, 10]; Thirdly, the quality of recording and playback devices are getting increasingly exquisite, making replay attack more and more serious.

To alleviate the risk of replay attack, a widely used approach is to employ the fact that humans can respond to queries dynamically. For example, the challenge-based approach [11] asks users to answer a random question, so that the recording and replay become more difficult. The widely adopted prompt-based system follows this idea, which asks users to repeat a short and simple string of some characters (e.g., digits) that are generated by the system randomly [12]. Another approach employs the intrinsic randomness of human speech. For example, the template-matching approach [13] checks the discrepancy between the test speech and the enrollment speech, and if the test speech is a replay of the enrollment speech, the discrepancy between the two speech signals would be less significant. These high-level approaches are application-specific and do not offer a principled and general solution to replay attacks. For example, if one has recorded the pronunciation of all the possible digits of the target speaker, then a simple concatenation method can deceive the prompt-based system.

A more interesting way is to *directly* detect whether a speech segment is from a real person (genuine) or a replayed version. This task is denoted by *replay detection*. Ideally, we hope there are some special patterns in the replayed speech signals, and these patterns can be exploited to distinguish genuine and replayed speech, by feature engineering and/or statistical modeling. Following this idea, numerous research has been conducted. Regarding the feature approach, various features have been demonstrated promising, including constant-Q cepstral coefficients (CQCC) [14], Mel frequency cepstral coefficients (MFCC), inverted Mel frequency cepstral coefficients (IMFCC) [15], and linear frequency cepstral coefficients (LFCC) [16]. The above cepstral features use carefully designed filters with different distributions or weights, however, the adaptive spectro-temporal resolution cepstral coefficients (AST-CC) [17] can adaptively scale the filters based on F-ratio analysis on the dataset. Other researchers focused on the shape of the filter, such as the rectangular frequency cepstral coefficients (RFCC) [18].

In addition to these magnitude features, phase-based features were also proposed, such as the modified group delay (MGD) function [19] and relative phase shift (RPS) feature [20]. Other features concern instant energy or instant phase, including those based on the Teager energy operator (TEO) [21], variable-length Teager energy operator [22], enhanced Teager energy operator [23], Hilbert transform (HT) [21], and energy separation algorithm (ESA) [21, 22]. Moreover, features based on the linear prediction analysis are also popular, such as the linear prediction cepstral coefficients (LPCC) [16], and the linear prediction residual (LPR) based features [24].

Regarding the modeling approach, the Gaussian mixture model (GMM) is among the most popular ones. Recently, more and more research focuses on deep neural nets. For instance, the light convolutional neural networks (LCNN) model [25] and ResNet model [26] have shown great potential. Other architectures were also be studied, such as Res2Net [27], DenseNet [28], LSTM [28], and attention-based neural model [29–31]. To perform the end-to-end learning, the models which can take the raw waveform as input were studied, such as SincNet [32–34] and RW-ResNet [35]. To learn an utterance-level representation, various pooling methods were studied, such as global averaged pooling (GAP) [26, 27], mean and standard deviation pooling [36], learnable dictionary encoding (LDE) [36], and attentive statistic pooling (ASP) [37]. Moreover, differentiable architecture search was also studied in [38, 39]. In addition to structure, many loss functions have been studied, such as large margin cosine loss [36], additive margin softmax (AM-softmax) loss [40], one-class softmax (OC-softmax) loss [40, 41], probability-to-similarity gradient (P2SGrad) loss [40, 42]. It is to be noticed that some of the methods mentioned above were designed for synthesis speech detection. However, it can also be used in replay detection since those architectures/losses are generic. More details about the recent progress on replay detection can be found in [4].

The ASVspoof challenge is perhaps the most popular benchmark for the ASV spoofing and anti-spoofing research [43–46]. Early challenges focused on attacks by speech synthesis and voice conversion [43]. Since 2017, replay detection became a separate track of the ASVspoof challenge series [44–46], in response to the increased popularity of ASV systems and the high risk of replay attack in practical applications. Thanks to this benchmark, numerous novel techniques have been fostered, including those have been mentioned above. Among these techniques, some were found to be highly effective, for example, the widely adopted CQCC feature [14].

In spite of the great success of the ASVspoof challenges, the results published by the participants have caused too many puzzles. In particular, techniques that have been demonstrated to be effective in one year may totally fail in another year. Especially in the latest ASVspoof 2021 challenge where only evaluation data were released, all the participants got much worse equal error rates (EERs) compared to the results released in the 2017 and 2019 challenges. Moreover, some quite effective models, especially those based on deep neural nets (DNNs), were found overfitting to some artifacts [47]. All these observations caused serious doubt about the existing techniques.

In this chapter, we firstly conduct a deep analysis of the distributions of genuine and replayed speech with a specifically designed database in Sects. 14.2 and 14.3, and then summarize the known artifacts in existing datasets in Sect. 14.4. Based on these analyses, we show the potential shortcomings of the two-class approach in both discrimination and generalization, and advocate the one-class approach to casting the replay detection task to an out-of-distribution detection task. As a demonstration, we present a novel one-class model based on enrollment-test divergence features for replay detection in Sect. 14.5. Experimental results show that our proposed one-class model can achieve promising results and demonstrates the great potential of the one-class approach.

14.2 PRAD: Dataset for Replay Analysis

Our primary goal is to analyze the properties of the recording and replay process. To support that analysis, we need a dataset that contains detailed information of the R&R process, including the important factors that we want to analyze, and the environmental settings. The important factors concerned in this study include (1) the speaker; (2) the ASV microphone (ASV Mic); (3) the attacker microphone (Att Mic); (4) the attacker replay device (Att Replay). The environmental settings include the acoustic ambient and the relative positions between speakers and devices. For genuine speech, it is the relative position between speakers and ASV Mics, and for replayed speech, it involves the relative position between speakers and Att Mics, as well as the relative position between Att Replays and ASV Mics.

So far, no existing datasets contain such detailed information. For example, the BTAS [48] dataset contained a limited number of ASV Mics and limited replay settings. The ASVspoof 2017 dataset [49] did not provide labels of ASV Mics (for genuine speech) or Att Mics (for replayed speech). The ASVspoof 2019 dataset [50] was constructed by simulation and only categorical labels of the simulation were provided, without detailed device information. The ASVspoof 2021 challenge [46] did not provide any training or development data, and the labels for the test data have not been released so far.

To support the research purpose, we constructed a new dataset called Parallel Replay Attack Digit (PRAD). The construction follows two principles: full combinatory and limited variation. For full combinatory, we mean a full coverage of the combinations of four important factors: speaker, ASV Mic, Att Mic, and Att Replay. Each combination of these four factors forms a *config*, and we intentionally let the dataset cover all possible configs. This full combinatory allows us to draw a complete picture of the impact of these factors. Moreover, the full combinatory prevents biased analysis caused by spurious correlations among the concerned factors. For example, if each speaker is recorded by only one particular Att Mic, there will be a spurious correlation between speaker and Att Mic, hence veiling the true impact of each individual factor. Covering the full combinatory configs will solve this problem.

For limited variation, we intentionally constrained the complexity in terms of environmental settings. For that purpose, we recorded all genuine speech in the same room and replayed it in only two rooms. Moreover, the relative positions between speakers and recording/replay devices were almost the same in all the recording sessions. Constraining the complexity in the recording settings allows us to focus on the most important factors. However, we still permit some variations in the environment. For instance, when recording replayed speech, the relative positions between Att Replays and ASV Mics were not fully fixed.

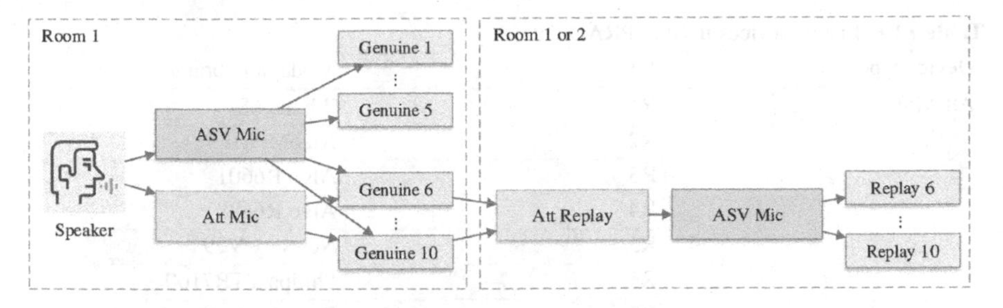

Fig. 14.1 Recording process of the PRAD dataset

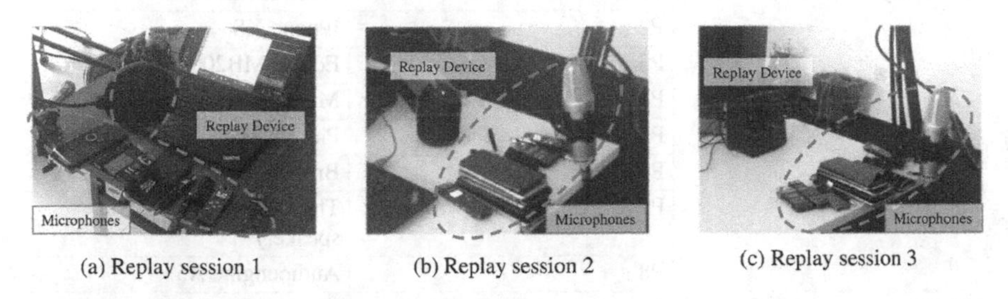

(a) Replay session 1 (b) Replay session 2 (c) Replay session 3

Fig. 14.2 Recording environment of the replay session 1–3

The recording process of the PRAD dataset is shown in Fig. 14.1. Firstly, 10 different sentences were read by speakers and were recorded by the ASV Mic and the Att Mic simultaneously in room 1. Each sentence contains 8 Chinese digits randomly sampled with a uniform distribution. Among the sentences recorded by the Att Mic, 5 sentences (ID 6–10) were replayed by Att Replay and captured by ASV Mic in room 1 or room 2. In total, there were 8 speakers, 7 Att Mics, 8 Att Replays, and 7 ASV Mics. There were totally 8 replay sessions, 3 in room 1 and 5 in room 2. Figure 14.2 shows the recording environment of the replay session 1–3. The details of the devices are shown in Table 14.1.

All the speakers, Att/ASV Mics, and Att Replays were crossly combined, forming full combinatory configs. Note that for each of the replayed speeches, we know the corresponding genuine speech. This enables us to investigate what has been changed during the R&R process, as shown in the next section.

In total, the number of genuine utterances in PRAD is $8 \times 7 \times 10 = 560$, and the total number of replayed utterances is $8 \times 7 \times 8 \times 7 \times 5 = 15,680$. As shown in Table 14.2, although PRAD is not large compared to other datasets such as ASVspoof 2017 and 2019, while it contains *full* combinatory to support the analysis of the recording and replay process.

Table 14.1 List of devices used in PRAD dataset

Device type	ID	Model and brand
Att Mic	R1	iPhone 6S
	R2	XiaoMi MI 5
	R3	Algo R6601
	R4	Algo R6688
	R5	Newsmy V29
	R6	Philips VTR7100
	R7	Rode NT1000
Att Replay	P1	Huawei P20 Pro
	P2	iPhone XS
	P3	Edifier MB200
	P4	Microlab B17
	P5	Philips SPA311
	P6	Brookstone Macaron 155579
	P7	Thinkpad E470 (built-in speaker)
	P8	Audioengine A5
ASV Mic	T1	Huawei mate 20 X
	T2	Huawei nova 3
	T3	Honor 6 H60-L01
	T4	JNN Q70
	T5	Amoi A1 USB
	T6	Sony UX560F
	T7	Philips VTR6900

Table 14.2 Profile of PRAD, ASVspoof 2017 V2 and ASVspoof 2019

Dataset	#Genuine	#Replayed	Factors			
			#Speaker	#Att Mic	#Att Replay	#ASV Mic
PRAD	560	15,680	8	7	8	7
ASVspoof 2017 V2 [49]	3,565	14,465	42	–	26	25
ASVspoof 2019 [50]	28,890	189,540	78	1	9[a]	1

[a] ASVspoof 2019 only contains the categorical labels without detailed device information.

14.3 Distribution Analysis

In this section, we will use the PRAD dataset to analyze the distribution profile of genuine speech and replayed speech, i.e., how they overlap with each other, and how they are impacted by important factors, including speaker, Att Mic, Att Replay, and ASV Mic.

14.3.1 Data Preparation

To perform the distribution analysis, we split the PRAD dataset into a training set and a test set. The training set contains speech data of 5 speakers, with the configs involving 4 Att Mics, 5 Att Replays, and 4 ASV Mics. The rest data are for the test. Note that some of the speakers and devices in the test set are seen in the training speech. For a test speech, if the speaker/device is seen in the training set, we call the speaker/device as *known*, otherwise, it is *unknown*. In our study, we ensure that at most one factor is unknown in the test data, so that we can analyze how the distribution changes when a particular factor is unknown, hence studying model generalizability with this factor. Moreover, the utterances in the training and test sets are non-overlapped: the training set only contains genuine utterances of ID 1–5 and replayed utterances of ID 6–7. The test set contains the remaining data, that is, genuine utterances of ID 6–10 and replayed utterances of ID 8–10. According to which factor is unknown, we designed 5 subsets for testing, as shown in Table 14.3.

14.3.2 Analysis on Overall Distributions

As the first step, we visualize the distributions of the first two dimensions of the LFCC feature for genuine and replayed speech. For each type of speech, we consider distributions of the training data, the test data with all factors known, and the test data with one or several factors unknown (union of the unknown conditions in Table 14.3).

Table 14.3 Training and test sets for distribution analysis

Dataset	Condition	#Genuine	#Replayed
Train	/	100	800
Test	All known	100	1200
	Unknown speaker	60	720
	Unknown Att Mic	0	900
	Unknown Att Replay	0	720
	Unknown ASV Mic	75	900

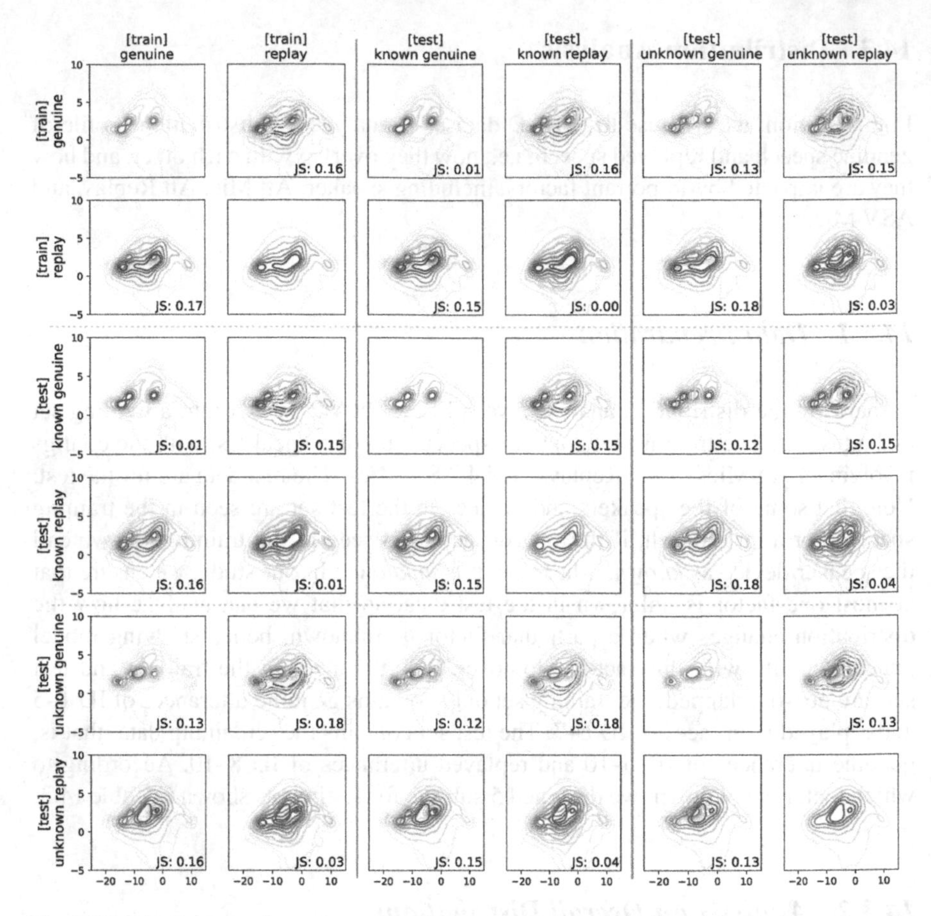

Fig. 14.3 Distributions of genuine and replayed speech. For each picture, the x-axis and y-axis represent the 1st and 2nd dimensions of the LFCC feature. The two distributions in each picture correspond to the two subsets of the data in the PRAD dataset corresponding to the labels of the row and column, respectively. 'JS' denotes the Jean-Shannon divergency between the two distributions in the picture

The results are shown in Fig. 14.3, where each picture plots the two distributions of the subsets corresponding to the row and column labels, and the Jean-Shannon (JS) divergence is also computed to show the quantitative discrepancy between the two distributions in the picture. Three main observations are: (1) The genuine speech has a smaller variance than replayed speech. This is not surprising, as the R&R process tends to cause additional distortion, hence leading to more uncertainty; (2) Unknown speakers/devices lead to a notable distribution shift. This is true for both genuine speech and replayed speech. It is clearly related to the weak generalization caused by limited training data. (3) Compared to replayed speech, genuine speech shows a larger distribution shift, measured by JS divergence.

14.3.3 Analysis on Important Factors

A key observation from Fig. 14.3 is that the distribution changes significantly when speakers or devices are changed, for both genuine and replayed speech. To gain more insight, Fig. 14.4 shows the distribution change when each of the four important factors is changed from known to unknown. It can be found that change in any individual factor leads to a distribution shift, for both genuine and replayed speech. Compared to other factors, ASV Mic causes the most notable shift, followed by Att Replay.

For more quantitative analysis, we trained a 64-component GMM with each subset defined in Table 14.3. We then compute Kullback-Leibler (KL) divergence and JS divergence between the GMM of the training data and the GMM of each test condition. For clarity, we test the divergence for genuine speech and replayed speech separately. The results are shown in Table 14.4. Note that this divergence is a little different from the divergence shown in Fig. 14.4: Fig. 14.4 shows the divergence between two *test* subsets, whereas Table 14.4 shows the divergence between the *training* set and each of the test subsets. Therefore, the divergence of Table 14.4 is more related to model generalization. We observe that the results in Table 14.4 are consistent with those in Fig. 14.4, i.e., ASV Mic is the most influential factor, and Att Replay is the second one. Speaker and Att Mic are relatively less significant.

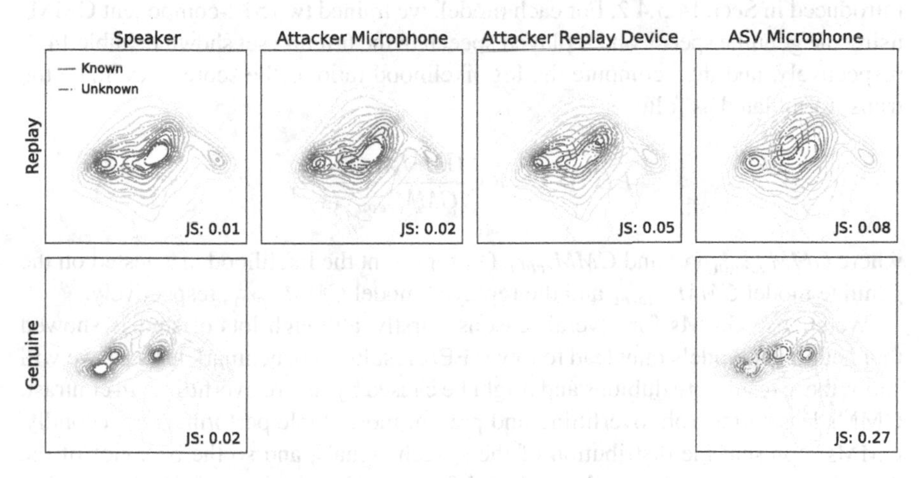

Fig. 14.4 Distribution of genuine and replayed speech, when each of the important factors is known or unknown. The x-axis and y-axis represent the 1st and 2nd dimensions of the LFCC feature

Table 14.4 Divergence between the GMMs of the training set and test subsets

Type	Test condition	KL divergence	JS divergence
Replayed	All known	0.98	0.17
	Unknown speaker	2.20	0.27
	Unknown Att Mic	2.70	0.36
	Unknown Att Replay	5.21	0.51
	Unknown ASV Mic	5.64	0.51
Genuine	All known	1.75	0.29
	Unknown speaker	3.61	0.40
	Unknown ASV Mic	9.25	0.60

14.3.4 Analysis on Discrimination and Generalization

In this section, we analyze how genuine speech and replayed speech can be discriminated, based on a materialized replay detection model. Compared to the distribution analysis, this analysis reflects the real performance of a replay detection system. We employed two detection systems, namely CQCC + GMM and LFCC + GMM, to perform the test. The two systems used CQCC and LFCC as the acoustic features, respectively. The details of the LFCC + GMM and CQCC + GMM systems are introduced in Sect. 14.5.4.2. For each model, we trained two 512-component GMMs using the genuine speech and replayed speech in the training set shown in Table 14.3, respectively, and then compute the log likelihood ratio as the score to evaluate the trials, formulated as follows:

$$LLR(x) = \log \frac{GMM_{genuine}(x)}{GMM_{replay}(x)},$$

where $GMM_{genuine}(x)$ and $GMM_{replay}(x)$ represent the likelihood of x tested on the genuine model $GMM_{genuine}$ and the replayed model GMM_{replay}, respectively.

We choose GMMs for several reasons. Firstly, although lots of studies showed that neural net models may lead to lower EER results on benchmark tests, as we will show, these results are dubious and might be caused by severe overfitting. In contrast, GMMs less suffer from overfitting and present more stable performance. Secondly, GMMs represent the distribution of the speech signals, and so the behavior of the detection system can be easily analyzed from the distributions of the data, as has been presented already. In contrast, analyzing neural nets is notoriously difficult.

The results are shown in Table 14.5, where the EER threshold[1] in the 'All known' condition is used to compute the FRR and FAR in all the conditions. It can be seen that both CQCC + GMM and LFCC + GMM systems achieve 0.0% FRR when the speakers and devices are all known, and the FAR is also low, demonstrating that the

[1] Due to the limited test data, the threshold that makes the FAR and FRR equal does not exist. Therefore, we use the threshold which makes FAR and FRR the closest.

Table 14.5 Performance of two systems under different test conditions defined in Table 14.3. The EER threshold of 'All known' condition is used for false alarm rate (FAR) and false reject rate (FRR) calculation. B1: CQCC + GMM; B2: LFCC + GMM

Condition	FRR(%)		FAR(%)	
	B1	B2	B1	B2
All known	0.00	0.00	0.75	0.17
Unknown speaker	0.00	0.00	0.15	0.42
Unknown Att Mic	/	/	2.33	2.00
Unknown Att Replay	/	/	13.89	12.36
Unknown ASV Mic	69.33	84.00	2.67	3.44

detection system works well.[2] When the speaker is unknown, both FRR and FAR are not notably impacted, indicating that the speaker factor does not cause much generalization problem. When devices are unknown, however, the performance is dropped. For genuine speech, the FRR increases significantly when ASV Mic is unknown. And for replay speech, when Att Replay is unknown, the FAR increases the most.

In order to gain more insight into the behavior of the detection system, we visualize the distributions of the log likelihood $GMM_{genuine}(x)$ and $GMM_{replay}(x)$ as well as the log likelihood ratio $LLR(x)$, where the test speeches are from subsets of different conditions. The results are shown in Fig. 14.5. In each picture, we show the distributions of both genuine speech and replayed speech.

Firstly focus on the first row, where all the speakers and devices are known. One can see that the genuine GMM produces a higher log likelihood for genuine speech than the replayed speech, demonstrating that it is discriminative. The GMM of the replayed speech, however, does not show any discrimination. In fact, it even assigns a higher log likelihood to genuine speech than replayed speech. This suggests that the low FRR and FAR reported in the first row of Table 14.5 are mainly due to the genuine model.

Now focus on the rest rows of Fig. 14.5. For each row, one factor is unknown so the generalization problem with a particular factor can be investigated. Note that for comparison, the distributions of the genuine and replayed speech in the 'All-known' condition (1st row) have been shown in the plots of other conditions (2nd-5th rows) as dotted curves. We have the following observations.

- When the speaker is unknown, the likelihood of both genuine and replayed speech is reduced, when tested on either the genuine model or the replayed model. This is understandable as new speakers lead to generalization problem. However, all the likelihood reductions are marginal, hence no notable performance reduction

[2] Note the reason that FAR is not equal to FRR (0.0%) in the 'All known' condition, is due to numerical inaccuracy.

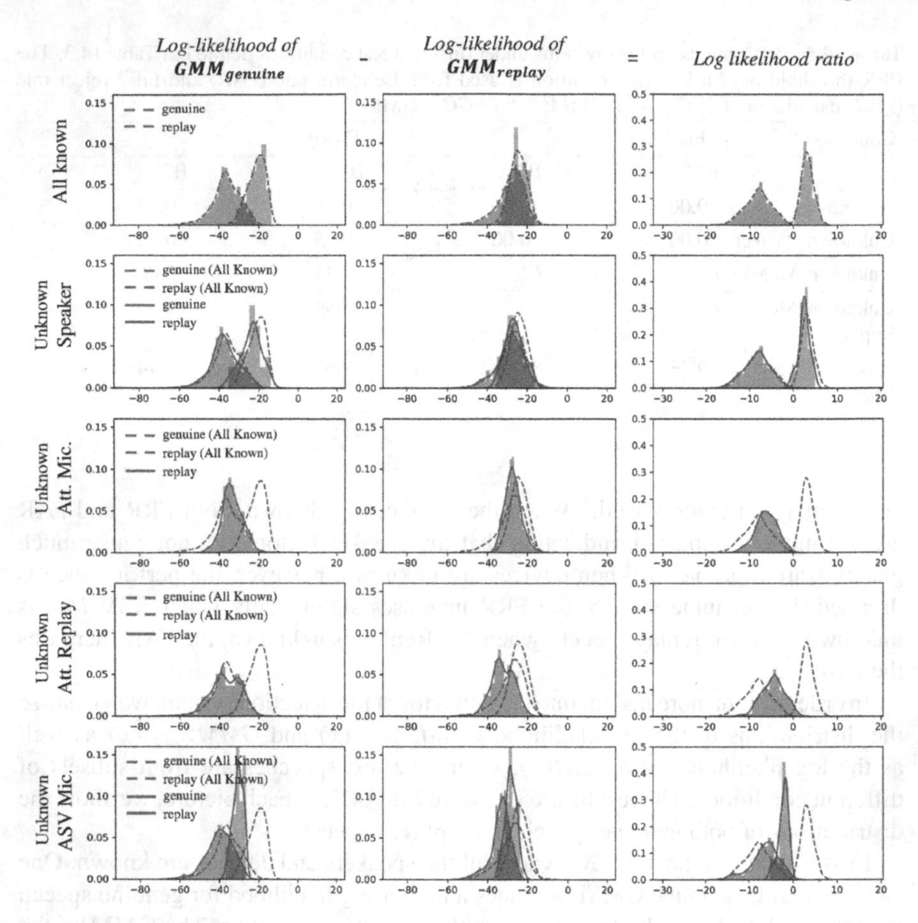

Fig. 14.5 Score distribution of LFCC + GMM. The three columns represent the log likelihood on the genuine model $GMM_{genuine}(x)$, the log likelihood on the replayed model $GMM_{replay}(x)$, and the log likelihood ratio $LLR(x)$, respectively. The rows show the results on different conditional sets defined in Table 14.3

is observed, in terms of both FRR (for genuine speech) and FAR (for replayed speech).

- When the Att Mic is unknown, the likelihoods of the replayed speeches on the replayed model are slightly reduced, leading to a marginal FAR increase. This is also expected, due to the same generalization problem. The likelihood of replayed speech on the genuine model seems to increase, however, we attribute the observation to randomness, as the genuine model was trained with genuine speech, so nothing to do with the change on Att Mic.
- When the Att Replay is unknown, the situation is similar to that with an unknown Att Mic. The only difference is that the replayed model seems more sensitive to the change on Att Replay compared to the change on Att Mic, and the likelihood

of replayed speech tested on the replayed model is severely reduced. Again, the change of the likelihood of the replayed speech on the genuine model should be regarded as randomness, as the genuine model is nothing to do with replayed speech, which means whether or not Att Replay is known does not matter.

- When the ASV Mic is unknown, the likelihood of both the genuine and replayed speech is reduced, on both the genuine model and the replayed model. Moreover, the reduction is more significant when the speech and model match (i.e., genuine speech on genuine model, and replayed speech on replayed model). This is expected as matched models suffer more from the generalization problem. More specifically, the likelihood reduction of genuine speech on the genuine model is much more significant than the likelihood reduction of genuine speech on the replayed model, leading to reduced LLR for genuine speech and hence a high FRR, as shown in Table 14.5. Similarly, the likelihood reduction of replayed speech on replayed model is more significant than the likelihood reduction of replayed speech on the genuine model, leading to the increased LLR, hence increased FAR, as shown in Table 14.5.

In summary, the results presented above indicate that the replayed model is not discriminative, and the detection performance is solely based on the genuine model in distinguishing between genuine and replayed speech. Regarding generalization, the two significant performance degradations in unknown conditions are (1) the increased FRR on genuine speech when ASV Mic is unknown, and (2) the increased FAR on replayed speech when Att Replay is unknown. In practice, the ASV Mic used in genuine trials is often the same as the one used in enrollment, indicating that if we assume that ASV Mic is the same for enrollment and genuine trials, the FRR problem can be largely avoided. In that case, the only problem that we need to solve is the high FAR associated with unknown R&R devices.

14.4 Dataset Analysis

We have analyzed the data distribution of genuine and replayed speech using the PRAD dataset. The main observation is that replay detection is very difficult, if not impossible. However, pretty low EER results were often reported in the literature. For example, in the ASVspoof 2017 challenge, the best EER was 6.73% [44] and in the ASVspoof 2019 challenge, the best EER was 0.39% [45]. However, post-evaluation analysis showed that these low EERs are partly due to some artifacts in dataset design, and thus can not reflect the true performance of present techniques.

In this section, we review the artifacts that have been identified in some public datasets. Some of the artifacts have been reported by multiple groups [19, 47, 51], and noticed by the ASVspoof organizers [49].

14.4.1 ASVspoof 2019 Physical Access Dataset

14.4.1.1 Introduction

The ASVspoof 2019 physical access (PA) dataset [50] was designed with carefully controlled acoustic and replay configurations. The whole dataset was constructed by simulation. The simulation consists of two parts: room simulation and playback device simulation. Room simulation was performed by Roomsimove,[3] which generated the finite impulse response (FIR) filter of the room based on several attributes: room size S, reverberation time T60, and distance D from the speaker to the microphone. The device was simulated by the generalized polynomial Hammerstein model [52], which can simulate not only linear distortions but also nonlinear distortions. The synchronized swept-sine tool [53] was used to measure the higher harmonic impulse responses (HHIRs) of devices. The training set and development set contain 14 devices, and the test set contains 9 devices. These HHIRs were finally transformed into the parameters of the generalized polynomial Hammerstein model.

The simulation process is shown in Fig. 14.6. Bonafide attempts were simulated by passing the source speech (from VCTK corpus) to the room simulator. Replay spoofing attempts were simulated by passing the source speech to a room simulator first (to simulate recording environments), followed by a device simulator (to simulate playback devices), and then another environment simulator (to simulate playback environments).

14.4.1.2 Artifacts

Due to the simulation tool, in the ASVspoof 2019 PA dataset, there are abnormal low-energy trailing silences in the generated speech, as shown in Fig. 14.7. This may be related to the room simulation process: the reverberation does not stop immediately when the signal stops. Therefore, simulation results in pure reverberation after the end of the speech signal. However, this does not happen in practice, because there are always noises in reality.

As shown in Fig. 14.8, replayed speech tends to contain longer trailing silence than genuine speech. When T60 increases, the difference is more obvious. Therefore, it is possible for a classifier to use the length of the trailing silence to determine whether the input is genuine or replay.

14.4.1.3 Impact

To investigate the influence of the artifacts on trailing silence, Cheng et al. [19] investigated the performance when the trailing silences on the training set and/or the

[3] http://homepages.loria.fr/evincent/software/Roomsimove_1.4.zip.

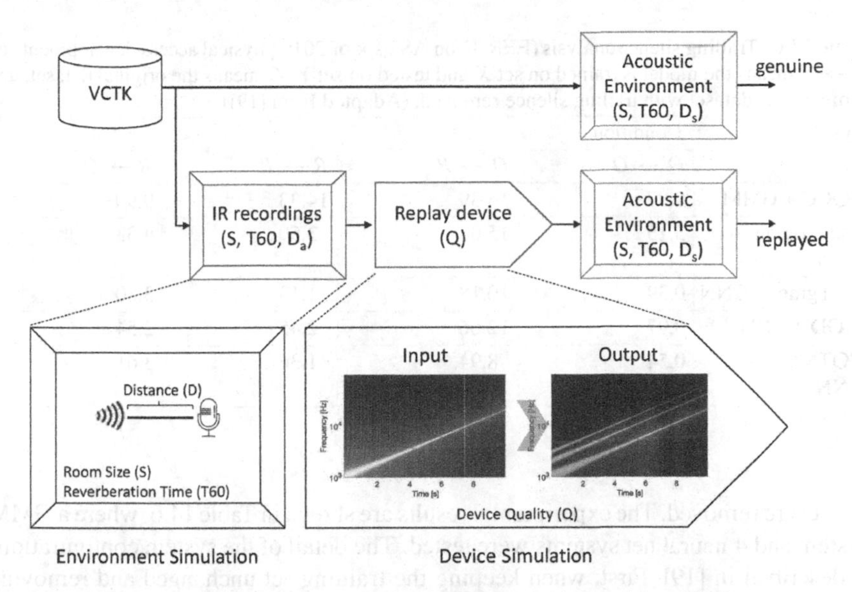

Fig. 14.6 Simulation process of ASVspoof 2019 PA dataset. (Adapted from [50])

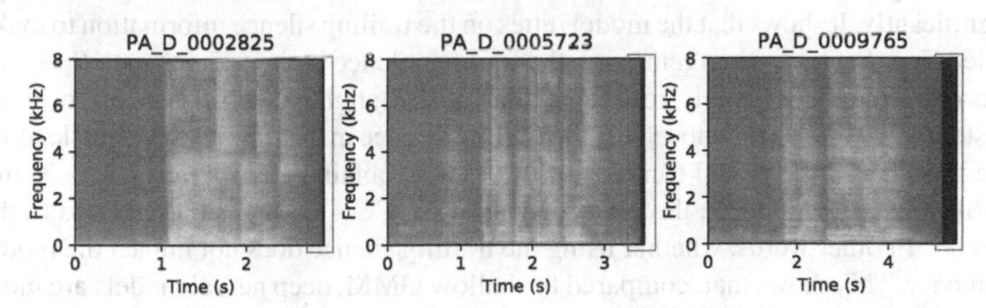

Fig. 14.7 Examples of trailing silence in ASVspoof 2019 PA dataset

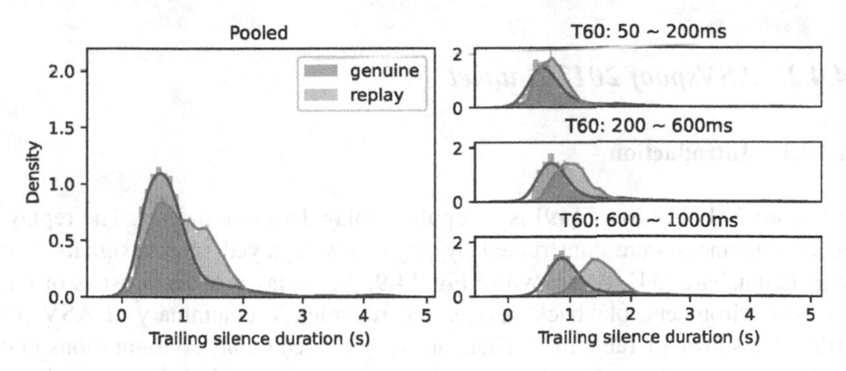

Fig. 14.8 Distribution of trailing silence duration on ASVspoof 2019 PA dataset. (Adapted from [19])

Table 14.6 Trailing silence analysis (EER%) on ASVspoof 2019 physical access development set. $X \rightarrow Y$ means the model is trained on set X and tested on set Y. O means the original dataset, and R means the dataset with trailing silence removed. (Adapted from [19])

System	Condition			
	$O \rightarrow O$	$O \rightarrow R$	$R \rightarrow R$	$R \rightarrow O$
CQCC + GMM	9.87	15.39	14.33	9.90
Spectrogram + CNN	3.15	15.05	3.59	4.33
CQTgram + CNN	0.39	10.18	1.13	3.80
MGD + CNN	0.97	12.66	2.45	2.54
CQTMGD + CNN	0.54	8.94	1.36	3.61

test set are removed. The experimental results are shown in Table 14.6, where a GMM system and 4 neural net systems were tested. The detail of the system configurations is described in [19]. First, when keeping the training set unchanged and removing all trailing silences in the test set ($O \rightarrow R$), the performance of neural models drops significantly. It shows that the model relies on the trailing silence information to make a decision. Second, when removing the trailing silence in the training set ($R \rightarrow R$), the performance is largely recovered, but is still slightly worse than the original system ($O \rightarrow O$). This shows that the trailing silence in the training set is misleading the model, and the model can use this artifact to obtain better performance. Third, unlike the neural models, the performance of CQCC + GMM is only related to the test set. In other words, whether using the trailing silence does not impact the model behavior. This shows that, compared to shallow GMM, deep neural models are more likely to be misled by artifacts.

14.4.2 ASVspoof 2017 Dataset

14.4.2.1 Introduction

ASVspoof 2017 version 2 [49] is a popular replay detection dataset. The replayed speech utterances were constructed by physically replayed speech signals in the Reddots database [54]. As shown in Fig. 14.9, the replay process consists of three factors: environment, playback device, and recorder. The summary of ASVspoof 2017 V2 is shown in Table 14.7. There are very limited replay configurations in the training set. Therefore, the detection system needs to be robust against unknown attacks.

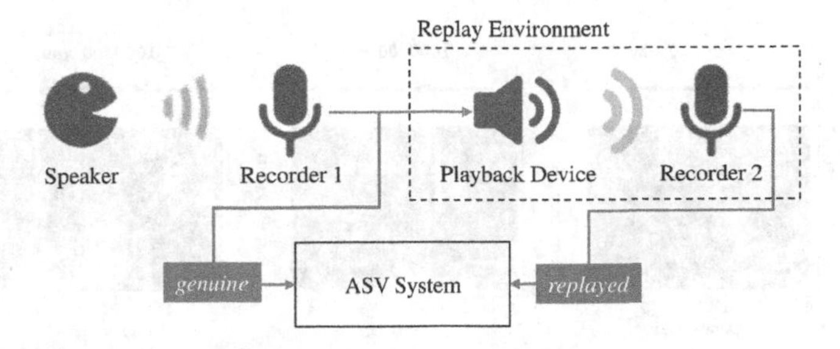

Fig. 14.9 An illustration of bonafide access and replay attack in ASVspoof 2017 dataset. (Adapted from [49])

Table 14.7 Summary of ASVspoof 2017 V2 dataset. A replay configuration is a triplet of environment, playback device, and recorder. Not all combinations are included in the dataset. (env: environment; pb: playback device; rec: recorder)

Subset	#Speaker	#Replay Configuration	#Genuine	#Replayed
Train	10	3 (2 env, 3 pb, 1 rec)	1,507	1,507
Dev	8	10 (6 env, 6 pb, 7 rec)	760	950
Eval	24	57 (24 env, 23 pb, 24 rec)	1,298	12,008
Total	42	62 (26 env, 26 pb, 25 rec)	3,565	14,465

14.4.2.2 Artifacts

Chettri et al. [47] reported some artifacts in ASVspoof 2017 V2 dataset:

- *Burst click sound (BCS)*: an abrupt click sound found at the start of audio recordings, as shown in Fig. 14.10a. It is usually found in genuine speech, but not in replayed speech.
- *Zero-valued silence*: there is some abnormal zero-valued silence without any noise at the start of the genuine speech, as shown in Fig. 14.10b. It does not appear in replayed speech.
- *Dual-tone multi-frequency signaling (DTMF) sound*: there are reverberation tails of DTMF at the start of the replayed speech, as shown in Fig. 14.10c. This is because DTMF is used as a separator between two adjacent replayed utterances and it is not completely cut out when splitting the replayed speech.
- *Non-speech segment in the first 0.3 s*: there are longer non-speech (noise, music, or silence) at the start of genuine speech compared with replayed speech.

Table 14.8 shows the proportion of the artifacts in the ASVspoof 2017 V2 dataset based on manual annotation. It can be observed that the proportions of these artifacts in the genuine and the replayed speech are different. These may lead to statistical bias that classifiers may make use of to discriminate between genuine and replayed speech.

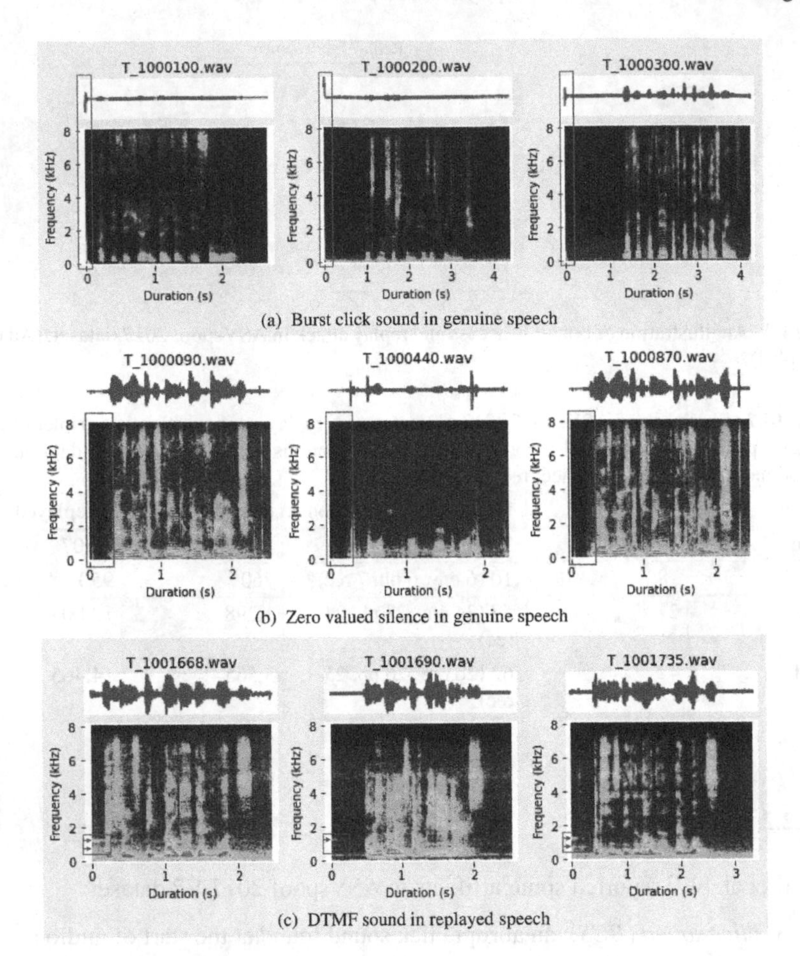

(a) Burst click sound in genuine speech

(b) Zero valued silence in genuine speech

(c) DTMF sound in replayed speech

Fig. 14.10 Examples of the artifacts in ASVspoof 2017 V2 dataset

Table 14.8 The proportion of artifacts occurrence in the ASVspoof 2017 V2 dataset. (Adapted from [47])

Artifact	Train		Dev	
	Genuine (%)	Replayed (%)	Genuine (%)	Replayed (%)
Burst click sound	36.36	2.45	41.06	0.00
Zero-valued silence	19.11	0.00	1.97	0.00
DTMF sound	0.00	45.58	0.00	16.63
Non-speech in first 300ms	60.45	31.26	73.55	58.95

Table 14.9 Artifact analysis (EER%) on ASVspoof 2017 V2 physical access development set. $X \rightarrow Y$ means the model is trained on set X and tested on set Y. O means the original dataset, and R means the dataset with artifacts removed

System	Condition			
	$O \rightarrow O$	$O \rightarrow R$	$R \rightarrow R$	$R \rightarrow O$
CQCC + GMM	29.66	34.98	33.81	28.20
LFCC + GMM	33.91	36.12	36.13	34.20
Spec + LCNN	11.40	13.24	12.47	12.42

14.4.2.3 Impact

To analyze the impact of the artifacts mentioned above, we retrained the replay detection models with these artifacts retained or removed, and tested the performance of these models on the test data with artifacts retained or removed. We use the endpoint annotation published in [47]. Note that Chettri et al. [47] has reported some detailed results on this topic; we reproduce some of the experiments in order to draw a similar picture as in Table 14.6 for ASVspoof 2017.

The experimental results are shown in Table 14.9, where we tested two GMM models and a neural model. One can observe that the trend of the performance change is similar to the results in the ASVspoof 2019 analysis shown in Table 14.6. Specifically, for the Spec + LCNN model trained on the original dataset, the performance degrades when it is tested on artifact-removed speech ($O \rightarrow R$). This shows that the model trained on O utilizes the artifact information to gain better performance. When the model is retrained with R and tested on artifact-removed speech ($R \rightarrow R$), the performance is better than the case $O \rightarrow R$. This is more close to the true performance of the model, as both the training and test data do not involve any artifacts. Another observation is that the $R \rightarrow R$ performance is worse than that of $O \rightarrow O$, indicating that the true performance of the LCNN model was over-estimated in the $O \rightarrow O$ condition. Moreover, the retrained model produces comparable performance when tested on speech with or without artifacts, demonstrating that the model is more robust by removing the artifacts from the training speech.

For the two GMM models, the performance is mostly determined by the condition of the test speech, and the artifact-retained test speech shows better performance. It implies that GMMs do not learn much bias. The performance reduction associated with artifact removal is perhaps due to the shortened speech utterances.

14.4.3 Cross Dataset Results

We have discussed some artifacts identified by researchers. However, some artifacts might be more *implicit*. These implicit artifacts can not be clearly identified, but they indeed lead to biased models, sometimes even very misleading. The artifacts,

Table 14.10 Cross-dataset performance (EER%). Since the AS19Real only contains an evaluation subset, we use the whole dataset either for training or for testing. (AS17: ASVspoof 2017 V2 [49]; AS19: ASVspoof 2019 PA [50]; AS19Real: ASVspoof 2019 Real PA [50])

System	Train	Test			
		PRAD	AS17	AS19	AS19Real
LFCC + GMM	PRAD	12.86	35.14	58.78	35.19
	AS17	45.71	33.91	81.22	47.59
	AS19	27.50	33.20	13.52	29.05
	AS19Real	30.72	35.52	20.10	/
CQCC + GMM	PRAD	14.31	37.98	36.76	26.83
	AS17	45.71	29.66	59.95	49.26
	AS19	29.28	37.98	11.25	12.57
	AS19Real	46.78	32.27	44.02	/
Spec + LCNN	PRAD	11.07	41.00	53.43	48.33
	AS17	36.43	11.40	65.45	47.41
	AS19	41.04	64.32	1.92	44.44
	AS19Real	33.22	54.01	42.24	/

no matter explicit or implicit, are clearly dataset dependent. Combining with data scarce, severe overfitting is often observed.

To gain more idea about the overfitting problem, we test the cross-dataset performance. The results are shown in Table 14.10, where we train and test two GMM-based models (LFCC + GMM and CQCC + GMM) and a neural model (Spec + LCNN) on four datasets: PRAD, ASVspoof 2017 V2, ASVspoof 2019 PA, and ASVspoof 2019 Real PA. It is clear to see that in all the test cases, the cross-dataset performance is much worse than the intra-dataset performance. This clearly demonstrates that existing techniques suffer from severe overfitting, and the performance reported on the benchmark test is highly suspectable. In particular, the neural model shows much more imbalanced performance compared to the two GMM models. For example, Spec + LCNN gains an EER 1.92% when trained and tested on AS19, but the number goes to 64.32% when it is tested on AS17. All these results seem to indicate that existing benchmark results and models are very doubtful.

14.5 One-Class Model

Inspired by the distribution analysis and dataset analysis, we found that the replay detection problem is more suitable to be solved by a one-class model, rather than a two-class model. Unfortunately, most popular techniques are based on two-class models. We will analyze the potential problem with the two-class approach, and present our recent study on a particularly designed one-class model.

14.5.1 Advocate One-Class Model

In the distribution analysis, we show that genuine and replayed speech are largely overlapped, and the discriminative capacity of replayed models is fairly low. Moreover, replayed models poorly generalize to unknown devices. These observations indicate that modeling replayed speech may produce more harm than benefit. A possible solution is to model genuine speech only, or a one-class approach [55, 56].

In the dataset analysis, we show that existing replay detection models tend to overfit artifacts. We argue that this learning-to-artifacts behavior is largely attributed to the two-class nature of these methods, i.e., treating genuine and replayed speech as two classes and building a classifier to discover and amplify the difference between them, which makes them easily overfit to artifacts. Powerful models (e.g., neural nets) tend to be more misled by artifacts. To solve the problem, one may consider modeling the underlying distribution of genuine speech, rather than the discrepancy between the two classes. Again, this is a one-class approach [55, 56].

We, therefore, advocate the one-class approach, i.e., model the distribution of genuine speech, and reject any speech whose likelihood tested on the genuine model is low. Essentially, it formulates replay detection as an out-of-distribution (OOD) detection problem, rather than a binary classification (genuine & replayed) problem. By this approach, it does not rely on special patterns of the R&R process, so can detect replayed speech of any form. On the other hand, since it does not model replayed speech, the generalization problem associated with unknown R&R devices and/or artifacts can be largely avoided.

14.5.2 Model Design

A straightforward way to implement the one-class approach is to build a probabilistic model for genuine speech, and then detect replayed speech by thresholding the likelihood of the test speech. Generally, any density model can be used, including models based on deep neural nets, such as variational auto-encoders (VAEs) and normalization flows (NFs). However, this 'general solution' did not work well in our investigation. As shown in Sect. 14.3, genuine speech and replayed speech are largely overlapped. For such overlapped classes, it is not possible to attain good discrimination, no matter with one-class model or two-class model.

We, therefore, address the problem in another way: rather than discriminate between genuine and replayed speech, we discriminate between genuine trials and replayed trials. The first glance is that these two tasks are the same, but they are not. The most significant difference is that by focusing on the trials, some prior knowledge of the authentication process can be utilized, hence boosting the detection performance. In this study, we heavily use the following prior to genuine trials: the speaker and ASV Mic do not change in enrollment and test. Specifically, we propose to compute the *divergence* of the test speech from the enrollment speech and model the density of the divergence. Under the prior condition that the speaker and ASV Mics remain unchanged, the divergence is ideally independent of speakers and ASV Mics. Hence, it is much easier to model.

We will first present features that we designed to reflect the enroll-test divergence, and then present the model architecture. For the feature side, we use utterance-level divergence rather than frame-level divergence, due to its low variation. For the model side, we choose GMM for the sake of simplicity and generalizability. This model was presented in [57], though more experiments and investigation will be presented in this paper.

14.5.2.1 Features Based on LTAS Divergence

We choose the long-term average spectrum (LTAS) [58] to produce the utterance-level representation, which is defined as:

$$\text{LTAS}_x(k) = \frac{1}{L}\sum_{l=1}^{L} log(|X(k,l)|), \tag{14.1}$$

where $X(k,l)$ is the spectrogram of a signal $x(n)$, k is the index of frequency bins, l is the index of frames, and L is the total number of frames. Then, the divergence feature can be estimated as:

$$d^{\text{LTAS}}(k) = \text{LTAS}_x(k) - \text{LTAS}_{enroll}(k), \tag{14.2}$$

where $enroll$ denotes the speech that concatenates all the enrollment utterances of the claimed speaker. We concatenate $d^{\text{LTAS}}(k)$ for all k, and denote it by *LTAS-based divergence feature*.

Since the discrete cosine transform (DCT) of the average of log magnitude spectrum is equivalent to the average of cepstral coefficients, we can extract the divergence feature from cepstral coefficients directly. Formally, we compute the LTAS-based LFCC divergence (LFCC-LTAS-D) as follows:

$$d^{\text{LFCC-LTAS}}(k) = \frac{1}{L}\sum_{l=1}^{L} LFCC_x(k,l) - \frac{1}{L'}\sum_{l=1}^{L'} LFCC_{enroll}(k,l),$$

where k is the index of LFCC coefficients, l is the index of frames, $LFCC_x$ and $LFCC_{enroll}$ is the LFCC of the test speech $x(n)$ and enrollment speech, respectively. The LTAS-based CQCC divergence (CQCC-LTAS-D) can be extracted in the same way by replacing LFCC with CQCC.

14.5.2.2 Phone-Attentive LTAS Divergence

The LTAS-based divergence feature may involve residual phonetic variation, especially when the contents of enrollment and test speech are different. We propose a phone-attentive approach to alleviate this problem. Formally, the phone-attentive average spectrum (PAAS) is designed as follows:

$$\text{PAAS}_x(q,k) = \frac{\sum_{l=1}^{L} P_x(q,l) log(|X(k,l)|)}{\sum_{l=1}^{L} P_x(q,l)}, \tag{14.3}$$

where $P_x(q,l)$ is the posterior of phone q at frame l for the signal $x(n)$. Then the divergence between enrollment speech and test speech is estimated by considering all the phone-dependent divergence:

$$d^{\text{PAAS}}(k) = \frac{\sum_q \omega_q(\text{PAAS}_x(q,k) - \text{PAAS}_{enroll}(q,k))}{\sum_q \omega_q}, \tag{14.4}$$

where ω_q is the prior weight of phone q, which is defined as:

$$\omega_q = \begin{cases} \sum_l P_x(q,l) + \sum_l P_{enroll}(q,l) & \text{if } \sum_l P_x(q,l) > \tau \text{ and } \sum_l P_{enroll}(q,l) > \tau \\ 0 & \text{else} \end{cases}, \tag{14.5}$$

where τ is the threshold to exclude low-frequency phones. We concatenate $d^{\text{PAAS}}(k)$ for all k, and denote it by *PAAS-based divergence feature*. Similar to the LTAS-based divergence feature, we extract the PAAS-based LFCC divergence (LFCC-PAAS-D) and the PAAS-based CQCC divergence (CQCC-PAAS-D) based on LFCC and CQCC, respectively.

14.5.2.3 Model Architecture

We simply model the divergence between enrollment and test speech in genuine trials by a Gaussian mixture model (GMM), parameterized by θ. The density of a divergence feature d is computed as follows:

$$P(d|\theta) = \sum_{i=1}^{K} w_i N(d|\mu_i, \Sigma_i) \tag{14.6}$$

where $N(d|\mu_i, \Sigma_i)$ is the i-th Gaussian component with mean vector μ_i and covariance matrix Σ_i; w_i is the mixture weight of the i-th component, and K is the total number of components. The expectation-maximization (EM) algorithm [59] can be used to estimate the parameters θ. Once the model is trained, for a trial whose divergence feature is d, the log likelihood $log(P(d|\theta))$ is used to decide whether it is a genuine trial or a replayed trial.

14.5.3 Related Work

Some researchers have proposed to use the genuine-only model in spoof countermeasure. For instance, Wu et al. [60] adopted an autoencoder to extract high-level features for replay detection. Some authors used a pre-trained model to extract more robust features, e.g., the wav2vec feature [61] and the speech2face feature [62]. All the models are trained with genuine speech only. However, the classifiers are still two-class models. Other researchers presented one-class model more similar to ours. Villalba et al. [63] extracted bottleneck feature from a deep neural net, and used them as a feature to train a one-class support vector machine (OC-SVM) for synthesis speech detection. They did not employ the genuine-trial prior, and so the OC-SVM has to deal with complex variation. Wang et al. [64] proposed another one-class approach. They hypothesized that a simple vocoder (e.g., WORLD) trained only with genuine speech could model the distribution of genuine speech. Therefore, the residual between a speech and the speech resynthesized by the vocoder can reflect the bias introduced by the replay process. They used this residual as features to train a one-class GMM. The residual feature is analog to our enroll-test divergence feature, where the vocoder-derived speech can be regarded as an estimation of the genuine speech. The advantage of this approach is that no enroll-test pairs are required during model training, but the vocoder needs to be carefully chosen.

14.5.4 Experiments

14.5.4.1 Dataset

The PRAD dataset, BTAS 2016 dataset [48], ASVspoof 2017 version 2 (AS17) dataset [49] and ASVspoof 2019 real physical access (AS19Real) dataset [45] were used for evaluation. We discarded the ASVspoof 2019 physical access dataset [45] since the replayed samples in this dataset are simulated by algorithms instead of replayed in reality. To evaluate the proposed one-class model, the enrollment data for each test utterance is necessary. The default partitions of the datasets mentioned above do not satisfy this request, so we firstly repartitioned the data as shown in Table 14.11. The detail is described as follows:

- *PRAD*. PRAD was used for both training and testing. The training set contains speech data of a full combination of 5 speakers, 4 Att Mics, 5 Att Replays, and 4 ASV Mics. And the eval set contains speech data of a full combination of the

Table 14.11 Distribution of the repartitioned databases

Database	Subset	#Speaker	#Genuine	#Replayed
PRAD	Train	5	200	2,000
	Eval-enroll	3	105	0
	Eval-test		105	5,877
BTAS-PA	Eval-enroll	16	80	0
	Eval-test		1,763	4,800
AS17	Train	10	1,507	1,507
	Dev	8	760	950
	Eval-enroll	17	170	0
	Eval-test		1,058	12,008
AS19Real	Eval-enroll	26	130	0
	Eval-test		260	1,860
THCHS30	Train	57	13,369	0

rest 3 speakers, all 7 Att Mics, 8 Att Replays, and 7 ASV Mics. The eval set is further split into two subsets, which are the eval-enroll set and the eval-test set. The eval-enroll set contains only genuine speech from ID 1 to 5. The eval-test set contains the genuine speech from ID 6 to 10 and the replayed speech from ID 6 to 10.

- *BTAS-PA*. BTAS was used for test only. The genuine and replayed speeches of BTAS were selected, which we called BTAS-PA. For the genuine speech, we selected utterances recorded by laptop. The first 5 genuine utterances with a content label 'read' in the 'sess1' session for each speaker were used as enrollment data (eval-enroll) and the rest utterances were used as test data (eval-test).
- *AS17*. AS17 was used for both training and testing. For model training, both training set and development set was used. For testing, speakers who do not have any spoofing data were removed. And for each speaker, onc genuine utterance per passphrase was selected as the enrollment data (eval-enroll). The rest genuine and replayed utterances were used as test data (eval-test). The partition of the AS17 evaluation dataset is the same as the one in [65].
- *AS19Real*. AS19Real was used for test only. For testing, we first removed all the utterances recorded by ASV Mic 'asv03' as their speaker labels are incorrect. Then the first 5 genuine utterances (sorted by filename) of each speaker were used as enrollment data (eval-enroll) and the rest utterances were used as the test data (eval-test).
- *THCHS30*. THCHS30 is a standard mandarin speech dataset that contains only genuine speech [66]. This dataset was used only for training the one-class models. Once the model was trained, they were tested on the eval-enroll and eval-test sets in the above-mentioned datasets. Note that training two-class models with genuine-only datasets is impossible.

14.5.4.2 Experiment Setup

Primary Baselines

We first built three primary baselines, namely LFCC + GMM, CQCC + GMM, Spec + LCNN. These models are essentially two-class models, as they model both the genuine and replayed speech. Moreover, these models do not employ the genuine-trial prior. The detail of the three baseline systems is shown below.

- *LFCC + GMM*. The configuration of LFCC was identical to the ASVspoof 2019 challenge baseline [45]. Specifically, the LFCC was extracted with a 20ms frame length, 10ms hop length, Hamming window, and 20 linear filter banks. Finally, a 60-dimension LFCC was formed by appending delta and delta-delta coefficients. Two 512-mixture GMMs were utilized to model the genuine and replayed data in the training set.
- *CQCC + GMM*. The configuration of CQCC was also identical to the ASVspoof 2019 challenge baseline [45]. The CQCC was extracted with the maximum frequency $f_{max} = 8kHz$, minimum frequency $f_{min} = f_{max}/2^9$, 96 bins per octave. Finally, a 90-dimension CQCC was formed by appending delta and delta-delta coefficients. Two 512-mixture GMMs were utilized to model the genuine and replayed data in the training set.
- *Spec + LCNN*. We adopted the CNN model published in [67]. It is a light CNN (LCNN) architecture, with the log power spectrogram as the input. The log power spectrogram (LPS) was extracted with 25ms frame length, 10ms hop length, Hannning window, 512 FFT points. The mean and variance normalization (MVN) per utterance was applied to the LPS. For the LCNN, the cross-entropy loss criterion was optimized by the stochastic gradient descent (SGD) optimizer with a learning rate of 0.001 and a momentum of 0.9. The learning rate was reduced by 10 times when the loss does not improve within 10 epochs. During the training, the batch size was set to 8. More detail on the LCNN can be found in [67].

MAP Baselines

We then built two maximum a posterior (MAP) adaptation-based baselines: LFCC + GMM + MAP and CQCC + GMM + MAP [65]. Firstly, a speaker-independent GMM was trained using the training set. Secondly, a *speaker-dependent* genuine GMM was attained for each speaker, by adapting from the speaker-independent GMM via MAP using the enrollment speech. During the MAP adaptation, the mean, variance, and weights were all adapted, by setting the relative factor to be 1. Finally, during test, the log likelihood ratio between the speaker-dependent genuine GMM and a universal replayed GMM was utilized to make the decision. These two baselines are still two-class models, however, the genuine-trial prior is employed.

One-Class Systems

For our one-class systems, we chose LFCC and CQCC as the frame-level features and used them to compute utterance-level LTAS/PAAS divergence features. During model training, some utterances of the training set were selected as the enrollment speech, and some others were selected as the test speech. Note that these enrollment and test utterances are all *training data*, the split is only for computing divergence features in model training. For the PRAD dataset, the first 5 genuine utterances for each speaker-terminal pair (ensure both speaker and terminal are the same) were selected as enrollment utterances, and the rest of the utterances of that speaker-terminal pair were used as the test utterances. For AS17 and THCHS30 databases, the first 10 utterances were selected as enrollment utterances, and the rest of the utterances of the same speaker were selected as test utterances. These utterance-level features were then used to train a *speaker-independent* GMM, by which we can test a trial of any speaker. The configurations of LFCC/CQCC extraction were the same as the two types of baseline systems, except that we used the static coefficients only.

To compute LFCC-PAAS-D and CQCC-PAAS-D features, a pre-trained automatic speech recognition (ASR) model is required to produce phone posteriors. For datasets BTAS-PA, AS17, and AS19Real, we used an English pre-trained model,[4] and for PRAD and THCHS30 datasets, we used a Chinese pre-trained model.[5] The threshold τ of the PAAS method was set to 1. Finally, all the one-class models involve 64 Gaussian components.

14.5.4.3 Result

Table 14.12 shows the performance of different systems trained on the PRAD dataset. Firstly, the primary baselines obtained poor performance. Spec + LCNN looks better on the PRAD test set (intra-dataset test), but the performance is still poor on other datasets. The two MAP baselines achieved much better performance than the primary baselines. We believe that the improvement comes from the prior information of the speaker-dependent genuine speech distribution, which leads to a more accurate genuine model. However, they still use a replayed model, leading to performance degradation on inter-dataset tests. Finally, the proposed one-class model achieved significantly better performance compared to all the baselines. Most significantly, the inter-dataset performance was substantially improved, confirming our argument that one-class model can largely alleviate the overfitting problem, by removing the variance on speakers and ASV Mics as well as discarding the replayed model. Comparing the two frame-level features, CQCC obtained better performance. Comparing the two utterance-level features, PAAS that considers phone dependency obtained better performance. All these results are in expectation.

---

[4] http://www.kaldi-asr.org/models/m13.

[5] http://www.kaldi-asr.org/models/m11.

Table 14.12 EER(%) results of systems trained on PRAD

Type	System	Test set			
		PRAD	BTAS-PA	AS17	AS19Real
Primary baseline	LFCC + GMM	10.61	24.50	34.99	34.67
	CQCC + GMM	11.62	23.31	40.26	25.38
	Spec + LCNN	5.67	26.10	43.28	46.98
MAP baseline	LFCC + GMM + MAP	3.81	11.90	19.28	13.07
	CQCC + GMM + MAP	6.86	5.61	20.15	20.68
One-class (Ours)	LFCC-LTAS-D + GMM	1.86	6.92	15.50	10.38
	CQCC-LTAS-D + GMM	1.78	5.00	9.27	6.47
	LFCC-PAAS-D + GMM	0.95	7.71	13.23	3.94
	CQCC-PAAS-D + GMM	0.87	5.56	8.13	2.69

Table 14.13 EER(%) results of systems trained on AS17

Type	System	Test set			
		PRAD	BTAS-PA	AS17	AS19Real
Primary baseline	LFCC + GMM	45.72	50.93	31.36	42.74
	CQCC + GMM	47.62	53.49	23.44	49.24
	Spec + LCNN	32.38	11.96	10.02	37.69
MAP baseline	LFCC + GMM + MAP	21.77	15.71	20.78	39.70
	CQCC + GMM + MAP	12.57	12.83	10.97	42.31
One-class (Ours)	LFCC-LTAS-D + GMM	2.86	10.77	12.31	11.14
	CQCC-LTAS-D + GMM	3.82	5.28	9.08	6.93
	LFCC-PAAS-D + GMM	2.99	8.79	10.51	8.45
	CQCC-PAAS-D + GMM	3.67	2.73	9.35	5.30

Table 14.13 shows the performance of the models trained on the AS17 dataset. Again, our one-class systems exhibit a clear advantage. Most significantly, comparable performance was observed on the intra-dataset and inter-dataset tests, indicating that the one-class model is fairly reliable.

Finally, we learn the one-class model using THCHS30, a dataset that contains genuine speech only. As shown in Table 14.14, the results are surprisingly good, even on datasets with a different language, i.e., BTAS-PA, AS17, and AS19Real. On AS17, the best performance is even better than the model trained with the AS17 training data. We attribute the good performance to the relatively large data volume for model training. These promising results once again demonstrated the potential of one-class models.

Table 14.14 EER(%) results of one-class systems trained on THCHS30

System	Test set			
	PRAD	BTAS-PA	AS17	AS19Real
LFCC-LTAS-D + GMM	1.91	6.24	13.61	7.69
CQCC-LTAS-D + GMM	1.91	5.09	10.58	4.24
LFCC-PAAS-D + GMM	1.91	6.58	11.15	3.45
CQCC-PAAS-D + GMM	1.16	5.27	8.04	1.60

14.6 Conclusions

This chapter conducted a deep analysis of the replay detection task. Based on these analyses, we showed the potential fragile of the conventional two-class methods and advocated the one-class approach. We also presented some experimental evidence to show that one-class models are quite promising.

Our study started by constructing a new dataset called PRAD. By a careful design, this new dataset permits us performing a deep investigation on the properties of replayed speech. Then, the distributions of genuine speech and replayed speech were analyzed. We found that the distributions of genuine class and replayed class are largely overlapped, and the replayed class is more dispersed than the genuine class. Moreover, in the case that Att Replay is unknown, the replayed model usually produces a low likelihood, leading to increased FAR. All the observations indicate that modeling replayed speech is less useful, and may lead to severe generalization problem. A one-class model is, therefore, more desirable.

Next, we performed a dataset analysis, by summarizing the artifacts in existing data sources. The subjective observation and objective experimental results showed that those artifacts did mislead replay detection models, in particular those two-class models. This again suggests a one-class model approach. We, therefore, advocated a one-class view for replay detection, by which only genuine speech is modeled, and replay detection is formulated as OOD detection.

As a demonstration, we presented a one-class model based on enrollment-test divergence features. Experimental results show that even using a simple GMM as the classifier, the one-class approach can achieve very promising results, and the advantage is more remarkable on inter-dataset tests. These results demonstrated the great potential of the one-class approach.

References

1. Campbell JP (1997) Speaker recognition: a tutorial. Proc IEEE 85(9):1437–1462. https://doi.org/10.1109/5.628714
2. Reynolds DA (2002) An overview of automatic speaker recognition technology. In: IEEE international conference on acoustics, speech and signal processing (ICASSP), vol 4. IEEE,

pp IV–4072. https://doi.org/10.1109/ICASSP.2002.5745552
3. Hansen J, Hasan T (2015) Speaker recognition by machines and humans: a tutorial review. IEEE Signal Process Mag 32(6):74–99. https://doi.org/10.1109/MSP.2015.2462851
4. Mittal A, Dua M (2021) Automatic speaker verification systems and spoof detection techniques: review and analysis. Int J Speech Technol. https://doi.org/10.1007/s10772-021-09876-2
5. Beranek B (2013) Voice biometrics: success stories, success factors and what's next. Biom Technol Today 2013(7):9–11. https://doi.org/10.1016/s0969-4765(13)70128-0
6. Wu Z, Evans N, Kinnunen T, Yamagishi J, Alegre F, Li H (2015) Spoofing and countermeasures for speaker verification: a survey. Speech Commun 66:130–153. https://doi.org/10.1016/j.specom.2014.10.005
7. Evans NW, Kinnunen T, Yamagishi J (2013) Spoofing and countermeasures for automatic speaker verification. In: Interspeech, pp 925–929. https://doi.org/10.21437/Interspeech.2013-288
8. Zhang W, Zhao S, Liu L, Li J, Cheng X, Zheng TF, Hu X (2021) Attack on practical speaker verification system using universal adversarial perturbations. In: ICASSP 2021. IEEE, pp 2575–2579. https://doi.org/10.1109/ICASSP39728.2021.9413467
9. Alegre F, Janicki A, Evans N (2014) Re-assessing the threat of replay spoofing attacks against automatic speaker verification. In: 2014 international conference of the biometrics special interest group (BIOSIG). IEEE, Darmstadt, Germany, pp 1–6. (10–12 Sept 2014)
10. Singh M, Mishra J, Pati D (2016) Replay attack: its effect on GMM-UBM based text-independent speaker verification system. In: 2016 IEEE Uttar Pradesh section international conference on electrical, computer and electronics engineering (UPCON). IEEE, Varanasi, India, pp 619–623. https://doi.org/10.1109/UPCON.2016.7894726
11. Baloul, M., Cherrier, E., Rosenberger, C.: Challenge-based speaker recognition for mobile authentication. In: 2012 BIOSIG-Proceedings of the International Conference of Biometrics Special Interest Group (BIOSIG), pp. 1–7. IEEE (2012)
12. Liu Y, He L, Zhang WQ, Liu J, Johnson MT (2018) Investigation of frame alignments for GMM-based digit-prompted speaker verification. In: 2018 Asia-Pacific signal and information processing association annual summit and conference (APSIPA ASC). IEEE, pp 1467–1472. https://doi.org/10.23919/apsipa.2018.8659790
13. Paul A, Das RK, Sinha R, Prasanna SM (2016) Countermeasure to handle replay attacks in practical speaker verification systems. In: 2016 international conference on signal processing and communications (SPCOM), pp 1–5. https://doi.org/10.1109/SPCOM.2016.7746646
14. Todisco M, Delgado H, Evans N (2017) Constant Q cepstral coefficients: a spoofing countermeasure for automatic speaker verification. Comput Speech Lang 45:516–535. https://doi.org/10.1016/j.csl.2017.01.001
15. Li L, Chen Y, Wang D, Zheng TF (2017) A study on replay attack and anti-spoofing for automatic speaker verification. In: Interspeech 2017. ISCA, Stockholm, Sweden, pp 92–96. https://doi.org/10.21437/Interspeech.2017-456
16. Witkowski M, Kacprzak S, Żelasko P, Kowalczyk K, Gałka J (2017) Audio replay attack detection using high-frequency features. In: Interspeech 2017. ISCA, Stockholm, Sweden, pp 27–31. https://doi.org/10.21437/Interspeech.2017-776
17. Liu M, Wang L, Lee KA, Chen X, Dang J (2021) Replay-attack detection using features with adaptive spectro-temporal resolution. In: ICASSP 2021—2021 IEEE international conference on acoustics, speech and signal processing (ICASSP), pp 6374–6378. https://doi.org/10.1109/ICASSP39728.2021.9414250
18. Font R, Espın JM, Cano MJ (2017) Experimental analysis of features for replay attack detection—Results on the ASVspoof 2017 challenge. In: Interspeech 2017, pp 7–11. https://doi.org/10.21437/Interspeech.2017-450
19. Cheng X, Xu M, Zheng TF (2020) A multi-branch ResNet with discriminative features for detection of replay speech signals. APSIPA Trans Signal Inf Process 9. https://doi.org/10.1017/ATSIP.2020.26
20. Kantheti S, Patil H (2018) Relative phase shift features for replay spoof detection system. In: 6th workshop on spoken language technologies for under-resourced languages (SLTU 2018). ISCA, pp 98–102. https://doi.org/10.21437/sltu.2018-21

21. Kamble MR, Tak H, Patil HA (2018) Effectiveness of speech demodulation-based features for replay detection. In: Interspeech 2018, pp 641–645. https://doi.org/10.21437/Interspeech.2018-1675
22. Patil HA, Kamble MR, Patel TB, Soni MH (2017) Novel variable length teager energy separation based instantaneous frequency features for replay detection. In: Interspeech 2017. ISCA, Stockholm, Sweden, pp 12–16. https://doi.org/10.21437/Interspeech.2017-1362
23. Guido RC (2019) Enhancing teager energy operator based on a novel and appealing concept: signal mass. J Franklin Inst 356(4):2346–2352. https://doi.org/10.1016/j.jfranklin.2018.12.007
24. Phapatanaburi K, Wang L, Nakagawa S, Iwahashi M (2019) Replay attack detection using linear prediction analysis-based relative phase features. IEEE Access 7:183,614–183,625. https://doi.org/10.1109/ACCESS.2019.2960369
25. Lavrentyeva G, Novoselov S, Tseren A, Volkova M, Gorlanov A, Kozlov A (2019) STC anti-spoofing systems for the ASVspoof2019 challenge. In: Interspeech 2019. ISCA, Graz, pp 1033–1037. https://doi.org/10.21437/Interspeech.2019-1768
26. Chen Z, Xie Z, Zhang W, Xu X (2017) ResNet and model fusion for automatic spoofing detection. In: Interspeech 2017, pp 102–106. https://doi.org/10.21437/Interspeech.2017-1085
27. Li X, Wu X, Lu H, Liu X, Meng H (2021) Channel-wise gated Res2Net: towards robust detection of synthetic speech attacks. arXiv:2107.08803 [cs, eess]
28. Huang L, Pun CM (2020) Audio replay spoof attack detection by joint segment-based linear filter bank feature extraction and attention-enhanced Densenet-BiLSTM network. IEEE/ACM Trans Audio, Speech, Lang Process 28:1813–1825. https://doi.org/10.1109/TASLP.2020.2998870
29. Gunendradasan T, Wickramasinghe B, Le PN, Ambikairajah E, Epps J (2018) Detection of replay-spoofing attacks using frequency modulation features. In: Interspeech 2018, pp 636–640. https://doi.org/10.21437/Interspeech.2018-1473
30. Lai CI, Abad A, Richmond K, Yamagishi J, Dehak N, King S (2019) Attentive filtering networks for audio replay attack detection. In: ICASSP 2019. IEEE, pp 6316–6320. https://doi.org/10.1109/ICASSP.2019.8682640
31. Ling H, Huang L, Huang J, Zhang B, Li P (2021) Attention-based convolutional neural network for ASV spoofing detection. In: Interspeech 2021. ISCV, pp 4289–4293. https://doi.org/10.21437/Interspeech.2021-1404
32. Zeinali H, Stafylakis T, Athanasopoulou G, Rohdin J, Gkinis I, Burget L, Černocký J et al (2019) Detecting spoofing attacks using VGG and SincNet: BUT-OMILIA submission to ASVspoof 2019 challenge. arXiv:1907.12908
33. Tak H, Patino J, Todisco M, Nautsch A, Evans N, Larcher A (2020) End-to-end anti-spoofing with RawNet2. arXiv:2011.01108 [eess]
34. Ren Y, Liu W, Liu D, Wang L (2021) Recalibrated bandpass filtering on temporal waveform for audio spoof detection. In: 2021 IEEE international conference on image processing (ICIP), pp 3907–3911. https://doi.org/10.1109/ICIP42928.2021.9506427
35. Ma Y, Ren Z, Xu S (2021) RW-Resnet: a novel speech anti-spoofing model using raw waveform. In: Interspeech 2021. ISCA, pp 4144–4148. https://doi.org/10.21437/Interspeech.2021-438
36. Chen T, Khoury E, Phatak K, Sivaraman G (2021) Pindrop labs' submission to the ASVspoof 2021 challenge. In: 2021 edition of the automatic speaker verification and spoofing countermeasures challenge. ISCA, pp 89–93. https://doi.org/10.21437/ASVSPOOF.2021-14
37. Tomilov A, Svishchev A, Volkova M, Chirkovskiy A, Kondratev A, Lavrentyeva G (2021) STC antispoofing systems for the ASVspoof2021 challenge. In: 2021 edition of the automatic speaker verification and spoofing countermeasures challenge. ISCA, pp 61–67. https://doi.org/10.21437/ASVSPOOF.2021-10
38. Ge W, Patino J, Todisco M, Evans N (2021) Raw differentiable architecture search for speech deepfake and spoofing detection. In: 2021 edition of the automatic speaker verification and spoofing countermeasures challenge. ISCA, pp 22–28. https://doi.org/10.21437/ASVSPOOF.2021-4
39. Ge W, Panariello M, Patino J, Todisco M, Evans N (2021) Partially-connected differentiable architecture search for Deepfake and spoofing detection. In: Interspeech 2021. ISCA, pp 4319–4323. https://doi.org/10.21437/Interspeech.2021-1187

40. Wang X, Yamagishi J (2021) A comparative study on recent neural spoofing countermeasures for synthetic speech detection. In: Interspeech 2021. ISCA, p 5. https://doi.org/10.21437/Interspeech.2021-702
41. Zhang Y, Jiang F, Duan Z (2021) One-class learning towards synthetic voice spoofing detection. arXiv:2010.13995 [cs, eess]
42. Zhang L, Wang X, Cooper E, Yamagishi J, Patino J, Evans N (2021) An initial investigation for detecting partially spoofed audio. In: Interspeech 2021. ISCA, pp 4264–4268. https://doi.org/10.21437/Interspeech.2021-738
43. Wu Z, Kinnunen T, Evans N, Yamagishi J, Hanilçi C, Sahidullah M, Sizov A (2015) ASVspoof 2015: the first automatic speaker verification spoofing and countermeasures challenge. In: Interspeech 2015. https://doi.org/10.21437/Interspeech.2015-462
44. Kinnunen T, Sahidullah M, Delgado H, Todisco M, Evans N, Yamagishi J, Lee KA (2017) The ASVspoof 2017 challenge: assessing the limits of replay spoofing attack detection. In: Interspeech 2017. ISCA, Stockholm, Sweden, pp 2–6. https://doi.org/10.21437/Interspeech.2017-1111
45. Todisco M, Wang X, Vestman V, Sahidullah M, Delgado H, Nautsch A, Yamagishi J, Evans N, Kinnunen T, Lee KA (2019) ASVspoof 2019: future horizons in spoofed and fake audio detection. In: Interspeech 2019. ISCA, pp 1008–1012. https://doi.org/10.21437/Interspeech.2019-2249
46. Yamagishi J, Wang X, Todisco M, Sahidullah M, Patino J, Nautsch A, Liu X, Lee KA, Kinnunen T, Evans N, Delgado H (2021) ASVspoof 2021: accelerating progress in spoofed and deepfake speech detection. In: 2021 edition of the automatic speaker verification and spoofing countermeasures challenge. ISCA, pp 47–54. https://doi.org/10.21437/ASVSPOOF.2021-8
47. Chettri B, Benetos E, Sturm BLT (2020) Dataset artefacts in anti-spoofing systems: a case study on the ASVspoof 2017 benchmark. IEEE/ACM Trans Audio, Speech, Lang Process 28:3018–3028. https://doi.org/10.1109/TASLP.2020.3036777
48. Korshunov P, Marcel, S, Muckenhirn H, Gonçalves AR, Souza Mello AG, Velloso Violato RP, Simoes FO, Neto MU, de Assis Angeloni M, Stuchi JA, Dinkel H, Chen N, Qian Y., Paul D, Saha G, Sahidullah M (2016) Overview of BTAS 2016 speaker anti-spoofing competition. In: 2016 IEEE 8th international conference on biometrics theory, applications and systems (BTAS), pp 1–6. https://doi.org/10.1109/BTAS.2016.7791200
49. Delgado H, Todisco M, Sahidullah M, Evans N, Kinnunen T, Lee KA, Yamagishi J (2018) ASVspoof 2017 version 2.0: meta-data analysis and baseline enhancements. In: Odyssey 2018—the speaker and language recognition workshop. https://doi.org/10.21437/Odyssey.2018-42
50. Wang X, Yamagishi J, Todisco M, Delgado H, Nautsch A, Evans NWD, Sahidullah M, Vestman V, Kinnunen T, Lee KA, Juvela L, Alku P, Peng Y-H, Hwang H-T, Tsao Y, Wang H-M, Le Maguer S, Becker M, Ling Z-H (2020) ASVspoof 2019: a large-scale public database of synthesized, converted and replayed speech. Comput Speech Lang 64:101114. https://doi.org/10.1016/j.csl.2020.101114. https://dblp.org/rec/journals/csl/WangYTDNESVKLJA20.bib
51. Chettri B, Sturm BL (2018) A deeper look at gaussian mixture model based anti-spoofing systems. In: 2018 IEEE international conference on acoustics, speech and signal processing (ICASSP). IEEE, pp 5159–5163. https://doi.org/10.1109/ICASSP.2018.8461467
52. Novak A, Simon L, Kadlec F, Lotton P (2010) Nonlinear system identification using exponential swept-sine signal. IEEE Trans Instrum Meas 59(8):2220–2229. https://doi.org/10.1109/tim.2009.2031836
53. Novak A, Lotton P, Simon L (2015) Synchronized swept-sine: theory, application, and implementation. J Audio Eng Soc 63(10):786–798. https://doi.org/10.17743/jaes.2015.0071
54. Kinnunen T, Sahidullah M, Falcone M, Costantini L, Hautamäki RG, Thomsen D, Sarkar A, Tan ZH, Delgado H, Todisco M et al (2017) Reddots replayed: a new replay spoofing attack corpus for text-dependent speaker verification research. In: ICASSP 2017. https://doi.org/10.1109/ICASSP.2017.7953187
55. Khan SS, Madden MG (2014) One-class classification: taxonomy of study and review of techniques. Knowl Eng Rev 29(3):345–374. https://doi.org/10.1017/S026988891300043X

56. Perera P, Oza P, Patel VM (2021) One-class classification: a survey. arXiv:2101.03064 [cs]
57. Cheng X, Xu M, Zheng TF (2021) Cross-database replay detection in terminal-dependent speaker verification. In: Interspeech 2021. ISCA, pp 4274–4278. https://doi.org/10.21437/Interspeech.2021-960
58. Kinnunen T, Hautamäki V, Fränti P (2006) On the use of long-term average spectrum in automatic speaker recognition. In: 5th Internat. Symposium on Chinese Spoken Language Processing (ISCSLP'06), Singapore, pp 559–567
59. Dempster AP, Laird NM, Rubin DB (1977) Maximum likelihood from incomplete data via the EM algorithm. J Roy Stat Soc: Ser B (Methodol) 39(1):1–22. https://doi.org/10.1111/j.2517-6161.1977.tb01600.x
60. Wu Z, Das RK, Yang J, Li H (2020) Light convolutional neural network with feature genuinization for detection of synthetic speech attacks. In: Interspeech 2020. ISCA, pp 1101–1105. https://doi.org/10.21437/Interspeech.2020-1810
61. Xie Y, Zhang Z, Yang Y (2021) Siamese network with wav2vec feature for spoofing speech detection. In: Interspeech 2021. ISCA, pp 4269–4273. https://doi.org/10.21437/Interspeech.2021-847
62. Xue J, Zhou H, Wang Y (2021) Physiological-physical feature fusion for automatic voice spoofing detection. arXiv:2109.00913 [eess]
63. Villalba J, Miguel A, Ortega A, Lleida E (2015) Spoofing detection with DNN and one-class SVM for the ASVspoof 2015 challenge. In: Interspeech 2015. ISCA, Dresden, Germany, pp 2067–2071. https://doi.org/10.21437/Interspeech.2015-468
64. Wang X, Qin X, Zhu T, Wang C, Zhang S, Li M (2021) The DKU-CMRI system for the ASVspoof 2021 challenge: vocoder based replay channel response estimation. In: 2021 edition of the automatic speaker verification and spoofing countermeasures challenge. ISCA, pp 16–21. https://doi.org/10.21437/ASVSPOOF.2021-3
65. Suthokumar G, Sriskandaraja K, Sethu V, Wijenayake C, Ambikairajah E, Li H (2018) Use of claimed speaker models for replay detection. In: 2018 Asia-Pacific signal and information processing association annual summit and conference (APSIPA ASC). IEEE, Honolulu, HI, USA, pp 1038–1046. https://doi.org/10.23919/APSIPA.2018.8659510
66. Wang D, Zhang X (2015) THCHS-30: a free Chinese speech corpus. arXiv:1512.01882 [cs]
67. Wang H, Dinkel H, Wang S, Qian Y, Yu K (2020) Dual-adversarial domain adaptation for generalized replay attack detection. In: Interspeech 2020. ISCA, pp 1086–1090. https://doi.org/10.21437/Interspeech.2020-1255

Chapter 15
Generalizing Voice Presentation Attack Detection to Unseen Synthetic Attacks and Channel Variation

You Zhang, Fei Jiang, Ge Zhu, Xinhui Chen, and Zhiyao Duan

Abstract Automatic Speaker Verification (ASV) systems aim to verify a speaker's claimed identity through voice. However, voice can be easily forged with replay, text-to-speech (TTS), and voice conversion (VC) techniques, which may compromise ASV systems. Voice presentation attack detection (PAD) is developed to improve the reliability of speaker verification systems against such spoofing attacks. One main issue of voice PAD systems is its generalization ability to unseen synthetic attacks, i.e., synthesis methods that are not seen during training of the presentation attack detection models. We propose one-class learning, where the model compacts the distribution of learned representations of bona fide speech while pushing away spoofing attacks to improve the results. Another issue is the robustness to variations of acoustic and telecommunication channels. To alleviate this issue, we propose channel-robust training strategies, including data augmentation, multi-task learning, and adversarial learning. In this chapter, we analyze the two issues within the scope of synthetic attacks, i.e., TTS and VC, and demonstrate the effectiveness of our proposed methods.

Y. Zhang (✉) · G. Zhu (✉) · Z. Duan (✉)
University of Rochester, Rochester, USA
e-mail: you.zhang@rochester.edu

G. Zhu
e-mail: ge.zhu@rochester.edu

Z. Duan
e-mail: zhiyao.duan@rochester.edu

F. Jiang (✉)
Tencent Technology Co., Ltd., Shenzhen, China
e-mail: faithjiang@tencent.com

X. Chen (✉)
Lehigh University, Bethlehem, USA
e-mail: xic721@lehigh.edu

S. Marcel et al. (eds.), *Handbook of Biometric Anti-Spoofing*, Advances in Computer Vision and Pattern Recognition, https://doi.org/10.1007/978-981-19-5288-3_15

15.1 Introduction

Automatic Speaker Verification (ASV) systems aim to verify whether a given utterance is from the target person. It is widely used in industries including financial services, law enforcement, online education, mobile workforce, healthcare, and IT security. ASV systems typically work in two stages: *enrollment* and *verification*, i.e., users first register their voices into the system, and then the system is able to verify their voices in various applications [1]. Existing ASV systems are expected to receive natural speech spoken by humans, i.e., *bona fide* speech, and they work well for these inputs. However, when fake speech, i.e., *presentation attacks*, are presented, these systems become vulnerable as they often cannot discern attacks from bona fide speech [2–4].

There are different kinds of presentation attacks. In a common categorization, as adopted in the ASVspoof 2019 challenge [5], *physical access (PA)* attacks refer to those pre-recorded utterances presented to the microphone of the ASV systems, whereas *logical access (LA)* attacks bypass the microphone and feed directly to the verification algorithm, utilizing synthetic utterances generated by text-to-speech (TTS) and voice conversion (VC) algorithms. As TTS and VC techniques are being developed at a fast pace, LA attacks pose an ever-growing threat to existing ASV systems. Some attacks are specially designed to spoof ASV systems with TTS and VC techniques [6, 7]. Some attacks can even deceive human perception [8–10]. Therefore, in this chapter, we present the voice presentation attack detection (PAD) issues in the context of LA attacks. Nevertheless, we believe that PA attacks could suffer from similar issues and the proposed techniques could be applied to address PA attacks as well.

The main challenge in the design of robust voice PAD systems is to ensure that such systems can generalize well to unknown conditions in real-world scenarios. One generalization is to unseen synthetic attacks, i.e., speech synthesis methods not seen in training the PAD models. With the development of speech synthesis technology, synthetic speech is causing increasing concerns for users. First, with more open-sourced speech synthesis algorithms, people have easier access to these technologies, but at the same time, these technologies are more likely to be applied to generate synthetic attacks on ASV systems. Second, as new speech synthesis methods are being developed, synthetic attacks used in training PAD models would never be complete to cover all possible attacks in the real world. As a result, it is important to improve the generalization ability to unseen attacks. As reviewed in Sect. 15.2, the recent development of PAD models has achieved reasonably good performance on seen attacks, however, their performance on unseen attacks still falls far behind.

Another generalization is to the variability of acoustic and telecommunication channel conditions. Due to limited conditions presented in the training data, the performance of PAD systems would degrade when exposed to channel variation. Under degraded conditions, even humans have worse performance [9] in detecting spoofing attacks. As reviewed in Sect. 15.3, this issue has received less attention. The performance degradation of state-of-the-art voice PAD systems has been revealed in some cross-dataset studies. Of all the factors presenting cross-dataset effects, we tested and verified the channel effect, especially the device impulse responses, as a reason for performance degradation of voice PAD systems in [11]. Without considering these effects, systems may fail to produce robust results under different channel variations.

This chapter will focus on LA attacks and elaborate on the techniques used to improve the generalization ability against unseen synthetic attacks and channel variation. We first introduce one-class learning to improve the generalization ability to unseen attacks in Sect. 15.2. We then introduce the training strategies for resolving the channel robustness in Sect. 15.3. The results indicate that with our proposed training strategies, the generalization ability can be improved. We also provide an outlook to open challenges that remain unaddressed and propose possible strategies to refine our method in the discussion of each section. Lastly, we conclude this chapter and point out some possible directions for further improving the generalization ability.

15.2 Generalize to Unseen Synthetic Attacks

Due to the fast development of speech synthesis technology, TTS and VC attacks, especially unseen attacks, impose increasing threats onto ASV systems. By unseen attacks, we mean that no examples of such *kinds* of attacks are available during the training of the ASV systems. As an analogy, suppose we are discriminating images of humans and animals, and the training set only contains dogs, cats, and birds for the animal class. If images of monkeys, however, are contained in the test set, then they would be considered unseen, and our classifier is likely to misclassify them as images of humans. This is different from a typical machine learning evaluation setup where only the data instances are unseen but the kinds or types are already included in the training set. For PAD, we are dealing with a case that the spoofing attacks have different distributions in training and testing since they are generated with different TTS and VC algorithms.

As an intrinsic problem to solve in voice PAD, much research has been attempted to improve the generalization ability to unseen attacks with front-end speech features and backend model architectures.

Some methods explore speech features that could reveal artifacts in spoofing attacks to distinguish them from bona fide speech, such as CFCCIF [12], LFCC [13], CQCC [14], and SCC [15]. Some methods use a data-driven approach and rely on deep learning model architectures, such as LCNN [16], ResNet [17], Res2Net [18], TDNN [19], capsule network [20], siamese network [21], and RawNet2 [22]. All of the methods have shown promising performance on seen attacks. Nevertheless, the performance on unseen attacks is still limited.

All of these methods treat PAD as a regular binary classification problem, however, such formulation assumes that the training and test data have similar distributions. For unseen attacks, this assumption is clearly incorrect, as the distribution of unseen attacks in testing can differ significantly from that of attacks in training. To improve the generalization ability, the idea of one-class learning with one-class Softmax as the loss function is proposed in [23]. The basic idea is to view and represent bona fide speech as the target class, but spoofing attacks as data points that may appear anywhere not close to the bona fide speech class in the embedding space. This is implemented by introducing two different margins in the loss function to compact the bona fide speech and to push away the spoofing attacks. Since the proposal of the one-class learning idea [23], the one-class Softmax loss function has been used in some recent works [24–27] and produced good results.

15.2.1 One-Class Learning

The concept of one-class classification has been used in various contexts. Broadly speaking, one-class classification aims to train a classifier with training data where the non-target class either lacks data or is not statistically representative [28, 29]. Given that the voice PAD problem suffers from unseen attacks that are statistically underrepresented, the one-class classification fits the problem formulation. The target class refers to bona fide speech and the non-target class refers to spoofing attacks.

Some methods deal with the case where data of only one class of interest, i.e., the target class, are accessible for training [30–33]. Our previous attempt showed that training the model only with bona fide speech does not work well. The possible reason is that synthetic speech generated by advanced methods shares many similar features with bona fide speech, and it is difficult for the model to discriminate between them without learning from spoofing attacks. This urges us to make use of data from both classes in the training set.

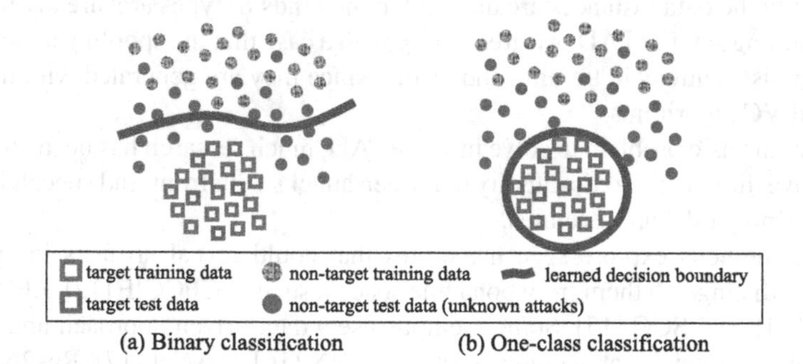

Fig. 15.1 Illustration of binary classification and one-class classification. The hollow squares denote the target data and the solid circles denote the non-target data. Training data are textured and colored in blue. As the non-target class is not well represented in the training set, it shows a different distribution in the test set. With binary classification, the classification boundary is learned to classify the two classes in the training set, but it often misclassifies non-target data in the test set. One-class classification, on the other hand, aims to learn a tight boundary around the target class, and often shows a better generalization ability to non-target data in the test set

In this setup, both one-class classification and binary classification train on both classes, but their concepts are different, as illustrated in Fig. 15.1. For binary classification, the decision boundary well separates the training data, but it lacks the ability to detect unseen non-target data due to its distribution mismatch between training and testing. The one-class classification boundary is much tighter around the target class, and successfully detects unseen distributions of non-target data in the test set.

For voice PAD, the potential of one-class classification was first explored in [34] where a one-class support vector machine (OC-SVM) [30] is employed based on local binary patterns of speech cepstrograms, without taking advantage of any information from spoofing attacks. To leverage information from spoofing attacks, a two-stage method was proposed in [35]. The speech feature embedding space is learned through binary classification with a deep neural network, and then an OC-SVM is used to fit a classification boundary close to the bona fide class. However, the two-stage training fails to optimize the learned speech embedding space for the OC-SVM to set up the boundary. We introduce the idea of one-class learning, where the embedding space is learned jointly with the one-class classification objective. In this way, the learned embedding space is better suited for one-class classification. We implement this idea by utilizing or designing loss functions to train the machine learning model. For bona fide speech, we encourage their embeddings to be concentrated, while for spoofing attacks, we push away their embeddings from the cluster of the bona fide speech. The loss function evaluates distances or similarities between learned features and the cluster center, where the distance can be Euclidean distance or cosine similarity in the embedding space. The illustration of the loss design is shown in Fig. 15.2. In this way, a tighter classification boundary is easier to draw around the target class (i.e., bona fide speech), and the generalization to unseen non-target-class data (i.e., unseen spoofing attacks) can be improved.

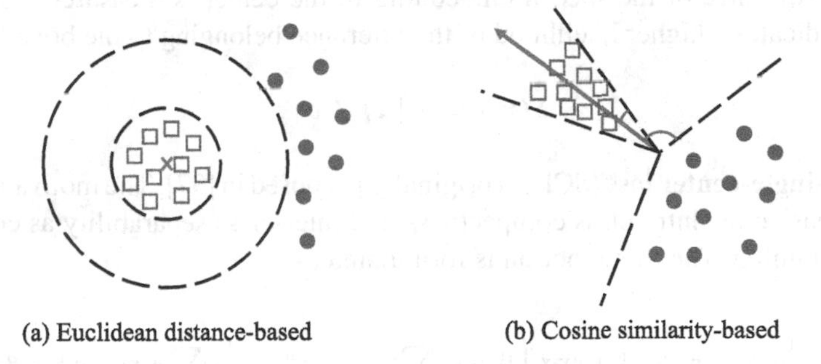

Fig. 15.2 Illustration of one-class learning with Euclidean distance-based loss and cosine similarity-based loss. The hollow squares represent the bona fide speech data and the solid circles represent the spoofing attacks in the embedding space

Notation. Given a mini-batch that has N samples $(\boldsymbol{x}_i, y_i)$, where $\boldsymbol{x}_i$ is the speech embedding from the output of the embedding network and $y_i \in \{0, 1\}$ is the label, we define $\Omega = \{i | y_i = 0\}$ as the set of bona fide speech. Then $\overline{\Omega} = \{i | y_i = 1\}$ is the set of spoofing attacks, and $|\Omega| + |\overline{\Omega}| = N$.

15.2.1.1 Euclidean Distance-Based One-Class Loss

As shown in Fig. 15.2a, the goal is to compact the target data distribution and to push away the non-target data. We introduce two loss functions originally proposed for image forgery detection and elaborate on the idea in the context of one-class learning.

We refer **isolate loss** (ISO) to the loss originally proposed in [36] for deepfake detection. The motivation was to induce the compactness for the embeddings of natural faces by moving them close to a center and separating them from the embeddings of manipulated faces. In the context of voice PAD, we compact bona fide speech and isolate spoofing attacks. The formula can be represented as

$$L_{ISO} = \frac{1}{|\Omega|}\sum_{\Omega} \max\left(0, \|\boldsymbol{x}_i - \boldsymbol{c}\| - r_0\right) + \frac{1}{|\overline{\Omega}|}\sum_{\overline{\Omega}} \max\left(0, r_1 - \|\boldsymbol{x}_i - \boldsymbol{c}\|\right), \tag{15.1}$$

where $\boldsymbol{c}$ is the target class center, and $\|\cdot\|$ denotes for the Euclidean distance. The radii r_0 and r_1 ($r_0 < r_1$) are pre-set hyperparameters (margins) for the target class and the non-target class, respectively. If a bona fide sample is outside the small circle with radius r_0, a loss will be incurred, and minimizing the loss will help pull back this sample. Similarly, if a spoofing sample is inside the large circle with a radius r_1, a loss will be incurred, and minimizing the loss will help push away the sample. The margin $r_1 - r_0$ ensures the discriminative ability of the model. For scoring, the negative distance of the speech embedding to the center is measured, so a higher score indicates a higher likelihood of the utterance belonging to the bona fide class.

$$S(\boldsymbol{x}_i) = -\|\boldsymbol{x}_i - \boldsymbol{c}\|. \tag{15.2}$$

The **single-center loss** (SCL) is originally proposed in [37]. The motivation is also to emphasize the intra-class compactness and inter-class separability as constraints during training. The loss function is formulated as:

$$L_{SCL} = \frac{1}{|\Omega|}\sum_{\Omega} \| \boldsymbol{x}_i - \boldsymbol{c} \| + \max\left(0, \frac{1}{|\Omega|}\sum_{\Omega} \| \boldsymbol{x}_i - \boldsymbol{c} \| - \frac{1}{|\overline{\Omega}|}\sum_{\overline{\Omega}} \| \boldsymbol{x}_i - \boldsymbol{c} \| + \beta\sqrt{D}\right). \tag{15.3}$$

The confidence score would be the same as in Eq. (15.2). Instead of using two radii, the margin is $m\sqrt{D}$ in this loss function, where D denotes the embedding dimension. A scale factor β controls the ratio with respect to the volume of the embedding space, which makes the hyperparameter setting easier. As stated in [37], the single-center loss needs to be trained together with the Softmax loss since updating the center point makes the model unstable. The global information of the dataset is carried with the Softmax loss and can guide the training of the center.

The curse of dimensionality states that data points tend to concentrate on the surface of the sphere with the increasing dimensionality [38]. As an induction, there is little difference in the Euclidean distances between different pairs of samples [39]. As such, the hyperparameters in the Euclidean distance-based loss functions are generally hard to tune and may fail to generalize to other feature embedding spaces. For instance, when the embedding dimension D changes, the radii r_0 and r_1 in Eq. (15.1) and the scale factor β in Eq. (15.3) need to be carefully adjusted to compensate for the volume change.

15.2.1.2 Cosine Similarity-Based One-Class Loss

With the assumption that data lie on a manifold, a hypersphere constraint can be added to the embedding space and encourage the model to learn angular discriminative features as in [40]. In such space, the embeddings are compared with a center vector using cosine similarity. As shown in Fig. 15.2b, a small angle (large cosine similarity) can make the target class concentrate around the center vector, whereas a relatively large angle (small cosine similarity) can push the non-target data away.

The **one-class Softmax** (OC-Softmax, OCS) was proposed in our previous work [23]. We introduced two different margins for better compacting the bona fide speech and pushing away the spoofing attacks. Derived from Softmax and AM-Softmax [41] while incorporating one-class learning, the OC-Softmax is defined as

$$L_{OCS} = \frac{1}{N} \sum_{i=1}^{N} \log\left(1 + e^{\alpha(m_{y_i} - \hat{\boldsymbol{w}}^T \hat{\boldsymbol{x}}_i)(-1)^{y_i}}\right), \tag{15.4}$$

where $\hat{\boldsymbol{w}}$ and $\hat{\boldsymbol{x}}_i$ are normalized to have unit L2 norm from the weight vector $\boldsymbol{w}$ and embedding vector $\boldsymbol{x}$, respectively. The normalized weight vector $\hat{\boldsymbol{w}}$ serves as the angular center, which refers to the optimization direction of the target class embeddings. Two margins ($m_0, m_1 \in [-1, 1], m_0 > m_1$) are introduced here for bona fide speech and spoofing attacks, respectively, to bound the angle θ_i between $\boldsymbol{w}$ and $\boldsymbol{x}_i$. When $y_i = 0$, m_0 is used to force θ_i to be smaller than $\arccos m_0$, whereas when $y_i = 1$, m_1 is used to force θ_i to be larger than $\arccos m_1$. The connection of OC-Softmax with Softmax and AM-Softmax is shown in Appendix. The confidence score is defined as:

$$S(\boldsymbol{x}_i) = \hat{\boldsymbol{w}}^T \hat{\boldsymbol{x}}_i. \tag{15.5}$$

Note that the Eq. (15.4) may be influenced by the data imbalance issue. For instance, the positive samples are 10 times less than the negative samples in the training set of ASVspoof 2019 LA. In that case, the negative samples would contribute more gradients when updating the model, leading to a bias toward pushing attacks out of the boundary. To address this imbalance issue, we can modify Eq. (15.4) by grouping the positive samples and negative samples and normalizing their losses as

$$L_{AISO} = \frac{1}{|\Omega|}\sum_{\Omega} \log\left(1 + e^{\alpha\left(m_0 - \hat{w}^T\hat{x}_i\right)}\right) + \frac{1}{|\overline{\Omega}|}\sum_{\overline{\Omega}} \log\left(1 + e^{\alpha\left(\hat{w}^T\hat{x}_i - m_1\right)}\right). \tag{15.6}$$

Thanks to the normalization factors $|\Omega|$ and $|\overline{\Omega}|$, now compacting the bona fide speech and pushing the attacks are equally important. Interestingly, we find that the form of Eq. (15.6) is very similar to the isolated loss in Eq. (15.1) and can be viewed as its angular version, so we name it **angular isolate loss** (A-ISO).

Compared with Euclidean distance-based losses in Sect. 15.2.1.1, our angular losses have advantages in two-fold. On one hand, rather than the radius that depends on the embedding dimension, we use the angle between feature vectors. This results in a more convenient hyperparameter tuning process. It is also shown in [40] that using angular margin is more effective for face recognition. On the other hand, the softplus function is utilized instead of the hinge function. The soft operation can assign a larger weight to a negative pair with a higher similarity, which is more informative to train the discrimination [42, 43]. The derivation is shown in Appendix.

15.2.2 *Experiments*

15.2.2.1 Experimental Setup

The **ASVspoof 2019 LA** sub-challenge provides a new database [10] split into training, development, and evaluation sets. The training and development set contains bona fide speech and 6 synthetic voice spoofing attacks (A01–A06), and the evaluation set includes 13 different attacks (A07–A19) that are unseen during training. We employ this dataset to benchmark the generalization ability of voice PAD systems to unseen synthetic attacks. The official protocol is followed and no data augmentation or model fusion is used to boost the performance.

The speech features, linear frequency cepstral coefficients (LFCC) for each utterance, are extracted to feed into the network. We randomly chop or use repeat padding to each utterance to obtain a 5 s segment for feature extraction. The evaluation metric we use is equal error rate (EER) and a minimum of the tandem detection cost function (min t-DCF) [44], which are the official metrics for the ASVspoof 2019 challenge. The lower is better for both metrics. EER is calculated by setting a threshold θ on the PAD confidence score such that the false alarm rate is equal to the miss rate, where the two rates are defined as:

$$P_{fa}(\theta) = \frac{\#\{\text{ spoof trials with score } > \theta\}}{\#\{\text{ total spoof trials }\}},$$
$$P_{\text{miss}}(\theta) = \frac{\#\{\text{ bona fide trials with score } \leq \theta\}}{\#\{\text{ total bona fide trials }\}}. \quad (15.7)$$

The lower the EER is, the better the PAD system is at detecting spoofing attacks. The t-DCF metric assesses the influence of PAD on the reliability of an ASV system. The ASV system is fixed to compare different PAD systems. The lower the minimum t-DCF is, the more reliable the ASV system is under the influence of the PAD system.

We use ResNet18 with attentive temporal pooling [17] as our backend model architecture to extract embeddings from speech features. Regarding the hyperparameters for each of the losses in Sect. 15.2.1, we set the embedding dimension D as 256 for all the experiments. The r_0 and r_1 in the isolate loss are 25 and 75, respectively. The scale factor β is 0.5 in the single-center loss. Note that the single-center loss needs to be additive to Softmax, we set the weight of SCL as 0.5. The scale factor α is 20, and the m_0 and m_1 are 0.5 and 0.2, respectively in the OC-Softmax and angular isolate loss. We train the network for 100 epochs using the Adam optimizer and select the model with the smallest validation loss for evaluation.

15.2.2.2 Results

To verify the effectiveness of one-class learning, we compare different loss functions with the same speech feature and the same backend model to produce the speech embedding. As observed in [25], the same model may perform differently when changing the random seed in training; Sometimes the differences between the best and worst trials of the same model could be even larger than those between different models. Therefore, with our experimental setup, we set different random seeds for each of the training-evaluation trials. We report both the best and average results of the trials in Table 15.1.

Table 15.1 Results on the evaluation set of the ASVspoof 2019 LA dataset using different loss functions. Except for the Softmax loss, all the other four losses are one-class losses. The best and average results across 4 trials with random training seeds are shown. All of the voice PAD systems are trained on the training set of ASVspoof 2019 LA and validated on its development set

Loss functions	Best		Average	
	EER (%) ↓	min t-DCF ↓	EER (%) ↓	min t-DCF ↓
Softmax	4.45	0.1096	4.74	0.1128
Isolate loss	2.45	0.0632	2.77	0.0686
Single-center loss	2.41	0.0619	2.97	0.0810
One-class softmax	2.11	0.0596	**2.63**	0.0685
Angular isolate loss	**1.96**	**0.0537**	2.74	**0.0675**

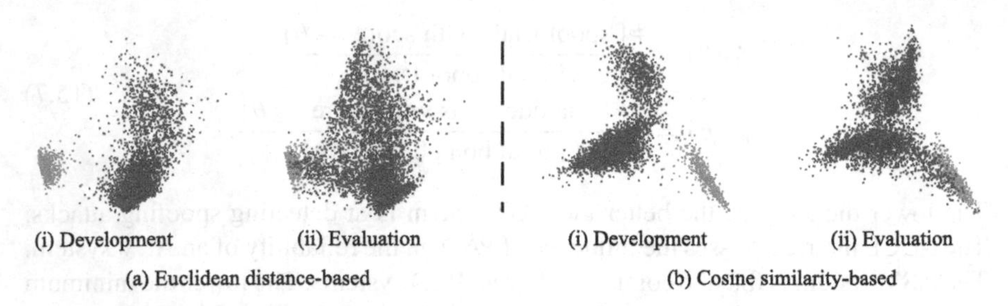

Fig. 15.3 Visualization of the learned feature embeddings with two kinds of loss functions. The light blue represent bona fide speech, while the red dots denote the spoofing attacks

From Table 15.1, the performance from the one-class learning losses is significantly better than the binary classification with the Softmax loss. Since all of the attacks in the evaluation set are unseen during training, this shows the superiority of one-class learning on the generalization ability to unseen synthetic attacks. Among the four one-class losses, they achieved similar results, and the cosine similarity-based losses show slightly better performance. Note that the cosine similarity-based losses are easier to set hyperparameters and are more robust to the curse of dimensionality. We believe these advantages make them better to integrate into other models and tasks.

To further investigate the one-class learning procedure, we applied Principal Component Analysis (PCA) dimensionality reduction to the learned speech embeddings of the development and evaluation sets of the ASVspoof 2019 LA dataset, as shown in Fig. 15.3. In other words, the visualizations of the two sets use the same coordinate systems.

In subfigure (a)(i), the embeddings of the bona fide speech are compact as a cluster, and a certain distance is injected between bona fide and spoofing attacks, verifying our hypothesis in Fig. 15.2a. Similarly, as shown in subfigure (b)(i), the bona fide speech is concentrated as a slender shape, and the spoofing attacks are isolated with an angular margin, verifying our hypothesis in Fig. 15.2b.

Comparing subfigures (i) and (ii) in both the Euclidean distance-based loss (a) and the cosine similarity-based loss (b), the bona fide speech has the same distribution in both sets, while unknown attacks in the evaluation set show different distributions from the known attacks in the development set. This suggests that the bona fide class is well characterized by the instances in the training data, but the spoofing attacks in the training data cannot form a statistical representation of the unknown attacks, hereby verifying our problem formulation with one-class learning. In both subfigures (ii), although the distributions of spoofing attacks are changed due to unseen attacks, the distance or angle between bona fide speech and spoofing attacks is still maintained, showing the effectiveness of our one-class losses.

15.2.3 Discussions

Our method for voice PAD with one-class learning shows promising results to enhance the robustness of the model against unseen spoofing attacks. With one-class learning, the model aims to learn a speech embedding space in which bona fide speech has a compact distribution while spoofing attacks reside outside by an (angular) margin. We believe it could also work in other biometric anti-spoofing applications, especially problems facing similar issues of generalization to unseen attacks.

Our proposed one-class learning method still has some limitations in voice PAD. Since we encourage the boundary for the bona fide speech to be tight, it might overfit the bona fide training utterances and might fail to generalize to their variations, for instance, utterances of different speakers, languages, and recording conditions. Future work could be exploring other ways of combining one-class classification and binary classification, for example, one-and-a-half class classification proposed in [45, 46], in which one of the two classifiers only models the target data to encourage compactness and the other learns to discriminate the two classes. Incorporating adversarial training in one-class learning [47] to generate out-of-distribution data may also help improve the performance.

15.3 Generalize to Channel Variation

Voice presentation attack detection (PAD) systems in real-world applications are expected to be exposed to diverse conditions such as reverberation and background noise in recording environments, frequency responses of recording devices, and transmission and compression-decompression algorithms in telecommunication systems. Some cross-dataset studies in LA have demonstrated that performance degradation exists in voice PAD [11, 48]. Some work also proposed methods to improve the robustness to noisy condition [49, 50], channel variation [11], and domain adaptation [51].

Recently, the LA sub-challenge of the ASVspoof 2021 challenge [52] takes care of the robustness to channel variability, especially the transmission codec variation. Prior to being processed by the voice PAD systems, the bona fide speech and LA spoofing attacks are usually transmitted through a public switched telephone network (PSTN) or voice over Internet protocol (VoIP) network using some particular codec [53]. Such transmission codec causes some degradation on the speech data due to bandwidths packet loss and different sampling frequencies, which should not be considered as artifacts that we aim to detect for discriminating spoofing attacks. Due to limited access to the channel variation of the training data, voice PAD systems may misregard the degradation as the artifacts of the speech data, hence being frail in practice. Without properly considering and compensating for these effects, voice PAD systems may overfit the limited channel effects presented in the training set and fail to generalize to unseen channel variation. For example, the bona fide speech utterances in ASVspoof 2019 LA were all from the VCTK corpus, and the TTS/VC systems used in generating LA attacks were all trained on the VCTK corpus. This may introduce strong biases to PAD systems on the limited channel variation. Recent attempts [27, 54–56] in the ASVspoof 2021 Challenge have focused on data augmentation and model fusion to achieve good results in the challenge.

In our previous work [11], we conducted cross-dataset studies on the existing LA-attack datasets and observed performance degradation. We hypothesized that the channel mismatch contributes to the degradation and verified that with a new channel-augmented dataset named ASVspoof 2019 LA-Sim. We then proposed several strategies (data augmentation, multi-task learning, and adversarial learning) to improve the robustness of voice PAD systems to channel variation, and obtained significant improvement in both in-domain and cross-dataset tests. Our one-class learning and channel-robust training strategies were also applied in our submitted PAD system [26] to ASVspoof 2021 challenge.

In this section, we introduce data augmentation techniques and three training strategies proposed in [11, 26] to improve the generalization ability of voice PAD systems to channel variation.

15.3.1 Channel-Robust Strategies

15.3.1.1 Data Augmentation

Data augmentation methods are shown effective to improve the cross-dataset performance of voice PAD systems [11]. To address the channel variability in the LA task, we design different augmentation methods. We augment the channel effects of the ASVspoof 2019 LA dataset through an open-source channel simulator [57]. The channel simulator provides three types of degradation processes including additive noise, telephone and audio codecs, and device/room impulse responses (IRs).

Device Augmentation. We choose 12 out of 74 different device IRs and apply each of them to all utterances of ASVspoof 2019 LA. This channel-augmented dataset is named *ASVspoof 2019 LA-Sim*, and the same train/dev/eval split is followed from ASVspoof 2019 LA. As shown in our previous study [11] where we applied device augmentation, it is beneficial to improve cross-dataset performances.

Transmission Augmentation. We apply landline, cellular, and VoIP codec groups, which are the most commonly used transmission channels. To be specific, the landline codec includes μ-law and A-law companding and adaptive-differential PCM. Cellular includes GSM and narrow-band and wide-band advance multi-rate (AMR-NB and AMR-WB). And VoIP includes ITU G.722, ITU G.729 standards besides SILK and SILK-WB. All of the above-mentioned codecs are equipped with different bitrate options. After going through these different transmission codecs, we resample all of the resulting audio files with 16 kHz for further processing. The codec parameters such as bit rate and packet loss are randomly sampled to handle different settings. This results in 60 transmission codec options, among which we randomly sampled and created 21 parallel transmission codec audio files for each utterance in the ASVspoof 2019 LA dataset.

Dual Augmentation. To combine the benefits from both the device augmentation and transmission augmentation introduced above, we introduce a dual combination, i.e., convolving the original audio data with recording device IRs after applying transmission augmentation. We randomly sampled 21 out of the 60 transmission codecs and 12 out of the 74 device IRs. For each utterance, we created 21 audio files with dual augmentation, where a random device augmentation and a random transmission augmentation are applied. We believe combining device augmentation with transmission augmentation would further improve the robustness against unseen transmission variation.

15.3.1.2 Training Strategies

We refer to the model trained only on the ASVspoof 2019 LA dataset as the Vanilla model, based on which three channel-robust training strategies were proposed in our previous work [11] to make use of the augmented data. The model structures of them are compared in 15.4. An embedding network, parameterized by θ_e, aims to learn a discriminative embedding from speech features, which is then classified by a fully connected (FC) layer as an anti-spoofing classifier, parameterized by θ_a, into bona fide speech or spoofing attacks.

Augmentation (AUG) uses the same model architecture as the Vanilla model, as shown in Fig. 15.4a, but the model is trained on the augmented dataset.

$$(\hat{\theta}_e, \hat{\theta}_a) = \underset{\theta_e, \theta_a}{\arg\min} L_a\left(\theta_e, \theta_a\right). \tag{15.8}$$

Multi-Task Augmentation (MT-AUG) adds a classifier, parameterized by θ_c, to the model architecture, as shown in Fig. 15.4b. The input is the speech embedding output from the embedding network. In a supervised learning fashion, the channel labels are provided to guide the training. The gradient from the channel classification cross-entropy loss is backpropagated to update the parameters in the embedding net-

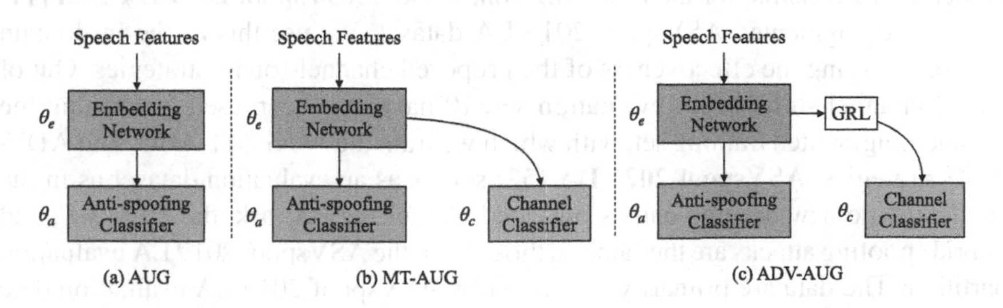

Fig. 15.4 Comparison of model structures of three channel-robust training strategies

work, which encourages the channel discrimination ability of the speech embeddings. The addition of this channel classifier informs the voice PAD system of the channel information during training, potentially leading to improved channel robustness.

$$(\hat{\theta}_e, \hat{\theta}_a, \hat{\theta}_c) = \underset{\theta_e,\theta_a,\theta_c}{\arg\min} L_a(\theta_e, \theta_a) + \lambda L_c(\theta_e, \theta_c). \tag{15.9}$$

Adversarial Augmentation (ADV-AUG) inserts a Gradient Reversal Layer (GRL) [58] between the embedding network and the channel classifier, as shown in Fig. 15.4c. The gradient from the channel classification loss is backpropagated through the GRL, with the sign reversed, to optimize the embedding network. Therefore, the embedding network aims to maximize the channel classification error while the channel classifier aims to minimize it, forming an adversarial training paradigm. When equilibrium is reached, the learned speech embeddings would be channel-agnostic, making the anti-spoofing classifier robust to channel variation. The overall training objective is as follows:

$$\begin{aligned}(\hat{\theta}_e, \hat{\theta}_a) &= \underset{\theta_e,\theta_a}{\arg\min} L_a(\theta_e, \theta_a) - \lambda L_c(\theta_e, \hat{\theta}_c),\\ (\hat{\theta}_c) &= \underset{\theta_c}{\arg\min} L_c(\hat{\theta}_e, \theta_c).\end{aligned} \tag{15.10}$$

15.3.2 *Experiments*

To verify the effectiveness of our channel-robust strategies, we conduct in-domain and out-of-domain tests to explore how the strategies contribute to generalizing channel variation.

15.3.2.1 Experimental Setup

ASVspoof 2019 LA [10], as introduced in Sect. 15.2.2.1, serves as the base dataset for data augmentation. The model trained only on ASVspoof 2019 LA (Vanilla model) is the baseline for the following comparison. **ASVspoof 2019 LA-Sim** [11] is a device augmented ASVspoof 2019 LA dataset. We use this as the in-domain test for verifying the effectiveness of the proposed channel-robust strategies. Out of the 12 channel shifts in the evaluation set, 10 have also been used in forming the channel-augmented training set, with which we train the AUG, MT-AUG, and ADV-AUG strategies. **ASVspoof 2021 LA** [52] serves as an evaluation dataset as in the challenge, no new training data is provided. As for the LA task, the TTS, VC, and hybrid spoofing attacks are the same as those from the ASVspoof 2019 LA evaluation partition. The data are primarily generated by ASVspoof 2019 LA evaluation data going through several unknown transmission codec channels, in total 181,566 trials.

The training and evaluation procedure is the same as in Sect. 15.2.2.1. For MT-AUG and ADV-AUG, λ is updated according to $\lambda_n = \frac{2}{1+\exp(-\gamma n)} - 1$, as suggested in [58], where n represents the epoch number at the current epoch. We set the γ as 0.001. This aims to alleviate the noisy signal from the channel classifier at the early stage of training.

15.3.2.2 Results

In the in-domain test, we test the training strategies on the evaluation set of our simulated ASVspoof 2019 LA-Sim dataset. We choose three widely used models that provided state-of-the-art performance in the literature, LCNN [16], ResNet [17], and RawNet [22]. We train them with our proposed OC-Softmax loss. LCNN and ResNet take LFCC feature as the input while RawNet takes raw waveform as the input. We train the models on the device augmented training set, containing the original ASVspoof 2019 LA training data and only 10 out of the 12 channel shifts of the ASVspoof 2019 LA-Sim training data, except the Vanilla model. We expect that the training strategies with augmentation would decrease the EER and min t-DCF compared with the Vanilla model, which would indicate the effectiveness of the strategies. The results are shown in Table 15.2.

The Vanilla model shows less promising results compared with their performance (2–5% EER) on the ASVspoof 2019 LA evaluation set. This indicates that the model trained only on ASVspoof 2019 LA cannot generalize well to channel variation. Comparing the performance achieved by the channel-robust training strategies with the Vanilla model, the significant decrease of EER and min t-DCF shows that the proposed strategies do improve the channel robustness of the voice PAD systems. These strategies work for different kinds of models, spectral-domain methods, and time-domain methods. Comparing the three strategies, they perform similarly. Surprisingly, the performance of the RawNet model decreases much less than the LCNN and ResNet model, showing the superiority and robustness of the raw waveform features.

Table 15.2 EER and min t-DCF performance on the evaluation set of the ASVspoof 2019 LA-Sim (device augmented ASVspoof 2019 LA dataset) using different training strategies

EER (%)	Training strategies			
min t-DCF	**Vanilla**	**AUG**	**MT-AUG**	**ADV-AUG**
LCNN	40.15	5.57	**4.22**	4.85
	0.9134	0.1373	**0.1089**	0.1165
ResNet	43.54	4.75	**4.32**	6.02
	0.9988	0.1132	**0.1131**	0.1319
RawNet	11.65	7.19	5.02	**4.72**
	0.3870	0.1874	0.1389	**0.1171**

Table 15.3 EER and min t-DCF performance on the evaluation set of the ASVspoof 2021 LA trained with data augmented ASVspoof 2019 LA using different data augmentation and training strategies

EER (%) / min t-DCF	Vanilla	Training strategies: AUG	MT-AUG	ADV-AUG
Device augmentation	21.78	11.71	14.33	**10.60**
	0.7485	**0.3608**	0.3891	0.4290
Transmission augmentation		7.67	10.23	**7.20**
		0.3159	0.3755	0.3190
Dual augmentation		7.06	9.17	**6.84**
		0.3447	0.4051	**0.3222**

In the out-of-domain test, we test the performance on ASVspoof 2021 LA. Since the ASVspoof 2021 LA contains some unknown transmission variation, we use our transmission augmentation, device augmentation, and dual augmentation, and compare them. Without loss of generality and for the sake of space, we only demonstrate these strategies on the ResNet PAD system trained with OC-Softmax. The results are shown in Table 15.3.

The Vanilla model achieves 21.78% EER on the ASVspoof 2021 LA dataset, showing that the vanilla model has limited generalization ability on channel variation. All the models trained using data augmentation and training strategies achieved much better performance than the Vanilla model, hereby verifying the effectiveness of the channel-robust training strategies. As for the three augmentation methods, the device augmentation can marginally improve the performance on ASVspoof 2021 LA. This suggests that the device IR is an important factor that affects the transmission variation. The transmission codec achieved reasonable performance as the ASVspoof 2021 LA evaluation set mainly contains transmission variation. The performance is further improved with dual augmentation, which shows that including both device and transmission variation in training is beneficial for generalizing to unknown transmission variability. Comparing the three training strategies, the MT-AUG shows worse performance than AUG, showing that explicitly learning channel classification as an auxiliary task may make the model overfit to known channels presented in the training set and hard to generalize to unseen channels. The ADV-AUG achieves the best results, showing that the channel-agnostic representation is useful to improve the robustness to channel variation.

Note that the best baseline system provided by the organizers achieved 8.90% EER, but the best in Table 15.3 achieves 6.84% EER, outperforming the baseline system. We further improved the model architecture and leveraged model fusion in [26] as our submitted system and achieved 5.46% EER.

15.3.3 Discussions

Channel-robust strategies, including data augmentation and training strategies (data augmentation, multi-task learning, and adversarial learning), are effective to improve the generalization ability of voice PAD systems to channel variation, and obtained significant improvement in both in-domain and cross-dataset tests.

One limitation would be that the training strategies rely on data augmentation and may be hard to generalize to a totally different dataset, e.g., with new speakers or languages. Other methods such as knowledge distillation [50] and domain adaptation [51] can be further utilized to improve such generalization ability.

The ASVspoof 2021 challenge also includes a new task speech deepfake (DF) to detect fake audio without ASV systems. For this task, developing systems that can be applied in multiple and unknown conditions such as audio compression is also critical. The evaluation set includes multiple compression algorithms. However, the participating teams did not achieve promising results (generally 20–30% EER), and the best team achieved 15.64% EER. Further investigations are needed after we have access to the complete information of the ASVspoof 2021 LA dataset, including the metadata of the channel information.

15.4 Conclusions and Future Directions

This chapter reviewed and extended our previous works on improving the generalization ability with respect to unseen synthetic attacks and channel variation. With one-class learning and channel-robust strategies, the voice PAD systems gain a better ability to generalize to unseen conditions.

As for further improving the generalization ability to unseen synthetic attacks and channel variation, some problems mentioned in the discussions of each section are worth investigating. Besides those, while one-class learning aims to encourage the intra-class compactness for the bona fide speech in the embedding space, channel-robust strategies aim to increase the model capacity to handle variance in the bona fide speech data. Combining these two would cause some contradiction and may lead to a compromised solution for the generalization ability. The mutual impact of one-class learning and channel variation robustness is worth more attention.

There are also other aspects of generalization ability for voice PAD systems, such as noisy environments and speaker and language variation. For different kinds of attacks, LA, PA, and adversarial attacks, a unified system to handle these would be promising. With respect to applications, from the end user's perspective, designing a speaker-dependent PAD system will be an interesting direction.

For more generalized biometrics anti-spoofing (such as face, iris, fingerprint, multimodal anti-spoofing), one-class learning methods can also be explored to help detect unseen attacks. Our methods for generalizing to channel variation can also be adapted to other domains like illumination variation in face PAD. We believe that the methods described in this chapter are helpful for improving the generalization ability for biometrics anti-spoofing in general.

Acknowledgements This work is supported by National Science Foundation grant No. 1741472, a New York State Center of Excellence in Data Science award, and funding from Voice Biometrics Group. You Zhang would like to thank the synergistic activities provided by the NRT program on AR/VR funded by NSF grant DGE-1922591. The authors would also like to thank Dr. Xin Wang and Prof. Junichi Yamagishi for the valuable discussions.

15.5 Appendix

Connections of OC-Softmax with the original Softmax and AM-Softmax

The original Softmax loss for binary classification can be formulated as

$$\begin{aligned} L_S &= -\frac{1}{N}\sum_{i=1}^{N}\log\frac{e^{\boldsymbol{w}_{y_i}^T\boldsymbol{x}_i}}{e^{\boldsymbol{w}_{y_i}^T\boldsymbol{x}_i}+e^{\boldsymbol{w}_{1-y_i}^T\boldsymbol{x}_i}} \\ &= \frac{1}{N}\sum_{i=1}^{N}\log\left(1+e^{(\boldsymbol{w}_{1-y_i}-\boldsymbol{w}_{y_i})^T\boldsymbol{x}_i}\right), \end{aligned} \tag{15.11}$$

where $\boldsymbol{x}_i \in \mathbb{R}^D$ and $y_i \in \{0, 1\}$ are the embedding vector and label of the i-th sample, respectively, $\boldsymbol{w}_0$, $\boldsymbol{w}_1 \in \mathbb{R}^D$ are the weight vectors of the two classes, and N is the number of samples in a mini-batch. The confidence score is as follows:

$$S_S(\boldsymbol{x}_i) = \frac{e^{\boldsymbol{w}_0^T\boldsymbol{x}_i}}{e^{\boldsymbol{w}_0^T\boldsymbol{x}_i}+e^{\boldsymbol{w}_1^T\boldsymbol{x}_i}}. \tag{15.12}$$

AM-Softmax [41] improves upon this by introducing an angular margin to make the embedding distributions of both classes more compact, around the weight difference vector's two directions:

$$\begin{aligned} L_{AMS} &= -\frac{1}{N}\sum_{i=1}^{N}\log\frac{e^{\alpha(\hat{\boldsymbol{w}}_{y_i}^T\hat{\boldsymbol{x}}_i-m)}}{e^{\alpha(\hat{\boldsymbol{w}}_{y_i}^T\hat{\boldsymbol{x}}_i-m)}+e^{\alpha\hat{\boldsymbol{w}}_{1-y_i}^T\hat{\boldsymbol{x}}_i}} \\ &= \frac{1}{N}\sum_{i=1}^{N}\log\left(1+e^{\alpha\left(m-(\hat{\boldsymbol{w}}_{y_i}-\hat{\boldsymbol{w}}_{1-y_i})^T\hat{\boldsymbol{x}}_i\right)}\right), \end{aligned} \tag{15.13}$$

where α is a scale factor, m is the margin for cosine similarity, and $\hat{\boldsymbol{w}}, \hat{\boldsymbol{x}}$ are normalized $\boldsymbol{w}$ and $\boldsymbol{x}$ respectively. The confidence score is as follows:

$$S_{AMS}(\boldsymbol{x}_i) = (\hat{\boldsymbol{w}}_0 - \hat{\boldsymbol{w}}_1)^T \hat{\boldsymbol{x}}_i. \tag{15.14}$$

According to the formulae of Softmax and AM-Softmax in Eq. (15.11) and Eq. (15.13), for both loss functions, the embedding vectors of the target and non-target class tend to converge around two opposite directions, i.e., $\boldsymbol{w}_0 - \boldsymbol{w}_1$ and $\boldsymbol{w}_1 - \boldsymbol{w}_0$, respectively. For AM-Softmax, the embeddings of both target and non-target classes are imposed with an identical compactness margin m.

The traditional Softmax and the AM-Softmax are used to produce the posterior probabilities for each class. For our proposed OC-Softmax, it is more natural to use cosine similarity (in [–1, 1]) between the speech embedding $\boldsymbol{x}_i$ and the weight vector $\boldsymbol{w}$ as the output confidence score. As such, we also change the output of Softmax and the AM-Softmax to be cosine similarity between the speech embedding $\boldsymbol{x}_i$ and the weight vector difference $\boldsymbol{w}_0 - \boldsymbol{w}_1$. If we would like to treat the output as probability, we can derive it as below in Eq. (15.15).

$$\begin{aligned} L_{OCS} &= \frac{1}{N} \sum_{i=1}^{N} \log\left(1 + e^{\alpha\left(m_{y_i} - \hat{\boldsymbol{w}}^T \hat{\boldsymbol{x}}_i\right)(-1)^{y_i}}\right) \\ &= \frac{1}{N}\left(\sum_{|\Omega|} \log\left(1 + e^{\alpha\left(m_0 - \hat{\boldsymbol{w}}^T \hat{\boldsymbol{x}}_i\right)}\right) + \sum_{|\overline{\Omega}|} \log\left(1 + e^{\alpha\left(\hat{\boldsymbol{w}}^T \hat{\boldsymbol{x}}_i - m_1\right)}\right)\right) \\ &= -\frac{1}{N}\left(\sum_{|\Omega|} \log \frac{1}{1 + e^{\alpha\left(m_0 - \hat{\boldsymbol{w}}^T \hat{\boldsymbol{x}}_i\right)}} + \sum_{|\overline{\Omega}|} \log \frac{1}{1 + e^{\alpha\left(\hat{\boldsymbol{w}}^T \hat{\boldsymbol{x}}_i - m_1\right)}}\right) \end{aligned} \tag{15.15}$$

where $\Omega = \{i \,|\, y_i = 0\}$ is the set of bona fide speech, and $\overline{\Omega} = \{i \,|\, y_i = 1\}$ is the set of spoofing attacks. $|\Omega| + |\overline{\Omega}| = N$. The two terms inside log in the last line would be the posterior probabilities for the bona fide and spoofing class, respectively.

Softplus Function and Hinge Function

In Sects. 15.2.1.1 and 15.2.1.2, we discussed the difference between the Euclidean distance-based one-class losses and our proposed cosine similarity-based one-class losses. One difference is how we penalize the data if the embedding does not fit the distribution in Fig. 15.2. We denote the distance of the embedding from the (angular) margin as t. Details of computing t can be referred in Eqs. (15.1) (15.3) (15.4) (15.6). The hinge function is formulated as:

$$L_{hinge} = \max(0, t). \tag{15.16}$$

The softplus function can be represented as:

$$L_{softplus} = \log(1 + e^t). \tag{15.17}$$

As can be found with the function curves, the softplus is a soft version of the hinge function. In terms of the gradient,

$$\frac{\partial L_{hinge}}{\partial t} = \begin{cases} 0, & \text{if } t \leq 0 \\ 1, & \text{if } t > 0 \end{cases}, \tag{15.18}$$

and

$$\frac{\partial L_{softplus}}{\partial t} = \frac{e^t}{1 + e^t}. \tag{15.19}$$

As $t > 0$ increases, the derivative does not change for the hinge function but becomes larger for the softplus function. During backpropagation, according to the chain rule, the gradient of t with respect to the parameters in the network is weighted based on the derivatives in Eqs. (15.18) and (15.19). Instead of a constant weight for margin violation in the hinge function, with softplus function, we can have larger weights for more difficult negative samples so that discriminative training can be more encouraged.

Resources

- Open-source code for experiments in this chapter is available at https://github.com/yzyouzhang/HBAS_chapter_voice3
- Datasets:
  - ASVspoof 2019 LA:
    https://doi.org/10.7488/ds/2555 Please check the LA subfolder.
  - ASVspoof 2019 LA-Sim:
    https://doi.org/10.5281/zenodo.5794671
  - ASVspoof 2021 LA:
    https://doi.org/10.5281/zenodo.4837263

References

1. Hannani AE, Petrovska-Delacrétaz D, Fauve B, Mayoue A, Mason J, Bonastre JF, Chollet G (2009) Text-independent speaker verification. Springer London, London, pp 167–211. https://doi.org/10.1007/978-1-84800-292-0_7
2. Ergünay SK, Khoury E, Lazaridis A, Marcel S (2015) On the vulnerability of speaker verification to realistic voice spoofing. In: Proceedings of the IEEE international conference on biometrics theory, applications and systems (BTAS), pp 1–6. https://doi.org/10.1109/BTAS.2015.7358783

3. Marcel S, Nixon M, Li S (eds) (2014) Handbook of biometric anti-spoofing. Springer https://doi.org/10.1007/978-1-4471-6524-8
4. Wu Z, Evans N, Kinnunen T, Yamagishi J, Alegre F, Li H (2015) Spoofing and countermeasures for speaker verification: a survey. Speech Commun 66:130–153. https://doi.org/10.1016/j.specom.2014.10.005
5. Nautsch A, Wang X, Evans N, Kinnunen TH, Vestman V, Todisco M, Delgado H, Sahidullah M, Yamagishi J, Lee KA (2021) ASVspoof 2019: spoofing countermeasures for the detection of synthesized, converted and replayed speech. IEEE Trans Biom, Behav, Identity Sci 3(2):252–265. https://doi.org/10.1109/TBIOM.2021.3059479
6. Yuan M, Duan Z (2019) Spoofing speaker verification systems with deep multi-speaker text-to-speech synthesis. arXiv:1910.13054
7. Tian X, Das RK, Li H (2020) Black-box attacks on automatic speaker verification using feedback-controlled voice conversion. In: Proceedings of the odyssey the speaker and language recognition workshop, pp 159–164
8. Müller NM, Markert K, Böttinger K (2021) Human perception of audio deepfakes. arXiv:2107.09667
9. Terblanche C, Harrison P, Gully AJ (2021) Human spoofing detection performance on degraded speech. In: Proceedings of the interspeech, pp 1738–1742. https://doi.org/10.21437/Interspeech.2021-1225
10. Wang X, Yamagishi J, Todisco M, Delgado H, Nautsch A, Evans N, Sahidullah M, Vestman V, Kinnunen T, Lee KA et al (2020) ASVspoof 2019: a large-scale public database of synthesized, converted and replayed speech. Comput Speech Lang 64:101,114
11. Zhang Y, Zhu G, Jiang F, Duan Z (2021) An empirical study on channel effects for synthetic voice spoofing countermeasure systems. In: Proceedings of the interspeech, pp 4309–4313. https://doi.org/10.21437/Interspeech.2021-1820
12. Patel T, Patil H (2015) Combining evidences from MEL cepstral, cochlear filter cepstral and instantaneous frequency features for detection of natural versus spoofed speech. In: Proceedings of the interspeech
13. Sahidullah M, Kinnunen T, Hanilçi C (2015) A comparison of features for synthetic speech detection. In: Proceedings of the interspeech, pp 2087–2091
14. Todisco M, Delgado H, Evans NW (2016) A new feature for automatic speaker verification anti-spoofing: constant Q cepstral coefficients. In: Proceedings of the odyssey, pp 283–290
15. Sriskandaraja K, Sethu V, Ambikairajah E, Li H (2016) Front-end for antispoofing countermeasures in speaker verification: scattering spectral decomposition. IEEE J Sel Top Signal Process 11(4):632–643
16. Lavrentyeva G, Tseren A, Volkova M, Gorlanov A, Kozlov A, Novoselov S (2019) STC antispoofing systems for the ASVspoof2019 challenge. In: Proceedings of the interspeech, pp 1033–1037
17. Monteiro J, Alam J, Falk TH (2020) Generalized end-to-end detection of spoofing attacks to automatic speaker recognizers. Comput Speech Lang 63:101,096. https://doi.org/10.1016/j.csl.2020.101096
18. Li X, Li N, Weng C, Liu X, Su D, Yu D, Meng H (2021) Replay and synthetic speech detection with res2net architecture. In: Proceedings of the IEEE international conference on acoustics, speech and signal processing (ICASSP), pp 6354–6358. https://doi.org/10.1109/ICASSP39728.2021.9413828
19. Alam J, Fathan A, Kang WH (2021) End-to-end voice spoofing detection employing time delay neural networks and higher order statistics. In: Proceedings of the international conference on speech and computer. Springer, pp 14–25
20. Luo A, Li E, Liu Y, Kang X, Wang ZJ (2021) A capsule network based approach for detection of audio spoofing attacks. In: Proceedings of the IEEE international conference on acoustics, speech and signal processing (ICASSP), pp 6359–6363. https://doi.org/10.1109/ICASSP39728.2021.9414670
21. Xie Y, Zhang Z, Yang Y (2021) Siamese network with wav2vec feature for spoofing speech detection. In: Proceedings of the interspeech, pp 4269–4273. https://doi.org/10.21437/Interspeech.2021-847

22. Tak H, Patino J, Todisco M, Nautsch A, Evans N, Larcher A (2021) End-to-end anti-spoofing with rawnet2. In: Proceedings of the IEEE international conference on acoustics, speech and signal processing (ICASSP), pp 6369–6373
23. Zhang Y, Jiang F, Duan Z (2021) One-class learning towards synthetic voice spoofing detection. IEEE Signal Process Lett 28:937–941. https://doi.org/10.1109/LSP.2021.3076358
24. Ling H, Huang L, Huang J, Zhang B, Li P (2021) Attention-based convolutional neural network for ASV spoofing detection. In: Proceedings of the interspeech, pp 4289–4293. https://doi.org/10.21437/Interspeech.2021-1404
25. Wang X, Yamagishi J (2021) A comparative study on recent neural spoofing countermeasures for synthetic speech detection. In: Proceedings of the interspeech, pp 4259–4263. https://doi.org/10.21437/Interspeech.2021-702
26. Chen X, Zhang Y, Zhu G, Duan Z (2021) UR channel-robust synthetic speech detection system for ASVspoof 2021. In: Proceedings of the 2021 edition of the automatic speaker verification and spoofing countermeasures challenge, pp 75–82. https://doi.org/10.21437/ASVSPOOF.2021-12
27. Kang WH, Alam J, Fathan A (2021) CRIM's system description for the ASVSpoof2021 challenge. In: Proceedings of the 2021 edition of the automatic speaker verification and spoofing countermeasures challenge, pp 100–106. https://doi.org/10.21437/ASVSPOOF.2021-16
28. Khan SS, Madden MG (2014) One-class classification: taxonomy of study and review of techniques. Knowl Eng Rev 29(3):345–374
29. Seliya N, Abdollah Zadeh A, Khoshgoftaar TM (2021) A literature review on one-class classification and its potential applications in big data. J Big Data 8(1):1–31
30. Schölkopf B, Williamson RC, Smola AJ, Shawe-Taylor J, Platt JC (2000) Support vector method for novelty detection. In: Advances in neural information processing systems, pp 582–588
31. Tax DM, Duin RP (2004) Support vector data description. Mach Learn 54(1):45–66
32. Ruff L, Vandermeulen R, Goernitz N, Deecke L, Siddiqui SA, Binder A, Müller E, Kloft M (2018) Deep one-class classification. In: Proceedings of the international conference on machine learning (ICML), pp 4393–4402
33. Alam S, Sonbhadra SK, Agarwal S, Nagabhushan P (2020) One-class support vector classifiers: a survey. Knowl-Based Syst 196:105,754. https://doi.org/10.1016/j.knosys.2020.105754
34. Alegre F, Amehraye A, Evans N (2013) A one-class classification approach to generalised speaker verification spoofing countermeasures using local binary patterns. In: Proceedings of the IEEE international conference on biometrics: theory, applications and systems (BTAS). IEEE, pp 1–8
35. Villalba J, Miguel A, Ortega A, Lleida E (2015) Spoofing detection with DNN and one-class SVM for the ASVspoof 2015 challenge. In: Proceedings of the interspeech
36. Masi I, Killekar A, Mascarenhas RM, Gurudatt SP, AbdAlmageed W (2020) Two-branch recurrent network for isolating deepfakes in videos. In: Proceedings of the European conference on computer vision (ECCV). Springer, pp 667–684
37. Li J, Xie H, Li J, Wang Z, Zhang Y (2021) Frequency-aware discriminative feature learning supervised by single-center loss for face forgery detection. In: Proceedings of the IEEE/CVF conference on computer vision and pattern recognition (CVPR), pp 6458–6467
38. Bellman RE (2015) Adaptive control processes. Princeton University Press
39. Beyer K, Goldstein J, Ramakrishnan R, Shaft U (1999) When is "nearest neighbor" meaningful? In: Proceedings of the international conference on database theory. Springer, pp 217–235
40. Liu W, Wen Y, Yu Z, Li M, Raj B, Song L (2007) Sphereface: deep hypersphere embedding for face recognition. In: IEEE/CVF conference on computer vision and pattern recognition (CVPR)
41. Wang F, Cheng J, Liu W, Liu H (2018) Additive margin softmax for face verification. IEEE Signal Process Lett 25(7):926–930. https://doi.org/10.1109/LSP.2018.2822810
42. Yi D, Lei Z, Liao S, Li SZ (2014) Deep metric learning for person re-identification. In: Proceedings of the international conference on pattern recognition (ICPR), pp 34–39. https://doi.org/10.1109/ICPR.2014.16

43. Wang X, Han X, Huang W, Dong D, Scott MR (2019) Multi-similarity loss with general pair weighting for deep metric learning. In: Proceedings of the IEEE/CVF conference on computer vision and pattern recognition (CVPR)
44. Kinnunen T, Lee KA, Delgado H, Evans N, Todisco M, Sahidullah M, Yamagishi J, Reynolds DA (2018) t-DCF: a detection cost function for the tandem assessment of spoofing countermeasures and automatic speaker verification. In: Proceedings of the odyssey the speaker and language recognition workshop, pp 312–319. https://doi.org/10.21437/Odyssey.2018-44
45. Biggio B, Corona I, He ZM, Chan PP, Giacinto G, Yeung DS, Roli F (2015) One-and-a-half-class multiple classifier systems for secure learning against evasion attacks at test time. In: International workshop on multiple classifier systems. Springer, pp 168–180
46. Barni M, Nowroozi E, Tondi B (2020) Improving the security of image manipulation detection through one-and-a-half-class multiple classification. Multimed Tools Appl 79(3):2383–2408
47. Perera P, Nallapati R, Xiang B (2019) OCGAN: one-class novelty detection using Gans with constrained latent representations. In: Proceedings of the IEEE/CVF conference on computer vision and pattern recognition (CVPR), pp 2898–2906
48. Das RK, Yang J, Li H (2020) Assessing the scope of generalized countermeasures for anti-spoofing. In: Proceedings of the IEEE international conference on acoustics, speech and signal processing (ICASSP), pp 6589–6593
49. Hanilci C, Kinnunen T, Sahidullah M, Sizov A (2016) Spoofing detection goes noisy: an analysis of synthetic speech detection in the presence of additive noise. Speech Commun 85:83–97
50. Liu P, Zhang Z, Yang Y (2021) End-to-end spoofing speech detection and knowledge distillation under noisy conditions. In: Proceedings of the 2021 international joint conference on neural networks (IJCNN). IEEE, pp 1–7
51. Himawan I, Villavicencio F, Sridharan S, Fookes C (2019) Deep domain adaptation for anti-spoofing in speaker verification systems. Comput Speech Lang 58:377–402. https://doi.org/10.1016/j.csl.2019.05.007
52. Yamagishi J, Wang X, Todisco M, Sahidullah M, Patino J, Nautsch A, Liu X, Lee KA, Kinnunen T, Evans N, Delgado H (2021) ASVspoof 2021: accelerating progress in spoofed and deepfake speech detection. In: Proceedings of the 2021 edition of the automatic speaker verification and spoofing countermeasures challenge, pp 47–54. https://doi.org/10.21437/ASVSPOOF.2021-8
53. Delgado H, Evans N, Kinnunen T, Lee KA, Liu X, Nautsch A, Patino J, Sahidullah M, Todisco M, Wang X, et al (2021) ASVspoof 2021: automatic speaker verification spoofing and countermeasures challenge evaluation plan. arXiv:2109.00535
54. Chen T, Khoury E, Phatak K, Sivaraman G (2021) Pindrop labs' submission to the ASVspoof 2021 challenge. In: Proceedings of the 2021 edition of the automatic speaker verification and spoofing countermeasures challenge, pp 89–93. https://doi.org/10.21437/ASVSPOOF.2021-14
55. Cáceres J, Font R, Grau T, Molina J (2021) The biometric VOX system for the ASVspoof 2021 challenge. In: Proceedings of the 2021 edition of the automatic speaker verification and spoofing countermeasures challenge, pp 68–74. https://doi.org/10.21437/ASVSPOOF.2021-11
56. Tomilov A, Svishchev A, Volkova M, Chirkovskiy A, Kondratev A, Lavrentyeva G (2021) STC antispoofing systems for the ASVspoof challenge. In: Proceedings of the 2021 edition of the automatic speaker verification and spoofing countermeasures challenge, pp 61–67. https://doi.org/10.21437/ASVSPOOF.2021-10
57. Ferras M, Madikeri S, Motlicek P, Dey S, Bourlard H (2016) A large-scale open-source acoustic simulator for speaker recognition. IEEE Signal Process Lett 23(4):527–531
58. Ganin Y, Lempitsky V (2015) Unsupervised domain adaptation by backpropagation. In: International conference on machine learning (ICML). PMLR, pp 1180–1189

Part V
Other Biometrics and Multi-Biometrics

Chapter 16
Introduction to Presentation Attacks in Signature Biometrics and Recent Advances

Carlos Gonzalez-Garcia, Ruben Tolosana, Ruben Vera-Rodriguez, Julian Fierrez, and Javier Ortega-Garcia

Abstract Applications based on biometric authentication have received a lot of interest in the last years due to the breathtaking results obtained using personal traits such as face or fingerprint. However, it is important not to forget that these biometric systems have to withstand different types of possible attacks. This chapter carries out an analysis of different Presentation Attack (PA) scenarios for on-line handwritten signature verification. The main contributions of this chapter are: *(i)* an updated overview of representative methods for Presentation Attack Detection (PAD) in signature biometrics; *(ii)* a description of the different levels of PAs existing in on-line signature verification regarding the amount of information available to the impostor, as well as the training, effort, and ability to perform the forgeries; and *(iii)* an evaluation of the system performance in signature biometrics under different scenarios considering recent publicly available signature databases, DeepSignDB (https://github.com/BiDAlab/DeepSignDB) and SVC2021_EvalDB (https://github.com/BiDAlab/SVC2021_EvalDB), (https://competitions.codalab.org/competitions/27295). This work is in line with recent efforts in the Common Criteria standardization community towards security evaluation of biometric systems.

C. Gonzalez-Garcia · R. Tolosana (✉) · R. Vera-Rodriguez (✉) · J. Fierrez (✉) · J. Ortega-Garcia
Biometrics and Data Pattern Analytics—BiDA Lab, Escuela Politecnica Superior, Universidad Autonoma de Madrid, 28049 Madrid, Spain
e-mail: ruben.tolosana@uam.es

R. Vera-Rodriguez
e-mail: ruben.vera@uam.es

J. Fierrez
e-mail: julian.fierrez@uam.es

C. Gonzalez-Garcia
e-mail: carlos.gonzalezgarcia@estudiante.uam.es

J. Ortega-Garcia
e-mail: javier.ortega@uam.es

S. Marcel et al. (eds.), *Handbook of Biometric Anti-Spoofing*, Advances in Computer Vision and Pattern Recognition, https://doi.org/10.1007/978-981-19-5288-3_16

16.1 Introduction

Signature verification systems have become very popular in many applications such as banking, e-health, e-education, and security in recent times [1]. This evolution has been motivated due to two main factors: (*i*) the technological evolution and the improvement of sensors quality, which has made general purpose devices (smartphones [2] and tablets [3]) more accessible to the general population, and therefore, the social acceptance has increased; and (*ii*) the evolution of biometric recognition technologies, especially through the use of deep learning techniques [4, 5]. However, it is important to highlight that this biometric systems have to endure different types of possible attacks [6, 7], some of them highly complex [8].

In this chapter we focus on the study of different Presentation Attack (PA) scenarios for on-line handwritten signature biometric verification systems due to the significant amount of attention received in the last years thanks to the development of new scenarios (e.g. device interoperability [9] and mobile scenarios [10, 11]) and writing tools (e.g. finger [2]). These new scenarios have grown hand in hand with the rapid expansion of mobile devices, such as smartphones, which allow the implementation of biometric-based verification systems far from the traditional office-like ones [12].

In general, two different types of impostors can be found in the context of signature verification: (1) *random (zero-effort* or *accidental)* impostors, the case in which no information about the signature of the user being attacked is known and impostors present their own genuine signature claiming to be another user of the system, and (2) *skilled* impostors, the case in which attackers have some level of information about the signature of the user to attack (e.g. global shape of the signature or signature dynamics) and try to forge the signature claiming to be that user in the system.

In [13], Galbally et al. discussed different approaches to report accuracy results in handwritten signature verification. They considered skilled impostors as a particular case of biometric PAs which is performed against a behavioural biometric characteristic (also referred as *mimicry*). There are important differences between PAs and mimicry: while traditional PAs involve the use of some physical artefacts such as fake masks and gummy fingers (and therefore, they can be detected in some cases at the sensor level), in the case of mimicry the interaction with the sensor is exactly the same followed in a genuine access attempt. In [13] a different nomenclature of impostor scenarios is proposed following the literature standard in the field of biometric Presentation Attack Detection (PAD) : the classical random impostor scenario is referred to as Bona Fide (BF) scenario, while the skilled impostor scenario is referred to as PA scenario. This nomenclature has also been used in this chapter.

If during the development of a biometric verification system those PAs are expected, it is possible to include specific modules for PAD, which in the signature verification literature are commonly referred to as forgery detection modules. A comprehensive study of these PAD methods is out of the scope of the chapter, but in Sect. 16.2 we provide a brief overview of some selected representative works in that area.

A different approach to improve the security of a signature verification system against attacks different from including a PAD module is template protection [14–20]. Traditional on-line signature verification systems work with very sensitive biometric data such as the X and Y spatial coordinates and store that information without any additional protection. This makes very easy for attackers to steal this information. If an attacker has the information of spatial coordinates along the time axis it would be very easy for him/her to generate very high quality forgeries. Template protection techniques involve feature transformation and the use of biometric cryptosystems. In [21], an extreme approach for signature template generation was proposed not considering information related to X, Y coordinates and their derivatives on the biometric system, providing therefore a much more robust system against attacks, as this critical information would not be stored anywhere. Moreover, the results achieved had error rates in the same range as more traditional systems which store very sensitive information. An interesting review and classification of different biometric template protection techniques for on-line handwritten signature application is conducted in [22].

The main contributions of this chapter are: *(i)* a brief overview of representative methods for PAD in signature biometrics ; *(ii)* a description of the different levels of PAs existing in on-line signature verification regarding the amount of information available to the impostor, as well as the training, effort and ability to perform the forgeries; and *(iii)* an evaluation of the system performance in signature biometrics under different scenarios following the recent SVC-onGoing competition[1] [23].

The remainder of the chapter is organized as follows. The introduction is completed with a short overview of PAD in signature biometrics (Sect. 16.2). After that, the main technical content of the chapter begins in Sect. 16.3, with a review of the most relevant features of all different impostor scenarios, pointing out which type of impostors are included in many different well-known public signature databases. Section 16.4 describes the on-line signature databases considered in the experimental work. Section 16.5 describes the experimental protocol and the results achieved. Finally, Sect. 16.6 draws the final conclusions and points out some lines for future work.

16.2 Review of PAD in Signature Biometrics

Presentation Attack Detection (PAD) in signature biometrics is a field that has been extensively studied since the late 70s to the present [24]. In this section we describe some state-of-the-art forgery detection methods.

Some of the studies that can be found in the literature are based on the *Kinematic Theory* of rapid human movements and its associated Sigma LogNormal model. In [25], the authors proposed a new scheme in which a module focused on the detection

[1] https://competitions.codalab.org/competitions/27295.

of skilled forgeries (i.e. PA impostors) was based on four parameters of the Sigma LogNormal writing generation model [26] and a linear classifier. That new binary classification module was supposed to work sequentially before a standard signature recognition system [27]. Good results were achieved using that approach for both skilled (i.e. PA) and random (i.e. BF) scenarios. In [28], Reillo et al. proposed PAD methods based on the use of some global features such as the total number of strokes and the signing time of the signatures. They acquired a new database based on 11 levels of PAs regarding the level of knowledge and the tools available to the forger. The results achieved in that work using the proposed PAD methods reduced the Equal Error Rate (EER) from a percentage close to 20.0% to below 3.0%.

In [29], authors proposed an off-line signature verification and forgery detection system based on fuzzy modelling. The verification of genuine signatures and detection of forgeries was achieved via angle features extracted using a grid method. The derived features were fuzzified by an exponential membership function, which was modified to include two structural parameters regarding variations of the handwriting styles and other factors affecting the scripting of a signature. Experiments showed the capability of the system in detecting even the slightest changes in signatures.

Brault and Plamondon presented in [30] an original attempt to estimate, quantitatively and a priori from the coordinates sampled during its execution, the difficulty that could be experienced by a typical imitator in reproducing both visually and dynamically that signature. To achieve this goal, they first derived a functional model of what a typical imitator must do to copy dynamically any signature. A specific difficulty coefficient was then numerically estimated for a given signature. Experimentation geared specifically to signature imitation demonstrated the effectiveness of the model. The ranking of the tested signatures given by the difficulty coefficient was compared to three different sources: the opinions of the imitators themselves, the ones of an expert document examiner, and the ranking given by a specific pattern recognition algorithm. They provided an example of application as well. This work was one of the first attempts of PAD for on-line handwritten signature verification using a special pen attached to a digitizer (Summagraphic Inc. model MM1201). The sampling frequency was 110 Hz, and the spatial resolution was 0.025 in.

Finally, it is important to highlight that new approaches based on deep learning architectures are commonly used in the literature [5, 23]. Several studies use Convolutional Neural Network (CNN) architectures in order to predict whether a signature is genuine or a forgery presented by a PA impostor [31–33]. Other state-of-the-art architectures are based on Recurrent Neural Networks (RNNs) such as the ones presented in [34]. Also, some recent works focus on analyzing the system performance against both BF and PA scenarios depending of the signature complexity [35, 36].

16.3 Presentation Attacks in Signature Biometrics

The purpose of this section is to clarify the different levels of skilled forgeries (i.e. PA impostors) that can be found in the signature biometrics literature regarding the amount of information provided to the attacker, as well as the training, effort and ability to perform the forgeries. In addition, the case of random forgeries (i.e. zero-effort impostors) is also considered although it belongs to the BF scenario and not to the PA scenario in order to review the whole range of possible attacks in on-line signature verification.

Previous studies have applied the concept of Biometric Menagerie in order to categorize each type of user of the biometric system as an animal. This concept was initially formalized by Doddington et al. in [37], classifying speakers regarding the ease or difficulty with which the speaker can be recognized (i.e. sheep and goats, respectively), how easily they can be forged (i.e. lambs) and finally, how adept/effective they are at forging/imitating the voice of others (i.e. wolves). Yager and Dunstone extended the Biometric Menagerie in [38] by adding four more categories of users (i.e. worms, chameleons, phantoms, and doves). Their proposed approach was investigated using a broad range of biometric modalities, including 2D and 3D faces, fingerprints, iris, speech, and keystroke dynamics. In [39], Houmani and Garcia-Salicetti applied the concept of Biometric Menagerie for the different types of users found in the on-line signature verification task proposing the combination of their personal and relative entropy measures as a way to quantify how difficult it is a signature to be forged. Their proposed approach achieved promising classification results on the MCYT database [40], where the attacker had access to a visual static image of the signature to forge.

In [41], some experiments were carried out to reach the following conclusions: (1) some users are significantly better forgers than others; (2) forgers can be trained in a relatively straight-forward way to become a greater threat; (3) certain users are easy targets for forgers; and (4) most humans are relatively poor judges of handwriting authenticity, and hence, their unaided instincts cannot be trusted. Additionally, in that work authors proposed a new metric for impostor classification more realistic to the definition of security, i.e., *naive*, *trained*, and *generative*. They considered naive impostors as random impostors (i.e. zero-effort impostors) in which no information about the user to forge is available whereas they referred to trained and generative impostors to skilled forgeries (i.e. PA impostors) when only the image or the dynamics of the signature to forge is available, respectively.

In [42], the authors proposed a software tool implemented on two different computer platforms in order to achieve forgeries with different quality levels (i.e. PA impostors). Three different levels of PAs were considered: (1) *blind forgeries*, the case in which the attacker writes on a blank surface having access just to textual knowledge (i.e. precise spelling of the user's name to forge); (2) *low-force forgeries*, where the attacker gets a blueprint of the signature projected on the writing surface (dynamic information is not provided), which they may trace; and (3) *brute-force forgeries*, in which an animated pointer is projected onto the writing pad showing the whole realization of the signature to forge. The attacker may observe the sequence and follow the pointer. The authors carried out an experiment based on the use of 82 forgery samples performed by four different users in order to detect how the False Acceptance Rate (FAR) is affected regarding the level of PA. They considered a signature verification system based on the average quadratic deviation horizontal and vertical writing signals. Results obtained for four different threshold values confirmed the requirement of strong protection of biometric reference data as it was proposed in [21].

16.3.1 Types of Presentation Attacks

Alonso-Fernandez et al. carried out an exhaustive analysis of the different types of forgeries found in handwritten signature verification systems [43]. In that work, authors considered random impostors and 4 different levels of PA impostors, classified regarding the amount of information provided and the tools used by the attacker in order to forge the signature:

- **Random or zero-effort forgeries**, in which no information of the user to forge is available and the attacker uses its own genuine signature (accidentally or not) claiming to be another user of the system.
- **Blind forgeries**, in which the impostor has access to a descriptive or textual knowledge of signatures to forge (e.g. the name of the subject to forge).
- **Static forgeries** (low-force in [42]), where the attacker has available a static image of the global shape of the signature to forge. In this case, there are two ways to generate the forgeries. In the first one, the attacker can train to imitate the signature with or without time restrictions and blueprint, and then forge it without the use of the blueprint, which leads to **static trained forgeries**. In the second one, the attacker uses a blueprint to first copy the genuine signature of the user to forge and then put it on the screen of the device while forging, leading to **static blueprint forgeries**, more difficult to detect as they have quite the same appearance as the original ones.
- **Dynamic forgeries** (brute-force in [42]), where the impostor has access to both the global image and also the whole realization process (i.e. dynamics) of the signature to forge. The dynamics can be obtained in the presence of the original writer or through the use of a video-recording. In a similar way as the previous category, we can distinguish first **dynamic trained forgeries** in which the attacker can use specific tools to analyze and train to forge the genuine signature, and second, **dynamic blueprint forgeries** which are generated by projecting on the acquisition area a real-time pointer that the forger only needs to follow.
- **Regained forgeries**, the case where the impostor has only available the static image of the signature to forge and makes use of a dedicated software to recover the signature dynamics [44], which are later analyzed and used to create dynamic forgeries.

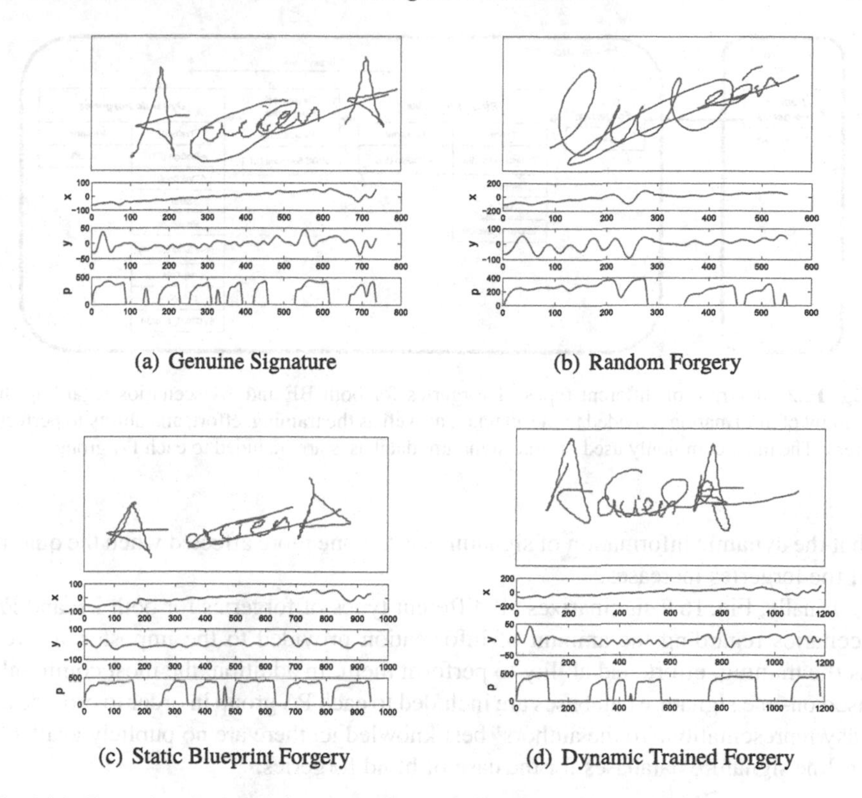

Fig. 16.1 Examples of one genuine signature and three different types of forgeries performed for the same user

As expected, dynamic forgeries are the forgeries with better quality (in some cases, very similar to the genuine signatures that are forged), followed by static forgeries. Random and blind forgeries are usually very different from the signature forged. Figure 16.1 shows examples of a genuine signature and three different types of forgeries (i.e. random, static blueprint and dynamic trained) performed for the same user. The image shows both the static and dynamic information with the X and Y coordinates and pressure.

Besides the forgery classification carried out in [43], Alonso-Fernandez et al. studied the impact of an incremental level of quality forgeries against handwritten signature verification systems. The authors considered off-line and on-line systems using the BiosecurID database [45]. For the off-line verification system, they considered a system based on global image analysis and a minimum distance classifier [46] whereas a system based on Hidden Markov Models (HMM) [47] was considered for the on-line system. The experiments carried out proved that the performance of the off-line approach is only degraded when the highest quality level of forgeries is used. The on-line system shows a progressive degradation of its performance when the quality level of the forgeries is increased. This lead the authors to the conclusion

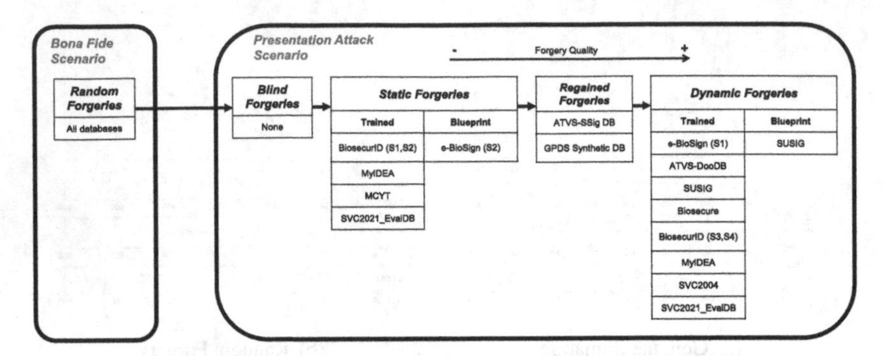

Fig. 16.2 Diagram of different types of forgeries for both BF and PA scenarios regarding the amount of information provided to the attacker, as well as the training, effort, and ability to perform them. The most commonly used on-line signature databases are included to each PA group

that the dynamic information of signatures is the one more affected when the quality of the forgeries increases.

Finally, Fig. 16.2 summarizes all different types of forgeries for both BF and PA scenarios regarding the amount of information provided to the impostor, as well as the training, effort, and ability to perform them. In addition, the most commonly used on-line signature databases are included to each PA group in order to provide an easy representation. To the authors' best knowledge, there are no publicly available on-line signature databases for the case of blind forgeries.

16.3.2 Synthetic Forgeries

On-line signature synthesis has become a very interesting research line due to, among other reasons, the lack of forgery signatures in real scenarios, which makes the development of robust signature verification systems difficult [4].

One of the most popular approaches in the literature for realistic handwriting generation was presented in [48]. In that study, the author presented a Long Short-Term Memory (LSTM) Recurrent Neural Network (RNN) architecture to generate complex sequences. The proposed architecture was tested on handwriting, achieving very good visual results. Currently, the Sigma LogNormal model is one of the most popular on-line signature synthesis approaches [49, 50], and has been applied to on-line signature verification systems, generating synthetic samples from a genuine signature, increasing the amount of information and improving the performance of the systems [51–53].

Other important contributions in this area are the following ones. In [44], Ferrer et al. proposed a system for the synthetic generation of dynamic information for both static and dynamic handwritten signatures based on the motor equivalence theory, which divides the action of human handwriting into an effector dependent cognitive level and an effector independent motor level, achieving good results. In [54], Tolosana et al. proposed DeepWriteSYN, a novel on-line handwriting signature synthesis approach based on deep short-term representations. The DeepWriteSYN architecture is composed by two modules: a first module which divides the signature in short-time strokes and a second module based on a sequence-to-sequence Variational Autoencoder (VAE) in charge of the synthesis of those short-time strokes. DeepWriteSYN is able to generate realistic handwriting variations of a given handwritten structure corresponding to the natural variation within a given population or a given subject. For more information, an exhaustive study of the evolution of synthetic handwriting is conducted in [55].

16.4 On-Line Signature Databases

The following two public databases are considered in the experiments reported here. Both of them are currently used in the popular SVC-onGoing on-line signature verification competition.[2]

16.4.1 DeepSignDB

The DeepSignDB[3] database [56] is composed by a total of 1,526 subjects from four different well known state-of-the-art databases: MCYT (330 subjects) [40], BiosecurID (400 subjects) [45], Biosecure DS2 (650 subjects) [57], e-BioSign (65 subjects) [2], and a novel on-line signature database composed by 81 subjects. DeepSignDB comprises more than 70 K signatures acquired using both stylus and finger writing inputs in both office and mobile scenarios. A total of 8 different devices were considered during the acquisition process (i.e., 5 Wacom devices and 3 Samsung general purpose devices). In addition, different types of impostors and number of acquisition sessions are considered along the database.

The available information when using the pen stylus as writing input is *X* and *Y* spatial coordinates and pressure. In addition, pen-up trajectories are also available. For the case of using the finger as writing input, the only available information is *X* and *Y* spatial coordinates.

[2] https://competitions.codalab.org/competitions/27295.

[3] https://github.com/BiDAlab/DeepSignDB.

16.4.2 SVC2021_EvalDB

The SVC2021_EvalDB[4] is a novel database specifically acquired for the ICDAR 2021 Signature Verification Competition [23] and then used as well for the SVC-onGoing Competition [23]. In this database, two scenarios are considered: office and mobile scenarios.

- **Office scenario:** on-line signatures from 75 subjects were collected using a Wacom STU-530 device with the stylus as writing input. It is important to highlight that all the acquisition took place in an office scenario under the supervision of a person with experience in the on-line signature verification field. The subjects considered in the acquisition of SVC2021_EvalDB database are different compared to the ones considered in the previous DeepSign database. All the signatures were collected in two different sessions separated by at least 1 week. For each genuine subject, a total of 8 genuine signatures (4 genuine signatures per session) and 16 skilled forgeries (8 static forgeries and 8 dynamic forgeries, performed by 4 different subjects in two different sessions) were collected. Regarding the skilled forgeries, static forgeries were collected in the first acquisition session and dynamic forgeries were considered in the second one. The following information is available for every signature: X and Y spatial coordinates, pressure, pen-up trajectories and timestamp.
- **Mobile scenario:** on-line signatures from a total of 119 subjects were acquired using the same acquisition framework considered in MobileTouchDB database [12]: an Android App was developed in order to work with unsupervised mobile scenarios. All users could download the application and use it on their own smartphones without any kind of supervision, simulating a real scenario (e.g., standing, sitting, walking, in public transport, etc.). As a result, a total of 94 different smartphone models from 16 different brands are available in the database. Regarding the acquisition protocol, between four and six separated sessions were acquired for every user with a time gap between first and last session of at least 3 weeks. The number and type of the signatures for every user is the same as on the office scenario. Timestamp and spatial coordinates X and Y are available for every signature.

16.5 Experimental Work

16.5.1 On-line Signature Verification System

We consider for the experimental analysis the state-of-the-art signature verification system presented in [5, 12] based on Time-Alignment Recurrent Neural Network (TA-RNN).

[4] https://github.com/BiDAlab/SVC2021_EvalDB.

For the input of the system, the network is fed with 23 time functions extracted from the signature [58]. Information related to the azimuth and altitude of the pen angular orientation is not considered in this case. The TA-RNN architecture is based on two consecutive stages: *(i)* time sequence alignment through DTW (*Dynamic Time Warping*) , and *(ii)* feature extraction and matching using a RNN. The RNN system comprises three layers. The first layer is composed of two Bidirectional Gated Recurrent Unit (BGRU) hidden layers with 46 memory blocks each, sharing the weights between them. The outputs of the first two parallel BGRU hidden layers are concatenated and serve as input to the second layer, which corresponds to a BGRU hidden layer with 23 memory blocks. Finally, a feed-forward neural network layer with a sigmoid activation is considered, providing an output score for each pair of signatures. This learning model was presented in [5] and was retrained for the SVC-onGoing competition [23] adapted to the stylus scenario by using only the stylus-written signatures of the development set of DeepSignDB (1,084 users). The best model has been then selected using a partition of the development set of DeepSignDB, leaving out of the training the DeepSignDB evaluation set (442 users).

16.5.2 Experimental Protocol

The experimental protocol has been designed to allow the study of both random forgeries (i.e. BF) and skilled forgeries (i.e. PA) scenarios on the system performance. Additionally, the case of using the stylus or the finger as writing tool is considered.

For the study of the writing input impact in the system performance, the same three scenarios considered in the SVC-onGoing competition [23] have been used:

- **Task 1:** analysis of office scenarios using the stylus as input.
- **Task 2:** analysis of mobile scenarios using the finger as input.
- **Task 3:** analysis of both office and mobile scenarios simultaneously.

For the development of the system, the training dataset of the DeepSignDB database (1084 subjects) has been used. This means that the system has been trained using only signatures captured with a stylus writing tool. This will have a considerable impact on the system performance, as will be seen in Sect. 16.5.3. It is also important to highlight that, in order to consider a very challenging impostor scenario, the skilled forgery comparisons included in the evaluation datasets (not in the training ones) of both databases have been optimised using machine learning methods, selecting only the best high-quality forgeries.

In addition, SVC-onGoing simulates realistic operational conditions **considering random and skilled forgeries simultaneously in each task.** A brief summary of the proposed experimental protocol used can be seen in Fig. 16.3. For more details, we refer the reader to [23].

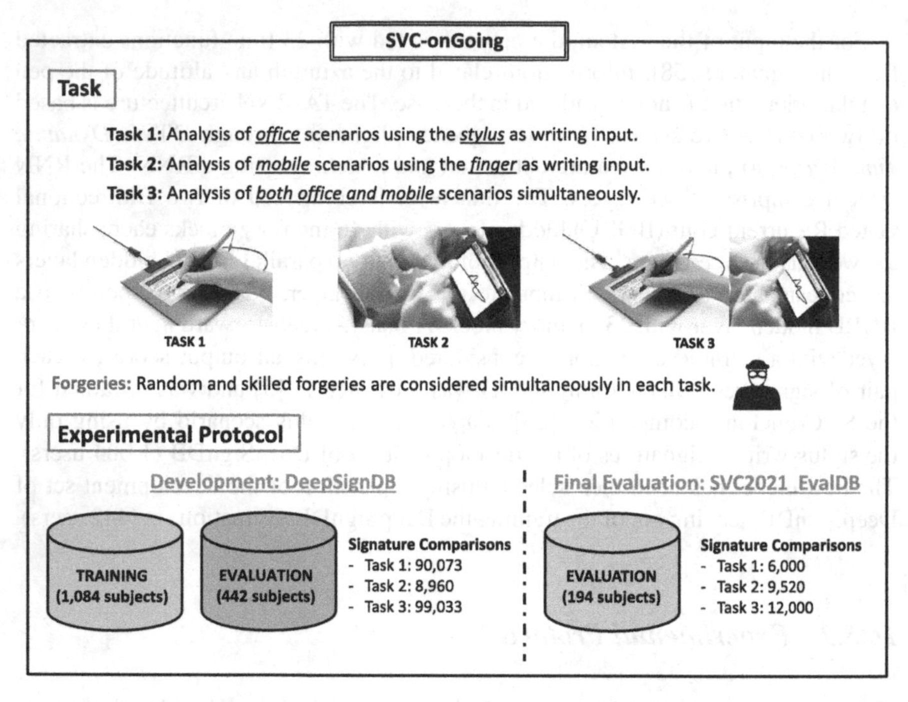

Fig. 16.3 Description of the tasks and experimental protocol details considered in SVC-onGoing Competition

16.5.3 Experimental Results

This section analyzes the results achieved in both DeepSignDB and SVC2021_EvalDB databases.

16.5.3.1 DeepSignDB

In this first case, the evaluation dataset (442 subjects) of DeepSignDB was used to evaluate the performance of both DTW and TA-RNN systems. Figure 16.4 shows the results achieved in each of the three tasks using Detection Error Tradeoff (DET) curves and considering both random and skilled forgeries simultaneously. A Baseline DTW system (similar to the one described in [59] based on X, Y spatial time signals, and their first- and second-order derivatives) is included in the image for a better comparison of the results. First, in all tasks we can see that the TA-RNN system has outperformed the traditional Baseline DTW. For Task 1, focused on the analysis of office scenarios using the stylus as writing input, the TA-RNN approach obtained a 4.31% EER. Regarding Task 2, focused on mobile scenarios using the finger as writing input, a considerable system performance degradation is observed compared

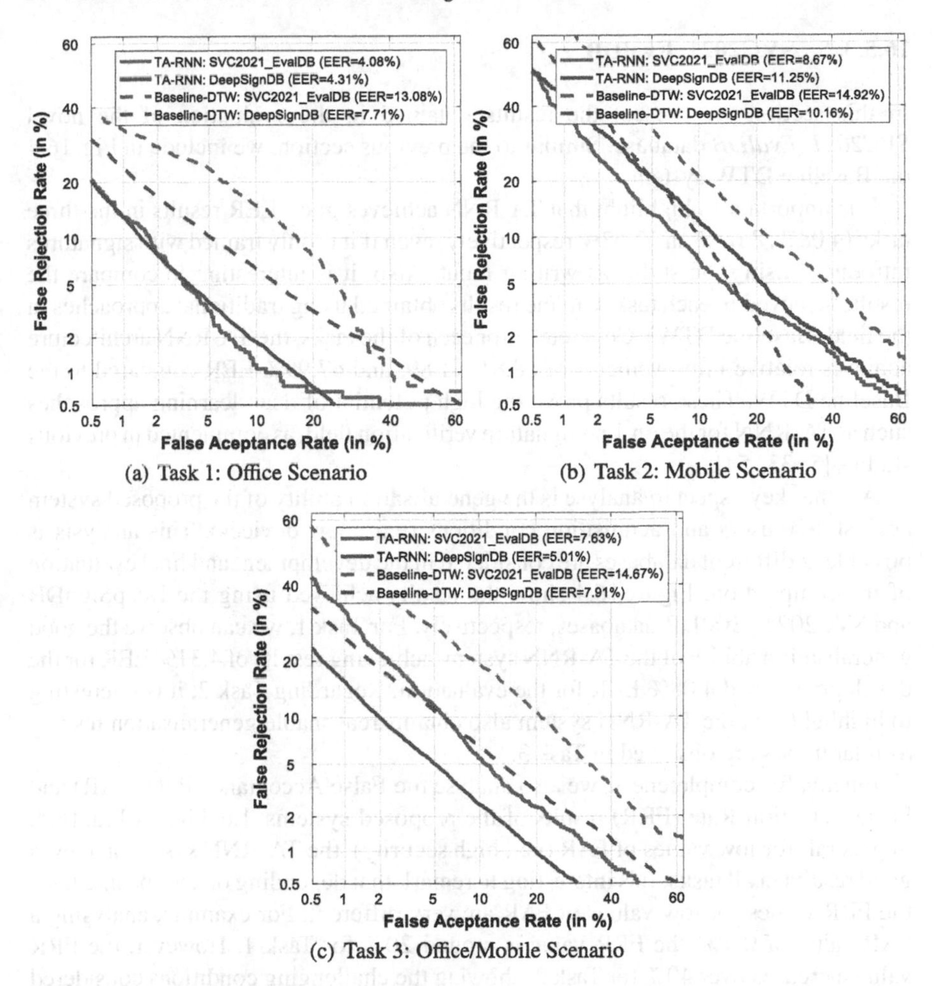

Fig. 16.4 Task Analysis: Results in terms of DET curves over the evaluation dataset of DeepSignDB and SVC2021_EvalDB for the three tasks considered

to the results of Task 1. In this case, the EER obtained was 11.25%. This result proves the bad generalisation of the stylus model (Task 1) to the finger scenario (Task 2) as the model considered was trained using only signatures acquired through the stylus, not the finger. Finally, good results are generally achieved in Task 3 taking into account that both office and mobile scenarios are considered together, using both stylus and finger as writing inputs. The system obtained an EER of 5.01%.

16.5.3.2 SVC2021_EvalDB

In this section we present the results obtained in the evaluation of the novel *SVC2021_EvalDB* database. Similar to the previous section, we include in Fig. 16.4 the Baseline DTW system.

It is important to highlight that TA-RNN achieves good EER results in the three tasks (4.08%, 8.67% and 7.63% respectively) even if it is only trained with signatures introduced using the stylus as writing input. Also, it is interesting to compare the results achieved in each task with the results obtained using traditional approaches in the field (Baseline DTW). Concretely, for each of the tasks, the TA-RNN architecture achieves relative improvements of 68.81, 41.89, and 47.99% EER compared to the Baseline DTW. These results prove the high potential of deep learning approaches such as TA-RNN for the on-line signature verification field, as commented in previous studies [5, 33, 54].

Another key aspect to analyse is the generalisation ability of the proposed system against new users and acquisition conditions (e.g., new devices). This analysis is possible as different databases are considered in the development and final evaluation of the competition. Figure 16.4 show the results achieved using the DeepSignDB and SVC2021_EvalDB databases, respectively. For Task 1, we can observe the good generalisation ability of the TA-RNN system, achieving results of 4.31% EER for the development, and 4.08% EER for the evaluation. Regarding Task 2, it is interesting to highlight that the TA-RNN system also obtains reasonable generalisation results. Similar trends are observed in Task 3.

Finally, for completeness, we also analyse the False Acceptance Rate (FAR) and False Rejection Rate (FRR) results of the proposed systems. Looking at Fig. 16.4, in general, for low values of FAR (i.e., high security), the TA-RNN system achieves good results in all tasks. It is interesting to remark that depending on the specific task, the FRR values for low values of FAR are very different. For example, analysing a FAR value of 0.5%, the FRR value is around 20% for Task 1. However, the FRR value increases over 40% for Task 2, showing the challenging conditions considered in real mobile scenarios using the finger as writing input. A similar trend is observed for low values of FRR (i.e., high convenience).

16.5.3.3 Forgery Analysis

This section analyzes the impact of the type of forgery in the proposed on-line signature verification system. In the evaluation of SVC-onGoing, both random and skilled forgeries are considered simultaneously in order to simulate real scenarios. Therefore, the winner of the competition was the system that achieved the highest robustness against both types of impostors at the same time [23]. We now analyse the level of security of the two systems considered for each type of forgery, i.e., random and skilled. Figure 16.5 shows the DET curves of each task and type of forgery, including also the EER results, over both DeepSignDB and SVC2021_EvalDB databases.

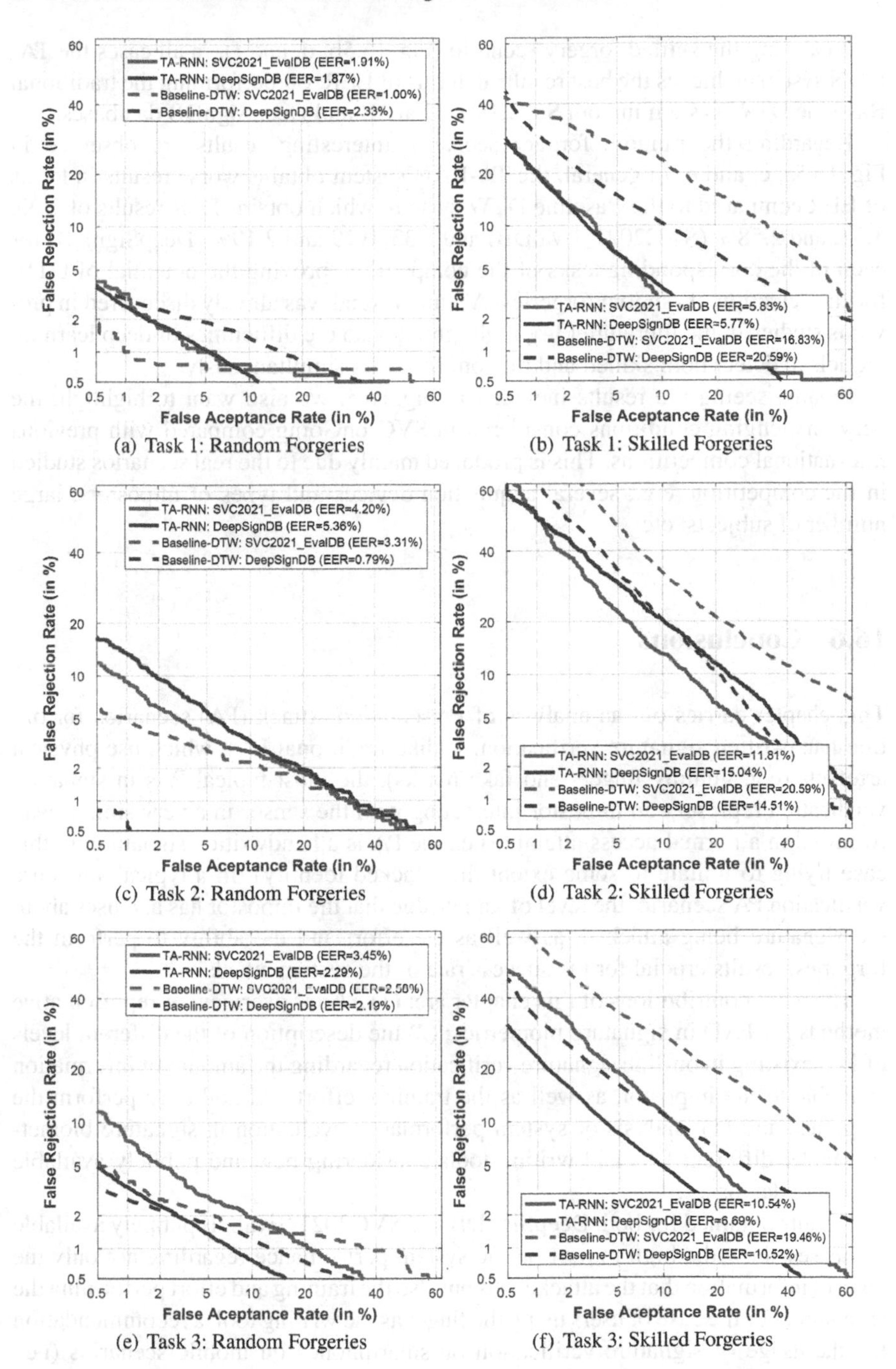

Fig. 16.5 Forgery Analysis: Results in terms of DET curves over the evaluation dataset of DeepSignDB and SVC2021_EvalDB for the three tasks and both types of forgeries separately

Analysing the skilled forgery scenario (Fig. 16.5b, d, and f), in all cases the TA-RNN system achieves the best results in terms of EER, outperforming the traditional Baseline DTW system in both SVC2021_EvalDB and DeepSignDB databases.

Regarding the random forgery scenario, interesting results are observed in Fig. 16.5a, c, and e. In general, the TA-RNN system obtains worse results in terms of EER compared to the Baseline DTW system, which obtains EER results of 1.00, 3.31, and 2.58% (SVC2021_EvalDB) and 2.33, 0.79 and 2.19% (DeepSignDB) for each of the corresponding tasks of the competition, proving the potential of DTW for the detection of random forgeries. A similar trend was already discovered in previous studies in the literature [34], highlighting also the difficulties of deep learning models to detect both skilled and random forgeries simultaneously.

Finally, seeing the results included in Fig. 16.5, we also want to highlight the very challenging conditions considered in SVC-onGoing compared with previous international competitions. This is produced mainly due to the real scenarios studied in the competition, e.g., several acquisition devices and types of impostors, large number of subjects, etc.

16.6 Conclusions

This chapter carries out an analysis of Presentation Attack (PA) scenarios for on-line handwritten signature verification. Unlike traditional PAs, which use physical artefacts (e.g. gummy fingers and fake masks), the most typical PAs in signature verification represent an impostor interacting with the sensor in a very similar way followed in a normal access attempt (i.e., the PA is a handwritten signature, in this case trying to imitate to some extent the attacked identity). In a typical signature verification PA scenario, the level of knowledge that the impostor has and uses about the signature being attacked, as well as the effort and the ability to perform the forgeries, results crucial for the success rate of the system attack.

The main contributions of this chapter are: (1) a brief overview of representative methods for PAD in signature biometrics; (2) the description of the different levels of PAs existing in on-line signature verification regarding the amount of information available to the impostor, as well as the training, effort and ability to perform the forgeries; and (3) analysis of system performance evaluation in signature biometrics under different PAs and writing tools considering new and publicly available signature databases.

Results obtained for both DeepSignDB and SVC2021_EvalDB publicly available databases show the high impact on the system performance regarding not only the level of information that the attacker has but also the training and effort performing the signature. For the case of users using the finger as the writing tool, a recommendation for the usage of signature verification on smartphones on mobile scenarios (i.e., sitting, standing, walking, indoors, outdoors, etc.) would be to protect themselves from other people that could be watching while performing their genuine signature, as this is more feasible to do in a mobile scenario compared to an office scenario.

This way skilled impostors (i.e. PA impostors) might have access to the global image of the signature but not to the dynamic information and system performance would be much better. This work is in line with recent efforts in the Common Criteria standardization community towards security evaluation of biometric systems, where attacks are rated depending on, among other factors: time spent, effort, and expertise of the attacker; as well as the information available and used from the target being attacked [60].

Acknowledgements The chapter update for the 3rd Edition of the book has received funding from the European Union's Horizon 2020 research and innovation programme under the Marie Sklodowska-Curie grant agreement No 860315 (PRIMA) and No 860813 (TRESPASS-ETN). Partial funding also from INTER-ACTION (PID2021-126521OB-I00 MICINN/FEDER), Orange Labs, and Cecabank.

References

1. Faundez-Zanuy M, Fierrez J, Ferrer MA, Diaz M, Tolosana R, Plamondon R (2020) Handwriting biometrics: applications and future trends in e-security and e-health. Cogn Comput 12(5):940–953
2. Tolosana R, Vera-Rodriguez R, Fierrez J, Morales A, Ortega-Garcia J (2017) Benchmarking desktop and mobile handwriting across COTS devices: the e-BioSign biometric database. PLOS ONE 1–17
3. Alonso-Fernandez F, Fierrez-Aguilar J, Ortega-Garcia J (2005) Sensor interoperability and fusion in signature verification: a case study using tablet PC. In: Proceedings of the IWBRS, LNCS, vol 3781. Springer, pp 180–187
4. Diaz M, Ferrer MA, Impedovo D, Malik MI, Pirlo G, Plamondon R (2019) A perspective analysis of handwritten signature technology. ACM Comput Surv (Csur) 51(6):1–39
5. Tolosana R, Vera-Rodriguez R, Fierrez J, Ortega-Garcia J (2021) DeepSign: deep on-line signature verification. IEEE Trans Biom Behav Ident Sci
6. Galbally J, Fierrez J, Ortega-Garcia J (2007) Vulnerabilities in biometric systems: attacks and recent advances in liveness detection. In: Proceedings of the Spanish workshop on biometrics, (SWB)
7. Rathgeb C, Tolosana R, Vera-Rodriguez R, Busch C (2021) Handbook of digital face manipulation and detection: from DeepFakes to morphing attacks. Advances in computer vision and pattern recognition. Springer
8. Hadid A, Evans N, Marcel S, Fierrez J (2015) Biometrics systems under spoofing attack: an evaluation methodology and lessons learned. IEEE Signal Process Mag 32(5):20–30. https://doi.org/10.1109/MSP.2015.2437652
9. Tolosana R, Vera-Rodriguez R, Ortega-Garcia J, Fierrez J (2015) Preprocessing and feature selection for improved sensor interoperability in online biometric signature verification. IEEE Access 3:478–489
10. Impedovo D, Pirlo G (2021) Automatic signature verification in the mobile cloud scenario: survey and way ahead. IEEE Trans Emerg Top Comput 9(1):554–568
11. Martinez-Diaz M, Fierrez J, Galbally J, Ortega-Garcia J (2009) Towards mobile authentication using dynamic signature verification: useful features and performance evaluation, pp 1 – 5
12. Tolosana R, Vera-Rodriguez R, Fierrez J, Ortega-Garcia J (2020) BioTouchPass2: touchscreen password biometrics using time-aligned recurrent neural networks. IEEE Trans Inf Forensics Secur 5:2616–2628

13. Galbally J, Gomez-Barrero M, Ross A (2017) Accuracy evaluation of handwritten signature verification: rethinking the random-skilled forgeries dichotomy. In: Proceedings of the IEEE international joint conference on biometrics, pp 302–310
14. Campisi P, Maiorana E, Fierrez J, Ortega-Garcia J, Neri A (2010) Cancelable templates for sequence based biometrics with application to on-line signature recognition. IEEE Trans Syst Man Cybernet Part A: Syst Humans 3:525–538
15. Delgado-Mohatar O, Fierrez J, Tolosana R, Vera-Rodriguez R (2019) Biometric template storage with blockchain: a first look into cost and performance tradeoffs. In: Proceedings of the IEEE/CVF conference on computer vision and pattern recognition workshops
16. Freire MR, Fierrez J, Ortega-Garcia J (2008) Dynamic signature verification with template protection using helper data. In: Proceedings of the IEEE international conference on acoustics, speech, and signal processing, ICASSP, pp 1713–1716
17. Gomez-Barrero M, Galbally J, Morales A, Fierrez J (2017) Privacy-preserving comparison of variable-length data with application to biometric template protection. IEEE Access 5:8606–8619
18. Gomez-Barrero M, Maiorana E, Galbally J, Campisi P, Fierrez J (2017) Multi-biometric template protection based on homomorphic encryption. Pattern Recogn 67:149–163
19. Nanni L, Maiorana E, Lumini A, Campisi P (2010) Combining local, regional and global matchers for a template protected on-line signature verification system. Expert Syst Appl 37(5):3676–3684
20. Ponce-Hernandez W, Blanco-Gonzalo R, Liu-Jimenez J, Sanchez-Reillo R (2020) Fuzzy vault scheme based on fixed-length templates applied to dynamic signature verification. IEEE Access 8:11152–11164
21. Tolosana R, Vera-Rodriguez R, Ortega-Garcia J, Fierrez J (2015) Increasing the robustness of biometric templates for dynamic signature biometric systems. In: Proceedings of the 49th annual international carnahan conference on security technology
22. Malallah FL, Ahmad SS, Yussof S, Adnan W, Iranmanesh V, Arigbabu O (2013) A review of biometric template protection techniques for online handwritten signature application. Int Rev Comput Softw 8(12):1–9
23. Tolosana R, Vera-Rodriguez R, Gonzalez-Garcia C, Fierrez J, Morales A, Ortega-Garcia J, Carlos Ruiz-Garcia J, Romero-Tapiador S, Rengifo S, Caruana M, Jiang J, Lai S, Jin L, Zhu Y, Galbally J, Diaz M, Angel Ferrer M, Gomez-Barrero M, Hodashinsky I, Sarin K, Slezkin A, Bardamova M, Svetlakov M, Saleem M, Lia Szcs C, Kovari B, Pulsmeyer F, Wehbi M, Zanca D, Ahmad S, Mishra S, Jabin S (2022) Svc-ongoing: signature verification competition. Pattern Recogn 127:108,609
24. Nagel R, Rosenfeld A (1977) Computer detection of freehand forgeries. IEEE Trans Comput C-26:895–905
25. Gomez-Barrero M, Galbally J, Fierrez J, Ortega-Garcia J, Plamondon R (2015) Enhanced on-line signature verification based on skilled forgery detection using sigma-lognormal features. In: Proceedings of the IEEE/IAPR international conference on biometrics, ICB, pp 501–506
26. O'Reilly C, Plamondon R (2009) Development of a sigma-lognormal representation for on-line signatures. Pattern Recogn 42(12):3324–3337
27. Fierrez J, Morales A, Vera-Rodriguez R, Camacho D (2018) Multiple classifiers in biometrics. Part 1: fundamentals and review. Inf Fus 44:57–64
28. Sanchez-Reillo R, Quiros-Sandoval H, Goicochea-Telleria I, Ponce-Hernandez W (2017) Improving presentation attack detection in dynamic handwritten signature biometrics. IEEE Access 5:20463–20469
29. Madasu V, Lovell B (2008) An automatic off-line signature verification and forgery detection system In: Verma B, Blumenstein M (eds), Pattern recognition technologies and applications: recent advances, IGI Global, pp 63–88
30. Brault J, Plamondon R (1993) A complexity measure of handwritten curves: modeling of dynamic signature forgery. IEEE Trans Syst Man Cybern 23:400–413
31. Wu X, Kimura A, Iwana BK, Uchida S, Kashino K (2019) Deep dynamic time warping: end-to-end local representation learning for online signature verification. In: 2019 international conference on document analysis and recognition (ICDAR). IEEE, pp 1103–1110

32. Vorugunti CS, Mukherjee P, Pulabaigari V et al (2019) OSVNet: convolutional siamese network for writer independent online signature verification. In: 2019 international conference on document analysis and recognition (ICDAR). IEEE, pp 1470–1475
33. Lai S, Jin L, Zhu Y, Li Z, Lin L (2021) SynSig2Vec: forgery-free learning of dynamic signature representations by sigma lognormal-based synthesis. IEEE Trans Pattern Anal Mach Intell
34. Tolosana R, Vera-Rodriguez R, Fierrez J, Ortega-Garcia J (2018) Exploring recurrent neural networks for on-line handwritten signature biometrics. IEEE Access 1–11
35. Caruana M, Vera-Rodriguez R, Tolosana R (2021) Analysing and exploiting complexity information in on-line signature verification. In: Proceedings of the international conference on pattern recognition workshops, ICPRw
36. Vera-Rodriguez R, Tolosana R, Caruana M, Manzano G, Gonzalez-Garcia C, Fierrez J, Ortega-Garcia J (2019) DeepSignCX: signature complexity detection using recurrent neural networks. In: Proceedings of the 15th international conference on document analysis and recognition, ICDAR
37. Doddington G, Liggett W, Martin A, Przybocki M, Reynolds D (1998) Sheeps, goats, lambs and wolves: a statistical analysis of speaker performance in the NIST 1998 speaker recognition evaluation. In: Proceedings of the international conference on spoken language processing
38. Yager N, Dunstone T (2010) The biometric menagerie. IEEE Trans Pattern Anal Mach Intell 32(2):220–230
39. Houmani N, Garcia-Salicetti S (2016) On hunting animals of the biometric menagerie for online signature. PLoS ONE 11(4):1–26
40. Ortega-Garcia J, Fierrez-Aguilar J, Simon D, Gonzalez J, Faundez-Zanuy M, Espinosa V, Satue A, Hernaez I, Igarza JJ, Vivaracho C et al (2003) MCYT baseline corpus: a bimodal biometric database. IEE Proc-Vis Image Signal Process 150(6):395–401
41. Ballard L, Lopresti D, Monroe F (2007) Forgery quality and its implication for behavioural biometric security. IEEE Trans Syst Man Cybernet Part B 37(5):1107–1118
42. Vielhauer C, Zöbisch F (2003) A test tool to support brute-force online and offline signature forgery tests on mobile devices. In: Proceedings of the international conference multimedia and expo, vol 3, pp 225–228
43. Alonso-Fernandez F, Fierrez J, Gilperez A, Galbally J, Ortega-Garcia J (2009) Robustness of signature verification systems to imitators with increasing skills. In: Proceedings of the 10th international conference on document analysis and recognition
44. Ferrer M, Diaz M, Carmona-Duarte C, Morales A (2017) A behavioral handwriting model for static and dynamic signature synthesis. IEEE Trans Pattern Anal Mach Intell 39(6):1041–1053
45. Fierrez J, Galbally J, Ortega-Garcia J et al (2010) BiosecurID: a multimodal biometric database. Pattern Anal Appl 13(2):235–246
46. Fierrez-Aguilar J, Alonso-Hermira N, Moreno-Marquez G, Ortega-Garcia J (2004) An off-line signature verification system based on fusion of local and global information. In: Proceedings of the European conference on computer vision, workshop on biometric authentication, BIOAW, LNCS, vol 3087. Springer, pp 295–306
47. Tolosana R, Vera-Rodriguez R, Ortega-Garcia J, Fierrez J (2015) Update strategies for HMM-based dynamic signature biometric systems. In: Proceedings of the 7th IEEE international workshop on information forensics and security, WIFS
48. Graves A (2013) Generating sequences with recurrent neural networks. arXiv:1308.0850
49. Ferrer MA, Diaz M, Carmona-Duarte C, Plamondon R (2018) iDeLog: Iterative dual spatial and kinematic extraction of sigma-lognormal parameters. IEEE Trans Pattern Anal Mach Intell 42(1):114–125
50. Vera-Rodriguez R, Tolosana R, Hernandez-Ortega J, Acien A, Morales A, Fierrez J, Ortega-Garcia J (2020) Modeling the complexity of signature and touch-screen biometrics using the lognormality principle. World Scientific, pp 65–86
51. Diaz M, Fischer A, Ferrer M, Plamondon R (2016) Dynamic signature verification system based on one real signature. IEEE Trans Cybernet 48:228–239
52. Galbally J, Fierrez J, Martinez-Diaz M, Ortega-Garcia J (2009) Improving the enrollment in dynamic signature verification with synthetic samples. In: Proceedings of the IAPR international conference on document analysis and recognition, ICDAR, pp 1295–1299

53. Lai S, Jin L, Lin L, Zhu Y, Mao H (2020) SynSig2Vec: learning representations from synthetic dynamic signatures for real-world verification. In: Proceedings of the AAAI conference on artificial intelligence, vol 34, pp 735–742
54. Tolosana R, Delgado-Santos P, Perez-Uribe A, Vera-Rodriguez R, Fierrez J, Morales A (2021) DeepWriteSYN: on-line handwriting synthesis via deep short-term representations. In: Proceedings AAAI conference on artificial intelligence
55. Carmona-Duarte C, Ferrer MA, Parziale A, Marcelli A (2017) Temporal evolution in synthetic handwriting. Pattern Recogn 68:233–244
56. Tolosana R, Vera-Rodriguez R, Fierrez J, Morales A, Ortega-Garcia J (2019) Do you need more data? The DeepSignDB on-line handwritten signature biometric database. In: Proceedings international conference on document analysis and recognition (ICDAR)
57. Houmani N, Mayoue A, Garcia-Salicetti S, Dorizzi B, Khalil M, Moustafa M, Abbas H, Muramatsu D, Yanikoglu B, Kholmatov A, Martinez-Diaz M, Fierrez J, Ortega-Garcia J, Alcobé JR, Fabregas J, Faundez-Zanuy M, Pascual-Gaspar J, Cardeñoso-Payo V, Vivaracho-Pascual C (2012) Biosecure signature evaluation campaign (BSEC'2009): evaluating on-line signature algorithms depending on the quality of signatures. Pattern Recogn 45(3):993–1003
58. Martinez-Diaz M, Fierrez J, Krish R, Galbally J (2014) Mobile signature verification: feature robustness and performance comparison. IET Biom 3(4):267–277
59. Martinez-Diaz M, Fierrez J, Hangai S (2015) Signature matching. In: Li SZ, Jain A (eds), Encyclopedia of biometrics. Springer, pp 1382–1387
60. Tekampe N, Merle A, Bringer J, Gomez-Barrero M, Fierrez J, Galbally J (2016) Toward common criteria evaluations of biometric systems. Technical Report BEAT Public Deliverable D6.5. https://www.beat-eu.org/project/deliverables-public/d6-5-toward-common-criteria-evaluations-of-biometric-systems

Chapter 17
Extensive Threat Analysis of Vein Attack Databases and Attack Detection by Fusion of Comparison Scores

Johannes Schuiki, Michael Linortner, Georg Wimmer, and Andreas Uhl

Abstract The last decade has brought forward many great contributions regarding presentation attack detection for the domain of finger and hand vein biometrics. Among those contributions, one is able to find a variety of different attack databases that are either private or made publicly available to the research community. However, it is not always shown whether the used attack samples hold the capability to actually deceive a realistic vein recognition system. Inspired by previous works, this study provides a systematic threat evaluation including three publicly available finger vein attack databases and one private dorsal hand vein database. To do so, 14 distinct vein recognition schemes are confronted with attack samples, and the percentage of wrongly accepted attack samples is then reported as the Impostor Attack Presentation Match Rate. As a second step, comparison scores from different recognition schemes are combined using score level fusion with the goal of performing presentation attack detection.

17.1 Introduction

Since most biometric data are inseparably linked, to a human individual, security considerations for biometric systems are of utter importance. However, as some biometric traits are left behind (e.g., latent fingerprint) or are always exposed to the public (e.g., gait, face), one potential attack scenario that has to be dealt with is the presentation of a replica of a biometric sample to the biometric reader with the

---

J. Schuiki (✉) · M. Linortner (✉) · G. Wimmer (✉) · A. Uhl (✉)
Department of Computer Science, University of Salzburg, Salzburg, Austria
e-mail: jschuiki@cs.sbg.ac.at

M. Linortner
e-mail: mlinortner@cs.sbg.ac.at

G. Wimmer
e-mail: gwimmer@cs.sbg.ac.at

A. Uhl
e-mail: uhl@cosy.sbg.ac.at

S. Marcel et al. (eds.), *Handbook of Biometric Anti-Spoofing*, Advances in Computer Vision and Pattern Recognition, https://doi.org/10.1007/978-981-19-5288-3_17

goal of either impersonate someone or not being recognized. This is known as the presentation attack and is defined by the ISO/IEC [1] as *presentation to the biometric data capture subsystem with the goal of interfering with the operation of the biometric system*.

Due to the circumstance that blood vessels are hard to detect with the human visual system or consumer cameras, biometric systems that employ vascular patterns in the hand region were often attributed with being insusceptible against such attacks. However, German hackers [2] demonstrated the use of a hand vein pattern which was acquired from a distance of a few meters to successfully deceive a commercial hand vein recognition device during a live session at a hacking conference in 2018. Besides that, they suggested that hand dryers in the restroom could be manipulated to secretly capture hand vein images. Also, one must never rule out the possibility of an already existing vein database being compromised by an attacker. Hence, plenty of studies were published that address the possibility of vein recognition systems getting deceived by forged samples over the course of the last decade. A comprehensive overview of countermeasures to such attacks (also known as presentation attack detection, *PAD*) can be found in Table 14.1 in [3].

In order to evaluate the effectiveness of such PAD algorithms, either private databases or publicly available databases are used. For the creation of such attack samples, the following strategies can be found in the literature:

One of the pioneer works was published in 2013 by Nguyen et al. [4], where one of the first successful attempts to fool a finger vein recognition algorithm was reported. They created a small-scale attack database from an existing finger vein database consisting of seven individuals. Their recipe for generating the presentation attacks was to print selected finger vein templates from an existing database onto two types of paper and on overhead projector film using a laser printer, thereby creating three different attack types.

Tome et al. [5] created presentation attacks using a subset from a publicly available finger vein recognition database, that was initially released at the same time. Likewise, as Nguyen et al. in 2013 [4], a laser printer was used for presentation attack creation, and additionally the contours of the veins were enhanced using a black whiteboard marker. One year later, in 2015, the first competition on countermeasures to finger vein spoofing attacks was held [6] by the same authors as in [5]. With this publication, their presentation attack database was extended such that every biometric sample from their reference database has a corresponding spoofed counterpart. This was the first complete finger vein presentation attack database publicly available for research purposes.

Instead of merely printing vein samples on paper, Qiu et al. [7] suggested to print the sample on two overhead projector films and sandwich a white paper in-between the aligned overhead films. The white paper is meant to reduce overexposure due to the transparency of the overhead films.

Another experiment was shown in [8]. Here dorsal hand vein images were acquired using a smartphone camera that had, without modification, a moderate sensitivity in the near-infrared region. The images were then shown on the smartphone display and presented to the biometric sensor.

A very different approach was published by Otsuka et al. [9] in 2016. They created a working example of what they call "wolf attack". This type of attack has the goal to exploit the working of a recognition toolchain to construct an attack sample that will generate a high similarity score with any biometric template stored. They showed that their master sample worked in most of the finger vein verification attempts, therefore, posing a threat to this particular recognition toolchain used.

In 2020, Debiasi et al. [10] created a variety of presentation attacks that employ wax and silicone casts. However, no successful comparison experiments with bona fide samples were reported. In [11], the attack generation recipe using casts of beeswax was reworked and finally shown to be functional.

In order to demonstrate the capability of actually deceiving a vein recognition algorithm, attack samples can be evaluated using a so-called "2 scenario protocol", which is described in Sect. 17.3.1. In the authors' previous works [11, 12], an extensive threat evaluation was carried out for the finger vein databases introduced in [6, 11, 13] by confronting twelve vein recognition schemes with the attack samples. The twelve recognition schemes can be categorized into three types of algorithms based on which features they extract from the vein images. Additionally, a presentation attack detection strategy was introduced that employs score level fusion of recognition algorithms. This was done by cross-evaluation of five fusion strategies along with three feature scaling approaches.

This study extends the authors' previous work by adding two feature extraction and comparison schemes that introduce a fourth algorithm category. Further, three additional feature normalization techniques and two additional fusion strategies are included in the experiments of this chapter. Besides that, this study transfers the threat analysis procedure to a private dorsal hand vein video database that was used in earlier studies [14, 15], but it was never shown whether it could actually deceive a vein recognition system.

The remainder of this chapter is structured as follows. Section 17.2 describes the databases used later in this study. Section 17.3 contains an evaluation of the threat that is emitted by the databases from Sect. 17.2 to 14 vein recognition schemes that can be categorized into 4 types of algorithms. In Sect. 17.4, the comparison scores from Sect. 17.3 are combined using score level fusion in order to carry out presentation attack detection. Finally, Sect. 17.5 includes a summary of this study.

17.2 Attack Databases

In total, four databases are included in this study which are described hereafter.

(A) *Paris Lodron University of Salzburg Palmar Finger Vein Spoofing Data Set (PLUS-FV3 Spoof)*[1]: The PLUS-FV3 Spoof data set uses a subset of the PLUS Vein-FV3 [16] database as bona fide samples. For the collection of presentation

[1] https://wavelab.at/sources/PLUS-FV3-PALMAR-Image-Spoof/.

attack artefacts, binarized vein images from 6 fingers (i.e., index, middle and ring finger of both hands) of 22 subjects were printed on paper and sandwiched into a top and bottom made of beeswax. The binarization was accomplished by applying Principal Curvature [17] feature extraction in two different levels of vessel thickness, named *thick* and *thin*. The original database was captured with two types of light sources, namely *LED* and *Laser*. Therefore, presentation attacks were created for both illumination variants. While the original database was captured in 5 sessions per finger, only three of those were reused for presentation attack generation. Summarized, a total of 396 (22*6*3) presentation attacks per light source (LED and Laser) and vein thickness (thick and thin) corresponding to 660 (22*6*5) bona fide samples are available. Every sample is of size 192×736.

(B) *The Idiap Research Institute VERA Fingervein Database (IDIAP VERA)*[2]: The IDIAP VERA finger vein database consists of 440 bona fide images that correspond to 2 acquisition sessions of left-hand and right-hand index fingers of 110 subjects. Therefore, these are considered as 220 unique fingers captured 2 times each. Every sample has one presentation attack counterpart. Presentation attacks are generated by printing preprocessed samples on high quality paper using a laser printer and enhancing vein contours with a black whiteboard marker afterwards. Every sample is provided in two modes named *full* and *cropped*. While the full set is comprised of the raw images captured with size 250×665, the cropped images were generated by removing a 50 pixel margin from the border, resulting in images of size 150×565.

(C) *South China University of Technology Spoofing Finger Vein Database (SCUT-SFVD)*[3]: The SCUT-SFVD database was collected from 6 fingers (i.e., index, middle and ring finger of both hands) of 100 persons captured in 6 acquisition sessions, making a total of 3600 bona fide samples. For presentation attack generation, each finger vein image is printed on two overhead projector films which are aligned and stacked. In order to reduce overexposure, additionally, a strong white paper ($200\,\mathrm{g/m^2}$) is put in-between the two overhead projector films. Similar to the IDIAP VERA database, the SCUT-SFVD is provided in two modes named *full* and *roi*. While in the full set every image sample has a resolution of 640×288 pixel, the samples from the roi set are of variable size. Since the LBP and the ASAVE matching algorithm cannot be evaluated on variable-sized image samples, a third set was generated for this study named *roi-resized* where all roi samples have been resized to 474×156 which corresponds to the median of all heights and widths from the roi set.

(D) *Paris Lodron University of Salzburg Dorsal Hand Vein Video Spoofing Data Set (PLUS-DHV Spoof):* This database was initially created by [14] in order to analyze a finger vein video presentation attack detection scheme proposed by [18]. The database consists of video samples from both hands of 13 participants, that were captured using two different illumination variants: *Reflected light*,

[2] https://www.idiap.ch/en/dataset/vera-fingervein.

[3] https://github.com/BIP-Lab/SCUT-SFVD.

where the illumination source is placed next to the imaging sensor in order to capture the light which is not absorbed but reflected by the user's hand, as well as *transillumination*, where the light source comes from the opposite side and goes through the user's hand. For every bona fide video sample, five different video attacks exist: (i) printed on paper using a laser printer (*Paper Still*), (ii) printed on paper with applied movement back and forth (*Paper Moving*), (iii) shown on a smartphone display (*Display Still*), (iv) shown on smartphone display with programmed sinusoidal translation oscillation along the x-axis (*Display Moving*) and (v) shown on smartphone display with programmed sinusoidal scaling oscillation in every direction (*Display Zoom*). For this study, 10 equally spaced frames were extracted throughout every video sequence, resulting in 260 (26*10) bona fide as well as attack samples, per attack type. Further, region of interest preprocessing was applied. Note, however, that unfortunately, this database is a private one.

17.3 Threat Analysis

This section follows the idea from previous publications [11, 12], in which databases (A)–(C) from Sect. 17.2 were subjected to experiments in order to evaluate the threat these attack samples emit to a variety of recognition algorithms. The goal of the experiments in this section is (i) to carry out the experiments from previous publications with a slightly different setting in the options as well as (ii) the transfer of this threat evaluation to the dorsal hand vein database (D) explained in Sect. 17.2. To do so, the threat evaluation protocol described in Sect. 17.3.1 is used for the experiments in this section. Altogether 14 distinct feature extraction and comparison schemes are used in this study that can be categorized into four classes of algorithms based on the type of feature they extract from the vein samples:

- *Binarized vessel networks*: Algorithms from this category work by transforming a raw vein image into a binary image where the background (and also other parts of the human body such as flesh) is removed and only the extracted vessel structures remain. The binarized image is then used as a featured image for the comparison step. Seven different approaches are included in this study that finally create such a binarized vein image. *Maximum Curvature (MC)* [19] and *Repeated Line Tracking (RLT)* [20] try to achieve this by looking at the cross-sectional profile of the finger vein image. Other methods such as *Wide Line Detector (WLD)* [21], *Gabor Filter (GF)* [22] and *Isotropic Undecimated Wavelet Transform (IUWT)* [23] also consider local neighbourhood regions by using filter convolution. A slightly different approach is given by *Principal Curvature (PC)* [17] which first computes the normalized gradient field and then looks at the eigenvalues of the Hessian matrix at each pixel. All so far described binary image extraction methods use a correlation measure to compare probe and template samples which are often referred to as *Miura-matching* due to its introduction in Miura et al. [20]. One more sophisticated vein pattern-based feature extraction and matching strategy is *Anatomy Structure Analysis-Based Vein Extraction (ASAVE)* [24], which includes two different techniques for binary vessel structure extraction as well as a custom matching strategy.

- *Keypoints*: The term keypoint is generally understood as a specific pixel or pixel region in a digital image that provides some interesting information to a given application. Every keypoint is stored by describing its local neighbourhood and its location. This research uses three keypoint-based feature extraction and matching schemes. One such keypoint detection method, known as *Deformation Tolerant Feature Point Matching (DTFPM)* [25], was especially tailored for the task of finger vein recognition. This is achieved by considering shapes that are common in finger vein structures. Additionally, modified versions of general purpose keypoint detection and matching schemes, *SIFT* and *SURF*, as described in [26] are tested in this research. The modification includes filtering such that only keypoints inside the finger are used while keypoints at the finger contours or even in the background are discarded.
- *Texture information*: Image texture is a feature that describes the structure of an image. Shapiro and Stockman [27] define image texture as something that gives information about the spatial arrangement of color or intensities in an image or selected region of an image. While two images can be identical in terms of their histograms, they can be very different when looking at their spatial arrangement of bright and dark pixels. Two methods are included in this work that can be counted as texture-based approaches. One of which is a *Local Binary Pattern* [28] descriptor that uses histogram intersection as a similarity metric. The second method is a *convolutional neural network (CNN)*-based approach that uses triplet loss as presented in [29]. Similarity scores for the CNN approach are obtained by computing the inverse Euclidean distance given two feature vectors corresponding to two finger vein samples.
- *Minutiae-based*: The term minutiae descends from the domain of fingerprint biometrics. Every fingerprint is a unique pattern that consists of ridges and valleys. The locations where such a pattern has discontinuities such as ridge endings or bifurcations are named “minutiae points”. This concept of finding such minutiae points was successfully transferred [30] to the vein biometrics domain by skeletonization of an extracted binarized vein image, such as described in the first category. This study now employs two schemes in order to perform vein recognition, and both use these extracted minutiae points. First, the proprietary software *VeriFinger SDK (VF)*[4] is used for comparison of these minutiae points. A second method, *Location-based Spectral Minutiae Representation (SML)* [31] uses the minutiae points as an input in order to generate a representation that can finally be compared using a correlation measure.

[4] https://www.neurotechnology.com/verifinger.html.

17.3.1 Threat Evaluation Protocol

To evaluate the level of threat exhibited by a certain database, a common evaluation scheme is used that employs two consecutive steps, hence often coined "2 scenario protocol" [5, 8, 32], is adopted in this study. The two scenarios are briefly summarized hereafter (description taken from [12]):

- **Licit Scenario (Normal Mode)**: The first scenario employs two types of users: Genuine (positives) and zero effort impostors (negatives). Therefore, both enrolment and verification are accomplished using bona fide finger vein samples. Through varying the decision threshold, the False Match Rate (FMR, i.e., the ratio of wrongly accepted impostor attempts to the number of total impostor attempts) and the False Non Match Rate (FNMR, i.e., the ratio of wrongly denied genuine attempts to the total number of genuine verification attempts) can be determined. The normal mode can be understood as a matching experiment which has the goal to determine an operating point for the second scenario. The operating point is set at the threshold value where the FMR = FNMR (i.e., Equal Error Rate).
- **Spoof Scenario (Attack Mode)**: The second scenario uses genuine (positives) and presentation attack (negatives) users. Similar to the first scenario, enrolment is accomplished using bona fide samples. Verification attempts are performed by matching presentation attack samples against their corresponding genuine enrolment samples or templates. Given the threshold from the licit scenario, the proportion of wrongly accepted presentation attacks is then reported as the Impostor Attack Presentation Match Rate (IAPMR), as defined by the ISO/IEC 30107-3:2017 [1].

17.3.2 Experimental Results

The experimental results are divided into two tables: Table 17.1 contains the outcomes from the threat evaluation on the finger vein databases (PLUS, IDIAP Vera and SCUT) and Table 17.2 contains the results from the experiments on the dorsal hand vein database. Note that for databases where multiple types of attacks exist for the same corresponding bona fide samples, the reported EER values are equal since the error rate calculation is solely based on bona fide data. The table that contains the threat analysis for the hand vein samples (Table 17.2) includes always two IAPMR values per attack type. The reason for this is illustrated in Fig. 17.1: Due to the fact that only one video sequence is available per subject to extract frames from, the samples are somewhat biased in terms of intra-class variability. Hence, genuine and impostor scores can often be separated perfectly such that a whole range of decision thresholds would be eligible options to set the EER (the range is illustrated in the figure as the region in-between the vertical dash-dotted lines). The dash-dotted lines represent the extreme cases where the EER would still be zero, i.e., perfect separation of genuine

Table 17.1 Vulnerability analysis using 2 scenario protocol for the finger vein DBs

	MC	PC	GF	IUWT	RLT	WLD	ASAVE	DTFPM	SIFT	SURF	LBP	CNN	SML	VF
PLUS LED Thick														
EER	0.60	0.75	1.06	0.68	4.62	1.14	2.35	2.35	1.06	3.33	3.86	2.90	2.79	0.80
IAPMR	72.81	69.93	36.99	79.08	40.65	69.15	24.31	16.34	0.00	0.00	0.00	0.67	4.02	5.16
PLUS LED Thin														
EER	0.60	0.75	1.06	0.68	4.62	1.14	2.35	2.35	1.06	3.33	3.86	2.90	2.79	0.80
IAPMR	89.65	79.80	59.72	89.77	32.83	83.96	19.07	15.78	0.00	0.00	0.25	0.35	5.87	5.24
PLUS Laser Thick														
EER	1.29	1.65	2.72	1.97	6.21	2.80	2.50	2.90	1.04	3.71	4.27	6.82	2.95	1.01
IAPMR	59.90	57.47	31.55	79.44	22.48	57.22	9.07	4.98	0.00	0.00	0.00	0.00	2.67	6.54
PLUS Laser Thin														
EER	1.29	1.65	2.72	1.97	6.21	2.80	2.50	2.90	1.04	3.71	4.27	6.82	2.95	1.01
IAPMR	76.52	66.92	52.65	84.34	16.79	77.78	2.02	5.18	0.13	0.00	0.00	0.05	3.60	5.62
IDIAP Vera FV Full														
EER	2.66	2.73	6.83	4.95	29.55	6.03	9.11	10.45	4.54	11.39	8.17	6.35	6.69	4.04
IAPMR	93.18	90.45	85.76	93.18	35.00	93.48	72.58	26.21	14.24	0.91	16.97	17.27	59.09	21.67
IDIAP Vera FV Cropped														
EER	5.52	6.91	11.36	6.36	27.19	9.09	19.10	6.69	5.43	11.62	5.80	9.93	17.27	11.11
IAPMR	92.27	89.24	81.97	91.82	38.64	90.45	68.79	81.97	44.55	14.24	69.09	7.84	64.55	21.97
SCUT-SFVD Full														
EER	4.01	4.79	9.40	6.29	14.01	7.60	11.56	8.90	2.37	5.49	8.78	0.74	7.63	1.41
IAPMR	86.33	84.67	54.90	74.06	40.36	74.21	74.98	73.75	30.43	4.18	45.43	68.65	59.76	32.90
SCUT-SFVD ROI-Resized														
EER	2.28	2.12	3.52	2.10	2.94	2.36	6.03	5.63	1.92	9.41	3.51	0.84	7.25	3.78
IAPMR	60.36	47.95	36.88	49.85	9.46	52.49	59.88	55.08	34.93	7.42	55.36	55.00	61.22	40.40

Table 17.2 Results of the threat analysis for the hand vein attacks

	MC	PC	GF	IUWT	RLT	WLD	ASAVE	DTFPM	SIFT	SURF	LBP	CNN	SML	VF
PLUS-DHV Transillumination Paper-Still														
EER	0.00	0.00	0.00	0.00	2.66	0.00	0.00	0.00	0.00	0.00	0.00	1.62	0.17	0.09
IAPMR-R	0.00	0.00	0.00	0.00	0.00	0.00	0.00	0.00	0.00	0.00	0.00	3.31	3.15	2.31
IAPMR-L	94.90	5.03	0.28	6.01	0.00	11.75	38.95	85.52	0.35	0.00	6.78	3.31	3.15	2.31
PLUS-DHV Transillumination Paper-Moving														
IAPMR-R	0.00	0.00	0.00	0.00	0.00	0.00	0.00	0.00	0.00	0.00	0.00	3.38	3.78	2.24
IAPMR-L	92.38	4.97	0.42	5.87	0.00	10.77	32.94	79.86	0.00	0.00	3.85	3.38	3.78	2.24
PLUS-DHV Transillumination Display-Still														
IAPMR-R	1.40	0.49	24.20	0.56	8.46	10.77	0.00	0.00	0.00	0.00	0.00	18.58	5.31	9.51
IAPMR-L	99.65	33.15	32.87	42.24	8.46	61.82	2.17	96.22	3.64	0.00	0.00	18.46	5.31	9.51
PLUS-DHV Transillumination Display-Moving														
IAPMR-R	0.63	0.28	23.01	1.19	7.13	10.84	0.00	0.00	0.00	0.00	0.00	18.54	5.59	10.77
IAPMR-L	98.95	34.83	29.93	40.70	7.13	54.69	5.66	95.03	3.43	0.00	0.00	18.50	5.59	10.77
PLUS-DHV Transillumination Display-Zoom														
IAPMR-R	0.14	0.21	22.87	0.21	9.02	10.56	0.00	0.00	0.00	0.00	0.00	18.54	6.08	9.72
IAPMR-L	99.65	33.22	31.47	43.29	9.02	57.55	2.87	96.50	3.43	0.00	0.00	18.54	6.08	9.72

(continued)

Table 17.2 (continued)

	MC	PC	GF	IUWT	RLT	WLD	ASAVE	DTFPM	SIFT	SURF	LBP	CNN	SML	VF
PLUS-DHV Reflected Light Paper-Still														
EER	0.00	0.00	0.10	0.00	3.25	0.00	0.00	0.00	0.00	0.18	0.00	2.62	0.34	0.69
IAPMR-R	0.00	0.00	2.66	0.00	0.00	13.50	0.00	0.00	0.00	0.00	0.00	0.00	1.47	0.42
IAPMR-L	92.80	0.00	2.66	0.07	0.00	13.50	17.90	8.18	0.00	0.00	0.00	0.00	1.47	0.42
IAPMR-R	0.00	0.00	2.52	0.00	0.00	10.56	0.00	0.00	0.00	0.00	0.00	0.00	1.75	0.42
IAPMR-L	91.96	0.00	2.52	0.63	0.00	10.56	19.65	23.43	0.00	0.00	0.00	0.00	1.75	0.42
PLUS-DHV Reflected Light Display-Still														
IAPMR-R	0.00	8.46	12.24	2.45	64.83	0.00	0.00	0.00	0.07	3.08	0.00	8.00	1.33	3.08
IAPMR-L	94.97	67.90	12.24	30.07	64.83	0.00	13.15	54.62	3.57	3.08	0.00	8.00	1.33	3.08
PLUS-DHV Reflected Light Display-Moving														
IAPMR-R	0.00	7.97	12.38	2.59	65.38	0.00	0.00	0.00	0.00	2.45	0.00	8.54	1.75	3.29
IAPMR-L	93.15	71.33	12.38	37.34	65.38	0.00	19.86	64.13	3.43	2.45	0.00	8.54	1.75	3.29
PLUS-DHV Reflected Light Display-Zoom														
IAPMR-R	0.49	9.58	11.96	2.73	59.16	0.00	0.00	0.00	0.00	2.38	0.00	6.42	1.54	3.15
IAPMR-L	95.31	73.43	11.96	36.92	59.16	0.00	12.66	58.39	3.36	2.38	0.00	6.42	1.54	3.15

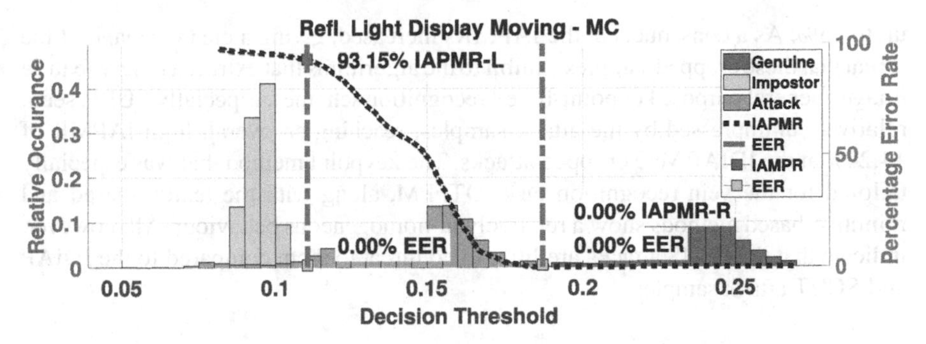

Fig. 17.1 Exemplary visualization of the outcomes from the two scenario protocol when the Maximum Curvature scheme was confronted with hand vein attack samples from type: reflected light display moving

and impostor scores. When looking at the two IAPMRs in the extreme cases, one can see that this rate varies (in this case) between 0.00 and 93.15%. Thus, the IAPMRs for the hand vein database is always reported as for the left-most (IAPMR-L) and right-most (IAPMR-R) extreme case. Note that for EERs $\neq 0$ both IAPMRs are identical.

The minutiae-based recognition schemes are evaluated using the *VeinPLUS+ Framework* [33]. The binary vessel network-based methods, the keypoint-based methods and the LBP scheme were evaluated using the *PLUS OpenVein Toolkit* [34]. The CNN-based approach was implemented in Python. Due to the fact that this is a learning-based approach and such learning is non-deterministic, the EERs and IAPMRs for the CNN method are calculated by taking the arithmetic mean over a 2-fold-cross validation. This is done for every database except for the PLUS LED and PLUS Laser, since there exists a second open source database [35] which descends from a similar imaging sensor that is used for training the CNN. For comparison, the "FVC" protocol, named after the Fingerprint Verification Contest 2004 [36] where it was introduced, was used. This protocol executes every possible genuine comparison but only compares the first sample of each subject with the first sample of all remaining subjects. Hence, it is ensured that every genuine comparison is made, but largely reduces the amount of impostor scores.

17.3.2.1 Finger Vein Attack Databases

When confronted with the finger vein attacks, binary vessel pattern-based schemes tend to produce IAPMRs up to 93.48% for the IDIAP Vera, 86.33% for the SCUT and 89.54% for the PLUS databases. Such IAPMRs indicate that roughly up to 9 out of 10 attack samples could wrongly be classified as a bona fide presentation. Compared to the reference paper [12], the "Miura matching" comparison step for the region of interest (roi/cropped) versions was also executed with parameters that allow for a certain translation invariance. By doing so, some EER results could be reduced by up to 15%. As a consequence, the IAPMRs increased, giving a clearer image of the threat that these cropped samples exhibit to the algorithms that extract a binary feature image. General purpose keypoint-based recognition schemes, especially SURF, seem relatively unimpressed by the attack samples, reaching an overall high IAPMR of 14.24% at the IDIAP Vera cropped attacks. The keypoint method that was especially tailored for the vein recognition task, DTFPM, along with the texture-based and minutiae-based methods show a relatively in-homogeneous behaviour. All, however, indicate that the PLUS attacks are little to no threat to them compared to the IDIAP and SCUT attack samples.

17.3.2.2 Dorsal Hand Vein Attack Database

When looking at the threat analysis for the dorsal hand vein attacks (Table 17.2), one is able to observe that (i) attack samples that were captured through transillumination are more likely to pose a threat to the recognition schemes and also (ii) attacks shown on a smartphone display tend to work better than the laser printed paper attacks. One exception constitutes the MC recognition scheme. It seems that, when treating both extreme cases IAPMR-R and IAPMR-L as equally valid, all attacks have an overall high potential to fool the MC algorithm. The same holds true for the DTFPM keypoint recognition scheme. For other binary vessel network schemes, as mentioned, only the display attacks seem to have a certain attack potential. The remaining general purpose keypoint schemes, SIFT and SURF, the Minutiae-based schemes as well as LBP seem to remain unaffected regardless of the attack type used. Interestingly, the display attacks seem to have at least some attack potential on the CNN recognition scheme.

17.4 Attack Detection Using Score Level Fusion

Using the observation that different recognition schemes vary in their behaviour and thus probably contain complementary information, this section includes experiments to combine multiple recognition schemes in order to achieve presentation attack detection. Note that throughout the following section, genuine scores (i.e., scores that descend from intra-subject comparisons) are viewed as bona fide scores and scores that descend from comparisons where an attack sample is compared to its bona fide counterpart are considered to be presentation attack scores. Solving the presentation attack problem via a combination of multiple similarity scores descending from distinct recognition schemes can be formally written as follows. Let s_{ij} be the ith comparison score out of m total eligible comparisons that was assigned by the jth recognition scheme out of n considered recognition schemes at a time, i.e., $i \in \{1, \ldots, m\}$ and $j \in \{1, \ldots, n\}$. Note that for an ith comparison, every recognition scheme needs to compare the same biometric samples. Further, let x_i be the the vector $x_i = (s_{i1}, \ldots, s_{in})$ that describes the ith comparison by concatenation of all n recognition schemes considered at a time. Let $\mathbf{X}$ be the set of all m eligible comparisons from a certain database $\mathbf{X} = \{x_1, \ldots, x_m\}$. Since some of the following fusion strategies demand training data, the set of all comparisons $\mathbf{X}$ is always divided into a train $\mathbf{X_{train}}$ and a test split $\mathbf{X_{test}}$ using k-fold cross validation. The combined score that remains after score level fusion for the ith comparison shall be denoted as S_i. The comparison can then be classified as bona fide or attack comparison by simple thresholding of S_i.

For the experiments in this study, three simple fusion strategies (*Min-Rule Fusion*, *Max-Rule Fusion* and *Simple Sum-Rule Fusion*) were adopted from [37] that can be seen in the following equations.

- *Min-Rule Fusion*

$$S_i = \min(x_i) \tag{17.1}$$

- *Max-Rule Fusion*

$$S_i = \max(x_i) \tag{17.2}$$

- *Simple Sum-Rule Fusion*

$$S_i = \sum_{j=1}^{n} s_{ij} \tag{17.3}$$

A slightly more sophisticated method is given by the *Weighted Sum-Rule Fusion*. In [38], this technique is also named as "Matcher Weighting" in order to distinguish the weighting strategy from the one where every user or individual would get different weights instead of every recognition scheme.

- *Weighted Sum-Rule Fusion*

$$S_i = \sum_{j=1}^{n} s_{ij} * w_j \tag{17.4}$$

In order to use the weighted sum rule fusion, a strategy has to be chosen on how to assign the weights w_j in Eq. 17.4. Snelick et al. [38] propose to choose weights indirectly proportional to the error rates. Since it is not intuitively clear which error rates to use in this study (since the goal is to use score level fusion of different recognition schemes to achieve presentation attack detection), the training split $\mathbf{X_{train}}$ is used to calculate an Equal Error Rate estimate EER_j per recognition scheme j. This decision can be justified since $\mathbf{X_{train}}$ consists of bona fide and attack comparisons and therefore gives a rough estimate on how much a particular recognition scheme contributes to the PAD-task.

$$w_j = \frac{\frac{1}{EER_j}}{\sum_{v=1}^{n} \frac{1}{EER_v}} \tag{17.5}$$

Additionally, three binary classifiers are evaluated that are trained using $\mathbf{X_{train}}$ in order to predict the test data $\mathbf{X_{test}}$ per fold: (i) *Fisher Linear Discriminant*, (ii) *Support Vector Machine with Linear Kernel* and (iii) *Support Vector Machine with Radial Basis Function Kernel*.

Similarity scores from distinct recognition algorithms, however, do not necessarily lie in the same range. Therefore, a variety of comparison score normalization strategies are applied to x_i in order to create a normalized comparison feature vector x_i' along with the option of applying no feature scaling at all (*No Norm:* $x_i' = x_i$). Note that calculations over the whole data (e.g., mean, min, max,...) are also conducted over the training split $\mathbf{X_{train}}$ for the normalization step as well in order to simulate a realistic scenario where the sample under test is not included in parameter determination. Together with the option of omitting score normalization, six strategies are included in this research.

Three popular score normalization techniques [39], *Min-Max Norm*, *Z-Score Norm* and *Tanh-Norm* are described in the following equations, where σ represents the operator for calculation of the standard deviation and μ stands for calculation of the arithmetic mean.

- *Min-Max Norm*

$$x_i' = \frac{x_i - \min(\mathbf{X_{train}})}{\max(\mathbf{X_{train}}) - \min(\mathbf{X_{train}})} \tag{17.6}$$

- *Tanh-Norm*

$$x_i' = 0.5 * \left(\tanh\left(0.01 * \frac{x_i - \mu(\mathbf{X_{train}})}{\sigma(\mathbf{X_{train}})}\right) + 1\right) \tag{17.7}$$

- *Z-Score Norm*

$$x_i' = \frac{x_i - \mu(\mathbf{X_{train}})}{\sigma(\mathbf{X_{train}})} \tag{17.8}$$

Another normalization technique was proposed by He et al. [40] named *Reduction of high-scores effect (RHE) normalization*. Here, $\mathbf{X_{train_{bf}}}$ indicates to use only the bona fide comparison scores for calculation of the mean and standard deviation.

- *Rhe-Norm*

$$x_i' = \frac{x_i - \min(\mathbf{X_{train}})}{\mu(\mathbf{X_{train_{bf}}}) + \sigma(\mathbf{X_{train_{bf}}}) - \min(\mathbf{X_{train}})} \tag{17.9}$$

Additionally, rescaling the feature vector x to unit length is tested. Note that for the cases where only a single recognition scheme is considered, the *Unit Length Norm* would result in ones (i.e., $\frac{x}{||x||} = 1$ for the case where x is a scalar). Therefore the normalization is omitted for these cases.

- *Unit Length Norm*

$$x_i' = \frac{x_i}{||x_i||} \tag{17.10}$$

The ISO/IEC 30107-3:2017 [1] defines decision threshold dependent metrics for presentation attack detection such as Attack Presentation Classification Error Rate (APCER) and Bona Fide Presentation Classification Error Rate (BPCER):

- *Attack Presentation Classification Error Rate (APCER)*: Proportion of attack presentations incorrectly classified as bona fide presentations in a specific scenario
- *Bona Fide Presentation Classification Error Rate (BPCER)*: Proportion of bona fide presentations incorrectly classified as presentation attacks in a specific scenario

The PAD performance is reported in Table 17.3 in terms of detection equal error rate D-EER (operating point where BPCER = APCER), BPCER20 (BPCER at APCER $<=$ 0.05) and BPCER100 (BPCER at APCER $<=$ 0.01).

17.4.1 Experimental Results

For the score level fusion experiments described above, results are obtained by conducting an exhaustive cross combination that includes all of the aforementioned fusion and normalization strategies. Per database, $2^{14} - 1 = 16\,383$ method constellations (since 14 different recognition schemes are used) exist, all of which are further evaluated using $6*7 = 42$ Norm-Fusion combinations. Doing so allows for a broader perspective on which score level fusion techniques tend to work well very often. The dorsal hand vein database was not included in these experiments since for every attack case there is at least one recognition scheme that achieves 0.00% IAPMR, meaning that at least one single method would suffice to separate bona fide from attack samples, thus annulling the reason of performing score level fusion.

Since the SCUT database has a lot more samples than the other databases, the experiments in this section only considered comparisons of samples with IDs 1 and 2 per subject in order to keep the experiments computationally feasible in a reasonable amount of time. In order to split the comparison scores into train and test data, 2-fold cross validation is used. The best working constellation per database can be seen in Table 17.3. It is worth noting, however, that for the PLUS LED and PLUS Laser database often multiple method constellations would be eligible while still achieving 00.00% D-EER. Hence, the reported constellations are chosen to require as few methods as possible and preferably use a computationally inexpensive norm-fusion combination. Each boxplot in Fig. 17.2 represents 16 383 method constellations for a particular normalization and fusion method, evaluated on a certain database. Every number on the x-axis corresponds to a certain norm-fusion-combination which can be deciphered using the lookup Table 17.4. One particular mode was chosen for every database since the overall trend is roughly the same for every sub-database. One can see that the score level fusion—presentation attack detection works well on the PLUS data. This observation coincides with the threat analysis, where the PLUS attacks had a hard time deceiving the keypoint, texture and minutiae recognition schemes, therefore all contributing valuable information for the PAD. The three peaks that can

Table 17.3 Selection of best working method constellations in terms of detection error rate. Note that for the PLUS databases, there are often multiple eligible options. This is only one constellation that uses as little recognition schemes as possible

Database	D-EER	BPCER20	BPCER100	Fusion	Norm	MC	PC	WLD	RLT	GF	IUWT	ASAVE	DTFPM	SURF	SIFT	LBP	CNN	SML	VF
PLUS-LED thick	0.00	0.00	0.00	svm-rbf	z-norm										✓		✓		
PLUS-LED thin	0.00	0.00	0.00	sum-rule	z-norm									✓	✓				
PLUS-Laser thick	0.00	0.00	0.00	sum-rule	no-norm										✓		✓		
PLUS-Laser thin	0.00	0.00	0.00	svm-lin	z-norm										✓	✓	✓		
IDIAP VERA full	1.14	0.91	1.82	fisher-lda	unit-length		✓	✓	✓	✓		✓		✓	✓	✓	✓		
IDIAP VERA cropped	3.18	2.27	13.18	svm-lin	rhe-norm	✓	✓	✓	✓	✓	✓	✓	✓	✓	✓		✓	✓	✓
SCUT-SFVD full	0.33	0.00	0.17	svm-lin	min-max				✓		✓	✓	✓	✓		✓	✓		
SCUT-SFVD roi-resized	0.76	0.67	0.67	svm-rbf	rhe-norm		✓	✓	✓	✓		✓		✓		✓	✓		✓

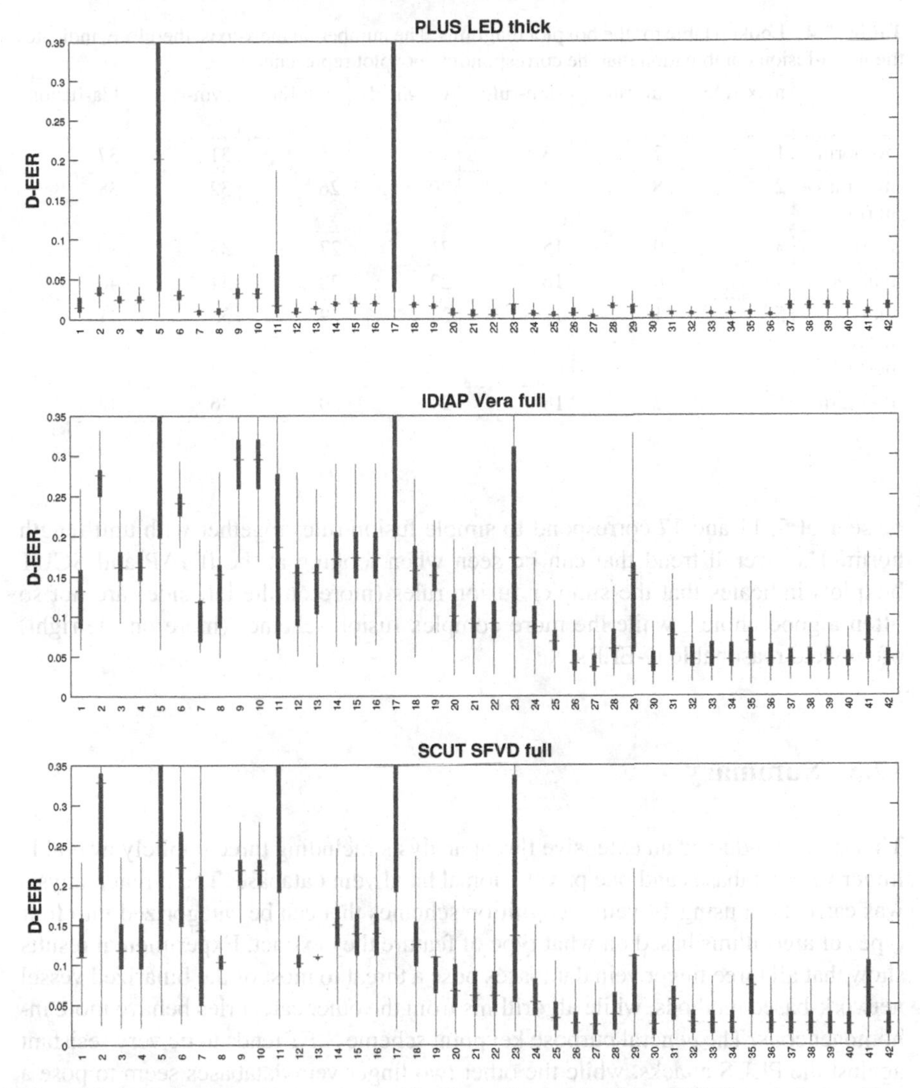

Fig. 17.2 Boxplots including cross combinations of normalization and fusion strategies. Note that every boxplot contains 16 383 values for $2^{14} - 1$ possible method constellations. To see what number on the x-axis corresponds to which norm-fusion-combination, Table 17.4 serves as a lookup table

Table 17.4 Lookup table for the boxplot diagrams. The number on the x-axis, therefore, indicates the norm-fusion combination that the corresponding boxplot represents

	max-rule	min-rule	sum-rule	weighted-sum	svm-lin	svm-rbf	lda-fusion
no-norm	1	7	13	19	25	31	37
min-max-norm	2	8	14	20	26	32	38
z-norm	3	9	15	21	27	33	39
tanh-norm	4	10	16	22	28	34	40
unit-length-norm	5	11	17	23	29	35	41
rhe-norm	6	12	18	24	30	36	42

be seen at 5, 11 and 17 correspond to simple fusion rules together with unit-length norm. The overall trend that can be seen when looking at the IDIAP and SCUT boxplots indicates that the simpler fusion rules (more on the left side) are not so often a good choice, while the more complex fusion schemes (more on the right) often yield reasonable D-EERs.

17.5 Summary

This study conducted an extensive threat analysis including three publicly available finger vein databases and one private dorsal hand vein database. The threat analysis was carried out using 14 vein recognition schemes that can be categorized into four types of algorithms based on what type of feature they extract. Experimental results show that all three finger vein databases pose a threat to most of the binarized vessel network-based methods, while algorithms from the other categories behave more in-homogeneous. The general purpose keypoint scheme SIFT tends to be very resistant against the PLUS attacks, while the other two finger vein databases seem to pose a threat, having IAPMRs ranging from 14 to 44%. The other general purpose keypoint recognition scheme, SURF, seems to be very unimpressed by the presented attacks overall. Similar to the SIFT recognition scheme, the texture-based methods LBP and CNN tend to be only susceptible to the IDIAP and SCUT attack samples. Minutiae-based methods as well as the DTFPM keypoint scheme show minor (5–16% IAPMR) receptiveness for the PLUS attacks and higher (21% and above) for the IDIAP and SCUT attacks. The second part of this research tries to use the in-homogeneity from the recognition schemes in order to perform presentation attack detection. To do so, comparison scores from different recognition schemes are combined using a range of different score level fusion techniques. After exhaustive cross combination of all recognition schemes together with normalization and fusion schemes, the best working method constellations show indeed that presentation attack detection can be achieved.

Acknowledgements This work has been partially supported by the Austrian Science Fund and the Salzburg State Government, FWF project no. P32201.

References

1. ISO/IEC JTC 1/SC 37 Biometrics (2017) Information technology — biometric presentation attack detection — part 3: testing and reporting. Standard ISO/IEC 30107-3:2017, International Organization for Standardization, Geneva, CH. https://www.iso.org/standard/67381.html
2. Krissler J, Julian (2018) Venenerkennung hacken - Vom Fall der letzten Bastion biometrischer Systeme. Chaos Computer Club e.V. https://doi.org/10.5446/39201. Accessed 21 Jan 2021
3. Kolberg J, Gomez-Barrero M, Venkatesh S, Ramachandra R, Busch C (2020) Presentation attack detection for finger recognition. Springer International Publishing, Cham, pp 435–463. https://doi.org/10.1007/978-3-030-27731-4_14
4. Nguyen DT, Park YH, Shin KY, Kwon SY, Lee HC, Park KR (2013) Fake finger-vein image detection based on fourier and wavelet transforms. Digit Signal Process 23(5):1401–1413. https://doi.org/10.1016/j.dsp.2013.04.001. www.sciencedirect.com/science/article/pii/S1051200413000808
5. Tome P, Vanoni M, Marcel S (2014) On the vulnerability of finger vein recognition to spoofing. In: 2014 international conference of the biometrics special interest group (BIOSIG), pp 1–10
6. Tome P, Raghavendra R, Busch C, Tirunagari S, Poh N, Shekar BH, Gragnaniello D, Sansone C, Verdoliva L, Marcel S (2015) The 1st competition on counter measures to finger vein spoofing attacks. In: 2015 international conference on biometrics (ICB), pp 513–518. https://doi.org/10.1109/ICB.2015.7139067
7. Qiu X, Kang W, Tian S, Jia W, Huang Z (2018) Finger vein presentation attack detection using total variation decomposition. IEEE Trans Inf Forensics Secur 13(2):465–477. https://doi.org/10.1109/TIFS.2017.2756598
8. Patil I, Bhilare S, Kanhangad V (2016) Assessing vulnerability of dorsal hand-vein verification system to spoofing attacks using smartphone camera. In: 2016 IEEE international conference on identity, security and behavior analysis (ISBA), pp 1–6
9. Otsuka A, Ohki T, Morita R, Inuma M, Imai H (2016) Security evaluation of a finger vein authentication algorithm against wolf attack. In: 37th IEEE symposium on security and privacy, San Jose, CA
10. Debiasi L, Kauba C, Hofbauer H, Prommegger B, Uhl A (2020) Presentation attacks and detection in finger- and hand-vein recognition. In: Proceedings of the joint Austrian computer vision and robotics workshop (ACVRW'20), Graz, Austria, pp 65–70. https://doi.org/10.3217/978-3-85125-752-6-16
11. Schuiki J, Prommegger B, Uhl A (2021) Confronting a variety of finger vein recognition algorithms with wax presentation attack artefacts. In: Proceedings of the 9th IEEE international workshop on biometrics and forensics (IWBF'21), Rome, Italy (moved to virtual), pp 1–6
12. Schuiki J, Wimmer G, Uhl A (2021) Vulnerability assessment and presentation attack detection using a set of distinct finger vein recognition algorithms. In: 2021 IEEE international joint conference on biometrics (IJCB), Shenzhen (China), pp 1–7. https://doi.org/10.1109/IJCB52358.2021.9484351
13. Qiu X, Tian S, Kang W, Jia W, Wu Q (2017) Finger vein presentation attack detection using convolutional neural networks. In: Zhou J, Wang Y, Sun Z, Xu Y, Shen L, Feng J, Shan S, Qiao Y, Guo Z, Yu S (eds) Biometric recognition. Springer International Publishing, Cham, pp 296–305

14. Herzog T, Uhl A (2020) Analysing a vein liveness detection scheme. In: Proceedings of the 8th international workshop on biometrics and forensics (IWBF'20), Porto, Portugal, pp 1–6
15. Schuiki J, Uhl A (2020) Improved liveness detection in dorsal hand vein videos using photoplethysmography. In: Proceedings of the IEEE 19th international conference of the biometrics special interest group (BIOSIG 2020). Darmstadt, Germany, pp 57–65
16. Kauba C, Prommegger B, Uhl A (2018) Focussing the beam - a new laser illumination based data set providing insights to finger-vein recognition. In: 2018 IEEE 9th international conference on biometrics theory, applications and systems (BTAS), Los Angeles, California, USA, pp 1–9. https://doi.org/10.1109/BTAS.2018.8698588
17. Choi JH, Song W, Kim T, Lee SR, Kim HC (2009) Finger vein extraction using gradient normalization and principal curvature. In: Image processing: machine vision applications II, vol 7251, pp 7251–7251–9. https://doi.org/10.1117/12.810458
18. Raghavendra R, Avinash M, Marcel S, Busch C (2015) Finger vein liveness detection using motion magnification. In: 2015 IEEE 7th international conference on biometrics theory, applications and systems (BTAS), pp 1–7
19. Miura N, Nagasaka A, Miyatake T (2007) Extraction of finger-vein patterns using maximum curvature points in image profiles. IEICE - Trans Inf Syst E90-D(8):1185–1194
20. Miura N, Nagasaka A, Miyatake T (2004) Feature extraction of finger-vein patterns based on repeated line tracking and its application to personal identification. Mach Vis Appl 15:194–203. https://doi.org/10.1007/s00138-004-0149-2
21. Huang B, Dai Y, Li R, Tang D, Li W (2010) Finger-vein authentication based on wide line detector and pattern normalization. In: 2010 20th international conference on pattern recognition, pp 1269–1272. https://doi.org/10.1109/ICPR.2010.316
22. Kumar A, Zhou Y (2012) Human identification using finger images. IEEE Trans Image Process 21(4):2228–2244. https://doi.org/10.1109/TIP.2011.2171697
23. Starck J, Fadili J, Murtagh F (2007) The undecimated wavelet decomposition and its reconstruction. IEEE Trans Image Process 16(2):297–309. https://doi.org/10.1109/TIP.2006.887733
24. Yang L, Yang G, Yin Y, Xi X (2018) Finger vein recognition with anatomy structure analysis. IEEE Trans Circuits Syst Video Technol 28(8):1892–1905. https://doi.org/10.1109/TCSVT.2017.2684833
25. Matsuda Y, Miura N, Nagasaka A, Kiyomizu H, Miyatake T (2016) Finger-vein authentication based on deformation-tolerant feature-point matching. Mach Vis Appl 27. https://doi.org/10.1007/s00138-015-0745-3
26. Kauba C, Reissig J, Uhl A (2014) Pre-processing cascades and fusion in finger vein recognition. In: Proceedings of the international conference of the biometrics special interest group (BIOSIG'14), Darmstadt, Germany
27. Stockman G, Shapiro LG (2001) Computer vision, 1st edn. Prentice Hall PTR, USA
28. Lee EC, Lee HC, Park KR (2009) Finger vein recognition using minutia-based alignment and local binary pattern-based feature extraction. Int J Imaging Syst Technol 19(3):179–186
29. Wimmer G, Prommegger B, Uhl A (2020) Finger vein recognition and intra-subject similarity evaluation of finger veins using the cnn triplet loss. In: Proceedings of the 25th international conference on pattern recognition (ICPR), pp 400–406
30. Linortner M, Uhl A (2021) Towards match-on-card finger vein recognition. In: Proceedings of the 2021 ACM workshop on information hiding and multimedia security, IH&MMSec '21. Association for Computing Machinery, New York, NY, USA, pp 87–92. https://doi.org/10.1145/3437880.3460406
31. Xu H, Veldhuis RNJ, Bazen AM, Kevenaar TAM, Akkermans TAHM, Gokberk B (2009) Fingerprint verification using spectral minutiae representations. IEEE Trans Inf Forensics Secur 4(3):397–409. https://doi.org/10.1109/TIFS.2009.2021692
32. Chingovska I, Mohammadi A, Anjos A, Marcel S (2019) Evaluation methodologies for biometric presentation attack detection. In: Marcel S, Nixon MS, Fierrez J, Evans N (eds) Handbook of biometric anti-spoofing: presentation attack detection. Springer International Publishing, Cham, pp 457–480. https://doi.org/10.1007/978-3-319-92627-8_20

33. Linortner M, Uhl A (2021) Veinplus+: a publicly available and free software framework for vein recognition. In: 2021 international conference of the biometrics special interest group (BIOSIG), pp 1–5. https://doi.org/10.1109/BIOSIG52210.2021.9548286
34. Kauba C, Prommegger B, Uhl A (2019) Openvein - an open-source modular multipurpose finger vein scanner design. In: Uhl A, Busch C, Marcel S, Veldhuis R (eds) Handbook of vascular biometrics, Chap 3. Springer Nature Switzerland AG, Cham, Switzerland, pp 77–111. https://doi.org/10.1007/978-3-030-27731-4_3
35. Sequeira AF, Ferryman J, Chen L, Galdi C, Dugelay JL, Chiesa V, Uhl A, Prommegger B, Kauba C, Kirchgasser S, Grudzien A, Kowalski M, Szklarski L, Maik P, Gmitrowicz P (2018) Protect multimodal db: a multimodal biometrics dataset envisaging border control. In: Proceedings of the international conference of the biometrics special interest group (BIOSIG'18), Darmstadt, Germany, pp 1–8. https://doi.org/10.23919/BIOSIG.2018.8552926
36. Maio D, Maltoni D, Cappelli R, Wayman JL, Jain AK (2004) Fvc 2004: third fingerprint verification competition. In: Zhang D, Jain AK (eds) Biometric authentication. Springer, Berlin, pp 1–7
37. Snelick R, Indovina M, Yen J, Mink A (2003) Multimodal biometrics: issues in design and testing. In: Proceedings of the 5th international conference on multimodal interfaces, ICMI '03. Association for Computing Machinery, New York, NY, USA, pp 68–72. https://doi.org/10.1145/958432.958447
38. Snelick R, Uludag U, Mink A, Indovina M, Jain A (2005) Large-scale evaluation of multimodal biometric authentication using state-of-the-art systems. IEEE Trans Pattern Anal Mach Intell 27(3):450–455. https://doi.org/10.1109/TPAMI.2005.57
39. Jain A, Nandakumar K, Ross A (2005) Score normalization in multimodal biometric systems. Pattern Recognit 38(12):2270–2285. https://doi.org/10.1016/j.patcog.2005.01.012. www.sciencedirect.com/science/article/pii/S0031320305000592
40. He M, Horng SJ, Fan P, Run RS, Chen RJ, Lai JL, Khan MK, Sentosa KO (2010) Performance evaluation of score level fusion in multimodal biometric systems. Pattern Recognit 43(5):1789–1800. https://doi.org/10.1016/j.patcog.2009.11.018. www.sciencedirect.com/science/article/pii/S0031320309004373

Chapter 18
Fisher Vectors for Biometric Presentation Attack Detection

Lazaro Janier Gonzalez-Soler, Marta Gomez-Barrero, Jose Patino, Madhu Kamble, Massimiliano Todisco, and Christoph Busch

Abstract Biometric systems have experienced a large development over the last years since they are accurate, secure and in many cases, more user convenient than traditional credential-based access control systems. In spite of their benefits, biometric systems are vulnerable to attack presentations, which can be easily carried out by a non-authorised subject without having a deep computational knowledge. This way, he/she can gain access to several applications where biometric systems are frequently deployed, such as bank accounts and smartphone unlocking. In order to mitigate such threats, we present in this work a study on the feasibility of using the Fisher Vector (FV) representation to spot unknown-attack presentations over different biometric modalities such as fingerprint, face and voice. By learning a common feature space from a set of local features, extracted from known samples, the FVs lead to the construction of reliable discriminative models which can successfully distinguish a bona fide presentation from an attack presentation. The experimental evaluation over publicly available databases (i.e. LivDets, CASIA-FASD, SiW-M and ASVspoof, among others) yields error rates outperforming most state-of-the-art algorithms for challenging scenarios where species, recipies or capture devices remain unknown.

L. J. Gonzalez-Soler (✉) · C. Busch (✉)
da/sec - Biometrics and Internet-Security Research Group, Hochschule Darmstadt, Darmstadt, Germany
e-mail: lazaro-janier.gonzalez-soler@h-da.de

C. Busch
e-mail: christoph.busch@h-da.de

M. Gomez-Barrero (✉)
Hochschule Ansbach, Ansbach, Germany
e-mail: marta.gomez-barrero@hs-ansbach.de

J. Patino · M. Kamble · M. Todisco (✉)
EURECOM, Biot, France
e-mail: massimiliano.todisco@eurecom.fr

J. Patino
e-mail: jose.patino@eurecom.fr

M. Kamble
e-mail: madhu.kamble@eurecom.fr

S. Marcel et al. (eds.), *Handbook of Biometric Anti-Spoofing*, Advances in Computer Vision and Pattern Recognition, https://doi.org/10.1007/978-981-19-5288-3_18

18.1 Introduction

Biometric systems have considerably been developed in recent years. Several studies [1] have shown that the extensive development of biometric systems has increased security and accuracy in many applications such as border controls, financial transaction authentication and mobile device unlocking. This is due to the fact that biometric characteristics such as fingerprint, face, iris or voice offer a high discriminative capability (i.e. they are 'unique') and cannot be forgotten or shared with other subjects [2].

In spite of their advantages, biometric systems are still vulnerable to different external attacks [4], as shown in Fig. 18.1. In particular, we focus on attacks on the capture device, known as 'Attack Presentations (APs)', which can be easily launched by any subject without having a vast expert knowledge. As a consequence of the wide development experienced by several social networks (e.g. Facebook, LinkedIn, Instagram, or YouTube) a non-authorised subject can learn from a video tutorial and create a copy of our biometric characteristics, denoted as Presentation Attack Instrument (PAI). Thus, she/he could gain access to those aforementioned applications. Furthermore, the free access to large-scale public databases together with the recent advances in the creation of very realistic fake contents or 'Deep Fakes' also pose a serious threat to biometric systems [5].

The risk posed by PAIs is not only reduced to an academic issue. APs were addressed for the first time in 1998 [6]. Willis and Lee showed how four out of six evaluated biometric systems were vulnerable to PAIs. In 2000, Zwiesele et al. [7] carried out a comparative study on biometric identification systems which revealed the high vulnerability of these systems to PAIs. Two years later, Matsumoto et al. [8] analysed the weakness of 11 commercial fingerprint-based biometric systems to gummy fingerprints. The experimental evaluation reported that 68–100% of the PAIs created with cooperative methods were falsely accepted as bona fide presentations (i.e. genuine). In 2009, Japan reported the use of PAIs in one of its airports, and in

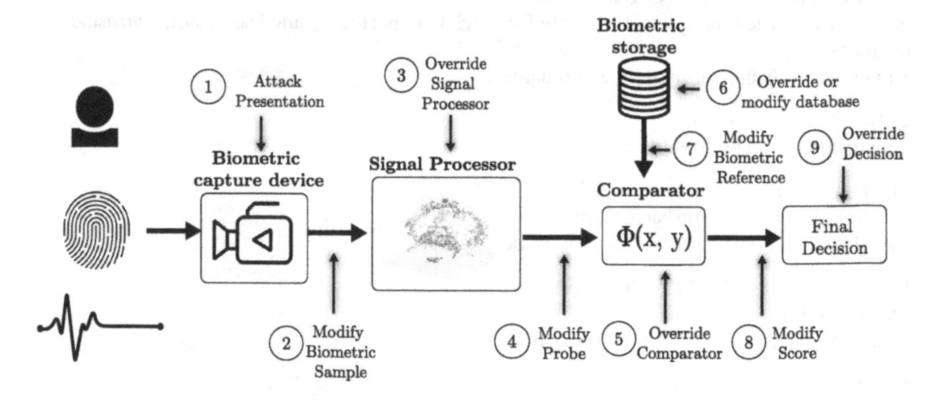

Fig. 18.1 Attack points on biometric systems derived from [3]

2013, a Brazilian doctor used artificial silicone fingerprints to tamper a biometric attendance system at the Sao Paulo hospital [9].

In order to address those security threats, several Presentation Attack Detection (PAD) techniques have been proposed in the literature [10]. Depending on how those PAD methods are integrated into biometric systems, they can be categorised as hardware- and software-based algorithms [10]. The former seek to spot PAIs by detecting the biological characteristics of the captured subject using a special sensor: electric resistance [11], temperature [11, 12] and blood pressure [13], among others [14–17].

In contrast to hardware-based approaches, software-based techniques are more interoperable, as they are not dependent on sensing devices and can therefore be deployed in many more applications than the former. These algorithms assume that properties in bona fide presentations (BPs) must be intrinsically opposed to the ones in attack presentations (APs) mainly due to capturing properties or the materials used in the fabrication of the artificial replicas or PAIs. In this context, different strategies have emerged in the last decades: textural handcrafted features such as Local Binary Patterns (LBP) [18, 19], Histogram of Oriented Gradients (HOG) [20], Local Phase Quantisation (LPQ) [20, 21], frequency domain analysis [22–24] and image quality assessment [16, 25, 26]. More recently, the success of deep learning techniques in several academic and industrial fields has led to the development of more sophisticated PAD approaches which considerably outperform earlier PAD algorithms [24, 27, 28].

In spite of the aforementioned and other efforts, current PAD algorithms struggle to generalise well beyond the PAI species (i.e. attack types) on which they were trained. Specifically, the best-performing deep learning-based techniques have reported a high detection performance for identifying PAIs when both PAI species (i.e. attack type) and acquisition conditions are known a priori. However, there are still some issues to be resolved:

- *Poor generalisation capabilities for unknown PAI species*: Current state-of-the-art techniques face difficulties to detect unknown PAI species (i.e. attack presentations created with a particular species different from those in the training set), thereby resulting in a degradation of the detection accuracy.
- *Poor generalisation capabilities across several datasets*: Most state-of-the-art PAD methods show a decreasing detection performance when evaluated over a new database. Since capture devices might age and will eventually be replaced, PAD methods must be able to successfully classify samples acquired with a new capture device. Therefore, generalisation across multiple datasets is of utmost importance.
- *Specialisation on a particular biometric characteristic*: The most sophisticated PAD algorithms have been developed to detect PAIs through a particular biometric characteristic (e.g. face, fingerprint, or voice). Therefore, their application across different modalities is not straightforward and could lead to a wide performance deterioration.

- *Large number of hyperparameters*:Deep learning-based approaches are usually based on dense CNNs having a large number of learnable parameters (exceeding 2.7 million [29]). Such models are not viable in mobile environments with limited resources.

In order to tackle those unresolved issues, we describe in this chapter the feasibility of using the Fisher Vectors for biometric PAD. This approach combines the generalisation capabilities of generative models with discriminative techniques to improve PAD generalisation. In essence, the FV computes in a single vector how the distribution of local input descriptors differs from those learned by a generative model. This vector can, in turn, be fed to any discriminative model (e.g. Support Vector Machine - SVM) for BP versus AP classification. The high generalisation capabilities provided by the FV representation have been recently confirmed in the last two editions of the Fingerprint Liveness Detection Competitions: first prize with an overall accuracy of 96.17% in LivDet 2019 [30] and the best-performing algorithm for detecting unknown PAI species in LivDet 2021 [31]. It should be noted that these datasets are not used in the evaluation of our method, as they are not available to the biometric research community.

To sum up, the main contributions of this work are:

- An extensive review of the state-of-the-art techniques employed for fingerprint, facial and voice PAD. We highlight those methods focused on the generalisation of unknown PAI species and databases.
- A thorough analysis confirming the high generalisation capabilities of the FV representation for detecting challenging unknown PAI species stemming from different biometric characteristics: fingerprint, face and voice.
- An extensive study over challenging cross-dataset scenarios frequently encountered on the fingerprint, face and voice PAD task. Those challenging scenarios include variations on capture devices and environmental conditions, among others.
- A thorough evaluation compliant with the ISO/IEC 30107-3 standard on biometric PAD [32] over well-established fingerprint, face and voice databases: LivDet 2011, 2013 and 2017 [33–35] for fingerprint; CASIA Face Anti-Spoofing [36], REPLAY-ATTACK [37], MSU-MFSD [38] and SiW-M [39] for face; and ASVspoof 2019 [40] for voice.

The remainder of this chapter is organised as follows: a review about different techniques applied for detecting APs stemming from fingerprint, face and voice characteristics is included in Sect. 18.2. Section 18.3 describes the generalisable PAD algorithm based on the FV representation. The experimental protocol is explained in Sect. 18.4, and the results benchmarking the performance of the FV representation with the top state-of-the-art approaches for the three biometric characteristics explored are discussed in Sect 18.5. Conclusions and future work directions are finally summarised in Sect. 18.6.

Table 18.1 Summary of some studies focused on PAD generalisation per biometric characteristic. The results are reported in terms of D-EER(%)

	Method	Known-attacks	Unknown-PAIs	Cross-database
Fingerprint	VGG [41]	3.87%	6.30%	30.70%
	TripleNet [42]	2.41%	5.86%	15.20%
	FSB-v1 [43]	1.70%	3.50%	18.90%
	FSB-v2 [44]	1.11%	2.93%	17.91%
	Proposed method	**1.74%**	**4.32%**	**9.15%**
Face	Textural fusion [45]	2.43%	–	21.30%
	DeepPixelBis [46]	0.42%	5.97%	–
	CDCN++ [47]	0.69%	11.95%	18.15%
	DR-UDA [48]	3.63%	–	17.93%
	Proposed method	**0.45%**	**11.45%**	**18.24%**
Voice	B01 [49]	9.78%	11.04%	–
	B02 [49]	11.96%	13.53%	–
	Proposed method	**3.68%**	**3.66%**	–

18.2 Related Work

See Table 18.1.

18.2.1 Hardware-Based Approaches

Hardware-based techniques integrate an extra sensor into the capture device to detect biological characteristics of a human body. Such living characteristics are, for instance, intrinsic properties (e.g. blood pressure [13], skin structure analysis through Optical Coherence Tomography (OCT) [50–54], electric resistance [11], reflectance [55, 56] or the combination of the two latter through impedance [17]), involuntary signals (e.g. thermal radiation stemming either from fingertips [11] or faces [12]), responses to external stimuli (e.g. motion estimation [57]) or articulatory gestures and oral airflow for speech [58, 59]. In general, those methods have reported a high detection performance to spot particular PAI species. However, the integration with an extra sensor can significantly increase their development cost (e.g. a thermal sensor for an iPhone exceeds EUR 250[1]). Moreover, their accuracy considerably decreases for the detection of unknown-attacks, as the sensing technology employed is designed for particular PAI species [16].

[1] https://amz.run/44Mp.

18.2.2 *Software-Based PAD Approaches*

Whereas hardware-based techniques are mostly expensive and not user-friendly, software-based approaches detect PAI species by analysing a single image or a set of frames acquired with the same capture device used for recognition purpose. They often provide a high security, efficiency and interoperability, thereby leading to a wide development in the last decade [10, 16, 60].

18.2.2.1 Handcrafted-Based Methods

Depending on the biometric modality, several properties have been analysed [10]. For fingerprint, skin distortions [61, 62] and perspiration produced by pores [63, 64] reported promising results one decade ago. However, they depend on the pressure applied by subjects on the capture device surface during the acquisition process.

In addition, several properties at different timeslot have been analysed for face PAD: involuntary gestures such as eye-blinking [65–68], face and head gestures (e.g. nodding, smiling, looking in different directions) [69–71]. In spite of those and other efforts, these approaches fail to spot PAI species such as printed attacks whose eye region is replaced by the attacker eyes. Furthermore, video replay attacks cannot be successfully detected.

To compensate for such weaknesses, a large number of studies have analysed texture properties. Handcraft-based approaches usually employ processing tools such as Fourier Spectrum to describe the global frequency of images [22, 23, 72], Gaussian [73] or Gabor [74] filters to extract specific frequency information, wavelet multiresolution analysis [75], statistical models to detect image noise [76] or traditional texture descriptors (e.g. Local Binary Pattern (LBP) [18, 19, 37], Histogram Oriented Gradients (HOG) [77], Binarised Statistical Image Features (BSIF) [78] and Local Phase Quantisation (LPQ) [21]).

A common approach for handcrafted speech features is to decompose one-dimensional voice signals into a number of orthogonal or quasi-orthogonal signals that convert them into a two-dimensional signal [79]. These methods include various techniques, the most popular of which are the short-time Fourier transform (STFT) and mel-frequency cepstral coefficients (MFCC), which have a great relevance in many tasks in audio processing [80, 81].

Numerous handcrafted methods exist in the literature that attempt to capture the artefacts that give away artificial/replayed speech [82]. Among the most successful ones are the constant Q cepstral coefficients [83] (CQCC). CQCCs are based on the constant-Q transform [84] (CQT), a perceptually inspired alternative to Fourier-based approaches for time–frequency analysis. As reported in the literature, CQCCs generalise across different databases (i.e. ASVspoof 2015 [85], ASVspoof [86] and RedDots replayed database [87]), delivering performance close to the state of the art in each case.

Alternative methods include those generating high-dimensional magnitude- and phase-based features, which have shown a good ability to discriminate between bona fide and attack presentations [88, 89]. Features extracted using linear sub-band processing were also explored, such as the Linear Frequency Cepstral Coefficients (LFCCs) [90], which have been shown to detect APs with high accuracy in the recent ASVspoof 2019 [91]. The basic motivation behind sub-band processing is that artefacts of converted speech occur differently in different sub-bands.

18.2.2.2 Deep Learning Methods

The high accuracy reported by Convolutional Neural Networks (CNNs) on numerous computer vision applications has led to the development of several sophisticated algorithms for PAD. These techniques have reported a high detection performance which outperforms most of the aforementioned handcrafted-based methods. In 2014, Yang et al. [92] fine-tuned the ImageNet pre-trained CaffeNet [93] and VGG-face [94] models for bi-class classification. Following this idea, Nogueira et al. [41] established a benchmark between three CNNs, achieving the best results in the LivDet 2015 competition with an overall accuracy of 95.5%. Pala and Bhanu [42] trained a triple-stream CNN fed with random patches extracted from images. Based on the fact that PAIs produce spurious minutiae on a fingerprint image, Chugh et al. [43, 44] proposed a framework for independently classifying minutiae-centred local patches extracted from a fingerprint image.

In the context of facial PAD, Xu et al. [95] combined Long Short-Term Memory (LSTM) units with CNNs to learn temporal features from face videos. The authors showed that the spatio-temporal features were helpful for facial PAD, thereby resulting in a reduction by half of the error rates reported by handcrafted feature baselines (5.93 vs. 10.00%). Keeping spatio-temporal features in mind, Gan et al. [96] proposed a 3D CNN for facial PAD, which, unlike traditional 2D CNNs, extracts the temporal and spatial dimension features from a frame sequence. Atoum et al. [97] also combined two-stream CNNs for extracting local features and depth estimation maps from facial images.

Deep residual learning [98], successfully used for image processing tasks, was adopted for voice PAD. In particular, ResNet has been introduced to avoid vanishing/exploding gradients earlier in deep CNN architectures. This model is successfully used along with image-like speech spectrogram (e.g. Mel spectrogram) for the detection of voice APs [99, 100]. Recently, end-to-end approaches, which make use of the raw voice waveform, have also employed residual networks [101]. In this case, the weights of the first 1D convolution layer can be learnable [102] or fixed [103] and forced to have a *sinc* curve. In both cases performances are comparable with the state of the art [91, 104].

In spite of the advances achieved for handcrafted- and deep learning-based approaches, they still struggle to identify PAIs when, (i) species employed in the fabrication of PAIs remain unknown in training (i.e. unknown PAI species) and (ii) samples in training and testing sets are acquired with different capture devices under different acquisition conditions (i.e. cross-dataset), thereby leading to a generalisation ability decrease. The aforementioned state-of-the-art limitations can be observed in the results reported in Table 18.1.

18.2.2.3 Anomaly Detection-Based Methods

In order to tackle those generalisation shortcomings, several anomaly detection-based PAD methods have been proposed. In 2013, de Freitas Pereira et al. [105] already reported poor generalisation capabilities to unknown-attacks of state-of-the-art face PAD methods based on LBP and SVMs. In particular, the error rates increased by at least 100%. Motivated by those findings, Arashloo et al. [106] experimented over several unknown-attack scenarios and concluded that anomaly detection approaches trained only on bona fide data can reach a detection performance comparable to two-class classifiers. However, the results are reported only in terms of the area under the Receiving Operating Characteristic curve (AUC), thus lacking a proper quantitative analysis compliant with the ISO/IEC 30107-3 standard on biometric PAD [32]. Following a similar idea, Rattani et al. [107] proposed an automatic adaptation of Weibull-calibrated support vector machines (SVMs) and evaluated it over the LivDet 2011 database. The experimental evaluation showed that D-EERs oscillated between 20 and 30% in the presence of unknown PAIs. On the other hand, Ding and Ross analysed an ensemble of one-class SVMs trained only on bona fide data in [108], which lowered the error rates to 10–22% over the same generalisation task.

More recently, Nikisins et al. [109] showed how a one-class Gaussian Mixture Model (GMM) can outperform two-class classifiers depending on the PAI species included in the test set. Following the same anomaly detection paradigm, Xiong and AbdAlmageed studied in [19] the detection performance of one-class SVMs and autoencoders in combination with LBP descriptors. In most of the scenarios tested, the detection rates increased with respect to common two-class classifiers. Liu et al. also analysed in [39] the performance of a Deep Tree Network (DTN) by clustering the PAI species into semantic sub-groups. The experimental evaluation focused on unknown PAI species over the challenging SiW-M database [39], reported a mean D-EER of 16% which is considerably higher than those for known-attacks. Chingovska and Dos Anjos [110] explored the feasibility of client-specific information for facial PAD. Finally, George and Marcel [111] also combined a one-class GMM with a Multi-Channel CNN (MCCNN), which fed with face samples acquired at different light spectra (i.e. RGB, thermal and infrared). Despite the fact that the experimental evaluation over the SiW-M dataset showed a performance improvement with respect to the DTN technique, its generalisation capability to detect unknown PAIs was still poor (i.e. a D-EER of 12.00%).

Anomaly detection for voice PAD was also explored in a one-class classification framework by Zhang et al. [112]. To avoid overfitting to know attacks, the authors introduced two different margins in the softmax loss function for better modelling the bona fide speech and isolating the PAIs.

18.2.2.4 Domain Adaptation-Based Methods

Generally, acquisition conditions such as appearance, illumination or capture devices vary between datasets. In order to overcome poor cross-dataset generalisation issues, new PAD approaches have explored Domain Adaptation to transfer the knowledge learned from a source domain to a target domain [113]. By assuming that the relationship between BP and AP face samples on a given subject can be modelled with a linear transformation, Yang et al. [114] proposed a subject domain adaptation method to synthesise virtual features. Following this idea, Li et al. [115] transformed knowledge learned from a labelled source domain to an unlabelled target domain by minimising the maximum mean discrepancy [116] for facial PAD. De Freitas Pereira [117] proposed a CNN-based method which builds a common feature space from face images, captured on different visual spectra domains, for improving face recognition. To transfer knowledge to the unlabelled target domain, Wang et al. [48, 118] proposed an unsupervised domain adaptation with disentangled representation, which builds a feature space shared between the source and target domains. Even if this common feature space appeared to be suitable to overcome cross-dataset issues, experimental results showed a poor detection performance over known-attack scenarios (i.e. D-EERs of 3.20, 6.00 and 7.20% for CASIA Face Anti-spoofing [36], MSU-MFSD [38] and Rose-Youtu [115] databases, respectively).

Regarding domain adaptation for fingerprint PAD, Gajawada et al. tried to tackle the dependency on the PAIs contained in the training set from a different perspective in [119]. They propose a so-called deep learning-based "Universal Material Translator" (UMT). Given a reduced number (e.g. five) of known AP samples, the UMT generated synthetic PAI samples by embedding the main appearance features of those PAIs with known bona fide samples. Those synthetic samples were reutilised for training its detector, thereby resulting in an improvement up to 17% over the baseline. Despite these promising results, it should be noted that this approach does require some PAI samples (i.e. five) which should be carefully selected.

Domain adversarial training for voice PAD has been explored in [120]. In this paper, the authors treated the cross-dataset scenario as a domain-mismatch problem and addressed it using a domain adversarial training framework. Same authors further proposed a dual-adversarial domain adaptation [121] framework to enable fine-grained alignment of attack and bona fide presentations separately by using two domain discriminators.

18.2.2.5 Generative-Based Methods

Nowadays, generative models are in the vanguard of unsupervised learning. Techniques such as Gaussian Mixture Models (GMM) [122], Boltzmann Machines [123], Variational Autoencoders [124] and Generative Adversarial Networks [125] have been successfully applied in numerous computer vision [126], speech recognition and generation [127] and natural language [128, 129] tasks. Those algorithms try to capture the inner data probabilistic distribution to generate new similar data [130].

However, to the best of our knowledge, a rather limited number of works have been employed for PAD. Engelsma and Jain [131] fed several GANs with bona fide samples acquired by a RaspiReader fingerprint capture device. The experimental results for high-security threshold over unknown-attacks showed a detection performance very sensitive to the training set.

In voice PAD, generative models have been purposely used to augment the data in the training phase. Recently, Wang et al. [112] have proposed a vocoder replay channel response estimation based on MelGAN [132] and HifiGAN [133] on the ASVspoof 2021 [104], the results of which showed a good generalisation ability.

18.3 Generalisable FV-Based PAD Approach

Building upon the generalisation capabilities provided by generative models, we build our proposed PAD subsystem. This is based on four main steps: (*i*) features are extracted from a regular grid of points along the whole input biometric sample (i.e. fingerprint, face, and voice); (*ii*) a generalisable common feature space is built by learning an unsupervised (GMM) model from the aforementioned features; (*iii*) the final descriptor, which represents the biometric sample at hand, is subsequently encoded by computing the differences of first- and second-order statistics with respect to the learned generative model parameters; and (*iv*) a BP or AP decision is finally taken by a linear SVM. Linear SVMs are helpful since they perform well in high-dimensional spaces, avoid overfitting, and have good generalisation capabilities. It is worth noting that our approach is rotation- and pose-invariant, as it is based on a coherence analysis of the fingerprint orientation map and a frequency study on the facial and voice data. Therefore, no preprocessing step is applied on input images (Fig. 18.2).

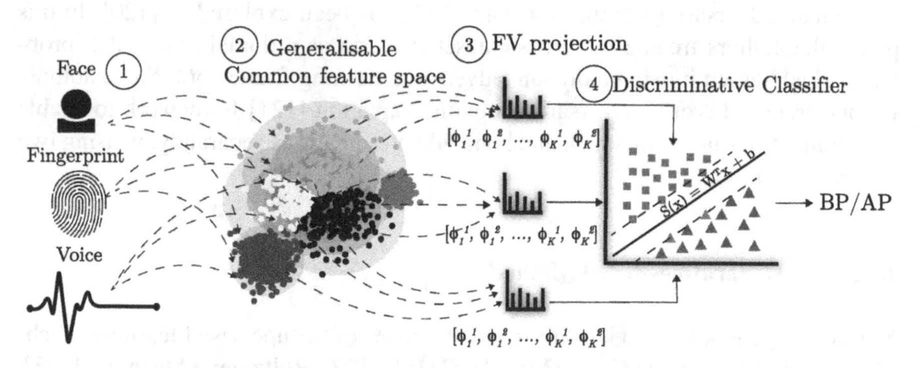

Fig. 18.2 General overview of our generalisable FV-based approach

18.3.1 Reliable Features for Different Biometric Characteristics

As mentioned in Sect. 18.1, we focus on three biometric characteristics, namely fingerprint, face and voice, each of which has intrinsic properties that make them different from each other. Whereas fingerprint comprises mainly ridges and valley [1], facial images are composed of facial aesthetic units [134]. In addition, voice data are generally represented by time-domain, frequency-domain or time–frequency-domain representations known as spectrograms [49]. Therefore, the creation of a single set of universal features to represent those characteristics would lead to a low biometric performance.

18.3.1.1 Dense Multi-scale Features

Since artefacts produced in the fabrication of PAIs might be located in any area of the input image, we follow in our approach the strategies in [135, 136] for the feature computation. Therefore, local descriptors are densely extracted at fixed points on a regular grid with an uniform spacing (e.g. 3 pixels). In addition, those artefacts might have different sizes. Hence, descriptors are computed over four circular patches with different radii σ. Thus, each point in the grid is represented by four descriptors, as depicted in Fig. 18.3.

18.3.1.2 Fingerprint Orientation Field-Based Features: SIFT

Scale Invariant Feature Transform (SIFT) [137] is a popular histogram-based descriptors whose features are invariant to changes in scale and rotation. In addition, they

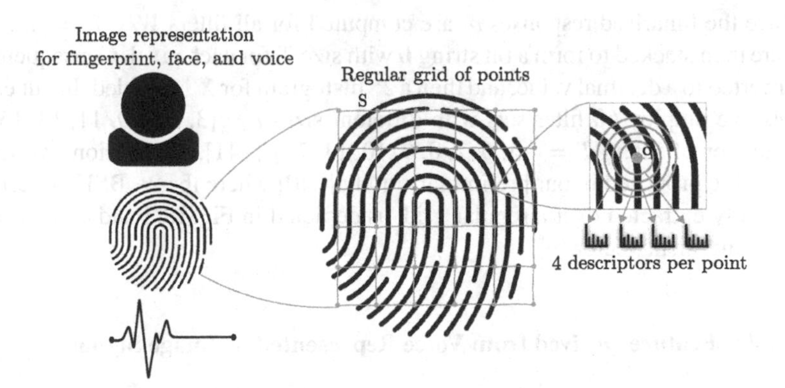

Fig. 18.3 Feature extraction over the biometric samples. Dense multi-scale features are computed for face and spectrogram-image data, as it is shown exemplary for fingerprint

have shown to provide robust recognition capabilities across different affine distortions, change in 3D viewpoints, addition of noise and illumination changes. This method involves four stages to generate the set of image features: (*i*) scale-space extrema detection, (*ii*) keypoint localisation, (*iii*) orientation assignment and (*iv*) keypoint descriptor. SIFT features can capture and describe the discontinuities produced over fingerprint ridges during the fabrication of PAIs [135]. Therefore, we utilise steps three and four in our implementation, as keypoints are fixed over a regular grid along the whole fingerprint image (see Sect. 18.3.1.1). To efficiently compute SIFT descriptors, we use the implementation provided in [138], which delivers a speed-up of up to 60x by i) exploiting the uniform sampling and overlapping between descriptors, and ii) using linear interpolation with integral image convolution.

18.3.1.3 Facial Texture Features

Whereas keypoint-based descriptors such as SIFT have shown to be appropriate for fingerprint samples [135, 136, 139], in which minutiae can be regarded as landmarks within the image, for facial images the textural information is more relevant than the geometric details related to landmarks [140]. For this reason, most facial PAIs include properties which can be successfully detected by convolving the input image with a particular kernel. BSIF [141] is a texture descriptor based on Independent Component Analysis (ICA) [142], which employs a set of pre-trained filters to obtain a meaningful representation of the data. Let X be an image patch with size $l \times l$ pixels and $W = \{W_1, W_2, \ldots, W_T\}$ a set of linear filters with the same size of X, the binarised response b_i for X can be computed as follows:

$$b_i = \begin{cases} 1 & \sum_{n,m} W_i(n, m)X(n, m) > 0 \\ 0 & \text{otherwise} \end{cases} \tag{18.1}$$

Once the binarised responses b_i are computed for all filters $W_i : i = 1 \ldots T$, they are then stacked to form a bit string **b** with size T for each pixel. Consequently, **b** is converted to a decimal value, and then a 2^T histogram for X is yielded. In our experiments, we employ 60 filter sets with different sizes $l = \{3, 5, 7, 9, 11, 13, 15, 17\}$ and number of filters $T = \{5, 6, 7, 8, 9, 10, 11, 12\}$ [141]. In addition, we follow the pipeline proposed in our preliminary study [140] where the 2^T BSIF descriptors are densely extracted over a regular grid, as depicted in Fig. 18.3 and compacted in a 128-component vector.

18.3.1.4 Feature Derived from Voice Represented as Image Signal

The visualisation of audio/speech signals is key for the analysis of several voice processing tasks. Usually, it involves some time-domain, frequency-domain or time–frequency-domain representations known as spectrograms, which show the signal

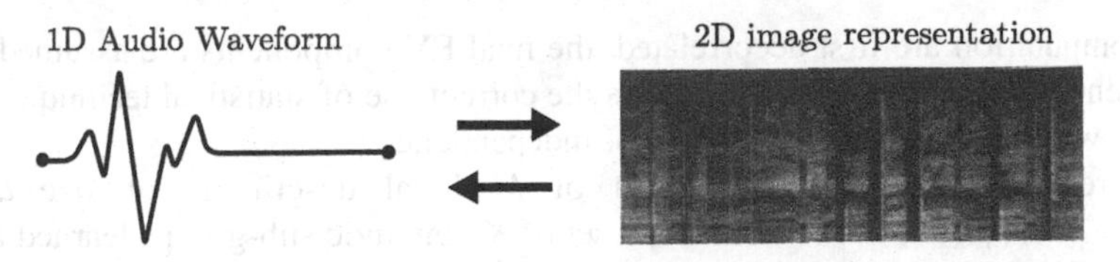

Fig. 18.4 Speech spectrogram to its corresponding 2D image representation

amplitude over time at a set of discrete frequencies. In order to transform a speech spectrogram to its corresponding 2D image representation, as shown in Fig. 18.4, various time–frequency techniques have been proposed, each with different characteristics. In particular, we focus on the best-performing recent time–frequency approach [49]: constant Q transform (CQT) [84], which is a perceptually motivated approach to time–frequency analysis.

By transforming a speech spectrogram to its corresponding 2D image representation, we can successfully apply well-known pattern recognition techniques. Especially, for PAD, we represent textural information of spectrogram-images by BSIF features, which are able to detect those time–frequency changes over the spatial domain [49].

18.3.2 Fisher Vector Encoding

The FV representation derives a kernel from the generative model parameters (i.e. GMM) [143]. This representation describes how the distribution of unknown features (e.g. features extracted from unknown PAI species) differs from the distribution depicting features from known APs and BPs. Hence, the final features are more robust to new samples, which might stem from unknown scenarios and thus be different from the samples used for training, as shown in a preliminary evaluation in [140].

Building upon the idea in [144], we train a diagonal-covariance GMM from a subsampling form the population features (i.e. compact dense-BSIF descriptors for face and voice data or dense-SIFT for fingerprint). In essence, a GMM based on K-components, which is represented by their mixture weights (π_k), means (μ_k) and covariance matrices (σ_k), with $k = 1, \ldots, K$, allows discovering or building semantic sub-groups (i.e. a common feature space) from known PAIs and BP samples, which could successfully enhance the detection of unknown-attacks. Since components in the RGB colour space are correlated with each other [45], we firstly decorrelated the input local descriptors through Principal Component Analysis (PCA) [145]. This decreases their size to $d = 64$ components while retaining 95% of the variance. Then, the FV representation computes the average first-order and second-order statistic differences between the local features and each semantic sub-group previously learned by the GMM [146]. Since local histograms used for training the GMM and for the

FV computation are first decorrelated, the final FV components are assumed to be independent of each other. This allows the correct use of statistical techniques (e.g. SVM) which rely upon assumptions of independence.

More formally, let $\mathbf{X}$ be a set of M local descriptors of size d and $G_K = \{(\pi_k, \mu_k, \sigma_k) : k = 1 \ldots K\}$ a set of K semantic sub-groups learned by the GMM. The FV representation for $\mathbf{X}$ is defined as the conditional probability:

$$FV_X = P(X|\mu_k, \sigma_k) \tag{18.2}$$

By applying Bayesian properties, we can rewrite the previous equation as

$$\phi_k^1 = \frac{1}{M\sqrt{\pi_k}} \sum_{i=1}^{d} \alpha_i(k) \left(\frac{X_i - \mu_k}{\sigma_k} \right), \tag{18.3}$$

$$\phi_k^2 = \frac{1}{M\sqrt{2\pi_k}} \sum_{i=1}^{d} \alpha_i(k) \left(\frac{(X_i - \mu_k)^2}{\sigma_k^2} - 1 \right), \tag{18.4}$$

where $\alpha_i(k)$ is the soft assignment weight of the i-th feature x_i to the k-th Gaussian. Finally, the FV representation that defines a biometric characteristic is obtained by stacking the differences $\phi_G = [\phi_1^1, \phi_1^2, \cdots, \phi_K^1, \phi_K^2]$, thereby resulting in a $2 \cdot d \cdot K = 2 \cdot 64 \cdot K$ sized vector. Following [144], ϕ_G is further enhanced by a signed square-rooting and L_2 normalisation. This allows the final features to be more discriminative for its use with linear classifiers [143].

18.3.3 Classification

Now, we should combine the FV representations with a discriminative model for the final BP versus AP decision. To that end, we utilise an SVM. Since the FV dimensionality ($2 \cdot 256 \cdot 64 \leq 2 \cdot K \cdot D \leq 2 \cdot 1024 \cdot 64$) is large with respect to the number of instances employed for training ($\leq$25 K), we link the SVM with a linear mapping. This has reported to have a similar performance with respect to a non-linear mapping, while considerably outperforming the latter in terms of efficiency [147].

In order to find the optimal hyperplane separating the bona fide from the attack presentations, the optimisation algorithm bounds the loss from below. Therefore, we have trained the linear SVM as follows: The SVM labels the bona fide samples as +1 and the presentation attacks as −1, thereby yielding the corresponding $\mathbf{W}'$ (weights) and $\mathbf{b}'$ (bias) classifier parameters. Thus, given an FV encoding $\mathbf{x}$, the final score $s_{\mathbf{x}}$, which estimates the category of the sample at hand, is computed as the confidence of such decision (i.e. the absolute value of the score is the distance to the hyperplane):

$$s_{\mathbf{x}} = \mathbf{W}' \cdot \mathbf{x} + \mathbf{b}' \tag{18.5}$$

18.4 Experimental Set-up

We design the experimental protocol with the aim of carrying out a fair evaluation of the detection performance of the proposed PAD approach for different scenarios, evaluating three aspects: (*i*) detection performance of the FV representation over traditional known-attack scenarios to spot PAIs of the adopted biometric characteristics, (*ii*) performance benchmark of our proposal with the top state-of-the-art approaches, and (*iii*) realistic and challenging scenarios with unknown-attacks, unknown capture devices and cross-database settings. Keeping these goals in mind, three different scenarios are defined:

- *Known-attacks*, which includes an analysis of all PAI species. In particular, samples included in training and testing stem from the same known set of PAI species (e.g. gelatine fingerprint, Printed photo, silicone mask, and so on) [10, 16].
- *Unknown PAI species*, in which the PAI species evaluated remain unknown for training the detector. Specifically, we use the leave-one-out testing protocol explained in [106] for facial evaluation. For fingerprint, we follow the protocol defined in [135] and for voice the one in [49].
- *Cross-database*, in which the datasets employed for testing are different from the databases used for training. Both datasets contain the same PAI species to ensure that the performance degradation is due to the dataset change and not to the lacking knowledge about the PAI species.

In the following, we present the detection performance for a limited set of datasets. For a more comprehensive analysis for more databases per biometric characteristics as well as parameter optimisation, the reader is referred to [49, 135, 148].

18.4.1 Databases

The experimental evaluation is carried out on the freely available fingerprint, face and voice databases: fingerprint—LivDet 2011 [33], 2013 [34] and 2017 [35]; face-CASIA Face Anti-Spoofing [36], REPLAY-ATTACK [37], MSU-MFSD [38] and SiW-M [39]; and for voice—ASVspoof 2019 [40]. We only selected the physical attack (PhA) partition from the ASVspoof database, which simulates attack presentations in reverberant conditions [40]. A summary of the database characteristics is presented in Table 18.2. It is important to highlight that the LivDet 2017 includes unknown PAI species in the test set and will therefore be used for the evaluation of unknown PAI species.

Table 18.2 A summary of databases used in our evaluation

Bio	DB	# Samples	Capture device	PAI species
Fingerprint	LivDet 2011	16,000	Digital P. 4000B	Gelatin, Latex, PlayDoh
			Sagem MSO300	Silicone, Wood Glue
			Biometrika FX2000	Gelatine, Latex, Silgum, Wood Glue
			Italdata ET10	EcoFlex (platinum-catalysed silicone)
	LivDet 2013	8,000	Biometrika FX2000	Gelatine, Latex, Modasil, Wood Glue
			Italdata ET10	EcoFlex (platinum-catalysed silicone)
	LivDet 2017	18,984	Digital P. U.are.U 5160	EcoFlex, Body Double, Wood Glue
			GreenBit DactyScan84C	Gelatine, Latex, Liquid EcoFlex
			Orcanthus Certis2 Image	
Face	CASIA FASD	600	Low-quality USB camera	Warped photo or Printed attacks
			Normal-quality USB camera	Cut photo, Video replay
			High-quality Sony NEX-5 camera	
	REPLAY-ATTACK	1,200	Low-quality 13-inch MacBook webcam	Printed, Digital replay, Video replay
	MSU-MFSD	440	Low-quality 13-inch MacBook webcam	Printed, Video replay
			Low-quality Google Nexus 5 camera	
	SiW-M	968	High-quality Logitech C920 webcam,	Printed, Video replay, Half mask, Silicone mask
			High-quality Canon EOS T6	Transparent, Papercraft, Mannequin, Obfuscation
				Impersonation, Cosmetic, Funny Eye
				Paper Glasses, Partial Paper
Voice	ASVspoof2019 (PhA)	218,430	Commercial grade simulated	Replay attacks created
			Audio devices	According to 9 different replay configurations

18.4.2 Evaluation Metrics

The evaluation is conducted in compliance with the methodology standardised in ISO/IEC 30107-3 [32], reporting the following metrics:

- Attack Presentation Classification Error Rate (APCER), which denotes the percentage of misclassified attack presentations for a fixed threshold.
- Bona Fide Presentation Classification Error Rate (BPCER), which refers to the percentage of wrongly classified bona fide presentations for a fixed threshold.

Along with the above metrics, we also report the corresponding Detection Error Trade-off (DET) curves between both APCER and BPCER. In addition, the BPCERs for a fixed APCER of 10% (BPCER10), 5% (BPCER20) and 1% (BPCER100) are included. Finally, in order to allow a fair benchmark with the available literature, we report the Detection Equal Error Rate (D-EER), which is defined as an operating point where APCER = BPCER.

18.5 Experimental Results

18.5.1 Detection of Known PAI Species

In the first set of experiments, we carry out an in-depth analysis of the detection performance of the FV representation to spot PAI species in a known-attack scenario. To that end, we select the best-performing key parameter to build de common feature space per biometric characteristics, i.e. $K = 512$ for fingerprint [135] and for voice [49] and $K = 1024$ for face [148]. Building upon the results reported by [148], we also choose $T = 10$ filters with size $l = 9$ for the BSIF computation. In Table 18.3, we report the D-EERs for different biometric characteristics. As it should be noted, our algorithm is capable of outperforming most state-of-the-art approaches, yielding D-EERs ranging from 0.00 to 3.68%, depending on the characteristic at hand. In particular, the proposed common feature space attains a mean D-EERs of 1.73% for fingerprint, 0.12% for face and 3.68% for voice, respectively.

18.5.2 Detection of Unknown PAI Species

In the second set of experiments, we evaluate challenging unknown PAI species over the analysed biometric characteristics. We follow for face the leave-one-out protocol in [106] over the SiW-M database. For fingerprint, the protocol in [41] is adopted where test set contains more than one PAI species. In addition, for voice PAD we selected the protocol proposed by the ASVspoof 2019 challenge [40]. Figure 18.5

Table 18.3 Benchmark with the state of the art in terms of D-EER(%) for known-attacks. The best results per dataset are highlighted in bold

Method	Fingerprint-LivDet 2011				Face			Voice
	Biometrika	Digital P.	Italdata	Sagem	CASIA	MSU	RA	PhA
VGG [41]	5.20	3.20	8.00	1.70	–	–	–	–
TripleNet [42]	5.15	1.85	5.10	1.23	–	–	–	–
FSB-v1 [43]	2.60	2.70	3.25	1.80	-	–	–	-
tinyFCNN [149]	**1.10**	1.10	4.75	1.56	-	-	-	–
FSB-v2 [44]	1.24	1.61	2.45	**1.39**	–	–	–	–
CSURF + FV [150]	–	–	–	–	2.80	0.10	3.70	–
Texture fusion [45]	–	–	–	–	4.60	1.50	1.20	–
shallowCNN-LE [151]	–	–	–	–	4.00	8.41	3.70	–
Depth CNNs [97]	–	–	–	–	2.67	0.35	0.72	–
DR-UDA [48]	–	–	–	–	3.30	6.30	1.30	–
B01 [49]	–	–	–	–	–	–	–	9.78
B02 [49]	–	–	–	–	–	–	–	11.96
Proposed method	2.80	**0.30**	**2.40**	1.42	**0.37**	**0.00**	**0.00**	**3.68**

shows the DET curves for unknown PAI scenarios over the three biometric characteristics. We observe that our approach reports a mean BPCER100 of 21.53% for the challenging mask attacks (i.e. Half mask, Silicone, Transparent, Papercraft and Mannequin). In contrast, the proposed method yields a BPCER of 0.00% for any APCER over the Papercraft masks. In addition, our algorithm also achieves a BPCER of 0.00% for any APCER over impersonation attacks. The latter, unlike other attacks (e.g. obfuscation attacks), have reported a biometric performance decrease for current deep face recognition systems (i.e. GMR = 51.2% @ FMR = 0.01% [152]).

In the context of fingerprint PAD (Fig. 18.5-b), it should be noted that all three DET curves present a similar behaviour. We obtain a mean BPCER100 of 12.06% ± 2.25 over the three selected dataset configurations. In particular, BPCER values for a high-security threshold (i.e. APCER = 1.00%) range from 10.58 to 14.65%. Therefore, the algorithm rejects at most 15 of 100 bona fide samples while one out of 100 unknown-attack presentations is accepted.

Subsequently to the above results, we observe in Fig. 18.5c that our technique also outperforms the state-of-the-art baselines for physical access, resulting in a BCER100 of 15.56%. Thus, it improves the results reported by B01 and B02 by a relative 81.67%, hence confirming the soundness of our proposed common feature space to generalise across unknown PAI species over voice data.

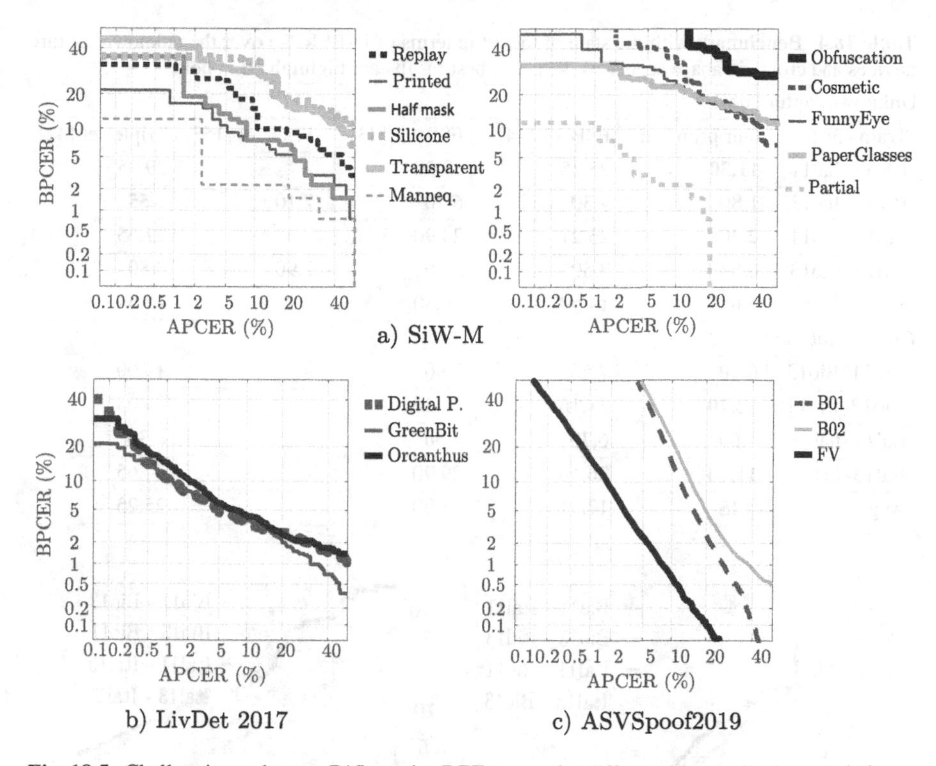

Fig. 18.5 Challenging unknown PAI species DET curves for different biometric characteristics. **a** It represents DET curves for facial data, **b** fingerprint and **c** voice

18.5.3 Cross-Database

Finally, we evaluate the proposed algorithm for challenging cross-database scenarios where different capture devices, illumination conditions and subjects for testing might be different from those employed for training the detector. For fingerprint, we also evaluate the change of data collection over the same capture device (e.g. train and test over the same capture device, but acquired for LivDet 2011 and LivDet 2013, respectively). It is important to highlight that the cross-database analysis for voice data has not been carried out due to a missing database for this purpose. Up to now, the ASVspoof 2021 database, which contains a reliable cross-database protocol with the ASVspoof 2019 database, has not been released yet.

18.5.3.1 Fingerprint

We first evaluate the soundness of our proposal in scenarios where different capture devices are used following the protocols in [41] (i.e. capture device inter-operability

Table 18.4 Benchmark with the state of the art in terms of D-EER(%) over the unknown capture devices and cross-database scenarios [41]. The best results are highlighted in bold

Train–Test	Our proposal	FSB-v2 [44]	FSB-v1 [43]	FLDNet [153]	TripleNet [42]
Unknown capture devices					
Bio11–Ital11	**11.30**	25.35	29.50	–	29.35
Bio13–Ital13	1.80	4.30	6.70	2.10	**1.55**
Ital11–Bio11	**2.40**	25.21	24.90	–	27.65
Ital13–Bio13	**0.80**	3.50	5.10	2.90	3.80
Avg.	**4.08**	14.59	16.60	–	15.59
Cross-database					
Bio11–Bio13	**6.80**	7.60	7.90	–	14.00
Bio13–Bio11	**12.70**	31.16	34.40	–	34.05
Ital11–Ital13	5.60	6.70	**3.30**	–	8.30
Ital13–Ital11	**11.50**	26.16	29.90	–	44.65
Avg.	**9.15**	17.91	18.90	–	25.25

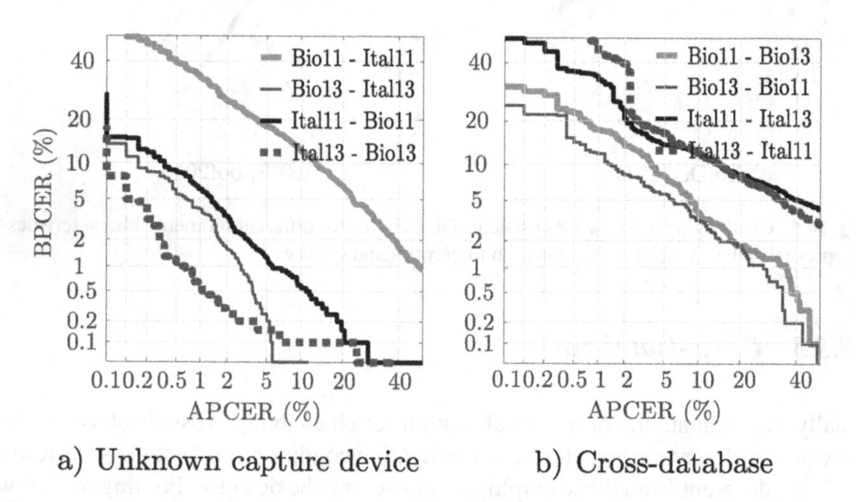

Fig. 18.6 Challenging evaluation on unknown capture devices and cross-database for fingerprint

analysis). Table 18.4 reports the detection performance of the proposed method and establishes a benchmark with the top state-of-the-art techniques. We note that FVs are capable of outperforming the results reported in the literature (i.e. a mean D-EER = 4.08 vs. 14.59% for FSB-v2 [44]) by up to 72% (relative).

We show in Fig. 18.6a the corresponding security analysis for the proposed approach. As it can be noted, the training of our detector using the Italdata subset attains a better detection performance than the training over Biometrika (black vs. orange and blue vs. red curves). In particular, our algorithm achieves a BPCER100 of 11.05%, which reduces by 78.96% the top state-of-the-art result

(BPCER100 = 52.52% for FSB-v2 [44]), thereby confirming its soundness for this challenging scenario.

As mentioned, we also evaluate the performance over the change of data collection over the same capture device (i.e. cross-database, herein train and test over the same capture device, but acquired for LivDet 2011 and LivDet 2013, respectively). In the last five rows of Table 18.4, we can observe similar results to the ones reported in the first part of this table: our approach outperforms the rest of PAD techniques (e.g. D-EER = 17.91% for FSB-v2 [44]) by up to a 64% relative improvement.

Subsequently, Fig. 18.6b confirms the results in Table 18.4: fingerprint samples acquired by the same capture device at different years lead to similar DET curves. Specifically, Biometrika 2011 and Biometrika 2013 report a joint BPCER10 = 3.60%, BPCER20 = 7.10% and BPCER100 = 4 15.10% with a standard deviation in the range (0.42–3.82%). In this train–test configuration, current state-of-the-art PAD algorithms achieve a poor BPCER100 of 65.06%. In contrast, for the Italdata 2011 and Italdata 2013 configurations, our method yields a joint BPCER10 = 12.55 ± 0.21, BPCER20 = 15.70% ± 1.98 and BPCER100 = 41.95 ± 9.40, thereby confirming its soundness once more for this challenging scenario.

18.5.3.2 Face

We report in Table 18.5 the D-EERs for different train–test configurations, and the corresponding DET curves are depicted in Fig. 18.7. As it may be observed, our approach achieves a D-EER of 18.24% on average for the best BSIF filter configuration. It should be noted that the best performing deep learning-based scheme for this scenario (i.e. DR-UDA [48]) attains a mean D-EER of 17.93%. This algorithm is fully focused on domain adaptation, which transfers the knowledge learned from a source domain to a target domain. Despite its results for cross-database, the DR-UDA approach is unable to achieve reliable error rates for known-attacks (i.e. D-EERs of 3.20, 6.00, and 7.20% for CASIA, MSU-MFSD and Rose-Youtu databases, respectively in Table 18.3). Thus, we may state that our method outperforms PAD techniques

We observe in Fig. 18.7 that our proposal suffers a detection performance decrease for high-security thresholds: a high average BPCER100 of 73.90% states the need for enhanced interoperable PAD schemes in order to improve their generalisation capabilities for this scenario without losing accuracy for the remaining scenarios.

18.5.4 Common Feature Space Visualisation

Finally, we show in Fig. 18.8 the t-SNE visualisation for the generalisable common feature space over three adopted biometric characteristics. It should be observed that most cases of the feature spaces representing PAIs appear to be closer with each other (intra-variability) than with those for BP samples (inter-variability). For

Table 18.5 Benchmark of our FVs with the state of the art in terms of the D-EER (%) for the cross-dataset scenario over facial data. The best results are highlighted in bold

Train	MSU		RA		CASIA		Avg.
Test	CASIA	RA	MSU	CASIA	MSU	RA	
Colour Texture [154]	46.00	33.90	34.10	37.70	24.40	30.30	34.40
Texture fusion [45]	29.20	16.20	21.40	31.20	19.90	**9.90**	21.30
DupGAN [155]	27.10	35.40	36.20	46.50	33.40	42.4	36.83
KSA [115]	**9.10**	33.30	34.90	**12.30**	15.10	39.30	24.00
ADA [156]	17.70	5.10	30.50	41.50	9.30	17.50	20.27
DR-UDA [48]	16.80	**3.00**	29.00	34.20	**9.00**	15.60	**17.93**
Proposed Method	24.67	6.57	**11.67**	29.33	12.86	24.36	18.24
Optimum BSIF filters	$T = 7$	$T = 7$	$T = 12$	$T = 10$	$T = 8$	$T = 6$	–
	$l = 7$	$l = 3$	$l = 11$	$l = 9$	$l = 13$	$l = 3$	

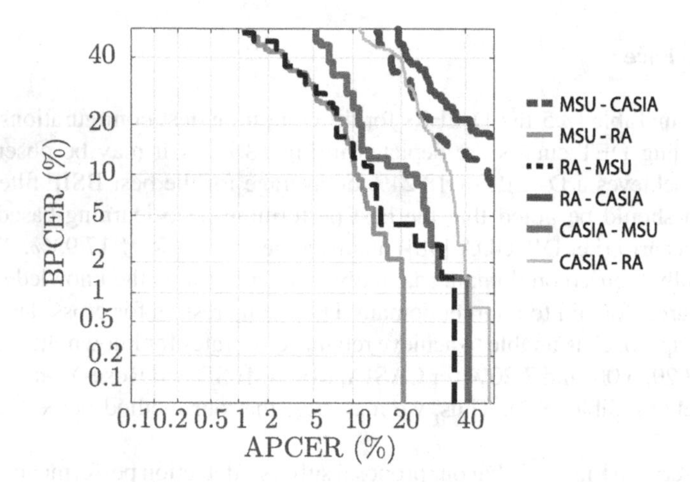

Fig. 18.7 Detection performance of FVs for the cross-dataset scenario over facial data

instance, fingerprint PAIs are successfully separated from the corresponding BPs (see Fig. 18.8-a). Similar results can be perceived for facial and voice data (see Fig. 18.8-b, c). Despite the results achieved, we can note an overlap of some PAI species such as AA, BA and CA with the BP features for voice data, hence indicating that the data distribution learned by a GMM still needs to be improved to obtain a better generalisable common feature space.

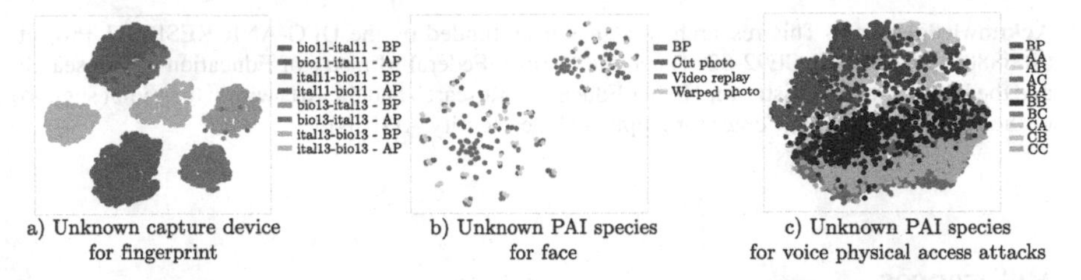

Fig. 18.8 t-SNE visualisation of our generalisable representation for different biometric characteristics taken from [49, 135, 140]. It should be noted that the t-SNE plot of the facial data **b** appears more scattered than the other plots due to the number of samples in the CASIA database (i.e. 600)

18.6 Conclusion

In this work, a novel generalisable common feature space to address challenging scenarios such as unknown PAI species and cross-database scenarios in the PAD task was proposed. In essence, this method projects local descriptors (e.g. SIFT or BSIF) into a new feature space which allows the definition of semantic feature subgroups from known samples in order to improve the generalisation capabilities of the module for different biometric characteristics.

The experimental evaluation over several freely available databases and three biometric characteristics confirmed the soundness of our proposal for spotting both known and unknown PAIs. In particular, high D-EERs close to 0.00% for most modalities over known-attacks were attained. Regarding the most challenging scenarios where PAI species remain unknown, we observed that the algorithm is capable of yielding a reliable performance for high-security thresholds, e.g. a BPCER of 0.00% for any APCER was reported for Papercraft and Impersonation attacks over facial data. In addition, a remarkable BPCER100 ranging from 0.00 to 4.60% was achieved for fingerprint. The experimental evaluation for voice data also showed a performance improvement with respect to the state of the art.

In the context of the cross-database evaluation, we noted that the proposed method considerably outperforms the top state-of-the-art fingerprint-based approaches by up to a relative 72%. Consequently, it showed a high generalisation ability for facial data, thereby resulting in a mean D-EER of 18.24%. It is important to highlight that a cross-database evaluation for voice data was not performed due to the lack of a proper evaluation protocol. Therefore, it will be analysed in future work. Finally, we showed the feature space visualisation which does confirm the soundness of our proposal to generalise across unknown PAI species and databases and shows that there is still some room for improvements. In future work, we plan to address the PAD task through multi-biometric solutions leading to increased security of the biometric system.

Acknowledgements This research work has been funded by the DFG-ANR RESPECT Project (406880674 / ANR-18-CE92-0024) and the German Federal Ministry of Education and Research and the Hessen State Ministry for Higher Education, Research and the Arts within their joint support of the National Research Center for Applied Cybersecurity ATHENE.

References

1. Jain A, Flynn P, Ross A (2007) Handbook of biometrics. Springer
2. Maltoni D, Maio D, Jain A, Prabhakar S (2009) Handbook of fingerprint recognition, 1st edn. Springer
3. ISO/IEC JTC1 SC37 Biometrics: ISO/IEC 30107-1 (2016) Information Technology—Biometric presentation attack detection—Part 1: Framework. International Organization for Standardization
4. Ratha NK, Connell JH, Bolle RM (2001) Enhancing security and privacy in biometrics-based authentication systems. IBM Syst J 40(3):614–634
5. Tolosana R, Vera-Rodriguez R, Fierrez J, Morales A, Ortega-Garcia J (2020) Deepfakes and beyond: a survey of face manipulation and fake detection. Inf Fusion 64:131–148
6. Willis D, Lee M (1998) Six biometric devices point the finger at security. Comput Secur 5(17):410–411
7. Zwiesele A, Munde A, Busch C, Daum H (2000) BioIS study-comparative study of biometric identification systems. In: 34th annual 2000 IEEE international Carnahan conferences on security technology (CCST). IEEE Computer Society, pp 60–63
8. Matsumoto T, Matsumoto H, Yamada K, Hoshino S (2002) Impact of artificial gummy fingers on fingerprint systems. In: Electronic imaging 2002. International Society for Optics and Photonics, pp 275–289
9. Schuckers S (2016) Presentations and attacks, and spoofs, oh my. Image Vis Comput 55:26–30
10. Marasco E, Ross A (2015) A survey on antispoofing schemes for fingerprint recognition systems. ACM Comput Surv (CSUR) 47(2):28
11. Drahansky M (2008) Experiments with skin resistance and temperature for liveness detection. In: International conferences on intelligent information hiding and multimedia signal processing. IEEE, pp 1075–1079
12. Sun L, Huang W, Wu M (2011) Tir/vis correlation for liveness detection in face recognition. In: International conferences on computer analysis of images and patterns. Springer, pp 114–121
13. Lapsley PD, Lee JA, Ferrin PD, Hoffman N (1998) Anti-fraud biometric scanner that accurately detects blood flow. US Patent 5,737,439
14. Drahansky M, Lodrova D (2008) Liveness detection for biometric systems based on papillary lines. In: International conferences on information security and assurance. IEEE, pp 439–444
15. Drahansky M, Notzel R, Funk W (2006) Liveness detection based on fine movements of the fingertip surface. In: Proceedings of the information assurance workshop. IEEE, pp 42–47
16. Galbally J, Marcel S, Fierrez J (2014) Biometric antispoofing methods: a survey in face recognition. IEEE Access 2:1530–1552
17. Kolberg J, Gläsner D, Breithaupt R, Gomez-Barrero M, Reinhold J, von Twickel A, Busch C (2021) On the effectiveness of impedance-based fingerprint presentation attack detection. Sensors 21(17) (2021)
18. Peng F, Qin L, Long M (2020) Face presentation attack detection based on chromatic co-occurrence of local binary pattern and ensemble learning. J Vis Commun Image Represent 66:102–746
19. Xiong F, AbdAlmageed W (2018) Unknown presentation attack detection with face rgb images. In: Proceedings of the international conferences on biometrics theory, applications and systems (BTAS), pp 1–9

20. Agarwal A, Yadav D, Kohli N, Singh R, Vatsa M, Noore A (2017) Face presentation attack with latex masks in multispectral videos. In: Proceedings of the international conferences on computer vision and pattern recognition workshops, pp 81–89
21. Raghavendra R, Venkatesh S, Raja K, Wasnik P, Stokkenes M, Busch C (2018) Fusion of multi-scale local phase quantization features for face presentation attack detection. In: Proceedings of the international conferences on information fusion (FUSION), pp 2107–2112
22. Jin C, Kim H, Elliott S (2007) Liveness detection of fingerprint based on band-selective fourier spectrum. In: Proceedings of the international conferences on information security and cryptology (ICISC). Springer, pp 168–179
23. Li J, Wang Y, Tan T, Jain AK (2004) Live face detection based on the analysis of fourier spectra. In: Biometric technology for human identification, vol 5404. International Society for Optics and Photonics, pp 296–303
24. Sahidullah M, Delgado H, Todisco M, Kinnunen T, Evans N, Yamagishi J, Lee K (2019) Introduction to voice presentation attack detection and recent advances. In: Handbook of biometric anti-spoofing. Springer, pp 321–361
25. Galbally J, Marcel S, Fierrez J (2013) Image quality assessment for fake biometric detection: application to iris, fingerprint, and face recognition. IEEE Trans Image Process 23(2):710–724
26. Jin C, Li S, Kim H, Park E (2011) Fingerprint liveness detection based on multiple image quality features. In: Proceedings of the international conferences on information security applications (ISA). Springer, pp 281–291
27. Chugh T, Jain AK (2020) Fingerprint spoof detector generalization. IEEE Trans Inf Forensics Secur (TIFS) 16:42–55
28. George A, Marcel S (2021) On the effectiveness of vision transformers for zero-shot face anti-spoofing. In: Proceedings of the international joint conferences on biometrics (IJCB). IEEE, pp 1–8
29. Deb D, Jain AK (2020) Look locally infer globally: a generalizable face anti-spoofing approach. IEEE Trans Inf Forensics Secur (TIFS) 16:1143–1157
30. Orrù G, Casula R, Tuveri P, Bazzoni C, Dessalvi G, Micheletto M, Ghiani L, Marcialis GL (2019) Livdet in action-fingerprint liveness detection competition 2019. In: Proceedings of the international conferences on biometrics (ICB). IEEE, pp 1–6
31. Casula R, Micheletto M, Orrù G, Delussu R, Concas S, Panzino A, Marcialis GL (2021) Livdet 2021 fingerprint liveness detection competition-into the unknown. In: Proceedings of the international joint conferences on biometrics (IJCB). IEEE, pp 1–6
32. ISO/IEC JTC1 SC37 Biometrics: ISO/IEC 30107-3. Information Technology—Biometric presentation attack detection—Part 3: Testing and Reporting. International Organization for Standardization (2017)
33. Yambay D, Ghiani L, Denti P, Marcialis GL, Roli F, Schuckers S (2012) LivDet 2011 finger-print liveness detection competition 2011. In: Proceedings of the international conferences on biometrics (ICB). IEEE, pp 208–215
34. Ghiani L, Yambay D, Mura V, Tocco S, Marcialis GL et al (2013) LivDet 2013 fingerprint liveness detection competition 2013. In: Proceedings of the international conferences on biometrics (ICB). IEEE, pp 1–6
35. Mura V, Orrù G, Casula R, Sibiriu A, Loi G, Tuveri P, Ghiani L, Marcialis GL (2018) Livdet 2017 fingerprint liveness detection competition 2017. In: Proceedings of the international conferences on biometrics (ICB). IEEE, pp 297–302
36. Zhang Z, Yan J, Liu S, Lei Z, Yi D, Li SZ (2012) A face antispoofing database with diverse attacks. In: Proceedings of the international conferences on biometrics (ICB), pp 26–31
37. Chingovska I, Anjos A, Marcel S (2012) On the effectiveness of local binary patterns in face anti-spoofing
38. Wen D, Han H, Jain AK (2015) Face spoof detection with image distortion analysis. IEEE Trans Inf Forensics Secur 10(4):746–761
39. Liu Y, Stehouwer J, Jourabloo A, Liu X (2019) Deep tree learning for zero-shot face anti-spoofing. In: Proceedings of the conferences on computer vision and pattern recognition (CVPR), pp 4680–4689

40. Todisco M, Wang X, Vestman V, Sahidullah M, Delgado H, Nautsch A, Yamagishi J, Evans N, Kinnunen T, Lee KA (2019) Asvspoof 2019: future horizons in spoofed and fake audio detection. In: Proceedings of the interspeech
41. Nogueira RF, de Alencar Lotufo R, Machado RC (2016) Fingerprint liveness detection using convolutional neural networks. IEEE Trans Inf Forensics Secur 11(6):1206–1213
42. Pala F, Bhanu B (2017) Deep triplet embedding representations for liveness detection. In: Deep learning for biometrics, pp 287–307
43. Chugh T, Cao K, Jain AK (2017) Fingerprint spoof detection using minutiae-based local patches. In: Proceedings of the international joint conferences on biometrics (IJCB), pp 581–589
44. Chugh T, Cao K, Jain AK (2018) Fingerprint spoof buster: use of minutiae-centered patches. IEEE Trans Inf Forensics Secur 13(9):2190–2202
45. Boulkenafet Z, Komulainen J, Hadid A (2018) On the generalization of color texture-based face anti-spoofing. Image Vis Comput 77:1–9
46. George A, Marcel S (2019) Deep pixel-wise binary supervision for face presentation attack detection. In: Proceedings of the international conferences on biometrics (ICB). IEEE, pp 1–8
47. Yu Z, Zhao C, Wang Z, Qin Y, Su Z, Li X, Zhou F, Zhao G (2020) Searching central difference convolutional networks for face anti-spoofing. In: Proceedings of international conference on computer vision and pattern recognition (CVPR), pp 5295–5305
48. Wang G, Han H, Shan S, Chen X (2020) Unsupervised adversarial domain adaptation for cross-domain face presentation attack detection. IEEE Trans Inf Forensics Secur 16:56–69 (2020)
49. Gonzalez-Soler LJ, Patino J, Gomez-Barrero M, Todisco M, Busch C, Evans N (2020) Texture-based presentation attack detection for automatic speaker verification. In: Proceedings of the international workshop on information forensics and security (WIFS), pp 1–6
50. Sousedik C, Breithaupt R, Busch C (2013) Volumetric fingerprint data analysis using optical coherence tomography. In: Proceedings of the international conferences of the biometrics special interest group (BIOSIG). IEEE Computer Society, pp 1–6
51. Sousedik C, Busch C (2014) Quality of fingerprint scans captured using optical coherence tomography. In: Proceedings of the international joint conferences on biometrics (IJCB). IEEE Computer Society (2014)
52. Auksorius E, Boccara A (2015) Fingerprint imaging from the inside of a finger with full-field optical coherence tomography. Biomed Opt Express 6(11)
53. Darlow L, Connan J, Singh A (2016) Performance analysis of a hybrid fingerprint extracted from optical coherence tomography fingertip scans. In: 2016 International conferences on biometrics (ICB)
54. Chugh T, Jain A (2019) OCT fingerprints: resilience to presentation attacks. arXiv:1908.00102
55. Kose,N, Dugelay J (2013) Reflectance analysis based countermeasure technique to detect face mask attacks. In: Proceedings of the international conferences on digital signal processing (DSP), pp 1–6
56. Wang Y, Hao X, Hou Y, Guo C (2013) A new multispectral method for face liveness detection. In: Proceedings of the Asian conferences on pattern recognition, pp 922–926
57. Kollreider K, Fronthaler H, Bigun J (2009) Non-intrusive liveness detection by face images. Image Vis Comput 27(3):233–244
58. Zhang L, Yang J (2021) A continuous liveness detection for voice authentication on smart devices. arXiv:2106.00859
59. Wang Y, Cai W, Gu T, Shao W, Li Y, Yu Y (2019) Secure your voice: an oral airflow-based continuous liveness detection for voice assistants. In: Proceedings of the ACM on interactive, mobile, wearable and ubiquitous technologies, vol 3(4)
60. Sequeira AF, Gonçalves T, Silva W, Pinto JR, Cardoso JS (2021) An exploratory study of interpretability for face presentation attack detection. IET Biomet 10(4):441–455
61. Antonelli A, Cappelli R, Maio D, Maltoni D (2005) A new approach to fake finger detection based on skin distortion. In: Advances in biometrics. Springer, pp 221–228

62. Zhang Y, Tian J, Chen X, Yang X, Shi P (2007) Fake finger detection based on thin-plate spline distortion model. In: Advances in biometrics. Springer, pp 742–749
63. Derakhshani R, Schuckers S, Hornak L, O'Gorman L (2003) Determination of vitality from a non-invasive biomedical measurement for use in fingerprint scanners. Pattern Recogn 36(2):383–396
64. Plesh R, Bahmani K, Jang G, Yambay D, Brownlee K, Swyka T, Johnson P, Ross A, Schuckers S (2019) Fingerprint presentation attack detection utilizing time-series, color fingerprint captures. In: Proceedings of the international conferences on biometrics (ICB), pp 1–8
65. Jee HK, Jung SU, Yoo JH (2006) Liveness detection for embedded face recognition system. Proc Intl J Biol Med Sci 1(4):235–238 (2006)
66. Pan G, Sun L, Wu Z, Lao S (2007) Eyeblink-based anti-spoofing in face recognition from a generic webcamera. In: Proceedings of the international conference on computer vision, pp 1–8
67. Kollreider K, Fronthaler, Bigun J (2008) Verifying liveness by multiple experts in face biometrics. In: Proceedings of the computer society conference on computer vision and pattern recognition workshops, pp 1–6
68. Patel K, Han H, Jain AK (2016) Cross-database face antispoofing with robust feature representation. In: Proceedings of the Chinese conference on biometric recognition, pp 611–619
69. Bigun J, Fronthale H, Kollreider K (2004) Assuring liveness in biometric identity authentication by real-time face tracking. In: Proceedings of the international conference on computational intelligence for homeland security and personal safety (CIHSPS), pp 104–111
70. Ali A, Deravi F, Hoque S (2012) Liveness detection using gaze collinearity. In: Proceedings of the international conference on emerging security technologies, pp 62–65
71. Tirunagari S, Poh N, Windridge D, Iorliam A, Suki N, Ho AT (2015) Detection of face spoofing using visual dynamics. IEEE Trans Inf Forensics Secur 104:762–777
72. Lee H, Maeng H, Bae Y (2009) Fake finger detection using the fractional fourier transform. In: Biometric ID management and multimodal communication. Springer, pp 318–324
73. Zhang Z, Yan J, Liu S, Lei Z, Yi D, Li SZ (2012) A face antispoofing database with diverse attacks. In: Proceedings of the international conference on biometrics (ICB). IEEE, pp 26–31
74. Tan B, Schuckers S (2008) New approach for liveness detection in fingerprint scanners based on valley noise analysis. J Electron Imaging 17(1):011,009–011,009
75. Abhyankar A, Schuckers S (2006) Fingerprint liveness detection using local ridge frequencies and multiresolution texture analysis techniques. In: Proceedings of the international conference on on image processing. IEEE, pp 321–324
76. Nguyen HP, Delahaies A, Retraint F, Morain-Nicolier F (2019) Face presentation attack detection based on a statistical model of image noise. IEEE Access 7:175,429–175,442
77. Dubey RK, Goh J, Thing VL (2016) Fingerprint liveness detection from single image using low-level features and shape analysis. IEEE transactions on information forensics and security, vol 11(7), pp 1461–1475
78. Arashloo SR, Kittler J, Christmas W (2015) Face spoofing detection based on multiple descriptor fusion using multiscale dynamic binarized statistical image features. IEEE transactions on information forensics and security, vol 10(11), pp 2396–2407
79. Audio Processing Systems (2008) Wiley
80. Kinnunen T, Li H (2010) An overview of text-independent speaker recognition: From features to supervectors. Speech Commun 52(1):12–40
81. Huang X, Deng L (2010) An overview of modern speech recognition
82. Sahidullah M, Kinnunen T, Hanilçi C (2015) A comparison of features for synthetic speech detection. In: Proceedings of the interspeech, pp 2087–2091
83. Todisco M, Delgado H, Evans N (2017) Constant Q cepstral coefficients: a spoofing countermeasure for automatic speaker verification. Comput Speech Lang 45:516–535
84. Brown J, Calculation of a constant Q spectral transform. J Acoust Soc Am 89(1):425–434
85. Todisco M, Delgado H, Evans N (2016) A new feature for automatic speaker verification anti-spoofing: constant Q cepstral coefficients. In: Proceedings of the Odyssey, vol 25, pp 249–252

86. Ergünay SK, Khoury E, Lazaridis A, Marcel S (2015) On the vulnerability of speaker verification to realistic voice spoofing. In: Proceedings of the international conference on biometrics theory, applications and systems (BTAS), pp 1–6
87. Kinnunen T, Sahidullah M, Falcone M, Costantini L, Hautamäki R, Thomsen D, Sarkar A, Tan Z, Delgado H, Todisco M, Evans N, Hautamäki V, Lee KA (2017) Reddots replayed: a new replay spoofing attack corpus for text-dependent speaker verification research. In: Proceedings of the international conference on acoustics, speech and signal processing (ICASSP), pp 5395–5399
88. Srinivas K, Das RK, Patil HA (2018) Combining phase-based features for replay spoof detection system. In: Proceedings of the international symposium on Chinese spoken language processing (ISCSLP), pp 151–155
89. Cheng X, Xu M, Zheng TF (2019) Replay detection using cqt-based modified group delay feature and resnewt network in asvspoof 2019. In: Proceedings of the Asia-Pacific signal and information processing association annual summit and conference (APSIPA ASC), pp 540–545
90. Tak H, Patino J, Nautsch A, Evans N, Todisco M (2020) Spoofing attack detection using the non-linear fusion of sub-band classifiers. In: Proceedings of the interspeech
91. Wang X, Yamagishi J, Todisco M, Delgado H, Nautsch A, Evans N, Sahidullah M, Vestman V, Kinnunen T, Lee K (2020) Asvspoof 2019: a large-scale public database of synthesized, converted and replayed speech. Comput Speech Lang 64:101–114
92. Yang J, Lei Z, Li S (2014) Learn convolutional neural network for face anti-spoofing. arXiv:1408.5601
93. Jia Y, Shelhamer E, Donahue J, Karayev S, Long J, Girshick R, Guadarrama S, Darrell T (2014) Caffe: convolutional architecture for fast feature embedding. In: Proceedings of the international conference on multimedia, pp 675–678
94. Parkhi O, Vedaldi A, Zisserman A (2015) Deep face recognition. In: Proceedings of the British machine vision conference (BMVC). British Machine Vision Association
95. Xu Z, Li S, Deng W (2015) Learning temporal features using LSTM-CNN architecture for face anti-spoofing. In: Proceedings of the Asian conference on pattern recognition (ACPR), pp 141–145
96. Gan J, Li S, Zhai Z, Liu C (2017) 3D convolutional neural network based on face anti-spoofing. In: Proceedings of the international conference on multimedia and image processing (ICMIP), pp 1–5
97. Atoum Y, Liu Y, Jourabloo A, Liu X (2017) Face anti-spoofing using patch and depth-based CNNS. In: Proceedings of the international joint conference on biometrics (IJCB), pp 319–328
98. He K, Zhang X, Ren S, Sun J (2016) Deep residual learning for image recognition. In: Proceedings of the international conference on computer vision and pattern recognition (CVPR), pp 770–778
99. Chen T, Kumar A, Nagarsheth P, Sivaraman G, Khoury E (2020) Generalization of audio deepfake detection. In: Proceedings of the Odyssey, pp 132–137
100. Tomilov A, Svishchev A, Volkova M, Chirkovskiy A, Kondratev A, Lavrentyeva G (2021) STC antispoofing systems for the asvspoof2021 challenge. In: Proceedings of the automatic speaker verification and spoofing countermeasures challenge, pp 61–67
101. Ma Y, Ren Z, Xu S (2021) Rw-resnet: a novel speech anti-spoofing model using raw waveform. arXiv:2108.05684
102. Ravanelli M, Bengio Y (2018) Speaker recognition from raw waveform with sincnet. In: Proceedings of the IEEE spoken language technology workshop (SLT), pp 1021–1028
103. Tak H, Patino J, Todisco M, Nautsch A, Evans N, Larcher A (2021) End-to-end anti-spoofing with rawnet2. In: Proceedings of the international conference on acoustics, speech and signal processing (ICASSP), pp 6369–6373
104. Yamagishi J, Wang X, Todisco M, Sahidullah MJ, Nautsch A, Liu X, Lee KA, Kinnunen T, Evans N, Delgado H (2021) Asvspoof 2021: accelerating progress in spoofed and deepfake speech detection. In: Proceedings of the automatic speaker verification and spoofing countermeasures challenge (ASVSpoof2021), pp 47–54

105. de Freitas Pereira T, Anjos A, Martino JD, Marcel S (2016) Can face anti-spoofing countermeasures work in a real world scenario? In: Proceedings of the international conference on biometrics (ICB), pp 1–8
106. Arashloo SR, Kittler J, Christmas W (2017) An anomaly detection approach to face spoofing detection: a new formulation and evaluation protocol. IEEE Access 5:13868–13882
107. Rattani A, Scheirer WJ, Ross A (2015) Open set fingerprint spoof detection across novel fabrication materials. IEEE transactions on information forensics and security, vol 10(11), pp 2447–2460
108. Ding Y, Ross A (2016) An ensemble of one-class SVMS for fingerprint spoof detection across different fabrication materials. In: Proceedings of the international workshop on information forensics and security (WIFS). IEEE, pp 1–6
109. Nikisins O, Mohammadi A, Anjos A, Marcel S (2018) On effectiveness of anomaly detection approaches against unseen presentation attacks in face anti-spoofing. In: Proceedings of the international conference on biometrics (ICB), pp 75–81
110. Chingovska I, Anjos AD (2015) On the use of client identity information for face antispoofing. IEEE Trans Inf Forensics Secur 10(4):787–796
111. George A, Marcel S (2020) Learning one class representations for face presentation attack detection using multi-channel convolutional neural networks. IEEE Trans Inf Forensics Secur 16:361–375
112. Zhang Y, Fei J, Duan Z (2021) One-class learning towards synthetic voice spoofing detection. IEEE Signal Process Lett 28:937–941
113. Ganin Y, Lempitsky V (2015) Unsupervised domain adaptation by backpropagation. In: Proceedings of the international conference on machine learning, pp 1180–1189
114. Yang J, Lei Z, Yi D, Li SZ (2015) Person-specific face antispoofing with subject domain adaptation. IEEE Trans Inf Forensics Secur **10**(4):797–809
115. Li H, Li W, Cao H, Wang S, Huang F, Kot AC (2018) Unsupervised domain adaptation for face anti-spoofing. IEEE Trans Inf Forensics Secur 13(7):1794–1809
116. Long M, Zhu H, Wang J, Jordan MI (2016) Unsupervised domain adaptation with residual transfer networks. In: Advances in neural information processing systems, pp 136–144
117. Pereira TDF (2019) Learning how to recognize faces in heterogeneous environments. Technical report, EPFL
118. Wang G, Han H, Shan S, Chen X (2020) Cross-domain face presentation attack detection via multi-domain disentangled representation learning. In: Proceedings of the international conference on computer vision and pattern recognition, pp 6678–6687
119. Gajawada R, Popli A, Chugh T, Namboodiri A, Jain A (2019) Universal material translator: towards spoof fingerprint generalization. In: Proceedings of the international conference on biometrics (ICB). IEEE, pp 1–8
120. Wang H, Dinkel H, Wang S, Qian Y, Yu K (2019) Cross-domain replay spoofing attack detection using domain adversarial training. In: Proceedings of the interspeech, pp 2938–2942
121. Wang H, Dinkel H, Wang S, Qian Y, Yu K (2020) Dual-adversarial domain adaptation for generalized replay attack detection. In: Proceedings of the interspeech 2020, pp 1086–1090. https://doi.org/10.21437/Interspeech.2020-1255
122. McLachlan GJ, Peel D (2004) Finite mixture models. Wiley
123. Fahlman SE, Hinton GE, Sejnowski TS (1983) Massively parallel architectures for al: Netl, thistle, and boltzmann machines. In: Proceedings of the national conference on artificial intelligence (AAAI)
124. Kingma DP, Welling M (2013) Auto-encoding variational bayes. arXiv:1312.6114
125. Goodfellow I, Pouget-Abadie J, Mirza M, Xu B, Warde-Farley D, Ozair S, Courville A, Bengio Y (2014) Generative adversarial nets. In: Advances in neural information processing systems, pp 2672–2680
126. Krizhevsky A, Sutskever I, Hinton GE (2012) Imagenet classification with deep convolutional neural networks. In: Advances in neural information processing systems, pp 1097–1105

127. Hinton GE, Deng L, Yu D, Dahl GE, Mohamed A, Jaitly N, Senior A, Vanhoucke V, Nguyen P, Sainath T (2012) Others: deep neural networks for acoustic modeling in speech recognition: the shared views of four research groups. IEEE Signal Process Mag 29(6):82–97
128. Klein D, Manning CD (2003) Fast exact inference with a factored model for natural language parsing. In: Advances in neural information processing systems, pp 3–10
129. Cotterell R, Eisner J (2018) A deep generative model of vowel formant typology. arXiv:1807.02745
130. Oussidi A, Elhassouny A (2018) Deep generative models: Survey. In: Proceedings of the international conference on intelligent systems and computer vision (ISCV), pp 1–8
131. Engelsma J, Jain AK (2019) Generalizing fingerprint spoof detector: learning a one-class classifier. In: Proceedings of the international conference on on biometrics (ICB), pp 1–8
132. Kong J, Kim J, Bae J (2020) Hifi-gan: generative adversarial networks for efficient and high fidelity speech synthesis
133. Kumar K, Kumar R, de Boissiere T, Gestin L, Teoh WZ, Sotelo J, de Brebisson A, Bengio Y, Courville A (2019) Melgan: generative adversarial networks for conditional waveform synthesis
134. Gonzalez-Ulloa M (1956) Restoration of the face covering by means of selected skin in regional aesthetic units. British J Plast Surg 9:212–221
135. Gonzalez-Soler LJ, Gomez-Barrero M, Chang L, Perez-Suarez A, Busch C (2021) Fingerprint presentation attack detection based on local features encoding for unknown attacks. IEEE Access 9:5806–5820. https://doi.org/10.1109/ACCESS.2020.3048756
136. Gonzalez-Soler LJ, Gomez-Barrero M, Kolberg J, Chang L, Perez-Suarez A, Busch C (2021) Local feature encoding for unknown presentation attack detection: an analysis of different local feature descriptors. IET Biometrics 10(4):374–391
137. Lowe GD, Distinctive image features from scale-invariant keypoints (2004) Int J Comput Vis 60:91–110
138. Vedaldi A, Fulkerson B (2008) VLFeat: an open and portable library of computer vision algorithms. http://www.vlfeat.org/
139. González-Soler LJ, Gomez-Barrero M, Chang L, Perez-Suarez A, Busch C (2019) On the impact of different fabrication materials on fingerprint presentation attack detection. In: Proceedings of the international conference on biometrics (ICB)
140. Gonzalez-Soler LJ, Gomez-Barrero M, Busch C (2020) Fisher vector encoding of dense-bsif features for unknown face presentation attack detection. In: Proceedings of the international conference on of the special interest group on biometrics (BIOSIG 2020), LNI, pp 1–11. GI
141. Kannala J, Rahtu E (2012) BSIF: binarized statistical image features. In: Proceedings of the international conference on pattern recognition (ICPR), pp 1363–1366
142. Hyvärinen A, Hurri J, Hoyer PO (2009) Natural image statistics: a probabilistic approach to early computational vision, vol 39. Springer Science & Business Media
143. Sánchez J, Perronnin F, Mensink T, Verbeek J (2013) Image classification with the fisher vector: theory and practice. Proc Intl J Comput VIs 105(3):222–245
144. Perronnin F, Sánchez J, Mensink T (2010) Improving the fisher kernel for large-scale image classification. In: Proceedings of the international European conference on computer vision (ECCV), pp 143–156
145. Jegou H, Perronnin F, Douze M, Sánchez J, Perez P, Schmid C (2012) Aggregating local image descriptors into compact codes. IEEE Trans Pattern Anal Mach Intell 34(9):1704–1716
146. Simonyan, K., Parkhi, O.M., Vedaldi, A., Zisserman, A.: Fisher vector faces in the wild. In: British Machine Vision Conf. (BMVC), vol. 2, p. 4 (2013)
147. Hsu C, Chang C, Lin C et al (2003) A practical guide to support vector classification. National Taiwan University, Taipei, Taiwan, Tech. rep
148. Gonzalez-Soler LJ, Gomez-Barrero M, Busch C (2021) On the generalisation capabilities of fisher vector based face presentation attack detection. IET Biometrics 10(5):480–496
149. Park, E., Cui, X., Nguyen, T.H., Kim, H.: Presentation attack detection using a tiny fully convolutional network. IEEE Trans. on Information Forensics and Security **14**(11), 3016–3025 (2019)

150. Boulkenafet Z, Komulainen J, Hadid A (2016) Face antispoofing using speeded-up robust features and fisher vector encoding. IEEE Signal Processing Letters 24(2):141–145
151. Qu X, Dong J, Niu S (2019) shallowcnn-le: a shallow CNN with laplacian embedding for face anti-spoofing. In: International conference on automatic face & gesture recognition, pp 1–8
152. Singh M, Chawla M, Singh R, Vatsa M, Chellappa R (2019) Disguised faces in the wild 2019. In: Proceedings of the international conference on computer vision workshops, p 0
153. Zhang Y, Pan S, Zhan X, Li Z, Gao M, Gao C (2020) Fldnet: light dense CNN for fingerprint liveness detection. IEEE Access 8:84141–84152
154. Boulkenafet Z, Komulainen J, Hadid A (2016) Face spoofing detection using colour texture analysis. IEEE Trans Inf Forensics Secur 11(8):1818–1830
155. Hu L, Kan M, Shan S, Chen X (2018) Duplex generative adversarial network for unsupervised domain adaptation. In: Proceedings of the international conference on computer vision and pattern recognition, pp 1498–1507
156. Wang G, Han H, Shan S, Chen X (2019) Improving cross-database face presentation attack detection via adversarial domain adaptation. In: Proceedings of the international conference on biometrics (ICB)

Chapter 19
Smartphone Multi-modal Biometric Presentation Attack Detection

Martin Stokkenes, Raghavendra Ramachandra, Amir Mohammadi, Sushma Venkatesh, Kiran Raja, Pankaj Wasnik, Eric Poiret, Sébastien Marcel, and Christoph Busch

Abstract Biometric verification is widely employed on smartphones for various applications, including financial transactions. In this work, we present a new multi-modal biometric dataset (face, voice, and periocular) acquired using a smartphone. The new dataset consists of 150 subjects captured in six different sessions reflecting real-life scenarios of smartphone authentication. A unique feature of this dataset is that it is collected in four different geographic locations representing a diverse population and ethnicity. Additionally, we also present a multi-modal presentation attack dataset using low-cost presentation attack Instruments such as print and electronic display attacks. The novel acquisition protocols and the diversity of the data subjects collected from different geographic locations will allow the development of novel

M. Stokkenes (✉) · R. Ramachandra (✉) · S. Venkatesh (✉) · K. Raja (✉) · P. Wasnik (✉) · C. Busch (✉)
Norwegian University of Science and Technology, Gjøvik, Norway
e-mail: martin.stokkenes@ntnu.no

R. Ramachandra
e-mail: raghavendra.ramachandra@ntnu.no

S. Venkatesh
e-mail: sushma.venkatesh@ntnu.no

K. Raja
e-mail: kiran.raja@ntnu.no

P. Wasnik
e-mail: pankaj.wasnik@ntnu.no

C. Busch
e-mail: christoph.busch@ntnu.no

A. Mohammadi (✉) · S. Marcel (✉)
IDIAP Research Institute, Martigny, Switzerland
e-mail: amir.mohammadi@idiap.ch

S. Marcel
e-mail: marcel@idiap.ch

E. Poiret (✉)
IDEMIA, Paris, France
e-mail: eric.poiret@idemia.com

S. Marcel et al. (eds.), *Handbook of Biometric Anti-Spoofing*, Advances in Computer Vision and Pattern Recognition, https://doi.org/10.1007/978-981-19-5288-3_19

algorithms for either uni-modal or multi-modal biometrics. Further, we also report results obtained on the newly collected dataset after conducting performance evaluation experiments using a set of baseline verification and presentation attack detection algorithms that will be described in further detail.

19.1 Introduction

Secure and reliable access control using biometrics are deployed in various applications including border control, banking transactions, financial services, attendance systems, and smartphone unlocking. Use of biometrics in access control applications not only improves reliability but also improves user experience by authenticating the user based on who they are. Thus, the user is not required to remember a passcode or possess a smart card to gain access in the control process. Based on ISO/IEC 2382-37, biometrics is defined as: 'automated recognition of individuals based on their behavioral and biological characteristics [1]'. Biometric data can be sampled from either a physical (face, iris, fingerprint, etc.) or a behavioral (keystroke, gait, etc.) characteristic, that can be used to recognize the data subject. Even though the utility of biometrics on smartphones brings several advantages, there exist a number of challenges in real-life applications that must be solved in order to take full advantage of the biometrics authentication process on consumer smartphone. Among the many challenges, vulnerability toward attacks and interoperability are key problems that limit reliable and secure applications of biometrics on smartphones.

The smartphone is an ideal device for biometric capture as it comes with a variety of sensors that can be used to measure an individual's biometric characteristic. Even though the manufacturers have started to introduce dedicated biometric sensors in their devices, these are often not suitable for more general purpose usage of biometrics. In most cases, they are locked down from a hardware perspective and a third party that would like to integrate such a biometric solution in their services will not be able to obtain certainty about the identity of the individual using the integrated biometric solution. Consider the scenario where a financial institution wants to onboard a new customer. They would require an Identity Document as a reference to establish the identity of the potential customer. In a digital onboarding process this could be done with the help of a smartphone to scan the ID and the capture a facial image of the person to match against the biometric reference from the ID. This process would not be possible using the integrated biometric systems from the device manufacturers.

From the perspective of a service provider the smartphone and the user is considered to be untrusted. The vulnerability of smartphones to both direct presentation attacks (PAs) and indirect attacks are well exemplified in the literature [2, 3]. Further, it is also demonstrated in [4] that, the biometric reference is retrievable from the smartphone hardware chip, which further indicates another vulnerability. Recent works based on master prints [5] demonstrate the vulnerability of the fingerprint recognition system itself, irrespective of the manufacturer. These factors motivated the research toward smartphone biometrics that can be independent of the devices. To this extent, several attempts have been made to develop both uni-modal and multi-modal biometric solutions. The most popular biometric characteristics investigated include face [6], visible iris [7], soft-biometrics [8], finger photo [9].

A crucial aspect to foster advances in the area of smartphone biometrics is the availability of suitable biometric datasets for benchmarking of newly developed algorithms and reproducible evaluation. This can help to identify shortcomings, new issues, or challenges that exist for the particular scenario. For example, biometric capture on a smartphone would be operated by the user, not a trained operator, and would be used in uncontrolled environments such as on the train or outdoors. These scenarios can introduce noise and degrading factors in the biometric data that could impact on the recognition performance of a biometric system. There exists a limited number of publicly available smartphone-based datasets particularly addressing biometric characteristics such as face [10], finger photo [9], iris [7], soft-biometrics [8] and multi-modality [11]. However, the collection of biometric data is a tedious process with additional efforts required in selecting the capture device, design of data collection protocols, post-processing of captured data, annotation of captured data, data anonymization, and the collection itself. Further, one also needs to consider the legal and ethical requirements related to the collection of biometric data.

This chapter presents a recently collected smartphone-based multi-modal biometric dataset within the framework of the SWAN project [12] with an accompanying multi-modal presentation attack dataset. The SWAN Multi-modal Biometric Dataset (SWAN-MBD) and SWAN multi-modal Presentation Attack Dataset (SWAN-MPAD) consists of data from 150 subjects with face, periocular, and voice biometric characteristics and presentation attack instruments (PAIs) such as high-quality print and display attack (using iPhone 6S and iPad PRO).

The following are the main contribution of this chapter:

- Overview of other related datasets
- New multi-modal dataset collected using smartphone in six different sessions from 150 subjects.
- New multi-modal presentation attack dataset collected using two different PAIs for face, periocular, and voice biometrics.
- Performance evaluation protocols to benchmark the accuracy of both verification and presentation attack detection (PAD) algorithms.
- Quantitative result of the baseline algorithms are reported using ISO/IEC SC 37 metrics on both verification and PAD.

The rest of the chapter is organized as follows: Sect. 19.2 presents the related work on the available multi-modal biometric datasets collected using smartphone, Sect. 19.3 presents the SWAN multi-modal biometric and presentation attack dataset, Sect. 19.4 presents the evaluation protocols to benchmark the algorithms, Section 19.5 discuss the different baseline algorithms used to benchmark performance on the SWAN multi-modal biometric dataset. Section 19.6 discuss the quantitative results of the baseline systems and Sect. 19.7 draws the conclusion.

19.2 Related Work

In this section, we discuss the publicly available smartphone-based multi-modal datasets. There are limited smartphones-based multi-modal biometric datasets currently available for researchers. The majority of these available datasets are based on the two modalities that can be captured together, for example, face (talking) and voice, iris, and periocular. Table 19.1 presents an overview of the different smartphone-based multi-modal datasets that are publicly available.

The **BioSecure** dataset (DS3) [18] is one of the earlier and large scale publicly available datasets (available upon license fee payment). The BioSecure dataset comprises four different faces, voice, signature, and fingerprint collected from 713 subjects in 2 sessions. Only face and voice are collected using the mobile device Samsung Q1.

Table 19.1 Publicly available smartphone-based multi-modal datasets

Dataset	Year	Devices	No. of subjects	Biometric	Availability
MOBIO [10]	2012	Nokia N93i Mac-book	152	Face, Voice	Free
CSIP [13]	2015	Sony Xperia Apple IPhone 4	50	Iris, Periocular	Free
FTV [14]	2010	HP iPAQ rw6100	50	Face, Teeth, Voice	Free
MobBIO [15]	2014	ASUS PAD TF 300	105	Voice, Face, periocular	Free
UMDAA [16]	2016	Nexus 5	48	Face, Behavior patterns	Free
MobiBits [17]	2018	Huawei Mate S Huawei P9 Lite CAT S60	53	Signature, Voice, Face, Periocular, Hand	Free
BioSecure-DS3 [18]	2010	Samsung Q1 Philips SP900NC Webcam HP iPAQ hx2790 PDA	713	Voice, Signature, Face, Fingerprint	Paid
SWAN dataset	2019	iPhone 6S	150	Face, Periocular, Multilingual Voice Presentation Attack dataset	Free

The **MOBIO** database [10] consists of data from 153 subjects where biometric characteristics (face and voice) are collected using Nokia N93i and MacBook laptop. The complete dataset is collected in 12 sessions by capturing the face (talking) together with voice. Voice samples are collected based on both pre-defined text and free text, and the face characteristics are recorded while the subject is talking.

The **CSIP** database [13] contains 50 subjects with iris and periocular biometric characteristics captured using four different smartphone devices. The entire dataset is collected from different backgrounds to reflect the real-life scenario.

The **FTV** dataset [14] consists of face, teeth, and voice biometric characteristics captured from 50 data subjects using a smartphone HP iPAQ rw6100. The face and teeth samples are collected using the smartphone camera while the microphone of smartphone is used to collect the voice samples.

The **MobBIO** dataset [15] is based on the three different biometric characteristics such as the face, voice, and periocular, captured using a tablet Asus Transformer Pad TF 300T. Voice samples are collected using a microphone in the Asus Transformer Pad TF 300T where a data subject was asked to read out 16 sentences in Portuguese. The face and periocular samples are collected using the 8MP camera from Asus Transformer Pad TF 300T. This dataset holds data from 105 data subjects collected in two different lighting conditions.

The **UMDAA** dataset [16] is collected using the Nexus 5 from 48 data subjects. This database has a collection of both physical and behavioral biometric characteristics. The data collection sensors include the front-facing camera, touchscreen, gyroscope, magnetometer, light sensor, GPS, Bluetooth, accelerometer, Wi-Fi, proximity sensor, temperature sensor, and pressure sensor. The entire dataset is collected for two months and provides face and other related behavior patterns suitable for continuous authentication.

The **MobiBits** dataset [17] consists of five different biometric characteristics namely voice, face, iris, hand, and signature which are collected from 55 data subjects. Three different smartphones are used to collect the biometric characteristics such as Huawei Mate smartphone is used to collect signature and periocular samples, CAT S60 smartphone is used to collect hand and face samples, Huawei P9 Lite smartphone is used to collect voice samples.

19.2.1 Features of the SWAN Multi-modal Biometric Dataset

SWAN Multi-modal Biometric Dataset (SWAN-MBD) is collected to complement the existing datasets and the data collection protocols are designed to meet real-life scenario such as banking transactions. The following are the main features of the newly introduced SWAN Multi-modal Biometric Dataset:

- The data collection protocol is designed to capture data in six different sessions. Each session reflects both time variation and capturing conditions (outdoor and indoor).

- The data collection application is developed to make data collection consistent with ease of use with regard to the user interaction such that self-capture is facilitated. This is a unique feature of this dataset in which the data is self-captured by the participants.
- The data collection is carried out in four different geographic locations such as India, Norway, France, and Switzerland with data subjects belonging to various ethnic groups.
- Three different biometric characteristics: face, periocular, and voice are captured from 150 data subjects in all 6 different sessions. The voice samples are captured based on the question and answers which is text dependent. The voice samples are collected in four different languages: English, Norwegian, French, and Hindi.
- The whole dataset is collected using iPhone 6S smartphone and iPad PRO.
- In addition, we also present a new SWAN Multi-modal Presentation Attack dataset (SWAN-MPAD) for all three biometric characteristics (face, periocular, and voice).

19.3 SWAN Multi-modal Biometric Dataset

19.3.1 Database Acquisition

To facilitate the data collection at different locations, a smartphone application is developed for the iOS platform (version 9) that can be installed in the data capture devices (both iPhone 6S and iPad Pro). This application has a Graphical User Interface that allows the data collection moderator to select session number, biometric characteristics, location ID, subject ID, and other relevant information for data collection. Figure 19.1 shows the graphical user interface (GUI) of SWAN data collection application while Figure 19.2 shows the example images during biometric data collection. Thus, the application is designed to make sure the data collection process can be easily carried out such that data subjects can use it seamlessly during the self-capture protocol. The data collection process is broadly divided into two phases as explained below.

19.3.2 SWAN Multi-modal Biometric Dataset

The SWAN multi-modal biometric dataset was jointly collected within SWAN project, between April 2016 till December 2017, by partners at four different geographic locations: Norway, Switzerland, France, and India. The dataset collected from 150 volunteer subjects such that 50 data subjects collected in Norway, 50 data subjects are collected in Switzerland, 5 data subjects in France, and 45 data subjects in India. The age distribution of the data subjects is between 20 and 60 years. The dataset was collected using Apple iPhone 6S and Apple iPad Pro (12.9-inch iPad

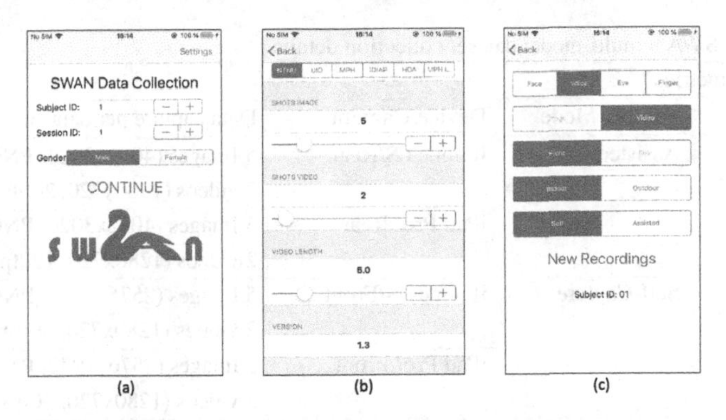

Fig. 19.1 Screenshot from SWAN multi-modal biometric data capture application

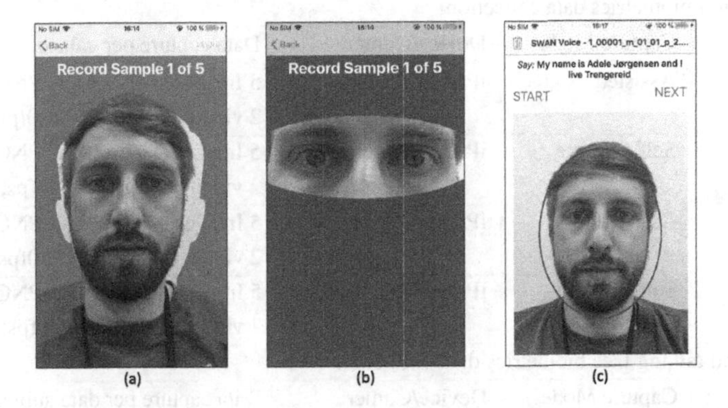

Fig. 19.2 Illustration of SWAN multi-modal biometric data capture application during **a** Face capture **b** Eye region capture **c** Voice and talking face capture

Pro). Three biometric modalities are collected: (1) Face: both Ultra HD images and HD (including slow motion) video recordings of faces from data subjects (2) Voice: HD audio-visual recordings of talking faces from data subjects (3) Eye: both Ultra HD images and HD (including slow motion) video recordings of eyes from data subjects. The whole dataset is collected in six different sessions such that, **Session 1** is captured in an indoor environment with uniform illumination of the face, quiet environment reflecting high-quality supervised enrollment. **Session 2** is captured indoors with illumination from ceiling lights and windows without any degrading shadow on the face and semi-quiet office environment. **Session 3** is captured outdoors with uncontrolled natural illumination, natural noise environment. **Session 4** is captured indoors with illumination e.g., side light from windows that creates shadow in the face, natural noise environment. **Session 5** is captured indoors with uncontrolled illumination, natural noise environment in crowded place. **Session 6** is captured indoors with uncontrolled illumination e.g., side light from windows, natural noise environment. The average time duration between sessions varies from 1 week to 3 weeks. The multi-modal biometric samples are captured in both assisted

Table 19.2 SWAN multi-modal dataset collection details

(a) Face biometrics

Session	Capture Mode	Device/Camera	Data capture per data subject
S1	Assisted	iPhone 6S/Rear	5 Images (4032x3024, PNG) 2 videos (1280x720, 240fps, 5s, MP4)
		iPad Pro/ Rear	5 Images (4032x3024, PNG) 2 videos (1280x720, 120fps, 5s, MP4)
	Self-Capture	iPhone 6S/Front	5 Images (2576x1932, PNG) 2 videos (1280x720, 30fps, 5s, MP4)
		iPad Pro/ Front	5 Images (2576x1932, PNG) 2 videos (1280x720, 30fps, 5s, MP4)
S2 - S6	Self-Capture	iPhone 6S/Front	2 videos (1280x720, 30fps, 5s, MP4)

(b) Eye region biometrics data collection

Session	Capture Mode	Device/Camera	Data capture per data subject
S1	Assisted	iPad Pro/Rear	5 Images (4032x3024, PNG) 2 videos (1280x720, 120fps, 5s, MP4)
	Self-Capture	iPad Pro/Front	5 Images (2576x1932, PNG) 2 videos (1280x720, 30fps, 5s, MP4)
S1-S6	Assisted	iPhone 6S/Rear	5 Images (4032x3024, PNG) 2 videos (1280x720, 240fps, 5s, MP4)
	Self-Capture	iPhone 6S/Front	5 Images (2576x1932, PNG) 2 videos (1280x720, 30fps, 5s, MP4)

(c) Voice and talking face biometrics data collection

Session	Capture Mode	Device/Camera	Data capture per data subject
S1-S6	Self-Capture	iPhone 6S/Front	8 videos: 4 videos in English and 4 videos in native language (1280x720, 30fps, variable length, MP4)

capture mode (where data subjects are assisted during capturing) and self-capture mode (here data subject control the capture process on their own). Multi-modal biometrics are collected using both rear and front cameras from iPhone 6S and iPad Pro. The iPhone 6S is used as the primary device to collect the dataset, while the iPad Pro is used to collect the data only in session 1 to capture a high-quality image that can be used to generate a PAI.

Table 19.2 indicates the data collection protocol and the sample collection (both images and videos) for the facial biometric characteristic. The biometric face capture, indicated in Table 19.2 A, corresponds to one data subject. iPhone 6S is used to capture the face data in all six sessions and iPad Pro is used only in session 1. During each acquisition of session 1, the face data is captured in both assisted and self-captured mode using rear and front camera from iPhone 6S and iPad Pro respectively. For session 2—session 6 it was decided to adopt only self-capture to reduce the overall

Table 19.3 SWAN presentation attack data collection details

Modality	PA name	PA source selection			PAI	Biometric capture sensor
		Session	Sensor	type		
Face	PA.F.1	1	iPhone 6S back camera	photos	Epson Expression Photo XP-860 Epson Photo Paper Glossy, PN: S041271	iPhone 6S front camera (video)
	PA.F.5	1	iPhone 6S front camera	videos	iPhone 6S display	iPhone 6S front camera (video)
	PA.F.6	1	iPad Pro front camera	videos	iPad-Pro display	iPhone 6S front camera (video)
Eye	PA.EI.1	1	iPhone 6S back camera	photos	Epson Expression Photo XP-860 Epson Photo Paper Glossy, PN: S041271	iPhone 6S front camera (video)
	PA.EI.4	1	iPad-Pro front camera	photos	iPad-Pro display	iPhone 6S front camera (video)
	PA.EI.5	1	iPhone 6S front camera	videos	iPhone 6S display	iPhone 6S front camera (video)
Voice	PA.V.4	1	iPad-Pro microphone	audio	Logitech high-quality loudspeaker	iPhone 6S microphone (audio)
	PA.V.7	1	iPhone 6S microphone	audio	iPhone 6S speakers	iPhone 6S microphone (audio)

time burden of the subject for a complete capture session. The data collected using the iPad Pro is used to generate presentation attacks while data collected from iPhone 6S is used to perform the biometric verification. Thus, in total there are: 150 subjects $\times 20$ images $= 3000$ images and 150 subjects $\times 18$ videos $= 2700$ videos. Table 19.2 B presents the data statistics corresponding to eye region biometric characteristic, which is in both self-capture and assisted mode. Since the goal of the eye region data collection is to get high-quality images that can be used for both periocular and visible iris recognition, we collected the eye region dataset in both capture modes across all 6 sessions. Therefore, the assisted mode is captured using the rear camera of iPhone 6S with 12mega pixels that provide good quality images for visible iris recognition. However, the images collected from the self-capture process using the frontal camera can be used to develop the periocular verification systems. Similar to the face capture process, the iPad Pro is used only in session 1 to capture a good quality eye region image that is used to generate a PAI, to be used against the periocular biometric system. The whole dataset consists of 150 subjects $\times 70$ images $= 10500$ image and 150 subjects $\times 28$ videos $= 4200$ videos.

Table 19.2 C indicates the statistics and protocol of the voice and talking face data collection. To cover both text independent and text dependent modes, the audio recordings (actually audio-video recordings) are captured with the data subjects pronouncing these 4 utterances in English followed by these 4 utterances in a national language depending on the site (Norwegian, French, and Hindi). The four sentences include: *Sentence #1*: "My name is `FAKE_FIRSTNAME FAKE_NAME` and I live `FAKE_ADDRESS`" the address is a short string that is limited to the street name. Thus, no street number, no zip code, no city, and no country information is recorded. *Sentence #2*: "My account number is `FAKE_ACCOUNTNUMBER`". The account numbers are presented by groups of digits (eg. "5354" "8745"). The data subject is free to pronounce the groups of digits the way he/she wants (digits by digits, as a single number, or as a combination). *Sentence #3:* "The limit of my card is 5000 euros". *Sentence #4:* "The code is 9 8 7 6 5 4 3 2 1 0". The 10 digits are presented one by one and the data subjects were asked to pronounce the digits separately. The audio-visual data with voice and talking faces are collected in the self-capture mode using the frontal camera of iPhone 6S in all 6 sessions. There is no dedicated assisted capture as the data subject is assisted by the capture application itself. Thus, $150\times$ Subjects $\times 48$ videos $= 7200$ videos corresponding to audio-visual data. Figure 19.3 illustrates the example images from SWAN multi-modal dataset collected in all six session indicating an inter-session variability.

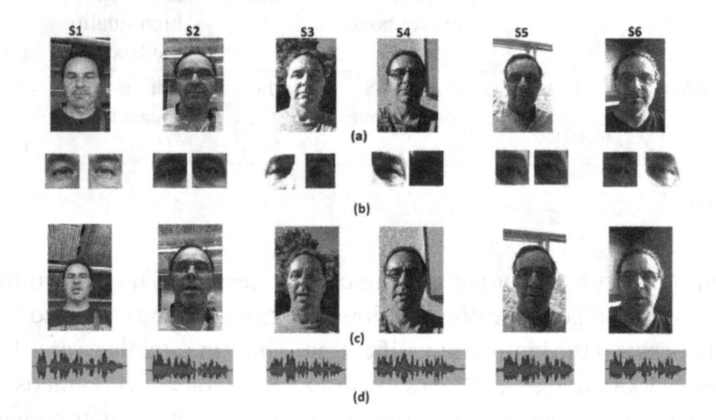

Fig. 19.3 Illustration of SWAN multi-modal biometric dataset images **a** Face capture **b** Eye region capture **c** Talking face capture (one only frame is included for the simplicity) **d** voice sample

19.3.3 SWAN-Presentation Attack Dataset

The SWAN presentation attack dataset contains three different types of presentation attacks that are generated for three different biometric characteristics, namely, face, eye, and voice. The presentation attack (PA) database generation generally requires obtaining biometric samples "PA source selection", generating artifacts, and presenting attack artifacts to the biometric capture sensor. Session 1 recordings from SWAN multi-modal biometric dataset are used to generate the presentation attack dataset. The artifact generation is carried out using five different PAIs such as (1) High-quality photo on paper generated using the photo printer (face and Eye). The print artifacts on face and eye are generated using Epson Expression Photo XP-860 with high-quality paper Epson Photo Paper Glossy (Product Number: S041271; Basis Weight: 52 lb.(200 g/m^2); Size: A4; Thickness: 8.1 mil.). (2) Electronic display of artifacts using iPad-Pro (face and Eye) (3) Electronic display of artifacts using iPhone 6S (face and Eye) (4) Logitech high-quality loudspeaker (Voice) (5) iPad-Pro loudspeaker (Voice). We have used these PAI by considering the cost for generation versus attack potential and thus selected PAIs, which are of low cost (reasonably) and at the same time indicates a high vulnerability for the biometric system under attack.

Table 19.3 presents the PAD data collection procedure with presentation attack source, PAI, and biometric capture sensor. All the presentation attack data for face and eye are collected using a frontal camera of iPhone 6S and all recordings were videos and at least 5 seconds long for each PA and in the *.mp4* format. Both the biometric capture device (iPhone 6S) and the PAI (paper or display) are mounted on stands. In the case of voice samples, the audio files were collected using iPhone 6S with the same compression specifications as bona fide samples. Figures 19.4 and 19.5 shows the example images for face and eye presentation attacks, respectively.

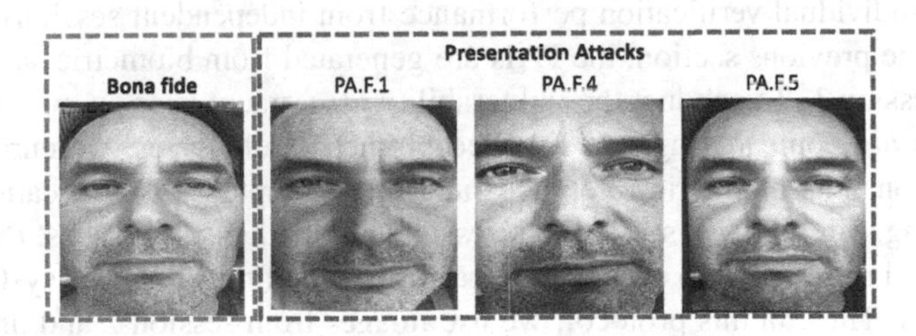

Fig. 19.4 Illustration of SWAN-Presentation Attack dataset **a** Bona fide **b** Presentation Attack (PA.F.1) **c** Presentation Attack (PA.F.4) **d** Presentation Attack (PA.F.5)

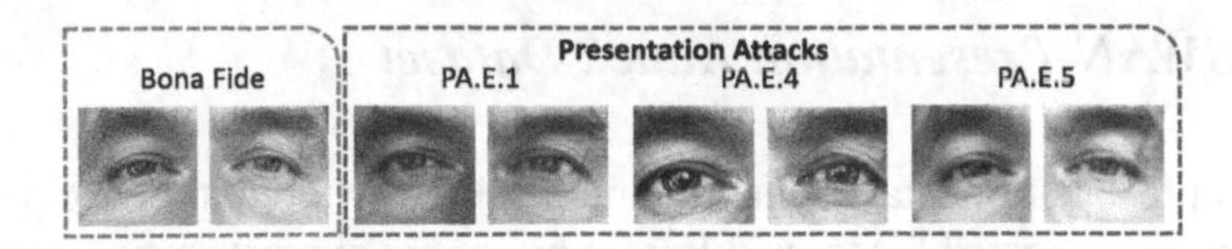

Fig. 19.5 Illustration of SWAN-Presentation Attack dataset **a** Bona fide **b** Presentation Attack (PA.E.1) **c** Presentation Attack (PA.E.4) (d) Presentation Attack (PA.E.5)

19.3.4 Dataset Distribution

The Idiap subset of the database will be distributed online on the Idiap dataset distribution portal (https://www.idiap.ch/dataset). We are also aiming to make the full database available through the BEAT platform[1] [19]. The BEAT platform is a European computing e-infrastructure solution for open access and scientific information sharing and re-use. Both the data and the source code of experiments are shared while protecting privacy and confidentiality. As the data on the BEAT platform is easily available for experimental research but cannot be downloaded, it makes BEAT an ideal platform for biometrics research while the privacy of users who participated in the data collection is maintained.

19.4 Experimental Performance Evaluation Protocols

In this section, we discuss the evaluation protocol that is used to report the performance of the baseline algorithm for both verification and PAD subsystems.

Biometric verification performance evaluation protocol: To evaluate the biometric verification performance, we propose two protocols: **Protocol-1:** designed to study the individual verification performance from independent sessions. As mentioned in the previous section, the PAIs are generated from biometric samples captured in session 1. To conduct the vulnerability assessment later on, where samples from PAIs are compared against enrolled biometric references, we enroll images from session 2 as biometric references and then compute the verification performance using samples from session 3 to session 6 individually as probes. **Protocol-2:** this protocol is designed to evaluate the performance of the biometric system across all sessions. Thus, in this protocol, we use images from session 2 and probe using samples from session 3 to session 6, to evaluate the verification performance. However, for simplicity, we have used only self-captured data to report the performance of the baseline verification algorithms.

Presentation Attack Detection protocol: To evaluate the performance of the baseline presentation attack detection algorithms we propose two different evaluation protocols. **Protocol-1**: This protocol is to evaluate the performance of the

[1] https://www.beat-eu.org/platform/.

PAD algorithms independently on each PAI, thus training and testing of the PAD algorithms are carried out independently on each PAI. **Protocol-2**: This protocol is designed to evaluate the performance of the PAD algorithms when trained and tested with all PAIs. To effectively evaluate the PAD algorithms, we divide the whole dataset into three independent partitions such that the training set has 90 subjects, the development set has 30 subjects, and the testing set has 30 subjects. Further, we perform cross-validation by randomizing the selection of training, development, and testing set for $N = 5$ times and average the results, which are reported with standard deviation.

19.5 Baseline Algorithms

In this section, we discuss baseline biometric algorithms used to benchmark the performance of biometric verification and presentation attack detection.

19.5.1 Biometric Verification

Face Biometrics: Two deep learning based face biometric algorithms are evaluated: VGG-Face[2] [20] and FaceNet [21, 22]. The choice of networks is based on the obtained recognition accuracy on the LFW dataset—challenging FR dataset [23] where FaceNet reported an accuracy of 99.2% and VGG-Face reported an accuracy of 98.95%.

Eye Biometrics: To evaluate the periocular biometric system, we have used five different algorithms that include: Coupled-Autoencoder [24], Collaborative representation of Deep-sparse Features [25], Deep Convolution Neural Network (DCNN) features from pre-trained networks such as AlexNet, VGG16, and ResNet50. These baseline algorithms are selected based on the reported verification performance, especially in the smartphone environment.

Audio-visual biometrics: Inter-Session Variability (ISV) [26, 27] and a deep learning based network (ResNet) [28] is used for evaluating the voice biometrics. ISV is a GMM-UBM [29] based system which explicitly models the session variability. We have used an extended ResNet implementation of [28] named dilated residual network (DRN) which is publicly available.[3] The DRN model is one of the state-of-the-art algorithms on the Voxceleb 1 database [30] evaluations achieving 4.8% EER on the dataset.

---

[2] Website: www.robots.ox.ac.uk/~vgg/software/vgg_face.

[3] https://www.idiap.ch/software/bob/docs/bob/bob.learn.pytorch/v0.0.4/guide_audio_extractor.html.

The audio-visual system is the result of score fusion between the FaceNet and the DRN models which are the best performing algorithm for face and voice subsystems, respectively.

19.5.2 Presentation Attack Detection Algorithms

Face and Eye Attack Instruments: We report the detection performance of five different PAD algorithms on both eye and face attack instruments. The baseline algorithms include Local Binary Pattern (LBP) [31], Local Phase Quantisation (LPQ) [31], Binarised Statistical Image Features (BSIF) [32], Image Distortion Analysis (IDA) [33] and Color texture features [34]. We use linear SVM as the classifier that is trained and tested with these features following different evaluation protocols as discussed in Sect. 19.4. We have selected these five PAD algorithms as the baseline algorithms to cover the spectrum of both micro-texture and image quality-based attack detection algorithms.

Audio-visual attack instruments: We use three features: MFCC, SCFC, and LFCC with two classifiers (SVM and GMM) [35] to develop voice PAD systems. The same PAD systems that were used for still faces were also used here. The final audio-visual PAD system is the result of score-level fusion of the best algorithms of face and audio PAD systems (Color Textures-SVM and LFCC-GMM).

19.6 Experimental Results

In this section, we present the baseline performance of algorithm evaluation for both biometric verification and presentation attack detection. The performance of the baseline algorithms is presented using the Equal Error Rate (EER %) following the experimental protocol presented in Section 19.4. The performance of the baseline PAD algorithms is presented using *Bona fide Presentation Classification Error Rate (BPCER) and Attack Presentation Classification Error Rate (APCER)*, in addition to the detection of equal error rate (D-EER). **BPCER** is defined as the proportion of bona fide presentations incorrectly classified as attacks while **APCER** is defined as the proportion of attack presentations incorrectly classified as bona fide presentations. In particular, we report the performance of the baseline PAD algorithms by reporting the value of BPCER while fixing the APCER to 5% and 10% according to the recommendation from ISO/IEC 30107-3 [36]. The PAD evaluation protocol is presented in Sect. 19.4.

19.6.1 Biometric Verification Results

Table 19.4 indicates the performance of the uni-modal biometric systems with eye, 2D face, and audio-visual. For the simplicity, we have used the self-capture images from each session to present the quantitative results on eye verification using baseline algorithms (see Section 19.4 for evaluation protocols). Based on the obtained results as presented in Table 19.4, the following can be observed (1) among the five different algorithms, the DeepSparse-CRC [25] method shows the marginal improvement over other algorithms in both Protocol-1 and Protocol-2 (c). (2) Both DeepSparse-CRC [25] and Deep Autoencoder [24] algorithms indicate a degraded performance for Protocol-2 when compared to that of the Protocol-1. This can be attributed to the capture data quality reflected from all four sessions (S3-S6).

Both face biometric systems showed similar performance across all sessions. The FaceNet system outperformed the VGG-Face system by a large margin. However, the same FaceNet system's performance was degraded when evaluated on audio-visual data. This can be attributed to the fact the camera was held at a reading position (lower down) during the audio-visual data capture process compared to the holding device higher up when taking still face videos. The DRN speaker verification system performed well compared to the ISV system (not shown for brevity, the ISV's EER on all sessions was 13.1%). The worst performance was achieved in session 3 which can be mainly attributed to the noise (especially from wind) outdoor. The final audio-visual biometric system was the average score of FaceNet and DRN systems. Performing this fusion improved the results (except for session 4) which showed that the information from face and voice were mainly complementary in performing verification.

19.6.2 Biometric Vulnerability Assessment

Figure 19.6 illustrates the vulnerability analysis on the uni-modal biometrics from the SWAN multi-modal biometric dataset. The vulnerability analysis is performed using the baseline uni-modal biometric in which, the bona fide samples are enrolled, and presentation attack samples are used as the probe. Finally, the comparisons scores obtained on the probe samples are compared against the operational threshold set, and finally, the quantitative results of the vulnerability are presented using an Impostor Attack Presentation Match Rate (IAPMR (%)).

For vulnerability analysis with an eye biometric system, we have used the DeepSparse-CRC [25] system, for 2D face biometrics we have used FaceNet (InceptionResNetV1) [22] and for audio-visual biometric system we have employed FaceNet (InceptionResNetV1) [22] for face and DRN [28] for voice whose scores are fused at comparisons core level. All three types of presentation attack samples are used as the probe samples to compute the IAMPR. As indicated in the Fig. 19.6, all three modalities have indicated IAMPR of 100% on both face and audio-visual bio-

Table 19.4 Baseline performance of uni-modal biometric algorithms

(a) Face biometrics

Algorithms	Enrollment	Probe	EER(%)
FaceNet (InceptionResNetV1) [22]	Session 2	Session 3	4.3
		Session 4	3.3
		Session 5	5.3
		Session 6	2.7
		All Session	4.2
VGG-Face [37]	Session 2	Session 3	17.2
		Session 4	17.3
		Session 5	16.6
		Session 6	17.2
		All Session	17.0

(b) Audio-Visual biometrics

Algorithms	Enrollment	Probe	EER(%)
FaceNet (InceptionResNetV1) [22]	Session 2	Session 3	14.2
		Session 4	13.1
		Session 5	13.9
		Session 6	13.0
		All Session	13.5
DRN	Session 2	Session 3	4.3
		Session 4	2.1
		Session 5	3.2
		Session 6	3.3
		All Session	3.2
FaceNet-DRN-score-mean-fusion	Session 2	Session 3	3.4
		Session 4	3.0
		Session 5	3.1
		Session 6	2.9
		All Session	3.1

(c) Eye biometrics

Algorithms	Enrollment	Probe	EER(%)
Deep Autoencoder [24]	S2	S3	25.59
		S4	23.88
		S5	27.26
		S6	23.07
		All	25.67
DeepSparse-CRC [25]	S2	S3	21.32
		S4	22.30
		S5	22.55
		S6	23.54
		All	22.41

(continued)

Table 19.4 (continued)

AlexNet features	S2	S3	24.23
		S4	28.72
		S5	27.21
		S6	21.46
		All	25.46
ResNet features	S2	S3	29.59
		S4	24.78
		S5	24.17
		S6	20.79
		All	24.77
VGG16 features	S2	S3	34.62
		S4	27.41
		S5	27.45
		S6	24.82
		All	28.62

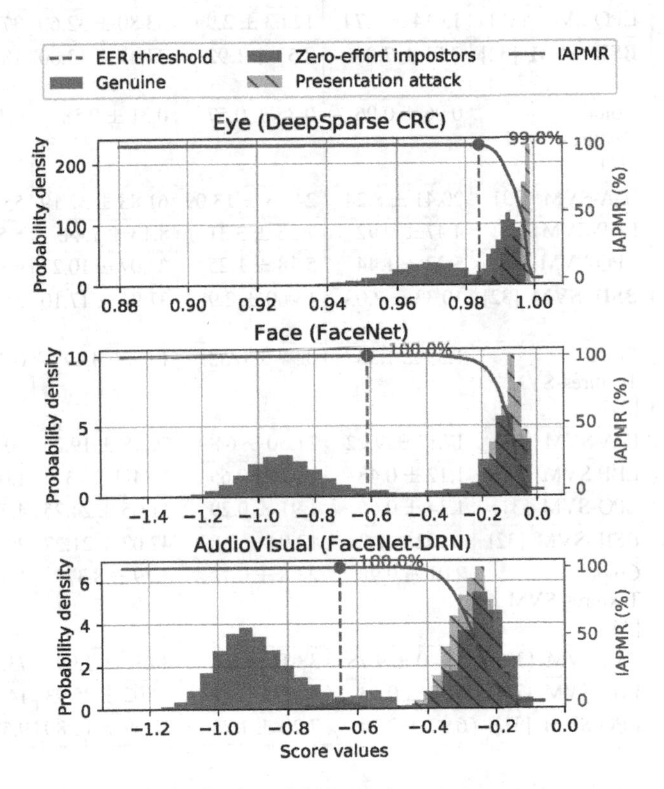

Fig. 19.6 Vulnerability analysis of uni-modal biometric systems

metrics and 99.8% on eye biometrics. The obtained results further justify the quality of the presentation attacks generated in this work and also the need for presentation attack detection algorithms to mitigate the presentation attacks.

Table 19.5 Baseline performance of PAD algorithms on Eye biometrics

Evaluation Protocol	Algorithms	Development database	Testing database		
		D-EER(%)	D-EER(%)	BPCER@ APCER	
				= 5(%)	= 10(%)
Protocol-1 (PA.E.1)	BSIF-SVM [32]	10.14 ± 1.16	11.78 ±2.62	57.95 ± 40.16	37.54 ± 29.97
	Color Textures-SVM [34]	3.65 ± 2.13	1.07 ± 1.47	17.59 ± 16.49	10.51±11.54
	IDA-SVM [33]	19.42 ± 17.46	24.23 ± 19.80	44.87 ± 26.86	37.77 ± 25.83
	LBP-SVM [31]	5.52 ± 1.17	9.55 ± 2.15	17.70 ± 14.90	16.86 ± 16.41
	LPQ-SVM [31]	13.14 ± 3.74	12.63 ± 2.99	53.80 ± 32.60	37.34 ± 28.74
Protocol-1 (PA.E.4)	**BSIF-SVM** [32]	7.51 ± 2.04	8.53 ± 1.92	34.41 ± 13.80	18.82 ± 5.59
	Color Textures-SVM [34]	0.06 ± 0.06	0.85 ± 0.57	0.31 ± 0.58	0.11+ 0.22
	IDA-SVM [33]	20.41 ± 8.24	24.35 ± 13.09	61.88 ± 32.19	55.44 ± 37.87
	LBP-SVM [31]	4.47 ± 0.92	7.26 ± 3.31	8.13 ± 7.44	5.51 ± 6.74
	LPQ-SVM [31]	5.22 ± 1.44	5.18 ± 1.25	22.07 ± 10.27	9.84 ± 6.54
Protocol-1 (PA.E.5)	**BSIF-SVM** [32]	10.33 ± 2.03	13.60 ± 2.98	31.96 ± 17.10	21.80 ± 11.40
	Color Textures-SVM [34]	1.35 ± 1.33	0.97 ± 0.52	1.64 ± 3.49	0.76 ± 1.66
	IDA-SVM [33]	17.81 ± 9.22	21.50 ± 6.89	29.18 ± 19.26	20.97 ± 14.33
	LBP-SVM [31]	1.12 ± 0.48	1.52 ± 0.65	2.14 ± 2.13	1.02 ± 1.24
	LPQ-SVM [31]	1.34 ± 0.65	1.91 ± 0.29	11.65 ± 24.25	4.76 ± 10.35
Protocol-2	BSIF-SVM [32]	10.02 ± 1.34	12.04 ± 2.89	42.02 ± 21.27	24.60 ± 10.51
	Color Textures-SVM [34]	2.01 ± 0.98	2.68 ± 0.72	4.79 ± 8.37	1.93 ± 3.50
	IDA-SVM [33]	35.40 ± 9.78	38.16 ± 8.37	80.35 ± 13.96	73.29 ± 16.75
	LBP-SVM [31]	6.34 ± 0.63	8.95 ± 2.11	22.02 ± 25.93	14.75 ± 19.41
	LPQ-SVM [31]	6.92 ± 2.77	7.20 ± 1.73	22.69 ± 12.80	9.57 ± 5.79

19.6.3 Biometric PAD Results

Table 19.5 indicates the quantitative performance of the baseline PAD algorithms on the eye recognition subsystem on Protocol-1 and Protocol-2. Based on the obtained results, the following can be observed: (1) Among the three different PAI, the detection of PA.E.4 (iPad Pro front camera) indicates an excellent detection accuracy on

Table 19.6 Baseline performance of PAD algorithms on face biometrics

Evaluation Protocol	Algorithms	Development database	Testing database		
		D-EER(%)	**D-EER(%)**	**BPCER@ APCER**	
				=5(%)	=10(%)
Protocol-1 (PA.F.1)	BSIF-SVM [32]	5.20±8.75	10.13±2.58	38.92±29.24	30.16± 24.20
	Color Textures-SVM [34]	2.20±1.25	8.71±1.28	24.87± 29.68	18.90±22.16
	IDA-SVM [33]	34.50±22.41	37.61±22.32	63.78±30.58	56.77±36.27
	LBP-SVM [31]	3.08±1.91	13.63±2.06	39.23±27.97	33.78±24.86
	LPQ-SVM [31]	5.67±2.67	13.87±2.69	30.52±12.52	22.54±9.47
Protocol-1 (PA.F.5)	BSIF-SVM [32]	17.16±5.11	20.30±1.19	67.48±30.14	50.32±23.73
	Color Textures-SVM [34]	11.57±4.37	12.46±3.21	45.76±19.67	27.96±9.17
	IDA-SVM [33]	24.55±3.11	29.08±13.40	56.53±26.66	46.87±28.95
	LBP-SVM [31]	15.83±3.65	17.13±4.38	42.05±18.31	30.96±16.13
	LPQ-SVM [31]	24.25±2.75	18.15±3.07	63.63±25.64	53.83±26.49
Protocol-1 (PA.F.6)	BSIF-SVM [32]	17.76±4.27	27.74±1.55	70.50±16.30	59.48±15.43
	Color Textures-SVM [34]	4.21±1.79	5.66±2.13	12.29±7.79	8.08±4.85
	IDA-SVM [33]	31.78±12.69	30.99±12.99	71.09±29.95	62.75±37.14
	LBP-SVM [31]	6.70±0.98	9.41±2.59	32.21±22.19	19.30±16.50
	LPQ-SVM [31]	17.07±3.18	17.04±2.41	51.63±26.95	38.36±25.62
Protocol-2	BSIF-SVM [32]	20.05±3.97	25.55±1.51	58.12±13.61	46.06±12.76
	Color Textures-SVM [34]	7.07±1.13	14.45±1.56	47.31±7.66	36.06±6.33
	IDA-SVM [33]	32.66±19.22	33.49±16.17	66.54±15.45	58.79±16.99
	LBP-SVM [31]	27.67±11.90	35.95±7.65	75.09±12.18	64.23±16.14
	LPQ-SVM [31]	18.47±3.93	20.23±3.84	48.28±10.95	39.05±9.33

Table 19.7 Baseline performance of PAD algorithms on Talking Faces biometrics

Evaluation protocol	Algorithms	Development database	Testing database		
		D-EER (%)	D-EER (%)	BPCER@ APCER	
				=5(%)	=10(%)
Protocol-1 (PA.F.5)	BSIF-SVM	23.25±5.08	27.32±3.91	97.92±1.77	91.88±6.18
	Color Textures-SVM	2.67±1.57	5.45±1.50	57.93±27.78	43.67±27.25
	IDA-SVM	35.00±8.99	37.87±9.20	53.79±44.49	50.84±43.82
	LBP-SVM	17.83±12.12	24.06±11.01	52.37±37.12	39.69±33.97
	LPQ-SVM	5.25±2.86	8.66±1.38	30.87±25.64	19.46±19.70
Protocol-1 (PA.F.6)	BSIF-SVM	21.74±4.55	29.34±3.67	81.30±11.92	68.47±19.95
	Color Textures-SVM	5.47±2.98	15.21±2.86	48.55±40.99	43.67±43.22
	IDA-SVM	37.88±17.72	41.69±9.97	69.82±35.84	66.42±35.48
	LBP-SVM	38.51±8.25	41.13±8.85	96.36±4.05	92.79±7.52
	LPQ-SVM	19.17±3.06	25.10±1.97	63.90±33.95	54.94±37.27
Protocol-2	BSIF-SVM	26.48±3.31	24.56±2.95	65.17±14.69	53.78±17.40
	Color Textures-SVM	4.31±1.50	5.18±1.58	9.72±7.24	5.22±4.61
	IDA-SVM	37.07±10.02	39.15±9.48	65.94±35.50	58.39±36.43
	LBP-SVM	34.00±7.64	31.35±6.93	52.11±38.95	45.56±41.87
	LPQ-SVM	16.72±1.98	16.86±3.68	60.44±21.89	51.61±25.25

all five baseline algorithms employed in this work. (2) Among five different baseline PAD algorithms, the PAD technique based on Color Textures-SVM [34] has indicated the best performance on both Protocol-1 and Protocol-2. (3) The performance of the PAD algorithms in Protocol-2 shows a degraded performance when compared to that of the Protocol-1.

Table 19.6 and 19.7 indicates the quantitative performance of the baseline PAD algorithms on 2D face modality on Protocol-1 and Protocol-2. Based on the obtained results, it is noted that (1) the performance of the PAD algorithms is degraded in the Protocol-2 when compared to Protocol-1. (2) Among five different baseline PAD algorithms, the PAD technique based on Color Textures-SVM [34] has indicated the best performance on both Protocol-1 and Protocol-2.

Table 19.8 indicates the quantitative performance of the baseline PAD algorithms on voice biometrics. In all protocols, the LFCC-GMM system showed well performance (0% error rates) in the detection of the presentation attacks. Among other systems, protocol 2 was the more challenging protocol in the MFCC-SVM and SCFC-GMM baselines.

Table 19.8 Baseline performance of PAD algorithms on Voice biometrics

Evaluation protocol	Algorithms	Development database	Testing database		
		D-EER (%)	D-EER (%)	BPCER@ APCER	
				=5(%)	=10(%)
Protocol-1 (PA.V.4)	MFCC-SVM	0.00±0.00	0.11±0.13	0.00±0.00	0.00±0.00
	LFCC-GMM	0.00±0.00	0.00±0.00	0.00±0.00	0.00±0.00
	SCFC-GMM	1.28±0.38	1.21±0.38	0.02±0.03	0.00±0.00
Protocol-1 (PA.V.7)	MFCC-SVM	0.00±0.00	0.12±0.10	0.00±0.00	0.00±0.00
	LFCC-GMM	0.00±0.00	0.00±0.00	0.00±0.00	0.00±0.00
	SCFC-GMM	1.58±0.49	1.17±0.21	0.08±0.06	0.00±0.00
Protocol-2	MFCC-SVM	1.52±0.43	1.67±0.17	0.14±0.16	0.03±0.04
	LFCC-GMM	0.00±0.00	0.00±0.00	0.00±0.00	0.00±0.00
	SCFC-GMM	2.09±0.41	1.48±0.07	0.04±0.02	0.00±0.00

Table 19.9 Baseline performance of PAD algorithms on Audio-Visual biometrics

Evaluation protocol	Algorithm:	Development database	Testing database		
		D-EER (%)	D-EER (%)	BPCER@ APCER	
				=5(%)	=10(%)
Protocol-1 Low-Quality (PA.F.5 and PA.V.7)	Fusion of Color Textures-SVM and LFCC-GMM	0.08±0.17	0.09±0.11	0.65±0.68	0.24±0.15
Protocol-1 High-Quality (PA.F.6 and PA.V.4)		21.74±4.55	29.34±3.67 8	1.30±11.92 6	8.47±19.95
Protocol-2 All Attacks		26.48±3.31	24.56±2.95 6	5.17±14.69 5	3.78±17.40

Finally, Table 19.9 reports the baselines from score mean fusion of Color Textures-SVM and LFCC-GMM (best performing systems of each modality) on Audio-Visual attacks. The low-quality attacks (PA.F.5 and PA.V.7) were detected with error rates less than 1%. However, the performance is degraded significantly when high-quality attacks are used with a D-EER of 29.34% on the testing database. This degradation in performance is also evident in protocol 2 where both low- and high-quality attacks are used.

19.7 Conclusions

Smartphone-based biometric verification has gained a lot of interest among researchers from the past few years. The availability of public datasets is crucial to driving the research forward so that new algorithms are developed and benchmarked with the existing algorithms. However, the collection of biometric datasets is resource consuming, especially in collecting multi-modal biometric dataset from different geographic locations. In this work, we present a new multi-modal biometric dataset captured using a smartphone together with the evaluation of the baseline algorithms. This dataset is a result of the collaboration between three European partners within the framework of the SWAN project sponsored by Research Council of Norway. This new dataset is captured in challenging conditions in four different geographic locations. The whole dataset is obtained from 150 data subjects in six different sessions that can simulate real-life scenarios. Besides, a new presentation attack dataset is also presented for multi-modal biometrics. The vulnerability presented for the verification algorithms indicates the need for a presentation attack component to complete the dataset and the necessity of presentation attack detection algorithms. A brief description of the performance evaluation protocols and the baseline biometric verification and presentation attack detection algorithms are benchmarked. Experimental findings using the baseline algorithms are highlighted for both biometric verification and presentation attack detection and show the potential for improvement for several of the biometric modalities.

Acknowledgements This work is carried out under the funding of the Research Council of Norway (Grant No. IKTPLUSS 248030/O70)

References

1. ISO/IEC JTC1 SC37 Biometrics (2017) ISO/IEC 2382-37:2017 Information technology - vocabulary - Part 37: biometrics. International Organization for Standardization (2017)
2. Ramachandra R, Busch C (2017) Presentation attack detection methods for face recognition systems: a comprehensive survey. ACM Comput Surv 50(1):8:1–8:37. https://doi.org/10.1145/3038924
3. Marcel S, Nixon MS, Li SZ (2018) Handbook of biometric anti-spoofing, vol 1. Springer
4. Zhang Y, Chen Z, Xue H, Wei T (2015) Fingerprints on mobile devices: abusing and leaking. In: Black hat conference (2015)
5. Roy A, Memon N, Ross A (2017) Masterprint: exploring the vulnerability of partial fingerprint-based authentication systems. IEEE Trans Inf Forensics Secur 12(9):2013–2025. https://doi.org/10.1109/TIFS.2017.2691658
6. Rattani A, Derakhshani R (2018) A survey of mobile face biometrics. Comput Electr Eng 72:39 – 52 https://doi.org/10.1016/j.compeleceng.2018.09.005. http://www.sciencedirect.com/science/article/pii/S004579061730650X Computers & Electrical Engineering **72**, 39 – 52 (2018). https://doi.org/10.1016/j.compeleceng.2018.09.005. http://www.sciencedirect.com/science/article/pii/S004579061730650X

7. De Marsico M, Nappi M, Riccio D, Wechsler H (2015) Mobile iris challenge evaluation (miche)-i, biometric iris dataset and protocols. Pattern Recognit Lett 57(C):17–23. https://doi.org/10.1016/j.patrec.2015.02.009
8. Patel VM, Chellappa R, Chandra D, Barbello B (2016) Continuous user authentication on mobile devices: Recent progress and remaining challenges. IEEE Signal Process Mag 33(4):49–61. https://doi.org/10.1109/MSP.2016.2555335
9. Sankaran A, Malhotra A, Mittal A, Vatsa M, Singh R (2015) On smartphone camera based fingerphoto authentication. In: 2015 IEEE 7th international conference on biometrics theory, applications and systems (BTAS), pp 1–7
10. McCool C, Marcel S, Hadid A, Pietikäinen M, Matejka P, Cernocký J, Poh N, Kittler J, Larcher A, Lévy C, Matrouf D, Bonastre J, Tresadern P, Cootes T (2012) Bi-modal person recognition on a mobile phone: Using mobile phone data. In: 2012 IEEE international conference on multimedia and expo workshops, pp 635–640
11. Bartuzi E, Roszczewska K, Białobrzeski R et al (2018) Mobibits: multimodal mobile biometric database. In: 2018 international conference of the biometrics special interest group (BIOSIG). IEEE, pp 1–5
12. NTNU (2019) Secure access control over wide area network (SWAN). https://www.ntnu.edu/iik/swan/. Accessed from 2019-08-09
13. Santos G, Grancho E, Bernardo MV, Fiadeiro PT (2015) Fusing iris and periocular information for cross-sensor recognition. Pattern Recognit Lett 57:52–59. https://doi.org/10.1016/j.patrec.2014.09.012, http://www.sciencedirect.com/science/article/pii/S0167865514003006. Mobile Iris CHallenge Evaluation part I (MICHE I)
14. Kim D, Chung K, Hong K (2010) Person authentication using face, teeth and voice modalities for mobile device security. IEEE Trans Consum Electron 56(4):2678–2685. https://doi.org/10.1109/TCE.2010.5681156
15. Sequeira AF, Monteiro JC, Rebelo A, Oliveira HP (2014) Mobbio: a multimodal database captured with a portable handheld device. In: 2014 international conference on computer vision theory and applications (VISAPP), vol 3, pp 133–139 (2014)
16. Mahbub U, Sarkar S, Patel VM, Chellappa R (2016) Active user authentication for smartphones: a challenge data set and benchmark results. In: 2016 IEEE 8th international conference on biometrics theory, applications and systems (BTAS), pp 1–8
17. Bartuzi E, Roszczewska K, Rokielewicz M, Białobrzeski R (2018) Mobibits: multimodal mobile biometric database. In: 2018 international conference of the biometrics special interest group (BIOSIG), pp 1–5. https://doi.org/10.23919/BIOSIG.2018.8553108
18. Ortega-Garcia J, Fierrez J, Alonso-Fernandez F, Galbally J, Freire MR, Gonzalez-Rodriguez J, Garcia-Mateo C, Alba-Castro J, Gonzalez-Agulla E, Otero-Muras E, Garcia-Salicetti S, Allano L, Ly-Van B, Dorizzi B, Kittler J, Bourlai T, Poh N, Deravi F, Ng MNR, Fairhurst M, Hennebert J, Humm A, Tistarelli M, Brodo L, Richiardi J, Drygajlo A, Ganster H, Sukno FM, Pavani S, Frangi A, Akarun L, Savran A (2010) The multiscenario multienvironment biosecure multimodal database (bmdb). IEEE Trans Pattern Anal Mach Intell 32(6):1097–1111. https://doi.org/10.1109/TPAMI.2009.76
19. Anjos A, El-Shafey L, Marcel S (2017) BEAT: An Open-Science Web Platform. In: International Conference on Machine Learning (ICML). https://publications.idiap.ch/downloads/papers/2017/Anjos_ICML2017_2017.pdf
20. Parkhi OM, Vedaldi A, Zisserman A (2015) Deep face recognition. In: British machine vision conference, vol. 1. BMVA Press, pp 41.1 – 41.12
21. Schroff F, Kalenichenko D, Philbin J (2015) FaceNet: a unified Embedding for Face Recognition and Clustering. In: Proceedings of IEEE conference on computer vision and pattern recognition (CVPR). IEEE, pp 815 – 823. 00297
22. Sandberg D (2017) facenet: face recognition using tensorflow. https://github.com/davidsandberg/facenet. Accessed from 2017-08-01
23. Huang GB, Ramesh M, Berg T, Learned-Miller E (2007) Labeled faces in the wild: a database for studying face recognition in unconstrained environments. Technical Report 07-49, University of Massachusetts, Amherst

24. Raghavendra R, Busch C (2016) Learning deeply coupled autoencoders for smartphone based robust periocular verification. In: 2016 IEEE international conference on image processing (ICIP), pp 325–329. https://doi.org/10.1109/ICIP.2016.7532372
25. Raja KB, Raghavendra R, Busch C (2016) Collaborative representation of deep sparse filtered features for robust verification of smartphone periocular images. In: 2016 IEEE international conference on image processing (ICIP), pp 330–334. https://doi.org/10.1109/ICIP.2016.7532373
26. Vogt R, Sridharan S (2008) Explicit modelling of session variability for speaker verification. Comput Speech Lang 22(1):17–38. https://doi.org/10.1016/j.csl.2007.05.003, http://www.sciencedirect.com/science/article/pii/S0885230807000277
27. McCool C, Wallace R, McLaren M, El Shafey L, Marcel S (2013) Session variability modelling for face authentication. IET Biom 2(3):117–129. https://doi.org/10.1049/iet-bmt.2012.0059
28. Le N, Odobez JM (2018) Robust and discriminative speaker embedding via intra-class distance variance regularization. In: Proceedings interspeech, pp 2257–2261
29. Reynolds DA, Quatieri TF, Dunn RB (2000) Speaker verification using adapted Gaussian mixture models. Digit signal Process 10(1):19–41
30. Nagrani A, Chung JS, Zisserman A (2017) VoxCeleb: a large-scale speaker identification dataset. arXiv:1706.08612 [cs]
31. Chingovska I, André A, Marcel S (2014) Biometrics evaluation under spoofing attacks. IEEE Trans Inf Forensics Secur (T-IFS) 9(12):2264–2276
32. Ramachandra R, Busch C (2014) Presentation attack detection algorithm for face and iris biometrics. In: 22nd European signal processing conference, EUSIPCO 2014, Lisbon, Portugal, pp 1387–1391
33. Wen D, Han H, Jain A (2015) Face spoof detection with image distortion analysis. IEEE Trans Inf Forensics Secur 10(99):1–16
34. Boulkenafet Z, Komulainen J, Hadid A (2015) Face anti-spoofing based on color texture analysis. In: IEEE international conference on image processing (ICIP). IEEE, pp 2636–2640
35. Sahidullah M, Kinnunen T, Hanilçi C (2015) A comparison of features for synthetic speech detection. In: Proceedings of INTERSPEECH, pp 2087–2091. Citeseer. http://citeseerx.ist.psu.edu/viewdoc/download?doi=10.1.1.709.5379&rep=rep1&type=pdf
36. ISO/IEC JTC1 SC37 Biometrics (2017) ISO/IEC 30107-3. Information Technology - Biometric presentation attack detection - Part 3: testing and Reporting. International Organization for Standardization
37. Simonyan K, Zisserman A (2014) Very deep convolutional networks for large-scale image recognition. arXiv:1409.1556

Part VI
Legal Aspects and Standards

Chapter 20
Legal Aspects of Image Morphing and Manipulation Detection Technology

Els J. Kindt and Cesar Augusto Fontanillo López

Abstract Manipulation Attack Detection (MAD) becomes increasingly important when relying on biometric information systems which serve a large variety of purposes, such as border checks, access control to digital or physical environments, and convenience, e.g., airline check-in. Besides technical aspects, legal aspects shall not be overlooked. When deploying MAD technology, responsible entities shall determine, *inter alia,* the objectives of the use of this technology (legitimacy and purpose limitation principle), the law and the legal ground on which the deployment of MAD technology will be based (legality and lawfulness), the information to be provided to individuals relating to data processing (transparency), and also how fair and accurate the processing is and shall be (fairness). Also, human oversight and intervention where automatic decision-making is in place is a must. This contribution delves into these principles and poses legal questions which arise in the context of MAD technology and practical applications.

20.1 Introduction

It should not surprise that in the information society in which we arrived, where data is representing information and is of great value, such data will be prone to malicious manipulation attacks. In some of these scenarios, the data is not removed or leaked as such, but rather modified or manipulated in a subtle manner in order not to alarm those relying on the data. Data manipulation is a Trojan horse awaiting to exploit the vulnerabilities of technology.

If such data is representing biometric information for verifying or identifying individuals, such as facial images, fingerprints, or voices, the gain for perpetrators manipulating the biometric data could be significant, not only in terms of financial

E. J. Kindt · C. A. Fontanillo López (✉)
KU Leuven – Law Faculty – Citip – iMec, Sint-Michielsstraat 6, 3000 Leuven, Belgium
e-mail: cesar.fontanillo@kuleuven.be

E. J. Kindt
e-mail: els.kindt@law.kuleuven.be

S. Marcel et al. (eds.), *Handbook of Biometric Anti-Spoofing*, Advances in Computer Vision and Pattern Recognition, https://doi.org/10.1007/978-981-19-5288-3_20

profit; it can facilitate the clandestine movement of criminals and terrorists wishing to ride on a false identity, and hence play a key role in their criminal activities. Manipulation of biometric data is a real threat in a wide variety of use cases, including crossing borders under false identities, unlawful access to restricted areas, buildings, and digital applications, such as for banking or health applications, and when biometric information is used for security purposes combined with convenience. This kind of manipulation of biometric data is not purely hypothetical and starts to spread fast. For example, German activists announced in 2018 that they enrolled a morphed image of an activist and of the former High Representative of the Union for Foreign Affairs and Security Policy and Vice-President of the European Commission Federica Mogherini for obtaining a German passport.[1] At the same time, Member States of the European Union or commercial players are not eager to disclose figures of discovered uses of these types of manipulation attacks.

It is obvious that the rise in the number of biometric systems and this type of manipulation poses threats not only to personal rights and safety but also to the confidence and dynamism of the economy, society, and democracy in the European Union.[2] In this context, the Union requires effective responses, such as to detect these manipulations to confront these kinds of security challenges.[3] Manipulation Attack Detection (MAD), of which detecting image morphing may be considered as a sub-category, has therefore attracted the interest of policymakers and become a new field of research, also in view of the increasing importance and utilization of

[1] R. Thelen and J. Horchert, *'Aktivisten schmuggeln Fotomontage in Reisepass'*, in der Spiegel, 22.09.2018, available at: https://www.spiegel.de/netzwelt/netzpolitik/biometrie-im-reisepass-peng-kollektiv-schmuggelt-fotomontage-in-ausweis-a-1229418.html. Other examples are mimicking someone's voice by creating—through deep learning—synthetic voice to mislead.

[2] See K. Crockett, S. Zoltán, J. O'shea, T. Szklarski, A. Malamou, and G. Boultadakis, *'Do Europe's borders need multi-faceted biometric protection?'*, in Biometric Technology Today, Volume 2017, Issue 7, 2017, pp. 5–8, ISSN 0969-4765, https://doi.org/10.1016/S0969-4765(17)30137-6. The European Commission has addressed these challenges as a 'priority societal challenges' under the umbrella of the Horizon 2020 Framework Programme as well as in the recent Security Union Strategy for the period 2020–2025. See European Commission, *Priority societal challenges*, 2014, available at: https://cordis.europa.eu/programme/id/H2020-EU.3. See also European Commission, *Communication from the Commission to the European Parliament, the European Council, the Council, the European Economic and Social Committee and the Committee of the Regions on the EU Security Union Strategy*, COM(2020) 605 final, Brussels, 24.7.2020, p. 7 *et seq.*, available at: https://eur-lex.europa.eu/legal-content/EN/TXT/PDF/?uri=CELEX:52020DC0605&from=EN.

[3] Such an undertaking plays a critical role for the overall preservation of the Schengen area and the respect for the fundamental rights and freedoms of individuals who transit through it. And it becomes even more notable due to the impending European harmonized security framework for issuance of identity and residence documents. See European Commission, *Technologies to enhance border and external security*, 2018, available at https://cordis.europa.eu/programme/id/H2020_SU-BES02-2018-2019-2020.

biometric systems.[4] EU research projects, such as Tabula Rasa[5] and BEAT,[6] have created awareness for security risks, such as spoofing biometric sensors (broadly referred to as presentation attacks), and have proposed first approaches and tools for these security threats, such as algorithms to increase the robustness of biometric systems and evaluation. Manipulation attacks and their detection have been further researched by other European Union and national research councils' initiatives, such as in ANANAS[7] and SOTAMD,[8] and are also at the core of the European Union's Horizon 2020 program funded project iMARS.[9] It is important that the legal aspects of MAD are looked into as well, in particular, whether the technology (to be) developed and used respects the data protection principles in addition to fundamental rights and freedoms of individuals and other applicable laws.

In this chapter, we focus on facial image manipulation and discuss first the need for and the advantages of a legal definition of image morphing and manipulation attack and detection. Furthermore, we summarize some important legal requirements from mainly the data protection domain when deploying MAD in systems and applications.[10] When doing so, we may refer to mainly three use cases: border control, access control to digital or physical environments, and convenience applications (for example, airline check-in). We also discuss briefly the proposed AI Act where relevant.

---

[4] See, e.g., C. Rathgeb, R. Tolosana Moranchel, R. Vera-Rodriguez, C. Bush, *Handbook of Digital Face Manipulation and Detection,* Springer, 2022, p. 484, available at https://www.springer.com/gp/book/9783030876630 ('Rathgeb, a.o., Handbook 2022).

[5] About the EU 7th Framework Programme Trusted Biometrics under Spoofing Attacks (Tabula Rasa), see http://www.tabularasa-euproject.org/.

[6] About the EU 7th Framework Programme Biometrics Evaluation And Testing (BEAT) project, see https://www.beat-eu.org/#:~:text=The%20BEAT%20platform%20is%20an,and%20testing%20in%20computational%20science.

[7] Anomaly Detection for Prevention of Attacks on Authentication Systems Based on Facial Images (ANANAS). German Federal Ministry of Education and Research BMBF. Available at https://www.bmbf.de/en/index.html.

[8] State of the art of Morphing Detection (SOTAMD). European Union's Internal Security Fund—Borders and Visa. https://www.ntnu.edu/iik/sotamd.

[9] The image Manipulation Attack Resolving Solutions (iMARS) project is funded under grant agreement No 883356. In this project, technical solutions to detect image morphing and manipulation (for example, of a facial image in an ePassport) are investigated and developed to address threats to travel and identity documents (such as ePassports and eIDs), authenticity, their use for verification and fraud detection, during issuance, renewal, and Schengen border checks, or where the identification of suspects is necessary for law enforcement purposes. For more information about this project, please see https://imars-project.eu/.

[10] For ethical and societal issues, see K. Laas-Mikko, T. Kalvet, R. Derevski, and M. Tiits, *Promises, Social, and Ethical Challenges with Biometrics in Remote Identity Onboarding,* 2021, p. 20, available on the iMars project website.

20.2 Should Image Morphing and Image Manipulation Attack Detection (MAD) have a Legal Definition?

Biometric systems consist of various steps or phases, usually enrolment and presentation of the biometric information, comparison, and decision. Over the successive stages of the processing, biometric systems can be disrupted through an ample range of attacks, both direct and indirect.

Over the past years, especially the improper *presentation* of biometric characteristics to a biometric system at enrolment or during presentation has been researched,[11] e.g., detecting when an artifact (such as a gummy finger or a face mask) is presented to a sensor or camera,[12] when using mutilated fingerprints to subvert biometric systems,[13] or for detecting reply.[14] This is often referred to as spoofing or 'presentation attacks' (PA), i.e. attempts to log into the system with a fake biometric characteristic or other presentation attack instruments. With new sophisticated imaging and processing tools, which have also become widely available on the market for all, more misleading attacks are, however, possible. We focus herein on facial image manipulation and, in particular, on the merging of several facial images into one picture ('image morphing') prior to submission of an identity document or passport picture to an enrolment agency. For such manipulation attack, the fraudster would also *intentionally replace or alter* the appearance or properties of an image to circumvent the biometric system. More precisely, for a morphing attack, the fraudster would intentionally *combine two or multiple* facial images that may resemble the contributing subjects to subvert the biometric system.[15] Image morphing could hence be seen as a sub-category of image manipulation. Such fraudulent activities result in modified or morphed facial images that can mislead biometric systems (and humans), fooling them to give a positive outcome as being the named individual in

---

[11] R. Raghavendra and C. Busch. 'Presentation attack detection methods for face recognition systems: A comprehensive survey', *ACM Comput. Suvrv.*, 50(1), 2017, p. 37 , available at https://dl.acm.org/doi/pdf/10.1145/3038924 ('Raghavendra, et al., PAD Survey, 2017'). For a useful suggested classification of presentation attacks, into basically two types, (1) artificial and (2) human (lifeless, altered, non-conformant, coerced, and conformant), see p.6; see also A. Anjos and S. Marcel, *Counter-measures to photo attacks n face recognition: a public database and a baseline, 2011,* available via *http://www.tabularasa-euproject.org/publications/* and N. Evans, J. Yamagishi and T. Kinnunen, *Spoofing and countermeasures for automatic speaker verification, in Proc. Interspeech* 2013, pp. 925–929. See also C. Busch et al., *What is a Presentation Attack? And how do we detect it ?*, 2018, presentation, available at https://www.christoph-busch.de/files/Busch-TelAviv-PAD-180116.pdf.

[12] See also R. Tolosana, M. Gomez-Barrero, C. Busch, and J. Ortega-Garcia, 'Biometric Presentation Attack Detection: Beyond the Visible Spectrum', in *IEEE Transactions on Information Forensics and Security*, vol. 15, pp. 1261–1275, 2020, doi: 10.1109/TIFS.2019.2934867 .

[13] J. Feng, A. K. Jain, A. Ross, *Fingerprint Alteration, MSU Technical Report*, MSU-CSE-09-30, Dec. 2009, available at: http://citeseerx.ist.psu.edu/viewdoc/summary?doi=10.1.1.389.4636.

[14] Replay attacks occur when an individual's information is captured while being sent along a network and used at a later time by a hacker, e.g., to fraudulently gain access.

[15] S. Venkatesh, R. Ramachandra, K. Raja, C. Busch, *Face Morphing Attack Generation and Detection: A Comprehensive Survey,* 2020, p. 1, available at : https://arxiv.org/pdf/2011.02045.pdf.

the identity or travel document, or wrongly rejecting them as being in a list. Facial image morphing attacks further merit special consideration as multiple individuals could rely on the resulting morphed image and be successfully verified with probe images from each of the contributing subjects, for instance, when passing an identity control. In other words, the morphed image can be used by a variety of fraudsters (as many as probe image contributors) to subvert the biometric recognition system.[16]

In order to have a common understanding of presentation and manipulation attacks and fraud detection by specific MAD technology within the European and national infrastructure, and to distinguish it from other *bona fide* manipulations, we hold that also a legal definition of unlawful misleading manipulation attacks and MAD should be proposed and agreed upon. On a regular basis, *technical definitions* concerning biometrics are formulated, *inter alia*, in the standards issued by the International Standardization Organization (ISO) and the International Electrotechnical Commission (IEC).[17] In this context, ISO/IEC has issued the International Standard ISO/IEC 30107-1:2016 focusing on techniques for the automated detection of so-called biometric Presentation Attacks (PA) undertaken by biometric capture subjects at enrolment and presentation of the relevant biometric characteristics. The term 'Presentation Attack' is defined as a 'presentation to the biometric data capture subsystem with the goal of interfering with the operation of the biometric system',[18] while "Presentation Attack Detection" (PAD) is defined as 'the automated determination of a presentation attack'.[19] For specific image manipulations and the detection thereof, and especially for MAD, definitions are still missing.[20] Definitions are, in our view, however important, both technically but also for legal purposes; the latter, especially, with the aim of detecting *unauthorized and/or fraudulent* manipulations, as not all manipulations may pose a problem or be with a fraudulent intent. Manipulation attacks may also be inspired or even be based on daily habits, such as putting

---

[16] As a result, morphing attacks disrupt *per se* the rule of single ownership, as the unique link of "one-person, one-passport" for the protection of minors and the material of the passport or travel document issued in a machine-readable format is challenged. See European Commission, Commission Decision C(2005) 409 of 28 February 2005 establishing the technical specifications on the standards for security features and biometrics in passports and travel documents issued by Member States Search, available at https://ec.europa.eu/home-affairs/sites/homeaffairs/files/e-library/documents/policies/borders-and-visas/document-security/docs/comm_decision_c_2005_409_fr.pdf.

[17] These international organizations encompass national bodies that participate in the development of international standards through technical committees established by the respective organizations to deal with particular fields of technical activity, e.g., in relation to the quality of biometric data or for harmonized vocabulary on biometrics. For an overview of relevant standardization for a, see also Revelin, D5.7: *Standards for security evaluation under spoofing attacks,* Tabula Rasa, 2014, available at http://www.tabularasa-euproject.org/project/pdf/d57.pdf.

[18] See Sect. 3.6 of ISO/IEC 30107-1:2016.

[19] See Sect. 3.5 of ISO/IEC 30107-1:2016.

[20] There is however a reference to manipulation detection under the general standard ISO/IEC 2382:2015(en) on information technology. Here, Section 2126392 offers a definition of manipulation detection or modification detection as "procedure that is used to detect whether data have been modified, either accidentally or intentionally" without contextualizing such definition within the biometrics realm.

make-up, or on legitimate interests of individuals to protect their facial image, for example, to protect their privacy.[21] Legal definitions could hence be useful to clarify and categorize in which cases specific manipulation should be considered (un)lawful and therefore detected in biometric and other AI systems.

In the iMARS project, the consortium partners are currently developing technical definitions for MAD. The proposed definitions are expected to be submitted to the standardization subcommittee ISO/IEC SC37 JTC1, which focuses on the standardization of generic biometric technologies.[22] Once such technical definitions are developed and approved, it shall hence be determined to what extent such definitions are relevant and can be transposed to a legal and regulatory context. This is all the more important as standards will play an increasing role in AI systems' regulation.[23] One shall overall carefully assess how and to what extent technical terms are relevant and are fit for or should be transposed in the legal framework.

20.3 Privacy and Data Protection Aspects of MAD

The section below contains a succinct analysis of some important legal principles from a data protection law point of view. It will be zooming in to the tripartite principle of the GDPR of lawfulness, fairness, and transparency, and the principle of purpose limitation. It will also briefly describe the need for human oversight.

The aforementioned tripartite principle requires MAD solutions to be developed and used in a *lawful, fair,* and *transparent* way, thus calling for the respect of an ample bundle of checks and balances between the rights and interests of data subjects and data controllers, regardless of the use case. The tripartite principle should be understood in practice as an equilateral triangle where each one of the sides represents the necessary conditions to be respected for the sustainment of a balanced relationship between controllers and data subjects. By these means, whereas fairness (axiological element) is the principle to which lawfulness (substantive element) and transparency (procedural element) are projected, lawfulness represents the substantive legal conditions that should be respected for the processing to happen, and transparency, the procedural circumstances enabling the rebalance of power asym-

[21] E.g., or if they would face illegitimate facial recognition. Some individuals use for this purpose sometime geometric lines on their faces, for example, during demonstrations.

[22] Even if standards are not legally binding as long as they are not incorporated in legal texts as an obligation, they can provide useful insights and even legal relevance in court proceedings to settle legal disputes, e.g., in product liability cases, where manufacturers need to exercise and demonstrate conformity and even care and due diligence where it should be determined whether a product is "fault free". It should be noted, however, that standards will become of utmost importance further to the AI Act proposal (see below).

[23] In the Proposed AIA 2021, compliance by high-risk AI systems with harmonized standards (or parts thereof) if available and sufficient shall result in the presumption of compliance and conformity with the obligations of such AI systems under Chap. 2 of the Proposed AIA 2021 such as maintaining a risk management system, data governance, record-keeping, and transparency, to the extent covered by such standards (Art. 40 Proposed AI Act 2021).

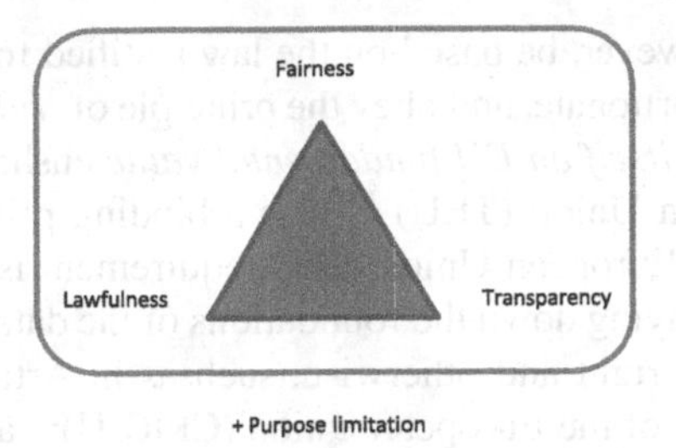

Fig. 20.1 The tripartite and purpose limitation principles—*Source* compiled by the authors

metry between controllers and data subjects for a fair processing. In addition to the tripartite principle, the principle of purpose limitation will also be discussed as this principle provides additional guarantees which underpin the baseline established by the tripartite principle. If the conditions set out by the tripartite principle establish the *minimum* requirements for the processing of personal data, the purpose limitation principle ensures that those minimum exigencies of lawful, fair, and transparent processing remain stable during the whole life cycle of the processing, thus providing a teleological safeguard which encloses, captures, and protects the legal relationship over time (Fig. 20.1).

First, the principle of lawfulness will be analyzed in relation to the rule of law and the legality assessment of MAD; such analysis remains relevant for any scenario. Then, we make some observations as to the relationship between the principle of fairness from a data protection law and technical perspective. Here, also some references to the recently proposed AI Act proposal will be made. Furthermore, we discuss the principle of transparency in relation to the obligations and rights that MAD controllers (and processors under the supervision of controllers) should respect for their personal biometric data processing to be *transparent*. Finally, the principle of purpose limitation will be also analyzed in relation to the deployment of MAD tools.

20.3.1 The 'Rule of Law' and Legality

20.3.1.1 'Rule of Law'

Despite enhancing the integrity and security of the Schengen area by preventing, *inter alia*, document fraud, MAD technologies may also affect the fundamental rights of individuals, such as the rights to dignity, freedom of movement or non-discrimination, and other protected values, such as autonomy or agency. While fundamental rights and freedoms are the highest priority in the hierarchy of our legal systems, to which all secondary laws shall conform, the fundamental rights and freedoms as proclaimed are mostly not absolute and interferences are possible, also in democratic societies. Such interferences shall, however, be based on the law justified for legitimate objectives, be necessary and proportionate, and obey the principle of *'rule of law'*. The principle of the rule of law *is in itself an EU fundamental value* enshrined in Article 2 of the Treaty of the European Union (TEU).[24] It is a binding principle applicable to all Member States of the European Union. This requirement is also mirrored in other laws, including those laying down the foundations of the data protection framework, both as a fundamental right and otherwise, such as in Article 8(2) of the Charter of Fundamental Rights of the European Union (CFREU)[25] and in Article 5.1 of the GDPR.[26]

As jurisdictions have *varying approaches and legal systems*, a common understanding of what the 'rule of law' implies is important.[27] In general, it refers to the relationship between governments and individuals and applies also to private actors in their relationship with individuals, as requiring *legality* (see legality requirement in the section below) and, in addition, some *democratic decision-making processes reflected in laws and procedures* as to if, when, and what for interferences by such governments and its officials and private actors could be acceptable. The regulatory objective is thus *to avoid arbitrary decisions or interventions*. Further, laws shall meet some *quality requirements*, such as the need to be publicly promulgated. Thirdly, the rule of law—which is debated in many domains recently as to understand its precise scope—is also *to be upheld by an independent and unbiased judiciary*, i.e., an independent judiciary in national courts and the Court of Justice of the European Union (CJEU).[28]

As a consequence, the rule of law requires the deployment and use of MAD tools to have a basis in laws and, foremost, in a legal system which respects this 'rule of law' principle as explained—in addition to fundamental rights and freedoms. For MAD tools, the law basis may stem from, at the European level, Article 77(1)(b) of the Treaty of Functioning of the European Union (TFEU) in relation to the Union's policy to carry out checks on persons and efficient monitoring of external borders while ensuring the free movement of persons, as further implemented by secondary law, provided that such law shall be sufficiently precise, transparent, accessible, and meet all other quality requirements (see the legality requirement in the section below). Hence, a specific legal basis will be needed as well as a 'rule of law' legal

---

[24] Article 2 TEU stipulates that *"[t]he Union is founded on (...) the rule of law and respect for human rights"*.

[25] Article 8(2) CFREU states that *"...data must be processed fairly for specified purposes and on the basis of the consent of the person concerned or some other legitimate basis laid down by law"*.

[26] Article 5(1)(a) GDPR requires personal data to be *"processed lawfully (...) in relation to the data subject"*.

[27] See also, e.g., R. Mańko, *Protecting the rule of law in the EU,* European Parliamentary Research Service, November 2019, 12 p. (summary), at https://www.europarl.europa.eu/RegData/etudes/BRIE/2019/642280/EPRS_BRI(2019)642280_EN.pdf.

[28] For some recent decisions in relation to the rule of law and the principle of an independent judiciary as a key part of the rule of law, see e.g., ECJ, *European Commission v Republic of Poland,* C-791/19, 15.07.2021, ECLI:EU:C:2021:596; see also E. Kindt, *Privacy and Data Protection Issues of Biometric Applications*, Springer, 2013, Sect. 309, pp. 459–462.

system, with due process and an independent judiciary, and such may also need to be established at the national level if necessary for the areas in which the deployment and use of the MAD tools would not fall under the Union's competences to regulate border management, asylum, and immigration.

20.3.1.2 Legality

The notion of legality as a principle under which the law shall govern any event of interference with fundamental rights and provide for the conditions of interference is part of the *broader notion of the 'rule of law'* as previously explained.[29] Under the general data protection legislation in the European Union, the principle of legality has been transposed as the requirement of 'lawfulness' in the GDPR, which, in addition and more precisely, also requires that any processing of personal data must have a *legal basis* or *legal ground* in order to be "lawful". Legal bases include not only the (explicit) consent of the data subject but also the necessity for a (balanced) legitimate interest of the controller, or for tasks in the public interest, as well as other legal bases described by the GDPR, and will vary depending on the use case.[30] A legal basis may also be found in the provisions laid down in other Union or Member State law.[31]

This is also applicable for using MAD tools and the resulting processing of personal data. In border control situations, the legality requirement is of utmost importance. Border and identity checks in combination with MAD tools may result in suspicion, investigation of criminal offenses, and/or a refusal to enter a given territory, and other (restrictive) measures which may have a significant effect for the individual concerned. As fundamental rights and freedoms are often at stake in these situations as well, any interference therewith shall be based on 'a law', as further defined and described in international court cases, including the European Court of Human Rights and the CJEU, and this in addition to the requirement of a 'legal basis' as put forward by the European data protection framework. This means, practically speaking, that if, for example, facial images would be checked for reasons of security or due to suspicion at Schengen border checks, Schengen-related regulations, such as the Schengen Border Code (see Regulation (EU) 2016/399), shall explicitly refer to and allow such by using MAD tools. The risk-based approach of the GDPR to data protection further imposes *higher standards of protection* where personal data which by their nature are particularly "sensitive" in relation to fundamental rights

[29] See also Article 8(2) of the Charter of Fundamental Rights of the European Union.

[30] Other legal bases, which may allow the processing of personal data, can not only be the necessity for the performance of a contract but also the necessity for compliance with a legal obligation and necessity for the exercise of official authority: see Article 6.1 of the GDPR. In the case of processing special categories of data (sensitive data), which includes biometric data processing for uniquely identifying (which could be argued if MAD tools are used), additional conditions, such as the necessity and reasons of substantial public interest and law, all as set out in Art. 9 GDPR, are applicable, and this to be exempted from the general prohibition of processing of personal 'sensitive' data.

[31] Recital 40 GDPR.

and freedoms are processed.[32] This implies that, in addition to a legal ground, other conditions will be necessary for the processing "sensitive data", including for "biometric data" if processed for uniquely identifying purposes.[33] Biometric data used for identifying purposes changes irrevocably the relation between body and identity in the sense that it makes the characteristics of the human body and its identity 'machine-readable'. One shall take into account that also in convenience scenarios, for instance, when MAD tools are used to allow or confirm the identity of an individual at airline check-in kiosks, as biometric data based on bodily characteristics deserves special protection.[34] As a result, the legislator imposes by way of the GDPR a general prohibition of the processing of biometric data for the purposes of uniquely identifying individuals,[35] unless one of the exceptions applies.[36]

Therefore, the principles of legality and lawfulness mandate the deployment of algorithmic solutions, such as the MAD tools dealt with in this chapter, to rely on one or more relevant legal bases in addition to the need for a law if and when interfering with fundamental rights, for example, if there is a risk of interference with the right to personal data protection.

20.3.1.3 Legality of MAD in Border Checks

As explained, the use of MAD tools within the border control infrastructure of the Member States needs to rely on law and a relevant legal basis if used for border check and security purposes. To our knowledge, there is no legislation yet within the *Schengen acquis* for the implementation of MAD tools, leave alone any performance standards or other requirements for such tools. Presently, the Schengen Border Code (SBC) for border checks states that the minimum requirements *necessary for the establishment of the identities* of EU citizens and their family members should be based on the *presentation of their travel documents* and this by a *minimum check*, which shall consist of a *"rapid and straightforward verification of the validity of the travel document for crossing the border, and of the presence of signs of falsification or counterfeiting, where appropriate by using 'technical devices'* (emphasis added)",

[32] Recital 51 GDPR.

[33] Article 4(14) GDPR. According to this definition, biometric data is: *"personal data resulting from a specific technical processing relating to the physical, physiological or behavioral characteristics of a natural person, which allow or confirm the unique identification of that natural person, such as facial images or dactyloscopic data"*.

[34] This is further justified by the fact that, in addition to the highly sensitive nature of biometric data, the possibilities of inter-linkage and the state of play of technology may produce unexpected and/or undesirable results for the data subject: see European Data Protection Supervisor (EDPS), *Opinion on a notification for prior checking received from the Data Protection Officer of the European Central Bank related to the extension of a pre-existing access control system by an iris scan technology for high-secure business areas*, 2008 (Case 2007-501), p. 3, available at: https://edps.europa.eu/sites/edp/files/publication/08-02-14_ecb_iris_scan_en.pdf.

[35] See this formulation in Art. 9(1) GDPR.

[36] There are ten different exceptions for the processing of biometric data in Art. 9(2) GDPR, which may be maintained or further developed by Member State law. See Art. 9(2) and Art. 9(4) GDPR.

which should lead to the confirmation of, inter alia, the authenticity and the validity of the travel document authorizing the legitimate holder to cross the border. In cases of doubt or possible threats, Union and national databases may be consulted, as well as biometric verification.[37] This text does not clearly refer to the use of MAD tools and technology based on AI algorithms for detecting image manipulation. At the same time, more stringent checking requirements are applicable for the identity checks of non-EU nationals, also referred to as Third Country Nationals (TCN), who shall undergo 'thorough checks' and 'a detailed examination covering (...) verification of the identity (...) and the authenticity and validity of the travel document', including by checking relevant databases and biometric verification (see art. 8.3 SBC). While this provision addresses identity verification more comprehensively, it could also be seen as vague in its acceptance as a legitimate legal basis for the use of MAD tools. To this extent, one could also note that information about the purposes and the procedures for the check shall be given to the travelers (see art. 8.5 SBC).

The integration of MAD algorithms within the SBC procedures of *minimum* and *thorough checks* raises in our view issues in relation to the legality principle. The aforementioned SBC articles are general and do not provide in our view an *explicit* legal provision for the deployment of MAD technologies. Hence, the legality of the processing of personal data with the deployment of MAD tools in these situations, if used, could raise concerns because there is not a sufficient precise law regulating the interference with fundamental rights. Indeed, the use of MAD tools may have consequences for the data subjects involved, such as raised suspicion, and because of possibly the recording of (additional) information in the context of various large EU IT systems for border management, including for their interoperability and for the biometric matching system.[38] In light of this, further research should be done on the interaction of the legality principle and the use of MAD tools in relation to the purposes of the processing as well as to the present EU information systems (e.g., EES and ETIAS) and databases for border management. It should be further noted that an internal border police code or guidelines which are easily *accessible* or *foreseeable* would not fulfill the requirement. This "quality" aspect of the law was also at the core of the discussion in the court case about the trial use of facial

[37] See Article 8.2 of the Schengen Border Code (SBC) (Regulation (EU) 2016/399, as updated). In addition to this, the relevant European and national databases are consulted to extract information relating to stolen, misappropriated, lost, and invalidated documents. See Article 8.2 SBC.

[38] We hereby point to the different European and national systems and interoperability infrastructure used for a variety of different purposes. Regulation (EU) 2019/818 of the European Parliament and of the Council of 20 May 2019 on establishing a framework for interoperability between EU information systems in the field of police and judicial cooperation, asylum, and migration and amending Regulations (EU) 2018/1726, (EU) 2018/1862 ,and (EU) 2019/816, available at: https://eur-lex.europa.eu/legal-content/EN/TXT/PDF/?uri=CELEX:32019R0818&from=en; Regulation (EU) 2019/817 of the European Parliament and of the Council of 20 May 2019 on establishing a framework for interoperability between EU information systems in the field of borders and visas and amending Regulations (EC) No 767/2008, (EU) 2016/399, (EU) 2017/2226, (EU) 2018/1240, (EU) 2018/1726, and (EU) 2018/1861 of the European Parliament and of the Council and Council Decisions 2004/512/EC and 2008/633/JHA, available at: https://eur-lex.europa.eu/legal-content/EN/TXT/PDF/?uri=CELEX:32019R0817&from=EN.

recognition software on members of the public by the South Wales Police.[39]

One shall further note that border checks will not only be governed by the aforementioned specific Schengen border legislation and large-scale IT systems regulations as specific legislation but may also be completed with the general provisions of the GDPR or the equivalent Regulation for EU Institutions (EUI),[40] and by the provisions of the Law Enforcement Directive (hereinafter, the "*LED*")[41] as implemented. This means that, where border control activities, particularly those relating to the exchange of information, are not specifically regulated, they will be referred to the GDPR or the GDPR for EUI,[42] unless they are covered by regime of the LED. In relation to the latter regime, the LED allows for the processing and for the free flow of personal data between the competent authorities while ensuring a high level of protection of personal data in the fields of judicial cooperation in criminal matters and police cooperation. In these cases as well, a legal basis is required for the lawfulness. This is particularly relevant for the processing of biometric data for the purpose of uniquely identifying a natural person by the border police, in which case the processing will be only allowed where strictly necessary,[43] subject to appropriate safeguards for the rights and freedoms of the data subject,[44] and only: (a) where

[39] See High Court of Justice (Divisional Court of Cardiff), 4.9.2019, EWCH 2341 (Admin) ('Bridges case'), available at https://www.judiciary.uk/wp-content/uploads/2019/09/bridges-swp-judgment-Final03-09-19-1.pdf, as appealed.

[40] European Parliament and the Council, Regulation (EU) 2018/1725 of the European Parliament and of the Council of 23 October 2018 on the protection of natural persons with regard to the processing of personal data by the Union institutions, bodies, offices, and agencies, and on the free movement of such data, and repealing Regulation (EC) No 45/2001 and Decision No 1247/2002/EC (Text with EEA relevance.).

[41] Directive (EU) 2016/680 of the European Parliament and of the Council of 27 April 2016 on the protection of natural persons with regard to the processing of personal data by competent authorities for the purposes of the prevention, investigation, detection, or prosecution of criminal offenses or the execution of criminal penalties, and on the free movement of such data, and repealing Council Framework Decision 2008/977/JHA.

[42] In the case of EU institutions, bodies, offices, and agencies carrying out the deployment of MAD tools for activities which fall within the scope of Chaps. 4 or 5 of Title V of Part Three of the TFEU to meet its objectives and tasks, the GDPR for EUIs should cumulatively apply. The scheme of both regulations is understood as equivalent according to the case law of the Court of Justice of the European Union. See Recital 5 of the GDPR for EUIs: *"[w]henever the provisions of this Regulation [GDPR for EUIs] follow the same principles as the provisions of Regulation (EU) 2016/679 [GDPR], those two sets of provisions should, under the case law of the Court of Justice of the European Union, be interpreted homogeneously, in particular because the scheme of this Regulation should be understood as equivalent to the scheme of Regulation (EU) 2016/679."*.

[43] According to Art. 8(1) LED, the necessity for processing biometric data for uniquely identifying purposes must be based on the performance of a task carried out in the public interest by competent authority based on Union or Member State law for the purposes of the prevention, investigation, detection, or prosecution of criminal offenses or the execution of criminal penalties, including the safeguarding against and the prevention of threats to public security.

[44] (Joint) controllers and processors processing biometric data for the purpose of unique identification shall comply with the obligations and security measures provided in Chap. IV of the LED.

authorized by Union or Member State law[45]; (b) to protect the vital interests of the data subject or of another natural person; *or* (c) where such processing relates to data which are manifestly made public by the data subject.[46] To this extent, consent is not to provide a legal ground for processing personal competent authorities, and the processing of personal data by MAD tools could only be possible if a public interest stemming from the task carried out by the competent authority or exercise of official authority based on Union or Member State law would lay down the legal basis for such processing to take place.[47] Still, even if the lawfulness would be fulfilled under the LED, GDPR, and/or EUI, the issue remains whether such authorities could use the MAD algorithm if the Schengen border Regulations or Code or any other applicable law would not be explicit about it (legality—see above).

20.3.1.4 Legality of MAD in Access and Convenience Applications

MAD technology is fit to be used by a wide variety of stakeholders, e.g., also airlines to allow seamless check-in procedures or boarding. If used by any private authority or controller, for example, for 'to know your customer checks', this shall be used in a *lawful and fair* manner as well.

Legality and lawfulness refer to the need—in accordance with the rule of law principle—to adopt adequate, accessible, and foreseeable laws with sufficient precision and sufficient safeguards whenever the MAD technology is used for identity checks for convenience purposes, e.g., for face recognition or any other biometric comparison, if such purposes could interfere with fundamental rights and freedoms. Consequently, the unproportionate and unnecessary use of MAD tools for convenience purposes may fail to fulfill the *rule of law* if such use is not covered by appropriate legislation. Furthermore, explicit consent or substantial public interest as the necessity requirement for lawfulness in addition to coverage by national or EU law for the use of MAD tools shall be considered.

In this realm, a more exhaustive analysis should be made depending on the specific scenario in which the MAD solutions could be employed. In such an analysis, it would remain important to determine which entities involved could be responsible as well as the purposes and means of the processing operations.

20.3.2 Fairness

Fairness is a principle which is at the core of many legal systems, including in the domain of data protection and also in the more recently proposed AI Act. In data

[45] To this extent, Member State law must specify at least the objectives of processing, the personal data to be processed, and the purposes of the processing, as required by Article 8(2) LED.

[46] Art. 10 LED.

[47] Rec. 35 LED.

protection, to understand fairness as a principle, fairness has to be juxtaposed with both the principles of lawfulness and transparency, as included in the GDPR (see also above). In relation to lawfulness, one must note that the lawfulness principle materializes the axiological character of fairness by requiring to establish legal basis, which are considered to be compliant with the fairness mandates in abstracto by the European legislator, prior to the processing of data.[48] In relation to the interdependecy between fairness and transparency, one should note that, by imposing certain information obligations to data (joint) controllers as well as by recognizing data subjects rights, transparency operationalizes fairness by creating the necessary conditions to dispel any possible obscurity in the processing of personal data. The interrelationships between fairness, lawfulness, and transparency are thus critical to conceptualize fairness both as a whole and as a part of the governing principles for the processing of personal data as well as to help modeling its modern doctrinal interpretation in the data protection law.[49] We explain this further below.

20.3.2.1 Explicit and Implicit Fairness

According to legal doctrine,[50] two dimensions of fairness could be distinguished: explicit and implicit fairness. *Explicit fairness* relates to a fair and transparent processing of personal data in terms of explicit *information requirements* for the data subjects.[51] These information requirements could result in explicit fairness being directly attributable to the consecution of a fair processing in relation to transparency, as it allows data subjects to make decisions relating to their personal data.[52] *Implicit fairness*, on the other side, points to a 'fair balance' regarding competing (fundamental) rights and interests. This is translated into the establishment of rights and obligations to compensate the asymmetric relationship between data subjects and controllers (and processors). While explicit fairness aims at creating fair and transparent procedures, implicit fairness aims at rebalancing the interests and reasonable

---

[48] See Recital 39 GDPR, which points out the intimate relationship of both principles by stating that *"[a]ny processing of personal data should be lawful and fair"*. About fairness and biometric data processing, see also E. Kindt, *Privacy and Data Protection Issues of Biometric Applications*, Springer, 2013, Sects. 256–278, pp. 426–444. About fairness in AI, see e.g., S. Wachter e.a., 'Why fairness cannot be automated: Bridging the gap between EU non-discrimination law and AI', C.L.S.R., 2021, 105567.

[49] See D. Clifford and J. Ausloos, *Data Protection and the Role of Fairness*, KU Leuven Centre for IT and IP Law, CiTiP Working Paper 29/2017, p. 9 (Clifford and Ausloos, Role of Fairness, 2017'). See also Recital 30 of the Proposal for a regulation of the European Parliament and of the Council on the protection of individuals with regard to the processing of personal data and on the free movement of such data (General Data Protection Regulation), COM/2012/011 final - 2012/0011 (COD), available at: https://eur-lex.europa.eu/legal-content/EN/TXT/PDF/?uri=CELEX:52012PC0011&from=EN. This Recital 30 later disappeared during the negotiations.

[50] Clifford and Ausloos, Role of Fairness, 2017, p. 11.

[51] For instance, the requirements for transparent information, communication, and modalities for allowing the exercise of the rights of the data subject as specified in Article 12 GDPR *et seq.*

[52] Clifford and Ausloos, Role of Fairness, 2017, p.37.

expectations of the data processing actors and the data subject. Moreover, one shall take into account that any conclusions as about what is fair may evolve over time.[53]

The incidence of the distinctive elements of fairness when deploying MAD tools would require that the controller communicates the data subject the information required under the transparency principle (see below).[54] As for implicit fairness, the fairness principle equally requires controllers of MAD tools to establish the conditions for lawful and fair processing including informing the data subject about the existence of their rights and to allow their exercise.[55]

Based on the two dimensions of fairness, the legal basis (legality) allowing the processing of personal data for MAD purposes in border control contexts should provide the required information as to comply with the fairness and transparency requirements and rights of data subjects. The same would apply if MAD is used for access control for security or convenience purposes, for example, in privacy and data protection policies.

20.3.2.2 Definitional Concerns of the Fairness Concept

Another important element of the discussion on fairness is the problem surrounding its conceptualization. It seems that there is a disparity in the understanding of fairness among the disciplines of knowledge.[56] Under data protection law, it relates, generally, to the achievement of a *balanced relationship between data subjects and controllers*.[57] Such interpretation significantly differs from the technical understanding of fairness, where fairness is rather understood as the absence of any prejudice

[53] L. A. Bygrave 'Core principles of data protection' in *Privacy Law and Policy Reporter* 7(9), 169, 2001 p. 11, available at http://www.austlii.edu.au/au/journals/PrivLawPRpr/2001/9.html.

[54] It requires, inter alia, information about the controller's identity and contact details as well as its DPO, where applicable; the purposes of the processing and the categories of personal data; the storage limitation period; the existence of automated decision-making; the recipients, if any; and any potential data transfer and its safeguards, where applicable. Art. 14(1) GDPR. In security contexts, this implies that individuals must be informed before being subjected to any automated biometric processing.

[55] This includes the rights to access, rectification, erasure, restriction, object, data portability, withdrawal of consent, and the right to lodge a complaint with a supervisory authority. A mention of the existence of such rights has to be equally done by MAD controllers in security contexts.

[56] See also D. Mulligan, J. Kroll, N. Kohli, and R. Wong, *'This Thing Called Fairness: Disciplinary Confusion Realizing a Value in Technology'*, in Proc. ACM Hum.-Comput. Interact. 3, CSCW, 36 p., available at https://papers.ssrn.com/sol3/papers.cfm?abstract_id=3459247.

[57] It implies a transparent data processing and a counterbalance of controllers and data subject rights in light of the power asymmetry arising from their interaction.

or favoritism towards an individual or a group based on their intrinsic or acquired traits in the context of decision-making.[58]

In this respect, one could argue that the notion of fairness in data protection law requires a *fair* system of checks and balances of rights and obligations rather than the consecution of *fair* algorithmic decisions. The first would ensure that personal data is processed *fairly*. The second would provide for *fair* algorithms. Different approaches to fairness might thus be at stake. While data protection law emphasizes more the *procedural* aspect of fairness in the way that it seeks to achieve a balanced legal relationship, computer science rather focuses on the *material* aspect of fairness, as it seeks to achieve fair algorithmic decisions which do not alter the balanced equilibrium. Data protection law does, in our view, not provide sufficient mechanisms to regulate the *material* component of fairness as it does not directly regulate, for example, the conditions for the achievement of unbiased or non-discriminatory algorithmic decision making.[59] Indeed, no specific rules are laid down in the GDPR to ensure algorithmic fairness from a technical point of view. At most, algorithmic fairness is regarded from the lenses of security and transparency, namely as the objective of appropriate measures to avoid the detrimental effects that the processing of personal data could have on the data subject in the case of undesirable events or accidents. Also, there is no general reference to key terms traditionally associated in the literature of algorithmic fairness, such as gender or demographic balance, as potential factors for algorithmic fairness.

[58] N. A. Saxena, K. Huang, E. DeFilippis, G. Radanovic, D. C. Parkes, and Y. Liu.2019, 'How Do Fairness Definitions Fare? Examining Public Attitudes Towards Algorithmic Definitions of Fairness' in *Proceedings of the 2019 AAAI/ACM Conference on AI, Ethics, and Society*. ACM, 99–106.

[59] In the GDPR, only few references to discrimination are made, while none of the references have been included as provisions in the GDPR. They have only been acknowledged in the recitals of the GDPR, which are not legally binding: Recital 71 GDPR refers to discrimination in the context of the security of personal data in relation to potential risks and on the basis of, inter alia, racial or ethnic origin. It seems from this that algorithmic fairness is contemplated more from the principle of integrity and confidentiality (security), i.e., for the protection against unauthorized or unlawful processing and against accidental loss, destruction, or damage, using appropriate technical or organizational measures. Recital 75 GDPR acknowledges a long list of risks to the rights and freedoms of natural persons, of varying likelihood and severity, which could lead to physical, material or non-material damage, including discrimination. However, no specific legal response is provided in the recital as to the ways in which algorithms should be developed to minimize discrimination. Finally, Recital 85 GDPR recognizes risks resulting in discrimination in relation to personal data breaches which are not addressed in an appropriate and timely manner, urging controllers to notify such breaches to the supervisory authority without undue delay. Discrimination is hence here thus treated from a transparency perspective, requiring controllers to promptly notify the supervisory authorities to limit the undesirable consequences of data breaches.

20.3.2.3 Could the Proposed AI Act Help?

The lack of precise pronouncement on fairness entails important consequences. Since only a few provisions address the *procedural* element of fairness,[60] and no reference is being made from a *material* point of view, it could be argued that the GDPR is not sufficient in itself for the regulation of fairness of MAD algorithms and tools. To this extent, other legal instruments focusing on the regulation of processes aimed at ensuring algorithmic fairness could therefore be needed.

Although there is no current legally binding legislation on algorithmic fairness in Europe, the European Commission issued in April 2021 a proposal for a regulation for an Artificial Intelligence Act (AIA).[61] Such piece of legislation is intended to regulate algorithms—and therefore also MAD tools—from a *hard law* perspective. Until its entry into force, regulatory approaches to algorithmic fairness are mainly to be sought in *soft law* instruments regulating algorithms from an ethical standpoint.[62]

Since the automated decisions resulting from the MAD algorithm could significantly impact the rights and freedoms of individuals, e.g., the right to freedom of movement, we argue there is an urgent need in legally assessing the fairness models under which these kinds of automated system are to be built on.[63] In this respect, as the development of MAD algorithms depends to a great extent on the data injected into the models, legal requirements should be imposed to safeguard the inclusion of data in algorithm's prediction models. Although no such requirements currently exist, the fore-mentioned AIA proposal contains provisions to ensure the use of high

[60] In particular, the communication to the data subject of the storage period of the processing, the existence of the data subject rights, the nature of the legal basis and the data subject obligations, and the existence of automated of automated decision-making with meaningful information about the logic involved, as well as the significance and the envisaged consequences of such processing for the data subject.

[61] EU Commission, Proposal for a Regulation of the European Parliament and the Council laying down harmonized rules on artificial intelligence (Artificial Intelligence Act) and amending certain union legislative acts, COM/2021/206 final, 2021, available at https://eur-lex.europa.eu/legal-content/EN/TXT/PDF/?uri=CELEX:52021PC0206&from=EN (Proposed AIA 2021).

[62] For instance, the European Commission's High-Level Expert Group on Artificial Intelligence (AI HLEG) has issued guidelines in pursuance of regulating fair algorithms: the Ethics Guidelines for Trustworthy AI. See AI HLEG, *Ethics Guidelines for Trustworthy Artificial Intelligence*, 2018, revised and updated in April 2019, available at: https://ec.europa.eu/newsroom/dae/document.cfm?doc_id=60419. The guidelines are based on four principles—respect for human autonomy, prevention of harm, fairness, and explicability—the AI HLEG came to seven ethical requirements which should be complied in the development, deployment, and use algorithms, which would also be included in the iMAD tools and systems. The fifth requirement on diversity, non-discrimination, and fairness, including the avoidance of unfair bias is particularly relevant. Moreover, the Ethics Guidelines on Trustworthy AI were enclosed with an Assessment List for AI practitioners to allow AI practitioners to evaluate their AI systems' compliance with the Ethics Guidelines in a practical manner.

[63] As noted by Pessach and Shmueli, it is important to consider the inherent trade-off between accuracy and fairness of algorithms. This means that any legal solution to this problem should consider the maximization of an acceptable degree of fairness while preserving the accuracy of the algorithms. See D. Pessach and E. Shmueli. *Algorithmic Fairness*. 2020, available at ArXiv, abs/2001.09784.

qualitative training, testing, and validation data in AI systems, *"with a view to ensure that high-risk AI systems perform as intended and safely and they do not become the source of discrimination prohibited by Union law"*.[64]

If the law is to provide regulatory approaches for artificial intelligence, further analysis should be made on the available mechanisms and solutions to minimize algorithmic bias. For instance, possible solutions to algorithmic fairness could be prospected in the realm of supervised learning and classification, in which methods to ensure parity or equality in relation to algorithmic decisions have been developed,[65] e.g. by including an external actor for reducing or eliminating any potential bias in the algorithmic decision.[66] Whereas this supervised control of algorithmic decisions is a very characteristic of the developmental stages, this idea can be further expanded in operational MAD contexts by integrating human oversight processes. For instance, civil servants or police officers in charge of border control should ultimately have the final say in relation to automated decision-making processes (see also below). The proposed AIA seems also to endorse this viewpoint for the purpose of preventing or minimizing the risks of AI. By these means, it puts additional onus on certain high-risk AI systems,[67] for which no action or decision can be taken on the basis of automatic processing unless verified and confirmed by at least *two* natural persons.[68]

20.3.3 Transparency

20.3.3.1 Importance

Along with lawfulness and fairness, transparency equally accounts as a fundamental pillar for the regulation of MAD under data protection law. The principle of transparency, as a unique feature of data protection law, contributes to the creation of

[64] Rec. 44 and Art. 10 Proposed AIA 2021. The literature also identified a set of elements which may lead to algorithmic unfairness. For instance, biases included in the datasets used for facial recognition systems such as demographic biases; biases caused my missing data, such as missing values on soft biometrics which may result in biometric datasets which are not representative of the captured subject's population; biases stemming from algorithmic configurations aiming at minimizing recognition errors and therefore benefiting certain demographic or gender groups over others; and biases caused by "proxy" attributes for sensitive attributes. See N. Mehrabi, F. Morstatter, N. Saxena, K. Lerman and A. Galstyan (2019). *A Survey on Bias and Fairness in Machine Learning*, available at ArXiv, abs/1908.09635.

[65] Y.R. Shrestha and Y. Yongjie. 2019. "Fairness in Algorithmic Decision-Making: Applications in Multi-Winner Voting, Machine Learning, and Recommender Systems" Algorithms 12, no. 9: 199. https://doi.org/10.3390/a12090199. See also C. Breger, *Criteria for algorithmic fairness metric selection under different supervised classification scenarios*, Master's thesis on Intelligent Interactive Systems Universitat Pompeu Fabra, 2020 available at: https://repositori.upf.edu/bitstream/handle/10230/46359/TFM_breger_2021.pdf?sequence=1&isAllowed=y.

[66] This approach seems to be in line with the wording of Art. 10 Proposed AIA 2021.

[67] High-risk AI systems referred to in point 1(a) of Annex III Proposed AIA 2021.

[68] Art. 14(5) Proposed AIA 2021.

trust in the data processing from a twofold perspective: it allows data subjects to be informed, understand and decide, if possible, about the processing of their personal data as well as enables them to challenge such processing, should that be necessary.

As for the principle of transparency, MAD controllers will have to provide the data subject, in the first place, with general information about the processing,[69] promote and facilitate the exercise of their rights,[70] and fulfill obligations in relation to data breaches.[71] Such duties have to be done in a specific way that enhances transparency. This means that the relevant information has to be conveyed, *inter alia*, in a concise, transparent, intelligible, and easily accessible way, as well as in a clear and plain language, in writing or by other means, and generally free of charge.[72] Extensive guidelines on the meaning of each one of the previous conditions have been issued by the Article 29 Working Party, which have been endorsed by the European Data Protection Board.[73]

20.3.3.2 Relevance for MAD

As inferred from the above-mentioned Guidelines, controllers deploying MAD, *e.g.*, at border crossing points, shall choose methods to provide the information to the data subjects which are *appropriate to the particular circumstances.*[74] For instance, if MAD tools are deployed in e-gates or kiosks, information requirements relating to transparency could be provided in electronic format if such devices have digital screens. In addition to that, visible boards containing the information, public signage, public information campaigns, or newspaper/media notices would be also advisable.[75] If icons or symbols are used to advert data subjects, it should be assessed how they are incorporated to effectively convey the informational duties of controllers.[76] It is encouraged that one shall investigate as to whether *standardized symbols/images* that are universally used and recognized across the EU could be deployed as shorthand for that information.[77] This should come in addition to the description and specification of the use of MAD tools in relevant legislation, such as for border control in the Schengen Border Code, as well as in the Schengen

[69] Cf. Art. 12 in relation to Arts. 13 and 14 GDPR.

[70] Cf. Art. 12 in relation to Arts. 15–22 GDPR.

[71] Cf. Art. 12 in relation to Art. 34 GDPR.

[72] See Art. 12 GDPR.

[73] Article 29 Working Party, Guidelines on transparency under Regulation 2016/679, 2018, p. 4, available at: https://ec.europa.eu/newsroom/article29/items/622227 ('A29WP, Guidelines on transparency').

[74] A29WP, *Guidelines on transparency*, p. 12.

[75] A29WP, *Guidelines on transparency*, p. 22.

[76] Arts. 13 and 14 GDPR.

[77] A29WP, *Guidelines on transparency*, p. 26.

Handbook[78] and the EUROSUR Handbook,[79] which propose guidelines, best practices, and recommendations relating to the performance of border guard dutics in the Schengen States. For other controllers deploying MAD for access control and even convenience, the relevant information shall be provided as well, including in the data processing policies, which are *clearly visible*, e.g., on websites.

The notification to the data subjects of the existence of *automated decision-making* by MAD would in our view be required as well. As MAD entails the use of algorithmic solutions which could significantly affect the fundamental rights of individuals, particularly the rights to dignity, freedoms of movement, and data protection, as well as diverse risks where automatic identification and/or verification does not provide clear results, data subjects should be informed about the existence of MAD together with meaningful information about the logic involved and the significant and envisaged consequences of the processing.[80] This is more important as data subjects may have legitimate reasons to have an altered (facial) appearance, such as for purposes of beautification or even to protect privacy and freedoms. In this regard, data subjects should not be taken by surprise by the processing of their personal data. Rules and safeguards, including the prior data protection impact assessments (DPIAs),[81] should be developed and adopted before MAD tools become operational by any (co)controller. As set out in the WP29 Guidelines on DPIAs,[82] data controllers may consider the publication of the DPIA for the purposes of fostering trust and demonstrating transparency and accountability. Adherence to a code of conduct may also go towards the demonstration of transparency.[83]

Apart from the GDPR requirements regarding transparency, the checklist issued by AI HLEG, which accompanies the Ethics Guidelines for Trustworthy AI,[84] requires that transparency would have to be checked and verified by the MAD controllers.[85] According to the Ethics Guidelines, transparency touches several elements of AI systems, including the data used, the system, and even the business models.

[78] European Commission, Annex to the Commission Recommendation establishing a common "Practical Handbook for Border Guards" to be used by Member States' competent authorities when carrying out the border control of persons and replacing Commission Recommendation C(2006) 5186 of 6 November 2006, Brussels, 8.10.2019 C(2019) 7131 final, available at: https://ec.europa.eu/home-affairs/system/files_en?file=2019-10/c2019-7131-annex.pdf.

[79] European Commission, Commission Recommendation of 15.12.2015 adopting the Practical handbook for implementing and managing the European Border Surveillance System (EUROSUR Handbook), Strasbourg, 15.12.2015 C(2015) 9206 final, available at: https://reliefweb.int/sites/reliefweb.int/files/resources/eurosur_handbook_en.pdf.

[80] Art. 22 GDPR.

[81] Art. 35.3 GDPR.

[82] Article 29 Data Protection Working Group, *Guidelines on Data Protection Impact Assessment (DPIA) and determining whether processing is "likely to result in a high risk" for the purposes of Regulation 2016/679*, WP 248 rev.1.

[83] Art. 40 GDPR.

[84] AI HLEG, *Ethics Guidelines for Trustworthy AI,* 2018, revised and updated in April 2019, available at https://ec.europa.eu/futurium/en/ai-alliance-consultation/guidelines#Top.

[85] See AI HLEG, *The Assessment List for Trustworthy AI—ALTAI,* available at and accompanied by a web based tool available at https://altai.insight-centre.org/.

The concept of transparency further includes (i) traceability, (ii) explainability, and (iii) communication. *Traceability* encompasses that, *inter alia*, the MAD algorithms are *documented to the best possible standards.* The reason and objective for this are to be able to trace the decisions made and to be able to know and identify why a decision was wrong. *Explainability* requires that not only the human decisions but also the technical processes of MAD are comprehensible. Finally, *communication* should ensure that humans are informed of their interaction with MAD tools. Communication entails that MAD solutions are identifiable and that there is an option to decide based on human interaction for alternatives to MAD's decisions. Although the fore-mentioned Ethical Guidelines are not binding, and should be regarded as soft law, they contain in our view useful interpretations of the legal requirement of transparency, as contained in Article 5 GDPR.

20.3.4 Purpose Limitation and Legitimate Processing

The principle of purpose limitation is another core tenet we briefly discuss in this chapter because of its importance. The principle delimits the scope of the processing of personal data,[86] and essentially requires that personal data may only be processed for the initial purpose as defined at the collection of the data,[87] and so long as processing is not incompatible with the original purpose.[88]

The first component of the principle of purpose limitation requires the controller *to specify* the purpose of the data processing and communicates such. This serves as an additional safeguard which underpins the tripartite principle of lawfulness, fairness, and transparency, as it ensures the fair processing of data also requiring the elucidation of a legal basis and the adequate standards of transparency over the life cyle of the processing. While specifying a legal basis for the processing of data, the specification of purposes provides the data subject with the *necessary information* on the data processing and serves as a tool to discover any potential risks that may cause harm to the data subject. Such level of transparency, in turn, helps in re-balancing the natural information asymmetry of the relationship data subject-controller and, consequently, materializes the fair processing.

The principle of purpose limitation becomes more prevalent where *the balancing* of colliding fundamental rights and freedoms of the data subject vis-à-vis the interests of the data controllers is at stake, for example, at border checks. The aim is to protect the individual's autonomy against potential risks and/or harms caused by the processing of its personal data where data is re-purposed, for example, as to store any information obtained by using the MAD tools in large scale IT systems, including

[86] See Article 29 Data Protection Working Group, Opinion 03/0213 on purpose limitation, pp. 4 and 6, available at https://ec.europa.eu/justice/article-29/documentation/opinion-recommendation/files/2013/wp203_en.pdf.

[87] Art. 5(1)(b) GDPR.

[88] See also Art. 6(4) GDPR.

the Biometric Matching System (BMS). But such purpose limitation is also required when used for security or convenience. MAD controllers cannot exceed the expectations of the data subjects as regards the re-purpose of their biometric information. This implies for border control, for example, that controllers of MAD cannot in principle further process (re-purposing; secondary processing) the biometric information extracted from ID documents and passports and from the facial captures for other purposes different to the identity control and to detect manipulation.

The processing for other than the initial purposes is only possible under certain conditions. The additional processing for another purpose could be deemed as compatible with the original purpose if a 'compatibility test' between purposes is conducted and provides a positive result. To this extent, a bundle of criteria have to be considered to delimit the compatibility of purposes.[89] The application of such a test has to be conducted in a strict way in order to avoid breaking the principles of lawfulness, fairness, and transparency. Only where MAD tools are researched and developed or personal data is processed for scientific research purposes, a presumption of compatibility with initial purposes is explicitly provided by the GDPR.[90] In these cases, appropriate technical and organizational safeguards, such as anonymization, if possible, or pseudonymization and access limitations, are needed.[91] For other uses, the test needs to be done as well. As such, the presumption is not a general authorization to further process data in all cases, as each case must be considered on its own *merits and circumstances*. It is thus advised that, in case of repurposing data, controllers of MAD tools still conduct a compatibility test, even where the re-purposing of biometric data would be done for scientific research.[92]

Finally, it is obvious that the use of the MAD technology should be restricted to well-defined legitimate purposes for which the processing is necessary. For use in a border control context, such purposes could include specific substantial public interests provided by Union or Member State law, such as preventing illegitimate migration, or detecting identity or document fraud, or for the protection of rights of others. These objectives, even in combination with biometric data processing, are usually however not the major obstacle when possible interferences are assessed by the courts.[93]

[89] For instance, the relationship between the original purpose and the further processing, the context of collection, the nature of the data and the impact caused by the later use on the individual, as well as the safeguards applied in order to prevent any undue impact are some of the criteria used for this assessment. See A29WP. *Opinion 03/2013 on purpose limitation*, 2013, pp. 23–27.

[90] Art. 5(1)(b) GDPR.

[91] Art. 89(1) GDPR.

[92] European Data Protection Supervisor, *A Preliminary Opinion on data protection and scientific research*, 2020, p. 23, available at https://edps.europa.eu/sites/edp/files/publication/20-01-06_opinion_research_en.pdf.

[93] See, e.g., ECJ, *Schwartz v. Bochum*, 17.10.2013, **Sect. 37**, ECLI:EU:C:2013:670: the Court of Justice herein agreed with the EU legislator that for taking and storing fingerprints for the ePassport *preventing illegal entry is a legitimate objective* of general interest recognized by the Union; see also E. Kindt, *Privacy and Data Protection Issues of Biometric Applications*, Springer, 2013, Sect. 320.

20.3.5 Human Oversight and Intervention

From a data protection perspective, decisions purely and solely based on automated processing, producing adverse legal effects or significantly affecting persons, are prohibited, unless authorized by law, and subject to appropriate safeguards, including at least human oversight and intervention.[94] However, human oversight as well as the possibility of having a final decision made by a human operator are also addressed in other bodies of law, such as the SBC,[95] and in soft law instruments as the Ethics Guidelines for Trustworthy AI and the Assessment List for Trustworthy AI (ALTAI).[96]

The provisions are all aligned in scope. They fundamentally have as a result that MAD should both support the user's agency and uphold fundamental rights. ALTAI further specifies that human supervision and intervention shall be guaranteed at each decision cycle, such as of the MAD algorithm, during the system design cycle and system performance monitoring. This would hence have implications for the overall activity of the MAD system, and at any given situation.[97]

The performance of tasks of preventing, investigating, detecting, or prosecuting criminal offenses institutionally conferred to law enforcement and border control authorities do not preclude the right of the data subject to not to be subject to a decision evaluating personal aspects relating to him or her which is based solely on automated processing and which produces adverse legal effects concerning, or significantly affects, him or her.[98] In any case, the processing of personal data by MAD solutions should be subject to suitable safeguards, including the provision of specific information to the data subject and the right to obtain human intervention, in particular to express his or her point of view, to obtain an explanation of the decision reached after such assessment or to challenge the decision.[99] Such requirements play a fundamental role in view of the benefits and risks that MAD technologies may pose to society as a whole.

The use of MAD tools should hence always be accompanied by human oversight and intervention. The accuracy of algorithms is not 100%, as they are only providing degrees of probability that the images were manipulated. Further investigation including questioning by human intervention will therefore remain very important.

---

[94] Arts. 22 GDPR and 11 LED; Recs. 71 GDPR and 38 LED.

[95] Art. 7 SBC.

[96] See AI HLEG, *The Assessment List for Trustworthy AI—ALTAI,* 17.07.2020, available at https://altai.insight-centre.org/.

[97] ALTAI p. 9.

[98] Art. 11 LED.

[99] Rec. 38 LED.

20.4 Conclusion

The present chapter discussed, in the first place, the need for legal definitions on MAD given the current lack of pronouncement in the area from a technical and legal perspective. This is needed for the purposes of creating legal certainty and for further regulating the use of this technology.

The chapter further elaborated on the most important privacy and data protection requirements. As statutory law, data protection holds a unique social value in steering the processing of personal information in all situations, including when done for border control, access control to digital or physical environments, and convenience applications. Although different data protection regimes may be applicable depending on the use and purpose of MAD tools, we hold that the respect for the core tenets of the data protection framework should be always ensured, as they may not significantly vary from use case to use case. To this extent, the tripartite principle of lawfulness, fairness and transparency were discussed in relation to the purpose limitation principle. The first unifying principle is giving the *essential* conditions for the processing of data, the latter providing for *stable channels* to ensure the happening of these *essential* conditions.

Most importantly, MAD tools must be based on a law and have a *legal basis* from a fundamental rights and data protection perspective. In addition, respect for the axiological principles of the data protection framework should be sought in order to rebalance the asymmetry of power between data controllers and data subjects. This comprises the communication to the data subject of the relevant information, allowing also the reconciliation between the concepts of fairness from a legal and technical standpoint as well as from both a *procedural* and *material* perspective. The *qualitative aspect* of the provision of information is also a point of attention. Specific recommendations as to the fashion in which the use of MAD tools are communicated to the public, e.g., at border controls, have been mentioned. Finally, the chapter highlighted the importance of not exceeding the expectations of the data subjects as regards the re-purpose of their biometric information and the need of human oversight.

While the assessment and compliance of these conditions remain fundamental requirements to justify the interference that these kinds of biometric systems may pose to individuals' fundamental and data protection rights, other data protection principles will remain important as well and must also be taken into account. Finally, an appropriate assessment of this new technology under these frameworks remains important as well, at all times, and this to fully ensure respect of the *rule of law*.

Acknowledgements This work is supported and partly funded by the European Commission under the H2020 Project image Manipulation Attack Resolving Solutions (iMARS), Grant agreement ID: 883356 and the ETN TReSPAsS under the Marie Skłodowska–Curie grant agreement No. 860813, as well as by the Cybersecurity Initiative Flanders—Strategic Research Program (CIF).

Chapter 21
Standards for Biometric Presentation Attack Detection

Christoph Busch

Abstract This chapter reports about the relevant international standardization activities in the field of biometrics and specifically describes standards on presentation attacks, that have established a framework including a harmonized taxonomy for terms in the field of liveness detection and spoofing attack detection, an interchange format for data records and moreover a testing methodology for presentation attack detection. The scope of the presentation attack detection multipart standard ISO/IEC 30107 is presented. Moreover standards regarding criteria and methodology for security evaluation of biometric systems are discussed.

21.1 Introduction

Biometric systems are characterized by two essential properties. On the one hand, functional components or subsystems are usually spatially separated. The enrolment may take place as part of an employment process with the personnel department, or as part of an ePassport application in the municipality administration. Subsequently, the biometric verification may take place at a different location, say, when the data subject (e.g., the staff member) approaches a certain enterprise gate (or any other physical border gate), or when the citizen uses the passport to cross an international border. On the other hand, although the biometric enrolment step is likely to be a supervised process (often also including guidance on capture device interaction from a biometric attendant), such supervision may not exist during the verification process. Furthermore, the verification is often conducted based on a probe sample that was generated in a unsupervised capture process. In consequence for the verification not only usability of the capture device and ease of human-machine interaction is essential but moreover a measure of confidence that the probe sample was indeed collected from the subject that was initially enrolled and not from an biometric artefact that was presented by an attacker with the intention to pretend the presence of

---

C. Busch (✉)
Hochschule Darmstadt and ATHENE (National Research Center for Applied Cybersecurity), Haardtring 100, 64295 Darmstadt, Germany
e-mail: christoph.busch@h-da.de

S. Marcel et al. (eds.), *Handbook of Biometric Anti-Spoofing*, Advances in Computer Vision and Pattern Recognition, https://doi.org/10.1007/978-981-19-5288-3_21

the enrollee. Such attacks are now becoming more relevant also for mobile biometric use cases and applications like biometric banking transactions. Thus industry stakeholders such as the FIDO-Alliance as well as financial institutions, which are in the process of implementing the security requirements of the European payment service directive (PSD2), are continuously concerned about unsupervised biometric authentication and robustness of biometric capture devices.

In most cases, the comparison of a probe biometric sample with the stored biometric reference will be spatially separated from the place of enrolment. Some applications store the reference in a centralized or de-centralized database. More prominent are token-based concepts like the ICAO ePassports [1] since they allow the subject to keep control of his/her personal biometric data. As individuals are travelling they decide themselves, whether and when they provide the token to the controlling instance. The same holds true for many smartphone-based biometric authentication procedures, where the reference is stored in secure memory.

The recognition task is likely to fail if the biometric reference is not readable according to a standardized format. Any open system concept, therefore, relies on an open standard, to allow biometrics components from different suppliers to be used for the recognition task. The prime purpose of a biometric reference is to represent a biometric characteristic. This representation must on the one hand allow a good biometric performance but on the other hand the encoding format must result in a compact record and fully support the interoperability requirements. Thus encoding of a biometric sample (i.e. fingerprint image or face image) according to the ISO/IEC Biometric data interchange format [2] became a prominent format structure for many applications. See Table 21.1.

For all data interchange formats, it is essential to store along with the representation of the biometric characteristic essential information (meta data) on the capture process and the generation of the sample. Meta data that is stored along with the biometric data includes information such as size and resolution of the image but also relevant data that impacted the data capturing process: examples are the capture device type ID, that identifies uniquely the device, which was used for the acquisition of the biometric sample and also the certification block data that reports the certification authority, which had tested the capture device and the corresponding certification scheme that was used for this purpose. These data fields are contained in standardized interchange records [2–7].

Table 21.1 International standards for biometric systems

Purpose	Standards	Notes
Taxonomy	2382-37, SC37 SD11	Harmonized vocabulary
Open systems	19794, 39794	Data interchange format
Interoperability	19784	BioAPI
Testing	19795, 30107-3	Testing and reporting
PAD	30107	PAD-framework, data, testing

Essential information that was further considered helpful for the verification capture process is a measure to describe the robustness of the capture device against presentation attacks (a.k.a. spoofing attacks). This becomes more pressing once the capture device is deployed remotely and is not directly connected to the decision subsystem. In recent years, ISO/IEC has developed a multipart standard covering this issue. We will discuss this standard in this chapter. The chapter will outline the strategy behind this standardization process, cover the framework architecture and taxonomy that was established and will discuss the constraints that had to be considered in the respective standardization projects.

The remainder of the chapter is structured as follows: Sect. 21.2 will introduce the international standardization activities. Then Sects. 21.3, 21.4 and 21.5 will detail the PAD multipart standard ISO/IEC 30107. Section 21.6 is devoted to testing and reporting and Sect. 21.7 is drawing the conclusions.

21.2 International Standards Developed in ISO/IEC JTC

International standardization in the field of information technology is driven by a Joint Technical Committee (JTC 1) formed by the International Organization for Standardization (ISO) and the International Electrotechnical Commission (IEC). An important part of the JTC1 is the Sub-Committee 37 (SC37) that was established in 2002. The initial batch of SC37 standards, adopted in 2005, quickly found wide acceptance. It is estimated that almost one billion ePassport implementations conforming to the SC37 standards have already been deployed in the field (at the time of writing). The standards and standing documents developed by SC37 include

- ISO/IEC 2382-37: defining a *Harmonized Biometric Vocabulary* (HBV), aimed at removing contradictions in biometrics terminology [8, 9],
- ISO/IEC SC37 SD11: defining of a General Biometric System that describes the distributed subsystems that comprise deployed systems [10],
- ISO/IEC 19784-1: defining a common biometric programming interface (BioAPI) that supports ease of integration of capture devices and SDKs [11],

Moreover over the years, the SC37 has concentrated on the development data interchange formats with the ISO/IEC 19794 family (in the first generation), which includes the following parts:

- Part 1: Framework
- Part 2: Finger minutiae data
- Part 3: Finger pattern spectral data
- Part 4: Finger image data
- Part 5: Face image data
- Part 6: Iris image data
- Part 7: Signature/Sign time series data
- Part 8: Finger pattern skeletal data

- Part 9: Vascular image data
- Part 10: Hand geometry silhouette data
- Part 11: Signature/Sign processed dynamic data
- Part 12: - void -
- Part 13: Voice data
- Part 14: DNA data
- Part 15: Palm crease image data

The framework part includes relevant information that is common to all subsequent modality specific parts such as an introduction of the layered set of SC37 standards and an illustration of a general biometric system with a description of its functional subsystems namely the capture device, signal processing subsystem, data storage subsystem, comparison subsystem and decision subsystem [2]. Furthermore this framework part illustrates the functions of a biometric system such as enrolment, verification and identification and explains the application context of biometric data interchange formats. Part 2 to Part 15 then detail the specification and provide modality-related data interchange formats for both image interchange and template interchange on feature level. The 19794-family gained relevance, as the International Civic Aviation Organization (ICAO) adopted image-based representations for finger, face and iris in order to store biometric references in Electronic Passports. Thus the corresponding ICAO standard 9303 [1] includes a normative reference to ISO/IEC 19794. The ICAO estimated that, by 2017, over 700 million ePassports had been issued by 112 member states of ICAO.

Meanwhile, the third generation of data interchange formats has been developed with the ISO/IEC 39794 family, consisting, at present, of four parts:

- Part 1: Framework
- Part 4: Finger image data
- Part 5: Face image data
- Part 6: Iris image data

A core concept followed by this third-generation multipart standard is that encoding of biometric data will be done in an extensible data structure, as it is needed for future ePassports.

In September 2020 ICAO has endorsed the 8th Edition of Doc 9303. Doc 9303 contains in its Part 9 on Deployment of Biometric Identification and Electronic Storage of Data in MRTDs the timeline for the transition from ISO/IEC 19794-X 1st Edition to ISO/IEC 39794-X as follows:

- ePassport reader equipment must be able to handle ISO/IEC 39794 data by 2025-01-01 after a five years preparation period starting 2020-01-01.
- Between 2025 and 2030, passport issuers can use the data formats specified in ISO/IEC 19794-X:2005 or in ISO/IEC 39794-X during a five years transition period. During this transition period, interoperability and conformity testing will be essential.
- From 2030-01-01 on, passport issuers must use ISO/IEC 39794-X for encoding biometric data.

21.3 Development of Presentation Attack Detection Standard ISO/IEC 30107

For almost two decades along with the enthusiasm for biometric technologies the insight into potential risks in biometrics systems was developed and is documented in the literature [12–14]. Within the context of this chapter, the risks of subversive attacks on the biometric capture device became a major concern in unsupervised applications. Over the years academic and industry lab researched on countermeasures to detect biometric presentation attacks that constitute a subversive activity. For a survey on attacks and countermeasures regarding face-, iris- and fingerprint recognition the reader is directed to [15–17].

With a general perspective, a presentation attack can be conducted from an outsider that interacts with a biometric capture device but could as well be undertaken from an experienced insider. However, the need to develop a harmonized perspective for presentation attacks that are conducted by biometric capture subjects became obvious. The motivation to develop a standard that is related to liveness detection and spoofing was supported from stakeholders of all three communities that are active in SC37 namely from industry (essentially representatives from vendors working on fingerprint-, vein-, face- and iris-capture-devices), from academia and research projects (e.g., European projects on liveness detection) as well as from governmental agencies (e.g., responsible for testing laboratories). The latter took the lead and have started the development. Since then experts from the biometric community as well as from the security community have intensively contributed to the multipart standard that is entitled 'ISO/IEC Information Technology—Biometric presentation attack detection' [18–20]. The intention of this standard is to provide a harmonized definition of terms and a taxonomy of attack techniques. Beyond that it defines a data format that can transport measures of robustness against said attacks and a testing methodology that can evaluate PAD mechanisms.

The objectives of a standardization project are best understood by analyzing the scope clause. For the Presentation Attack Detection (PAD) standard the scope indicates that it aims at establishing beyond the taxonomy, terms and definitions a specification and characterization of PAD methods. A further objective is to develop a common data format devoted to presentation attack assessments and a third objective is to standardize principles and methods for performance assessment of PAD algorithms. This field of standardization work becomes sharpened, when topics that are outside of the scope are defined: Outside of this standardization project are definitions of specific PAD detection methods as well as detailed information about countermeasures that both are commonly valuable IPR of the industrial stakeholders.

21.4 Taxonomy for Presentation Attack Detection

Literature and science tends to struggle in general, and specifically in a multidisciplinary community as biometrics, with a clear and non-contradicting use and understanding of its terms. Therefore, ISO/IEC has developed a HBV [9] that includes terms and definitions useful also in the context of presentation attacks. Without going into detail of the terminology definition process it is important to note that biometric *concepts* are always discussed in context (e.g., of one or multiple biometric subsystems) before a *term* and its *definition* for said concept can be developed. Thus, terms are defined in groups and overlap of groups ("concept clusters"). The interdependencies among these groups often lead to revisions of previously accepted definitions. The result of this work is published as ISO/IEC 2382-37:2017 [8] and is also available online [9]. It is of interest to consider here definitions in the HBV, as they are relevant for the taxonomy and terminology defined in ISO/IEC 30107 [18–20]. The following list contains definitions of interest[1]:

- **biometric characteristic**: biological and behavioural characteristic of an individual from which distinguishing, repeatable biometric features can be extracted for the purpose of biometric recognition (37.01.02)
- **biometric feature**: numbers or labels extracted from biometric samples and used for comparison (37.03.11)
- **biometric capture subject**: individual who is the subject of a biometric capture process (37.07.03)
- **biometric capture process**: collecting or attempting to collect a signal(s) from a biometric characteristic, or a representation(s) of a biometric characteristic(s,) and converting the signal(s) to a captured biometric sample set (37.05.02)
- **impostor**: subversive biometric capture subject who attempts to being matched to someone else's biometric reference (37.07.13)
- **identity concealer**: subversive biometric capture subject who attempts to avoid being matched to their own biometric reference (37.07.12)
- **subversive biometric capture subject**: biometric capture subject who attempts to subvert the correct and intended policy of the biometric capture subsystem (37.07.17)
- **subversive user**: user of a biometric system who attempts to subvert the correct and intended system policy (37.07.18)
- **uncooperative biometric capture subject**: biometric capture subject motivated to not achieve a successful completion of the biometric acquisition process (37.07.19)
- **uncooperative presentation**: presentation by a uncooperative biometric capture subject (37.06.19)

In order to formulate a common understanding of attacks on biometric systems the list of above terms was expanded with the following concepts that provided in

[1] The six-digit number at the end of each term refer to the clause in the standard, which can be looked up on the ISO/IEC website: https://www.iso.org/obp/ui/#iso:std:iso-iec:2382:-37:ed-2:v1:en.

ISO/IEC 30107-1 Biometric presentation attack detection—Part 1: Framework [18] and in ISO/IEC 30107-3 Biometric presentation attack detection—Part 3: Testing and reporting [19]. The use of the above terms is strongly recommended. The use of alternative terms, such as *fake*, is deprecated, notwithstanding the intensive previous use of such terms in the literature.

- **presentation attack/attack presentation**: presentation to the biometric data capture subsystem with the goal of interfering with the operation of the biometric system
- **bona fide presentation**: interaction of the biometric capture subject and the biometric data capture subsystem in the fashion intended by the policy of the biometric system
- **presentation attack instrument (PAI)**: biometric characteristic or object used in a presentation attack
- **PAI species**: class of presentation attack instruments created using a common production method and based on different biometric characteristics
- **artefact**: artificial object or representation presenting a copy of biometric characteristics or synthetic biometric patterns
- **presentation attack detection (PAD)**: automated determination of a presentation attack

In the development of ISO/IEC 30107 a framework was defined to understand PA characteristics and also detection methods. Figure 21.1 illustrates the potential targets in a generic biometric system [10] that could be attacked. In this chapter (and moreover, in this book) we concentrate only on presentation attacks, that is, attacks that occur at the capture device.

The framework defined in the ISO/IEC 30107-1 standard [18] considers two types of attacks: the Active Impostor Presentation Attack and the Identity Concealer Attack.

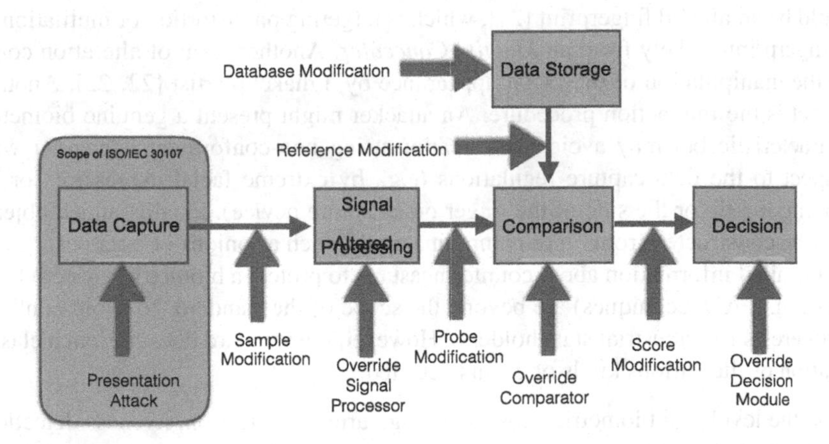

Fig. 21.1 Examples for points of attacks (following [18])

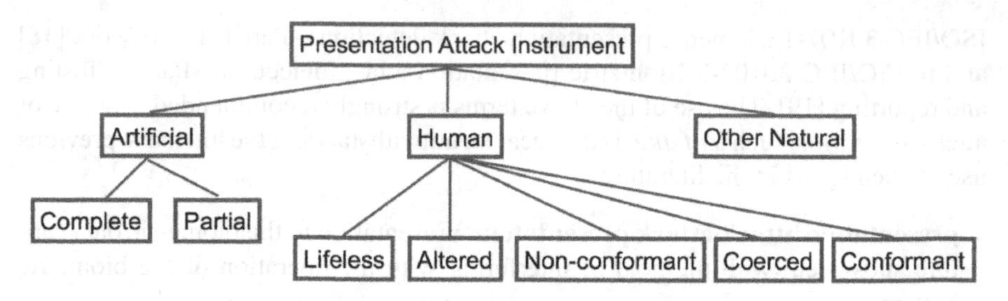

Fig. 21.2 A biometrics system may be attacked at various points. The figure shows a schematic architecture of a generic biometrics system. Each red arrow in the figure identifies a different kind of attack that may be performed against a biometrics system. (Following [18])

- The *Active Impostor Presentation Attack* attempts to subvert the correct and intended policy of the biometric capture subsystem. Here, the attacker aims to be recognized as a specific data subject known to the system (i.e., an impersonation attack).
- In an *Identity Concealer Presentation Attack*, on the other hand, the attacker aims to avoid being matched to his/her own biometric reference in the system.

To perpetrate an attack, the attacker will use an object that interacts with the capture device. Moreover, the potential of the attack (to succeed) will depend on the attacker's knowledge, the window of opportunity, and other factors, discussed in Sect. 21.6. The object that is employed in the attack can be of manifold nature: from *gummy fingers* to categories of alterations of the biometric characteristic itself. Figure 21.2 illustrates that, besides artificial objects (i.e., $artefacts$) natural material could be used to constitute a PA. When the expected biometric characteristic from an enrollee is absent and replaced by an *attack presentation characteristic* (i.e., the attack presentation object) this could be a human tissue from a deceased person (i.e., a cadaver part) or it could be an altered fingerprint [21], which is targeting on distortion or mutilation of a fingerprint—likely from an *Identity Concealer*. Another form of alteration could be the manipulation of the facial appearance by a makeup artist [22, 23]. Another aspect is the interaction procedure. An attacker might present a genuine biometric characteristic but may avoid identification using non-conformant behaviour with respect to the data capture regulations (e.g., by extreme facial expression, or by placing the tip or the side of the finger on a capture device). Finally, attack objects may be constructed from other natural materials, such as onions or potatoes.

Detailed information about countermeasures to protect a biometrics system from PAs (i.e., PAD techniques) are beyond the scope of the standard (to avoid conflicts of interests for industrial stakeholders). However, the standard does present a classification of the various kinds of events detected:

- on the level of a biometric subsystem (e.g., artefact detection, liveness detection, alteration detection, non-conformance detection) and
- through detection of non-compliant interaction in violation with system security policies (e.g., geographic or temporal exception).

Table 21.2 Selected data fields included in a PAD record according ISO/IEC 30107-2

Field name	Valid values	Notes
PAD decision	0,1	Optional
PAD score	0 to 100	Mandatory
Level of supervision	1 to 5	

21.5 Data Formats

One of the objectives of the ISO/IEC 30107 multipart standard is to transport information about the PAD results from the capture device to subsequent signal processing or decision subsystems. The container to transmit such information is the open data interchange format (DIF) according to the ISO/IEC 19794 series [24] and the new extensible data format specified in the ISO/IEC 39794 series [5]. This section outlines the conceptual data fields that are defined for a PAD record in ISO/IEC 30107-2 [19]. A selection of fields is illustrated in Table 21.2. It indicates that the result of the PAD functionality should be encoded as a scalar value in the range of 0 to 100 in analogy to the encoding of the sample quality assessment that is potentially also conducted by the capture device and stored as a quality score according to ISO/IEC 29794-1 [25].

The PAD decision is encoded with abstract values **NO_ATTACK** or **ATTACK**. The PAD score shall be rendered in the range between 0 and 100 provided by the attack detection techniques. *Bonafide* presentations (e.g., presentations in the intended fashion) shall tend to generate lower scores. Attack presentations shall tend to generate higher scores. The abstract value **FAILURE_TO_COMPUTE** shall indicate that the computation of the PAD score has failed. The level of supervision expresses the surveillance during the capture process. Possible abstract values are **UNKNOWN, CONTROLLED, ASSISTED, OBSERVED, UNATTENDED**.

In the absence of standardized assessment methods a PAD score would be encoded on a range of 0 (i.e., indicative of an attack) to 100 (i.e. indicative of a genuine capture attempt). This score can describe the trustworthiness of the transmitted biometric sample. Any decision based on this information is at the discretion of the receiver. The described PAD data record became an integral part of the representation block in ISO/IEC 39794-1 [5].

21.6 Testing and Reporting

One way to affirm the reliability of a transmitted PAD record is to include, in the interchange record, the performance-testing results of the capture device established at an independent testing laboratory. Test procedures as such are well known since the biometric performance testing standards ISO/IEC 19795-1 was established in

2006 [26] and has recently been revised in 2021 [27]. The framework for Biometric Performance Testing and Reporting was developed on the basis of established concepts such as the *Best Practices in Testing and Reporting Performance of Biometric Devices* [28]. It defines in which way algorithm errors such as false-match-rate (FMR) and false-non-match-rate (FNMR) as well as system errors ()such as false-accept-rate (FAR) and false-reject-rates (FRR)) must be reported. Such concepts are not applicable for testing the PAD performance of a biometrics system. Therefore, during the development of the ISO/IEC 30107 standard, various approaches were proposed by the different national nationalization delegations participating in the development of the standard.

A further evaluation should determine whether artefacts with abnormal properties are accepted by the traditional biometric system and can result in higher-than-normal acceptance against $bonafide$ enrolled references. This is quantified by the Impostor Attack Presentation Match Rate (IAPMR): defined, in a full-system evaluation of a verification system, as the proportion of impostor attack presentations using the same PAI species in which the target reference is matched.

Moreover, the ISO/IEC 30107-3 introduces three levels of PAD evaluation, namely:

- noindent(1) PAD subsystem evaluation: this level evaluates only a PAD system, which is either hardware or software based
- noindent(2) Data Capture subsystem evaluation: this will evaluate the data capture subsystem that may or may not include the PAD algorithms but is focused more on the biometric capture device itself
- noindent(3) Full system evaluation: providing the end-to-end system evaluation.

The PAD subsystem are evaluated using two different metrics namely [20]: (1) Attack Presentation Classification Error Rate (APCER): defined as the proportion of presentation attacks incorrectly classified as *bona bide* presentations (2) Bona Fide Presentation Classification Error Rate (BPCER): defined as the proportion of *bona fide* presentations incorrectly classified as presentation attacks.

The APCER can be calculated as follows:

$$APCER = \frac{1}{N_{PAIS}} \sum_{i=1}^{N_{PAIS}} (1 - RES_i) \tag{21.1}$$

Where, N_{PAIS} is the number of attack presentations for the given PAI [18]. RES_i takes the value 1 if the ith presentation is classified as an attack presentation and value 0 if classified as *bona fide* presentation.

While the BPCER can be calculated as follows:

$$BPCER = \frac{\sum_{i=1}^{N_{BF}} RES_i}{N_{BF}} \tag{21.2}$$

where N_{BF} is the number of *bona fide* presentations. RES_i takes the value 1 if the ith presentation is classified as an attack presentation and value 0 if classified as *bona fide* presentation.

21.7 Conclusion and Future Work

This chapter introduces the standardization work that began around 2010 in the area of PAD. Now that metrics are well established the published standards can contribute to mature taxonomy of PAD terms in future literature. The encoding details of the PAD interchange record are ready to be deployed in large-scale applications and will play a relevant role in the new ISO/IEC 39794 interchange standard.

One challenge with the now established testing and reporting methodology is that, unlike for biometric performance testing, (also known as 'technology testing'), a large corpus of test samples (i.e. PAIs) is often not available to testing labs. Top national laboratories are in possession of no more than 60 PAI species for a fingerprint recognition system. In such cases it is essential that the proportion is computed not to a large number of samples all derived from one single PAI species, that have similar material properties or stem from the same biometric source. At least the denominator should be defined by the number of *PAI species*. Note that one single artefact species would correspond to the set of fingerprint artefacts all made with the same *recipe* and the same materials but with different friction ridge patterns from different fingerprint instances. The APCER and BPCER are complementary measures, and should both be specified in every evaluation.

An essential difference of PAD testing is that obviously there is beyond the mere statistical observations as expressed by APCER and BPCER metrics the need to categorize the attack potential itself. Such methodology is well established in the scope of Common Criteria testing that developed the Common Methodology for Information Technology Security Evaluation [29]. It may be desirable to replace the indication of a defined level of difficulty by an attack potential attribute of a biometric PA. This new attribute would express the effort expended in the preparation and execution of the attack in terms of elapsed time, expertise, knowledge about the capture device being attacked, window of opportunity and equipment. The attack potential could be graded as *no rating*, *minimal*, *basic*, *enhanced-basic*,*moderate* or *high*. Such gradings are established in Common Criteria testing and would allow a straightforward understanding of a PAD result for security purposes.

Due to separation of work tasks in ISO/IEC JTC1, discussion of topics related to security is not in scope of the ISO/IEC 30107 standard. However, the Common Criteria concept of attack potential should be seen as both a good categorization for the criticality of an attack and the precondition to conduct later a security evaluation based on the results of a ISO/IEC 30107 metric. This work has been addressed in the ISO/IEC 19989 multipart standard [30].

The discussions on PAD standards are ongoing and the reader may consider contributing to this process via his or her national standardization body.

References

1. International Civil Aviation Organization NTWG (2015) Machine readable travel documents – part 4 – specifications for machine readable passports (MRPs) and other TD3 size MRTDs. http://www.icao.int/publications/Documents/9303_p4_cons_en.pdf
2. ISO/IEC JTC1 SC37 Biometrics (2011) ISO/IEC 19794-1:2011 information technology - biometric data interchange formats - part 1: framework. International Organization for Standardization
3. ISO/IEC JTC1 SC37 Biometrics (2011) ISO/IEC 19794-4:2011 information technology – biometric data interchange formats – part 4: finger image data. International Organization for Standardization
4. ISO/IEC JTC1 SC37 Biometrics (2011) ISO/IEC 19794-5:2011. Information technology - biometric data interchange formats - part 5: face image data. International Organization for Standardization
5. ISO/IEC JTC1 SC37 Biometrics (2019) ISO/IEC 39794-1:2019 information technology - extensible biometric data interchange formats - part 1: framework. International Organization for Standardization
6. ISO/IEC JTC1 SC37 Biometrics (2019) ISO/IEC 39794-4:2019 information technology - extensible biometric data interchange formats - part 4: finger image data. International Organization for Standardization
7. ISO/IEC JTC1 SC37 Biometrics (2019) ISO/IEC 39794-5:2019 information technology - extensible biometric data interchange formats - part 5: face image data. International Organization for Standardization
8. ISO/IEC JTC1 SC37 Biometrics (2017) ISO/IEC 2382-37 harmonized biometric vocabulary. International Organization for Standardization
9. ISO/IEC JTC1 SC37 Biometrics (2017) ISO/IEC 2382-37 harmonized biometric vocabulary. http://www.christoph-busch.de/standards.html
10. ISO/IEC JTC1 SC37 Biometrics (2008) ISO/IEC SC37 SD11 general biometric system. International Organization for Standardization
11. ISO/IEC TC JTC1 SC37 Biometrics (2006) ISO/IEC 19784-1:2006. Information technology – biometric application programming interface – part 1: BioAPI specification. International Organization for Standardization
12. Zwiesele A, Munde A, Busch C, Daum H (2000) Biois study - comparative study of biometric identification systems. In: 34th Annual 2000 IEEE international Carnahan conference on security technology (CCST). IEEE Computer Society, pp 60–63
13. Matsumoto T, Matsumoto H, Yamada K, Yoshino S (2002) Impact of artificial "Gummy" fingers on fingerprint systems. In: SPIE conference on optical security and counterfeit deterrence techniques IV, vol 4677, pp 275–289
14. Schuckers S (2002) Spoofing and anti-spoofing measures. Information security technical report 4, Clarkson University and West Virginia University
15. Raghavendra R, Busch C (2017) Presentation attack detection methods for face recognition systems: a comprehensive survey. ACM Comput Surv 50(1):8:1–8:37. https://doi.org/10.1145/3038924
16. Galbally J, Gomez-Barrero M (2016) A review of iris anti-spoofing. In: 4th International Conference on Biometrics and Forensics (IWBF), pp 1–6
17. Sousedik C, Busch C (2014) Presentation attack detection methods for fingerprint recognition systems: a survey. Biometrics IET 3(1):1–15
18. ISO/IEC JTC1 SC37 Biometrics (2016) ISO/IEC 30107-1. Information technology - biometric presentation attack detection - part 1: framework. International Organization for Standardization
19. ISO/IEC JTC1 SC37 Biometrics (2017) ISO/IEC 30107-2. Information technology - biometric presentation attack detection - part 2: data formats. International Organization for Standardization

20. ISO/IEC JTC1 SC37 Biometrics (2017) ISO/IEC 30107-3. Information technology - biometric presentation attack detection - part 3: testing and reporting. International Organization for Standardization
21. Ellingsgaard J, Busch C (2017) Altered fingerprint detection. Springer International Publishing, Cham, pp 85–123. https://doi.org/10.1007/978-3-319-50673-9_5
22. Rathgeb C, Drozdowski P, Busch C (2020) Detection of makeup presentation attacks based on deep face representations. In: Proceedings of the international conference on pattern recognition (ICPR). IEEE, pp 3443–3450
23. Rathgeb C, Drozdowski P, Busch C (2020) Makeup presentation attacks: review and detection performance benchmark. IEEE Access 8:224,958–224,973
24. ISO/IEC JTC1 SC37 Biometrics (2011) ISO/IEC 19794-1:2011 information technology - biometric data interchange formats - part 1: framework. International Organization for Standardization
25. ISO/IEC JTC1 SC37 Biometrics (2016) ISO/IEC 29794-1:2016 information technology - biometric sample quality - part 1: framework. International Organization for Standardization
26. ISO/IEC JTC1 SC37 Biometrics (2017) ISO/IEC 19795-1:2017. Information technology – biometric performance testing and reporting – part 1: principles and framework. International Organization for Standardization and International Electrotechnical Committee
27. ISO/IEC JTC1 SC37 Biometrics (2021) ISO/IEC 19795-1:2021. Information technology – biometric performance testing and reporting – part 1: principles and framework. International Organization for Standardization
28. Mansfield T, Wayman J (2002) Best practices in testing and reporting performance of biometric devices. CMSC 14/02 Version 2.01, NPL
29. Criteria C (2012) Common methodology for information technology security evaluation - evaluation methodology. http://www.commoncriteriaportal.org/cc/
30. ISO/IEC JTC1 SC37 Biometrics (2020) ISO/IEC 19989-1:2020. Information technology – criteria and methodology for security evaluation of biometric systems – Part 1: framework. International Organization for Standardization

Glossary

A-PBS: Attention-based Pixel-wise Binary Supervision 162
ACE: Average Classification Error 62
ACER: Average Classification Error Rate 226
ACER: Average Classification Error Rate 241
AFR: Automatic face recognition 236
AIA: Artificial Intelligence Act 456
ALTAI: Assessment List for Trustworthy AI 462
AM-Softmax: Additive Margin Softmax 317
Anti-spoofing (term to be deprecated): countermeasure to spoofing, *see* PAD 2
AP: Attack Presentation 398
APCER: Attack Presentation Classification Error Rate 136
APCER: Attack Presentation Classification Error Rate 162
APCER: Attack Presentation Classification Error Rate 226
APCER: Attack Presentation Classification Error Rate 241
APCER: Attack Presentation Classification Error Rate 433
APCER: Attack Presentation Classification Error Rate 475
APCER: Attack Presentation Classification Error Rate 50
APCER: Attack Presentation Classification Error Rate 82
AS17: ASVspoof 2017 Version 2 Dataset 334
AS19: ASVspoof 2019 PA Dataset 334
AS19Real: ASVspoof 2019 Real PA Dataset 334

S. Marcel et al. (eds.), *Handbook of Biometric Anti-Spoofing*, Advances in Computer Vision and Pattern Recognition, https://doi.org/10.1007/978-981-19-5288-3

ASP: Attentive Statistic Pooling 317
ASR: Automatic Speech Recognition 340
AST-CC: Adaptive Spectro-temporal Resolution Cepstral Coefficients 316
ASV Mic: ASV Microphone 318
ASV: Automatic Speaker Verification 315
ASV: Automatic Speaker Verification 346
ASVspoof: Automatic Speaker Verification Spoofing and Countermeasures Challenge 317
Att Mic: Attacker Microphone 318
Att Replay: Attacker Replay Device 318
BCE: Binary Cross-entropy 162
BCE: Binary Cross-entropy 225
BCE: Binary Cross-entropy 250
BCS: Burst Click Sound 331
BD: Body Double 44
BEAT: Biometrics Evaluation and Testing. 8
BEAT: Biometrics Evaluation and Testing. 95
BEAT: Biometrics Evaluation and Testing 431
BMS: Biometric Matching System 461
Bona fide: refers to real or genuine 194
BP: Bona fide Presentation 398
BPCER: Bona fide Presentation Classification Error Rate 136
BPCER: Bona fide Presentation Classification Error Rate 162
BPCER: Bona fide Presentation Classification Error Rate 226
BPCER: Bona fide Presentation Classification Error Rate 241
BPCER: Bona fide Presentation Classification Error Rate 433
BPCER: Bona fide Presentation Classification Error Rate 475
BPCER: Bona fide Presentation Classification Error Rate 50
BPCER: Bona fide Presentation Classification Error Rate 82
BSIF: Binarized statistical image features 213
BSIF: Binarized Statistical Image Features 42
BSIF: Binary Statistical Image Features 82
CBAM: Convolutional Block Attention Mechanism 162
CCD: Charge-coupled Device 82
CDC: Central Difference Convolution 246
CDL: Contrastive Depth Loss 246

CE: Cross-entropy 250
CFCCIF: Cochlear Filter Cepstral Coefficients with Instantaneous Frequency 302
CFREU: Charter of Fundamental Rights of the European Union 447
CI: Cadaver Iris 133
CJEU: Court of Justice of the European Union 448
Clarkson: Clarkson University, USA 123
CMFL: Cross-modal Focal Loss 225
CMOS: Complementary Metal Oxide Semiconductor 82
CNN: Convolutional Neural Network 105
CNN: Convolutional Neural Network 211
CNN: Convolutional Neural Network 213
CNN: Convolutional Neural Network 250
CNN: Convolutional Neural Network 368
CNN: Convolutional Neural Network 398
CNN: Convolutional Neural Network 51
CNN: Convolutional Neural Network 61
CNN: Convolutional Neural Network 9
CQCC-LTAS-D: LTAS-based CQCC Divergence Feature 336
CQCC-PAAS-D: PAAS-based CQCC Divergence Feature 337
CQCC: Constant Q Cepstral Coefficients 302
CQCC: Constant Q Cepstral Coefficients 316
CQT: Constant Q Transform 302
CVPR: Conference on Computer Vision and Pattern Recognition 123
D-EER: Detection Equal Error Rate 433
DCT: Discrete Cosine Transformation 336
DHI: DualHeadInception 75
DHR: DualHeadResNet 75
DNN: Deep Neural Network 317
DPIA: Data Protection Impact Assessment 459
DSLR: Digital Single-Lens Reflex 259
DTMF: Dual-Tone Multi-frequency Signaling 332
DTW: Dynamic Time Warping 374
EC: Ecoflex 44
ED: Electronic Display 133
EDPS: European Data Protection Supervisor 450
EER: Equal Error Rate 162
EER: Equal Error Rate 249
EER: Equal Error Rate 317
EER: equal error rate 352
EER: Equal Error Rate 368
EER: Equal Error Rate 433

EM:	Expectation–Maximization 337
ESA:	Energy Separation Algorithm 317
EUI:	European Union Institutes 451
F-PAD:	Fingerprint PAD 9
FAR:	False Alarm Rate 324
FAS:	Face Anti-spoofing 236
FDR:	Fisher Discriminant Ratio 163
FDR:	Fisher Discriminant Ratio 261
FIR:	Finite Impulse Response 327
FMR:	False Match Rate 50
FMR:	False Match Rate 82
FN:	False Negative 242
FNMR:	False Non-match Rate 50
FNMR:	False Non-match Rate 82
FP:	False Positive 242
FPAD:	Fingerprint Presentation Attack Detection 60
FPAD:	Fingerprint Presentation Attack Detector 41
FPR:	False Positive Rate 242
FRR:	False Reject Rate 324
FRVT:	Face Recognition Vendor Test 123
FV:	Fisher Vectors 398
FVC:	Fingerprint Verification Competition 62
GAP:	Global Averaged Pooling 317
GDPR:	General Data Protection Regulation 139
GDPR:	General Data Protection Regulation 446
GEL:	Gelatine 44
GMM:	Gaussian Mixture Model 317
GRL:	Gradient Reversal Layer 357
GRU:	Gated Recurrent Unit 108
GUI:	Graphical User Interface 426
HHIRs:	Higher Harmonic Impulse Responses 328
HLEG:	High-Level Expert Group 456
HQ-WMCA:	High-Quality Wide Multi-channel Attack 226
HT:	Hilbert Transform 317
HTER:	Half Total Error Rate 162
HTER:	Half Total Error Rate 213
IAPMR:	Impostor Attack Presentation Match Rate 435
IAPMR:	Impostor Attack Presentation Match Rate 50
ICCV:	International Conference on Computer Vision 123
ID:	Identity Document 422
IEC:	International Electrotechnical Commission 445
IIIT:	Indian Institutes of Information Technology 123
IIITD:	Indian Institutes of Information Technology, Delhi, India 130

IM: Integrated Matching 50
IMFCC: Inverted Mel Frequency Cepstral Coefficients 316
IQM: Image Quality Measures 213
IREX: Iris Exchange 123
IRS: Iris Recognition System 82
ISO/IEC: International Organization for Standardization/ International Electrotechnical Commission 50
ISO: International Standards Organization 445
JS: Jensen–Shannon Divergence 321
KL: Kullback–Leibler Divergence 323
LA: Logical Access 346
LAT: Latex 44
LBP: Local Binary Patterns 82
LBP: Local Binary Pattern 42
LCNN: Light Convolutional Neural Network 317
LDE: Learnable Dictionary Encoding 317
LE: Liquid Ecoflex 44
LED: Law Enforcement Directive 451
LFC: Light Field Camera 82
LFCC-LTAS-D: LTAS-based LFCC Divergence Feature 336
LFCC-PAAS-D: PAAS-based LFCC Divergence Feature 337
LFCC: Linear Frequency Cepstral Coefficients 302
LFCC: Linear Frequency Cepstral Coefficients 316
LFCC: linear frequency cepstral coefficients 352
LivDet: Fingerprint Liveness Detection Competition 65
LivDet: Liveness Detection 41
LLR: Log Likelihood Ratio 325
LOO: Leave-one-out 226
LPCC: Linear Prediction Cepstral Coefficients 317
LPQ: Local Phase Quantization 82
LPR: Linear Prediction Residual 317
LSTM: Long Short-term Memory 106
LSTM: Long Short-term Memory 317
LTAS: Long-term Average Spectrum 335
MAD: Manipulation Attack Detection 441
MAP: Maximum a Posteriori 340
MCC: Minutia Cylinder-Code 70
MFCC: Mel Frequency Cepstral Coefficients 316
MGD: Modified Group Delay 316
MLF: Multi-layer feature Fusion 162
MOD: Modasil 44
MS-LBP: Multi-scale Local Binary Pattern 199
MSA: Micro-stripe Analyses 162
NAS: Neural Architecture Search 274

NF: Normalization Flow 335
NIR: Near Infrared 125
NIR: Near Infrared 211
NIR: Near Infrared 236
NIST: National Institute of Standards and Technology 64
Notre Dame: University of Notre Dame, USA 123
OC-Softmax: One-Class Softmax 317
OC-SVM: One-Class Support Vector Machine 337
OC-SVM: one-class support vector machine 348
OOD: Out-of-Distribution 335
OS: Operating System 44
P2SGrad: Probability-to-Similarity Gradient 317
PA: Physical Access 327
PA: Physical Access 346
PA: Presentation Attack 162
PA: Presentation Attack 211
PA: Presentation Attack 236
PA: Presentation Attack 2
PA: Presentation Attack 366
PA: Presentation Attack 422
PA: Presentation Attack 82
PAAS: Phone-attentive Average Spectrum 336
PAD: Presentation Attack Detection 123
PAD: Presentation Attack Detection 162
PAD: Presentation Attack Detection 211
PAD: Presentation Attack Detection 214
PAD: Presentation Attack Detection 236
PAD: Presentation Attack Detection 2
PAD: Presentation Attack Detection 346
PAD: Presentation Attack Detection 366
PAD: Presentation Attack Detection 398
PAD: Presentation Attack Detection 423
PAD: Presentation Attack Detection 82
PAD: Presentation Attack Detection 9
PAI: Presentation Attack Instrument 236
PAI: Presentation Attack Instrument 398
PAI: Presentation Attack Instrument 423
PAI: Presentation Attack Instrument 43
PAI: Presentation Attack Instrument 471
PAI: Presentation Attack Instrument 82
PCA: Principal Component Analysis 354
PD: Play-Doh 44
PixBiS: Pixel-wise Binary Supervision 231
PP: Paper Printout 127

PPCL: Prosthetic/Printed eyes with Contact Lenses 133
PRAD: Parallel Replay Attack Digit Dataset 318
PSTN: public switched telephone network 355
RBF: Radial Basis Function 213
RF30: RFast30 44
RFCC: Rectangular Frequency Cepstral Coefficients 316
RNN: Recurrent Neural Network 105
RNN: Recurrent Neural Network 213
RNN: Recurrent Neural Network 368
ROC: Receiver Operating Characteristic 242
rPPG: Remote Photoplethysmography 265
RPS: Relative Phase Shift 316
R&R : Recording and Replay 316
RTV: room-temperature-vulcanizing 44
SBC: Schengen Border Code 450
SE: Squeeze-and-Excitation 242
SGD: Stochastic Gradient Descent 340
SIFT: Scale Invariant Feature Transform 82
SIFT: Scale Invariant Feature Transform 82
SIL: Silicone 44
Spoofing (term to be deprecated): attempt to impersonate a biometric system, *see* PA 2
SVM: Support Vector Machines 213
SVM: Support Vector Machines 398
SWAN-MBD: SWAN Multi-modal Biometric Dataset 423
SWAN-MPAD: SWAN Multi-modal Presentation Attack Dataset 423
SWIR: Short-wave Infrared 211
SWIR: Short-Wave Infrared 276
t-DCF: tandem detection cost function 352
TA-RNN: Time Alignment Recurrent Neural Network 374
TABULA RASA: Trusted Biometrics under Spoofing Attacks. 8
TABULA RASA: Trusted Biometrics under Spoofing Attacks. 95
TAR: True Acceptance Rate 62
TCL: Textured Contact Lenses 127
TCN: Third Country Nationals 451
TEO: Teager Energy Operator 317
TEU: Treaty of the European Union 447
TFEU: Treaty of Functioning of the European Union 448
TN: True Negative 242
TP: True Positive 242
TPR: True Positive Rate 242

TTS: text-to-speech 346
VAE: Variational Auto-encoder 335
VC: voice conversion 288
VC: voice conversion 346
VoIP: voice over Internet protocol 355
Warsaw: Warsaw University of Technology, Poland 123
WG: Wood glue 44
WVU: West Virginia University, USA 130

Index

S. Marcel et al. (eds.), *Handbook of Biometric Anti-Spoofing*, Advances in Computer Vision and Pattern Recognition, https://doi.org/10.1007/978-981-19-5288-3

Printed in the United States
by Baker & Taylor Publisher Services

Communications in Computer and Information Science 1654

More information about this series at https://link.springer.com/bookseries/7899

Constantine Stephanidis · Margherita Antona ·
Stavroula Ntoa · Gavriel Salvendy (Eds.)

HCI International 2022 – Late Breaking Posters

24th International Conference on Human-Computer Interaction
HCII 2022, Virtual Event, June 26 – July 1, 2022
Proceedings, Part I

Editors
Constantine Stephanidis
University of Crete and Foundation for Research and Technology – Hellas (FORTH)
Heraklion, Crete, Greece

Margherita Antona
Foundation for Research and Technology
Hellas (FORTH)
Heraklion, Crete, Greece

Stavroula Ntoa
Foundation for Research and Technology
Hellas (FORTH)
Heraklion, Crete, Greece

Gavriel Salvendy
University of Central Florida
Orlando, FL, USA

ISSN 1865-0929 ISSN 1865-0937 (electronic)
Communications in Computer and Information Science
ISBN 978-3-031-19678-2 ISBN 978-3-031-19679-9 (eBook)
https://doi.org/10.1007/978-3-031-19679-9

This Springer imprint is published by the registered company Springer Nature Switzerland AG
The registered company address is: Gewerbestrasse 11, 6330 Cham, Switzerland

Foreword

Human-computer interaction (HCI) is acquiring an ever-increasing scientific and industrial importance, as well as having more impact on people's everyday life, as an ever-growing number of human activities are progressively moving from the physical to the digital world. This process, which has been ongoing for some time now, has been dramatically accelerated by the COVID-19 pandemic. The HCI International (HCII) conference series, held yearly, aims to respond to the compelling need to advance the exchange of knowledge and research and development efforts on the human aspects of design and use of computing systems.

The 24th International Conference on Human-Computer Interaction, HCI International 2022 (HCII 2022), was planned to be held at the Gothia Towers Hotel and Swedish Exhibition & Congress Centre, Göteborg, Sweden, during June 26 to July 1, 2022. Due to the COVID-19 pandemic and with everyone's health and safety in mind, HCII 2022 was organized and run as a virtual conference. It incorporated the 21 thematic areas and affiliated conferences listed on the following page.

A total of 5583 individuals from academia, research institutes, industry, and governmental agencies from 88 countries submitted contributions, and 1276 papers and 275 posters were included in the proceedings that were published just before the start of the conference. Additionally, 296 papers and 181 posters are included in the volumes of the proceedings published after the conference, as "Late Breaking Work". The contributions thoroughly cover the entire field of human-computer interaction, addressing major advances in knowledge and effective use of computers in a variety of application areas. These papers provide academics, researchers, engineers, scientists, practitioners, and students with state-of-the-art information on the most recent advances in HCI. The volumes constituting the full set of the HCII 2022 conference proceedings are listed in the following pages.

I would like to thank the Program Board Chairs and the members of the Program Boards of all thematic areas and affiliated conferences for their contribution and support towards the highest scientific quality and overall success of the HCI International 2022 conference; they have helped in so many ways, including session organization, paper reviewing (single-blind review process, with a minimum of two reviews per submission) and, more generally, acting as good-will ambassadors for the HCII conference.

This conference would not have been possible without the continuous and unwavering support and advice of Gavriel Salvendy, Founder, General Chair Emeritus, and Scientific Advisor. For his outstanding efforts, I would like to express my appreciation to Abbas Moallem, Communications Chair and Editor of HCI International News.

July 2022 Constantine Stephanidis

HCI International 2022 Thematic Areas and Affiliated Conferences

Thematic Areas

- HCI: Human-Computer Interaction
- HIMI: Human Interface and the Management of Information

Affiliated Conferences

- EPCE: 19th International Conference on Engineering Psychology and Cognitive Ergonomics
- AC: 16th International Conference on Augmented Cognition
- UAHCI: 16th International Conference on Universal Access in Human-Computer Interaction
- CCD: 14th International Conference on Cross-Cultural Design
- SCSM: 14th International Conference on Social Computing and Social Media
- VAMR: 14th International Conference on Virtual, Augmented and Mixed Reality
- DHM: 13th International Conference on Digital Human Modeling and Applications in Health, Safety, Ergonomics and Risk Management
- DUXU: 11th International Conference on Design, User Experience and Usability
- C&C: 10th International Conference on Culture and Computing
- DAPI: 10th International Conference on Distributed, Ambient and Pervasive Interactions
- HCIBGO: 9th International Conference on HCI in Business, Government and Organizations
- LCT: 9th International Conference on Learning and Collaboration Technologies
- ITAP: 8th International Conference on Human Aspects of IT for the Aged Population
- AIS: 4th International Conference on Adaptive Instructional Systems
- HCI-CPT: 4th International Conference on HCI for Cybersecurity, Privacy and Trust
- HCI-Games: 4th International Conference on HCI in Games
- MobiTAS: 4th International Conference on HCI in Mobility, Transport and Automotive Systems
- AI-HCI: 3rd International Conference on Artificial Intelligence in HCI
- MOBILE: 3rd International Conference on Design, Operation and Evaluation of Mobile Communications

Conference Proceedings – Full List of Volumes

1. LNCS 13302, Human-Computer Interaction: Theoretical Approaches and Design Methods (Part I), edited by Masaaki Kurosu
2. LNCS 13303, Human-Computer Interaction: Technological Innovation (Part II), edited by Masaaki Kurosu
3. LNCS 13304, Human-Computer Interaction: User Experience and Behavior (Part III), edited by Masaaki Kurosu
4. LNCS 13305, Human Interface and the Management of Information: Visual and Information Design (Part I), edited by Sakae Yamamoto and Hirohiko Mori
5. LNCS 13306, Human Interface and the Management of Information: Applications in Complex Technological Environments (Part II), edited by Sakae Yamamoto and Hirohiko Mori
6. LNAI 13307, Engineering Psychology and Cognitive Ergonomics, edited by Don Harris and Wen-Chin Li
7. LNCS 13308, Universal Access in Human-Computer Interaction: Novel Design Approaches and Technologies (Part I), edited by Margherita Antona and Constantine Stephanidis
8. LNCS 13309, Universal Access in Human-Computer Interaction: User and Context Diversity (Part II), edited by Margherita Antona and Constantine Stephanidis
9. LNAI 13310, Augmented Cognition, edited by Dylan D. Schmorrow and Cali M. Fidopiastis
10. LNCS 13311, Cross-Cultural Design: Interaction Design Across Cultures (Part I), edited by Pei-Luen Patrick Rau
11. LNCS 13312, Cross-Cultural Design: Applications in Learning, Arts, Cultural Heritage, Creative Industries, and Virtual Reality (Part II), edited by Pei-Luen Patrick Rau
12. LNCS 13313, Cross-Cultural Design: Applications in Business, Communication, Health, Well-being, and Inclusiveness (Part III), edited by Pei-Luen Patrick Rau
13. LNCS 13314, Cross-Cultural Design: Product and Service Design, Mobility and Automotive Design, Cities, Urban Areas, and Intelligent Environments Design (Part IV), edited by Pei-Luen Patrick Rau
14. LNCS 13315, Social Computing and Social Media: Design, User Experience and Impact (Part I), edited by Gabriele Meiselwitz
15. LNCS 13316, Social Computing and Social Media: Applications in Education and Commerce (Part II), edited by Gabriele Meiselwitz
16. LNCS13317, Virtual, Augmented and Mixed Reality: Design and Development (Part I), edited by Jessie Y. C. Chen and Gino Fragomeni
17. LNCS 13318, Virtual, Augmented and Mixed Reality: Applications in Education, Aviation and Industry (Part II), edited by Jessie Y. C. Chen and Gino Fragomeni

18. LNCS 13319, Digital Human Modeling and Applications in Health, Safety, Ergonomics and Risk Management: Anthropometry, Human Behavior, and Communication (Part I), edited by Vincent G. Duffy
19. LNCS 13320, Digital Human Modeling and Applications in Health, Safety, Ergonomics and Risk Management: Health, Operations Management, and Design (Part II), edited by Vincent G. Duffy
20. LNCS 13321, Design, User Experience, and Usability: UX Research, Design, and Assessment (Part I), edited by Marcelo M. Soares, Elizabeth Rosenzweig and Aaron Marcus
21. LNCS 13322, Design, User Experience, and Usability: Design for Emotion, Well-being and Health, Learning, and Culture (Part II), edited by Marcelo M. Soares, Elizabeth Rosenzweig and Aaron Marcus
22. LNCS 13323, Design, User Experience, and Usability: Design Thinking and Practice in Contemporary and Emerging Technologies (Part III), edited by Marcelo M. Soares, Elizabeth Rosenzweig and Aaron Marcus
23. LNCS 13324, Culture and Computing, edited by Matthias Rauterberg
24. LNCS 13325, Distributed, Ambient and Pervasive Interactions: Smart Environments, Ecosystems, and Cities (Part I), edited by Norbert A. Streitz and Shin'ichi Konomi
25. LNCS 13326, Distributed, Ambient and Pervasive Interactions: Smart Living, Learning, Well-being and Health, Art and Creativity (Part II), edited by Norbert A. Streitz and Shin'ichi Konomi
26. LNCS 13327, HCI in Business, Government and Organizations, edited by Fiona Fui-Hoon Nah and Keng Siau
27. LNCS 13328, Learning and Collaboration Technologies: Designing the Learner and Teacher Experience (Part I), edited by Panayiotis Zaphiris and Andri Ioannou
28. LNCS 13329, Learning and Collaboration Technologies: Novel Technological Environments (Part II), edited by Panayiotis Zaphiris and Andri Ioannou
29. LNCS 13330, Human Aspects of IT for the Aged Population: Design, Interaction and Technology Acceptance (Part I), edited by Qin Gao and Jia Zhou
30. LNCS 13331, Human Aspects of IT for the Aged Population: Technology in Everyday Living (Part II), edited by Qin Gao and Jia Zhou
31. LNCS 13332, Adaptive Instructional Systems, edited by Robert A. Sottilare and Jessica Schwarz
32. LNCS 13333, HCI for Cybersecurity, Privacy and Trust, edited by Abbas Moallem
33. LNCS 13334, HCI in Games, edited by Xiaowen Fang
34. LNCS 13335, HCI in Mobility, Transport and Automotive Systems, edited by Heidi Krömker
35. LNAI 13336, Artificial Intelligence in HCI, edited by Helmut Degen and Stavroula Ntoa
36. LNCS 13337, Design, Operation and Evaluation of Mobile Communications, edited by Gavriel Salvendy and June Wei
37. CCIS 1580, HCI International 2022 Posters - Part I, edited by Constantine Stephanidis, Margherita Antona and Stavroula Ntoa
38. CCIS 1581, HCI International 2022 Posters - Part II, edited by Constantine Stephanidis, Margherita Antona and Stavroula Ntoa

39. CCIS 1582, HCI International 2022 Posters - Part III, edited by Constantine Stephanidis, Margherita Antona and Stavroula Ntoa
40. CCIS 1583, HCI International 2022 Posters - Part IV, edited by Constantine Stephanidis, Margherita Antona and Stavroula Ntoa
41. LNCS 13516, HCI International 2022 - Late Breaking Papers: Design, User Experience and Interaction, edited by Masaaki Kurosu, Sakae Yamamoto, Hirohiko Mori, Marcelo M. Soares, Elizabeth Rosenzweig, Aaron Marcus, Pei-Luen Patrick Rau, Don Harris and Wen-Chin Li
42. LNCS 13517, HCI International 2022 - Late Breaking Papers: Interaction in New Media, Learning and Games, edited by Gabriele Meiselwitz, Abbas Moallem, Panayiotis Zaphiris, Andri Ioannou, Robert A. Sottilare, Jessica Schwarz and Xiaowen Fang
43. LNCS 13518, HCI International 2022 - Late Breaking Papers: Interacting with eXtended Reality and Artificial Intelligence, edited by Jessie Y. C. Chen, Gino Fragomeni, Helmut Degen and Stavroula Ntoa
44. LNCS 13519, HCI International 2022 - Late Breaking Papers: Multimodality in Advanced Interaction Environments, edited by Masaaki Kurosu, Sakae Yamamoto, Hirohiko Mori, Dylan D. Schmorrow, Cali M. Fidopiastis, Norbert A. Streitz and Shin'ichi Konomi
45. LNCS 13520, HCI International 2022 - Late Breaking Papers: HCI for Today's Community and Economy, edited by Matthias Rauterberg, Fiona Fui-Hoon Nah, Keng Siau, Heidi Krömker, June Wei and Gavriel Salvendy
46. LNCS 13521, HCI International 2022 - Late Breaking Papers: HCI for Health, Well-being, Universal Access and Active Aging, edited by Vincent G. Duffy, Qin Gao, Jia Zhou, Margherita Antona and Constantine Stephanidis
47. LNCS 13522, HCI International 2022 - Late Breaking Papers: Ergonomics and Product Design, edited by Vincent G. Duffy and Pei-Luen Patrick Rau
48. CCIS 1654, HCI International 2022 - Late Breaking Posters (Part I), edited by Constantine Stephanidis, Margherita Antona, Stavroula Ntoa and Gavriel Salvendy
49. CCIS 1655, HCI International 2022 - Late Breaking Posters (Part II), edited by Constantine Stephanidis, Margherita Antona, Stavroula Ntoa and Gavriel Salvendy

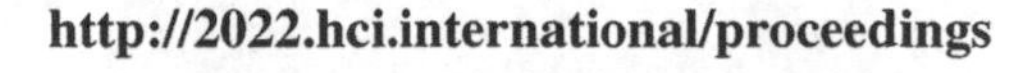

24th International Conference on Human-Computer Interaction (HCII 2022)

The full list with the Program Board Chairs and the members of the Program Boards of all thematic areas and affiliated conferences is available online at:

http://www.hci.international/board-members-2022.php

HCI International 2023

The 25th International Conference on Human-Computer Interaction, HCI International 2023, will be held jointly with the affiliated conferences at the AC Bella Sky Hotel and Bella Center, Copenhagen, Denmark, 23–28 July 2023. It will cover a broad spectrum of themes related to human-computer interaction, including theoretical issues, methods, tools, processes, and case studies in HCI design, as well as novel interaction techniques, interfaces, and applications. The proceedings will be published by Springer. More information will be available on the conference website: http://2023.hci.international/.

General Chair
Constantine Stephanidis
University of Crete and ICS-FORTH
Heraklion, Crete, Greece
Email: general_chair@hcii2023.org

http://2023.hci.international/

Contents – Part I

HCI Theory and Practice

Accessibility, Usability, and UX Design

HCI Research and Design Across Cultures

Cultural Heritage Experience Design

HCI for Health and Wellbeing

Design Case Studies

Contents – Part II

Robots, Agents, and Intelligent Environments